Fuel Cells Compendium

Elsevier Internet Homepage

http://www.elsevier.com

Consult the Elsevier homepage for full catalogue information on all books, journals and electronic products and services.

Elsevier Titles of Related Interest

HYDROGEN AND FUEL CELLS
Bent Sørensen
2005; ISBN: 0126552819
www.elsevier.com/locate/isbn/0126552819

HIGH-TEMPERATURE SOLID OXIDE FUEL CELLS:
FUNDAMENTALS, DESIGN AND APPLICATIONS
S.C. Singhal
2003; ISBN: 1856173879
www.elsevier.com/locate/isbn/1856173879

WORLD FUEL CELLS – AN INDUSTRY PROFILE WITH MARKET PROSPECTS TO 2010
G. Weaver
2002; ISBN: 1856173976
www.elsevier.com/locate/isbn/1856173976

Related Journals
The following journals can all be found at: http://www.sciencedirect.com

Fuel Cells Bulletin
Fuel
Journal of Power Sources
International Journal of Hydrogen Energy
Refocus
Electrochimica Acta
Energy Conversion and Management
Renewable & Sustainable Energy Reviews
Energy
Catalysis Today
Chemical Engineering Science
Renewable Energy
Journal of Membrane Science

To Contact the Publisher
Elsevier welcomes enquiries concerning publishing proposals: books, journal special issues, conference proceedings, etc. All formats and media can be considered. Should you have a publishing proposal you wish to discuss, please contact, without obligation, the publishing editor responsible for Elsevier's energy publishing programme:

Heather Johnstone
Publishing Editor, Materials Science & Engineering
Elsevier Limited
The Boulevard, Langford Lane
Kidlington, Oxford
OX5 1GB, UK

Tel.: +44 1865 84 3887
Fax: +44 1865 84 3920
E-mail: h.johnstone@elsevier.com

General enquiries including placing orders, should be directed to Elsevier's Regional Sales Offices – please access the Elsevier internet homepage for full contact details.

Fuel Cells Compendium

This volume contains 30 previously published papers from the journals

Journal of Power Sources
Solid State Ionics
Applied Catalysis B: Environmental
Catalysis Today
Current Opinion in Solid State and Materials Science
International Journal of Hydrogen Energy
Materials Science and Engineering
Surface Science Reports
Progress in Polymer Science
Current Opinion in Colloid and Interface Science
Materials Today
Electrochimica Acta
Chemical Engineering and Processing

Edited by

Nigel P. Brandon
Department of Chemical Engineering and
Chemical Technology, Imperial College, London, UK

Dave Thompsett
Johnson Matthey Technology Centre,
Sonning Common, Reading, UK

2005

ELSEVIER

AMSTERDAM • BOSTON • HEIDELBERG • LONDON • NEW YORK
OXFORD • PARIS • SAN DIEGO • SINGAPORE • SYDNEY • TOKYO

ELSEVIER B.V.
Radarweg 29
P.O. Box 211, 1000
AE Amsterdam
The Netherlands

ELSEVIER Inc.
525 B Street, Suite 1900
San Diego, CA 92101-4495
USA

ELSEVIER Ltd
The Boulevard, Langford
Kidlington, Oxford OX5 1GB
UK

ELSEVIER Ltd
84 Theobalds Road
London WC1X 8RR
UK

First edition 2005

ISBN: 0-08-044696-5

Typeset by Charon Tec Pvt. Ltd, Chennai, India
www.charontec.com
Printed in Great Britain

Contents

Foreword ix
Contributors xi

1 US distributed generation fuel cell program 1
M.C. Williams, J.P. Strakey and Subhash C. Singhal

2 From curiosity to "power to change the world®" 13
Charles Stone and Anne E. Morrison

3 A review of catalytic issues and process conditions for renewable hydrogen and alkanes by aqueous-phase reforming of oxygenated hydrocarbons over supported metal catalysts 29
R.R. Davda, J.W. Shabaker, G.W. Huber, R.D. Cortright and J.A. Dumesic

4 Fuel processing for low- and high-temperature fuel cells: challenges, and opportunities for sustainable development in the 21st century 53
Chunshan Song

5 Review of fuel processing catalysts for hydrogen production in PEM fuel cell systems 91
Anca Faur Ghenciu

6 Conversion of hydrocarbons and alcohols for fuel cells 107
Finn Joensen and Jens R. Rostrup-Nielsen

7 An assessment of alkaline fuel cell technology 117
G.F. McLean, T. Niet, S. Prince-Richard and N. Djilali

8 Molten carbonate fuel cells 147
Andrew L. Dicks

9 Phosphoric acid fuel cells: fundamentals and applications 155
Nigel Sammes, Roberto Bove and Knut Stahl

10 International activities in DMFC R&D: status of technologies and potential applications 167
R. Dillon, S. Srinivasan, A.S. Aricò and V. Antonucci

11 Transport properties of solid oxide electrolyte ceramics: a brief review 189
V.V. Kharton, F.M.B. Marques and A. Atkinson

12 A review on the status of anode materials for solid oxide fuel cells 215
W.Z. Zhu and S.C. Deevi

13 Advances, aging mechanisms and lifetime in solid-oxide fuel cells 235
Hengyong Tu and Ulrich Stimming

14 Components manufacturing for solid oxide fuel cells 249
F. Tietz, H.-P. Buchkremer and D. Stöver

15 Engineered cathodes for high performance SOFCs 261
R.E. Williford and P. Singh

16 Surface science studies of model fuel cell electrocatalysts 275
N.M. Marković and P.N. Ross, Jr.

17 Proton-conducting polymer electrolyte membranes based on hydrocarbon polymers 375
M. Rikukawa and K. Sanui

18 Advanced materials for improved PEMFC performance and life 411
Dennis E. Curtin, Robert D. Lousenberg, Timothy J. Henry, Paul C. Tangeman and Monica E. Tisack

19 Polymer–ceramic composite protonic conductors 425
Binod Kumar and J.P. Fellner

20 Recent developments in high-temperature proton conducting polymer electrolyte membranes 433
Patric Jannasch

21 PEM fuel cell electrodes 443
S. Litster and G. McLean

22 Review and analysis of PEM fuel cell design and manufacturing 469
Viral Mehta and Joyce Smith Cooper

23 Aging mechanisms and lifetime of PEFC and DMFC 503
Shanna D. Knights, Kevin M. Colbow, Jean St-Pierre and David P. Wilkinson

24 Materials for hydrogen storage 517
Andreas Züttel

25 Fuel economy of hydrogen fuel cell vehicles 531
Rajesh K. Ahluwalia, X. Wang, A. Rousseau and R. Kumar

26 PEMFC systems: the need for high temperature polymers as a consequence of PEMFC water and heat management 545
Ronald K.A.M. Mallant

27 Portable and military fuel cells 555
K. Cowey, K.J. Green, G.O. Mepsted and R. Reeve

28 Microfabricated fuel cells 561
J.S. Wainright, R.F. Savinell, C.C. Liu and M. Litt

29 Electro-catalytic membrane reactors and the development of bipolar membrane technology 573
J. Balster, D.F. Stamatialis and M. Wessling

30 Compact mixed-reactant fuel cells 593
Michael A. Priestnall, Vega P. Kotzeva, Deborah J. Fish and Eva M. Nilsson

Subject Index *607*

Foreword

This volume seeks to bring together a wide-ranging set of peer-reviewed literature which, in the opinion of the Editors, best represents the breadth of work ongoing in the field of fuel cell science and engineering in recent years.

Fuel cell technology is a multi-disciplinary and fast changing field, and it is the intent of the volume to bring together key review articles for the attention of those working in the field. In this regard the volume is particularly aimed at technology developers who do not have access to the broad range of literature from which these articles are drawn.

To that end papers are included from across 13 peer-reviewed Journals. They address all the main fuel cell technologies (alkaline, phosphoric acid, direct methanol, solid oxide, polymer and molten carbonate fuel cells), together with balance of plant issues such as fuel processing and hydrogen storage, and aspects of underlying science including catalysis, membranes and electrode materials.

While no one monograph can hope to include all the work from the many researchers active in this field, we hope that this collection of articles stimulates readers to delve more deeply into the literature in this fascinating and important area.

Professor Nigel P. Brandon
Shell Chair in Sustainable Development in Energy, Imperial College, London

Dr Dave Thompsett
Johnson Matthey Technology Centre

Contributors

Rajesh K. Ahluwalia
Argonne National Laboratory
9700 South Cass Avenue
Argonne, IL 60439, USA

V. Antonucci
Institute CNR-ITAE
Via Salita S. Lucia sopra Contesse 5,
98126-S. Lucia, Messina, Italy

A.S. Aricò
Institute CNR-ITAE
Via Salita S. Lucia sopra Contesse 5,
98126-S. Lucia, Messina, Italy

A. Atkinson
Department of Materials, Imperial College
Exhibition Road, London SW7 2AZ, UK

J. Balster
Membrane Technology Group
Faculty of Science and Technology
University of Twente
Postbus 217, 7500 AE Enschede
The Netherlands

Roberto Bove
Mechanical Engineering Department
University of Connecticut
44 Weaver Road,
Storrs, CT 06269, USA

H.-P. Buchkremer
Forschungszentrum Jülich
Institute for Materials and Processes in Energy Systems (IWV-1)
D-52425 Jülich, Germany

Kevin M. Colbow
Ballard Power Systems
9000 Glenlyon Parkway
Burnaby, BC, Canada V5J 5J9

Joyce Smith Cooper
Department of Mechanical Engineering
University of Washington
Seattle, WA 98195, USA

R.D. Cortright
Virent Energy Systems
3571 Anderson Street
Madison, WI 53704, USA

K. Cowey
QinetiQ Ltd.
Building 442
Cody Technology Park
Ivley Road
Hampshire, GU14 0LX, UK

Dennis E. Curtin
DuPont Fuel Cells
22828 NC Highway 87 W
Fayetteville, NC, USA

R.R. Davda
Department of Chemical and Biological Engineering
University of Wisconsin
1415 Engineering Drive
Madison, WI 53706, USA

S.C. Deevi
Research and Development Center
Chrysalis Technologies Incorporated
7801 Whitepine Road
Richmond, VA 23237, USA

Andrew L. Dicks
ARC Centre for Functional Nanomaterials
University of Queensland
Brisbane, Queensland 4072, Australia

R. Dillon
Center for Energy and Environmental Studies
Princeton University
Princeton, NJ 08544, USA

N. Djilali
University of Victoria
P.O. Box 3055, STN CSC Victoria,
BC, Canada V8W 3P6

J.A. Dumesic
Department of Chemical and Biological Engineering
University of Wisconsin
1415 Engineering Drive
Madison, WI 53706, USA

J.P. Fellner
Propulsion Directorate
Air Force Research Laboratory
Wright-Patterson Airforce Base
Dayton, OH 45433, USA

Deborah J. Fish
Scientific Generics Ltd.
Harston Mill, Harston
Cambridge CB2 5GG, UK

Anca Faur Ghenciu
Johnson Matthey Fuel Cells
1397 King Road
West Chester, PA 19380, USA

K.J. Green
QinetiQ Ltd.
Batteries, Fuel Cells & Power Systems
Room F11, ES Building
Haslar Road
Gosport, PO 12 2AG, UK

Timothy J. Henry
DuPont Fuel Cells
22828 NC Highway 87 W
Fayetteville, NC, USA

G.W. Huber
Department of Chemical and Biological Engineering
University of Wisconsin
1415 Engineering Drive
Madison, WI 53706, USA

Patric Jannasch
Department of Polymer Science and Engineering
Lund University
P.O. Box 124
SE-221 00, Lund, Sweden

Finn Joensen
Haldor Topsøe A/S
Nymøllevej 55,
DK-2800, Lyngby, Denmark

V.V. Kharton
Department of Ceramics and Glass Engineering
CICECO, University of Aveiro
3810-193 Aveiro, Portugal

Shanna D. Knights
Ballard Power Systems
9000 Glenlyon Parkway
Burnaby, BC, Canada V5J 5J9

Vega P. Kotzeva
Scientific Generics Ltd.
Harston Mill, Harston
Cambridge CB2 5GG, UK

Binod Kumar
University of Dayton Research Institute
300 College Park, KL 501
Dayton, OH 45469, USA

R. Kumar
Argonne National Laboratory
9700 South Cass Avenue
Argonne, IL 60439, USA

M. Litt
Department of Macromolecular Science and
Yeager Center for Electrochemical Sciences
Case Western Reserve University
Cleveland, OH 44106, USA

S. Litster
Angstrom Power Inc.
106 980 W. 1st-St
North Vancouver, BC, Canada V7P 3N4

C.C. Liu
Department of Chemical Engineering and
Electronics Design Center
Case Western Reserve University
Cleveland, OH 44106, USA

Robert D. Lousenberg
DuPont Fuel Cells
22828 NC Highway 87 W
Fayetteville, NC, USA

Ronald K.A.M. Mallant
Energy Research Centre of the Netherlands (ECN)
P.O. Box 1
1755 ZG, Petten, The Netherlands

N.M. Marković
Lawrence Berkeley National Laboratory
Materials Sciences Division
University of California
Berkeley, CA 94720, USA

F.M.B. Marques
Department of Ceramics and Glass Engineering
CICECO, University of Aveiro
3810-193 Aveiro, Portugal

G. McLean
Angstrom Power Inc.
106 980 W. 1st-St
North Vancouver, BC, Canada V7P 3N4

G.F. McLean
University of Victoria
P.O. Box 3055, STN CSC Victoria,
BC, Canada V8W 3P6

Viral Mehta
Department of Mechanical Engineering
University of Washington
Seattle, WA 98195, USA

G.O. Mepsted
QinetiQ Ltd.
Batteries, Fuel Cells & Power Systems
Room F11, ES Building
Haslar Road
Gosport, PO 12 2AG, UK

Anne E. Morrison
Ballard Power Systems Inc.
9000 Glenlyon Parkway
Burnaby, BC, Canada V5J 5J9

T. Niet
University of Victoria
P.O. Box 3055, STN CSC Victoria,
BC, Canada V8W 3P6

Eva M. Nilsson
Scientific Generics Ltd.
Harston Mill, Harston
Cambridge CB2 5GG, UK

Michael A. Priestnall
Scientific Generics Ltd.
Harston Mill, Harston
Cambridge CB2 5GG, UK

R. Reeve
QinetiQ Ltd.
Batteries, Fuel cells & Power Systems
Room F11, ES Building
Haslar Road
Gosport, PO 12 2AG, UK

S. Prince-Richard
University of Victoria
P.O. Box 3055, STN CSC Victoria,
BC, Canada V8W 3P6

M. Rikukawa
Department of Chemistry
Sophia University
7-1 Kioi-cho, Chiyoda-ku
Tokyo 102, Japan

P.N. Ross Jr
Lawrence Berkeley National Laboratory
Materials Sciences Division
University of California
Berkeley, CA 94720, USA

Jens R. Rostrup-Nielsen
Haldor Topsøe A/S
Nymøllevej 55
DK-2800, Lyngby, Denmark

A. Rousseau
Argonne National Laboratory
9700 South Cass Avenue
Argonne, IL 60439, USA

Jean St-Pierre
Ballard Power Systems
9000 Glenlyon Parkway
Burnaby, BC, Canada V5J 5J9

Nigel Sammes
Mechanical Engineering Department
University of Connecticut
44 Weaver Road, Storrs
CT 06269, USA

K. Sanui
Department of Chemistry
Sophia University
7-1 Kioi-cho, Chiyoda-ku
Tokyo 102, Japan

R.F. Savinell
Department of Chemical Engineering and
Yeager Center for Electrochemical Sciences
Case Western Reserve University
Cleveland, OH 44106, USA

J.W. Shabaker
Department of Chemical and Biological Engineering
University of Wisconsin
1415 Engineering Drive
Madison, WI 53706, USA

P. Singh
Pacific Northwest National Laboratory
P.O. Box 999
Mail Stop K2-44
Richland, WA 99352, USA

Subhash C. Singhal
Pacific Northwest National Laboratory
902 Battelle Boulevard
P.O. Box 999
Richland, WA 99352, USA

Chunshan Song
Clean Fuels and Catalysis Program
The Energy Institute
Department of Energy and Geo-Environmental Engineering
The Pennsylvania State University
209 Academic Projects Building
University Park, PA 16802, USA

S. Srinivasan
Center for Energy and Environmental Studies
Princeton University
Princeton, NJ 08544, USA

Knut Stahl
RWE Fuel Cells GmbH
Gutenbergstrasse 3
D-45128 Essen, Germany

D.F. Stamatialis
Membrane Technology Group
Faculty of Science and Technology
University of Twente
Postbus 217
7500 AE Enschede, The Netherlands

Ulrich Stimming
Physik-Department E19
Technische Universität München
James-Franck-Strasse 1
D-85748, Garching, Germany

Charles Stone
Ballard Power Systems Inc.,
9000 Glenlyon Parkway
Burnaby, BC, Canada V5J 5J9

D. Stöver
Forschungszentrum Jülich
Institute for Materials and Processes in Energy Systems (IWV-1)
D-52425 Jülich, Germany

J.P. Strakey
National Energy Technology Laboratory
U.S. Department of Energy
Pittsburgh, PA, USA

Paul C. Tangeman
DuPont Fuel Cells
22828 NC Highway 87 W
Fayetteville, NC, USA

F. Tietz
Forschungszentrum Jülich
Institute for Materials and Processes in Energy Systems (IWV-1)
D-52425 Jülich, Germany

Monica E. Tisack
DuPont Fuel Cells
22828 NC Highway 87 W
Fayetteville, NC, USA

Hengyong Tu
Physik-Department E19
Technische Universität München
James-Franck-Strasse 1
D-85748, Garching, Germany

J.S. Wainright
Department of Chemical Engineering and
Yeager Center for Electrochemical Sciences
Case Western Reserve University
Cleveland, OH 44106-7217, USA

X. Wang
Argonne National Laboratory
9700 South Cass Avenue
Argonne, IL 60439, USA

M. Wessling
Membrane Technology Group
Faculty of Science and Technology
University of Twente
Postbus 217
7500 AE Enschede, The Netherlands

David P. Wilkinson
Ballard Power Systems
9000 Glenlyon Parkway
Burnaby, BC, Canada V5J 5J9

M.C. Williams
National Energy Technology Laboratory
U.S. Department of Energy
Morgantown, WV, USA

R.E. Williford
Pacific Northwest National Laboratory
P.O. Box 999
Mail Stop K2-44
Richland, WA 99352, USA

W.Z. Zhu
Research and Development Center
Chrysalis Technologies Incorporated
7801 Whitepine Road
Richmond, VA 23237, USA

Andreas Züttel
Physics Department
University of Fribourg
Pérolles
CH-1700 Fribourg, Switzerland

Chapter 1
US distributed generation fuel cell program

M.C. Williams, J.P. Strakey and Subhash C. Singhal

Abstract

The Department of Energy (DOE) is the largest funder of fuel cell technology in the US. The Department of Energy – Office of Fossil Energy (FE) is developing high-temperature fuel cells for distributed generation. It has funded the development of tubular solid oxide fuel cell (SOFC) and molten carbonate fuel cell (MCFC) power systems operating at up to 60% efficiency on natural gas. The remarkable environmental performance of these fuel cells makes them likely candidates to help mitigate pollution. DOE is now pursuing more widely applicable SOFCs for 2010 and beyond. DOE estimates that a 5 kW SOFC system can reach $400 per kW at reasonable manufacturing volumes. SECA – the Solid State Energy Conversion Alliance – was formed by the National Energy Technology Laboratory (NETL) and the Pacific Northwest National Laboratory (PNNL) to accelerate the commercial readiness of planar and other SOFC systems utilizing 3–10 kW size modules by taking advantage of the projected economies of production from a "mass customization" approach. In addition, if the modular 3–10 kW size units can be "ganged" or "scaled-up" to larger sizes with no increase in cost, then commercial, microgrid and other distributed generation markets will become attainable. Further scale-up and hybridization of SECA SOFCs with gas turbines could result in penetration of the bulk power market. This paper reviews the current status of the SOFC and MCFC in the US.

Keywords: Stationary power; Distributed generation; Cogeneration; Solid oxide fuel cell; Molten carbonate fuel cell; Hybrids

Article Outline

1. Introduction . . . 1
2. Solid oxide fuel cells (SOFCs) . . . 2
   2.1. Siemens Westinghouse Power Corporation's Tubular SOFC Program . . . 2
   2.2. Solid State Energy Conversion Alliance (SECA) SOFC Programs . . . 7
3. Molten carbonate fuel cells (MCFCs) . . . 8
4. FutureGen . . . 9
5. Summary . . . 10
Acknowledgements . . . 10
References . . . 11

1. INTRODUCTION

Fuel cells have high efficiency, low environmental impact, potential low-cost even in small size units, and are easy to site. Because of these factors, together with the interest for distributed power generation, the

Fuel Cells Compendium

US Department of Energy – Office of Fossil Energy (FE) is funding several major programs for the development of fuel cell-based power generation systems [1]. These programs include both solid oxide and molten carbonate fuel cells, and can be categorized as follows:

- Solid oxide fuel cells (SOFCs):
  - Siemens Westinghouse Power Corporation Tubular SOFC Program
  - Solid State Energy Conversion Alliance (SECA) Programs
- Molten carbonate fuel cells (MCFCs):
  - FuelCell Energy (FCE), Inc. Direct Fuel Cell (DFC) Program.

In addition, a new concept termed FutureGen was initiated early this year to produce electricity and hydrogen from coal in a virtually emission-free plant; this concept will also employ fuel cells. This paper discusses the status of the various fuel cell programs.

2. SOLID OXIDE FUEL CELLS (SOFCs)

2.1. Siemens Westinghouse Power Corporation's Tubular SOFC Program

Siemens Westinghouse (formerly Westinghouse Electric Corporation) has been developing tubular SOFCs since late 1970s. In their latest tubular design, the cell components are deposited in the form of thin layers on a cathode (air electrode) tube, closed at one end [2]. Figure 1 schematically illustrates the design of the Siemens Westinghouse tubular cell [3,4]. The lanthanum manganite-based cathode tube (2.2 cm diameter, 2.2 mm wall thickness, about 180 cm length) is fabricated by extrusion followed by sintering to obtain about 30–35% porosity. Electrolyte, zirconia doped with about 10 mol% yttria (YSZ), is deposited in the form of about 40-μm-thick layer by an electrochemical vapor deposition (EVD) process [5,6]. In this process, chlorides of zirconium and yttrium are volatilized in a predetermined ratio and passed along with hydrogen and argon over the outer surface of the porous air electrode tube. Oxygen mixed with steam is passed inside the cathode tube. In the first stage of the reaction, molecular diffusion of oxygen, steam, metal chlorides, and hydrogen occurs through the porous cathode and these react to fill the pores in the cathode with the yttria-stabilized zirconia. During the second stage of the reaction after the pores in the air electrode are closed, electrochemical transport of oxide ions occurs through the already deposited yttria-stabilized zirconia in the pores from the high oxygen partial pressure side (oxygen/steam) to the low oxygen partial

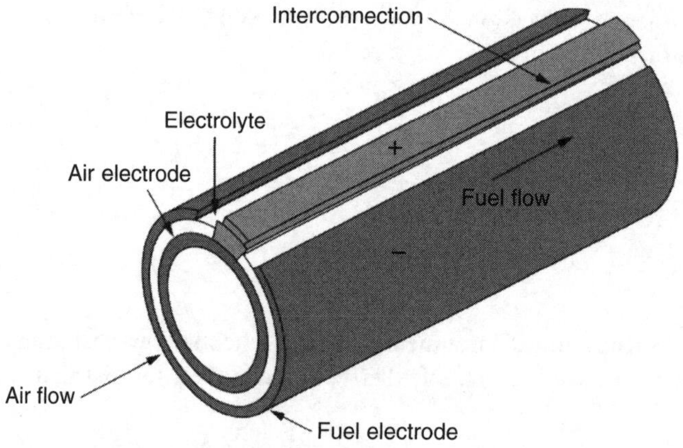

Fig. 1. Schematic illustration of a Siemens Westinghouse tubular SOFC [3]

pressure side (chlorides). The oxide ions, upon reaching the low oxygen partial pressure side, react with the metal chlorides and the electrolyte film grows in thickness. The ratio of yttrium chloride to zirconium chloride is so chosen that the electrolyte deposited contains about 10 mol% yttria.

The EVD technique deposits a very high quality, 100% dense, uniformly thick electrolyte film. However, this technique to deposit the electrolyte is complex, capital-cost intensive, and requires vacuum equipment that makes scaling it up to a cost-effective, continuous manufacturing process for high-volume SOFC production difficult. Fabrication of the YSZ electrolyte films by a more cost-effective non-EVD technique, such as plasma spraying followed by sintering, is presently being investigated to reduce cell manufacturing cost.

The Ni/YSZ anode, 100–150 μm thick, is deposited over the electrolyte by a two-step process. In the first step, nickel powder slurry is applied over the electrolyte. In the second step, YSZ is grown around the nickel particles by the same EVD process as used for depositing the electrolyte. Deposition of a Ni–YSZ slurry over the electrolyte followed by sintering has also yielded anodes that are equivalent in performance to those fabricated by the EVD process. Deposition of the anode by a thermal spraying method is also being investigated. Use of these non-EVD processes should result in a substantial reduction in the cost of manufacturing SOFCs.

Doped lanthanum chromite interconnection is deposited in the form of about 85 μm thick, 9 mm wide strip along the air electrode tube length by plasma spraying followed by densification sintering [7].

The cell tube is closed at one end. For cell operation, oxidant (air or oxygen) is introduced through an alumina injector tube positioned inside the cell. The oxidant is discharged near the closed end of the cell and flows through the annular space formed by the cell and the coaxial injector tube. Fuel flows on the outside of the cell from the closed end and is electrochemically oxidized while flowing to the open end of the cell generating electricity. At the open end of the cell, the oxygen-depleted air exits the cell and is combusted with the partially depleted fuel. Typically, 50–90% of the fuel is utilized in the electrochemical cell reaction. Part of the depleted fuel is recirculated in the fuel stream and the rest combusted to preheat incoming air and/or fuel.

A large number of tubular cells have been electrically tested over the years, some for times as long as 8 years; typical performance of these cells is illustrated in Fig. 2. These cells perform satisfactorily for extended periods of time under a variety of operating conditions with less than 0.1% per 1000 h performance degradation [3]. The tubular SOFCs have also shown the ability to be thermally cycled to room temperature from 1000°C over 100 times without any mechanical damage or electrical performance loss. This ability to sustain thermal cycles is essential for any SOFC generator to be commercially viable.

To construct an electric generator, individual cells are connected in both electrical parallel and series to form a semi-rigid bundle that becomes the basic building block of a generator [3]. Nickel felt is used to provide soft, mechanically compliant, low electrical resistance connections between the cells. This material bonds to the nickel particles in the fuel electrode and the nickel plating on the interconnection for the series connection, and to the two adjacent cell fuel electrodes for the parallel connection; such a series–parallel arrangement provides improved generator reliability. A three-in-parallel by eight-in-series cell bundle is shown in Fig. 3. The individual cell bundles are arrayed in series to build voltage and form generator modules.

Since 1984, Siemens Westinghouse has designed, built, and tested almost a dozen fully integrated power systems of successively increasing sizes. More recently, a 100 kW size atmospheric pressure power system, employing 1152 tubular cells in 48 bundles of 24 cells each, was built as shown in Fig. 4. It operated very successfully for over 20 000 h in the Netherlands and Germany on natural gas at an efficiency of 46% with no detectable performance degradation. Such atmospheric systems are ideal for combined heat and power (CHP) generation in distributed applications.

A scaled-up 250 kW size atmospheric pressure system, shown in Fig. 5, employing 2292 tubular cells, with heat extraction (CHP), has also been built and is now operating at a Kinetrics Facility in Toronto, Canada. Similar 250 kW size systems are planned for operation in 2004 at Stadwerke Hannover in Germany

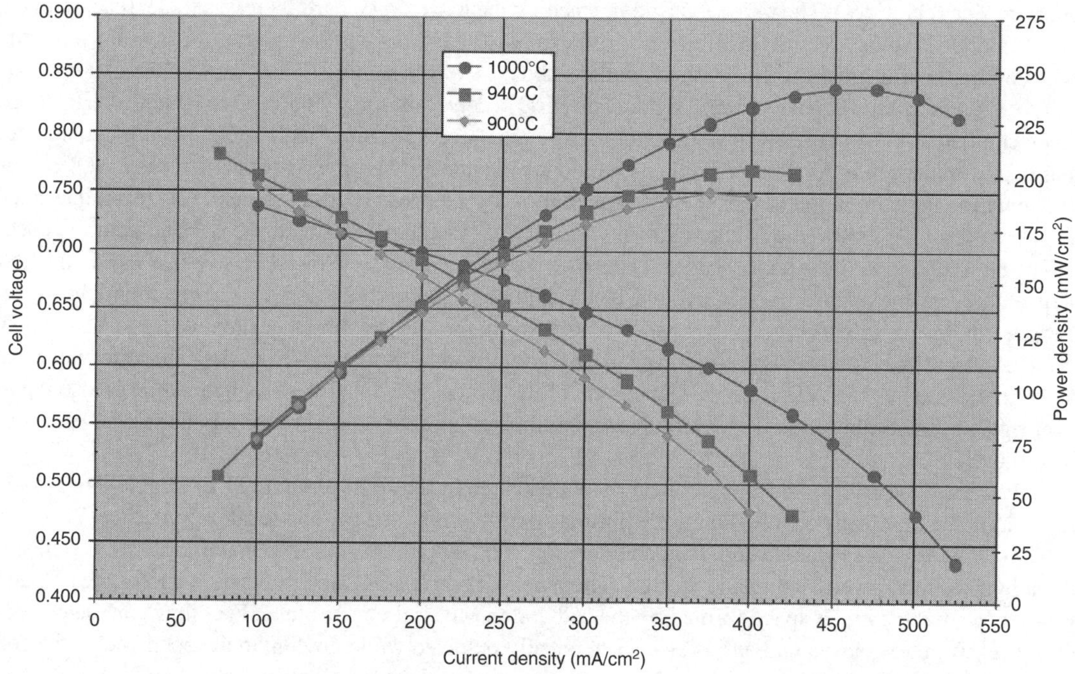

Fig. 2. Voltage–current density and power–current density plots of a commercial prototypical tubular SOFC. (Courtesy of Siemens Westinghouse Power Corporation)

Fig. 3. A three-in-parallel by eight-in-services tubular cell bundle [3]

and at BP America in Alaska. Such atmospheric CHP products with electrical efficiencies in the 45–50% range are expected to be the Siemens Westinghouse's initial commercial offering commencing in 2006.

Siemens Westinghouse has also tested tubular SOFCs at pressures up to 15 atm on hydrogen and natural gas fuels [3]. Figure 6 shows the effect of pressure on cell power output for a 2.2 cm diameter, 150 cm active length cell at 1000°C. Operation at elevated pressures yields a higher cell power at any current density due to increased Nernst potential and reduced cathode polarization, and thereby permits higher stack efficiency

Fig. 4. Siemens Westinghouse's 100 kW size atmospheric pressure CHP system

Fig. 5. Siemens Westinghouse's 250 kW size atmospheric pressure CHP system

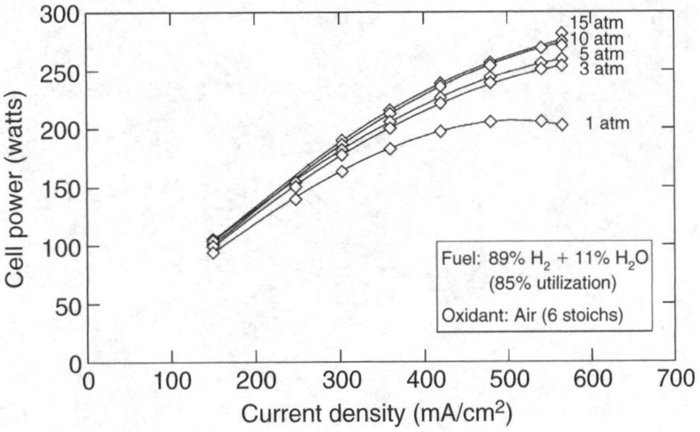

Fig. 6. Effect of pressure on the power of a tubular SOFC [3]

and greater power output. With pressurized operation, SOFCs can be successfully used as replacements for combustors in gas turbines for SOFC/turbine hybrid systems.

Siemens Westinghouse has designed, fabricated and tested a pressurized SOFC/gas turbine hybrid system for enhanced efficiency. The initial 200 kW hybrid system (PH200), shown in Fig. 7, underwent proof-of-concept testing at the National Fuel Cell Research Center in Irvine, CA, for about 3000 h, and the unit achieved 52% electrical efficiency. This first-of-a-kind system provided many useful lessons and demonstrated excellent emissions and efficiency performance. It demonstrated that SOFC/turbine hybrid systems are feasible, with promise of unparalleled efficiency, but stack and gas turbine development, system capacity scale-up, and validation must occur before commercialization. A scaled-up 330 kW size hybrid system (PH300) (Fig. 8) is presently undergoing factory test in Pittsburgh, PA, before its eventual shipment to a German utility, RWE, in Essen, Germany. This system is expected to achieve an electrical efficiency of 58%. Further demonstrations of hybrid systems in coming years are planned at utilities in US, Germany, and Italy.

Siemens Westinghouse's efforts are now focused on reducing the cost of SOFC power systems. A manufacturing facility with a production capacity of 15 MW per year has been built for start-up in late 2004; this facility will provide information on the rate of reduction in the cost of SOFC power systems as production volume increases.

Fig. 7. Siemens Westinghouse's 200 kW pressurized SOFC/gas turbine hybrid system

Fig. 8. Siemens Westinghouse 330 kW pressurized SOFC/turbine hybrid system for RWE, Germany

Siemens Westinghouse Power Corporation's (DOE) program is scheduled to wind down in FY2004. It is in a sense the precursor program to the SECA program. It was through the tubular SOFC program that many of the attributes of SOFCs have been first demonstrated.

2.2. Solid State Energy Conversion Alliance (SECA) SOFC Programs

The SECA program is the main thrust of the Department of Energy – Fossil Energy's distributed generation fuel cell program. SECA SOFC program supports Climate Change, FutureGen, Clear Skies, and Homeland Security initiatives. SECA is also recognized as part of the overall US Hydrogen program. Achieving SECA goals should result in the wide deployment of SOFC technology in high-volume markets. This means that benefits to the nation are large but the cost must be low. This is the SECA goal – less expensive materials, simple stack and system design, and high-volume markets. These criteria must be met to compete in today's energy market. Near-zero emissions, fuel flexibility, high efficiency, and simple CO_2 capture will provide a national payoff that gets bigger as these markets get larger.

The SECA program is dedicated to developing innovative, effective, low-cost ways to commercialize SOFCs [8]. The program is designed to move fuel cells out of limited niche markets into widespread market applications by making them available at a cost of $400/kW or less by 2010 through the mass customization of common SOFC modules. SECA fuel cells will operate on today's conventional fuels such as natural gas, gasoline and diesel, as well as the fuels of tomorrow – coal gas and hydrogen. The program will provide a bridge to the hydrogen economy beginning with the introduction of SECA fuel cells for stationary (for both central station and distributed power generation applications) and transportation's auxiliary power applications.

The SECA program is currently structured to include competing industry teams supported by a cross-cutting core technology program. SECA has six industry teams working on designs that can be mass-produced at costs that are almost 10-fold less than current costs. The SECA core technology program is made up of researchers from industrial suppliers and manufacturers as well as from universities and national laboratories, all working towards addressing key science and technology gaps to provide breakthrough solutions to critical issues facing SOFCs.

The SECA industry teams collectively are making very good progress. Delphi, in partnership with Battelle, is developing a compact, gasoline-fueled, 5 kW unit utilizing planar anode-supported SOFCs operating at 700–800°C for distributed generation and auxiliary power unit (APU) markets. Their first prototype APU, shown in Fig. 9, was installed in the trunk of a luxury automobile and tested successfully [9,10]. Delphi is expert at system integration and high-volume manufacturing and cost reduction. They are focused on making a very compact and light-weight system suitable for auxiliary power generation in transportation applications.

General Electric (GE) is initially developing a natural gas-fueled 5 kW system, also utilizing planar, anode-supported SOFCs, for residential power markets. GE is evaluating several stack designs and is especially interested in extending planar SOFCs to large hybrid systems. Presently, they are working on a radial design that can simplify packaging by minimizing the need for seals. GE has made good progress in achieving high fuel utilization with improved anode performance using standard materials.

Cummins and SOFCo team is developing a 10 kW product initially for recreational vehicles (RVs) that would run on propane using a catalytic partial oxidation (CPOX) reformer. The team has produced a conceptual design for a multilayer SOFC stack assembled from low-cost "building blocks." A thin electrolyte layer (50–75 μm) is fabricated by tape casting. Anode ink is screen-printed onto the one side of the electrolyte tape, and cathode ink onto the other. The printed cell is sandwiched between layers of a dense ceramic that accommodates reactant gas flow and electrical conduction. The assembly is then co-fired to form a single repeat unit.

Siemens Westinghouse, in addition to its ongoing tubular SOFC program for larger systems, is developing smaller, 5–10 kW size products in the SECA program to satisfy multiple markets. They have developed a

Fig. 9. Delphi/Battelle 5 kW auxiliary power unit [9,10]

new flattened, ribbed cell design for these smaller units that retains all the advantages of the cylindrical, cells such as not requiring seals, yet provides higher power density. These cells also make possible more efficient manufacturing, bundle assembly, and provide higher volumetric power density. These cells will be initially incorporated in 5 kW size systems being developed by Fuel Cell Technologies of Canada for residential and other distributed power applications.

Two additional industry teams, one led by FCE and the other by Acumentrics, recently initiated work under the SECA program. FCE will utilize lower-temperature planar anode-supported cells for distributed power systems, whereas Acumentrics plans to use microtubular cells for fast-starting small systems. Overall, the six industry teams are pursuing several design alternatives that enhance the prospects of success of SECA fuel cells for a broader market.

3. MOLTEN CARBONATE FUEL CELLS (MCFCs)

Department of Energy – Office of Fossil Energy has been funding molten carbonate-based Direct FuelCell® (DFC®) development at FuelCell Energy, Inc. for stationary power plant applications. FCE (Danbury, CT) is a world-recognized leader for the development and commercialization of high-efficiency fuel cells that can generate clean electricity at power stations or in distributed locations near the customer, including hospitals, schools, universities, hotels and other commercial and industrial applications. FCE has designed and is beginning to commercialize three different fuel cell power plant models (DFC300, DFC1500, and DFC3000). Rated output and footprint of these plants are provided in Fig. 10.

These power plants offer significant advantages compared to existing power generation technology – higher fuel efficiency, significantly lower emissions, quieter operation, flexible siting and permitting requirements, and scalability. Also, the exhaust heat can be used for cogeneration applications such as high-pressure steam, district heating, and air conditioning. Because hydrogen is generated directly within the fuel cell module from readily available fuels such as natural gas and waste water treatment gas, DFC power plants are ready today and do not require the creation of a hydrogen infrastructure.

FCE's products are based on its patented DFC technology [11,12]. Several DFC sub-megawatt power plants are currently operating in Europe, Japan, and the US. Accomplishments to date include over 17 million kWh generated with 12 million kWh at customer sites. FCE has also developed manufacturing and testing capabilities to produce 50 MW per year. Additional DFC power plants are scheduled for delivery in Europe, Japan,

DFC300
Output: 250 kW
Footprint: 10.5′ × 28′

DFC1500
Output: 1000 kW
Footprint: 42′ × 39′

DFC3000
Output: 2000 kW
Footprint: 42′ × 57′

Fig. 10. FuelCell Energy's DFC power plants for stationary application

Fig. 11. FuelCell Energy's DFC/T hybrid system

and the US over the next 12 months, including its first DFC1500 and DFC3000 units. In parallel, FCE is also developing technology for coal gas, logistic fuels, and other fossil and renewable fuels such as coal mine methane gas and anaerobic digester gas from municipal and industrial wastewater treatment facilities.

FCE is also developing a ultra-high-efficiency hybrid system, the patented Direct FuelCell/Turbine® (DFC/T®), a power plant designed to use the heat generated by the fuel cell to drive a unfired gas turbine for additional electricity [13]. During 2002, FCE completed successful proof-of-concept testing of a DFC/T power plant (Fig. 11) based on a 250 kW DFC integrated with a 30 kW modified microturbine. This proof-of-concept demonstration has provided information for the continued design of a 40 MW DFC/T power plant that is expected to approach 75% efficiency as well as to serve as a platform for high-efficiency DFC/T systems in smaller sizes. FCE is currently continuing its proof-of-concept testing of the DFC/T power plant with a 60 kW microturbine.

4. FUTUREGEN

FutureGen is a major new Presidential initiative to produce electricity and hydrogen from coal. It is a $1 billion government/industry partnership to design, build and operate a nearly emission-free, coal-fired electricity, and hydrogen production plant. The 275 MW prototype plant will serve as a large-scale engineering

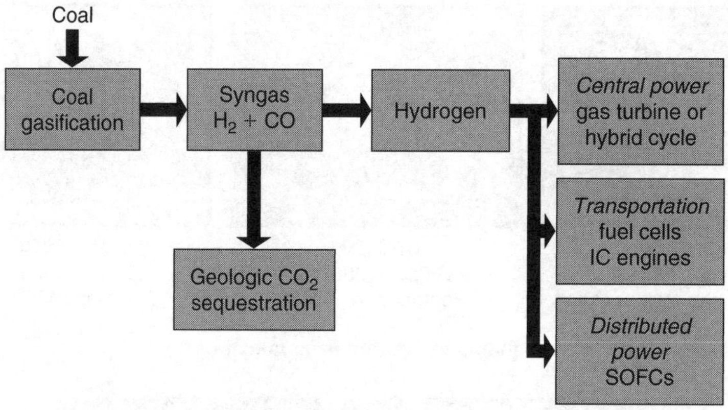

Fig. 12. Schematic illustration of the FutureGen concept

laboratory for testing new clean power generation, carbon capture, and coal-to-hydrogen technologies, and will establish the technical and economic feasibility of producing hydrogen and electricity from coal, the lowest cost and most abundant domestic energy resource. It will be the cleanest fossil fuel-fired power plant in the world. Virtually every aspect of the prototype plant will employ cutting-edge technology. As shown schematically in Fig. 12, rather than using traditional coal combustion technology, the plant will be based on coal gasification, which produces synthesis gas consisting of hydrogen and carbon monoxide. Advanced technology will be used to react the synthesis gas with steam to produce hydrogen and a concentrated stream of CO_2. Initially, the hydrogen will be used as a clean fuel for electricity production either in turbines, fuel cells, or fuel cell/turbine hybrids. The hydrogen could also be supplied as a feedstock for refineries. Later on, the plant could be a source of transportation-grade hydrogen fuel.

The captured CO_2 will be separated from the hydrogen, perhaps by novel membranes currently under development. It would then be permanently sequestered in a geologic formation.

The project will require 10 years to complete and will be led by an industrial consortium representing the coal and power industries.

5. SUMMARY

US Department of Energy – Office of Fossil Energy is funding several major programs for clean and efficient power generation; these include high-temperature solid oxide and molten carbonate fuel cells and the combinations of these with gas turbines in highly efficient hybrids. Significant progress has been made in these programs and 250 kW to few MW size systems are nearing commercialization. The biggest hurdle to large-scale commercialization is their rather high cost and the manufacturers are now concentrating on reducing the cost of fuel cell-based power systems. The SECA program is expected to provide SOFC systems that cost about $400/kW by 2010. Beyond 2010, SECA solid oxide fuel cells are expected to be incorporated into the FutureGen type plants to produce electricity and hydrogen in a virtually emission-free plant.

ACKNOWLEDGEMENTS

The authors thank Siemens Westinghouse Power Corporation and FuelCell Energy, Inc. for supplying data and photographs for this paper.

REFERENCES

1. M.C. Williams and J.P. Strakey. In: S.C. Singhal and M. Dokiya (Eds.), *Solid Oxide Fuel Cells VIII*, Vol. PV2003-07. The Electrochemical Society, Inc., Pennington, NJ, 2003, pp. 3–8.
2. S.C. Singhal. In: M. Dokiya, O. Yamamoto, H. Tagawa and S.C. Singhal (Eds.), *Solid Oxide Fuel Cells IV*, Vol. PV95-1. The Electrochemical Society, Inc., Pennington, NJ, 1995, pp. 195–207.
3. S.C. Singhal. *MRS Bull.* **25(3)** (2000) 16–21.
4. S.C. Singhal. *Solid State Ionics* **135** (2000) 305–313.
5. A.O. Isenberg. In: J.D.E. McIntyre, S. Srinivasan and F.G. Will (Eds.), *Electrode Materials and Processes for Energy Conversion and Storage*, Vol. PV77-6. The Electrochemical Society, Inc., Princeton, NJ, 1977, p. 572.
6. U.B. Pal and S.C. Singhal. *J. Electrochem. Soc.* **137** (1990) 2937.
7. L.J.H. Kuo, S.D. Vora and S.C. Singhal. *J. Am. Ceram. Soc.* **80** (1997) 589.
8. W.A. Surdoval, S.C. Singhal and G.L. McVay. In: H. Yokokawa and S.C. Singhal (Eds.), *Solid Oxide Fuel Cells VII*, Vol. PV2001-16. The Electrochemical Society, Inc., Pennington, NJ, 2001, p. 53.
9. S. Mukerjee, M.J. Grieve, K. Haltiner, M. Faville, J. Noetzel, K. Keegan, D. Schumann, D. Armstrong, D. England, J. Haller and C. DeMinco. In: H. Yokokawa and S.C. Singhal (Eds.), *Solid Oxide Fuel Cells VII*, Vol. PV2001-16. The Electrochemical Society, Inc., Pennington, NJ, 2001, p. 173.
10. S.C. Singhal. *Solid State Ionics* **152–153** (2002) 405–410.
11. M. Farooque and H. Ghezel-Ayagh. In: W. Vielstich, A. Lamm and H. Gasteiger (Eds.), *Handbook of Fuel Cells, Fundamentals, Technology and Applications: Fuel Cell Technology and Applications*, Part 2, Vol. 4. Wiley, New York, 2003, pp. 942–968 (Chapter 68).
12. H. Ghezel-Ayagh, A.J. Leo, H. Maru and M. Farooque. Overview of direct carbonate fuel cell technology and products development. In: *Proceedings ASME First International Conference on Fuel Cell Science, Engineering and Technology*, Rochester, NY, April 21–23, 2003.
13. H. Ghezel-Ayagh, J. Daly and Z. Wang. Advances in direct fuel cell/gas turbine power plants. In: *Proceedings of the ASME/IGTI Turbo Expo 2003*, ASME Paper GT2003-38941, 2003.

Chapter 2

From curiosity to "power to change the world®"

Charles Stone and Anne E. Morrison

Abstract

Proton exchange membrane fuel cells (PEMFCs) experienced a dramatic renaissance in the 1980s following their initial growth to everyday prominence in the 1960s through the activities of General Electric. However, it is the potential of PEMFC products to ultimately replace the internal combustion engine that has captured the greatest attention and provided the strongest impetus for technological expansion. Given that, it is quite clear that the impact of fuel cell technology will stretch much further than automotive applications and will offer innovative and practical products to meet the needs of stationary and portable electricity generation, as well as battery replacement.

A brief history of the development of fuel cell technology up to the early 1960s is provided, with a more detailed review of key events that culminated in the exponential growth of PEMFC technology, thereafter. The proliferation of demonstration programs is discussed, along with the development of businesses dedicated to the commercialization of this important technology. The beneficial far-reaching impact of this technology in addressing the insatiable needs of the consumer for reliable, low-cost, portable electricity supply, without sacrifice to environmental needs, is also explored. The proliferation of alliances, both end-user and supplier-based is discussed to emphasize its importance in accelerating PEMFC products into commercial production. The achievements of the technology development in increased performance, power density, reliability and lowered costs are outlined. As well, a clear description is given of the remaining areas where further advancements must be made to achieve successful commercialization. Finally, the factors that will define the rate of growth of PEMFC products into the marketplace and the required profitability for the companies that produce them are given as comment in the closing section.

Keywords: PEM fuel cells; History; Power generation; Alternative power technology; Environment-friendly
PACS classification codes: 84.60.D

Article Outline

1. Introduction 14
2. The technology development path to the PEMFC (1838–1960s) 14
3. PEMFC development from the 1970s into the new millennium 16
4. Key business and supplier alliances in PEMFCs for the 21st century 20
5. Why the world needs PEMFC products 22
6. The outlook for the future 22
Acknowledgements 23
References 23

Fuel Cells Compendium

1. INTRODUCTION

Through Edison's many inventions, which included the incandescent light bulb [1], the storage battery [2,3] and the phonograph [4,5], he was able to demonstrate the immense power and versatility of electricity to meet both the practical and recreational needs of a rapidly developing world. This was really the beginning of high-volume consumer products that utilized or produced electricity. Since then, there have been many incredible and ingenious ways in which humanity has conceived the means to produce and harness the power that is electricity. Going even further back than Edison, it was in 1802 that Sir Humphrey Davy observed that a galvanic cell could be used to produce hydrogen and oxygen [6]. It followed from there that the reversibility of this process should permit the production of electricity through the electrochemical combination of these two reactant gases. Almost 200 years later, the proton exchange membrane fuel cell (PEMFC) has been demonstrated by both academic [7–10] and industrial [11,12] entities to be a practical and environmentally desirable option in the production of electrical energy for various applications. The end use that has received the greatest attention has been the operation of PEMFC technology in transportation applications. While this is clearly a multi-billion dollar market, the use of PEMFC technology in both portable and stationary applications has equal potential and offers high-volume sales possibilities that can be realized in the early part of this decade.

The use of ion-exchange membrane-based fuel cell technology in battery-replacement type applications, such as cellular phones [13,14], laptop computers [15] and hand-held devices [16–19] is a market ripe for a product that can free users from the aggravation of repetitive recharging and limited usage cycles. This application may also provide an early market opportunity for a modified PEMFC in which the direct oxidation of a liquid fuel, methanol, obviates any concerns with the near-term availability of hydrogen. The direct methanol fuel cell (DMFC) [20,21] technology has been demonstrated by Ballard [22], the Los Alamos National Laboratory [23] and others [24–26] to possess many beneficial characteristics appropriate to commercial applications.

Driven by the need to generate power in a more efficient, user- and environmentally friendly manner, the enabling technology, that is, PEMFC, will become an integral part of everyone's daily life. The time span for this great and important change will depend on many factors. Significant investment in this technology has already taken place, resulting in truly substantial progress in reliability, increased power density, improved manufacturing technologies, product demonstrations and materials cost reduction. To fully appreciate a market-driven growth for PEMFC products, further advances in raw materials and components cost reduction will be required. This responsibility lies squarely on the shoulders of three key players. Firstly, there is the innovative supplier base for proton exchange membranes, electrocatalysts and carbon-based materials. Secondly, the PEMFC developers must continue to produce even simpler stacks and operating systems, that promote cost reduction through component and function integration, combined with ease of manufacturing and the most efficient use of materials. Finally, governments also have a role to play by ensuring adequate funding of activities that enhance progress of this technology into products through actions such as targeted legislative and policy initiatives, fuel infrastructure development, and consumer tax incentives for the purchase of PEMFC products. Together, these efforts will combine to facilitate the market entry and growth of PEMFC-related products to meet humanity's insatiable thirst for energy, without further sacrifice to the health of our planet.

2. THE TECHNOLOGY DEVELOPMENT PATH TO THE PEMFC (1838–1960s)

The development of the fuel cell as an electrochemical device to produce electricity actually dates before the inventions of the four-stroke internal combustion engine by Nikolaus August Otto in 1876 [27] and the compression-ignition engine by Rudolf Diesel in 1892 [28]. In fact, we have to look back to the work of

Christian Friedrich Schoenbein and his discovery of the "fuel cell effect" in 1838 [29], and from there to the work of his friend William Robert Grove [30], the inventor of the "gas battery" or fuel cell. Grove identified what was much later to be recognized as one of the most important areas for development in PEMFCs; namely, control of the interfaces between the electrolyte, electrocatalyst and reactant gas [31]. In 1889, Ludwig Mond and Charles Langer [32] developed a three-dimensional porous electrode structure to increase the interfacial surface area. They noted that although this material facilitated reactant gas distribution to the electrocatalyst, flooding problems related to product water formation quickly negated its effectiveness. These workers also observed that electricity was effectively produced in a fuel cell through the use of hydrogen, from steam reforming of coal, and oxygen from the air. They further identified the ability to increase power output from the fuel cell by "stacking" cells and by manifolding the reactant gases into the stack. William W. Jacques suggested in 1896 [33] that the fuel cell could be used to propel trains to great speed without the undesirable emission of smoke from conventional steam-powered engines. Further, Jacques identified the potential of fuel cells to meet the energy needs of household and marine applications.

Near the turn of the century, in 1910, Emil Baur et al. [34] plotted a polarization curve for a fuel cell. They observed a decrease in cell voltage as a function of increasing current density, thereby elucidating the means by which future performance improvements would be mapped. From the early 1930s, Francis T. Bacon researched fuel cells predominantly for their potential as energy storage devices [35]. It was not until August of 2000 [36] that a National Power subsidiary, Innogy, launched an energy storage device, Regenesys™, based on Bacon's proposal. Interestingly, it was Bacon who proposed that hydrogen produced by electrolysis from "cheap" electricity, rather than by the use of reformed fuels, would prove to be the most effective means of powering fuel cells. Bacon further cautioned that the use of high temperatures for fuel cells would require a careful choice of materials to avoid performance degradation. It was also likely Bacon who ran the first fuel cell product demonstration programs in 1959, when he powered a forklift truck, and energized other devices (up to 5 kW power output) to perform various tasks around his laboratory [37]. In the same year, Dr Harry Karl Ihrig of Allis-Chalmers Mfg. demonstrated a fuel cell-powered tractor having 1008 cells, split into 112 stacks, with a total output of 15 kW [38]. This interesting and early motive application for PEMFCs is shown in Fig. 1. Ultimately, it was Bacon's significant efforts in fuel cell development that fed into a commercial business for what was to become part of United Technologies

Fig. 1. Dr Harry Karl Ihrig driving the Allis-Chalmers' fuel cell-powered tractor (Photo reproduced from W. Mitchell (Ed.), Fuel Cells, Academic Press, New York, 1963.)

(UTC). Building on Bacon's original technology, UTC was able to produce fuel cells for the Apollo Lunar Missions that served as power sources for on-board applications.

In 1959, Willard Grubb, of General Electric (GE) received a US patent [39] related to the design of a fuel cell which, for the first time, incorporated an ion-exchange resin as the electrolyte (a solid electrolyte). With this intellectual property in hand, GE began an intensive program of research and development to produce PEMFC products, predominantly for space applications [40]. This effort led to GE receiving a US$9 million subcontract [41] to design and develop PEMFCs for the Gemini Space Program. The PEMFC approach won out over solar cells and other fuel cell approaches, based on its perceived simplicity and weight advantages, combined with optimum compatibility with the Gemini Program requirements.

Initial chemistries chosen for PEMs included blends of inert polymer with highly cross-linked polystyrene-based ionomers [42], sulfonated phenol–formaldehyde [42], and heterogeneous sulfonated divinylbenzene-cross-linked polystyrene [43]. These materials were prone to chemical and, in certain cases, mechanical degradation when used as PEMs in fuel cells. Efforts to mitigate the degradation [44] included the use of antioxidants, and the addition of Teflon® to the electrode materials. The first major advance in PEMFC lifetime came with the development of PEMs based on sulfonated poly(α,β,β-trifluorostyrene) [45], a material developed by GE to enhance chemical stability. However, poor membrane mechanical properties [46] remained as an issue. A solution finally came in the mid-1960s through the joint efforts of GE and DuPont [47]. This work resulted in the development of what is still today the PEM of choice, namely, DuPont's ubiquitous Nafion® membrane.

3. PEMFC DEVELOPMENT FROM THE 1970s INTO THE NEW MILLENNIUM

Ultimately, it was the development of the perfluorosulfonic acid proton exchange membrane, Nafion®, and its demonstrated reliability [48] that brought to life the possibility that PEMFCs would find use as products in widespread terrestrial applications. Once again, GE was the company that pioneered the work on PEMFCs for use in transportation products, citing that this technology had the potential to deliver clean power with high efficiency [48]. One key technical area that GE identified for further development was fuel cell performance and the related parameter, power density, that is, kilowatts per kilogram and kilowatts per liter for the PEMFC stack. In this endeavor, the elevation of stack operating temperature, control of water management issues (especially maintenance of moisture levels within the PEM) and decrease in PEM resistance were all identified as areas that would significantly increase the performance of the PEMFC.

In the late 1970s, GE's interest in PEMFCs, and indeed their appeal in general for this technology, began to diminish significantly. At this time, DuPont's Nafion® membranes were finding use in water electrolysis and chlor-alkali market segments, with studies indicating that these businesses would provide a steeper and more stable return on investment for the membrane producer. Even for stationary applications where the fuel cell's most attractive attribute of high efficiency of operation over a dynamic electrical output range was of optimum value, there was a developing preference for phosphoric acid (PAFC), molten carbonate (MCFC) and solid oxide (SOFC) technologies over that of PEMFC [49].

Given the relatively low-power density available from the PEMFC technology of the early 1980s, most investigators concluded [49] that unless the power density could be at least doubled, there would be no market acceptance in the transportation arena. The same study further concluded that given the high catalyst loading and expensive perfluorosulfonic acid membranes, PEMFCs would find great difficulty in meeting the cost requirements that would promote commercialization. With limited market opportunities for near-term revenue generation, GE decided to sell its PEMFC technology to International Fuel Cells, a division of UTC [50].

Meanwhile, in Canada the utility companies [51] and the Canadian Government [52] were promoting the continued development of PEMFC technology. For the Canadian utility companies, the driver was the production of cheap hydrogen through electrolysis, taking advantage of low electricity prices from hydro and nuclear generating plants. For the government, it was a desire to reduce the reliance on hydrocarbon fuels for electricity generation through the implementation of a hydrogen economy, as a long-term policy decision.

It was in this more positive atmosphere that, in 1983, Ballard Technologies, later to become Ballard Power Systems, was awarded a 3-year contract through two departments of the Canadian Government [53] to develop low-cost PEMFCs. By the end of this contract, Ballard had:

- developed single cell and stack PEMFC hardware that operated efficiently on either air or pure oxygen, with either pure hydrogen or synthetic reformate fuel.
- achieved efficient PEMFC operation using synthetic reformate fuel by "cleaning" the gas mixture of carbon monoxide (a well-known contaminant for platinum catalysts) through a process of selective oxidation [53].
- confirmed the findings of Niedrach et al. [54] that carbon monoxide content, to certain concentrations, could be tolerated in reformed fuels if a platinum/ruthenium anode catalyst was employed. This was achieved by operating a fuel cell, under CO-containing hydrogen, for greater than 1000 h, wherein only 5% of the starting voltage was lost, compared to a 13% performance drop when no ruthenium was used in the anode catalyst layer [53].
- demonstrated that the baseline performance of a cell operated using CO-containing fuel, with a platinum-only catalyst, could be immediately recovered by injection of air into the anode chamber [53].

The key contributor to the next substantial advancement in PEMFC technology came through the use of a novel perfluorosulfonic acid membrane developed by the Dow Chemical [55]. The membrane was received by Ballard in 1986 and, when evaluated, it provided a four-fold increase in electrical power output as compared to the standard Nafion® membrane [56]. The Dow membrane possessed a higher sulfonic acid concentration, higher water content at a given temperature, and was substantially thinner (relative to the standard Nafion® 117 which was ~180 μm). By 1990, using a combination of developments in PEMFC technology, a substantial improvement in performance was demonstrated at a current density of 5000 A/ft^2 at 0.5 V. This performance was maintained for greater than 2000 h at 500 A/ft^2 [57].

Further effort involved reductions in the quantities of electrocatalyst used in the membrane electrode assembly (MEA). Work performed at the Los Alamos National Laboratory [58] in 1986 demonstrated that a high surface area, supported electrocatalyst in contact with a proton conductor provided opportunity for dramatic catalyst loading reduction, without loss of performance. This was followed in 1988 by the reporting of Srinivasan et al. [59] that the use of heat and pressure in the preparation of MEAs, where the temperature was above the glass transition point of the PEM, resulted in establishing a more effective component interface. These combined approaches provided comparable PEMFC performance at 0.35 mg/cm^2 compared to the more conventional catalyst loading of 4 mg/cm^2. These workers also suggested that humidification of reactant gases at 5–10°C above the cell operating temperature would be required to maintain long-term operating stability.

In 1991, a US patent was issued to Ballard [60] describing the significant benefits to PEMFC performance that could be achieved through knowledgeable design of flow field patterns on the reactant gas carrier plates of the unit cell. One of the described benefits of a "serpentine" design for the flow channels was its ability to facilitate water removal from the cathode, thereby improving performance through enhanced oxidant gas distribution to the electrocatalyst. Having successfully enhanced PEMFC performance through design, attention quickly turned to the materials as a means of further reducing the cost and enhancing the performance of the unit cell. Machined graphite/resin composites were employed as reactant gas flow plates in place of the very expensive niobium plates used in the NASA fuel cell [56].

As early as 1989, Perry Energy Systems had demonstrated the first commercial PEMFC-powered submarine using a 2 kW Mk4 Ballard® fuel cell stack [57]. International Fuel Cells [61] and Ballard [62]

Fig. 2. PEMFC prototype stacks and end products (clockwise from upper left: Manhattan Scientifics, Plug Power, H Power, Ballard Power Systems, De Nora and Toyota)

identified the great value of demonstration programs in furthering the acceptance of PEMFCs as real products capable of meeting the needs of consumers in various markets. About this time, in 1991, a survey identified that a total of six commercial entities were involved in the development of PEMFC technology [63]. Two of these companies, International Fuel Cells [64] and Siemens [65], had purchased their base technology from GE. These companies were focusing their efforts on meeting the needs of niche markets applications where the, then, relatively high cost of PEMFC products would be viable.

The late 1980s and the 1990s saw a significant increase in demonstration programs for PEMFC prototype stacks and end products, some of which are shown below in Fig. 2. Daimler-Benz was evaluating a 20-cell Mk5 hydrogen/air stack, which produced 2 kW of power, as a feasibility study for the use of PEMFCs in transportation applications [56]. A significant milestone for mass transportation demonstration programs was achieved in 1990, when Ballard entered into an agreement with the Government of British Columbia to develop the first zero-emission bus powered by PEMFCs [62]. The initial phase of this program was completed in 1992 when a Ballard® PEMFC-based engine, installed in the Phase I bus, provided the same performance as a corresponding diesel engine, but with pure water vapour as the only emission [66]. Phase II of this program (1993–1995) commissioned the development of a higher power density stack for a full-size commercial bus, which required a PEMFC to produce 200 kW of power without reducing the number of passengers that could be transported [66]. The fuel cell was designed to produce all on-board power, that is, drive train, lighting and air conditioning. The objectives of both Phases I and II were achieved within all key requirements.

The success of the first two phases led to the all-important Phase III of the program, which was to provide two small fleets of zero-emission commercial-size buses for revenue demonstration programs run by the transit authorities of Chicago and Vancouver. The Chicago Transit Authority noted that all three of its PEMFC buses performed very well and had been enthusiastically received by their customers. Beyond the zero-emission aspect of these buses, passengers noted that they were substantially lower in noise pollution [67], operating at a full 20 dB quieter than corresponding diesel buses. Working with their automotive alliance partners, Ballard developed a pre-commercial PEMFC engine for a Phase IV bus. This engine embodied significant advances over the Phase III engine in regard to both volumetric and gravimetric power density, ease of maintenance and design for volume manufacture. A comparison of the two engine designs is shown in Fig. 3.

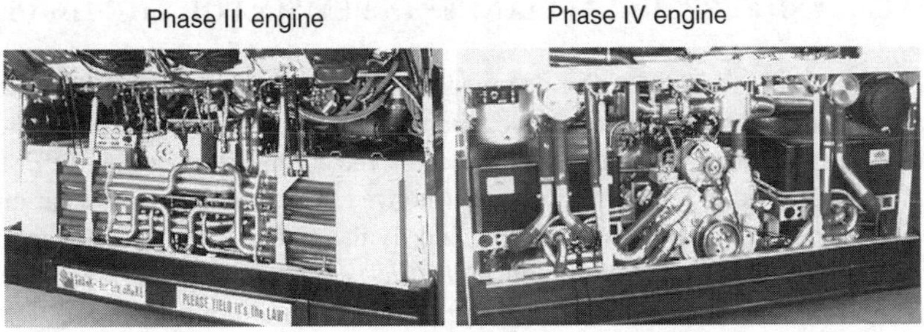

	Phase III engine	Phase IV engine
Weight:	4,850 kg	2,850 kg
Total volume:	2.7 m^3	2.2 m^3
Parts count:		
– Stacks	20	8
– Inverters	5	1
– Electric motor	11	1

Fig. 3. Comparison of Ballard's Phase III and Phase IV bus engine designs

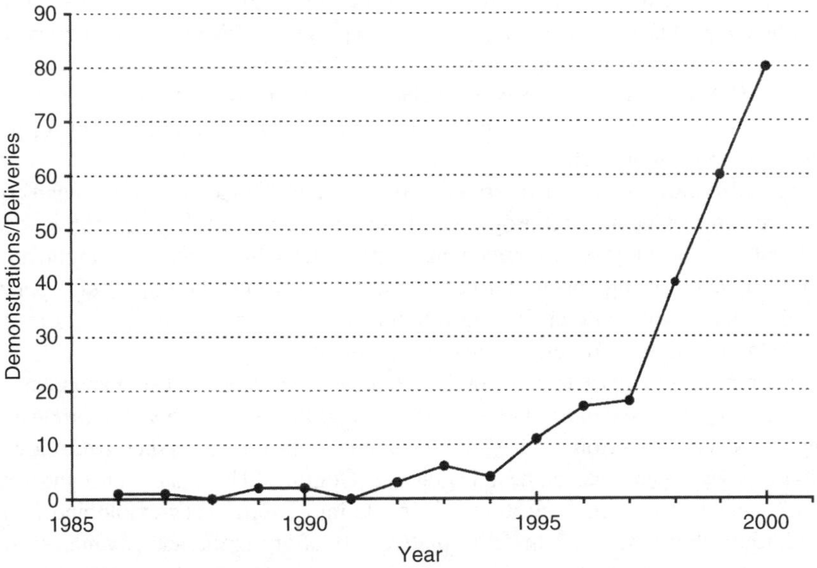

Fig. 4. Growth in publicly disclosed industrial PEMFC activity (demonstration projects and delivered prototypes)

This successful approach to demonstration programs was repeated in both automotive applications as well as in Ballard's stationary product applications [66,68]. Other PEMFC developers such as Siemens [69], De Nora [70] and Plug Power [71] have also embarked on demonstration programs as a means of strengthening consumer awareness and acceptance of PEMFC technology. A graphic visualization for this industry-wide exponential increase in the publicly disclosed product demonstrations and sales, for both PEMFC and DMFC, is shown in Fig. 4.

4. KEY BUSINESS AND SUPPLIER ALLIANCES IN PEMFCs FOR THE 21st CENTURY

Building on the success from the automotive demonstration programs of the late 1980s and 1990s, a decision was taken by Ballard in 1996 to form a close alliance with Daimler-Benz. Daimler-Benz paid C$450 million in return for its share in a venture to be focused on the commercialization of PEMFC engines for mobile applications [72]. Within 8 months, this was followed by an equally significant investment in the commercialization of PEMFC by Ford Motor [73]. Shortly thereafter, Toyota [74] announced that it was undertaking an aggressive R&D effort to produce electric vehicles that made practical use of PEMFC technology. By this time, PEMFC stacks were being developed and tested by almost every major automotive OEM in the world. Like Toyota, a number of OEMs had their own PEMFC stack development programs (e.g. Honda, Nissan and GM), but almost all of the major automotive OEMs were involved in evaluation programs that included a Ballard® fuel cell stack. This was aptly demonstrated at the opening of the California Fuel Cell Partnership [75] in November 2000, where 11 of the 14 vehicles exhibited were powered by Ballard® fuel cell stack hardware.

In the same timeframe, strategic alliances were forming to promote PEMFC products in the stationary power market. For example, Ballard Generation Systems entered in alliances, development agreements or supply agreements with six international corporations including GPU International, GEC Alstrom, EBARA, Tokyo Gas, Coleman Powermate and Matsushita Electric Works [76–81] – thereby creating global alliances for the manufacture, sale and distribution of stationary and portable power PEMFC products. This growing network of PEMFC stack and system developers was by no means restricted to Ballard and its partners. By the beginning of the new millennium, more than 400 universities, research institutes, private and public companies had entered into the race to progress PEMFC technology into a commercial reality. Among these groups, alliances and joint development programs are becoming more commonplace. For example, the MEA developer and producer Celanese Ventures formed significant alliances with both Honda and Plug Power to determine the viability of their high-temperature materials for both transportation and stationary applications [82].

Alliances and collaborations have also formed between PEMFC stack developers and component suppliers, to accelerate cost reduction and progress to high-volume manufacturing. Active areas of research have included materials development, component simplification, high-volume manufacturing processes, real-life PEMFC product testing and performance and reliability validation, all efforts directed toward decreasing total systems cost while enhancing reliability. Leveraging these advances in technology has resulted in a substantial increase in power density (see Fig. 5).

A significant contributor to this achievement has been the development, design and manufacture of new, low-cost materials. The once standard material, machined graphite plates, used to produce bipolar plates was determined to be too costly from both a material and manufacturing perspective. A low-cost graphitic material [83,84] was developed based on flexible graphite (Grafoil®). This material is light-weight, amenable to high-volume manufacturing and can be readily made into strong, yet flexible and thin bipolar plate material. In addition, thin embossed metallic plates [85] offer significant advantages over machined graphite plates, but require further development to address concerns over electrical conductivity and resistance to corrosion [86]. Sealing for gas-tight operation in a PEMFC stack is an area that has seen significant development, including a move toward injection-molded fluoropolymer seals [87] and integration of the sealing function into the design and componentry of the MEA [88].

For the gas diffusion layer (GDL) component, there are three key approaches being investigated by materials developers; namely, carbon-fibre paper [89], cloth materials [90] and non-woven materials [91]. For cost reduction and ease of high-volume manufacturing, the GDL must be processible as a rolling goods material. To contribute in the achievement of this objective, Ballard Materials Products have developed and commercialized a continuous carbon-fibre based gas diffusion material for use by all PEMFC manufacturers.

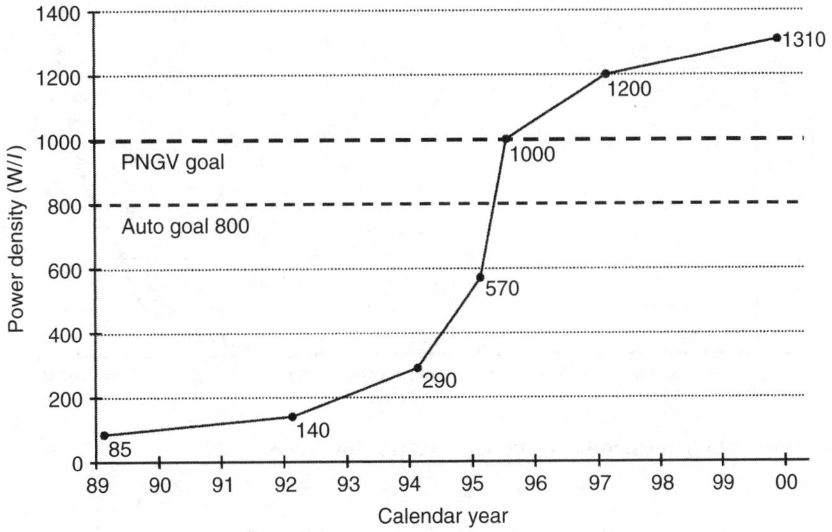

Fig. 5. Ballard's increasing stack power density (W/*l*)

Electrocatalyst development is fundamental to performance enhancement and cost reduction; all major noble metal catalyst producers, such as Johnson Matthey [92], Degussa Metals Catalyst Cerdec (dmc^2) [93], Tanaka Kikinzoku Kogyo (TKK) [94] and Engelhard [95], have some activity in this area. New alloys are being developed [96–98] to reduce cost, enhance performance and operational flexibility (cold start-up, CO tolerance, cell-reversal tolerance, etc.). There is also substantial academic research activity directed toward the development of non-noble metal catalysts for PEMFCs [99,100]. Positive results from these efforts would go a long way to achieving dramatic cost reduction, assuming no sacrifice to performance. Processing technologies for catalyst application are expanding to include decal application [101], dry spraying [102], electrodeposition [103] and chemical combustion vapour deposition (CCVD) [104].

Of all the areas for cost reduction and performance enhancement, the proton exchange membrane component stands out as an area of great importance. Today, there is essentially only one commercial membrane type, namely, the perfluorosulfonic acid PEMs sold by the DuPont, Asahi Glass and Asahi Chemical companies. There are, however, significant numbers of development activities underway in both academic and commercial entities [105]. These efforts are focused on addressing improved performance and reliability, dramatic cost reduction, ease of manufacture, optimization for use in specific applications, operation under reduced or zero external humidification, high-temperature operation and low-methanol crossover (e.g. in DMFC applications). Through Ballard Advanced Materials, significant effort and resources have been invested over the last 10 years to develop proprietary ionomers [106] and membrane processing technologies that will hasten the implementation of low-cost PEMs. This has recently culminated in an alliance with Victrex to develop new ionomer materials and pilot manufacturing capability [107].

In addition to materials development for the fuel cell stack, PEMFC systems development has seen quite revolutionary advances [108–111]. Modeling activities have increased the speed and reliability for systems development [112,113]. Recently, some exciting modeling work has been published [114,115] that should find use in facilitating the design of advanced unit cell components. Finally, developments in analytical techniques (SEM, TEM, STEM, CV, impedance, WAXS, SANS, current and voltage mapping, etc.) that probe the microstructure, function, interaction and performance of PEMFC components will continue to advance the characterization and optimization of materials.

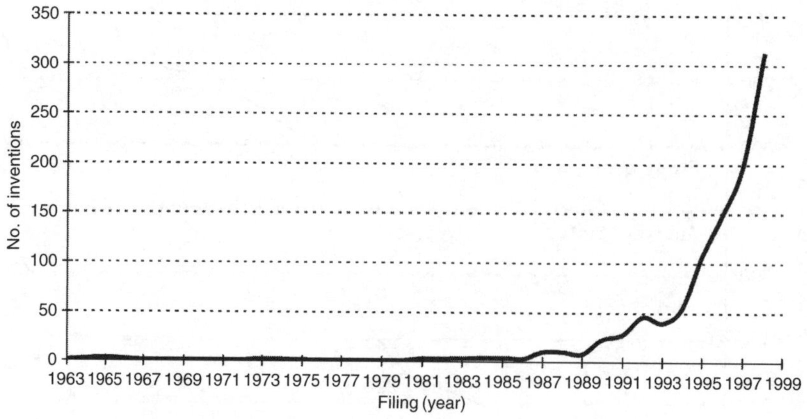

Fig. 6. Worldwide growth in the number of PEMFC-related inventions

Reflecting this enormous growth in technological advancement has been an equally impressive growth in worldwide patent activity. Figure 6 illustrates the exponential increase in patent applications for PEMFC-related inventions that has occurred in recent years.

5. WHY THE WORLD NEEDS PEMFC PRODUCTS

What is driving this vast investment of resources to bring PEMFC-based products to the marketplace? There are, as one would expect, multiple reasons. There is the high efficiency of the fuel cell reaction, the ability to place control of electricity generation into the hands of the individual consumer, concerns over conventional fuel sources [116–119], concerns over global warming [120,121], the environment [122,123], and noise pollution [124], to name but a few. The possibility of creating a product for which the most attractive fuel, hydrogen [125,126], is at once plentiful, renewable and, when used in a PEMFC product, produces no emissions, is truly compelling. PEMFC devices are lightweight, easily portable and scaleable to meet the need of a broad range of power generation, from watts to hundreds of kilowatts [122,127]. PEMFC stacks have no moving parts and, as such, are inherently more reliable and require less maintenance than conventional engines and generators. PEMFC products provide power directly at the site of use and avoid costly losses through energy distribution from a centralized power plant [128]. In addition, PEMFC products for home and office use are ideally suited to the highly energy efficient co-generation of electricity and heat [128].

Through the many effective PEMFC demonstration products put forth by both academic and industrial entities, the consumer is becoming a key player in the acceleration toward commercial PEMFC products. Our technology-driven society is finding more and more ways of using electricity to enhance and advance our everyday life experiences. While embracing these developments, the consumer is insisting that such benefits not be exploited at the expense of a healthy ecosystem. Before billions of consumers in the developing world become full partners in global energy production and consumption, we need to strike the right balance between power usage and safe, clean energy production. PEMFC products will play a leading role in achieving this delicate balance.

6. THE OUTLOOK FOR THE FUTURE

The chairman of Ford Motor, William C. Ford, Jr tells us that by the end of the first quarter of this new millennium, hydrogen-powered PEMFCs could be the predominant automotive power source [129]. Given the huge amount of resources directed toward making this technology a commercial reality, there will surely be other beneficial and profitable markets that will adopt PEMFC products. The micro-fuel cell, portable and stationary applications [130] all have the potential to drive a near-term revolutionary growth in the manufacture and sale of PEMFC products. To meet this potential, there is still significant work to be done and successes to be achieved. Further dramatic cost reduction is required, along with greater liberalization of existing power markets. Acceleration to a hydrogen economy and infrastructure would also be greatly beneficial, as would be an increased simplification in both stack and system components, design and fabrication.

We are in an era of constant and accelerating change; how we choose to manage this change will be a measure of our creativity and humanity. Technological change can only benefit the consumer if it offers solutions to unmet needs or provides an economic or ecological advantage. The products that proliferate from PEMFC technology will meet these requirements; however, the mass commercialization of these products will require further creativity, commitment and significant expenditures by all involved and interested parties. Beyond the PEMFC developers and manufacturers, governments, key component suppliers and developers, as well as the consumers themselves, must work together to ensure that PEMFC products fulfill and embody the promise of this enabling and revolutionary technology.

ACKNOWLEDGEMENTS

The authors acknowledge Dr Alfred E. Steck for the leadership and vision he has provided in over 17 years of work in PEMFCs with Ballard. The many individuals, universities, industrial entities and government bodies that have advanced fuel cell technology, and in particular PEMFC technology, to its current state of development are gratefully acknowledged. The dedication, commitment and sheer talent of the staff members of the Ballard Group of Companies and those of our Alliance partners are also recognized. Finally, the authors acknowledge and thank the Customer, who will surely be the final arbiter of success for PEMFC technology.

REFERENCES

1. T.A. Edison. US Patent 223 898 (1880).
2. T.A. Edison. *Edison Storage Battery*. US Patent 879 612 (1908).
3. T.A. Edison. *Edison Storage Battery*. US Patent 896 812 (1908).
4. T.A. Edison. US Patent 200 521 (1878).
5. T.A. Edison. US Patent 386 974 (1888).
6. H. Davy. *Nicholson's Mag.* **1** (1802) 144.
7. L. Crumbley. *Virginia Tech. Res. Mag.* (2001).
8. Case Western Reserve University press release, CWRU Researchers Develop Prototype of Miniature Fuel Cell, April 28, 2000.
9. V. Harri, P. Erni and S. Egger. In: F.N. Buchi (Ed.), *Proceedings of the Portable Fuel Cells, European Fuel Cell Forum*, Oberrohrdorf, Switzerland, 1999, p. 245.
10. Schatz Energy Research Center press release, Clean, Quiet, Cool – and Ready for the Road, April 22, 1998.
11. K. Dircks. In: D. Roller (Ed.), *Proceedings of the Clean Power Sources and Fuels, 31st ISATA Conference, Düsseldorf Trade Fair*, 1998, p. 77.
12. A.C. Lloyd. *Sci. Am.* **281(1)** (1999) 80.
13. R. Hockaday and C. Navas. *Fuel Cells Bull.* **10** (1999) 9.
14. Motorola press release, Motorola Researchers Report Progress in Miniaturizing Fuel Cell Power Source for Consumer Electronic Devices, September 26, 2000.
15. A. Heinzel, M. Zedda, A. Heitzler, T. Meyer and H. Schmidt. In: F.N. Buchi (Ed.), *Proceedings of the Portable Fuel Cells, European Fuel Cell Forum*, Oberrohrdorf, Switzerland, 1999, p. 55.

16. DCH Technology press release, DCH Technology Successfully Demonstrates Hydrogen Fuel Cell, April 8, 1999.
17. A. Jansen, S. van Leeuwen and A. Stevels. In: *Proceedings of the 2000 IEEE: International Symposium on Electronics and the Environment*. IEEE Publishers, Piscataway, NJ, 2000, p. 155.
18. Manhattan Scientifics press release, Manhattan Scientifics to Develop Fuel Cell Powered Vacuum Cleaner Prototype with Electrolux and Lunar Design, January 24, 2001.
19. S.G. Ehrenberg, J.M. Serpico, B.M. Sheikh-ali, T.N. Tagredi, E. Zador and G. Wnek. In: O. Savadogo and P.R. Roberge (Eds.), *Proceedings of the 2nd International Symposium on New Materials for Fuel Cell and Modern Battery Systems*. Ecole Polytechnique de Montreal, Montreal, Canada, 1997, p. 828.
20. A. Hamnett. *Catal. Today* **38** (1997) 445.
21. V. Antonucci. *Fuel Cells Bull.* **7** (1999) 6.
22. J. Zhang, K.M. Colbow and D.P. Wilkinson. *Ballard Power Systems*. US Patent 6 187 467, (2001).
23. X. Ren, P. Zelenay, S. Thomas, J. Davey and S. Gottesfeld. *J. Power Sources* **86** (2000) 111.
24. S.R. Narayanan, T.I. Valdez, A. Kindler, C. Witham, S. Surampudi and H. Frank. In: R.S.L. Das (Ed.), *Proceedings of the 15th Annual Battery Conference*. IEEE Publishers, Piscataway, NJ, 2000. p. 33.
25. M. Pien, S. Lis, A. Arkin, B. Taylor and R. Jalan. In: F.N. Buchi (Ed.), *Proceedings of the Portable Fuel Cells, European Fuel Cell Forum*, Oberrohrdorf, Switzerland, 1999, p. 101.
26. R. Hockaday and C. Navas. In: F.N. Buchi (Ed.), *Proceedings of the Portable Fuel Cells, European Fuel Cell Forum*, Oberrohrdorf, Switzerland, 1999, p. 45.
27. N.A. Otto. US Patent 194 047 (1877).
28. R. Diesel. German Patent 67207 (1892).
29. C.F. Schoenbein. *Philos. Mag. S.3* **14** (1839) 43.
30. W.R. Grove. *Philos. Mag. S.3* **14(86)** (1839) 127.
31. W.R. Grove. *Philos. Mag. S.3* **21(140)** (1842) 417.
32. L. Mond and C. Langer. *Proc. Roy. Soc. Lond.* **46** (1889) 296.
33. W.W. Jacques. *Harper's Mag.* **94** (1896) 144.
34. H.A. Liebhafsky and E.J. Cairns. In: *Fuel Cells and Fuel Batteries*. Wiley, New York, NY, 1968, pp. 34–42 and references cited within.
35. F.T. Bacon. *Int. J. Hydrogen Energ.* **10(7/8)** (1985) 423.
36. Innogy Technology Ventures press release, Innogy Commercialises Energy Storage on Both Sides of the Atlantic, August 21, 2000.
37. Anon. *Business Week* (September 19, 1959) 33.
38. Anon. *Business Week* (October 17, 1959) 68.
39. W.T. Grubb. *General Electric*. US Patent 2 913 511 (1959).
40. B.C. Hacker and C.C. Alexander. *On the Shoulders of Titans: A History of Project Gemini*. Scientific and Technical Information Office, NASA, Washington, DC, 1977 (Chapter 5).
41. J.M. Grimwood and B.C. Hacker. *Project Gemini: Technology and Operations; A Chronology*. Part I. Scientific and Technical Information Office, NASA, Washington, DC, 1969.
42. W.T. Grubb. *J. Electrochem. Soc.* **106(4)** (1959) 275.
43. H.A. Liebhafsky and E.J. Cairns. *Fuel Cells and Fuel Batteries*. Wiley, New York, NY, 1968 (Chapter 14).
44. B.C. Hacker and C.C. Alexander. *On the Shoulders of Titans: a History of Project Gemini*. Scientific and Technical Information Office, NASA, Washington, DC, 1977 (Chapter 8).
45. R.B. Hodgdon Jr., J.F. Enos, and E.J. Aiken. *General Electric*. US Patent 3 341 366 (1967).
46. A.J. Appleby and F.R. Foulkes. *Fuel Cell Handbook*. Van Nostrand Reinhold, New York, NY, 1989 (Chapter 10).
47. L.E. Chapman. In: *Proceedings of the 7th Intersociety Energy Conservation Engineering Conference*, American Nuclear Society, Hinsdale, IL, 1972, p. 466.
48. J.F. McElroy. In: *Proceedings of the Fuel Cells in Transportation Applications Workshop*, Los Alamos Scientific Laboratory, Los Alamos, NM, 1977, p. 53.
49. S. Srinivasan. In: T.N. Veziroglu and J.B. Taylor (Eds.), *Proceedings of the 5th World Hydrogen Energy Conference* Vol. 4. Pergamon, New York, NY, 1984, p. 1718.
50. A.J. Appleby and E.B. Yeager. *Energ. Int. J.* **11** (1986) 137.
51. W.J. Brown, P.B. Britton, L. Rucker and A.P. Scriven. In: *Hydrogen. A Challenging Opportunity*, Vol. 2. Ontario Hydrogen Energy Task Force, Toronto, Canada, 1981.

52. House of Commons Select Committee on Alternative Energy and Oil Substitution. In: *Energy Alternatives: Report of the Special Committee on Alternative Energy and Oil Substitution to the Parliament of Canada*. House of Commons. Ottawa, Canada, 1981, p. 230.
53. D. Watkins, K. Dircks, D. Epp and A. Harkness. In: *Proceedings of the 32nd International Power Sources Symposium*. IEEE Publications, Piscataway, NJ, 1986, p. 590.
54. L.W. Niedrach, D.W. McKee, J. Paynter and E.F. Danzig. *Electrochem. Technol.* **5(7/8)** (1967) 318.
55. B.R. Ezzell, W.P. Carl and W.A. Mod. *Dow Chemical*. US Patent 4 358 545 (1982).
56. K. Prater. *J. Power Sources.* **29** (1990) 239.
57. D. Watkins, K. Dircks, D. Epp and J. Blair. In: *Proceedings of the 5th Annual Battery Conference on Applications and Advances*. Electrochemical Society, Pennington, NJ, 1990.
58. I.D. Raistrick. In: R.E. White, K. Kinoshita, J.W. Van Zee and H.S. Burney (Eds.), *Proceedings of the Symposium on Diaphragms, Separators and Ion Exchange Membranes*, Vol. **86-13**, 1986, p. 172.
59. S. Srinivasan, E.A. Ticianelli, C.R. Derouin and A. Redondo. *J. Power Sources* **22** (1988) 359.
60. D.S. Watkins, K.W. Dircks and D.G. Epp. *Her Majesty the Queen as represented by the Minister of National Defence of Her Majesty's Canadian Government*. US Patent 4 988 583 (1991).
61. A.P. Meyer, J.V. Clausi and J.C. Trocciola. In: *Proceedings of the 33rd International Power Sources Conference*. Electrochemical Society, Pennington, NJ, 1988, p. 1.
62. K.B. Prater. In: *Proceedings of the 5th Canadian Hydrogen Workshop, Canadian Hydrogen Association*. Electrochemical Society, 1992, p. 216.
63. S. Gottesfeld and T.A. Zawodzinski. In: R.C. Alkire, H. Gerischer, D.M. Kolb and C.W. Tobias (Eds.), *Advances in Electrochemical Science and Engineering*, Vol. 5. Wiley-VCH, New York, NY, 1997 (Chapter 4).
64. G.A. Hards. *Platinum Metals Rev.* **35(1)** (1991) 17.
65. K. Straber. *Ber. Beusenges. Phys. Chem.* **94** (1990) 1000.
66. K.B. Prater. *J. Power Sources* **61** (1996) 105.
67. M. Jacoby. *Chem. Eng. News.* **77(4)** (1999) 31.
68. J.R. Huff and D.S. Watkins. In: *Proceedings of the 9th Annual Battery Conference on Applications and Advances*. Electrochemical Society, Pennington, NJ, 1994.
69. K.V. Schaller and C. Gruber. *Fuel Cells Bull.* **27** (2000) 9.
70. Anon. *Hydrog. Fuel Cell Lett.* **14(11)** (1999) 1.
71. Anon. *Fuel Cell News* **17(1)** (2000) 8.
72. F. Panik. *J. Power Sources* **71** (1998) 36.
73. Ballard Power Systems press release, Ford, Daimler-Benz and Ballard Complete Agreement to Develop Fuel-Cell Technology for Future Vehicles, April 7, 1998.
74. S. Kawatsu. *J. Power Sources* **71** (1998) 150.
75. A.C. Lloyd. *J. Power Sources* **86** (2000) 57.
76. S.A. Weiner. *J. Power Sources* **71** (1998) 61.
77. Ballard Power Systems press release, Ballard and GEC ALSTROM Complete C$110 Million Transaction To Commercialize Ballard Stationary Power Plants, May 29, 1998.
78. Ballard Power Systems press release, Ballard and EBARA Complete C$47.7 Million Transaction to Commercialize Ballard Stationary Power Plants, December 1, 1998.
79. Ballard Power Systems press release, Ballard, Tokyo Gas to Develop Fuel Processor for Residential Fuel Cell Generator, January 13, 2000.
80. Ballard Power Systems press release, Ballard, Coleman Powermate to Collaborate on Portable Fuel Cell Power Generators, January 16, 2000.
81. Ballard Power Systems press release, Ballard and Matsushita Electric Works Sign Fuel Cell Supply Agreement for Portable Power Generators, October 24, 2000.
82. Celanese press release, Honda and Celanese Sign Agreement to Improve Automotive Fuel Cell Systems, April 12, 2000.
83. P.R. Gibb. *Ballard Power Systems*. PCT WO/0041260, 2000.
84. R.A. Mercuri and J.J. Gough. *UCAR Carbon Technology*. US Patent 6 037 074 (2000).
85. D.P. Davies, P.L. Adcock, M. Turpin and S.J. Rowen. *J. Appl. Electrochem.* **30(1)** (2000) 101.
86. L. Ma, S. Warthesen and D.A. Shores. *J. New Mater. Electrochem. Syst.* **3** (2000) 221.

87. R.H. Barton, P.R. Gibb, J.A. Ronne and H.H. Voss. *Ballard Power Systems*. US Patent 6 057 054 (2000).
88. D.P. Wilkinson, J. Stumper, S.A. Campbell, M.T. Davis and G.J. Lamont. *Ballard Power Systems*. US Patent 5 976 726 (1999).
89. M. Inoue. *Toray Industries*. PCT WO/9962134, 1999.
90. M. DeMarinis, E.S. De Castro, R.J. Allen, K. Shaikh and De Nora. US Patent 6 103 077 (2000).
91. S.A. Campbell, J. Stumper, D.P. Wilkinson and M.T. Davis. *Ballard Power Systems*. US Patent 5 863 673 (1999).
92. Ballard Power Systems press release, Ballard Power Systems and Johnson Matthey Sign Collaboration and Supply Agreement, October 27, 1998.
93. R. Brand, A. Freund, J. Lang, T. Lehmann, J. Ohmer, T. Tacke, G. Heinz, R. Schwartz and Degussa. US Patent 5 489 563 (1996).
94. P. Stonehart. *Tanaka Kikinzoku Kogyo*. US Patent 5 593 934 (1997).
95. Plug Power press release, Plug Power Signs Agreement with Engelhard to Develop Advanced Catalysts for Fuel Cells, June 6, 2000.
96. M. Watanabe, H. Igarashi and T. Fujino. *Electrochemistry* **67(12)** (1999) 1194.
97. H. Bonnemann, R. Brinkmann, P. Britz, U. Endruschat, R. Mortel, U.A. Paulus, G.J. Feldmeyer, T.J. Schmidt, H.A. Gasteiger and R.J. Behm. *J. New Mater. Electrochem. Syst.* **3** (2000) 199.
98. M.K. Min, J. Cho, K. Cho and H. Kim. *Electrochim. Acta* **45** (2000) 4211.
99. R.W. Reeve, P.A. Christensen, A.J. Dickinson, A. Hamnett and K. Scott. *Electrochim. Acta* **45** (2000) 4237.
100. M. Pattabi, P.J. Sebastian and X. Mathew. *J. New Mater. Electrochem. Syst.* **4** (2001) 7.
101. M.S. Wilson and S. Gottesfeld. *J. Appl. Electrochem.* **22** (1992) 1.
102. E. Gulzow, M. Schulze, N. Wagner, T. Kaz, A. Schneider and R. Reissner. *Fuel Cells Bull.* **15** (1999) 8.
103. E.J. Taylor, E.B. Anderson and N.R.K. Vilambi. *J. Electrochem. Soc.* **139(5)** (1992) L45.
104. A.T. Hunt. *Microcoating Technologies*. PCT WO/0072391, 2000.
105. O. Savadogo. *J. New Mater. Electrochem. Syst.* **1** (1998) 47.
106. J. Wei, C. Stone and A.E. Steck. *Ballard Power Systems*. US Patent 5 422 411 (1995).
107. P. Charnock, D.J. Kemmish, P.A. Staniland and B. Wilson. *Victrex Manufacturing*. PCT WO/015691, 2000.
108. C. Zawodzinski, S. Moller-Holst, D.M. Webb and M.S. Wilson. In: *Proceedings of the Annual National Laboratory R&D Meeting of the DOE Fuel Cells for Transportation Program*, US DOE Office of Advanced Automotive Technologies, Washington, DC, 1999.
109. T. Moser, J.R. Rao and D. Grecksch. In: *Proceedings of the 13th International Electric Vehicle Symposium*, Japan Electric Vehicle Association, Tokyo, Japan, 1996, p. 680.
110. D.S. Watkins, K.W. Dircks, D.G. Epp, R.D. Merritt and B.N. Gorbell. *Ballard Power Systems*. US Patent 5 200 278 (1993).
111. G. Hornburg and Xcellsis. US Patent 6 190 791 (2001).
112. R. Kumar, S. Ahmed, M. Krumpelt and M. Myles. *Reformers for the Production of Hydrogen from Methanol and Alternative Fuels for Fuel Cell Powered Vehicles*, A Report Prepared for the DOE Office of Propulsion Systems Electric and Hybrid Propulsion Division. US DOE Office of Propulsion Systems Electric and Hybrid Propulsion Division, Washington, DC, 1992.
113. T.L. Reitz, S. Ahmed, M. Krumpelt, R. Kumar and H.H. Kung. *J. Mol. Catal.* **162** (2000) 275.
114. V. Gurau, F. Barbir and H. Liu. *J. Electrochem. Soc.* **147(7)** (2000) 2468.
115. T. Zhou and H. Liu. *Int. J. Transp. Phenom.* (in press).
116. J.J.J. Louis. *Fuel Cell Power for Transportation 2001*. Society of Automotive Engineers, Warrendale, PA, 2001.
117. General Motors, Argonne National Laboratory, BP amoco, ExxonMobil and Shell. Draft Final Report. *Well-to-Wheel Energy Use and Greenhouse Gas Emissions of Advanced Fuel/Vehicle Systems – North American Analysis*, Vol. 1. US DOE Office of Transportation Technologies, Washington, DC, 2001.
118. Anon. *Economist* **358(8208)** (2001) S13.
119. International Energy Agency. *World Energy Outlook 2000*. IEA Publications, Paris, France, 2000.
120. Kyoto Protocol to the United Nations Framework Convention on Climate Change, December 1997.
121. Intergovernmental Panel on Climate Change. In: J.T. Houghton (Ed.), *Report of Working Group 1 of the Intergovernmental Panel on Climate Change, Climate Change 2001: The Scientific Basis*, Summary for Policymakers. Cambridge University Press, New York, NY, 2001.

122. American Public Power Association. *Notice of Market Opportunities for Fuel Cells.* American Public Power Association, Washington, DC, 1988.
123. J. van der Veer. *Plenary speech, 16th World Petroleum Congress*, Calgary, June 13, 2000.
124. J.F. Contadini. In: *Proceedings of the IECEC 2000: 35th Intersociety Energy Conversion Engineering Conference and Exhibit*, American Institute of Aeronautics and Astronautics, Reston, VA, 2000, p. 1341.
125. L.D. Burns. *21st Century Vehicles: A Speech in Traverse City*, MI, August 10, 2000, General Motors news release, 2000.
126. J.S. Wallace and C.A. Ward. *Int. J. Hydrogen Energ.* **8(4)** (1983) 255.
127. G. Prentice. *Chemtech* **14** (1984) 684.
128. A.U. Dufour. *J. Power Sources* **71** (1998) 19.
129. W.C. Ford Jr. *Speech during the 5th Annual Greenpeace Business Conference*, London, October 5, 2000, Ford Motor news release, 2000.
130. F. Baentsch. *J. Power Sources* **86** (2000) 84.

Chapter 3

A review of catalytic issues and process conditions for renewable hydrogen and alkanes by aqueous-phase reforming of oxygenated hydrocarbons over supported metal catalysts

R.R. Davda, J.W. Shabaker, G.W. Huber, R.D. Cortright and J.A. Dumesic

Abstract

We have recently developed a single-step, low-temperature process for the catalytic production of fuels, such as hydrogen and/or alkanes, from renewable biomass-derived oxygenated hydrocarbons. This paper reviews our work in the development of this aqueous-phase reforming (APR) process to produce hydrogen or alkanes in high yields. First, the thermodynamic and kinetic considerations that form the basis of the process are discussed, after which reaction kinetics results for ethylene glycol APR over different metals and supports are presented. These studies indicate Pt-based catalysts are effective for producing hydrogen via APR. Various reaction pathways may occur, depending on the nature of the catalyst, support, feed and process conditions. The effects of these various factors on the selectivity of the process to make hydrogen versus alkanes are discussed, and it is shown how process conditions can be manipulated to control the molecular weight distribution of the product alkane stream. In addition, process improvements that lead to hydrogen containing low concentrations of CO are discussed, and a dual-reactor strategy for processing high concentrations of glucose feeds is demonstrated. Finally, various strategies are assembled in the form of a composite process that can be used to produce renewable alkanes or fuel-cell grade hydrogen with high selectivity from concentrated feedstocks of oxygenated hydrocarbons.

Keywords: Hydrogen production; Fuel cells; Reforming; Renewable energy; Supported metal catalysts; Hydrocarbon fuels; Renewable alkanes

Article Outline

1. Introduction . . . 30
2. Aqueous-phase reforming . . . 31
 2.1. Basis for aqueous-phase reforming process . . . 31
  2.1.1. Thermodynamic considerations . . . 31
  2.1.2. Kinetic considerations . . . 33
 2.2. Factors controlling selectivity for aqueous-phase reforming . . . 35
  2.2.1. Nature of the catalyst . . . 35
   2.2.1.1. Catalytic metal components . . . 35

Fuel Cells Compendium

2.2.1.2. Catalyst supports . . . 36
2.2.1.3. Modified nickel catalysts . . . 36
2.2.2. Reaction conditions . . . 38
2.2.3. Reaction pathways . . . 38
2.2.4. Nature of the feed . . . 39
2.3. Factors favoring production of heavier alkanes . . . 41
2.4. Producing hydrogen containing low levels of CO: ultra-shift . . . 42
2.5. Hydrogen from concentrated glucose feeds . . . 44
3. Discussion and overview . . . 45
4. Conclusions . . . 51
Acknowledgements . . . 51
References . . . 51

1. INTRODUCTION

Fuel cells have emerged as promising devices for meeting future global energy needs. In particular, fuel cells that consume hydrogen are environmentally clean, quiet, and highly efficient devices for electrical power generation. While hydrogen fuel cells have a low impact on the environment, current methods for producing hydrogen require high-temperature steam reforming of non-renewable hydrocarbon feedstocks [1] and [2]. The full environmental benefit of generating power from hydrogen fuel cells is achieved when hydrogen is produced from renewable sources such as solar power and biomass. Biomass and biomass in wastes are promising sources for the sustainable production of hydrogen in an age of diminishing fossil fuel reserves. However, conversion of biomass to hydrogen remains a challenge, since processes such as enzymatic decomposition of sugars, steam reforming of bio-oils, and gasification suffer from low hydrogen production rates and/or complex processing requirements [3] and [4].

The production of hydrogen for fuel cells and other industrial applications from renewable biomass-derived resources is a major challenge as global energy generation moves towards a 'hydrogen society'. Conversion of biomass in to hydrogen involves an extraction step to produce an aqueous carbohydrate feed stream from biomass, which can then be processed in a reformer to produce H_2 and CO_2. The CO_2 greenhouse gas can then be recycled back into the environment where it is consumed to grow more biomass, and the H_2 can be used for various applications (e.g., fed to a fuel cell, used in a hydrogenation process, consumed in an internal combustion engine, etc.). The development of an efficient reforming process is imperative to make the overall process feasible.

In the present review paper, we show that carbohydrates such as sugars (e.g., glucose) and polyols (e.g., methanol, ethylene glycol, glycerol and sorbitol) can be efficiently converted with water in the aqueous phase over appropriate heterogeneous catalysts at temperatures near 500 K to produce primarily H_2 and CO_2. Our aqueous-phase reforming (APR) process provides a route to generate hydrogen as a value-added chemical from aqueous-phase carbohydrates found in waste-water from biomass processing (e.g., cheese whey, beer brewery waste-water, sugar processing), from carbohydrates streams extracted from agricultural products such as corn and sugar beets, and from aqueous carbohydrates extracted by steam-aqueous fractionation of lower-valued hemicellulose from biomass [5–7]. The resulting hydrogen can be purified, if necessary, and utilized as:

1. a chemical feedstock for production of ammonia and fertilizers;
2. a chemical reagent for the future hydrogenation of carbohydrates to produce glycols;
3. a hydrogen-rich gas stream that augments the gas stream from biomass gasification units utilized for the production of liquid fuel via the Fischer–Tropsch process;
4. a future renewable fuel source for proton exchange membrane (PEM) fuel cells.

Hydrogen production using APR of carbohydrates has several advantages over existing methods of producing hydrogen via the steam reforming of hydrocarbons:

1. APR eliminates the need to vaporize both water and the oxygenated hydrocarbon, which reduces the energy requirements for producing hydrogen.
2. The oxygenated compounds of interest are nonflammable and non-toxic, allowing them to be stored and handled safely.
3. APR occurs at temperatures and pressures where the (WGS) reaction is favorable, making it possible to generate hydrogen with low amounts of CO in a single chemical reactor.
4. APR is conducted at pressures (typically 15–50 bar) where the hydrogen-rich effluent can be effectively purified using pressure-swing adsorption or membrane technologies, and the carbon dioxide can also be effectively separated for either sequestration or use as a chemical.
5. APR occurs at low temperatures that minimize undesirable decomposition reactions typically encountered when carbohydrates are heated to elevated temperatures.
6. Production of H_2 and CO_2 from carbohydrates may be accomplished in a single-step, low-temperature process, in contrast to the multi-reactor steam reforming system required for producing hydrogen from hydrocarbons.

Important selectivity challenges govern the production of H_2 by APR, because the mixture of H_2 and CO_2 formed in this process is thermodynamically unstable at low temperatures with respect to the formation of methane. Accordingly, the selective formation of H_2 represents a classic problem in heterogeneous catalysis and reaction engineering: the identification of catalysts and the design of reactors to maximize the yields of desired products at the expense of undesired byproducts formed in series and/or parallel reaction pathways. We show how the hydrogen selectivity can be controlled by altering the nature of catalytically active metal and metal–alloy components, and by choice of catalyst support. We also show how the reforming process can be operated at moderately high feed concentrations of oxygenated hydrocarbons (e.g., 10 wt.%), and we indicate how the APR processes can be conducted to achieve low levels of CO in the gaseous effluent (e.g., lower than 100 ppm). We also demonstrate how the APR process can be designed to favor the formation of heavier alkanes (e.g., hexane) from biomass-derived oxygenated compounds (e.g., glucose). This variation of the APR process provides a means to produce clean alkane streams from renewable resources, thereby providing an application for APR in the interim period until hydrogen fuel cells become more economical.

2. AQUEOUS-PHASE REFORMING

2.1. Basis for aqueous-phase reforming process

2.1.1. Thermodynamic considerations

Reaction conditions for producing hydrogen from hydrocarbons are dictated by the thermodynamics for the steam reforming of the alkanes to form CO and H_2 (reaction (1)) and the WGS reaction to form CO_2 and H_2 from CO (reaction (2)):

$$C_nH_{2n+2} + nH_2O \rightleftharpoons nCO + (2n+1)H_2 \tag{1}$$

$$CO + H_2O \rightleftharpoons CO_2 + H_2 \tag{2}$$

Figure 1 shows the changes in the standard Gibbs free energy ($\Delta G°/RT$) associated with reaction (1) for a series of alkanes (CH_4, C_2H_6, C_3H_8, C_6H_{14}), normalized per mole of CO produced. It can be seen that the

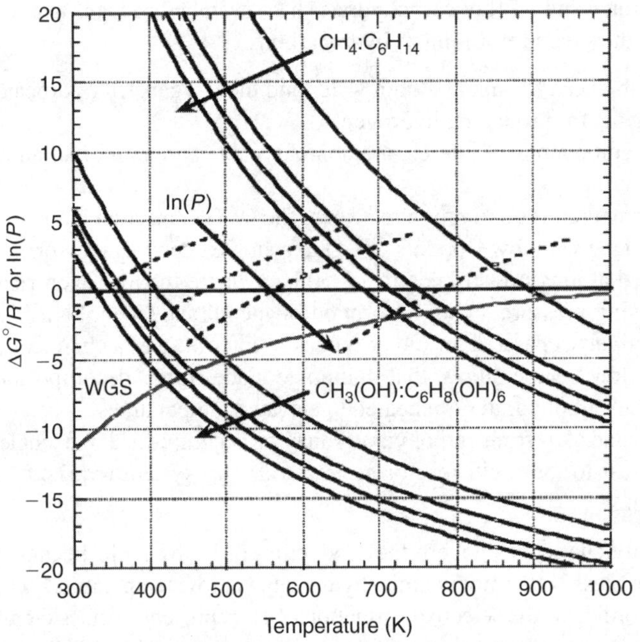

Fig. 1. $\Delta G°/RT$ vs. temperature for production of CO and H_2 from vapor-phase reforming of CH_4, C_2H_6, C_3H_8 and C_6H_{14}; $CH_3(OH)$, $C_2H_4(OH)_2$, $C_3H_5(OH)_3$ and $C_6H_8(OH)_6$; and water–gas shift. Dotted lines show values of ln(*P*) for the vapor pressures vs. temperature of $CH_3(OH)$, $C_2H_4(OH)_2$, $C_3H_5(OH)_3$, and $C_6H_8(OH)_6$ (pressure in units of atm)

steam reforming of alkanes is thermodynamically favorable (i.e., negative values of $\Delta G°/RT$) only at temperatures higher than 675 K (and higher than 900 K for methane reforming). Carbohydrates are oxygenated hydrocarbons having a C:O ratio of 1:1, and these compounds produce CO and H_2 according to reaction (3):

$$C_nH_{2y}O_n \rightleftharpoons nCO + yH_2 \tag{3}$$

Relevant oxygenated hydrocarbons having a C:O ratio of 1:1 are methanol (CH_3OH), ethylene glycol ($C_2H_4(OH)_2$), glycerol ($C_3H_5(OH)_3$), and sorbitol ($C_6H_8(OH)_6$). Importantly, sorbitol is produced via the hydrogenation of glucose ($C_6H_6(OH)_6$). Figure 1 shows that steam reforming of these oxygenated hydrocarbons to produce CO and H_2 is thermodynamically favorable at significantly lower temperatures than those required for alkanes with similar number of carbon atoms. Accordingly, the steam reforming of oxygenated hydrocarbons having a C:O ratio of 1:1 would offer a low-temperature route for the formation of CO and H_2. Figure 1 also shows that the value of $\Delta G°/RT$ for WGS of CO to CO_2 and H_2 is more favorable at lower temperatures. Therefore, it might be possible to produce H_2 and CO_2 from steam reforming of oxygenated compounds utilizing a single-step catalytic process, since the WGS reaction is favorable at the same low temperatures at which steam reforming of carbohydrates is possible.

The steam reforming of hydrocarbons typically takes place in the vapor phase. However, vapor-phase steam reforming of oxygenated hydrocarbons at low temperatures may become limited by the vapor pressures of these reactants. Figure 1 shows the plots of the logarithm of the vapor pressures (in atm) of CH_3OH, $C_2H_4(OH)_2$, $C_3H_5(OH)_3$, and $C_6H_8(OH)_6$ versus temperature. It is apparent that the vapor-phase reforming of methanol, ethylene glycol, and glycerol can be carried out at temperatures near 550 K, since the values of $\Delta G°/RT$ are favorable and the vapor pressures of these oxygenated reactants are higher than 1 atm at this

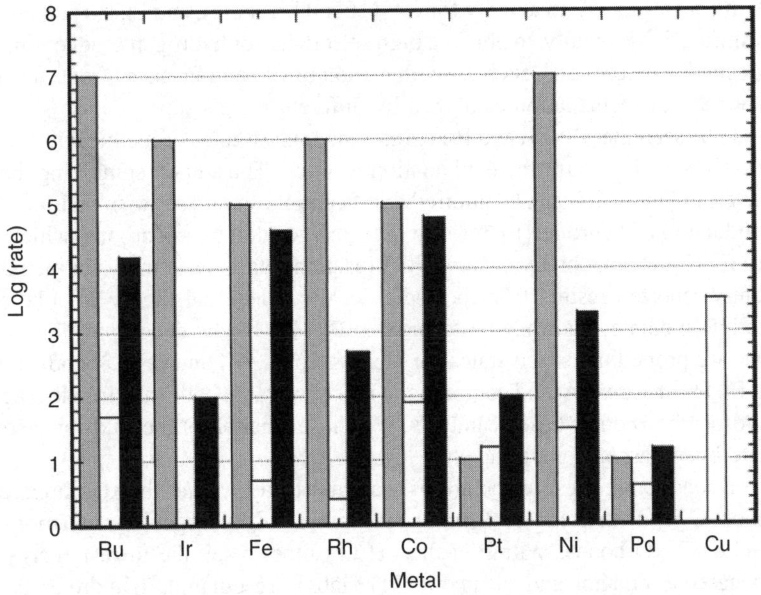

Fig. 2. Relative rates of C—C bond breaking reaction by Sinfelt (grey), WGS reaction by Grenoble, et al. (white), methanation reaction by Vannice (black). The rate of a particular reaction can be compared for the different metals; however, for a specific metal, the absolute rates of the three different reactions cannot be compared relative to each other. Adapted from Ref. [8]

temperature. In contrast, vapor-phase reforming of sorbitol must be carried out at temperatures near 750 K. Importantly, reforming of oxygenated hydrocarbons, if carried out in the liquid phase, would make it possible to produce H_2 from carbohydrate-derived feedstocks (e.g., sorbitol and glucose) that have limited volatility, thereby taking advantage of single-reactor processing at lower temperatures.

2.1.2. Kinetic considerations

Important reaction selectivity issues must be addressed if APR reactions are to be used for the production of H_2 from renewable biomass resources, since the H_2 and CO_2 produced at low temperatures are thermodynamically unstable with respect to alkanes and water. Alkanes (especially CH_4) can be formed from the subsequent reaction of H_2 and CO/CO_2 via methanation and Fischer–Tropsch reactions [12–15]. For example, the equilibrium constant at 500 K for the conversion of CO_2 and H_2 to methane ($CO_2 + 4H_2 \leftrightarrow CH_4 + 2H_2O$) is of the order of 10^{10} per mole of CO_2. Accordingly, selective hydrogen production via APR of oxygenated hydrocarbons would require an efficient catalyst that promotes reforming reactions (C—C scission followed by water-gas shift) and inhibits alkane-formation reactions (C—O scission followed by hydrogenation).

The catalytic activities of different metals for C—C bond breaking during ethane hydrogenolysis have been studied by Sinfelt and Yates [16], and the relative rates for the different metals are shown in Figure 2. It can be seen that Pt shows reasonable catalytic activity for C—C bond cleavage, although not as high as metals such as Ru, Ni, Ir and Rh. However, an effective catalyst for reforming of ethylene glycol must not only be active for cleavage of the C—C bond, but it must also be active for the WGS reaction to remove CO from the metal surface at the low temperatures of the reforming reaction. In this respect, Grenoble et al. [17] have reported the relative catalytic activities for WGS over different metals supported on alumina, and these data are also shown in Figure 2. It can be seen that Cu exhibits the highest water-gas shift rates among all the metals (although this metal shows no activity for C—C bond breaking), and Pt, Ru and Ni also show appreciable water-gas shift activity. Finally, to obtain a high selectivity for hydrogen production, the catalyst must not facilitate undesired side reactions, such as methanation of CO and Fischer–Tropsch synthesis. Figure 2 shows the relative rates of methanation catalyzed by different metals supported on silica, as reported by Vannice [15]. It can be seen that Ru, Ni and Rh exhibit the highest rates of methanation, whereas Pt, Ir and Pd show lower catalytic activities for the methanation reaction. Thus, upon comparing the catalytic activities of various metals in Figure 2, it can be inferred that Pt and Pd should show suitable catalytic activity and selectivity for production of hydrogen by reforming of oxygenated hydrocarbons, which requires reasonably high activity for C—C bond breaking and WGS reactions, and low activity for methanation.

We have recently reported results from periodic density functional theory calculations to probe the nature of surface intermediates that may be formed on Pt(111) by the decomposition of ethanol [18]. In these calculations, we probed transition states for cleavage of C—C and C—O bonds in these intermediates adsorbed on Pt(111) to identify and compare the most favorable pathways for cleavage of these bonds in oxygenated hydrocarbons on Pt-based catalysts. We chose ethanol for the study since this molecule is a simple oxygenated hydrocarbon containing a C—C bond.

The strategy for conducting these calculations was first to determine the stabilities of all 24 species (with stoichiometry C_2H_xO) that can be formed by removal of hydrogen atoms from ethanol, without cleavage of C—C or C—O bonds. Within each C_2H_xO isomeric set, the lowest-energy surface species (with respect to gaseous ethanol and clean Pt(111) slabs) are ethanol, 1-hydroxyethyl (CH_3CHOH), 1-hydroxyethylidene (CH_3COH), acetyl (CH_3CO), ketene (CH_2CO), ketenyl (CHCO), and CCO species. The next step in these calculations was to determine the stabilities of all possible reaction products resulting from C—C or C—O bond cleavage (i.e., stabilities of adsorbed O, OH, C_2H_x, and CH_xO species). From the results of these calculations it was then possible to determine the energy changes for all possible C—C and C—O cleavage reactions. We then selected those C—C and C—O cleavage reactions with the most favorable energy changes and identified their transition states. In general, the computational time to identify stable adsorbed species is shorter than to identify transition states. Thus, we started with rigorous DFT calculations for transition states corresponding to those reactions with the most favorable energy changes, since these reactions are expected to lead to transition states of low energy. Accordingly, we investigated with rigorous DFT calculations 14 transitions states from among the 48 transition states that are possible by cleavage of C—C and C—O bonds in the 24 species that can be formed by dehydrogenation of ethanol. The 1-hydroxyethylidene (CH_3COH) species has the lowest-energy transition state for C—O bond cleavage, and the ketenyl (CHCO) species has the lowest-energy transition state for C—C bond cleavage. Based on the results from DFT calculations for reactions with favorable energy changes, we generated a linear Brønted – Evans – Polanyi correlation for the energies of the transition states for C—C and C—O bond cleavage in terms of the energies of the adsorbed products for these reactions. We then used this correlation to estimate the energies of the transition states for the 34 remaining C—C and C—O bond cleavage reactions having less favorable energy changes, from which it was concluded that the energies of these remaining transition states were all significantly higher than the values of the lowest-energy transition energies that we had identified from our rigorous DFT calculations.

Figure 3 shows a simplified potential energy diagram of the stabilities and reactivities of dehydrogenated species derived from ethanol on Pt(111). Only the most stable species within each isomeric set and the most stable transition states for C—O and C—C bond cleavage are consolidated in this schematic potential energy diagram. Views of adsorbed species and transition states are shown in the insets. It can be seen from Figure 3 that C—O bond cleavage occurs on more highly hydrogenated species compared to C—C bond cleavage. Importantly, it can be seen that cleavage of the C—C bond in species derived from ethanol should be faster than cleavage of the C—O bond on Pt(111), since the energies with respect to ethanol of the lowest transition states for these reactions are equal to 4 and 42 kJ/mol, respectively. Furthermore, it appears that cleavage of the C—C bond in species derived from ethanol should be faster than cleavage of

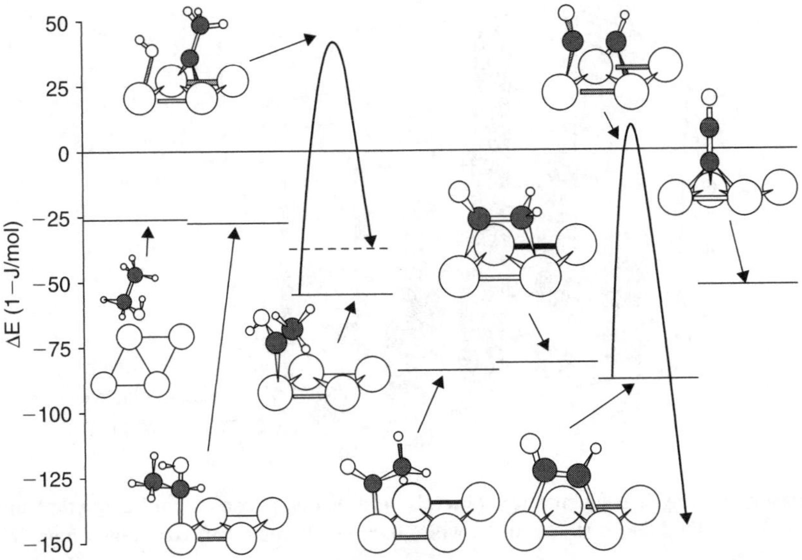

Fig. 3. Reaction energy diagram for ethanol reactions on Pt(111). The reference state is gas-phase ethanol and clean slab(s). Removed H atoms and bond cleavage products are each adsorbed on separate slabs. Solid curves represent bond cleavage reactions. Insets show views of stable and transition state species. The large white circles represent Pt atoms, grey medium circles represent C atoms, white medium circles represent O atoms, and the small white circles represent H atoms. Adapted from Ref. [18]

the C—C bond in ethane on Pt(111). In particular, the electronic energy associated with the formation of this transition state (and adsorbed H-atoms) from ethane is equal to 125 kJ/mol (ref), and this value is significantly higher than the energy of 4 kJ/mol for the transition state controlling C—C cleavage in ethanol.

2.2. Factors controlling selectivity for aqueous-phase reforming

2.2.1. Nature of the catalyst

2.2.1.1. Catalytic metal components

It is possible to produce hydrogen by steam reforming of methanol over copper-based catalysts at temperatures near 550 K [2] and [19]. However, copper-based catalysts are not effective for steam reforming of heavier oxygenated hydrocarbons, since these catalysts show low activity for cleavage of C—C bonds [16]. Therefore, it is more likely that effective catalysts for reforming of oxygenated hydrocarbons would be based on Group VIII metals, since these metals generally show higher activities for breaking C—C bonds [16] and [20]. Aqueous-phase reforming of ethylene glycol over silica-supported Group VIII metal catalysts was conducted in a fixed bed reactor [8, 21] Figure 4 summarizes the results of these studies at 483 K, and results at 498 K show similar trends. The rate of ethylene glycol reforming (as measured by the rate of CO_2 production) decreases in the following order:

$$Pt \sim Ni > Ru > Rh \sim Pd > Ir$$

The catalysts are also compared based on their selectivity for hydrogen production, which is defined [22] as the number of moles of H_2 in the effluent gas normalized by the number of moles of H_2 that would be present if each mole of carbon in the effluent gas had participated in the ethylene glycol reforming reaction to give 5/2 mol of H_2. In addition, the alkane selectivity is defined as the moles of carbon in the

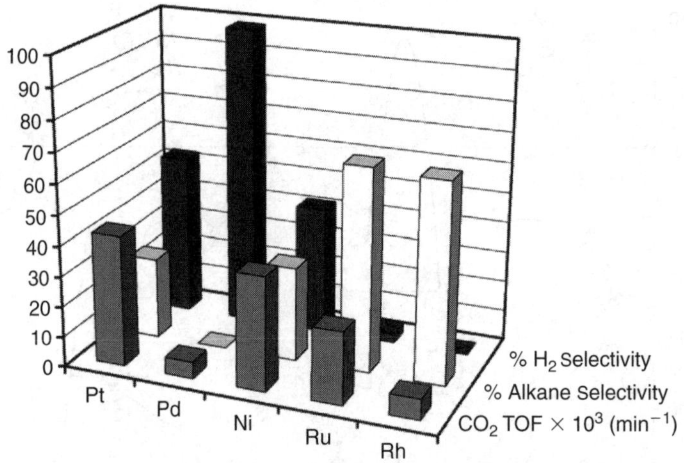

Fig. 4. Comparison of catalytic performance of metals for aqueous-phase reforming of ethylene glycol at 483 K and 22 bar (grey bar: CO_2 TOF $\times$ 10^3 (min^{-1}); white bar: % alkane selectivity; black bar: % H_2 selectivity). Adapted from Ref. [8]

gaseous alkane products normalized by the total moles of carbon in the gaseous effluent stream. Silica supported Rh, Ru and Ni show low selectivity for production of H_2 and high selectivity for alkane production. Unfortunately, Ni/SiO_2 deactivated under reaction conditions at 498 K.

While Pt, Ni and Ru exhibit relatively high activities for the reforming reaction, only Pt and Pd also show relatively high selectivity for the production of H_2. These trends suggest that active catalysts for APR reactions should possess high catalytic activity for the WGS reaction and sufficiently high catalytic activity for cleavage of C—C bonds. Moreover, Pt and Pd catalysts exhibit low activity for the C—O scission reactions and the series methanation and Fischer–Tropsch reactions between the reforming products, CO/CO_2 and H_2. On this basis, Pt-based catalysts were identified as promising systems for further study. Due to their low cost and good catalytic activity, Ni-based catalysts are also attractive despite their tendency to produce alkanes.

2.2.1.2. Catalyst supports

Various supported platinum catalysts were prepared to test the effect of the support on the activity and selectivity for production of hydrogen by aqueous reforming of ethylene glycol [10]. As shown in Figure 5A, turnover frequencies for production of hydrogen are the highest over Pt-black and Pt supported on TiO_2, carbon, and Al_2O_3 (i.e., 8–15 min^{-1} at 498 K for 10 wt.% ethylene glycol), while moderate catalytic activity for production of hydrogen is demonstrated by Pt supported on SiO_2—Al_2O_3 and ZrO_2 (near 5 min^{-1}). Lower turnover frequencies are exhibited by Pt supported on CeO_2, ZnO, and SiO_2 (less than about 2 min^{-1}), which may be due to deactivation caused by hydrothermal degradation of these support materials. As shown in Figure 5B, catalysts consisting of Pt supported on carbon, TiO_2, SiO_2—Al_2O_3 and Pt-black also lead to the production (about 1–3 min^{-1}) of gaseous alkanes and liquid-phase compounds that would lead to alkanes at higher conversions (e.g., ethanol, acetic acid, acetaldehyde). Thus, we conclude that Pt/Al_2O_3, and to a lesser extent Pt/ZrO_2 and Pt/TiO_2, are active as well as selective catalysts for the production of H_2 from liquid-phase reforming of ethylene glycol. By testing a sintered version of our highly dispersed Pt/Al_2O_3 catalyst (with a dispersion of only 31%), we conclude that effect of support on reforming activity and selectivity is greater than the effect of metal dispersion.

2.2.1.3. Modified nickel catalysts

A Sn-promoted Raney-Ni catalyst can be used to achieve good activity, selectivity, and stability for production of hydrogen by APR of biomass-derived oxygenated hydrocarbons [11] and [23]. This

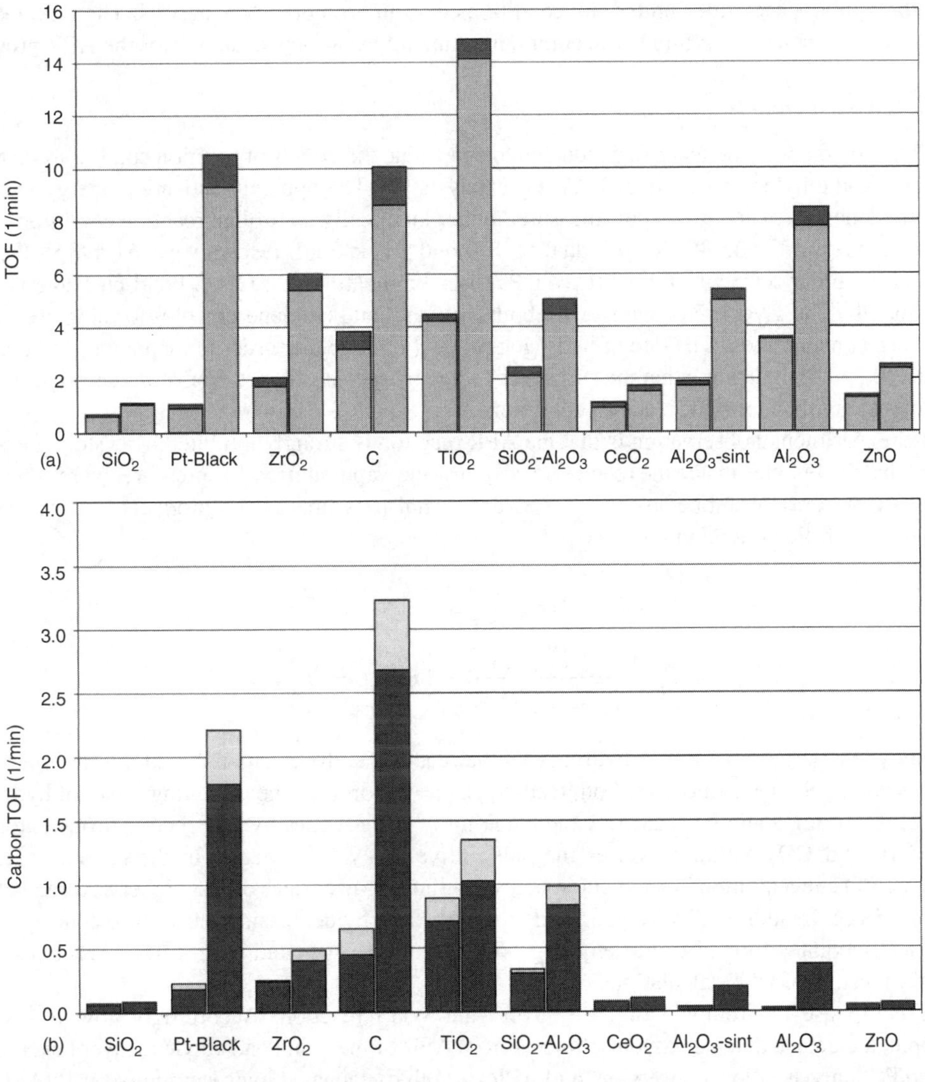

Fig. 5. Production rates for reforming of 10 wt.% aqueous ethylene glycol at 483 (left columns) and 498 K (right columns) over supported Pt catalysts. (a) H_2 production rate (gray) and the amount of hydrogen that could be generated by total reforming of the methanol byproduct (black). (b) Alkane (methane and ethane) production rate (gray) and production of alkane precursors (acetaldehyde, ethanol, and acetic acid) (black). Al_2O_3-sint indicates a Pt/Al_2O_3 catalyst that has been subjected to treatment at elevated temperature in H_2. Adapted from Ref. [10]

inexpensive material has catalytic properties that are comparable to those of Pt/Al_2O_3 for production of hydrogen from small oxygenate hydrocarbons, such as ethylene glycol, glycerol, and sorbitol. Rates of hydrogen production by APR of ethylene glycol over Raney-Ni—Sn catalysts with Ni—Sn atomic ratios of up to 14:1 are comparable to 3 wt.% Pt/Al_2O_3, based on reactor volume. The addition of Sn to Raney-Ni catalysts significantly decreases the rate of methane formation from series recombination of CO or CO_2 with H_2, while maintaining high rates of C—C cleavage necessary for production of H_2. However, it is necessary to operate the reactor near the bubble-point pressure of the feed and at moderate space–times to achieve high selectivities for production of H_2 over Raney-Ni—Sn catalysts, whereas it is impossible to achieve these high selectivities under any conditions over unpromoted Ni catalysts. These Ni—Sn catalysts illustrate the potential of bimetallic compounds and alloys as new catalysts for the APR process.

2.2.2. *Reaction conditions*

Reaction kinetic measurements were conducted to determine the effects of reaction conditions on the APR of methanol and ethylene glycol over Pt/Al_2O_3 catalysts [9]. The apparent activation energy barriers for APR of ethylene glycol and methanol (measured under kinetically controlled reaction conditions) at temperatures between 483 and 498 K are equal to 100 and 140 kJ/mol, respectively. At 498 K, these oxygenates have similar reactivity for APR over Pt/Al_2O_3, indicating that C—C bond cleavage is not rate limiting for ethylene glycol reforming. Also, both methanol and ethylene glycol reforming are fractional order in feed concentration. The rate of hydrogen production is higher order in methanol (0.8) than ethylene glycol (0.3–0.5), indicating that the surface coverage by species derived from ethylene glycol is higher than from methanol under APR reaction conditions.

Another experimental observation is that the APR reaction is strongly inhibited by system pressure. By assuming that the bubbles inside the reactor consist of water vapor at its vapor pressure, increasing the system pressure at constant temperature increases the partial pressures of the products (i.e., H_2 and CO_2) according to the following relations:

$$P_{\text{system}} \approx P_{\text{bubble}} = \sum_{\text{Feed}} P_i + \sum_{\text{products}} P_j,$$

$$P_j = \frac{P_{j,\text{diluted}}}{\sum_{\text{Products}} P_{j,\text{diluted}}} \left(P_{\text{bubble}} - \sum_{\text{Feed}} P_i \right)$$

Accordingly, the partial pressure of hydrogen in the reactor can be calculated, and a weak inhibition by hydrogen can thereby be deduced for both feed-stocks (−0.5 order). The inhibiting effect of hydrogen on the rate of reforming could be caused by the blocking of surface sites by adsorbed hydrogen atoms. The increased H_2 and CO_2 partial pressures may also drive the WGS reaction in the reverse direction to increase the CO concentration in the reactor, hence leading to lower rates due to higher coverage of CO on the metal surface. In addition, hydrogen could inhibit the rate by decreasing the surface concentrations of reactive intermediates formed from dehydrogenation of the oxygenated hydrocarbon reactants, as suggested from results of DFT calculations.

The liquid-phase environment of APR favors the WGS reaction. Accordingly, low levels of CO (<300 ppm) are detected in the gaseous effluents from APR of methanol and ethylene glycol over alumina-supported Pt catalysts at low conversion, and APR of both oxygenated hydrocarbons over Pt/Al_2O_3 leads to nearly 100% selectivity for the formation of H_2 (compared to the formation of alkanes). Since the selectivity for hydrogen production from methanol and ethylene glycol is essentially independent of conversion, it appears that the series hydrogenation of CO/CO_2 to alkanes is not significant over Pt/Al_2O_3.

2.2.3. *Reaction pathways*

Figure 6 shows a schematic representation of reaction pathways that we suggest are involved in the formation of H_2 and alkanes from an oxygenated hydrocarbon (e.g., ethylene glycol) over a metal catalyst. Ethylene glycol first undergoes reversible dehydrogenation steps to give adsorbed intermediates, prior to cleavage of C—C or C—O bonds. The adsorbed species can be formed on the metal surface either by formation of metal–carbon bonds and/or metal–oxygen bonds. On a metal catalyst such as platinum, the adsorbed species bonded to the surface by the formation of Pt—C bonds is more stable than the species involving Pt—O bonds [18]. However, Pt—O bonds may also be formed, since activation energy barriers for cleavage of O—H and C—H bonds are similar on Pt. Subsequent to formation of adsorbed species on

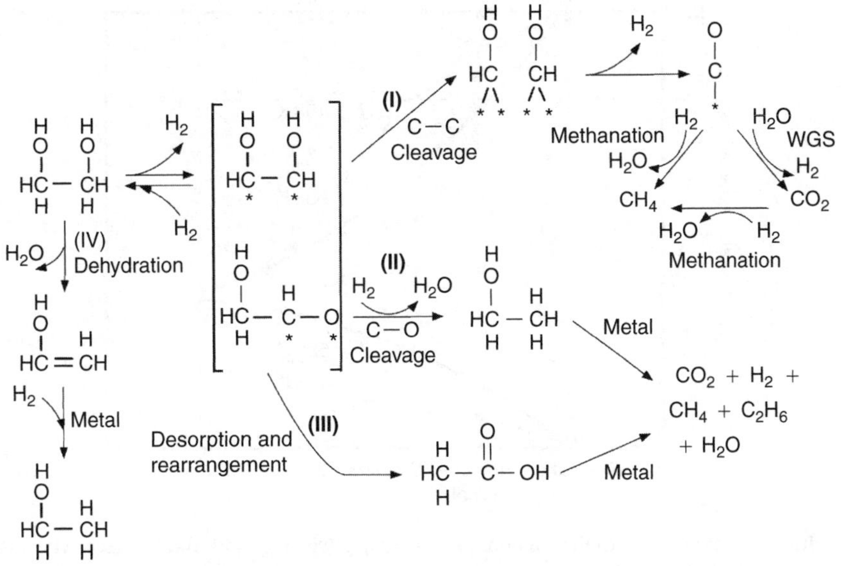

Fig. 6. Reaction pathways and selectivity challenges for production of H_2 from reactions of ethylene glycol with water (* represents a surface metal site). Adapted from Ref. [8]

the metal surface, three reaction pathways can occur, indicated as (I), (II) and (III) in Figure 6. Pathway I involves cleavage of the C—C bond leading to the formation of CO and H_2, followed by reaction of CO with water to form CO_2 and H_2 by the water-gas shift (WGS) reaction [17] and [24]. Further reaction of CO and/or CO_2 with H_2 (e.g., on metals such as Ni, Rh and Ru) leads to alkanes and water by methanation and Fischer–Tropsch reactions [12], [13], [14] and [15], and this degradation in the production of H_2 represents a series-selectivity challenge. Pathway II leads to the formation of an alcohol on the metal catalyst by cleavage of the C—O bond, followed by hydrogenation. The alcohol can undergo further reaction on the metal surface (adsorption, C—C cleavage, C—O cleavage) to form alkanes (CH_4, C_2H_6), CO_2, H_2 and H_2O. This degradation in the production of H_2 (by alkane formation) represents a parallel-selectivity challenge. Pathway III involves desorption of species from the metal surface followed by rearrangement (which may occur on the catalyst support and/or in the aqueous phase) to form an acid, which can then undergo surface reactions (adsorption, C—C cleavage, C—O cleavage) to form alkanes (CH_4, C_2H_6), CO_2, H_2 and H_2O. This pathway represents an additional parallel-selectivity challenge.

The catalyst support can affect the selectivity for H_2 production by catalyzing parallel dehydration pathways [25] and [26] that lead to the formation of alkanes. For example, the selectivity observed for the production of H_2 by APR of ethylene glycol over silica-supported Pt is significantly lower than we have observed for APR over alumina-supported Pt [8] and [22]. Thus, it appears that the higher acidity of silica compared to alumina, as reflected by the lower isoelectric point of silica [27], can facilitate acid-catalyzed dehydration reactions of ethylene glycol, represented by pathway IV in Figure 6, followed by hydrogenation on the metal surface to form an alcohol. The alcohol can subsequently undergo surface reactions, as in pathway II, to form alkanes (CH_4, C_2H_6), CO_2, H_2 and H_2O. This bi-functional dehydration/hydrogenation pathway consumes H_2, leading to decreased hydrogen selectivity and increased alkane selectivity.

2.2.4. Nature of the feed

Experimental results for the APR of reforming of glucose, sorbitol, glycerol, ethylene glycol and methanol are illustrated in Figure 7 [22]. Reactions were carried out over a Pt/Al_2O_3 catalyst at 498 and 538 K. Figure 7

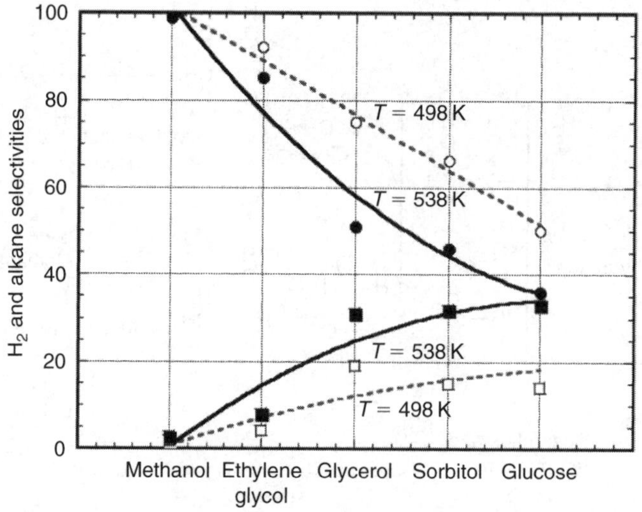

Fig. 7. Selectivities vs. oxygenated hydrocarbon. H_2 selectivity (circles) and alkane selectivity (squares) from aqueous-phase reforming of 1 wt.% oxygenated hydrocarbons over 3 wt.% Pt/Al_2O_3 at 498 K (open symbols and dashed curves) and 538 K (filled symbols and solid curves). Adapted from Ref. [22]

indicates that the selectivity for H_2 production improves in the order glucose < sorbitol < glycerol < ethylene glycol < methanol. The Figure also shows that lower operating temperatures result in higher H_2 selectivities, although this trend is in part due to the lower conversions achieved at lower temperatures. The selectivity for alkane production follows the opposite trend to that exhibited by the H_2 selectivity. The highest hydrogen yields are obtained when using sorbitol, glycerol and ethylene glycol as feed molecules for APR. Although these molecules can be derived from renewable feedstocks [28–31], the reforming of less reduced and more immediately available compounds such as glucose would be highly desirable. Unfortunately, as seen in Figure 7, the hydrogen yield from APR of glucose is lower than for these other more reduced compounds. Furthermore, while the high hydrogen yields for reforming of the polyols are insensitive to the concentration of the aqueous feed (e.g., from 1 to 10 wt.%), the hydrogen yield for reforming of glucose decreases further as the feed concentration increases to about 10 wt.%. The lower H_2 selectivities for the APR of glucose, compared to that achieved using the other oxygenated hydrocarbon reactants, are at least partially due to homogeneous reactions of glucose in the aqueous-phase at the temperatures employed in APR [32–35].

Reforming reactions are fractional order in the feed concentration [9], whereas glucose decomposition studies have shown first-order dependence on the feed concentration [34]. Thus, the rate of homogeneous glucose decomposition relative to the reforming rate increases with an increase in the glucose concentration from 1 to 10 wt.%. Accordingly, to collect the data presented in Figure 7 [22], low concentrations (i.e., 1 wt.%) of all feed molecules were employed to minimize effects from homogeneous decomposition reactions and thereby compare the selectivities under conditions where the conversion of reactant is controlled by the Pt/Al_2O_3 catalyst. This low feed concentration corresponds to a molar ratio H_2O/C of 165. Processing such dilute solutions is economically not practical, even though reasonably high hydrogen yields are achieved. However, the undesirable homogeneous reactions observed with glucose pose less of a problem when using sorbitol, glycerol, ethylene glycol and methanol, which makes it possible to generate high yields of hydrogen by the APR of more concentrated solutions containing these compounds (e.g., aqueous solutions containing 10 wt.% of these oxygenated hydrocarbon reactants, corresponding to molar ratios H_2O/C equal to 5).

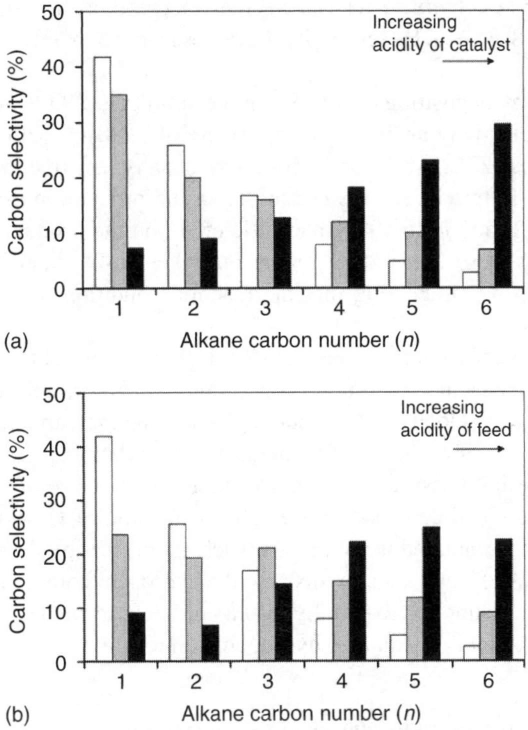

Fig. 8. Alkanes carbon selectivities for APR of 5 wt.% sorbitol at 538 K and 57.6 bar vs. (a) addition of solid acid (SiO_2–Al_2O_3) to 3 wt.% Pt/Al_2O_3 [Pt/Al_2O_3 (white), mixture 2: Pt/Al_2O_3 (3.30 g) and SiO_2–Al_2O_3 (0.83 g) (grey), and mixture 1: Pt/Al_2O_3 (1.45 g) and SiO_2–Al_2O_3 (1.11 g) (black)] and (b) addition of mineral acid (HCl) in feed over 3 wt.% Pt/Al_2O_3 [pH_{feed} = 7 (white), pH_{feed} = 3 (grey), and pH_{feed} = 2 (black)]. Adapted from Ref. [36]

2.3. Factors favoring production of heavier alkanes

Aqueous-phase reforming of sorbitol can be tailored to selectively produce a clean stream of heavier alkanes consisting primarily of butane, pentane and hexane. The conversion of sorbitol in to alkanes plus CO_2 and water is an exothermic process that retains approximately 95% of the heating value and only 30% of the mass of the biomass-derived reactant. This reaction takes place by a bi-functional pathway involving first the formation of hydrogen and CO_2 on the appropriate metal catalyst (such as Pt) and the dehydration of sorbitol on a solid acid catalyst (such as silica–alumina) or a mineral acid. These initial steps are followed by hydrogenation of the dehydrated reaction intermediates on the metal catalyst. When these steps are balanced properly, the hydrogen produced in the first step is fully consumed by hydrogenation of dehydrated reaction intermediates, leading to the overall conversion of sorbitol in to alkanes plus CO_2 and water. The selectivities for production of alkanes can be varied by changing the catalyst composition, the pH of the feed, the reaction conditions, and modifying the reactor design [36].

Figure 8 shows the effect of increasing the solid or mineral acidity on the alkane selectivity. As solid acid (SiO_2—Al_2O_3, containing 25 wt.% Al_2O_3 from Grace Davison) is added to Pt/Al_2O_3, the selectivity to heavier alkanes increases, as shown in Figure 8A. The alkanes formed are straight-chain compounds with only minor amounts of branched isomers (less than 5%). The H_2 selectivity decreases from 43 to 11% for the Pt/Al_2O_3 catalyst upon adding the solid acid SiO_2—Al_2O_3, indicating that the majority of the H_2 produced by the reforming reaction is consumed by the production of alkanes when the catalyst contains a sufficient number of acid sites. Similarly, the selectivity to heavier alkanes increases as a mineral acid (HCl) is added to the feed, as shown in Figure 8(B). The H_2 selectivity decreases from 43 to 6% as the pH of the feed decreases from 7 to 2.

A catalyst was prepared by depositing 4 wt.% Pt on the solid acid SiO_2—Al_2O_3 support [36], and this catalyst exhibited similar selectivity as the physical mixture of Pt/Al_2O_3 with SiO_2—Al_2O_3, for a similar ratio of Pt to solid acid sites [22]. The H_2 selectivity was usually less than 5% for the Pt/SiO_2—Al_2O_3, indicating that most of the H_2 produced was consumed in the production of the alkanes. Changing the temperature of the reactor from 538 to 498 K had little effect on the product selectivity. In contrast, the hexane selectivity increases from 21 to 40% for the Pt/SiO_2—Al_2O_3 catalyst at 498 K as the pressure increases from 25.8 to 39.6 bar, indicating that the reaction conditions can be used to manipulate the alkane selectivity.

As another option to produce heavier alkanes by APR, H_2 can be co-fed to the reactor with the aqueous feed. When H_2 is co-fed with an aqueous solution containing 5 wt.% sorbitol over the Pt/SiO_2—Al_2O_3 catalyst at 498 K and 34.8 bar, the selectivity to pentane plus hexane increases from 55 to 78% [36]. Under these conditions, approximately 90% of the effluent gas phase carbon is present as alkanes. Accordingly, increasing the hydrogen partial pressure in the reactor increases the rate of hydrogenation compared to C—C bond cleavage on the metal catalyst surface. The co-feeding of H_2 with the aqueous feed allows bi-functional catalysts to be formulated using metals (such as Pd) that by themselves show low activities for hydrogen production by APR reactions. In this case the hydrogen required for the formation of alkanes by the bi-functional catalyst is supplied externally instead of being formed in the reactor by APR, and the sole role of the metal component is to catalyze hydrogenation reactions.

2.4. Producing hydrogen containing low levels of CO: ultra-shift

Production of hydrogen containing low levels of CO is critical for energy generation using hydrogen PEM fuel cells. APR of oxygenates takes place over metal catalysts to produce CO and H_2. Adsorbed CO undergoes water-gas shift, which increases the amount of H_2 produced and removes CO from the catalyst surface [37]. The lowest partial pressure of CO that can be achieved depends on the thermodynamics of the WGS reaction and the operating conditions as given by

$$P_{CO} = \frac{P_{CO_2} P_{H_2}}{K_{WGS} P_{H_2O}}$$

where K_{WGS} is the equilibrium constant for the vapor-phase WGS and P_j the partial pressures. Since H_2, CO_2, and small amounts of alkanes (primarily CH_4) are produced by APR, gas bubbles are formed within the liquid-phase flow reactor. As noted above with respect to the effects of system pressure on reaction kinetics, the pressure in these bubbles can be approximated to be equal to the system pressure. Accordingly, the partial pressures of water vapor and the reaction products are dictated by the feed concentrations, system pressure and temperature, as outlined below [38].

For dilute product concentrations and system pressures above the saturation pressure of water, gaseous bubbles contain water vapor at a pressure equal to its saturation pressure at the reactor temperature, and the remaining pressure is the sum of the partial pressures of the product gases. The extent of vaporization, y, is defined as the percent of water in the vapor phase relative to the total amount of water flowing in the reactor. In contrast, for systems operated at pressures that are near the saturation pressure of water, all the liquid water may vaporize and the composition of the bubble is dictated by the stoichiometry of the feed stream. At this condition, the partial pressure of water is below its saturation pressure because the water vapor is diluted by the reforming product gases. Higher concentrations of ethylene glycol lead to lower partial pressures of water because of greater dilution from H_2 and CO_2 produced by reforming reactions. As the system pressure is increased, the partial pressure of water vapor increases until it reaches the saturation pressure of water, at which point any further increase in the system pressure leads to partial condensation of liquid water.

The above arguments indicate that the conditions which favor the lowest levels of CO from reforming of oxygenates are those which lead to the lowest partial pressures of H_2 and CO_2 in the reforming gas bubbles; and, these conditions are achieved by operating at system pressures that are near the saturation pressure of water and at low ethylene glycol feed concentrations. As the system pressure increases and extent of vaporization decreases below 100%, the partial pressures of H_2 and CO_2 in the bubble increase, thereby leading to higher equilibrium concentrations of CO. Similarly, as the ethylene glycol concentration in the feed increases, higher partial pressures of H_2 and CO_2 are developed, even for the case of complete vaporization, again leading to higher equilibrium CO concentrations.

Figure 9 shows the results for APR of 2 wt.% ethylene glycol over a Pt/Al_2O_3 catalyst contained in an upflow reactor, which was divided into two separately heated reaction zones [38]. Reforming reactions were carried out in the lower section (denoted as the reforming-zone), maintained at 498 K. The temperature of the top section (denoted as the shift-zone), system pressure, feed concentration, and feed flow rate were variables of the system. The observed CO concentration in the effluent gas and the corresponding equilibrium concentration are reported in Figure 9 for system pressures of 25.8, 32.0 and 36.2 bar, with the shift-zone of the reactor maintained at the same temperature as the reforming temperature of 498 K. Since the saturation pressure of water at 498 K is equal to 25.1 bar, liquid water is completely vaporized at a system pressure of 25.8 bar and the H_2 pressure in the bubble is calculated to be 0.77 bar, leading to a low equilibrium CO concentration of 66 ppm in the reactor effluent. At system pressures of 32 and 36.2 bar, the H_2 pressures are 4.60 and 7.53 bar, respectively, with only 18 and 11% vaporization of water occurring in each case. These conditions lead to higher equilibrium CO concentrations of 380 and 617 ppm, respectively.

Although the lowest levels of CO are obtained when the reactor is operated near the saturation pressure of water, this condition is not feasible for larger, non-volatile feeds such as glucose, which undergo undesirable decomposition reactions when vaporized [32–35]. It thus becomes imperative to operate the reformer at system pressures above the saturation pressure of water, to maintain liquid phase conditions. In these cases, we propose a process, which we denote as "ultra-shift", to achieve very low levels of CO in the product gas from APR of oxygenated hydrocarbons, conducted at pressures above the saturation pressure of water. This

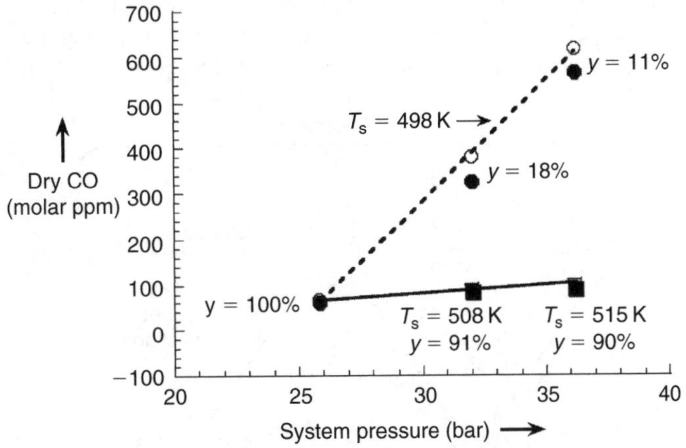

Fig. 9. Effect of process variables on CO concentration for 2% EG. Circles: CO concentration vs. system pressure at shift temperature T_s = 498 K, squares: ultra-shift at higher shift temperatures. Open symbols represent equilibrium CO and filled symbols represent observed CO. Adapted from Ref. [38]

process involves vaporization of liquid water (in the shift-zone) to dilute the H_2 and CO_2 in the bubbles, thereby favoring increased conversion of the WGS reaction and leading to lower CO concentrations.

The feasibility of the ultra-shift process is depicted in Figure 9 [38]. It is seen that at a system pressure of 32 bar, when the temperature of the shift-zone was increased from 498 to 508 K, 91% of the water in this zone vaporized, thereby decreasing the H_2 pressure to 1.05vbar, and lowering the measured CO concentration from 380 to 84 ppm. In the case where the system pressure was 36.2 bar, increasing the temperature of the shift-zone from 498 to 515 K decreased the H_2 pressure from 7.53 to 1.14 bar and the measured CO concentration from 564 to 89 ppm. At the higher temperatures, the effect of diluting the H_2 and CO_2 pressures with water vapor supersedes the less favorable thermodynamics, leading to ultra-shift.

2.5. Hydrogen from concentrated glucose feeds

Aqueous-phase reforming of glucose ($C_6O_6H_{12}$) is of particular interest for biomass utilization, because this sugar comprises the major energy reserves in plants and animals. As discussed earlier, the hydrogen selectivity from reforming of glucose decreases as the liquid concentration increases from 1 to 10 wt.% because of undesired hydrogen-consuming side reactions that occur in the liquid phase [34]. This decrease in selectivity is an important limitation, because processing dilute aqueous solutions involves the processing of excessive amounts of water.

Figure 10 depicts various reaction pathways that take place during APR of glucose and sorbitol [39]. Production of H_2 and CO_2 from glucose (G) and sorbitol (S) takes place on metal catalysts such as Pt or Ni—Sn alloys (pathways G1 and S1) via cleavage of C—C bonds followed by WGS. As discussed earlier, undesired alkanes can be formed by cleavage of C—O bonds on the metal catalyst and dehydration processes on acidic catalyst supports (pathways G2, S2). In the case of glucose, undesired reactions can also take place in the aqueous phase to form organic acids, aldehydes and carbonaceous deposits (pathway G3). These undesirable homogeneous decomposition reactions are first order in the glucose concentration, whereas the desirable reforming reactions on the catalyst surface are fractional order; therefore, high concentrations of glucose lead to low hydrogen selectivities. The hydrogenation of glucose to sorbitol (pathway G—S) also takes place on metal catalysts with high selectivity at low temperatures (e.g., 400 K) and high H_2 pressures.

The rates of pathways G1, G2 and G3 increase more rapidly with temperature than does the rate of pathway G—S. Accordingly, a strategy for improving the hydrogen selectivity from APR of glucose is to employ a dual-reactor system involving a low-temperature hydrogenation step followed by a higher-temperature reforming process. In this respect, we have conducted studies in which the partial pressures of the gas-phase products in the reactor were varied by co-feeding a gas stream (N_2 or H_2) with the liquid feed (10 wt.%

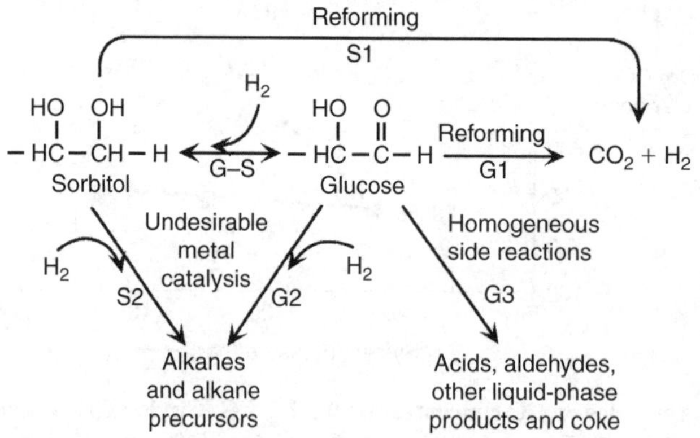

Fig. 10. Reaction pathways involved in glucose and sorbitol reforming. Adapted from Ref. [39]

glucose/sorbitol) at the inlet of the reactor. The performance of the reactor in this mode can be then compared to the case where only liquid was fed to the reactor and N_2 sweep gas was combined with the effluent stream at the exit of the reactor. It was found that the hydrogen selectivities observed for APR of glucose (~10–13%) were much lower compared to reforming of sorbitol (~60%) under similar conditions. Also higher alkane selectivities of 47–50% were observed for reforming of glucose, indicating that APR of glucose, even with hydrogen co-fed with the liquid reactant stream at the inlet of the reactor, is not selective for production of H_2. These results indicate that co-feeding hydrogen with liquid reactants into the reforming reactor at 538 K (and 52.4 bar) does not lead to rapid hydrogenation of glucose into sorbitol and its subsequent reforming to give high H_2 selectivities. Similarly, sorbitol does not undergo rapid dehydrogenation to glucose under conditions of APR where H_2 is not co-fed with the liquid reactant stream to the reforming reactor, a desirable result.

Experiments were conducted by co-feeding gaseous H_2 with aqueous solutions containing 10 wt.% sorbitol or glucose into a dual-reactor system consisting of a hydrogenation reactor at 393 K followed by a reforming reactor at 538 K [39]. The presence of the hydrogenation reactor did not affect the APR of sorbitol. However, when 10 wt.% glucose was co-fed with gaseous H_2 to the dual-reactor system, then high hydrogen selectivity (62.4%) and low alkane selectivity (21.3%) were obtained, these values being similar to the selectivities obtained from the reforming of sorbitol. These results indicate that glucose is first completely hydrogenated to sorbitol before being sent to the reformer in which the sorbitol is then converted with high selectivity in to H_2 and CO_2.

3. DISCUSSION AND OVERVIEW

Figure 11 summarizes the thermodynamic and kinetic aspects that form the foundation of the APR process. In contrast to alkane reforming which is favorable only at high temperatures, the reforming of oxygenated hydrocarbons (C:O ratio of 1:1) to form CO and H_2 is thermodynamically favorable at relatively lower temperatures of 400 K. Also, the subsequent WGS reaction, necessary to convert the carbon monoxide

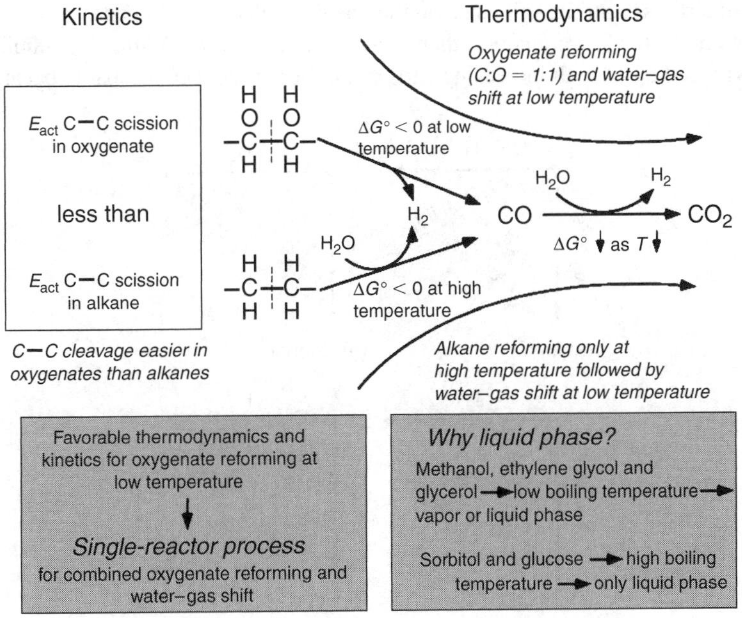

Fig. 11. Summary of thermodynamic and kinetic considerations that form the basis of the APR process

generated by reforming in to carbon dioxide, becomes increasingly favorable at lower temperatures. In addition, results from DFT calculations indicate that the activation energy required to break the C—C bond in oxygenated hydrocarbons is lower than that required for alkanes, indicating that C—C cleavage is easier in oxygenates than in alkanes. These results suggest that it is possible to design a process in which oxygenated hydrocarbons can be converted in to H_2 and CO_2 in a single-step process, wherein both the reforming and WGS reactions are conducted in the same low-temperature reactor. Although the more volatile oxygenates such as methanol, ethylene glycol and glycerol can be processed in the vapor as well as liquid phases, processing less volatile oxygenates such as glucose and sorbitol under vapor-phase conditions would require high-temperature reforming, followed by WGS at lower temperatures, thus obviating the advantages of a single-reactor low-temperature process.

Figure 12 summarizes the effects of various factors on the selectivity of hydrogen production, and also suggests process conditions that may lead to the selective production of alkanes, if desired, from oxygenated hydrocarbons. It is seen from Figure 12a, that metals such as Pt, Pd and Ni—Sn alloys show high selectivity for hydrogen production and very low tendency for formation of alkanes. In comparison, supported Ni catalysts tend to make more alkanes and also show some deactivation with time, which may be due to sintering of the metal particles, leading to lower dispersions. Figure 12a also shows that metals such as Ru and Rh are very active for alkane formation and make very little hydrogen.

The support plays an important role in the selectivity of the APR process, as summarized in Figure 12b. The more acidic supports (e.g., silica–alumina) lead to high selectivities for alkane formation, whereas the more basic/neutral supports (e.g., alumina) favor hydrogen production. Supports with mild acidity fall within the spectrum of the two extremes, as seen for titania-based catalysts. The acidity of the solution also affects the performance of the aqueous-phase reformer in a similar way. Depending on the nature of the byproduct/intermediate compounds formed in the reactor, the aqueous solution in contact with the catalyst can be acidic, neutral or basic. Also gaseous carbon dioxide dissolved in the solution at high pressures makes a slightly acidic solution (pH = 4–5). Acidic solutions (pH = 2–4) promote alkane formation, due to acid-catalyzed dehydration reactions that occur in solution (followed by hydrogenation on the metal). In contrast, neutral and basic solutions lead to high hydrogen selectivities and low alkane selectivities, as shown in Figure 12c.

The type of feed also has a strong influence on the reaction selectivity. In general, polyols (e.g., sorbitol) have a higher selectivity for H_2 production than sugars (e.g., glucose). Within the family of polyols, the hydrogen selectivity decreases with increasing carbon number of the feed, probably because the number of

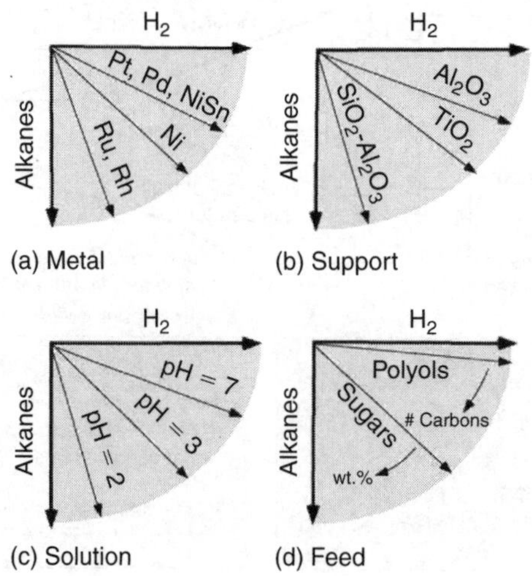

Fig. 12. Factors controlling the selectivity of the aqueous-phase reforming process

undesired hydrogen-consuming side reactions increases accordingly. Also, processing higher feed concentrations of glucose leads to lower hydrogen selectivities. For example, as the glucose feed concentration increases from 1 to 10 wt.%, the alkane selectivity increases from 30 to 50%, and the hydrogen selectivity decreases accordingly. This change with concentration is caused by undesired homogeneous decomposition reactions associated with sugars. These different feed effects have been summarized in Figure 12d.

Figure 13 outlines the various reaction pathways that can take place in the APR reactor (for polyols as feed), resulting in multiple selectivity challenges. The oxygenated hydrocarbon feedstock can undergo reforming via C—C cleavage on the metal surface to make the desired H_2 and CO_2. The CO_2 formed can undergo undesired series methanation/Fischer–Tropsch synthesis reactions to form alkanes. Alternately, some metals have a tendency to favor C—O bond scission, followed by hydrogenation to make alcohol intermediates, which can then further react on the metal surface (C—C/C—O cleavage) to form alkanes. These reactions constitute parallel-selectivity challenges on the metal surface. In addition to metal catalyzed reactions, undesired bi-functional catalysis can occur by combinations of metal, support and solution, as shown in Figure 13. Reaction pathways such as dehydration–hydrogenation leading to alcohols, and dehydrogenation-rearrangement to form acids can occur in solution or on the support aided by metal catalysis. These intermediates can then react further in solution or on the catalyst to make more alkanes. To obtain high selectivities for hydrogen production, the metal catalyst should show high rates of C—C cleavage, low rates

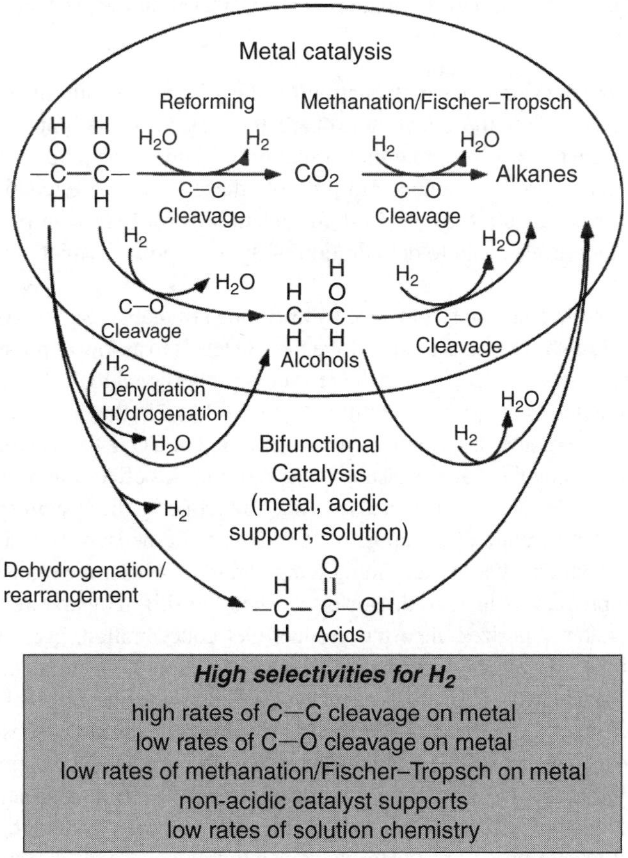

Fig. 13. Summary of reaction pathways involved in the APR of oxygenated hydrocarbons leading to multiple selectivity challenges

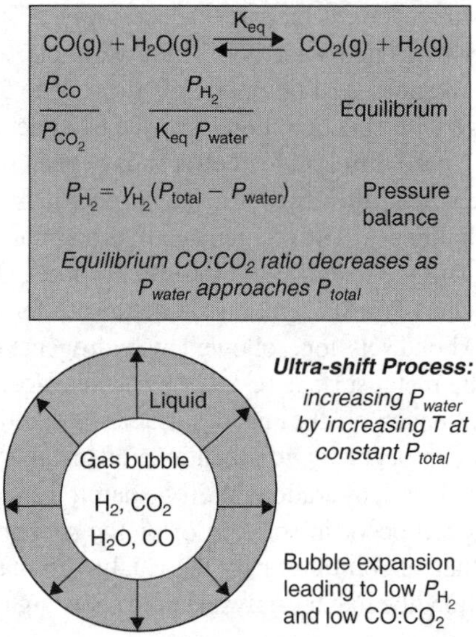

Fig. 14. Concept of the ultra-shift process for obtaining low CO levels in the product gas

of C—O cleavage and low rates of series methanation/Fischer–Tropsch synthesis reactions. Metals such as Pt and Ni—Sn alloy systems show these favorable characteristics. Also the support should not favor dehydration/C—O cleavage reactions or promote the formation of acids in aqueous solution. Acidic supports such as silica–alumina have shown a tendency to promote these undesirable reactions. Hence, non-acidic supports such as alumina are required to obtain high selectivities for hydrogen production via APR reactions. Finally, acidic solutions also lead to dehydration of the feed polyol, which ultimately leads to alkane formation.

Low CO levels in the product gas can be obtained by operating at suitable reaction conditions. Figure 14 outlines the concept of the ultra-shift process used to drive the CO levels to as low as possible. Based on the WGS equilibrium, the partial pressure of CO (P_{CO}) in the reactor is proportional to the partial pressure of hydrogen (P_{H_2}), which in turn decreases as the partial pressure of water (P_{water}) in the system approaches the total system pressure (P_{total}). Thus, to reduce the equilibrium concentration of CO in the reactor, the hydrogen pressure should be lowered by increasing the water pressure in the system. This effect can be achieved using the ultra-shift process, wherein the water pressure is increased either by reducing the system pressure in an isothermal reactor, or increasing the temperature of the upper ultra-shift-zone of the isobaric reactor. For example, when the shift temperature is maintained at the reforming temperature of 498 K, the effluent concentration of CO increases as the system pressure is increased. However, when the shift temperature is increased such that a majority of the liquid water vaporizes, then the product CO concentration, even at the higher pressures, decreases to below 100 ppm for 2% ethylene glycol. Thus, the addition of the ultra-shift zone to the reforming reactor leads to low CO levels while allowing for processing of non-volatile feedstocks such as glucose in the lower aqueous-phase reformer. Importantly, when the effluent from the reactor is subsequently cooled to condense liquid water, then the pressure of the non-condensable gases increases and approaches the system pressure. This process of ultra-shift, involving initial vaporization followed by condensation of liquid water, thus leads to the desirable production of fuel-cell-grade H_2 at high pressures and containing very low levels of CO.

The basis for the dual-reactor process for handling concentrated glucose feeds is summarized in Figure 15. One of the important limitations of the APR of glucose solutions is that the hydrogen selectivity decreases when higher concentration feeds are used. This limitation is specific to glucose feedstocks, and a

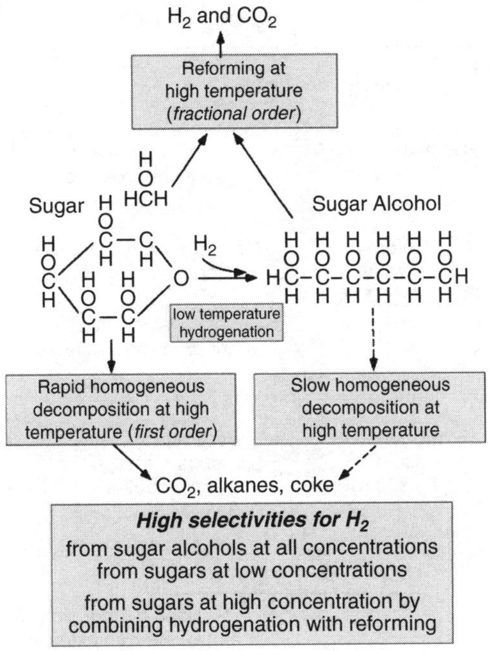

Fig. 15. Basis for the dual-reactor system employed in the processing of concentrated glucose feedstocks

sugar–alcohol such as sorbitol does not suffer this restriction. Starting with glucose, the main desirable reaction that can occur is the catalytic reforming at high temperatures to form H_2 and CO_2. In addition, undesired homogeneous decomposition reactions can occur at the same reaction conditions as reforming, to produce alkanes and coke. The rate of reforming increases more slowly with increasing feed concentration, as compared to the decomposition rates, leading to lower hydrogen selectivities at higher feed concentrations. Also, glucose can undergo hydrogenation (using H_2 produced by reforming or supplied externally) to form sorbitol, which can then undergo reforming on the catalyst surface to produce the desired H_2 and CO_2. Since sorbitol decomposition rates in aqueous solution are much slower relative to the sorbitol reforming rates, high hydrogen selectivities can be obtained at all concentrations of the feed. The activation energy barrier for glucose hydrogenation is lower than the activation energies for glucose reforming and glucose decomposition. Therefore, glucose hydrogenation becomes dominant at lower temperatures, and almost complete conversion of glucose in to sorbitol can be achieved, without any conversion in to H_2 and CO_2. The relative rates of the different reactions involved in glucose processing thus indicate that for the efficient extraction of hydrogen from glucose feedstocks, it is necessary to utilize an initial hydrogenation reactor that converts the glucose in to sorbitol, followed by a reforming reactor to produce hydrogen with high selectivities, as described in Figure 15.

We now outline how the APR process can be manipulated to obtain a product with the desired specifications. Sugar alcohols such as sorbitol give higher hydrogen selectivities than the corresponding sugar, i.e., glucose. Thus polyols (including sugar–alcohols) can be processed directly in the aqueous-phase reformer to obtain relatively high hydrogen selectivities, without any upstream processing. However, as shown in Figure 16, co-feeding hydrogen with the aqueous feed and employing a low-temperature hydrogenation reactor upstream from the reformer are used for processing sugars at high concentrations to obtain hydrogen in high selectivities. To obtain alkanes from concentrated sugars, a similar hydrogenation to sorbitol is first carried out, followed by reaction over an acidic catalyst in the presence of a hydrogen

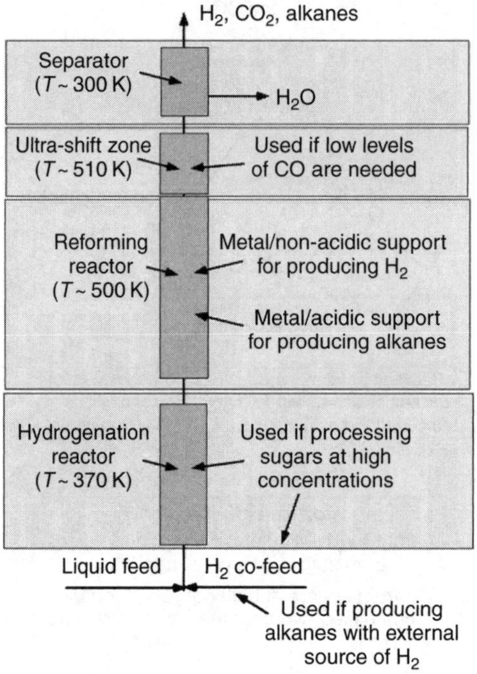

Fig. 16. Summary of the process conditions employed to obtain a product of the desired specifications using the APR process

co-feed. For applications such as PEM fuel cells wherein very low levels of CO in the hydrogen fuel are necessary for efficient operation, an ultra-shift zone, downstream of the reformer may be used to lower the outlet CO levels in the gas.

A novel aspect of the APR process is that a beneficial synergy is formed by operating the hydrogenation reactor, the reforming reactor, the ultra shift-zone of the reactor, and the gas–liquid separator (situated downstream of the reactors) at different temperatures while maintaining the total pressure of the system at a constant value, as depicted schematically in Figure 16. An aqueous solution of glucose can be co-fed with gaseous H_2 to the hydrogenation reactor, which is operated at a relatively low temperature ($T \sim 370\,K$) to minimize glucose decomposition reactions in the liquid phase. At the low vapor pressure of water corresponding to this hydrogenation temperature, the partial pressure of hydrogen in this reactor is high, and hence favorable for the conversion of glucose in to sorbitol. The aqueous solution of sorbitol and gaseous H_2 are then fed to the reforming (or alkane formation) reactor, which is operated at the higher temperature ($T \sim 500\,K$), necessary to convert sorbitol in to H_2 and CO_2 (or alkanes). If low levels of CO are required in the product gas, an ultra-shift-zone ($T \sim 510\,K$) downstream of the reforming reactor can be employed. Finally, the liquid and gaseous effluents from the reactor are cooled and sent to a separator, which is maintained at a low temperature ($T \sim 300\,K$), and the sum of the H_2 and CO_2 (or alkanes) pressures is essentially equal to the total pressure of the system. In the case of a hydrogen-rich reformate, this high pressure facilitates further removal of CO_2 from H_2 by pressure-swing adsorption or membrane separation. A fraction of the purified H_2 at high pressure can then be recycled to the hydrogenation reactor, and the remaining hydrogen may be directed to a fuel cell for generation of electrical power.

An important advantage of the APR process is that it can be tailored to make renewable fuels such as H_2 or alkanes in high selectivity. Another advantageous consequence of making H_2 and alkanes via the APR process is that the energy required for the endothermic reforming of oxygenates may be produced internally, by allowing a fraction of the oxygenated compound to form alkanes through exothermic reaction pathways. In this respect, the formation of a mixture of hydrogen and alkanes from APR of oxygenates may be essentially neutral energetically, and little additional energy is required to drive the reaction. The gas effluent from the reformer could then be utilized as a feed to an internal combustion engine or suitable fuel cell (such as a solid oxide fuel cell).

4. CONCLUSIONS

Aqueous-phase reforming of biomass-derived oxygenated hydrocarbons is a new process to produce renewable fuels consisting of hydrogen and alkanes. Catalysts based on Pt and Ni—Sn alloys are promising materials for hydrogen production. Various competing reaction pathways are involved in the reforming process, leading to parallel and series-selectivity challenges for production of hydrogen. The selectivity of the reforming process depends on various factors such as nature of the catalytically active metal, support, solution pH, feed and process conditions. By manipulating these factors, the APR process can be tailored to produce either H_2 or alkanes. Moreover, the molecular weight distribution of the alkane product stream can also be controlled, for example, to favor the formation of heavier alkanes using bi-functional catalysts containing metal and acidic components. Also, it is possible to operate the APR process to achieve very low CO levels in the product by using an ultra-shift zone downstream of the reactor. Aqueous-phase reforming is best conducted using polyol feed molecules (e.g., sorbitol), because the product distribution obtained is relatively insensitive to the aqueous feed concentration (e.g., from 1 to 10 wt.%, and probably higher). In contrast, processing high concentrations of sugars (e.g., glucose) is accompanied by undesirable decomposition reactions. However, in these cases, a dual-reactor approach can be used for processing high concentrations of glucose, in which glucose is hydrogenated to sorbitol in the first reactor, and sorbitol is passed to the APR reactor for selective production of hydrogen or alkanes.

We note, in closing, that aqueous-phase reforming is a remarkably flexible process that can be tailored to selective produce hydrogen or to produce an alkane stream with a desired molecular weight distribution. Accordingly, this process offers exciting research opportunities for developing new generation of heterogeneous catalysts using metals, metal alloys, supports, and promoters that are stable under aqueous-phase reaction conditions.

ACKNOWLEDGEMENTS

This work was supported by the US Department of Energy (DOE), Office of Basic Energy Sciences, Chemical Science Division, and by the National Science Foundation (NSF) through a STTR grant. Funding has also been provided by Conoco-Phillips and by Daimler-Chrysler. We would like to thank Kyle Allen, Dan Current, Andrew Richardson, Bret Wagner, Jonathan Tomshine, David Wishnik, Sibongile Nkosi, Dante Simonetti, Nicole Otto, Juben Chheda, Nitin Agarwal and Shampa Kandoi for assistance in catalyst preparation and reaction kinetic measurements, Marco Sanchez-Castillo and Joseph Napier for assistance with analysis of reaction products. We also thank Manos Mavrikakis for valuable discussions.

REFERENCES

1. J.R. Rostrup-Nielsen. *Steam Reforming Catalysts*. Danish Technical Press, Copenhagen, 1975.
2. J. Rostrup-Nielsen. *Phys. Chem. Chem. Phys.* **3** (2001) 283.
3. J. Woodward, M. Orr, K. Cordray and E. Greenbaum. *Nature* **405** (2000) 1014.
4. L. Garcia, R. French, S. Czernik and E. Chornet. *Appl. Catal. A Gen.* **201** (2000) 225.

5. K. Belkacemi, G. Turcotte and P. Savoie. *Ind. Eng. Chem. Res.* **41** (2002) 173.
6. S. Czernik, R. French, S. Feik and E. Chornet. *Ind. Eng. Chem. Res.* **41** (2002) 4209.
7. S.E. Jacobsen and C.E. Wyman. *Ind. Eng. Chem. Res.* **41** (2002) 1454.
8. R.R. Davda, J.W. Shabaker, G.W. Huber, R.D. Cortright and J.A. Dumesic. *Appl. Catal. B Environ.* **43** (2002) 13.
9. J.W. Shabaker, G.W. Huber, R.R. Davda, R.D. Cortright and J.A. Dumesic. *J. Catal.* **215** (2003) 344.
10. J.W. Shabaker, G.W. Huber, R.R. Davda, R.D. Cortright and J.A. Dumesic. *Catal. Lett.* **88** (2003) 1.
11. G.W. Huber, J.W. Shabaker and J.A. Dumesic. *Science* **300** (2003) 2075.
12. R.S. Dixit and L.L. Taviarides. *Ind. Eng. Chem. Proc. Des. Dev.* **22** (1983) 1.
13. E. Iglesia, S.L. Soled and R.A. Fiato. *J. Catal.* **137** (1992) 212.
14. C.S. Kellner and A.T. Bell. *J. Catal.* **70** (1981) 418.
15. M.A. Vannice. *J. Catal.* **50** (1977) 228.
16. J.H. Sinfelt and D.J.C. Yates. *J. Catal.* **8** (1967) 82.
17. D.C. Grenoble, M.M. Estadt and D.F. Ollis. *J. Catal.* **67** (1981) 90.
18. R. Alcalá, M. Mavrikakis and J.A. Dumesic. *J. Catal.* **218** (2003) 178.
19. B. Lindstrom and L.J. Pettersson. *Int. J. Hydrogen Energy* **26** (2001) 923.
20. G.A. Somorjai. *Introduction to Surface Chemistry and Catalysis*. Wiley, New York, (1994).
21. R.R. Davda, R. Alcala, J. Shabaker, G. Huber, R.D. Cortright, M. Mavrikakis and J.A. Dumesic. *Stud. Surf. Sci. Catal.* **145** (2003) 79.
22. R.D. Cortright, R.R. Davda and J.A. Dumesic. *Nature* **418** (2002) 964.
23. J.W. Shabaker, G.W. Huber and J.A. Dumesic. *J. Catal.* **222** (2004) 180.
24. S. Hilaire, X. Wang, T. Luo, R.J. Gorte and J. Wagner. *Appl. Catal. A Gen.* **215** (2001) 271.
25. S.P. Bates and R.A. Van Santen. *Adv. Catal.* **42** (1998) 1.
26. B. Gates. *Catalytic Chemistry*, Wiley, New York, (1992).
27. G.A. Parks. *Chem. Rev.* **65** (1965) 177.
28. H. Li, W. Wang and J.F. Deng. *J. Catal.* **191** (2000) 257.
29. B. Blanc, A. Bourrel, P. Gallezot, T. Haas and P. Taylor. *Green Chem.* **2** (2000) 89.
30. R. Narayan, G. Durrence and G.T. Tsao. *Biotech. Bioeng. Symp.* **14** (1984) 563.
31. E. Tronconi, N. Ferlazzo, P. Forzatti, I. Pasquon, B. Casale and L. Marini. *Chem. Eng. Sci.* **47** (1992) 2451.
32. G. Eggleston and J.R. Vercellotti. *J. Carbohyd. Chem.* **19** (2000) 1305.
33. B.M. Kabyemela, T. Adschiri, R.M. Malaluan and K. Arai. *Ind. Eng. Chem. Res.* **36** (1997) 1552.
34. B.M. Kabyemela, T. Adschiri, R.M. Malaluan and K. Arai. *Ind. Eng. Chem. Res.* **38** (1999) 2888.
35. A.R. Sapronov. *Khlebopek. Konditer. Prom.* **13** (1969) 12.
36. G.H. Huber, R.D. Cortright and J.A. Dumesic. *Angew. Chem. Int. Edit.* **43** (2004) 1549.
37. J. Novakova and L. Kubelkova. *Appl. Catal. B Environ.* **14** (1997) 273.
38. R.R. Davda and J.A. Dumesic. *Angew. Chem. Int. Edit.* **42** (2003) 4068.
39. R.R. Davda and J.A. Dumesic. *Chem. Commun.* Jan 7 (1) (2004) 36.

Chapter 4

Fuel processing for low- and high-temperature fuel cells: challenges, and opportunities for sustainable development in the 21st century

Chunshan Song

Abstract

This review paper first discusses the needs for fundamental changes in the energy system for major efficiency improvements in terms of global resource limitation and sustainable development. Major improvement in energy efficiency of electric power plants and transportation vehicles is needed to enable the world to meet the energy demands at lower rate of energy consumption with corresponding reduction in pollutant and CO_2 emissions. A brief overview will then be given on principle and advantages of different types of low-and high-temperature fuel cells. Fuel cells are intrinsically much more energy-efficient, and could achieve as high as 70–80% system efficiency (including heat utilization) in electric power plants using solid oxide fuel cells (SOFC, versus the current efficiency of 30–37% via combustion), and 40–50% efficiency for transportation using proton-exchange membrane fuel cells (PEMFC) or solid oxide fuel cells (versus the current efficiency of 20–35% with internal combustion (IC) engines). The technical discussions will focus on fuel processing for fuel cell applications in the 21st century. The strategies and options of fuel processors depend on the type of fuel cells and applications. Among the low-temperature fuel cells, PEMFC require H_2 as the fuel and thus nearly CO and sulfur-free gas feed must be produced from fuel processor. High-temperature fuel cells such as solid oxide fuel cells can use both CO and H_2 as fuel, and thus fuel processing can be achieved in less steps. Hydrocarbon and alcohol fuels can both be used as fuels for reforming on-site or on-board. Alcohol fuels have the advantages of being ultra-clean and sulfur-free and can be reformed at lower temperatures, but hydrocarbon fuels have the advantages of existing infrastructure of production and distribution and higher energy density. Further research and development on fuel processing are necessary for improved energy efficiency and reduced size of fuel processor. More effective ways for on-site or on-board deep removal of sulfur before and after fuel reforming, and more energy-efficient and stable catalysts and processes for reforming hydrocarbon fuels are necessary for both high and low-temperature fuel cells. In addition, more active and robust (non-pyrophoric) catalysts for water–gas-shift (WGS) reactions, more selective and active catalysts for preferential CO oxidation at lower temperature, more CO-tolerant anode catalysts would contribute significantly to development and implementation of low-temperature fuel cells, particularly proton-exchange membrane fuel cells. In addition, more work is required in the area of electrode catalysis and high-temperature membrane development related to fuel processing including tolerance to certain components in reformate, especially CO and sulfur species.

Keywords: Fuel processing; Reforming; Sulfur removal; Water–gas-shift; H_2; Fuel cell; Catalyst; Catalysis; Energy efficiency; Sustainable development

Fuel Cells Compendium

Article Outline

1. Introduction . . . 54
2. Sustainable development of energy . . . 55
2.1. Supply-side challenge of energy balance . . . 55
2.2. Sustainable development of energy . . . 57
2.3. Vision for efficient utilization of hydrocarbon resources . . . 57
3. Principle and advantages of fuel cells . . . 59
3.1. Concept of fuel cells . . . 59
3.2. Efficiency of fuel cells . . . 59
3.3. Types of fuel cells . . . 60
3.3.1. Proton-exchange membrane fuel cell . . . 61
3.3.2. Phosphoric acid fuel cell . . . 62
3.3.3. Alkaline fuel cell . . . 63
3.3.4. Molten carbonate fuel cell . . . 63
3.3.5. Solid oxide fuel cell . . . 64
3.4. Advantages of fuel cells compared to conventional devices . . . 65
4. Fuel processing for fuel cell applications . . . 65
4.1. Fuel options for fuel cells . . . 65
4.2. Fuel cells for electric power plants . . . 67
4.3. Fuel cells for transportation . . . 68
4.4. Fuel cells for residential and commercial sectors . . . 69
4.5. Fuel cells as portable power sources . . . 69
5. Challenges and opportunities for fuel processing research . . . 70
5.1. Sulfur removal from hydrocarbon fuels before/after reforming . . . 73
5.2. Fuel reforming for PEMFC and PAFC . . . 74
5.2.1. Reforming of alcohol fuels . . . 75
5.2.2. Reforming of hydrocarbon fuels . . . 76
5.3. Carbon formation during reforming . . . 78
5.4. Catalytic WGS . . . 78
5.5. Deep removal of CO . . . 79
5.6. Fuel processing for SOFC and MCFC . . . 79
5.7. Electrode catalysis related to fuel processing . . . 81
5.8. Direct oxidation of methanol and methane in fuel cells . . . 82
6. Concluding remarks . . . 82
Acknowledgments . . . 83
References . . . 83

1. INTRODUCTION

As the world moved into the first decade of the 21st century, a global view is due for energy consumption in the last century and the situations around energy supply and demand of energy and fuels in the future. The world of the 20th century is characterized by growth. Table 1 shows the changes in worldwide energy use in the 20th century, including consumption of different forms of energy in million tonnes of oil equivalent (MTOE), world population, and per capita energy consumption comparing the years 1900 and 1997, which are based on recent statistical data [1–3]. The rapid development in industrial and transportation sectors and improvements in living standards among residential sectors correspond to the dramatic growth in energy consumption from 911 MTOE in 1900 to 9647 MTOE in 1997. This is also due in part

Table 1. Worldwide energy use in million tonnes of oil equivalent (MTOE), world population and per capita energy consumption in the 20th century

	1900		1997	
Energy source	**MTOE**	**%**	**MTOE**	**%**
Petroleum	18	2	2940	30
Natural gas	9	1	2173	23
Coal	501	55	2122	22
Nuclear	0	0	579	6
Renewable	383	42	1833	19
Total	911	100	9647	100
Population (million)	1762		5847	
Per capita energy use (TOE)	0.517		1.649	
Global CO_2 emission (MMTC)[a]	534		6601	
Per capita CO_2 emission (MTC)	0.30		1.13	
CO_2 (ppmv)[b]	295		364	
Life expectancy (years)[c]	47		76	

[a] Global CO_2 emissions from fossil fuel burning, cement manufacture, and gas flaring; expressed in million metric tonnes of carbon (MMTC).
[b] Global atmospheric CO_2 concentrations expressed in parts per million by volume (ppmv).
[c] Life expectancy is based on the statistical record in the US [2,3].

to the rapid increase in population from 1762 million in 1900 to 5847 million in 1997, as can be seen from Table 1.

Table 1 also shows the data on combined global CO_2 emissions from fossil fuel burning, cement manufacture, and gas flaring expressed in million metric tonnes of carbon (MMTC) in 1990 and 1997 [4]. It is clear from Table 1 that global CO_2 emissions increased over 10 times, from 534 MMTC in 1900 to 6601 MMTC in 1997, in proportion with the dramatic increase in worldwide consumption of fossil energy. The emissions of enormously large amounts of gases from combustion into the atmosphere has caused a rise in global concentrations of greenhouse gases, particularly CO_2. Table 1 also includes data on the global atmospheric concentrations of greenhouse gas CO_2 in 1900 and in 1997, where the 1900 data was determined by measuring ancient air occluded in ice core samples [5], and that for 1997 was from actual measurement of atmospheric CO_2 in Mauna Loa, Hawaii [6]. The increase in atmospheric concentrations of CO_2 has been clearly established and can be attributed largely to increased consumption of fossil fuels by combustion. To control greenhouse gas emissions in the world, several types of approaches will be necessary, including major improvement in energy efficiency, the use of carbon-less (or carbon-free) energy, and the sequestration of carbon such as CO_2 storage in geologic formations.

2. SUSTAINABLE DEVELOPMENT OF ENERGY

2.1. Supply-side challenge of energy balance

Figure 1 shows the energy supply and demand (in quadrillion Btu) in the US in 1998 [7]. The existing energy system in the US and in the world today is largely based on combustion of fossil fuels – petroleum,

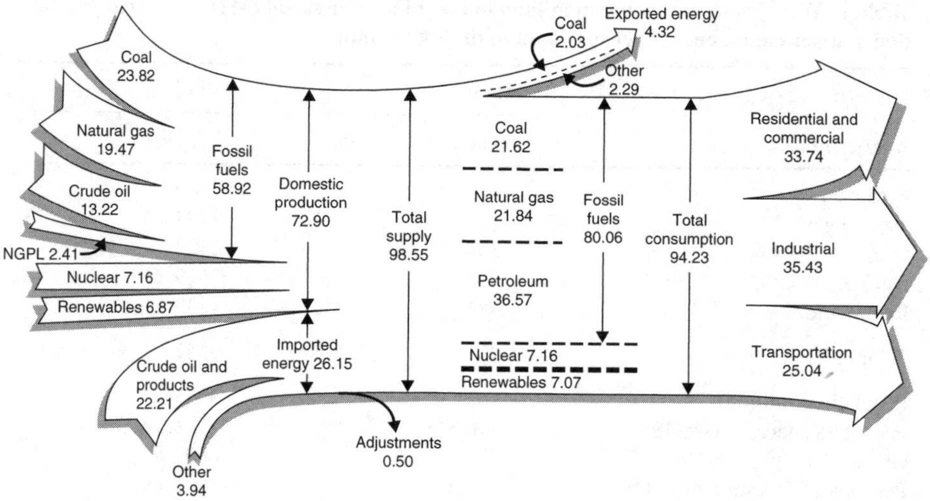

Fig. 1. Energy flow (quadrillion Btu) in the US, in 1998 [7]

natural gas, and coal – in stationary and mobile devices. It is clear from Fig. 1 that petroleum, natural gas, and coal are the three largest sources of primary energy consumption in the US. Renewable energies are important but small parts (6.87%) of the US energy flow, although they have potential to grow.

Figure 2 illustrates the energy input and the output of electricity (in quadrillion Btu) from power plants in the US in 1998 [7]. As is well known, electricity is the most convenient form of energy in industry and in daily life. The electric power plants are the largest consumers of coal. Great progress has been made in the electric power industry with respect to pollution control and generation technology with certain improvements in energy efficiency. What is not apparent in the energy supply–demand pictures is the following. The energy input into electric power plants represents 36.9% of the total primary energy supply in the US, but the majority of the energy input into the electric power plants, over 65%, is lost and wasted as conversion loss in the process, as can be seen from Fig. 2 for the electricity flow in the US including electric utilities and non-utility power producers. The same trend of conversion loss is also applicable for the fuels used in transportation, which represents 25.4% of the total primary energy consumption. This energy waste is largely due to the thermodynamic limitations of heat engine operations dictated by the maximum efficiency of the Carnot cycle.

How much more fossil energy resources are there? The known worldwide reserves of petroleum (1033.2 billion barrels in 1999) [8] would be consumed in about 39 years, based on the current annual consumption of petroleum (26.88 billion barrels in 1998). On the same basis, the known natural gas reserves in the world (5141.6 trillion cubic feet in 1999) would last for 63 years at the current annual consumption level (82.19 trillion cubic feet in 1998) [8]. While new exploration and production technologies will expand the oil and gas resources, two experts in oil industry, Campbell and Laherrere [9], have indicated that global production of conventional oil will begin to decline sooner than most people think and they have compellingly alluded to the end of cheap oil early in this century. Worldwide coal production and consumption in 1998 were 5042.7 and 5013.5 million short tonnes, respectively [7]. The known world recoverable coal reserves in 1999 are 1087.19 billion short tonnes [8], which is over 215 times the world consumption level in 1998. Thus, coal has great potential as a future source of primary energy, although environmental pressures may militate against expanded markets for coal as an energy source. However, even coal resources are limited. Prof. George Olah, the winner of Nobel Prize in chemistry in 1994, pointed out in 1991 that "Oil and gas resources under the most optimistic scenarios won't last much longer than through the next century. Coal reserves are more abundant, but are also limited. … I suggest we

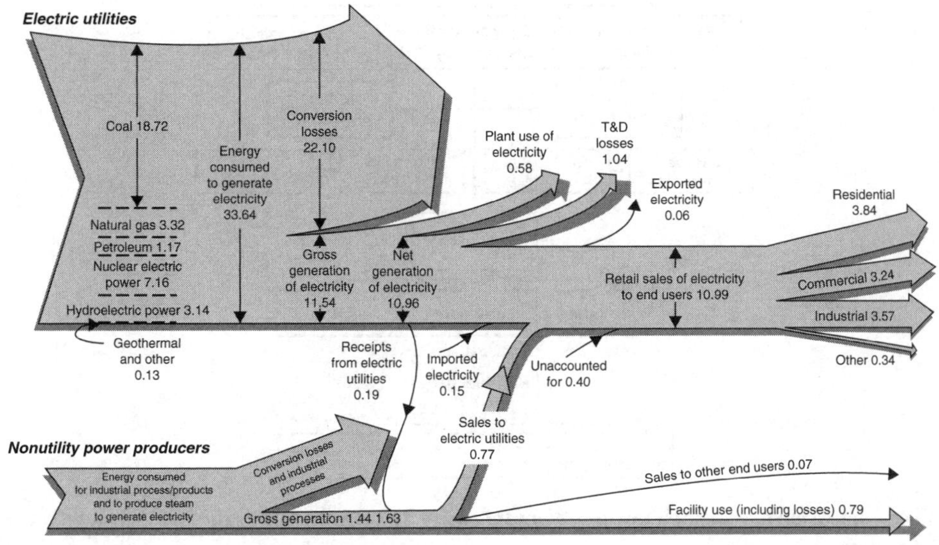

Fig. 2. Electricity flow (quadrillion Btu) in the US, in 1998 [7]

should worry much more about our limited and diminishing fossil resources" [10]. In this context, it is important to recognize the limitations of non-renewable hydrocarbon resources in the world.

2.2. Sustainable development of energy

Can the world sustain itself by continuously using the existing energy system based on combustion of fossil resources in the 21st century? Petroleum, natural gas and coal are important fossil hydrocarbon resources that are non-renewable. Sustainable development may have different meanings to different people, but a respected definition from the report "Our Common Future" [11], is as follows: "Sustainable development is development that meets the needs of the present without compromising the ability of future generations to meet their own needs" [12]. Sustainable development of the energy system focuses on improving the quality of life for all of the Earth's citizens by developing highly efficient energy devices and utilization systems that are cleaner and more environmentally friendly. This requires meeting the needs of the current population with a balanced clean energy mix while minimizing unintentional consequences caused by increases in atmospheric concentrations of greenhouse gases due to a rapid rise in global consumption of carbon-based energy. Ultimately, human society should identify and establish innovative ways to satisfy the needs for energy and chemical feedstocks without increasing the consumption of natural resources beyond the capacity of the globe to supply them indefinitely. Sustainable development requires an understanding that inaction has consequences and that we must find innovative ways to change institutional structures and influence individual behavior [12]. Sustainable development is not a new idea since many cultures over the course of human history have recognized the need for harmony between the environment, society and economy. What is new is an articulation of these ideas in the context of a global industrial and information society [12].

2.3. Vision for efficient utilization of hydrocarbon resources

Figure 3 presents a vision on directions and important issues in research on effective and comprehensive utilization of hydrocarbon resources that are non-renewable. It has been developed by the author for directing

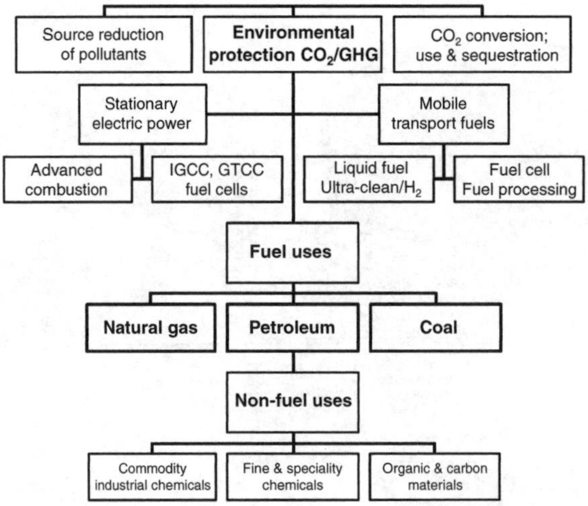

Fig. 3. A personal vision for research towards comprehensive and effective utilization of hydrocarbon resources in the 21st century

future research in our laboratory on clean fuels, chemicals, and catalysis. There are three fundamental elements in this vision: fuel uses, non-fuel uses, and environmental issues of energy and resources. This is a personal view reflecting my judgments and prejudices for future directions. It is helpful to us for seeing future directions and for promoting responsible and sustainable development in research on energy and fuels for the 21st century. Fundamentally, all fossil hydrocarbon resources are non-renewable and precious gifts from nature, and thus it is important to explore more effective and efficient ways of comprehensive utilization of all the fossil energy resources for sustainable development. The new processes and new energy systems should be much more energy-efficient, and also more environmentally benign.

Considering sustainable development seriously today is about being proactive and about taking responsible actions. The principle applies to all the nations in the world, but countries at different stages of economic development can take different but sustainable strategies. As indicated in "The Human Development Report" by the United Nations, Developing countries face a fundamental choice [13]. They can mimic the industrial countries and go through a development phase that is dirty and wasteful and creates an enormous legacy of pollution. Or they can leapfrog over some of the steps followed by industrial countries and incorporate efficient technologies [13]. It is therefore very important for "the present in the world" to make major efforts toward more efficient, responsible, comprehensive and environmentally benign use of the valuable fossil hydrocarbon resources, toward sustainable development.

Does the world really need new conversion devices in addition to internal combustion (IC) engines and heat engines for energy system? The fundamental answer to this question is yes, because the efficiencies of existing energy systems are not satisfactory since over 60% of the energy input is simply wasted in most power plants and in most vehicles for transportation. From an environmental standpoint, many of the existing processes in energy and chemical industries that rely on post-use clean-up to meet environmental regulations should be replaced by more benign processes that do not generate pollution at the source. For example, the current power plants use post-combustion SO_x and NO_x reduction system, but the future system should preferably eliminate or minimize SO_x and NO_x formation at the source. The current diesel fuels contain polycyclic sulfur and aromatic compounds that form SO_x and soot upon combustion in the diesel engines that would require exhaust gas treatment. In the future, ultra-clean fuels could be made at the source, the refinery, which will eliminate or minimize such pollutants before the fuel use in either current engines or future vehicles that may be based on fuel cells. Fuel cells are promising candidates as truly energy-efficient conversion devices [14].

3. PRINCIPLE AND ADVANTAGES OF FUEL CELLS

3.1. Concept of fuel cells

The principle of fuel cell was first discovered in 1839 by Sir William R. Grove, a British jurist and physicist, who used hydrogen and oxygen as fuels catalyzed on platinum electrodes [15,16]. A fuel cell is defined as an electrochemical device in which the chemical energy stored in a fuel is converted directly into electricity. A fuel cell consists of an electrolyte material which is sandwiched between two thin electrodes (porous anode and cathode). Specifically, a fuel cell consists of an anode – to which a fuel, commonly hydrogen, is supplied – and a cathode – to which an oxidant, commonly oxygen, is supplied. The oxygen needed by a fuel cell is generally supplied by feeding air. The two electrodes of a fuel cell are separated by an ion-conducting electrolyte. All fuel cells have the same basic operating principle. An input fuel is catalytically reacted (electrons removed from the fuel elements) in the fuel cell to create an electric current. The input fuel passes over the anode (negatively charged electrode) where it catalytically splits into electrons and ions, and oxygen passes over the cathode (positively charged electrode). The electrons go through an external circuit to serve an electric load while the ions move through the electrolyte toward the oppositely charged electrode. At the electrode, ions combine to create by-products, primarily water and CO_2. Depending on the input fuel and electrolyte, different chemical reactions will occur.

The main product of fuel cell operation is the DC electricity produced from the flow of electrons from the anode to the cathode. The amount of current available to the external circuit depends on the chemical activity and amount of the substances supplied as fuels and the loss of power inside the fuel cell stack. The current-producing process continues for as long as there is a supply of reactants because the electrodes and electrolyte of a fuel cell are designed to remain unchanged by the chemical reactions. Most individual fuel cells are small in size and produce between 0.5 and 0.9 V of DC electricity. Combination of several or many individual cells in a "stack" configuration is necessary for producing the higher voltages more commonly found in low and medium voltage distribution systems. The stack is the main component of the power section in a fuel cell power plant. The by-products of fuel cell operation are heat, water in the form of steam or liquid water, and CO_2 in the case of hydrocarbon fuel.

3.2. Efficiency of fuel cells

A simplified way to illustrate the efficiency of energy conversion devices is to examine the theoretical maximum efficiency [14]. The efficiency limit for heat engines such as steam and gas turbines is defined by Carnot cycle as maximum efficiency = $(T_1 - T_2)/T_1$, where T_1 is the maximum temperature of fluid in a heat engine, and T_2 the temperature at which heated fluid is released. All the temperatures are in Kelvin (K = 273 + degree Celsius), and therefore the lower temperature T_2 value is never small (usually >290 K). For a steam turbine operating at 400°C, with the water exhausted through a condenser at 50°C, the Carnot efficiency limit is (673 − 323)/673 = 0.52 = 52%. (The steam is usually generated by boiler based on fossil fuel combustion, and so the heat transfer efficiency is also an issue in overall conversion.) For fuel cells, the situation is very different. Fuel cell operation is a chemical process, such as hydrogen oxidation to produce water, and thus involves the changes in enthalpy or heat (ΔH) and changes in Gibbs free energy (ΔG). It is the change in Gibbs free energy of formation that is converted into electrical energy [14]. The Gibbs free energy is related to the fuel cell voltage via $\Delta G = -nF\Delta U_0$, where n is the number of electrons involved in the reaction, F the Faraday constant, and ΔU_0 the voltage of the cell for thermodynamic equilibrium in the absence of a current flow which can be derived by $\Delta U_0 = (-\Delta G)/(nF)$ [17]. For the case of H_2—O_2 fuel cell, the equilibrium cell voltage is 1.23 V corresponding to the ΔG of −237 kJ/mol for the overall reaction ($H_2 + (1/2)\ O_2 = H_2O$) at standard conditions (25°C).

The maximum efficiency for fuel cell can be directly calculated based on ΔG and ΔH as maximum fuel cell efficiency $= \Delta G/(-\Delta H)$. The ΔH value for the reaction is different depending on whether the product water is in vapor or in liquid state. If the water is in liquid state, then $(-\Delta H)$ is higher due to release of heat of condensation. The higher value is called higher heating value (HHV), and the lower value is called lower heating value (LHV). If this information is not given, then it is likely that the LHV has been used because this will give a higher efficiency value [14].

3.3. *Types of fuel cells*

On the basis of the electrolyte employed, there are five types of fuel cells. They differ in the composition of the electrolyte and are in different stages of development. They are alkaline fuel cells (AFC), phosphoric acid fuel cells (PAFC), proton-exchange membrane fuel cells (PEMFC), molten carbonate fuel cells (MCFC), and solid oxide fuel cells (SOFC). In all types there are separate reactions at the anode and the cathode, and charged ions move through the electrolyte, while electrons move round an external circuit. Another common feature is that the electrodes must be porous, because the gases must be in contact with the electrode and the electrolyte at the same time.

Table 2 lists the main features of the four main types of fuel cells summarized based on various recent publications [14,18–21]. Each of them has advantages and disadvantages relative to each other. Different types of fuel cells are briefly discussed below, which will pave the ground for further discussions on fuel

Table 2. Types of fuel cells and their features

Features	Fuel cell type			
Name	Polymer electrolyte	Phosphoric acid	Molten carbonate	Solid oxide
Electrolyte	Ion-exchange membrane	Phosphoric acid	Alkali carbonates mixture	Yttria-stabilized zirconia
Operating temperature (°C)	70–90	180–220	650–700	800–1000
Charge carrier	H^+	H^+	CO_3^{2-}	O^{2-}
Electrolyte state	Solid	Immobilized liquid	Immobilized liquid	Solid
Cell hardware	Carbon- or metal-based	Graphite-based	Stainless steel	Ceramic
Catalyst, anode	Platinum (Pt)	Platinum (Pt)	Nickel (Ni)	Nickel (Ni)
Fuels for cell	H_2	H_2	Reformate or CO/H_2	Reformate or CO/H_2 or CH_4
Reforming	External or direct MeOH	External	External or internal	External or internal, or direct CH_4
Feed for fuel processor	MeOH, natural gas, LPG, gasoline, diesel, jet fuel	Natural gas, MeOH, gasoline, diesel, jet fuel	Gas from coal or biomass, natural gas, jet fuel	Gas from coal or biomass, natural gas, gasoline, diesel, jet fuel
Oxidant for cell	O_2/air	O_2/air	CO_2/O_2/air	O_2/air
Co-generation heat	None	Low quality	High	High
Cell efficiency (% LHV)	40–50	40–50	50–60	50–60

processing for fuel cell applications. Detailed description on these fuel cells can be found in comprehensive references [14,20].

3.3.1. Proton-exchange membrane fuel cell

The PEMFC uses a solid polymer membrane as its electrolyte (Scheme 1). This membrane is an electronic insulator, but an excellent conductor of protons (hydrogen cations). The ion-exchange membrane used to date is fluorinated sulfonic acid polymer such as Nafion resin manufactured by Du Pont, which consist of a fluorocarbon polymer backbone, similar to Teflon, to which are attached sulfonic acid groups. The acid molecules are fixed to the polymer and cannot "leak" out, but the protons on these acid groups are free to migrate through the membrane. The solid electrolyte exhibits excellent resistance to gas cross-over [20]. With the solid polymer electrolyte, electrolyte loss is not an issue with regard to stack life. Typically the anode and cathode catalysts consist of one or more precious metals, particularly platinum (Pt) supported on carbon. Because of the limitation on the temperature imposed by the polymer and water balance, the operating temperature of PEMFC is less than 120°C, usually between 70 and 90°C.

PEMFC system, also called solid polymer fuel cell (SPFC), was first developed by General Electric in the US in the 1960s for use by NASA on their first manned space vehicle Germini spacecraft [14]. However, the water management problem in the electrolyte was judged to be too difficult to manage reliably and for Apollo vehicles NASA selected the "rival" alkali fuel cell; General Electric did not pursue commercial development of PEMFC [14]. Today PEMFC is widely considered to be a most promising fuel cell system that has widespread applications. The significant advances in PEMFC in the 1980s and early 1990s were due largely to major development efforts by Ballard Power Systems of Vancouver, Canada, and Los Alamos National Laboratory in the US [14]. The developments on solid polymer fuel cells at Ballard have been summarized by Prater [22]. PEMFC performance has improved over the last several years. Current densities of 850 A/ft^2 are achieved at 0.7 V per cell with hydrogen and oxygen at 65 psi, and over 500 A/ft^2 is obtained with air at the same pressure [18]. The PEMFC technology is primarily suited for residential/commercial (business) and transportation applications [21]. PEMFC offers an order of magnitude higher power density than any other fuel cell system, with the exception of the advanced aerospace AFC, which has comparable performance [18]. The use of a solid polymer electrolyte

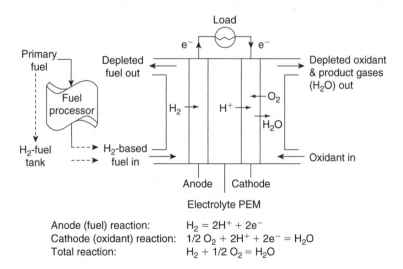

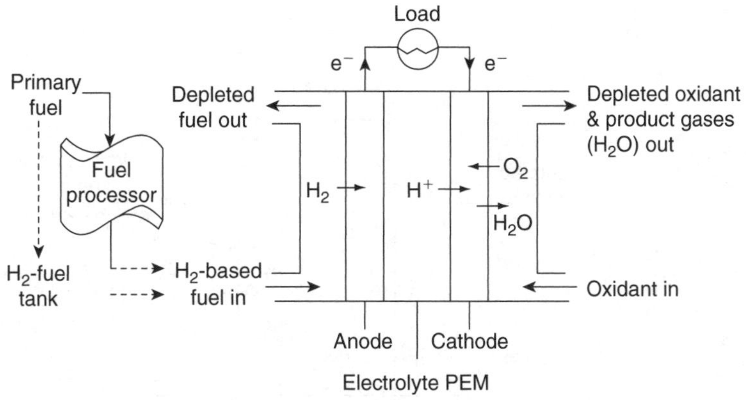

Anode (fuel) reaction: $H_2 = 2H^+ + 2e^-$

Cathode (oxidant) reaction: $1/2\ O_2 + 2H^+ + 2e^- = H_2O$

Total reaction: $H_2 + 1/2\ O_2 = H_2O$

Scheme 1. Concept of PEMFC system using on-board or on-site fuel processor, or on-board H_2 fuel tank

eliminates the corrosion and safety concerns associated with liquid electrolyte fuel cells. Its low operating temperature provides instant start-up and requires no thermal shielding to protect personnel. Recent advances in performance and design offer the possibility of lower cost than any other fuel cell system [18].

In addition to pure hydrogen, the PEMFC can also operate on reformed hydrocarbon fuels without removal of the by-product CO_2. However, the anode catalyst is sensitive to CO, partly because PEMFC operates at low temperatures. The traces of CO produced during the reforming process must be converted into CO_2 by a catalytic process such as selective oxidation process before the fuel gas enters the fuel cell. Higher loadings of Pt catalysts than those used in PAFCs are required in both the anode and the cathode of PEMFC [20]. CO must be reduced to $<$10 ppm, and the CO removal is typically a catalytic process which can be integrated into a fuel processing system. Water management is critical for PEMFC; the fuel cell must operate under conditions where the by-product water does not evaporate faster than it is produced because the membrane must be hydrated [20].

3.3.2. Phosphoric acid fuel cell

The PAFC uses liquid, concentrated phosphoric acid as the electrolyte (Scheme 2). The phosphoric acid is usually contained in a Teflon-bonded silicon carbide matrix. The small pore structure of this matrix preferentially keeps the acid in place through capillary action. Some acid may be entrained in the fuel or oxidant streams and addition of acid may be required after many hours of operation. Platinum supported on porous carbon is used on both the anode (for the fuel) and cathode (for the oxidant) sides of the electrolyte. PAFC operates at 180–220°C, typically around 200°C. The relative stability of concentrated phosphoric acid is high compared to other common acids, which enables PAFC operation at the high end of the acid temperature range of up to 220°C [20]. In addition, the use of concentrated acid of nearly 100% minimizes the water vapor pressure and therefore water management in PAFC is not difficult, unlike PEMFC.

PAFC power plant designs can achieve 40–45% fuel-to-electricity conversion efficiencies on a lower heating value basis (LHV) [23]. PAFC has a power density of 160–175 W/ft^2 of active cell area [18]. Turnkey 200 kW plants are now available and have been installed at more than 70 sites in the United States, Europe, and Japan [21]. Operating at about 200°C (400°F), the PAFC plant also produces heat for domestic hot water and space heating. PAFC is the most mature fuel cell technology in terms of system development and is already in the first stages of commercialization. It has been under development for more than 20 years and has received a total worldwide investment in the development and demonstration

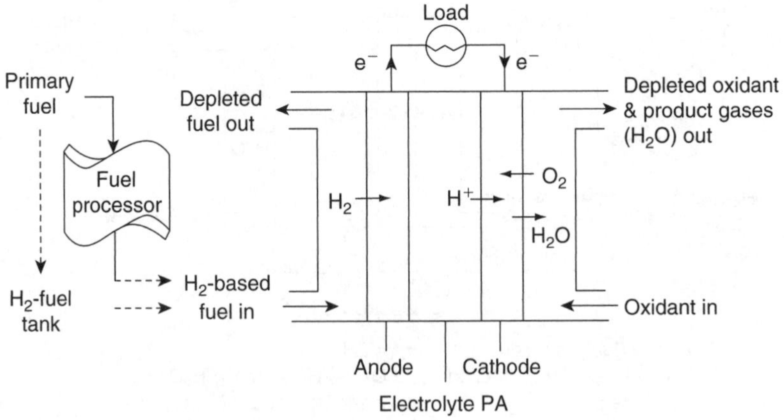

Scheme 2. Concept of PAFC system using on-board or on-site fuel processor

of the technology in excess of \$500 million [18]. The PAFC was selected for substantial development a number of years ago because of the belief that, among the low-temperature fuel cells, it was the only technology which showed relative tolerance for reformed hydrocarbon fuels and thus could have widespread applicability in the near term [18].

3.3.3. Alkaline fuel cell

AFC uses aqueous solution of potassium hydroxide (KOH) as its electrolyte. The electrolyte is retained in a solid matrix (usually asbestos), and a wide range of electrocatalysts can be used, including nickel, metal oxides, spinels, and noble metals electrode [20]. The operating temperatures of AFC can be higher than PAFC by using concentrated KOH (85%) for high-temperature AFC at up to 250°C, or lower by using less concentrated KOH (35–50%) for low-temperature AFC at <120°C. The fuel supply for AFC is limited to hydrogen; CO is a poison; and CO_2 reacts with KOH to form K_2CO_3, thus changing the electrolyte [20].

AFC concept has been described since 1902 in a US patent but they were not demonstrated till the 1940s and 1950s by Francis T. Bacon at Cambridge, England [14]. Since 1960s AFC has been used in space applications that took man to the moon with the Apollo missions [14]. However, the requirement of pure H_2 and the sensitivity to CO_2 appear to be among the major factors limiting the widespread application of AFC. The alkaline fuel cell is being phased out in the US where its only use has been in space vehicles [20]. However, it should be noted that AFC has its advantages of being simple in design and less expensive (electrolyte materials), and may have some applications where its disadvantages (require pure H_2, sensitive to CO_2) are not an issue such as with regenerative fuel cells involving water [14].

3.3.4. Molten carbonate fuel cell

The MCFC uses a molten carbonate salt mixture as its electrolyte (Scheme 3). The composition of the electrolyte varies, but usually consists of lithium carbonate and potassium carbonate (Li_2CO_3—K_2CO_3). At the operating temperature of about 650°C (1200°F), the salt mixture is liquid and a good ionic conductor. The electrolyte is suspended in a porous, insulating and chemically inert ceramic ($LiAlO_2$) matrix [18]. At the high operating temperatures in MCFCs, noble metals are not required for electrodes; nickel (Ni) or its alloy with chromium (Cr) or aluminum (Al) can be used as anode, and nickel oxide (NiO) as cathode [20]. The cell performance is sensitive to operating temperature. A change in cell temperature

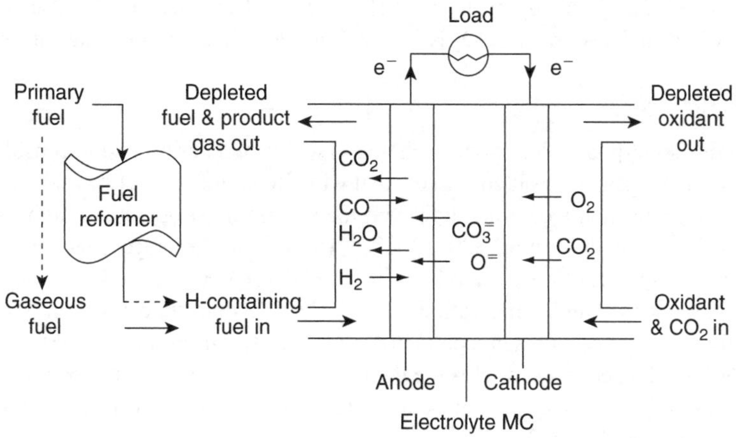

Scheme 3. Concept of MCFC using on-site external fuel reformer. The external reformer can be integrated to the fuel cell chamber directly or indirectly because of the sufficiently high operating temperatures of MCFC

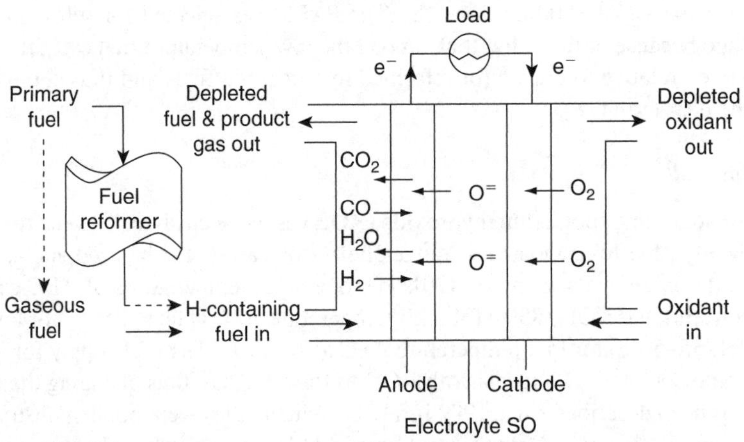

Anode (fuel) reaction: $H_2 + O^= \rightarrow H_2O + 2e^-$

$CO + O^= \rightarrow CO_2 + 2e^-$

Cathode (oxidant) reaction: $O_2 + 2e^- \rightarrow 2\,O^=$

Total reaction: $H_2 + CO + O_2 \rightarrow H_2O + CO_2$

Scheme 4. Concept of SOFC system using on-site or on-board external reformer of primary fuel (natural gas, gasoline, diesel, jet fuel, alcohol fuels, bio-fuels, etc.). The external reformer can be integrated to the fuel cell chamber directly or indirectly because of the higher operating temperatures of SOFC

from 1200°F (650°C) to 1110°F (600°C) results in a drop in cell voltage of almost 15% [18]. The reduction in cell voltage is due to increased ionic and electrical resistance and a reduction in electrode kinetics.

MCFCs evolved from work in the 1960s aimed at producing a fuel cell which would operate directly on coal [18]. While direct operation on coal seems less likely today, operation on coal-derived fuel gases or natural gas is viable. MCFCs are now being tested in full-scale demonstration plants and thus offer higher fuel-to-electricity efficiencies, approaching 50–60% (LHV) fuel-to-electricity efficiencies [23]. Because MCFCs operate at higher temperatures, around 650°C (1200°F), they are candidates for combined-cycle applications, in which the exhaust heat is used to generate additional electricity. When the waste heat is used, total thermal efficiencies can approach 85% [21]. The disadvantages of MCFC are that the electrolyte is corrosive and mobile, and a source of CO_2 is required at the cathode to form the carbonate ion [20].

3.3.5. Solid oxide fuel cell

SOFC uses a ceramic, solid-phase electrolyte (Scheme 4) which reduces corrosion considerations and eliminates the electrolyte management problems associated with the liquid electrolyte fuel cells. To achieve adequate ionic conductivity in such a ceramic, however, the system must operate at high temperatures in the range of 650–1000°C, typically around 800–1000°C (1830°F) in the current technology. The preferred electrolyte material, dense yttria (Y_2O_3)-stabilized zirconia (ZrO_2), is an excellent conductor of negatively charged oxygen (oxide) ions at high temperatures. The SOFC is a solid state device and shares certain properties and fabrication techniques with semiconductor devices [18]. The anode of SOFC is typically a porous nickel–zirconia (Ni—ZrO_2) cermet (cermet is the ceramic–metal composite) or cobalt–zirconia (Co—ZrO_2) cermet, while the cathode is typically magnesium (Mg)-doped lanthanum manganate or strontium (Sr)-doped lanthanum manganate $LaMnO_3$ [18,20].

At the operating temperature of 800–1000°C, internal reforming of most hydrocarbon fuels should be possible, and the waste heat from such a device would be easily utilized by conventional thermal electricity generating plants to yield excellent fuel efficiency. On the other hand, the high operating temperature of SOFC has its own drawbacks due to the demand and thermal stressing on the materials including the sealants and the longer start-up time [20]. Because the electrolyte is solid, the cell can be cast into various shapes such as tubular, planar, or monolithic [20]. SOFCs are currently being demonstrated in a 160 kW plant [21]. They are considered to be state-of-the-art fuel cell technology for electric power plants and offer the stability and reliability of all-solid-state ceramic construction. Operation up to 1000°C (1830°F) allows more flexibility in the choice of fuels and can produce better performance in combined-cycle applications [21]. Adjusting air and fuel flows allows the SOFC to easily follow changing load requirements. Like MCFCs, SOFCs can approach 50–60% (LHV) electrical efficiency, and 85% (LHV) total thermal efficiency [21].

3.4. Advantages of fuel cells compared to conventional devices

In general, all the fuel cells operate without combusting fuel and with few moving parts, and thus they are very attractive from both energy and environmental standpoints. A fuel cell can be two to three times more efficient than an IC engine in converting fuel into electricity [24]. A fuel cell resembles an electric battery in that both produce a direct current by using an electrochemical process. A battery contains only a limited amount of fuel material and oxidant, which are depleted with use. Unlike a battery, a fuel cell does not run down or require recharging; it operates as long as the fuel and an oxidizer are supplied continuously from outside the cell.

The general advantages of fuel cells are reflected by the following desirable characteristics: (1) high-energy conversion efficiency; (2) extremely low emissions of pollutants; (3) extremely low noise or acoustical pollution; (4) effective reduction of greenhouse gas (CO_2) formation at the source compared to low-efficiency devices; and (5) process simplicity for conversion of chemical energy to electrical energy. Depending on the specific types of fuel cells, other advantages may include fuel flexibility and existing infrastructure of hydrocarbon fuel supplies; co-generation capability; modular design for mass production; relatively rapid load response.

Therefore, fuel cells have great potential to penetrate into markets for both stationary power plants (for industrial, commercial, and residential home applications) and mobile power plants for transportation by cars, buses, trucks, trains and ships, as well as man-portable micro-generators. As indicated by US DOE, fuel cells have emerged in the last decade as one of the most promising new technologies for meeting the US energy needs well into the 21st century for power generation [21,25], and for transportation [26,27]. Unlike power plants that use combustion technologies, fuel cell plants that generate electricity and usable heat can be built in a wide range of sizes – from 200 kW units suitable for powering commercial buildings, to 100 MW plants that can add base-load capacity to utility power plants [21].

The disadvantages or challenges to be overcome include the following factors. The costs of fuel cells are still considerably higher than conventional power plants per kW. The fuel hydrogen is not readily available and thus on-site or on-board H_2 production via reforming is necessary. There are no readily available and affordable ways for on-board or on-site desulfurization of hydrocarbon fuels and this presents a challenge for using hydrocarbon fuels [28,29]. The efficiency of fuel processing affects the over system efficiency.

4. FUEL PROCESSING FOR FUEL CELL APPLICATIONS

4.1. Fuel options for fuel cells

Figure 4 illustrates the general concepts of processing gaseous, liquid, and solid fuels for fuel cell applications. For a conventional combustion system, a wide range of gaseous, liquid and solid fuels may be used, while hydrogen, reformate (hydrogen-rich gas from fuel reforming), and methanol are the primary fuels

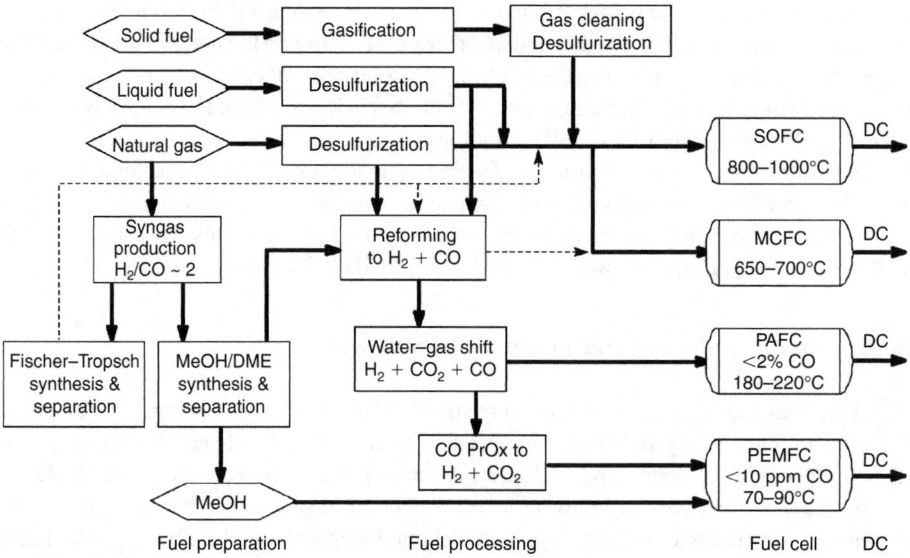

Fig. 4. The concepts and steps for fuel processing of gaseous, liquid and solid fuels for high- and low-temperature fuel cell applications

available for current fuel cells. The sulfur compounds in hydrocarbon fuels poison the catalysts in fuel processor and fuel cells and must be removed. Syngas can be generated from reforming. Reformate (syngas and other components such as steam and carbon dioxide) can be used as the fuel for high-temperature fuel cells such as SOFC and MCFC, for which the solid or liquid or gaseous fuels need to be reformulated. Hydrogen is the real fuel for low-temperature fuel cells such as PEMFC and PAFC, which can be obtained by fuel reformulation on-site for stationary applications or on-board for automotive applications. When natural gas or other hydrocarbon fuel is used in a PAFC system, the reformate must be processed by water – gas-shift (WGS) reaction. A PAFC can tolerate about 1–2% CO [20]. When used in a PEMFC, the product gas from WGS must be further processed to reduce CO to <10 ppm. Synthetic ultra-clean fuels can be made by Fischer–Tropsch synthesis [30] or methanol synthesis using the synthesis gas produced from natural gas or from coal gasification, as shown in Fig. 4, but the synthetic cleanness is obtained at the expense of extra cost for the extra conversion and processing steps.

Hohlein et al. [31] made a critical assessment of power trains for automobiles with fuel cell systems and different fuels including alcohols, ether and hydrocarbon fuels, and they indicated that hydrogen as PEFC fuel has to be produced on-board. H_2 can be obtained by catalytic steam reforming of methanol [32] and ethanol [33,34]. Methanol can also be used for direct electrochemical conversion to H_2 using direct methanol fuel cell (DMFC). Synthetic methanol has the advantage of being ultra-clean and easy to reform at lower temperatures. On the other hand, lower energy density and lack of infrastructure for methanol distribution and environmental concerns are some drawbacks for methanol. The advantages of existing infrastructures of worldwide production and distribution of natural gas, gasoline, diesel and jet fuels have led to active research on hydrocarbon-based fuel processors. Therefore, hydrogen production by processing conventional hydrocarbon fuels is considered by many researchers to be a promising approach [35–37].

It is increasingly recognized that the fuel processing subsystem can have a major impact on overall fuel cell system costs, particularly as ongoing research and development efforts result in reduction of the basic cost structure of stacks which currently dominate system costs [38]. The general processing schemes for syngas and H_2 production through steam reforming of hydrocarbons have been discussed by Gunardson

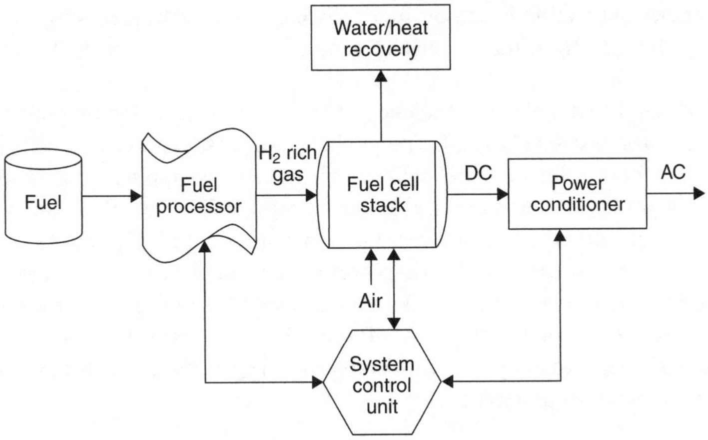

Fig. 5. The components of fuel cell systems for electric power plants

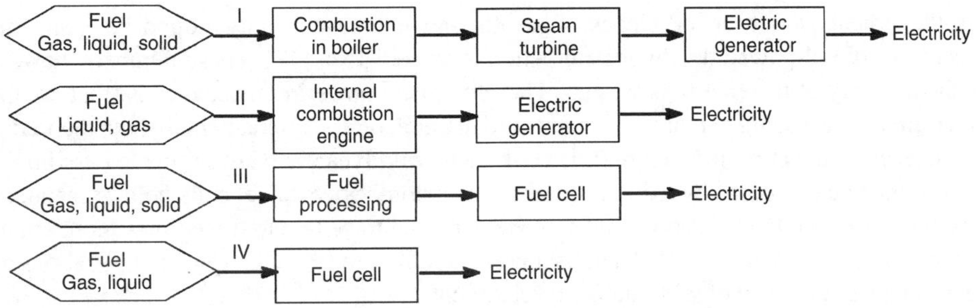

Fig. 6. Different paths of electricity generation from hydrocarbon-based solid, liquid and gaseous fuels

[39], Rostrup-Nielsen [40] and Armor [41] for stationary H_2 plants in the gas industry, and by Clarke et al. [42], Dicks [43], and Privette [44] for fuel cell applications.

4.2. Fuel cells for electric power plants

Figure 5 shows the components of fuel cell systems for electric power plants. Fuel cell systems can be grouped into three sections: fuel processor, generator (fuel cell stack), and power conditioner (DC/AC inverter). In the fuel processor, a fuel such as natural gas or gasoline is processed in several steps to produce hydrogen. The hydrogen-rich fuel and oxygen (air) are then fed into the generator section to produce DC electricity and reusable heat. The generator section includes a fuel cell stack which is a series of electrode plates interconnected to produce the required quantity of electrical power. The output DC electricity from fuel cell is then converted into AC electricity in the power conditioning section where it also reduces voltage spikes and harmonic distortions. The power conditioner can also regulate the voltage and current output from the fuel cells to accommodate variations in load requirements [45].

Figure 6 illustrates different paths of electricity generation from hydrocarbon-based solid, liquid and gaseous fuels by conventional technologies and new technologies based on fuel cells. As shown in Fig. 1 and Fig. 2 [7], a large amount of primary energy is consumed for electricity generation, and most of this electric power is generated via path I in Fig. 6 for fossil fuel-based power plants, and later half of path I for nuclear power plants. The efficiencies of the current electric power plants are about 30–37% in the US.

Path III is the electricity generation based on fuel cells including fuel processing, which is expected be more efficient than path I. Ideally, direct electricity generation based on path IV shown in Fig. 6 would be the most efficient.

Fuel cells have potential to double the efficiency of fossil fuel-based electric power generation, with a resultant slashing of CO_2 emissions [21,25]. The goals for the 21st century fuel cells program of the US DOE include development of solid state fuel cells with installed cost approaching \$400/kW (from current fuel cell cost of about \$4000/kW) and efficiencies up to 80% (LHV) by 2015, and applications include those in distributed power, central station power, and transportation [23,25]. SOFC and MCFC are promising for stationary applications such as electric power plants. Gasification of coal or other carbon-based fuels can be coupled to solid oxide-based or molten carbonate-based fuel cells for more efficient power generation. An extensive review on development of fuel cell technologies in the US, Europe and Japan up to 1995 has been published by Appleby [46] with emphasis on systems, economics and commercialization of fuel cells for stationary power generation.

4.3. Fuel cells for transportation

Currently, the typical overall fuel efficiency of gasoline-powered cars is only around 12%, and the overall fuel efficiency of diesel-powered vehicles are better, at around 15% [47]. These numbers, however, indicate that the majority of the energy is wasted. Therefore, new powering mechanisms (that are more efficient and clean) are also being explored by many auto manufacturers. Fundamentally, the theoretical upper limit of efficiency in the current IC engines is set by a thermodynamic (Carnot) cycle based on combustion, and this must be overcome by using different conversion devices. Fuel cells hold tremendous potential in this direction [48]. Fuel cell-powered cars are expected to be two to three times more efficient than the gasoline and diesel engines [153]. There is a great potential for the widespread applications and there is a fundamental need in view of sustainable development.

The consumption of transportation fuels is increasing worldwide. The total US consumption of petroleum products reached an all-time high of 18.68 million barrels per day (MBPD) in 1998. Of the petroleum consumed, 8.20 MBPD was used as motor gasoline, 3.44 MBPD as distillate fuels (including diesel and industrial fuels), 1.57 MBPD as jet fuels, 0.82 MBPD as residual fuel oil, and 1.93 MBPD as liquefied petroleum gas (LPG), and 2.72 MBPD for other uses [7]. Among the distillate fuels, about 2.2 MBPD of diesel fuel is consumed in the US road transportation market [49]. Due to the high demand and low domestic production in the US, crude oil and petroleum products were imported at the all-time high rate of 10.4 MBPD in 1998, while exports measured only 0.9 MBPD [7]. Between 1985 and 1998, the rate of net importation of crude oil and refinery products more than doubled from 4.3 to 9.5 MBPD [7], largely as a result of increasing demand for transportation fuels in the US. The demand for diesel fuels is increasing faster than the demand for other refined petroleum products and at the same time diesel fuel is being reformulated [50]. According to a recent analysis, diesel fuel demand is expected to increase significantly in the early part of the 21st century and both the US and Europe will be increasingly short of this product [51]. While the world will continue to rely on liquid fuels for transportation in the foreseeable future, the way the world uses liquid fuels in the future – sometime in the 21st century – may be significantly different from today.

PEM-based fuel cells seem to be promising for energy-efficient transportation in the 21st century. The power density that can be achieved with PEMFC is roughly a factor of 10 greater than that observed for the other fuel cell systems which represents a great potential for a significant reduction in stack size and cost over that possible for other systems [18]. The PEMFC typically operates at 70°C (160°F) to 85°C (185°F). About 50% of maximum power is available immediately at room temperature. Full operating power is available within about 3 min under normal conditions. The low temperature of operation also reduces or eliminates the need for thermal insulation to protect personnel or other equipment [18]. There is also hope for using SOFC for automotive applications using hydrocarbon fuels.

The transportation fuel cell program of the US DOE has been introduced in an overview by Milliken [27]. There is a cooperative research program called Partnership for a New Generation of Vehicles (PNGV) between the US federal government and the auto manufacturers including Daimler Chrysler, Ford Motor, and General Motors [52]. The review by Chalk et al. [53] described the status of the PNGV program and the key role and technical accomplishments of the DOE program on transportation fuel cells. A recent NRC report summarized the progress and the current status of fuel processor for automotive applications [52]. The PNGV program for automotive fuel cell applications aimed at creating an 80 miles per gallon PEMFC-powered car [53]. Fuel cells have potential to double the efficiency of energy utilization for transportation, and as an example, the transportation fuel cell program of US DOE has year 2004 target efficiencies up to 48% for gasoline-based vehicles [27]. In January 2002, the US government announced a new program called Freedom CAR (CAR stands for Cooperative Automotive Research), which replaces the PNGV program [54, 55]. The strategic objective of Freedom CAR seems to be directed at developing hydrogen-based fuel cells to power the cars of future [55].

In March 1999, Daimler Chrysler AG unveiled its newest fuel cell vehicle, Necar 4 (new electric car). This is the first time fuel cell system was mounted in the floor of the car. H_2 is the fuel for the fuel cell, and Necar 4 is powered by liquid hydrogen. Recently, Necar 5 has been announced by Daimler Chrysler, which uses the on-board methanol reformer for the fuel cell car; the first long-range fuel cell car test drive was conducted on Necar 5 in May 2002 starting from Sacramento in CA to Washington, DC for a driving distance of about 3000 miles [154]. In April 1999, a large number of companies and California state agencies formed the "California Fuel Cell Partnership" to advance further automotive fuel cell technology. The partnership plans to place 50 fuel cell cars and buses on the road between 2000 and 2003. Ogden et al. [56] made a comparison of hydrogen, methanol and gasoline as fuels for fuel cell vehicles, and discussed their implications for vehicle design and infrastructure development.

4.4. *Fuel cells for residential and commercial sectors*

While centralized electric utilities will continue to be the major generators of electricity in the near future, there are application markets where small fuel cells can serve as convenient generators for residential homes and commercial buildings. The general advantages for such applications include high energy efficiency, low noise, low emissions of pollutants, and low greenhouse gas emissions. For this type of applications using PEMFC, however, catalytic fuel processing should consider non-pyrophoric catalysts for the WGS reaction, as indicated recently [57]. The general principle of fuel processing is the same for most applications, and the fuel processor typically include the components of fuel reforming, WGS, and CO clean-up. The fuels, however, would preferably be those that have existing infrastructure in the distribution network such as natural gas [36]. For residential applications, in addition to natural gas, propane gas or LPG is also a potential fuel for on-site reforming for fuel cells [58].

4.5. *Fuel cells as portable power sources*

So far, the direct methanol fuel cell is the only option as the portable fuel cell. This type of fuel cell uses direct electrochemical oxidation of methanol without fuel reforming. Recently, research efforts have begun on developing miniaturized liquid hydrocarbon-based fuel processor as well as micro-reformer using methanol for micro-fuel cells, for use as man-portable electrical power sources. The advantage of liquid hydrocarbons is the higher energy density compared to methanol for micro-fuel processor development, which should preferably have at least an order of magnitude longer time of effective use without fuel replacement, as compared to batteries.

5. CHALLENGES AND OPPORTUNITIES FOR FUEL PROCESSING RESEARCH

The concepts and steps of fuel processing are illustrated in Fig. 4. There are challenges and opportunities for research and development on fuel processing for fuel cells. The progress in commercial development of fuel cells is faster than many people have predicted a few years ago. Fuel cells have become more promising and increasingly more important in the past few years, perhaps due to a combination of several factors [21,24–27,59–61,155] that stimulate investment in this area: (1) more stringent environmental regulations on controls of pollutant emissions such as EPA Tier II and California ZEV; (2) deregulation of electric power industry and the potential market for distributed generation; (3) intrinsically higher energy efficiency and environmentally friendly nature of fuel cells; (4) advances and successful demonstration of the technology by leading fuel cell companies (such as Ballard Power Systems Inc. in Canada, International Fuel Cells, Siemens Westinghouse, and the Fuel Cell Energy) and financially powerful alliances between fuel cell companies and large auto manufacturers (such as Daimler-Benz, Ford, General Motors and Toyota Motors) and various organizations including US DOE, US DOD, and NEDO in Japan; (5) potential to reduce CO_2 emissions while meeting the energy demands; (6) the potential to double the fuel efficiency in electric power plants by SOFC and MCFC and the potential to triple the fuel efficiency for transportation by PEMFC and SOFC development. Based on the report from Arthur D. Little Inc. in 1998, there are fuel processor technology paths, whose manufacturing cost analyses are consistent with fuel processor subsystem costs of under \$150/kW in stationary applications and \$30/kW in transport applications [38].

Table 3 summarizes the general fuel requirements of fuel cells and impacts of gas components on five different types of fuel cells [14,20,62]. Table 4 lists some of the performance targets for stationary and transport fuel cell applications according to the US DOE [23,25,27]. The US DOE is supporting research and development to address some of the biggest remaining challenges, which include fuel processing and lowering the cost of transportation fuel cell systems [26,27], and the development of more advanced fuel cell systems such as 21st century fuel cells with efficiency up to 70–80% [23,25].

Figure 7 shows the steps and current options for on-site and on-board processing to produce H_2 for low-temperature fuel cells such as PEMFC and PAFC. For catalytic research, needs and opportunities exist in

Table 3. The fuel requirements of fuel cells and impacts of gas components

Species	PEMFC	AFC	PAFC	MCFC	SOFC
Operating temperature (°C)	70–90	70–200	180–220	650–700	800–1000
H_2	Fuel	Fuel	Fuel	Fuel	Fuel
CO	Poison (>10 ppm)	Poison	Poison (>0.5%)	Fuel[a]	Fuel[a]
CH_4	Diluent	Diluent	Diluent	Diluent[a,b]	Diluent[a,b]
CO_2 and H_2O	Diluent	Poison[c]	Diluent	Diluent	Diluent
Sulfur (as H_2S and COS)	Poison (>0.1 ppm)	Unknown	Poison (>50 ppm)	Poison (>0.5 ppm)	Poison (>1 ppm)

[a] CO can react with H_2O to produce H_2 and CO_2 by shift reaction; CH_4 reacts with H_2 and CO faster than reacting as a fuel at the electrode.

[b] A fuel in the external or internal reforming MCFC and SOFC.

[c] CO_2 is a poison for AFC which more or less rules out its use with reformed fuels. Sources: From Refs. [14,20,62].

Table 4. Performance targets for stationary and transport fuel cell applications

Program and application	Parameters	Target values
Second generation fuel cell for stationary applications[a,b]	Efficiency(% LHV)	50–60
	Cost ($/kW)	1000–1500
	Target year	2003
21st century fuel cell for stationary applications[a,b]	Efficiency (% LHV)	70–80
	Cost ($/kW)	400
	Target year	2015
Transportation[c] targets in Partnership for New Generation Vehicle (PNGV) Program	50 kW gasoline fuel processor	
	Energy efficiency (%)	80
	Power density (W/l)	750
	Specific power (W/kg)	750
	CO tolerance (ppm)	10 (CO); 0 (sulfur)
	Emissions	<EPA Tier II
	Start to full power (min)	0.5
	Life time (h)	>5000
	Cost ($/kW)	10
	50 kW reformate fuel cell Subsystem	
	Efficiency	60% at 25% peak power
	Platinum loading (g Pt/kW)	0.2
	Start to full power (min)	0.5
	Cost ($/kW)	40
	Power density (W/l)	500
	CO tolerance (ppm)	100 (CO)
	Life time (h)	>5000
	Target year	2004
	50 kW gasoline-based fuel cell system by 2004	
	Energy efficiency	48% at 25% peak power
	Specific power (W/kg)	300
	Start-up to full power (min)	0.5
	Transient response (s)	1
	Cost ($/kW)	50

[a] See Ref. [23].
[b] See Ref. [25].
[c] See Ref. [27].

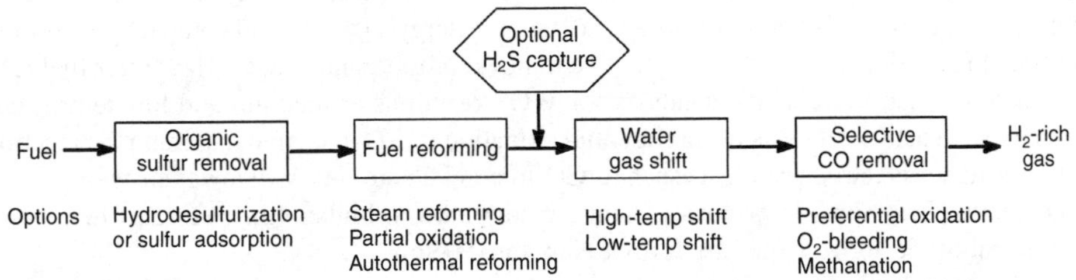

Fig. 7. Steps and current options for on-site and on-board processing liquid and gaseous hydrocarbon fuels and alcohol fuels to produce H_2-rich gas for low-temperature fuel cells (PEMFC)

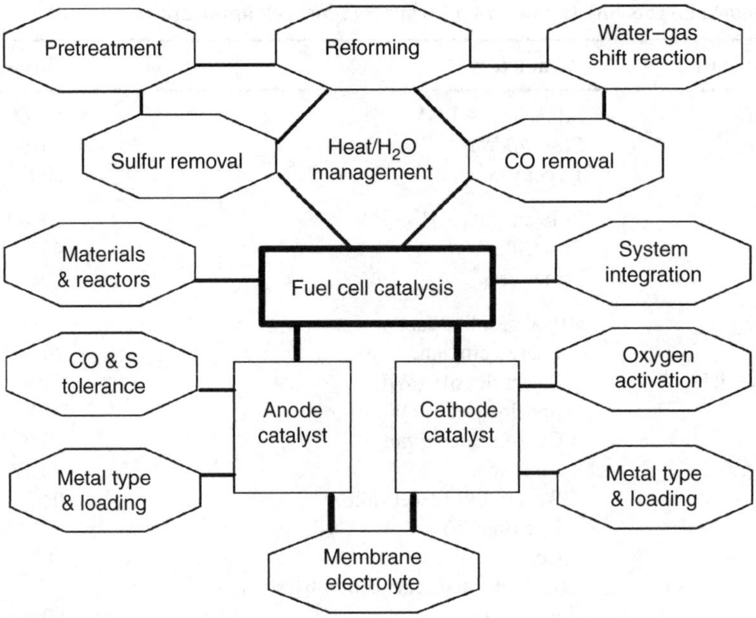

Fig. 8. Some key issues for research and development on fuel processor for fuel cells

several aspects in the area of fuel processing and electrode catalysis related to fuel processing, which involve one or more of the following aspects: catalytic materials development and application, process development, reactor development, system development, sensor and modeling development.

Gasoline, diesel fuels and jet fuels as well as natural gas are potential candidate fuels that all have existing infrastructure of manufacture and distribution, for hydrogen production for fuel cell applications either for stationary or mobile devices. Alcohol fuels such as methanol are among the candidate fuels. The reforming of alcohols can be done at lower temperatures. The processing sequence of hydrogen production from hydrocarbon fuels may involve several steps including fuel deep desulfurization, reforming (partial oxidation, steam reforming, autothermal reforming), WGS (high-temperature shift, low-temperature shift), CO clean-up (by either preferential CO oxidation or CO methanation), followed by feeding into fuel cells or feeding after some gas separation depending on the needs of purity of hydrogen and the impacts of impurities for the specific applications.

Based on the studies reported in literature and conducted in our laboratory, some key issues can be summarized, as shown in Fig. 8, for fuel processing for fuel cells. Based on a preliminary analysis of current situations, it appears necessary for further research to develop (1) effective ways for ultra-deep removal of sulfur from hydrocarbon fuels before reforming; (2) more energy-efficient and compact processors for on-site or on-board fuel reforming; (3) more effective removal of inorganic sulfur (H_2S) after fuel reforming; (4) non-pyrophoric, and more active catalysts for WGS reactions at medium and low temperatures; (5) highly selective and active catalysts for preferential oxidation of CO to enable maximum production of H_2; (6) high-performance electrode catalysts such as CO-tolerant electrodes with lower costs or lower loading of precious metals, and suitable proton-exchange membranes at higher (than current) temperatures for PEMFC. Several of the above issues are further elaborated below.

It should be noted that by using fuel processing for hydrogen production in multiple steps, the net efficiency of the fuel cell system is reduced, and its efficiency advantage is consequently reduced, although such an indirect fuel cell system would still display a significant efficiency advantage. In this context, it is important to develop highly efficient and compact fuel processor for fuel cell applications. While this review focuses on fuel processing, there are other important aspects of fuel cell system such as computational and experimental fluid dynamics [63,64].

5.1. Sulfur removal from hydrocarbon fuels before/after reforming

For conventional transportation fuels used in IC engines, catalytic research on clean fuels involve the following three aspects: (1) fuel processing for improved performance, (2) fuel refining for meeting environmental regulations such as deep removal of sulfur and reduction of aromatics, and (3) pollution control using the exhaust gas treatment system. For fuel cell applications, ultra-clean fuels are needed [28,62], and thus most of the recent discussions in literature on desulfurization of conventional refinery streams to make clean fuels [65–67] also apply to the fuels for fuel cells. Deep hydrodesulfurization of diesel fuels has been discussed in several recent reviews [68–70]. New types of catalysts for conventional hydrodesulfurization of diesel [71] and jet fuels [72] and low-temperature hydrotreating [73] as well as a new integrated system [152] are being explored in our laboratory for deep removal of sulfur from diesel and jet fuels.

However, even with the so-called ultra-low sulfur clean fuels which only contain <30 ppmw sulfur in gasoline and <15 ppmw sulfur in diesel fuels that would meet the EPA Tier II specifications for year 2006, the sulfur contents are still too high for fuel cell applications [28,29]. Some treatments for the sulfur removal is still necessary either before the fuel reforming, with possibly additional polishing for H_2S removal [74] after reforming before the reformate flows to the WGS reactor which also serves to protect PEMFC anode catalyst from sulfur poisoning [29,62]. This will be especially necessary for using petroleum-based gasoline, diesel fuels and jet fuels, because they will inevitably contain sulfur species, mostly in the two to three-ring polycyclic structures [75,76]. Different approaches for treating sulfur may be needed as a part of the fuel processor system [62,77].

A recent study explored the selective adsorption for removing sulfur (SARS) as a new process for on-site or on-board removal of organic sulfur species from hydrocarbon fuels for fuel cell systems [78,79]. It is advantageous to use the selective adsorption for sulfur removal from fuels before the reformer for fuel cells, since this approach can be used at ambient temperatures without using hydrogen [29,62,78]. As indicated by Bellows of International Fuel Cells [77], sulfur is a severe poison for catalysts in fuel processors for fuel cells, especially downstream of reformer; some developers are using sulfur traps before or after the reformer, but "other developers ignore sulfur removal and simply assume that when fuel cells are commercialized the refineries will produce sulfur-free or ultra-low-sulfur fuels" [77]. The selective adsorption [78] can be applied as organic sulfur trap for sulfur removal from fuels before the reformer for fuel cells on-board or on-site, and it may be applied in a periodically replaceable form such as a cartridge. Further improvement in adsorption capacity is desired.

Reformate from autothermal reforming of hydrocarbon fuels such as gasoline, diesel and jet fuels may contain H_2S at ppm levels, which can deactivate the catalysts for subsequent processing such as WGS and also poison the anode catalysts based on platinum. On-site or on-board sulfur removal from such reformate may be necessary. Some recent studies examined solid adsorbent such as ZnO to capture H_2S from reformate before it enters the WGS reactor [20,44,80,81]. The pre-desulfurization of sulfur-containing liquid fuels is also used prior to the catalytic autothermal in the multi-fuel processor being developed by McDermott Technology [82]. Capturing the organic sulfur with solid adsorbent by SARS before fuel reforming is an alternative approach [28,29,79] to conventional hydrodesulfurization. Any organic sulfur species will be converted into H_2S during reforming which produces a reducing atmosphere due to H_2-rich gas. For H_2S capture using ZnO, the morphology of ZnO in the adsorbent is important for effective sulfur removal [74]. More effective adsorbent materials for either organic sulfur or H_2S would be needed for more efficient deep sulfur removal for fuel cell applications.

5.2. Fuel reforming for PEMFC and PAFC

Various fuel cell systems and general fuel reforming methods have been reviewed by Larminie and Dicks [14], Hirschenhofer [20]. Privette [44], and Farrauto and Heck [59] have indicated recently that the PEMFC will be a major focus for research in catalytic fuel processing to make hydrogen from hydrocarbons. PEMFC require hydrogen but not necessarily pure H_2 as the fuel. PEMFC is sensitive to CO because CO poisons the precious metal in the anode at the PEMFC operating temperature. Hydrogen from on-site or on-board fuel processing is an important part of most PEM-based fuel cell systems. The fuel processor converts the hydrocarbon or alcohol fuels into hydrogen-rich gas in several steps. There are three common methods of processing hydrocarbon fuels to create the hydrogen required by the fuel cells. They are steam reforming, partial oxidation, and autothermal reforming or oxidative steam reforming, and the fuels include alcohols and hydrocarbons.

The following equations represent the possible reactions in different processing steps involving three representative fuels: natural gas (CH_4) and liquefied propane gas (LPG) for stationary applications, and liquid hydrocarbon fuels (C_mH_n) and methanol (MeOH) and other alcohols for mobile applications. Most reactions (Eqs (1)–(10), and (15)–(17)) require specific catalysts and process conditions in the current system. Some reactions (Eqs (11)–(14) and (18)) are undesirable but may occur under certain conditions. Trimm and Onsan [83] published a review for on-board fuel conversion and concluded that a combination of oxidation and steam reforming or direct partial oxidation are the most promising processes.

- *Steam reforming*

$$CH_4 + H_2O = CO + 3H_2 \tag{1}$$

$$C_mH_n + mH_2O = mCO + \left(m + \left(\tfrac{1}{2}\right)n\right)H_2 \tag{2}$$

$$CH_3OH + H_2O = CO_2 + 3H_2 \tag{3}$$

- *Partial oxidation*

$$CH_4 + O_2 = CO + 2H_2 \tag{4}$$

$$C_mH_n + \left(\tfrac{1}{2}\right)mO_2 = mCO + \left(\tfrac{1}{2}\right)nH_2 \tag{5}$$

$$CH_3OH + \tfrac{1}{2}O_2 = CO_2 + 2H_2 \tag{6}$$

$$CH_3OH = CO + 2H_2 \tag{7}$$

- *Autothermal reforming*

$$CH_4 + \tfrac{1}{2}H_2O + \tfrac{1}{2}O_2 = CO + \tfrac{5}{2}H_2 \tag{8}$$

$$C_mH_n + \left(\tfrac{1}{2}\right)mH_2O + \left(\tfrac{1}{4}\right)mO_2 = mCO + \left(\left(\tfrac{1}{2}\right)m + \left(\tfrac{1}{2}\right)n\right)H_2 \tag{9}$$

$$CH_3OH + \tfrac{1}{2}H_2O + \tfrac{1}{4}O_2 = CO_2 + 2.5H_2 \tag{10}$$

- *Carbon formation*

$$CH_4 = C + 2H_2 \tag{11}$$

$$C_mH_n = xC + C_{m-x}H_{n-2x} + xH_2 \tag{12}$$

$$2CO = C + CO_2 \tag{13}$$

$$CO + H_2 = C + H_2O \tag{14}$$

- *Water–gas-shift*

$$CO + H_2O = CO_2 + H_2 \tag{15}$$

$$CO_2 + H_2 = CO + H_2O\ (\text{RWGS}) \tag{16}$$

- *CO oxidation*

$$CO + O_2 = CO_2 \tag{17}$$

$$H_2 + O_2 = H_2O \tag{18}$$

5.2.1. Reforming of alcohol fuels

Production of H_2 from alcohol fuels can be achieved by steam reforming of methanol [84,85] and ethanol [33,35,86]. Peppley et al. [87] have reported on the reaction network for steam reforming of methanol on Cu/ZnO/Al_2O_3. Their experimental results showed that, in order to explain the complete range of observed product compositions, one need to include rate expressions for all three reactions (methanol–steam reforming, WGS and methanol decomposition) in the kinetic analysis. The same group has reported on surface mechanisms for steam reforming of methanol over Cu/ZnO/Al_2O_3 catalysts, which account for all three of the possible overall reactions: methanol and steam reacting directly to form H_2 and CO_2, methanol decomposition to H_2 and CO and the WGS reaction [88]. For practical application, Wiese et al. [89] reported on methanol steam reforming in a fuel cell drive system. Peters et al. [90] reported their study on a methanol reformer concept and they considered the particular impact of dynamics and long-term stability for use in a fuel cell-powered passenger car.

Because steam reforming is an endothermic reaction, one processing approach is to create nearly autothermal system by incorporating oxidation into steam reforming, as in the case of hydrocarbon fuel reforming. There are several recent reports on oxidative steam reforming of methanol [32,91–93,156]. Johnson Matthey has recently developed the HotSpot™ methanol processor which combines the steam reforming with catalytic partial oxidation in a single catalyst bed [156], followed by CO removal for on-board hydrogen generation [94]. Reitz et al. [32,92] reported some recent results on steam reforming of methanol over CuO/ZnO under oxidizing conditions. Recently, Velu et al. [93] reported on oxidative steam reforming of methanol over CuZnAl(Zr)-oxide catalysts for the selective production of hydrogen for fuel cells. Fierro [91] presented an overview on both partial oxidation and steam reforming involved in the oxidative steam reforming of methanol for the selective production of hydrogen. Fierro [91] covered studies on activity and effects of operating conditions as well as some mechanistic and kinetic aspects on oxidative steam reforming and partial oxidation of methanol over Cu/ZnO and Pd/ZnO catalysts. Steam reforming and oxidative steam reforming of alcohol are easier than that of a hydrocarbon, but WGS is still necessary for CO removal in general.

5.2.2. *Reforming of hydrocarbon fuels*

One of the recent focus areas is fuel processing for H_2 production from hydrocarbon fuels such as gasoline and diesel fuels for transportation as well as natural gas for stationary applications using low-temperature fuel cells, particularly PEMFC [35–37]. In addition to low cost, transport applications require a fuel processor that is compact and can start rapidly. The fuel processing subsystem for PEMFC, which is the focus of transport applications, includes the reforming, WGS, and deep CO removal [38].

There are two types of metals as candidate catalysts for reforming. The first is non-precious metal (base metal), and the second is precious metal (noble metal) catalyst. Typical base metal catalyst is nickel (Ni) supported on Al_2O_3, with or without alkali promoters. Typical precious metal that has been widely studied is platinum (Pt). While Al_2O_3 is still the widely used support material, various new support materials are being studied. Methods for hydrocarbon reformation include steam reforming, partial oxidation, and autothermal reforming. Steam reforming is widely used in industry for making H_2 and syngas [39–41]. Steam reforming generally give higher H_2/CO ratios (=3) compared to partial oxidation for a given feed, but steam reforming is endothermic and thus requires external heating. Direct partial oxidation (POX) of CH_4 to produce syngas [95, 96] and partial combustion of CH_4 for energy-efficient autothermal syngas production [97] are being explored. Liquid fuel can be reformed by partial oxidation; all the commercial partial oxidation reactors employ non-catalytic partial oxidation of the feed stream by oxygen in the presence of steam with flame temperatures of about 1300–1500°C [20]. These reactions are important but the catalytic partial oxidation is more difficult to control. The major operating problems in catalytic partial oxidation include the over-heating or hot spots due to the exothermic nature of the reactions, and coking problem.

Consequently, coupling the partial oxidation with endothermic steam reforming could lead to a more efficient catalytic autothermal reforming. Many papers and reviews have been published in the recent past on syngas production using autothermal reforming as a part of gas-to-liquids (GTL) research and development efforts worldwide, although such studies were directed for stationary syngas production [30, 97–100]. Trimm and Onsan [83] reported that indirect partial oxidation, which involves combustion of part of the fuel to produce sufficient heat to drive the endothermic steam reforming reaction, is the preferred process for on-board reforming of all fuels including methanol, methane, propane and octane.

The principle of autothermal reforming is applicable to both stationary syngas or H_2 plants and mobile fuel processors. However, non-catalytic POX or non-catalytic combustion, is not suitable for on-board autothermal reforming for fuel cells for mobile applications which prefer compact fuel processors where all the individual steps including reforming, WGS and CO clean-up are carried out inside one enclosure. Some technical issues for H_2 production by reforming of hydrocarbon fuels for PEMFC have been discussed by Bellows [101] and Krumpelt [102]. Several studies have been presented at recent conferences, including fuel processing research at Epyx [103], fuel-flexible processing system at Hydrogen Burner Technology [81], compact fuel processor for fuel cell vehicles being developed jointly by McDermott Technology and Catalytica [82], fuel processors for small-scale stationary PEMFC systems at Northwest

Power Systems [104], and reformate gas processing at Los Alamos National Laboratory [105]. Epyx Corp. (a subsidiary of Arthur D. Little) in the US merged with De Nora Fuel Cells in Italy to form a new company called Nuvera Fuel Cells in April 2000, which will produce fuel cell systems for applications in the stationary power and transportation markets.

Reformation of liquid or gaseous fuels may become an important process for hydrogen production for on-site stationary fuel cell or on-board mobile fuel cell applications, until the direct electrochemical conversion of fuels (or other more efficient conversion routes) become practically feasible. Consequently, there are considerable research interests and commercial developments in hydrocarbon-based fuel processing for transportation [31,38] using gasoline [157] or diesel fuel [158]. Fuel processing studies at US national laboratories have been presented at a recent conference. For example, studies reported by Argonne National Laboratory include catalytic autothermal reforming [106], alternative WGS catalysts [57], effects of fuel contaminants on reforming catalyst performance and durability [107], integrated fuel processor development for PEMFC-based vehicle applications [108], and sulfur removal from reformate [80]. Studies presented by Los Alamos National Laboratory include fuel processing with emphasis on the effects of fuel and fuel constituents on fuel processor performance and catalyst durability [109], and CO clean-up development by preferential oxidation (PrOx) [110]. Researchers from Pacific Northwest National Laboratory reported a compact fuel reforming reactor system based on micro-channel fuel processing [111].

Chalk et al. [26] discussed the challenges for fuel cells in transport applications. Fuel processing for H_2 production by on-board reforming of hydrocarbon fuels for cars and trucks may become important in the early part of next century. It is expected that in the near future a significant fraction of newer vehicles may be hybrid vehicles which use conventional fuels as well as the existing fuel handling and distribution system. An example is a fuel cell vehicle that uses conventional hydrocarbon fuels and performs on-board steam reforming to convert the hydrocarbon fuels into hydrogen and carbon monoxide, followed by the WGS and preferential oxidation to convert CO and water into H_2 and CO_2. In this case, clean hydrocarbon fuels that are extremely low in sulfur and aromatics will be needed. As the new technologies develop further and gain widespread acceptance, vehicles that do not use conventional fuels may penetrate more into road transportation.

A new approach for fuel reforming is to use H_2-selective membrane. The use of supported palladium membrane reactor for steam reforming has also been reported by Kikuchi [112] for membrane development and by Lin and Rei [113] for process development. Kikuchi [112] and Kikuchi et al. [114] have created a composite membrane consisting of thin palladium layer deposited on the outer surface of porous ceramics. By using electroless plating, the palladium layer could completely cover the surface, so that only hydrogen could permeate through the membrane with a 100% selectivity, and such membrane has been incorporated in a steam reformer being developed by Tokyo Gas and Mitsubishi Heavy Industries for the PEMFC system [112]. In a related study, Prabhu and Oyama [115] reported on the preparation and application of hydrogen-selective ceramic membranes for CO_2 reforming of methane.

The work on catalytic fuel reforming in our laboratory is directed toward energy-efficient oxidative steam reforming in carbon-free regions for hydrocarbon fuels as well as methanol for stationary and mobile applications. This is based on our prior work and on-going study for reforming of natural gas to produce syngas and H_2 under various conditions including high-pressure regime [29,65,75,116–119]. There are also fundamentally interesting issues and concepts in literature on hydrocarbon reforming such as oxygen spillover. Based on the report by Maillet et al. [120], oxygen species (OH, O) can be transferred from a Rh/Al_2O_3 catalyst to pure oxides such as ceria, and OH groups stored on the support migrate to the metal particles where the reaction with CH_x fragments from the activation of C_3H_8 (feed molecule) can occur. Rostrup-Nielsen and Alstrup [121] reported that the rates of steam reforming and hydrogenolysis are closely correlated indicating common rate-controlling steps.

It should also be noted that hydrogen production itself is an important subject. In addition to fuel cells, hydrogen production, storage and transportation have other potential applications as a clean energy, as a reactant for chemical processing such as hydrogenation, as well as for fuel processing such as hydrodesulfurization for making ultra-clean fuels. One could also envision some new developments that could result in high-capacity materials for safe storage and transportation of hydrogen that could also be released readily in a safe manner.

5.3. Carbon formation during reforming

Carbon formation is a problem in reforming of hydrocarbon fuels in the stationary syngas plants [122], particularly for hydrocarbon fuels with two or more carbon atoms in the main chain. The same problem can occur also during fuel reforming for fuel cell applications. For example, Sone et al. [159] recently reported on carbon deposition in fuel cell system. Partial oxidation, steam reforming and autothermal reforming can be used for converting liquid hydrocarbon fuels such as gasoline, jet fuel, and diesel fuels into synthesis gas, followed by WGS reaction and preferential CO oxidation to produce H_2-rich gas for use in fuel cells. Heavier hydrocarbons in the jet fuels and diesel fuels can form carbon deposits even at relatively lower temperatures such as 450°C due to fuel pyrolysis [123,124]. More aromatic fuels such as diesel fuels will have a higher tendency of carbon formation.

It is important to clarify the carbon-free conditions and to design effective reforming processes for stable and selective synthesis gas production, which also depend on the type and nature of catalysts. Computational analysis can be carried out to predict the thermodynamically carbon-free region of reforming operations, as has been shown for natural gas reforming under various conditions in our laboratory [116,117]. This is a complicated problem, because there are regions of reaction conditions with certain catalysts where thermodynamics predict no carbon formation but on some catalyst surface carbon is formed. The problem of carbon formation during reforming is also being studied in our laboratory using a tapered element oscillating microscope [118,125].

The high risk of carbon formation problem in steam reforming of liquid hydrocarbons (naphtha) and the importance of using a pre-reformer for reforming of liquid hydrocarbons have been discussed by Rostrup-Nielsen et al. [126]. There are two ways to suppress carbon formation, one is by changing process conditions such as steam/carbon ratio, and the other is by using carbon-resistant catalysts. Higher steam/carbon ratio is useful for minimizing carbon formation, but the use of more steam also increases energy cost. Modification of Ni catalyst by using Mg is beneficial for decreasing carbon formation, and the possible reasons are inhibition of dehydrogenation of adsorbed CH_x species, and enhanced steam adsorption [126]. Another approach is to use noble metal catalyst. It has been reported that the whisker carbon, which is frequently observed on Ni catalyst, does not form on noble metals because these metals do not dissolve carbon [126]. It is well known that essentially all hydrocarbon feeds contain sulfur at different concentrations, and sulfur is the main force for deactivation of pre-reforming and reforming catalysts [126].

5.4. Catalytic WGS

WGS is one of the major steps for H_2 production from gaseous, liquid and solid hydrocarbons or alcohols. WGS is already commercially practiced in the gas industry for syngas and H_2 production, for which the state-of-the-art has been summarized by Gunardson [39] and Armor [41]. For PEM-based fuel cell applications, CO is a poison to the Pt-based anode catalyst and thus deep removal of CO to the ppm level is necessary. On the other hand, the activity of existing commercial WGS catalysts is generally low, and as a result, the largest fraction of the reactor volume is occupied by the WGS part of the fuel processor for H_2 production. Development of more active catalysts would be necessary for a more efficient WGS step in the fuel processing train. Examples of some recent studies are mentioned below.

Thompson and coworkers have recently found that molybdenum carbide catalysts are more active than a commercial Cu—Zn—Al shift catalyst for the WGS reaction at 220–295°C under atmospheric pressure

[160]. They also noted that Mo_2C did not catalyze the methanation reaction, and is a promising candidate for new WGS catalyst. Li et al. [127] reported on a low-temperature WGS reaction over Cu- and Ni-loaded cerium oxide catalysts. Tabakova et al. [128] examined supported gold catalysts on various supports for the WGS reaction, and they concluded that the catalytic activity of the gold/metal oxide catalysts depends strongly not only on the dispersion of the gold particles but also on the state and the structure of the supports. Recently, Utaka et al. [129] made an attempt on CO removal by an oxygen-assisted WGS reaction over supported Cu catalysts. Cu/Al_2O_3—ZnO demonstrated an excellent activity for catalytic removal of CO by oxygen-assisted WGSR, and the equilibrium concentration obtained from thermodynamic data indicates that the reaction is desirable at lower temperatures [129]. New ways of catalyst preparation could lead to more active or more selective catalysts, and this is applicable to but not limited to WGS reaction. Shen and Song [130] reports a new method to prepare highly active Cu—ZnO—Al_2O_3 catalyst that can minimize CO formation and is active at lower temperature. Chandler et al. [131] reported on the preparation and characterization of supported bimetallic Pt—Au and Pt—Cu catalysts from bimetallic molecular precursors.

From the reactor engineering side, Tonkovich et al. [132] reported on a different approach to WGS using micro-channel reactors. Micro-channel reactors reduce heat and mass transport limitations for reactions, and thus facilitate exploiting fast intrinsic reaction kinetics, i.e. high effectiveness factors [132].

5.5. *Deep removal of CO*

For PAFC, the above-mentioned WGS is usually sufficient for producing H_2-rich fuel gas, because the anode catalyst in PAFC can tolerate about $<2\%$ CO. The anode catalyst for PEMFC is usually made of Pt/C, which is more sensitive to CO because PEMFC operates at lower temperatures at which CO can deactivate Pt metal. Usually, CO in the fuel must be reduced to <10 ppm. Even with Ru addition to modify Pt for improved CO tolerance by using Pt—Ru/C anode, CO in the H_2-rich gas should be reduced to <30 ppm. It is difficult for WGS to reach this level of CO reduction. Three processes can be used to further reduce CO in the feed, preferential or selective oxidation, methanation, and membrane separation [14]. In the PrOx, a small amount of air (usually about 2%) is added to the gas (fuel) stream from WGS, which then passes over a precious metal catalyst. This catalyst preferentially adsorbs CO, rather than H_2, where CO reacts with oxygen (from air). After WGS, selective oxidation of CO may be performed, preferably inside a compact unit [133]. Rohland and Plzak [134] reported on CO oxidation using Fe_2O_3-Au catalyst system and achieved relatively high oxidation rate at 1000 ppm CO and 5% "air bleed" at 80°C that could enable a PEMFC-integrated CO oxidation. Dudfield et al. [135] reported on a modeling study for a CO-selective oxidation reactor for solid polymer fuel cell automotive applications.

Methanation is the hydrogenation of CO using the H_2 that is already present in the feed stream. Methanation reaction is the opposite of steam reforming of methane. The methanation approach avoids the oxygen addition, and thus avoids the process complication. The methane produced does not poison the electrode, and only act as a diluent. However, the disadvantage of the method is the consumption of hydrogen [14].

Membrane approach is generally designed for separation and purification. Membrane can be used for separating hydrogen from gas mixtures. Palladium membrane has been studied more extensively than other types of membranes for hydrogen separation, but it is still very expensive for use in fuel cell system [14]. If a membrane is used, then the selective removal of hydrogen in a membrane reactor enables the hydrogen production by steam reforming at lower reaction temperatures than conventional processes.

5.6. *Fuel processing for SOFC and MCFC*

For the low-temperature fuel cells (PEMFC and PAFC) using hydrocarbon fuels, in addition to fuel reforming, several steps of fuel processing are required to convert the CO because H_2 is a fuel but CO is not a fuel and CO can deactivate the anode catalyst. These add to the cost and complexity of the fuel processing system when compared to those needed for the high-temperature fuel cells (SOFC and MCFC). Therefore, another focus area is fuel processing for high-temperature fuel cells including SOFC and MCFC, which could use either internal reforming (since the internal temperature is high enough for fuel conversion) or external reforming, or both. The fuel reforming discussed above for PEMFC is also applicable to the external reformer for SOFC and MCFC, which could use steam reforming, or partial oxidation, or autothermal reforming. However, WGS reaction and preferential CO oxidation will not be necessary when SOFC and MCFC are used.

When hydrocarbon fuels are to be used, high-temperature fuel cells (SOFC and MCFC) have an efficiency advantage over the PEMFC at the system level. One of the major problems of fuel reforming at high temperatures is carbon formation, as already discussed in the previous section. Fuel reforming for SOFC or MCFC is an active research subject. For example, recently, Peters et al. [90] reported on pre-reforming of natural gas in SOFC systems. Finnerty et al. [136] described a SOFC system with integrated catalytic fuel processing. The higher temperature waste heat of these systems (in the case of the SOFC and MCFC) can be used to assist in the reforming of hydrocarbon fuels, to drive air compressors, and to produce steam for thermal electric generation or other thermal load [18].

The catalytic aspects for the internal fuel reforming have been discussed recently [42,43]. There are two approaches in internal reforming, direct internal reforming fuel cell where both fuel reformation and electrochemical reaction takes place in anode chamber, and indirect internal reforming fuel cell where the fuel reformation and electrochemical reaction takes place on the two sides of the wall for anode gas chamber. Fuel Cell Energy is developing an externally manifolded, internally reforming molten carbonate fuel cell [25] called the direct fuel cell or DFC [137]. They have completed a 4000 h, 250 kW stack test in Danbury, CT [25]. Because the fuel is reformed to hydrogen-rich gas internally in the stack, this design eliminates the external fuel processing unit required by PAFC and PEMFC. The advantages of internal reforming include: (1) separate equipment to process the fuel externally is eliminated; (2) equipment count is lower, leading to simpler operation and higher reliability; and (3) efficiency of the system is increased [20,137].

One of the problems with internal reforming is anode deactivation. Coe et al. [138] reported on a kinetic study for the removal of surface carbon formed by methane decomposition following high-temperature reforming on a nickel/zirconia SOFC anode using methods based on temperature-programmed oxidation. They observed that the addition of small quantities of lithium to the anode resulted in a significant lowering of the activation energy for surface carbon removal by about 50 kJ/mol [138].

External fuel reforming is also being explored for high-temperature fuel cells. The advantage of external reforming is the flexibility of fuel processor design which is not limited by fuel cell stack design [20]. In the case of using liquid fuels (gasoline, diesel, and jet fuel) and gaseous fuels (natural gas, propane gas), a catalytic reformer can be placed adjacent to the anode gas chamber. In the design with external reformer, the reformer can be operated at elevated pressures although the fuel cell stack may be operated under atmospheric pressure. The reformer and the fuel cell do not have a direct physical effect on each other, other than heat transfer. This is an advantage also because this design eliminate the problem of deactivation of electrode catalyst due to carbon formation via fuel decomposition, which may be an important concern in the internal reforming fuel cells. Cavallaro and Freni [139] have discussed the feasibility and the overall process economy of an integrated system of autothermal reformer (ATR) and MCFC. M-C Power is developing an internally manifolded, externally reforming molten carbonate fuel cell, and they have completed two 75 kW stack tests in California [25].

The advantage of external reforming is also important when considering the SOFC or MCFC-type fuel cell-based electric power plants using coal gasification. Production of gaseous fuel (syngas) from coal gasification requires solid-gas reactions and involves both ash/residue disposal and hot gas clean-up before its use in fuel cell stack. It is better for such a process to be carried out outside the fuel cell stack. Detailed discussions on coal gasification are available in literature [46]. For SOFC and MCFC, CO does not act as a poison and can be used directly as a fuel. The SOFC is also the most tolerant of any fuel cell type to sulfur. It can tolerate several orders of magnitude more sulfur than other fuel cells. However, the 900–1000°C operating temperature of the SOFC requires a significant start-up time. The cell performance is very sensitive to operating temperature. A 10% drop in temperature results in ~12% drop in cell performance due to the increase in internal resistance to the flow of oxygen ions [18].

For both internal and external reforming, the reaction process needs to be carried out in the thermodynamically carbon-free region using suitable steam/carbon ratios. Even in such a region catalyst may still suffer from deactivation due to carbon formation (and in the case of MCFC, contamination of catalyst surface by other metal compounds). The development of catalysts that resist coke formation and other deactivation would be desirable. Increasing steam/carbon ratios for the reforming could reduce carbon formation, but higher steam/carbon ratios also result in lower energy efficiency, because vaporization and heating of water consume significant amount of energy. In industrial steam reformer, steam/carbon ratios of around 3 are used, but much lower ratios would be desirable for fuel cell processors [29].

SOFC offers inherently high efficiency up to 60–70% in individual systems and up to 80% in staged or hybrid systems [140]. The development of SOFC has been mainly directed toward large-scale power generation with early commercial devices of 1–3 MW [61]. The high temperature of SOFC also demands that the system include significant thermal shielding to protect personnel and to retain heat. While such requirements are acceptable in a utility application, they are not consistent with the demands of most transportation applications nor do they lend themselves to small, portable or transportable applications [18]. When the operating temperatures of SOFC can be lowered from 900–1000 to 600–800°C with improved cell performance, SOFC would be more promising for both stationary and mobile applications. Siemens Westinghouse is developing tubular SOFC under the support of US DOE, and they have tested multiple tubular SOFC for 70 000 h [25]. A recent study involved thermodynamic model and parametric analysis of a tubular SOFC module [161]. Recent advances have made SOFC seem also promising for transport applications in addition to stationary applications [140]. SOFC can handle liquid fuels by on-board reforming relatively easily because of the high cell operating temperatures enabling reformation of hydrocarbon fuels such as gasoline and diesel fuels [140]. There are also on-going efforts for developing hybrid fuel cell/turbine systems, for which the fuel-to-electricity efficiency goal is 70% [25].

For MCFC, the need for CO_2 in the oxidant stream requires that CO_2 from the spent anode gas be collected and mixed with the incoming air stream. Before this can be done, any residual hydrogen in the spent fuel stream must be burned. Future systems may incorporate membrane separators to remove the hydrogen for recirculation back to the fuel stream [18].

5.7. Electrode catalysis related to fuel processing

Precious metals are used for electrode catalysts in low-temperature fuel cells such as PEMFC and PAFC. It is important to reduce the amount of precious metal loadings such as Pt by using effective techniques [141]. The anode catalysts for PEM-based fuel cells are generally based on Pt, which is active at the low operating temperature of PEMFC but is sensitive to CO poisoning. Progress has been made in improving CO tolerance of the anode catalysts [27], and one of the ways to increase CO resistance is to use Pt-Ru as anode catalyst. Paulus et al. [142] used organometallic compounds for preparing Pt-Ru alloy colloids as precursors for fuel cell catalysts. Some recent studies have been reported on the characteristics of platinum-based electrocatalysts and more CO-tolerant Pt catalysts for mobile PEMFC applications [143]. Understanding the key factors affecting CO tolerance of the metal and alloy catalysts and the development of more CO-tolerant anode metal or alloy catalysts can reduce the need for deep CO removal or even reduce the need for extremely high CO conversion by the less-efficient WGS. This can lead to higher fuel cell system efficiency. New ways of reducing the minimum loading of precious metals or the use of non-precious metals, if successful, will be beneficial to the cost reduction of fuel cell system. Electrocatalysis for direct methanol oxidation (without reforming) has some problems due to conflicting demands.

The anode catalysts for high-temperature direct internal reforming fuel cells (SOFC, MCFC) need to be active and more resistant to carbon formation involving hydrocarbon fuels. The chemical composition and nature of anode catalysts also affect carbon formation in addition to the internal reforming reaction conditions. There are recent reports on internal reforming on anode catalyst in SOFC. For example, Finnerty et al. [136] have conducted internal steam reforming of methane using nickel/zirconia fuel reforming anodes in SOFC and observed that addition of small quantities of molybdenum leads to a significant reduction in the amount of carbon deposited, whilst having little effect on the reforming activity or cell performance. Based on temperature-programmed oxidation, Finnerty et al. [136] found that three types of carbon are formed on the anodes of SOFC during high-temperature steam reforming of methane. They also observed that as current is drawn from the cell, increased methane conversion occurs together with reduced carbon deposition, through reaction via partial oxidation and oxidative coupling with the flux of oxygen ions through the solid electrolyte. Nakagawa et al. [144] studied the catalytic activity of Ni—YSZ—CeO_2 anode for the steam reforming of natural gas in a direct internal reforming SOFC.

5.8. Direct oxidation of methanol and methane in fuel cells

All of the above-mentioned fuel cells require some fuel reformation when the fuels are hydrocarbons or alcohols. As indicated already, fuel processing for reformulation also reduces the efficiency of the overall energy conversion. It should be mentioned that the concepts of direct fuel-electricity conversion without fuel reforming are being explored, although the details are beyond the scope of this paper on fuel processing. Two principal approaches are DMFC based on PEMFC, and direct methane fuel cell (DMeFC) or direct hydrocarbon fuel cell based on SOFC.

Direct methanol fuel cell operate at relatively low temperatures in the range of ambient temperature to 150°C; it is promising for portable devices such as fuel cell for laptop computer and mobile telephone. Recent studies on direct methanol fuel cells and advances in this topic area have been summarized in several reviews [145–147]. The challenges include the development of more efficient anode and cathode catalysts, and the more effective membranes that resist methanol cross-over and can operate at higher temperatures.

Direct methane fuel cell based on SOFC requires much higher operating temperatures compared to direct methanol fuel cell. Direct electrochemical methane oxidation have been studied for some time. Further development of science and technology for SOFC could make it possible to directly use hydrocarbons such as CH_4 to generate electricity. More recent studies have been reported on direct electrochemical oxidation of hydrocarbons (beyond methane) in SOFC without fuel reforming [148–150]. These recent developments made direct oxidation based on SOFC more promising, but there are also major problems to be solved [151].

6. CONCLUDING REMARKS

Development and utilization of more efficient energy conversion devices are necessary for sustainable and environmentally friendly development in the 21st century.

Fuel cells are fundamentally much more energy-efficient, and can achieve as high as 70–80% system efficiency in integrated units including heat utilization, because fuel cells are not limited by the maximum efficiency of heat engines or IC engines dictated by the Carnot cycle.

Hydrogen would be an ideal fuel for fuel cells but due to the lack of infrastructure for distribution and storage, processing of fuels is necessary for producing H_2 on-site for stationary applications or on-board for mobile applications.

Hydrocarbons and alcohols can both be used as fuels for reforming on-site or on-board. Hydrocarbon fuels have the advantages of existing infrastructure of production and distribution, while alcohol fuels can be reformed at substantially lower temperatures.

Further research and development are necessary on fuel processing for improved energy efficiency and size reduction, and on electrode catalysis related to fuel processing such as tolerance to CO and sulfur components in reformate. For example, more effective ways of deep removal of sulfur before and after reforming, energy-efficient and stable autothermal reforming catalyst and processing scheme, more active and non-pyrophoric catalysts for WGS, more CO-tolerant anode catalysts or membrane electrode operating at higher temperature would contribute significantly to implementation of low-temperature fuel cells, particularly PEMFC.

ACKNOWLEDGMENTS

The author was motivated to write this review based on the discussion with Dr. Mark Williams, the Fuel Cell Product Manager, and Dr. Randy Gemmen, Scientist, at the National Energy Technology Laboratory, US Department of Energy, during their visit to our laboratory on October 5, 2000. The author gratefully acknowledges helpful general discussions on fuel cells with his colleagues at Penn State and experts in industry, particularly Dr. Chao-Yang Wang, Dr. Andre' L. Boehman, Dr. Serguei Lvov, Dr. Robert Santoro and Dr. Frank Rusinko of Penn State, as well as Dr. Robert Farrauto of Engelhard Corp., Dr. Sai Katikaneni of Fuel Cell Energy Corp., and Dr. John Armor and Dr. Robert Miller of Air Products and Chemicals Inc. The author is particularly grateful to Dr. Robert Farrauto of Engelhard Corp. for very helpful discussions on general challenges in fuel processing during his visit to our laboratory in September 2000. The author also wishes to thank his coworkers (X. Ma, M. Sparague, S. Velu, I. Novochinskii, J. Shen, and S.T. Srinivas, W. Pan) who read the draft of the manuscript and provided helpful comments.

REFERENCES

1. C. Flavin and S. Dunn. Reinventing the energy system. In: *State of the World*. Worldwatch Institute, Washington, DC, 1999, p. 22 (Chapter 2).
2. Statistical Abstract of the United States (SAUS) 1998, 118th edn. US Department of Commerce, 1999.
3. USCB, Historical Estimates of World Population, US Census Bureau, 1999.
4. G. Marland, T.A. Boden and R.J. Andres. Global, regional, and national CO_2 emissions. In: *Trends: A Compendium of Data on Global Change*. Carbon Dioxide Information Analysis Center, Oak Ridge National Laboratory, US Department of Energy, Oak Ridge, TN, 2000.
5. H. Friedli, H. Lötscher, H. Oeschger, U. Siegenthaler and B. Stauffer. Ice core record of 13C/12C ratio of atmospheric CO_2 in the past two centuries. *Nature* **324** (1986) 237–238.
6. C.D. Keeling and T.P. Whorf. Atmospheric CO_2 records from sites in the SIO air sampling network. In: *Trends: A Compendium of Data on Global Change*. Carbon Dioxide Information Analysis Center, Oak Ridge National Laboratory, US Department of Energy, Oak Ridge, TN, 2000.
7. EIA/AER, *Annual Energy Review* (1998). Energy Information Administration, US Department of Energy, Washington, DC, 1999.
8. EIA/IEA, *International Energy Annual*. Energy Information Administration, US Department of Energy, Washington, DC, 2000.
9. C.J. Campbell and J.H. Laherrere. The end of cheap oil. *Sci. Am.* **278**(**3**) (1998) pp. 78–83.
10. G.A. Olah. Nonrenewable fossil fuels. *Chem. Eng. News* **11** (1991) 50–51.
11. World Commission on Environment and Development (WCED). *Our Common Future*. Oxford University Press, Oxford, 1987, p. 43.
12. International Institute for Sustainable Development (IISD). Introduction to Sustainable Development (http://sdgateway.net/introsd), viewed 25 October 2000.

13. The United Nations (UN). *The Human Development Report*, Oxford University Press, Oxford, 1998.
14. J. Larminie and A.L. Dicks. *Fuel Cell Systems Explained.* Wiley, New York, 2000, 308 pp.
15. W.R. Grove. *Philos. Mag.* **14** (1839) 127.
16. W. Vielstich, *Fuel Cells*. Wiley Interscience, London, 1965, 501 pp.
17. L. Carrette, K.A. Friedrich and U. Stimming. Fuel cells – fundamentals and applications. *Fuel Cells* **1** (2001) 5–39.
18. Dodfuelcell. Overview of Fuel Cells, On-line publication (http://www.dodfuelcell.com/fcdescriptions.html), viewed 25 October 2000.
19. Encyclopedia Britannica. Fuel Cell, Encyclopædia Britannica On-line (http://www.eb.com:180/bol/topic?eu=108544&sctn=1&pm=1), viewed 18 October 2000.
20. J.H. Hirschenhofer, D.B. Stauffer, R.R. Engleman and M.G. Klett. *Fuel Cell Handbook*, 4th edn. DOE/FETC-99/1076, US Department of Energy, Federal Energy Technology Center, Morgantown, WV, November 1998.
21. M.C. Williams. Advanced Fuel Cell Power Systems (http://www.fe.doe.gov/coal_power/fuel_cells/fc_sum.html#top), viewed 25 October 2000.
22. K.B. Prater. Solid polymer fuel cell developments at Ballard. *J. Power Sources* **37(1–2)** (1992) 181–188.
23. V. Der and M.C. Williams. Advanced Generation Fuel Cells, Office of Fossil Energy, US DOE (http://www.fe.doe.gov/coal_power/fuelcells/index.html), viewed on-line 2 November 2000.
24. S. Thomas and M. Zalbowitz. Fuel Cells, Green Power, Publication No. LA-UR-99-3231, Los Alamos National Laboratory, Los Alamos, NM, 2000.
25. R.A. Baujura. Fuel cells: simple solutions in a complicated world. In: *Proceedings of the Joint DOE/EPRI/GRI Review Conference on Fuel Cell Technology*, Chicago, IL, August 3–5, 1999.
26. S.G. Chalk, J.F. Miller and F.W. Wagner. Challenges for fuel cells in transport applications. *J. Power Sources* **86**(1) (2000) 40–51.
27. J. Milliken. Transportation fuel cell program. In: *Proceedings of the Annual National Laboratory R&D Meeting of DOE Fuel Cells for Transportation Program*, Tri-Cities, WA, June 7–8, 2000.
28. C. Song. Keynote: catalysis and chemistry for deep desulfurization of gasoline and diesel fuels. An overview. In: *Proceedings of the fifth International Conference on Refinery Processing, AIChE 2002 Spring National Meeting*, New Orleans, March 11–14, 2002, pp. 3–12.
29. C. Song. Catalytic fuel processing for low- and high-temperature fuel cell applications. Challenges and opportunities. In: *Proceedings of the Topical Conference on Fuel Cell Technology, AIChE 2002 Spring National Meeting*, New Orleans, March 11–14, 2002, pp. 125–135.
30. M.M.G. Senden, A.D. Punt and A. Hoek. Gas-to-liquids processes: current status and future prospects. *Stud. Surf. Sci. Catal.* **119** (1998) 961–966.
31. B. Hohlein, S. von Andrian, T. Grube and R. Menzer. Critical assessment of power trains with fuel cell systems and different fuels. *J. Power Sources* **86(1)** (2000) 243–249.
32. T.L. Reitz, S. Ahmed, M. Krumpelt, R. Kumar and H.H. Kung. Oxidative methanol reforming over CuO/ZnO for H_2 production. In: *Proceedings of the Technical Program of 16th Meeting of North American Catalysis Society*, Boston, May 30–June 4, 1999, p. 107.
33. I. Fishtik, A. Alexander and R. Datta. Hydrous ethanol reforming for fuel cell applications: catalysis, thermodyanamics, mechanics and kinetics. In: *Proceedings of the Technical Program of 16th Meeting of North American Catalysis Society*, Boston, May 30–June 4, 1999, p. 108.
34. T. Ioannides. Thermodynamic analysis of ethanol processors for fuel cell applications. *J. Power Sources* **92(1)** (2001) 17–25.
35. P.S. Chintawar, C. Papile and W.L. Mitchell. Catalytic processes in fuel processors for fuel cells in automotive applications. In: *Proceedings of the Technical Program of 16th Meeting of North American Catalysis Society*, Boston, May 30– June 4, 1999, p. 109.
36. R. Farrauto. The generation of H_2 for fuel cells. In: *Proceedings of the Symposium on Clean Processes and Environment: The Catalytic Solution*, Lyon, France, December 6–8, 1999.
37. W.P. Teagan, B.M. Barnett and R.S. Weber. Catalyst opportunities in fuel cell systems. In: *Proceedings of the Technical Program of 16th Meeting of North American Catalysis Society*, Boston, May 30–June 4, 1999, p. 6.
38. W.P. Teagan, J. Bentley and B. Barnett. Cost reductions of fuel cells for transport applications: fuel processing options. *J. Power Sources* **71(1–2)** (1998) 80–85.
39. H. Gunardson. *Industrial Gases in Petrochemical Processing*. Marcel Dekker, New York, 1998, 283 pp.
40. J.R. Rostrup-Nielsen. Production of synthesis gas. *Catal. Today* **18(4)** (1993) 305–324.

41. J.N. Armor. The multiple roles for catalysis in the production of H_2. *Appl. Catal. A Gen.* **176** (1999) 159–176.
42. S.H. Clarke, A.L. Dicks, K. Pointon, T.A. Smith and A. Swann. Catalytic aspects of the steam reforming of hydrocarbons in internal reforming fuel cells. *Catal. Today* **38(4)** (1997) 411–423.
43. A.L. Dicks. Advances in catalysts for internal reforming in high temperature fuel cells. *J. Power Sources* **71(1)** (1998) 111–122.
44. R.M. Privette. Fuel processing technology. In: *Proceedings of the Fuel Cell Tutorial at 25th International Technical Conference on Coal Utilization and Fuel Systems*, Clearwater, FL, March 6, 2000.
45. F.H. Holcomb. How a Fuel Cell Operates, On-line publication (http://www.dodfuelcell.com/paper2.html), viewed 25 October 2000.
46. A.J. Appleby. Fuel cell technology: status and future prospects. *Energy* **21(7)** (1996) 521–653.
47. M. Jones. Hybrid vehicles – the best of both worlds. *Chem. Ind.* **15** (1995) 589–592.
48. T. Ford. Fuel cell – vehicles offer clean and sustainable mobility for the future. *Oil Gas J.* **97(50)** (1999) 130–133.
49. L.E. Bensabat. U.S. fuels mix to change in the next 2 decades. *Oil Gas J.* **97(28)** (1999) 46–53.
50. UOP. *Diesel Fuel. Specifications and Demand for the 21st Century*. UOP, 1998.
51. A. Brady. Global refining margins look poor in short term, buoyant later next decade. *Oil Gas J.* **97(46)** (1999) 75–80.
52. National Research Council (NRC). *Review of the Research Program of the Partnership for a New Generation of Vehicles*. Sixth Report. National Academy Press, Washington, DC, 2000, 114 pp.
53. S.G. Chalk, J. Milliken, J.F. Miller and S.R. Venkateswaran. The US department of energy – investing in clean transport. *J. Power Sources* **71(1–2)** (1998) 26–35.
54. J. Ball. Bush shifts gears on car – research priority. *The Wall Street J.* January 9, 2002.
55. N. Banerjee. White House shifts strategy on future fuel for vehicles. *The New York Times*. January 9, 2002.
56. J.M. Ogden, M.M. Steinbugler and T.G. Kreutz. A comparison of hydrogen, methanol and gasoline as fuels for fuel cell vehicles: implications for vehicle design and infrastructure development. *J. Power Sources* **79(2)** (1999) 143–168.
57. D. Myers, J. Krebs, T. Krause and M. Krumpelt. Alternative water – gas shift catalysts. In: *Proceedings of the Annual National Laboratory R&D Meeting of DOE Fuel Cells for Transportation Program*, Tri-Cities, WA, June 7–8, 2000.
58. K. Ledjeff-Hey, T. Kalk, F. Mahlendorf *et al.* Portable PEFC generator with propane as fuel. *J. Power Sources* **86(1)** (2000) 166–172.
59. R.J. Farrauto and R.M. Heck. Environmental catalysis into the 21st century. *Catal. Today* **55(1–2)** (2000) 179–187.
60. Fuel Cell 2000, Technology Updates, October 2000.
61. B.C.H. Steele. Fuel cell technology running on natural gas. *Nature* **400(6745)** (1999) 619–621.
62. C. Song. Catalytic fuel processing for fuel cell applications. Challenges and opportunities. *Am. Chem. Soc. Div. Fuel Chem. Prep.* **46(1)** (2001) 8–13.
63. S. Um, C.Y. Wang and C.S. Chen. Computational fluid dynamics modeling of proton exchange membrane fuel cells. *J. Electrochem. Soc.* **147(12)** (2000) 4485–4493.
64. Z.H. Wang, C.Y. Wang and K.S. Chen. Two-phase flow and transport in the air cathode of proton exchange membrane fuel cells. *J. Power Sources* **94(1)** (2001) 40–50.
65. C. Song, S. Murata, S.T. Srinivas, L. Sun and A.W. Scaroni. CO_2 reforming of CH_4 over zeolite-supported Ni catalysts for syngas production. *Am. Chem. Soc. Div. Petrol Chem. Prep.* **44(2)** (1999) 160–164.
66. C. Song, M.T. Klein, B. Johnson and J. Reynolds (Eds.). *Catalysis in Fuel Processing and Environmental Protection. Catal. Today* **50(1)** (1999).
67. C. Song. Tri-reforming: a new process concept for effective conversion and utilization of CO_2 in fuel gas from electric power plants. *Am. Chem. Soc. Div. Fuel Chem. Prep.* **45(4)** (2000) 772–776.
68. B.C. Gates and H. Topsoe. Reactivities in deep catalytic hydrodesulfurization: challenges, opportunities, and the importance of 4-methyldibenzothiophene and 4,6-dimethyl-dibenzothiophene. *Polyhedron* **16** (1997) 3213.
69. H. Topsoe, K.G. Knudsen, L.S. Byskov, J.K. Norskov and B.S. Clausen. Advances in deep desulfurization. *Stud. Surf. Sci. Catal.* **121** (1999) 13–22.
70. D.D. Whitehurst, T. Isoda and I. Mochida. Present state of the art and future challenges in the hydrodesulfurization of polyaromatic sulfur compounds. *Adv. Catal.* **42** (1998) 345.
71. C. Song and K.M. Reddy. Mesoporous molecular sieve MCM-41 supported Co-Mo catalyst for hydrodesulfurization of dibenzothiophene in distillate fuels. *Appl. Catal. A Gen.* **176(1)** (1999) 1–10.
72. U. Turaga and C. Song. Novel mesoporous Co-Mo/MCM-41 catalysts for deep hydrodesulfurization of jet fuels. In: *Proceedings of the North American Catalysis Society Meeting*, Canada, June, 2001.

73. C. Song. Designing sulfur-resistant, noble-metal hydrotreating catalysts. *ChemTech* **29**(**3**) (1999) 26–30.
74. I. Novochinskii, X. Ma, C. Song, J. Lampert, L. Shore and R. Farrauto. A ZnO-based sulfur trap for H_2S removal from reformate of hydrocarbons for fuel cell applications. In: *Proceedings of the Topical Conference on Fuel Cell Technology, AIChE Spring 2002 National Meeting*, New Orleans, March 10–14, 2002, pp. 98–105.
75. C. Song, S.T. Srimat, L. Sun and J.N. Armor. Comparison of high-pressure and atmospheric-pressure reactions for CO_2 reforming of CH_4 over Ni/Na-Y and Ni/Al_2O_3 catalysts. *Am. Chem. Soc. Div. Petrol Chem. Prep.* **45**(**1**) (2000) 143–148.
76. C. Song, C.S. Hsu and I. Mochida (Eds.). *Chemistry of Diesel Fuels*, Taylor & Francis, New York, 2000, 294 pp.
77. R. Bellows. Conventional and less conventional approaches to fuel processing for PEM fuel cells. *Am. Chem. Soc. Div. Fuel Chem. Prep.* **46**(**2**) (2001) 650–651.
78. X. Ma, L. Sun, Z. Yin and C. Song. New approaches to deep desulfurization of diesel fuel, jet fuel, and gasoline by adsorption for ultra-clean fuels and for fuel cell applications. *Am. Chem. Soc. Div. Fuel Chem. Prep.* **46**(**2**) (2001) 648–649.
79. X. Ma, M. Sprague, L. Sun and C. Song. Deep desulfurization of liquid hydrocarbons by selective adsorption for fuel cell applications. *Am. Chem. Soc. Div. Petrol Chem. Prep.* **47**(**1**) (2002) 48–49.
80. T. Krause, R. Kumar and M. Krumpelt. Sulfur removal from reformate. In: *Proceedings of the Annual National Laboratory R&D Meeting of DOE Fuel Cells for Transportation Program*, Tri-Cities, WA, June 7–8, 2000.
81. R. Woods. Fuel-Flexible, Fuel-processing subsystem development. In: *Proceedings of the Joint DOE/EPRI/GRI Review Conference on Fuel Cell Technology*, Chicago, IL, August 3–5, 1999.
82. R.M. Privette, T.J. Flynn, M.A. Perna, K.E. Kneidel, D.L. King and M. Cooper. Compact fuel processor for fuel cell powered vehicles. In: *Proceedings of the Joint DOE/EPRI/GRI Review Conference on Fuel Cell Technology*, Chicago, IL, August 3–5, 1999.
83. D.L. Trimm and Z.I. Onsan. Onboard fuel conversion for hydrogen-fuel-cell-driven vehicles. *Catal. Rev.Sci. Eng.* **43** (2001) 31–84.
84. J.P. Breen and J.R.H. Ross. Methanol reforming for fuel-cell applications: development of zirconia-containing Cu-Zn-Al catalysts. *Catal. Today* **51**(**3–4**) (1999) 521–533.
85. P.J. de Wild and M.J.F.M. Verhaak. Catalytic production of hydrogen from methanol. *Catal. Today* **60**(**1**) (2000) 3–10.
86. S. Cavallaro and S. Freni. Ethanol steam reforming in a molten carbonate fuel cell. a preliminary kinetic investigation. *Int. J. Hydrogen Energy* **21**(**6**) (1996) 465–469.
87. B.A. Peppley, J.C. Amphlett, L.M. Kearns and R.F. Mann. Methanol-steam reforming on Cu/ZnO/Al_2O_3. Part 1. The reaction network. *Appl. Catal. A Gen.* **179**(**1**) (1999) 21–29.
88. B.A. Peppley, J.C. Amphlett, L.M. Kearns and R.F. Mann. Methanol-steam reforming on Cu/ZnO/Al_2O_3. Part 2. A comprehensive kinetic model. *Appl. Catal. A Gen.* **179**(**1–2**) (1999) 31–49.
89. W. Wiese, B. Emonts and R. Peters. Methanol steam reforming in a fuel cell drive system. *J. Power Sources* **84**(**2**) (1999) 187–193.
90. R. Peters, H.G. Dusterwald and B. Hohlein. Investigation of a methanol reformer concept considering the particular impact of dynamics and long-term stability for use in a fuel-cell-powered passenger car. *J. Power Sources* **86**(**1–2**) (2000) 507–514.
91. J.L.G. Fierro. Oxidative methanol reforming reactions for the production of hydrogen. *Stud. Surf. Sci. Catal.* **130** (2000) 177–186.
92. T.L. Reitz, S. Ahmed, M. Krumpelt, R. Kumar and H.H. Kung. Methanol reforming over CuO/ZnO under oxidizing conditions. *Stud. Surf. Sci. Catal.* **130** (2000) 3645–3650.
93. S. Velu, K. Suzuki, M. Okazaki, M.P. Kapoor, T. Osaki and F. Ohashi. Oxidative steam reforming of methanol over CuZnAl(Zr)-oxide catalysts for the selective production of hydrogen for fuel cells: catalyst characterization and performance evaluation. *J. Catal.* **194**(**2**) (2000) 373–384.
94. I. Carpenter, N. Edwards, S. Ellis, J. Frost, S. Golunski, N. van Keulen, M. Petch, J. Pignon and J. Reinkingh. On-Board hydrogen generation for PEM fuel cells in automotive applications. *SAE Paper* No. 1999-01-1320, 1999, pp. 173–178.
95. D. Dissanayake, M.P. Rosynek, K.C.C. Kharas and J.H. Lunsford. Partial oxidation of methane to carbon-monoxide and hydrogen over a Ni/Al_2O_3 catalyst. *J. Catal.* **132**(**1**) (1991) 117–127.
96. D.A. Hickman, E.A. Haupfear and L.D. Schmidt. Synthesis gas formation by direct oxidation of methane over Rh monoliths. *Catal. Lett.* **17**(**3**) (1993) 223–237.

97. M.A. Pena, J.P. Gomez and J.L.G. Fierro. New catalytic routes for syngas and hydrogen production. *Appl. Catal. A Gen.* **144(1–2)** (1996) 7–57.
98. J. Shen, E. Schmertz, G.J. Kawalkin, J.C. Winslow, R.P. Noceti and B.J. Tomer. Ultra clean transportation fuels for the 21st century: The Fischer–Tropsch option – an overview. *Am. Chem. Soc. Div. Petrol Chem. Prep.* **45(2)** (2000) 190–193.
99. V.K. Venkataraman, H.D. Guthrie, R.A. Avellanet *et al.* Overview of US DOE's natural gas-to-liquids RD & D program and commercialization strategy. *Stud. Surf. Sci. Catal.* **119** (1998) 913–918.
100. Y.-Q. Zhang and B.H. Davis. Indirect coal liquefaction – where do we stand? In: *Proceedings of the 15th Annual International Pittsburgh Coal Conference*, Pittsburgh, September 14–18, 1998, Paper No. 26-6.
101. R.J. Bellows. Technical challenges for hydrocarbon fuel reforming. In: *Proceedings of the Joint DOE/ONR Fuel Cell Workshop*, Baltimore, October 6–8, 1999.
102. M. Krumpelt. Fuel processing session summary. In: *Proceedings of the Joint DOE/ONR Fuel Cell Workshop*, Baltimore, October 6–8, 1999.
103. W. Mitchell. Next Millennium ™ fuel processor for transportation and stationary fuel cell power. In: *Proceedings of the Joint DOE/EPRI/GRI Review Conference on Fuel Cell Technology*, Chicago, IL, August 3–5, 1999.
104. D. Edlund. Fuel processors for small-scale stationary PEMFC systems. In: *Proceedings of the Joint DOE/EPRI/GRI Review Conference on Fuel Cell Technology*, Chicago, IL, August 3–5, 1999.
105. N.E. Vanderborgh. Reformate gas processing. In: *Proceedings of the Joint DOE/ONR Fuel Cell Workshop*, Baltimore, October 6–8, 1999.
106. M. Krumpelt, T. Krause, J. Kopasz, R. Wilkenhoener and S. Ahmed. Catalytic autothermal reforming. In: *Proceedings of the Annual National Laboratory R&D Meeting of DOE Fuel Cells for Transportation Program*, Tri-Cities, WA, June 7–8, 2000.
107. J. Kopasz, D. Applegate, L. Ruscic, S. Ahmed and M. Krumpelt. Effects of fuels/contaminants on reforming catalyst performance and durability. In: *Proceedings of the Annual National Laboratory R&D Meeting of DOE Fuel Cells for Transportation Program*, Tri-Cities, WA, June 7–8, 2000.
108. S. Ahmed, S.H.D. Lee, E. Doss, C. Pereira, D. Colombo and M. Krumpelt. Integrated fuel processor development. In: *Proceedings of the Annual National Laboratory R&D Meeting of DOE Fuel Cells for Transportation Program*, Tri-Cities, WA, June 7–8, 2000.
109. R. Borup, M. Inbody, J. Hong, B. Morton and J. Tafoya. Fuel processing: fuel and fuel constituents effects on fuel processor and catalyst durability and performance. In: *Proceedings of the Annual National Laboratory R&D Meeting of DOE Fuel Cells for Transportation Program*, Tri-Cities, WA, June 7–8, 2000.
110. M. Inbody, R. Borup, J. Tafoya, J. Hedstrom and B. Morton. CO clean-up development. In: Proceedings of the Annual National Laboratory R&D Meeting of DOE Fuel Cells for Transportation Program, Tri-Cities, WA, 7–8 June 2000.
111. L. Pederson, W. TeGrotenhuis, G. Whyatt and B. Wegeng. Microchannel fuel processing: development of an efficient, compact fuel processor. In: *Proceedings of the Annual National Laboratory R&D Meeting of DOE Fuel Cells for Transportation Program*, Tri-Cities, WA, June 7–8, 2000.
112. E. Kikuchi. Membrane reactor application to hydrogen production. *Catal. Today* **56(1–3)** (2000) 97–101.
113. Y.M. Lin and M.H. Rei. Process development for generating high purity hydrogen by using supported palladium membrane reactor as steam reformer. *Int. J. Hydrogen Energ.* **25(3)** (2000) 211–219.
114. E. Kikuchi, Y. Nemoto, M. Kajiwara *et al.* Steam reforming of methane in membrane reactors: comparison of electroless-plating and CVD membranes and catalyst packing modes. *Catal. Today* **56(1)** (2000) 75–81.
115. A.K. Prabhu and S.T. Oyama. Highly hydrogen selective ceramic membranes: application to the transformation of greenhouse gases. *J. Membr. Sci.* **176(2)** (2000) 233–248.
116. W. Pan and C. Song. Computational analysis of energy aspects of CO_2 reforming and oxy-CO_2 reforming of methane at high pressure. *Am. Chem. Soc. Div. Petrol Chem. Prep.* **45(1)** (2000) 168–171.
117. W. Pan, S.T. Srinivas and C. Song. CO_2 Reforming and steam reforming of methane at elevated pressures: a computational thermodynamic study. In: *Proceedings of the 16th International Pittsburgh Coal Conference*, Pittsburgh, October 11–15, 1999, Paper No. 26-2.
118. S.T. Srimat, W. Pan, C. Song and J.N. Armor. Dynamic characterization of carbon formation during CO_2 reforming and steam reforming of CH_4 using oscillating microbalance. *Am. Chem. Soc. Div. Fuel Chem. Prep.* **46(1)** (2001) 92–93.

119. S.T. Srinivas and C. Song. Effects of pressure on CO_2 reforming of CH_4 over Rh catalysts. *Am. Chem. Soc. Div. Petrol Chem. Prep.* **45(1)** (2000) 153–156.
120. T. Maillet, Y. Madier, R. Taha, J. Barbier and D. Duprez. Spillover of oxygen species in the steam reforming of propane on ceria-containing catalysts. *Stud. Surf. Sci. Catal.* **112** (1997) 267–275.
121. J.R. Rostrup-Nielsen and I. Alstrup. Innovation and science in the process industry – steam reforming and hydrogenolysis. *Catal. Today* **53(3)** (1999) 311–316.
122. J.R. Rostrup-Nielsen. Carbon limits in steam reforming. In: *Fouling Science and Technology.* NATO ASI Series, Series E, Vol. **45**, 1988, pp. 405–424.
123. C. Song, S. Eser, H.H. Schobert and P.G. Hatcher. Pyrolytic degradation studies of a coal-derived and a petroleum-derived aviation jet fuel. *Energ. Fuels* **7(2)** (1993) 234–243.
124. C. Song, W.C. Lai and H.H. Schobert. Condensed-phase pyrolysis of *n*-tetradecane at elevated pressures for long duration product distribution and reaction mechanisms. *Ind. Eng. Chem. Res.* **33(3)** (1994) 534–547.
125. W. Pan, S.T. Srimat, C. Song and J.N. Armor. Carbon formation in CO_2 reforming of CH_4 at elevated pressures. Dynamic characterization using oscillating microbalance. In: *Proceedings of the Poster Program of 17th North American Catalysis Society Meeting*, Toronto, Canada, June 3–8, 2001, p. 84.
126. J.R. Rostrup-Nielsen, T.S. Christensen and I. Dybkjaer. Steam reforming of liquid hydrocarbons. *Stud. Surf. Sci. Catal.* **113** (1998) 81–95.
127. K. Li, Q. Fu and M. Flytzani-Slephanopoulos. Low-temperature water – gas shift reaction over Cu- and Ni-loaded cerium oxide catalysts. *Appl. Catal. B Environ.* **27(3)** (2000) 179–191.
128. T. Tabakova, V. Idakiev, D. Andreeva and I. Mitov. Influence of the microscopic properties of the support on the catalytic activity of Au/ZnO, Au/ZrO_2, Au/Fe_2O_3, Au/Fe_2O_3-ZnO, Au/Fe_2O_3-ZrO_2 catalysts for the WGS reaction. *Appl. Catal. A Gen.* **202(1)** (2000) 91–97.
129. T. Utaka, K. Sekizawa and K. Eguchi. CO removal by oxygen-assisted water gas shift reaction over supported Cu catalysts. *Appl. Catal. A Gen.* **194** (2000) 21–26.
130. J.-P. Shen and C. Song. Influence of preparation method on performance of Cu/Zn-based catalysts for low-temperature steam reforming and oxidative steam reforming of methanol for H_2 production for fuel cells, *Catal. Today* **77(1–2)** (2002) 89–98.
131. B.D. Chandler, A.B. Schabel and L.H. Pignolet. Preparation and characterization of supported bimetallic Pt – Au and Pt – Cu catalysts from bimetallic molecular precursors. *J. Catal.* **193(2)** (2000) 186–198.
132. A.Y. Tonkovich, J.L. Zilka, M.J. LaMont, Y. Wang and R.S. Wegeng. Microchannel reactors for fuel processing applications. I. Water gas shift reactor. *Chem. Eng. Sci.* **54(13–4)** (1999) 2947–2951.
133. C.D. Dudfield, R. Chen and P.L. Adcock. A compact CO selective oxidation reactor for solid polymer fuel cell powered vehicle application. *J. Power Sources* **86(1–2)** (2000) 214–222.
134. B. Rohland and V. Plzak. The PEMFC-integrated CO oxidation – a novel method of simplifying the fuel cell plant. *J. Power Sources* **84(2)** (1999) 183–186.
135. C.D. Dudfield, R. Chen and P.L. Adcock. Evaluation and modelling of a CO selective oxidation reactor for solid polymer fuel cell automotive applications. *J. Power Sources* **85(2)** (2000) 237–244.
136. C.M. Finnerty, N.J. Coe, R.H. Cunningham and R.M. Ormeroda. Carbon formation on and deactivation of nickel-based/zirconia anodes in solid oxide fuel cells running on methane. *Catal. Today* **46(2–3)** (1998) 137–145.
137. FCE Fuel Cell Energy-Research to Reality. On-line publication of the Fuel Cell Commercialization Group, 2000 (http://www.ttcorp.com/fccg/slideerc/slider01.htm), viewed 27 October 2000.
138. N.J. Coe, R.H. Cunningham and R.M. Ormerod. Calculation of the metal – carbon bond strength of surface carbon deposited on solid oxide fuel cell nickel/zirconia fuel reforming anodes. *Catal. Lett.* **49(3)** (1997) 189–192.
139. S. Cavallaro and S. Freni. Syngas and electricity production by an integrated autothermal reforming/molten carbonate fuel cell system. *J. Power Sources* **76(2)** (1998) 190–196.
140. M.C. Williams and W. Surdoval. Solid State Energy Conversion Alliance, National Energy Technology Laboratory, US Department of Energy, July 2000.
141. M.S. Wilson and S. Gottesfield. High performance catalyzed membranes of ultra-low Pt loadings for polymer electrolyte fuel cells. *J. Electrochem. Soc.* **139(2)** (1992) L28–L30.
142. U.A. Paulus, U. Endruschat, G.J. Feldmeyer, T.J. Schmidt, H. Bönnemann and R.J. Behm. New PtRu alloy colloids as precursors for fuel cell catalysts. *J. Catal.* **195(2)** (2000) 383–393.
143. K.A. Starz, E. Auer, T. Lehmann and R. Zuber. Characteristics of platinum-based electrocatalysts for mobile PEMFC applications. *J. Power Sources* **84**(2) (1999) 167–172.

144. N. Nakagawa, H. Sagara and K. Kato. Catalytic activity of Ni—YSZ—CeO_2 anode for the steam reforming of methane in a direct internal-reforming solid oxide fuel cell. *J. Power Sources* **92(1)** (2001) 88–94.
145. B.D. McNicol, D.A.J. Rand and K.R. Williams. Direct methanol – air fuel cells for road transportation. *J. Power Sources* **83(1)** (1999) 15–31.
146. X.M. Ren, P. Zelenay, S. Thomas, J. Davey and S. Gottesfeld. Recent advances in direct methanol fuel cells at Los Alamos National Laboratory. *J. Power Sources* **86(1–2)** (2000) 111–116.
147. S. Wasmus and A. Kuver. Methanol oxidation and direct methanol fuel cells: a selective review. *J. Electroanal. Chem.* **461(1–2)** (1999) 14–31.
148. E.P. Murray, T. Tsai and S.A. Barnett. A direct-methane fuel cell with a ceria-based anode. *Nature* **400(6745)** (1999) 649–651.
149. S. Park, J.M. Vohs and R.J. Gorte Direct oxidation of hydrocarbons in a solid-oxide fuel cell. *Nature* **404(6775)** (2000) 265–267.
150. S. Park, R.J. Gorte and J.M. Vohs. Applications of heterogeneous catalysis in the direct oxidation of hydrocarbons in a solid-oxide fuel cell. *Appl. Catal. A Gen.* **200(1–2)** (2000) 55–61.
151. K. Kendall. Hopes for a flame-free future. *Nature* **404(6775)** (2000) 233–235.
152. X. Ma, L. Sun and C. Song. A new approach to deep desulfurization of gasoline, diesel fuel and jet fuel by selective adsorption for ultra-clean fuels and for fuel cell applications. *Catal. Today* **77(1–2)** (2002) 107–116.
153. T.U.S. Chang. Consumption of alternative road fuels growing. *Oil Gas J.* **97(28)** (1999) 37–39.
154. News-FC Car, Fuel-cell car makes tracks across country. *USA Today*, Section B, May 29, 2002.
155. C. Hanisch. Powering tomorrow's cars. *Environ. Sci. Technol.* **33(21)** (1999) 458A–462A.
156. N. Edwards, S.R. Ellis, J.C. Frost, S.E. Golunski, A.N.J. van Keulen, N.G. Lindewald and J.G. Reinkingh. On-board hydrogen generation for transport applications: the HotSpot ™ methanol processor. *J. Power Sources* **71(1–2)** (1998) 123–128.
157. A. Doctor and A. Lamm. Gasoline fuel cell systems. *J. Power Sources* **84(2)** (1999) 194–200.
158. J.C. Amphlett, R.F. Mann, B.A. Peppley, P.R. Roberge, A. Rodrigues and J.P. Salvador. Simulation of a 250 kW diesel fuel processor PEM fuel cell system. *J. Power Sources* **71(1–2)** (1998) 179–184.
159. Y. Sone, H. Kishida, M. Kobayashi and T. Watanabe. A study of carbon deposition on fuel cell power plants – morphology of deposited carbon and catalytic metal in carbon deposition reactions on stainless steel. *J. Power Sources* **86(1–2)** (2000) 334–339.
160. J. Patt, D.J. Moon, C. Phillips and L. Thompson. Molybdenum carbide catalysts for water–gas shift. *Catal. Lett.* **65(4)** (2000) 193–195.
161. S. Campanari. Thermodynamic mode and parametric analysis of a tubular SOFC module. *J. Power Sources* **92(1–2)** (2001) 26–34.

Chapter 5

Review of fuel processing catalysts for hydrogen production in PEM fuel cell systems

Anca Faur Ghenciu

Abstract

The rapid development in recent years of the proton-exchange membrane (PEM) fuel cell technology has stimulated research in all areas of fuel processor catalysts for hydrogen generation. The principal aim is to develop more active catalytic systems that allow for the reduction in size and increase the efficiency of fuel processors. The overall selectivity in generating a low CO content hydrogen stream as needed by the PEM fuel cell catalyst is dependent on the efficiency of the catalysts in each segment of the fuel processor. This article reviews the advances achieved during the past few years in the development of catalytic materials for hydrogen generation through fuel reforming,[1] water-gas shift and carbon monoxide preferential oxidation, as used or aimed to be of use in fuel processing for PEM fuel cell systems.

Keywords: Fuel processing catalysts; Autothermal reforming (ATR); Water-gas shift (WGS); CO preferential oxidation (PROX); Platinum-group metals (PGM)

Article Outline

1. Introduction 92
2. Reformer catalysts 92
    2.1. Methane and hydrocarbon steam reforming and partial oxidation 93
    2.2. Gasoline/hydrocarbon autothermal reforming 94
    2.3. Methanol reforming 95
3. Water gas shift 96
    3.1. Non-precious metal catalysts 97
    3.2. Precious metal catalysts 98
4. CO clean-up through preferential oxidation 100
5. Conclusions 101
Acknowledgements 102
References 102

Fuel Cells Compendium

1. INTRODUCTION

Fuel cells are a viable alternative for clean energy generation [1–3]. At present the major commercial markets are in residential applications and public or private transportation. The variety of applications for fuel cell technology ranges from portable/micro power and transportation through to stationary power for buildings and distributed generation, applications that will be in large numbers worldwide [4–7]. Over the past few years, fuel cell and automotive companies have announced new technologies or prototype vehicles adopting fuel cells in an effort to reduce atmospheric pollution. A variety of fuel cells for different applications is under development [8–10]: solid polymer fuel cells (SPFC), also known as proton-exchange membrane (PEM) fuel cells operating at ~80°C, alkaline fuel cells (AFC), operating at ~100°C, phosphoric acid fuel cells (PAFC) for ~200°C operation, molten carbonate fuel cells (MCFC) at ~650°C, solid oxide fuel cells (SOFC) for high temperature operation, 800–1100°C. PEM fuel cells possess a series of advantageous features that make them leading candidates for mobile power applications (vehicles) or for small stationary power units: low operating temperature, sustained operation at high current density, low weight, compactness, potential for low cost and volume, long stack life, fast start-ups, suitability to discontinuous operation.

Except for the case of direct methanol fuel cells (DMFC), the ideal fuel for PEM fuel cells is pure hydrogen, with less than 50 ppm carbon monoxide, as dictated by the poisoning limit of the Pt fuel cell catalyst. Size, weight, cost and technical limitations make it difficult to store hydrogen in necessary quantity and density, therefore the hydrogen gas will likely be generated on site and on demand, by reforming available fuels such as natural gas (NG), gasoline, propane (LPG), methanol. Practical efficiencies of existent reformer/fuel cell systems-based on the lower heating value (LHV) of the fuel-are in the range of 35–50%. While this appears to be only slightly better than the theoretical efficiency of the internal combustion engine (ICE), the net advantage over ICE is the limited production of harmful pollutants. If operated with pure hydrogen, the local emissions are truly zero. If hydrogen is generated from hydrocarbon fuels, the emissions can be limited to CO_2 and perhaps small amounts of hydrocarbons generated during start up. In general, due to the relatively low temperatures in fuel processing on active catalysts, pollutants such as NO_x and particulates can be avoided. Further advantages over ICE include the load-independent efficiency, and quiet energy production with no moving parts.

The method used to produce the hydrogen for the fuel cell plays a decisive role in the design of fuel processor catalysts. For high temperature fuel cells hydrogen is produced through internal reforming [11], with the temperature being high enough for the endothermic reforming process to occur within the cell. Low-temperature PEM fuel cells utilize the hydrogen produced by external reforming using steam, air or a combination of both. Depending on the reformer type, operating conditions and fuel, the outlet stream from the reformer contains 3–10% CO, which is further converted into hydrogen and carbon dioxide via the water-gas shift WGS reaction. These two catalytic processes in series produce a hydrogen-rich stream containing 40–75% H_2, 0.5–1% CO, 15–25% CO_2, 15–30% H_2O, and 0–25% N_2. The CO content is further reduced to <50 ppm in a CO clean-up system.

Some of the current technical goals and challenges for fuel processor catalysts are the development of very active, poison-resistant materials that will result in small catalytic volumes and reduced start-up times, durability under steady-state and transient conditions at the required temperatures, cost reductions, and versatility to variations in fuel/feed composition. In the following sections, some of the latest achievements and trends in fuel processor catalyst development for PEM fuel cell systems are reviewed and discussed.

2. REFORMER CATALYSTS

The demands placed on all fuel processing catalysts are exacting, but particularly so for the front-end reforming catalyst. It must be active for the fuel of choice, withstand high temperatures and transients, be resistant to poisons and coking and have mechanical resistance to shock or large variations of temperature. In addition, it must not introduce a high pressure drop into the system. Catalysts are usually deposited on ceramic or metallic monoliths, foams, or other structured inert supports. Depending on the application and fuel load, electrically heated catalysts can be used to facilitate start-up.

Several routes for reforming fuels to produce syngas have been investigated, analyzed and are employed today in fuel processing: steam reforming, catalytic partial oxidation (CPO), and autothermal reforming (ATR) [4–6,12]. The choice of reforming route is based on the type of fuel cell, the demands and volume of the system, and heat management strategy. ATR combines oxidation and steam reforming in one single unit, with the exothermic partial oxidation driving the endothermic steam reforming.

ATR is practical for small or medium size fuel processors because it reduces the size and heat transfer limitation of the steam reformer while achieving high H_2 concentrations with less coking and with faster start-up. Heat management can be finely tuned through the steam-to-carbon (S/C) and air-to-fuel (A/F) ratios. Eqs. (1)–(6) describe the main reactions taking place in autothermal reforming of hydrocarbons. ATR can achieve high reforming efficiencies with no complicated subsystem for heat management for a large variety of fuels [13] such as NG [14], LPG, gasoline [6], diesel [15] or methanol [3,16–19]:

$$\text{Steam reforming: } C_nH_m + n\text{H}_2\text{O} \rightarrow n\text{CO} + (n + m/2)\text{H}_2,\quad \Delta H^0_{298} > 0 \tag{1}$$

$$\text{Partial oxidation: } C_nH_m + n/2\ \text{O}_2 \rightarrow n\text{CO} + m/2\ \text{H}_2,\ \Delta H^0_{298} < 0 \tag{2}$$

$$\text{Water-gas shift: CO} + \text{H}_2\text{O} \rightarrow \text{CO}_2 + \text{H}_2,\ \Delta H^0_{298} = -41.2\ \text{kJ/mol} \tag{3}$$

$$\text{Methanation: CO} + 3\text{H}_2 \rightarrow \text{CH}_4 + \text{H}_2\text{O}\ \Delta H^0_{298} = -206.2\ \text{kJ/mol} \tag{4}$$

$$\text{Product oxidation: CO} + \frac{1}{2}\text{O}_2 \rightarrow \text{CO}_2\ \Delta H^0_{298} = -283.6\ \text{kJ/mol} \tag{5}$$

$$\text{H}_2 + \frac{1}{2}\text{O}_2 \rightarrow \text{H}_2\text{O}\ \Delta H^0_{298} = -243.5\ \text{kJ/mol} \tag{6}$$

The choice of fuel for a given application is complex and can depend on local economic or political factors, as well as technical criteria.

2.1. Methane and hydrocarbon steam reforming and partial oxidation

Traditional methane steam reforming catalysts (Eq. (1)) for industrial production of hydrogen and synthesis gas are based on nickel/nickel oxide or cobalt compositions on refractory alumina or supports such as magnesium alumina spinel [20,21], often promoted with alkali or alkali-earth compounds to accelerate carbon removal [12,20,22], precious metals (Rh, Ru, Pt, Pd, Re) on alumina [20], or on rare-earth oxides, particularly ceria [23]. The effect of alkali in steam reforming has been the subject of investigations and yet is not completely understood. Used for activity improvement and for their ability to reduce methanation or to facilitate coke gasification, alkali promoters seem to suffer from the drawback of having increased "volatility" in high temperature steam environments [20]. This can potentially lead to catalyst deactivation by pore blocking [11], an effect that is likely to be dependent on the type of support and catalyst preparation.

Partial oxidation (Eq. (2)) is a much faster reaction than steam reforming, offering therefore the advantage of smaller reactors and higher throughputs. A comprehensive review has been written by Bharadwaj and Smith [12]. Catalysts include: supported nickel (NiO—MgO), nickel-modified hexa-aluminates [24], platinum group metals (PGM) Pt, Rh, Pd/alumina [25,26], on ceria-containing supports, or on titania [27], as well as supported metallic clusters [28]. Some of these materials present only academic interest for the time being. PGM catalysts are in general more active, however more expensive and sensitive to sulfur; based on the desired compactness of the reformer they tend to find more employment in the development of very active catalysts.

Catalysts for the oxidative reforming of methane based on ceria involve group VIII metals, especially PGM deposited on ceria-containing mixed oxides; for example, Ni/ZrO_2, Ni/CeO_2, $Ni/CeO_2—ZrO_2$ [29], Pt- or Ca-modified $Pt/CeO_2—ZrO_2$ [30,31], $Ru/CeO_2—ZrO_2$ [30]. The role and high activity of ceria and solid solutions of ceria-mixed oxides have been much documented and exploited in automotive, emission control catalysts. The improved activity obtained in hydrocarbon and CO activation on ceria-containing supports has been explained based on the oxygen storage capacity (OSC) of ceria under oxidizing–reducing conditions [32]. This feature is enhanced for ceria–zirconia mixed oxides [33] by the metal–support interaction which results in active catalytic sites for hydrocarbon activation.

2.2. Gasoline/hydrocarbon autothermal reforming

The design of ATR catalysts can be challenging, particularly for gasoline reforming due to the complex and ill-defined nature of the fuel. ATR catalysts have to be active for both steam reforming (Eq. (1)) and partial oxidation (Eq. (2)), be robust at high temperature and resistant to sulfur and coke formation, especially in the catalytic zone that runs oxygen limited. Early autothermal reformers used two separate catalytic units, a burner and a steam reformer located for example on either side of a heat exchanger. Newly developed ATR technologies combine the two segments into one single compact unit using an active catalyst that has both steam reforming and partial oxidation function [1–3]. Full conversion of all hydrocarbon components is desired; low levels of non-methane hydrocarbons, such as aromatics and olefins can undergo transformations, particularly hydrogenation on the downstream WGS catalyst, thus consuming hydrogen and decreasing the overall efficiency of the fuel processor.

Catalyst formulations for ATR fuel processors depend on the fuel choice and operating temperature. For methanol, Cu-based formulations similar to commercial methanol synthesis catalysts and low temperature WGS catalysts can be used. For higher hydrocarbons the catalyst typically comprises of metals such as Pt, Rh, Ru and Ni deposited or incorporated into carefully engineered oxide supports such as ceria-containing oxides [34]. These can be further promoted or doped with other elements for improved thermal robustness or better activity. US-DOE Argonne National Laboratory has developed sulfur-tolerant ATR catalysts for a large variety of hydrocarbon fuels [35,36] using catalyst formulations derived from solid oxide fuel cell technology and structure-formed into shapes (spiral microchannel configurations) to minimize diffusional resistances [35]. With the non-noble metal formulations such as Fe, Co, Ni supported on ion-conducting doped ceria substrates, activities similar to PGM-containing catalysts have been reported. The highest H_2 selectivity in reforming iso-octane has, however, been obtained for a Pt formulation, $Pt/Ce_{0.8}Sm_{0.15}Gd_{0.05}O_2$. Long-term tests of Pt/doped ceria supports have demonstrated improved durability for ATR of a benchmark liquid fuel, for 1000 h time on stream and with a low total hydrocarbon breakthrough [35]. Similar bifunctional catalysts, in cermet form, containing PGM as a dehydrogenation component and a selective oxidation function provided by an oxygen-conduction support (CeO_2, ZrO_2, Bi_2O_3, $BiVO_4$, $LaGaO_3$, $Ce_{0.8}Gd_{0.2}O_{1.9}$) have been developed by the same group as partial oxidation [37,38] or steam reforming catalysts [39].

Work carried out at Johnson Matthey (JM) on fuel processing catalysts has led to the development of improved supported PGM formulations that are able to effectively convert a range of C_1–C_{12} straight-chained,

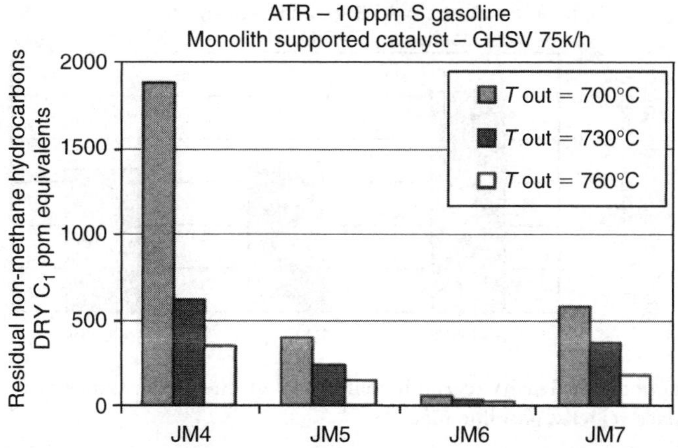

Fig. 1. Performance of Johnson Matthey formulations for ATR of sulfur-containing gasoline

branched, saturated, unsaturated, oxygenated and aromatic hydrocarbons including complex commercial gasolines. One main focus has been to develop materials with high thermal stability and mechanical robustness that can cope with the high temperatures seen in the oxidizing front zone while maintaining high activities in the lower temperature steam reforming regions of the ATR catalyst. These catalysts also include the WGS function, such that equilibrium levels of CO exit from the reformer. These formulations are optimized for specific operating regimes, S/C and A/F ratios through the choice of the active metal, careful tailoring of the support, promoter choice and loading. For example, textural modifiers can help maintain high metal and support surface areas at high temperatures, while the structural promoters can be selectively added to lower carbon formation. Specific catalyst formulations have been developed for high sulfur tolerance, these being able to cope with impurities in excess of 100 ppm sulfur. Fig. 1 shows the performance of several JM formulations differentiated based on residual hydrocarbons in reformate at several outlet temperatures, for catalyst deposited on a monolithic support. The very active formulations, such as JM6, yielded less than 200 ppm residual hydrocarbon (CH_4 equivalent) in the reformate composition, even at space velocities higher than 80 000 h^{-1}.

Another desirable characteristic of ATR catalysts for fuel processor reformer is resistance to intermittent operation and cycles, particularly on start-up and shut-down. Johnson Matthey has developed new formulations with excellent partial oxidation performance, low light-off temperature and thermal resistance such that the system can be started up in the absence of water; the A/F ratio can then be tuned as the S/C ratio improves and the system can be run at optimum efficiency. The latest formulations, when washcoated onto ceramic monoliths, demonstrate essentially complete conversions at space velocities in excess of 125 000 h^{-1}, with no degradation over several hundred hours. At these short contact time values, presentation of the reactants on the monolithic support has a significant influence as mass transport limitations begin to dominate the kinetics. This can be alleviated by using high cell density monoliths (Fig. 2), but still remains an issue for removing the last few ppm of residual non-methane hydrocarbons from the gas stream.

2.3. Methanol reforming

Typical catalysts for methanol steam reforming are based on $Cu—ZnO—Al_2O_3$. Albeit an old formulation, improvement of this catalyst has continued especially during the last decade, for the achievement of better activity, CO_2 selectivity, and durability [19,40–42], or to elucidate the catalytic mechanism [43].

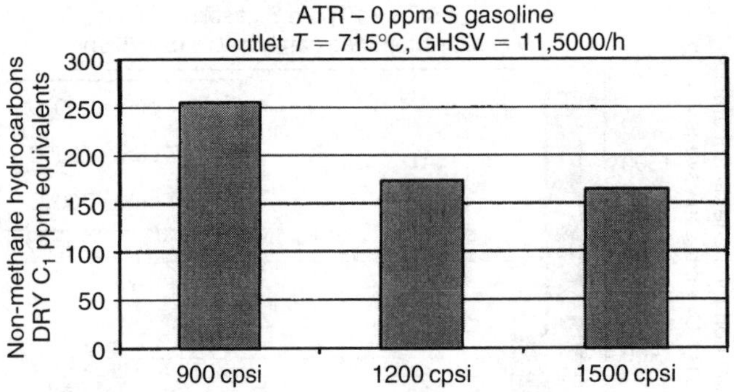

Fig. 2. Influence of support cell density (ceramic monoliths) on the performance of a Johnson Matthey ATR formulation at high space velocity, gasoline fuel

Steam reforming of methanol can theoretically achieve 75% hydrogen concentration, at 100% CO_2 selectivity. Practical H_2 outlet concentrations on various catalysts are usually greater than 70%.

ATR reforming of methanol in the presence of air and steam provides high H_2 yield at near complete conversion, at 250–330°C. It can use similar catalysts as for methanol steam reforming, or supported Cu formulations [17], with the concern of balancing the temperature profile in the reactor in order to maintain the active phase of the Cu—ZnO—Al_2O_3 catalyst [2,3,16,19]. Some of the efforts in improving CuZnO catalysts for methanol reforming include modification of Cu—Zn—alumina catalysts with zirconium [19, 40,41] for increased activity or improved carbon dioxide selectivity, or with titanium [42] for better durability and prevention of the demonstrated Cu crystallite sintering [44] compared to the unpromoted formulations. These catalysts are extremely sensitive to oxidizing environments, which causes issues in operating a real system.

3. WATER GAS SHIFT

The WGS reaction is a critical step in fuel processors for preliminary CO clean up and additional hydrogen generation prior to the CO preferential oxidation or methanation step. WGS units are placed downstream of the reformer to further lower the CO content and improve the H_2 yield (Eq. (3)). Ideally, the WGS stage(s) should reduce the CO level to less than 5000 ppm. For this equilibrium outlet CO to be obtained from reformate, the WGS catalyst has to be active at low temperatures, 200–280°C, depending on the inlet concentrations in reformate. The reaction is moderately exothermic, with low CO levels resulting at low temperatures, however with favorable kinetics at higher temperatures. Under adiabatic conditions, conversion in a single bed is thermodynamically limited, but improvements in conversion are achieved by using subsequent stages with inter-bed cooling [22]. As the flow contains CO, CO_2, H_2O, H_2, additional reactions can occur, depending on the H_2O/CO ratio [45] and favored at high temperatures: methanation (Eq. (4)), CO disproportionation or decomposition [22]. A catalyst that is active at low temperatures is sought. In industrial applications under continuous operation, the classical catalyst formulations employed are Fe—Cr oxide for the first stage (high-temperature shift (HTS)), and Cu—ZnO—Al_2O_3 for subsequent stages (low-temperature shift (LTS)) [22], for good performance under steady state conditions.

A large number of studies have been published on the classical WGS catalysts, focusing on catalyst preparation [42,46], kinetics or reaction mechanism [22b,47,48]. Most of the debate resides in the mechanistic area, and, particularly for the CuZnO catalyst, on whether the mechanism is associative taking place through intermediates such as formates and also associated with the formation of surface species such as carbonates, hydroxycarbonates of which participation to the mechanism is yet under question, or regenerative, via redox reactions involving special forms of copper. A generally accepted mechanism for WGS yet fails to exist, and this is not surprising considering the large variety of catalytic surfaces, having different crystal structures for the range of experimental conditions chosen in such studies. Combined mechanisms seem probable [48], depending on the reaction conditions or induced surface reconstruction in some cases [43]. Raney Cu—ZnO [49] formulations or, as in the case of methanol reforming catalysts, modification with zirconium or titanium [19,42] appear to bring better durability and activity under steady state conditions. Improved activity of Cu—ZnO formulations has also been reported upon promotion with alkali [50,51], however in all cases, Cu—ZnO catalysts suffer from the drawback of being pyrophoric and highly susceptible to poisons [22,52].

The use of Fe—Cr and Cu—ZnO formulations in fuel processors poses a series of disadvantages: the low activity of the HTS catalyst and its thermodynamic limitations at high temperatures, the sensitivity to air or temperature excursions of the CuZnO catalyst, the lengthy pre-conditioning of such catalysts for intermittent operation (pre-reduction/passivation), or the large reactor volume dictated by the slow WGS kinetics of the Cu—ZnO catalyst at low temperatures. The FeCr combined with CuZnO setting is therefore unsuitable for use in either residential or automotive applications, where fast start-ups dictate that low catalyst volume (fewer stages) and non-pyrophoric catalysts be used.

While research efforts to develop alternative formulations are under way, experimental and early generation fuel processors use HTS–LTS classical formulations in adiabatic reactors with interstage cooling. Shift reactors, loaded with current, commercial WGS formulations occupy ca. 30–50% of the catalyst volume and weight of the fuel processor. For typical reformate (8–10% inlet CO, wet basis), the first stage (HTS) – operating at near equilibrium at 350–400°C – reduces the CO to ~3–5%, while the LTS reactor(s), at 180–240°C can achieve 0.3–1% CO, depending on the amount of catalyst/number of stages. Improvements in operating WGS Cu—ZnO systems aimed to reducing the size of the LTS reactor by injecting air in the tail section [53] and to lowering the CO content that enters to CO clean-up system [54] exist.

The design of new WGS catalysts to be used for fuel processors must address the above mentioned deficiencies. Amongst the requirements imposed on WGS catalysts for use in fuel processors are: high activity at relatively low temperature (<300°C), stability under typical reformate outlet to insure low, steady outlet CO concentration within minimal catalyst volume, non-pyrophoric formulations in order to eliminate pre-conditioning steps, durability under steady state and transient conditions, mechanical integrity under shock or temperature excursions, stability to condensation and to poisons such H_2S, chlorine, high selectivity imposed by a range of H_2O/CO ratios, with no side reactions, particularly methanation that would consume valuable hydrogen, resistance to the adverse effects of potential hydrocarbons, particularly olefins and aromatics from the upstream reformate.

In trying to find the best compromise between activity and cost, two trends can be distinguished in the development of new, non-pyrophoric WGS catalysts for fuel processors: non-precious metal catalysts, usually active at high temperatures – with some formulations, however, being reported as possessing low-temperature activity, and precious metal formulations based on PGM or gold, possessing high activity over a larger temperature interval.

3.1. Non-precious metal catalysts

Non-precious metal formulations for WGS are sought based on their low cost compared to PGM catalysts.

Haldor–Topsoe has recently developed a series of alkaline-promoted, sulfur-resistant basic oxide HTS formulations based on Mg, Al (Mg aluminate spinel), La, Nd, Ce, Pr, Mn, Co—$MgAl_2O_4$, K-ZSM5, Mg—ZrO_2 [55]. These formulations are active at >400°C, at low CO conversion (~25–30%), however the reported benefit over other high temperature formulations such as Co- or Ni-promoted Mo, V, W oxides [56] is the absence of methanation. Cobalt–molybdenum or nickel–molybdenum sulfide [57] catalysts or their alkaline-promoted forms are active, sulfur-tolerant HTS formulations; in a feed containing only CO and H_2O, they have been reported to give reasonable CO conversions (>40%) at low space velocities (<8000 h^{-1}) [58].

The activity of the above formulations at high temperature, and their acceptable CO conversions at relatively low space velocities (<20 000 h^{-1}) would dictate that, for their application in fuel processors, additional downstream WGS reactors charged with more active catalysts for operation at lower temperature be used.

In researching for alternatives for the commercial CuZnO, Argonne National Laboratory has developed a series of cobalt–vanadium binary oxides [59]; these materials have low surface areas and therefore display specific activities (normalized to surface area) higher than CuZnO. Patt et al. [60], studying the feasibility of high surface area molybdenum carbide for the WGS reaction, reported that Mo_2C is an active LTS formulation with activity comparable or higher than that of Cu—ZnO—Al_2O_3. These nanocrystalline materials, known as active hydrotreating catalysts, are however, sensitive to oxygen, and their performance under real fuel processing conditions has yet to be studied. Non-pyrophoric, base metal WGS catalysts with no methanation activity were also reported by Ruettinger et al. [61].

Cerium oxide-containing WGS formulations have attracted interest based on the OSC of ceria [32] and based on the cooperative effect of ceria–metal leading to highly active sites. Although ceria or ceria-promoted formulations are mainly reported in conjunction with precious metals, non-PGM–ceria WGS catalysts have also been developed as potentially better alternatives to Cu—ZnO. Li et al. reported that copper or nickel deposited on high surface area $Ce(La)O_x$ supports prepared by urea precipitation–gelation [62] displayed good LTS activity at high space velocities, when tested under low CO concentrations (2%). The high activity was interpreted based on the enhanced reducibility of ceria in the presence of the metal. Durability performance was not reported for these formulations.

3.2. Precious metal catalysts

Pt/CeO_2 has raised interest for WGS reaction since the early 1980s, when in connection with the development of the three-way catalysts (TWC), it was discovered that "ceria is the best non-noble metal oxide promoter for a typical Pt-Pd-Rh TWC supported on alumina largely because it enhances the WGS reaction" (see reference 2 in Ref. [32]). In 1985, the methanation and WGS activity of Pt/CeO_2 were reported by Mendelovich and Steinberg [63,64]. NexTech Materials has developed a Pt/CeO_2 catalyst that has already been studied for high temperature WGS reaction for several years [59,65]. This Pt/CeO_2 formulation has been reported as active and non-pyrophoric, with activity higher than that of conventional WGS catalysts in the medium-temperature range (300–400°C), thus a potential candidate for use in fuel processor WGS reactors.

Academic research on Pt, Pd, Rh, Ni, Fe, Co–ceria for WGS involved mechanistic studies [66,67] and the implication of OSC on WGS activity. The regenerative (redox) mechanism and a high OSC are generally claimed to be operative and, respectively, responsible for the high activity when metal–ceria interactions are established. At least in some cases, the WGS activity is reported to depend not only on PGM–support interaction, but also on the degree of surface hydration [68] or CO_2 coverage [67]. The preparation method plays an important role in establishing the metal–support interaction, with effect on the low-temperature activity [69,70]. Non-pyrophoric, precious metal–HTS catalysts further promoted to suppress methanation were also reported by Engelhard [61]. Ruthenium deposited on α-Fe_2O_3 has been mentioned in the literature as giving promising WGS conversions with no methanation activity [71].

Work performed at JM on Pt—CeO_2 catalysts indicated that despite the high initial activity obtained in the medium–high temperature range (325–400°C, Fig. 3), the catalyst loses activity under synthetic and real reformate tests. The deactivation can be explained by several mechanisms, including surface coverage with in situ formed carbonate-like species, and partial loss of the re-oxidizing ability in the highly reducing CO/H_2 environment. An initial decrease in metal dispersion and total BET surface area has been seen after

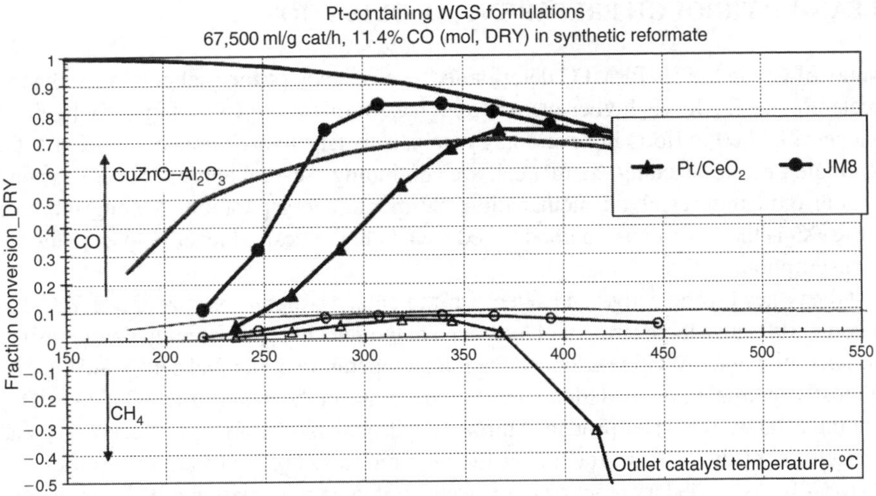

Fig. 3. Performance of a JM Pt-containing WGS catalyst compared to Pt—CeO_2 at the same metallization, synthetic reformate containing 11.4% CO (mol DRY)

the first hours of operation. At extended time-on-stream in reformate, ceria crystallite size slowly increases, leading to a further gradual decrease in the total BET surface area and to the occlusion of Pt particles in the support. These behaviors may be indicative of multiple operating mechanisms in addition to the redox process generally claimed, depending on the temperature and inlet concentrations. Consequently, multiple deactivation pathways are also available. The overall deactivation using typical reformate tests leads to the partial loss of WGS activity to levels that would require over-designing of the WGS reactors for long-term operation, as also reported by others [72]. In addition, methanation takes place on Pt/CeO_2 [64], also seen in Fig. 3 at temperatures 375°C.

New WGS catalysts developed at JM include non-pyrophoric PGM formulations with improved durability and no methanation activity over a large range of temperatures (200–500°C). Figure 3 shows the performance of one of the Pt-containing catalysts having the same metal content as the Pt—CeO_2 catalyst shown in the same graph. As it can be observed from Fig. 3, this formulation (JM8) has a much higher WGS activity while not producing methane over the entire temperature range of interest [the methanation conversion in Fig. 3 was calculated based on the 1.4% CH_4 (mol, dry) inlet concentration]. Further improved, also non-pyrophoric PGM catalysts with no methanation activity have been developed for even lower temperature operation. One of the novel WGS catalysts has been tested in Johnson Matthey's fuel processor for stationary applications at 2–10 kW load (electric) with reformate generated from methane using one of the new ATR formulations, under both steady state and transient conditions, with no decrease in activity at 250–260°C for more than 1500 h. With this performance and durability, the WGS operation of Johnson Matthey's fuel processor can now be conducted in one single stage, and translates into a large reduction factor in reactor volume compared to the case when commercial Cu—ZnO was used.

Gold catalysts show rapidly growing interest for WGS due to their high activity for CO oxidation at low temperature [69,73,74]. Gold-particles supported on metal oxides: CeO_2 [75,76], TiO_2 [77], Fe_2O_3 [78] have been reported as active for WGS, with the improved activity at low temperature being explained as due to the synergism of gold–metal oxide. Gold catalysts, however, are sensitive to the preparation conditions, the desired properties of the final material depending on dispersion, gold particle size, and the intimate metal–support contact [69]. The gold particle size can easily change during the reaction and strongly impact the activity. Improved stability has been recently reported [75], however it appears that more development is needed for these catalysts to be candidates for the demanding conditions in fuel processing applications.

4. CO CLEAN-UP THROUGH PREFERENTIAL OXIDATION

Trace amounts of CO in the H_2-PEMFC deteriorate the efficiency of the fuel cell Pt catalyst via CO poisoning, accelerated at CO levels higher than 50 ppm CO. While more CO-tolerant fuel cell catalysts are being developed [9,79,80], efforts in developing catalysts to selectively remove the 0.5–1% CO from the H_2-rich reformate prior to reaching the fuel cell are continuing. Several approaches are currently applied: CO preferential oxidation, catalytic methanation, Pd-membrane separation. Among these, preferential oxidation (PROX) is the lowest cost method to reduce CO to the desired level [1–3,81] without excessive hydrogen consumption.

PROX catalysts need to be active and selective; the catalyst should oxidize 0.5–1% CO to less than 50 ppm without oxidizing a large amount of hydrogen at the selected process temperature, usually between the outlet temperature of the WGS reactor and the inlet temperature of the fuel cell (~80°C). The reformate containing mostly hydrogen, its oxidation leads to a decrease in the overall fuel efficiency. Depending on the nature of the catalyst, the water produced through hydrogen oxidation may affect catalyst activity. The lower the selectivity of the process, the higher the required ratio O_2/CO has to be in order to completely oxidize CO to CO_2. As secondary reactions, reversed WGS and methanation of CO can occur, depending on temperature, ratio O_2/CO, and contact time. For an inlet CO of 1%, the overall CO conversion has to be higher than 99.5% for a reduction of the CO level to less than 50 ppm. As the oxidation is exothermic, multistage PROX systems with interstage cooling and/or water injections between stages (Demonox™) [1,3] can be used, with the catalyst usually being coated on monolithic supports. Series of catalyzed heat exchangers [82] or catalyzed microchannel heat exchangers [83] are good approaches to assure closer to isothermal operation, therefore better catalyst utilization.

The low-temperature CO oxidation has been reviewed and it is documented in the literature, however, the number of publications on the selective CO oxidation in the presence of H_2 has only recently increased as a result of the interest in PEM fuel cell technology. Early patents on the subject belong to Engelhard; catalysts were comprised of Pt deposited on supports like alumina, silica, kieselguhr, and diatomaceous earths [84].

Based on the high activity required to remove CO while maintaining a high CO oxidation selectivity, catalyst formulations used for PROX involve, in general, a high PGM loading on high surface area supports, and are operated at relatively low temperature, 80–200°C. Formulations comprise of Pt or promoted Pt, Ru, Pd, alloys of Pt—Sn or Pt—Ru, or Rh on alumina or on molecular sieves, or, more recently, Au catalysts. Among these, Pt catalysts appear to offer the best results over a larger temperature interval and have been the most studied. Copper catalysts on alternative supports such as ceria, ceria–samaria, or other ceria-promoted supports are also being developed in an attempt to provide selective surface oxygen for CO oxidation at low temperatures.

Korotkikh and Farrauto [85,86] investigated the CO selective oxidation at low temperature in a H_2 stream in the absence of CO_2, on a Fe-promoted, Pt catalyst on a monolithic support. For the same metallization, the Fe-promoted catalyst performed at higher conversion than the unpromoted formulation, 5% $Pt/\gamma\text{-}Al_2O_3$, while maintaining the selectivity at 90–150°C [85]. The role of the Fe promoter was interpreted based on its ability to provide additional sites for O_2 adsorption/dissociation. When in close contact with Pt, Fe is believed to induce electron-rich Pt surfaces and provide, by means of a dual-site mechanism, the readily available oxygen for selective CO oxidation [86].

Kahlich et al. [87] investigated the kinetics of CO oxidation on Pt/Al_2O_3 under integral and differential conditions and found that on 0.5% Pt/Al_2O_3, the presence of hydrogen actually increases the rate of CO oxidation. In trying to find a correlation between CO surface coverage and selectivity, the authors proposed a mechanism similar to that of CO oxidation in the absence of H_2, "low-rate branch" vs. "high-rate branch" [87]. According to this mechanism, the decrease in CO conversion at temperatures higher than 250°C is a consequence of the loss in selectivity due to the decrease in the "protective" CO coverage. In a later study, the same group confirmed that the reaction kinetics and selectivity depend on the steady-state coverage of the metal with CO in the presence of H_2, and were able to explain the differences between Pt/γ-Al_2O_3 and on Au/α-Fe_2O_3 catalysts [88]; the decrease in the selectivity of Au/α-Fe_2O_3 with temperature compared to that of Pt/γ-Al_2O_3 was interpreted based on the difference in the CO surface coverage as a function of temperature on the two surfaces [88]. This behavior explains the larger range of operation for Pt catalysts in multistage PROX reactors, whereas Au formulations remain attractive for low temperature operation.

Mechanistic studies using in situ DRIFTS (diffuse reflectance infrared Fourier transform spectroscopy) on Pt–alumina catalysts indicated the existence of formate surface species, of which generation is enhanced by hydroxylated surfaces [89]. The results obtained by Manasilp and Gulari [90] on sol–gel Pt/alumina, having the 1–2% metal incorporated in the oxide lattice, indicated that for this type of catalyst morphology, water has a beneficial effect.

Research published by a number of laboratories on Au catalysts for several applications [73,91–93] suggested that gold catalysts are potential candidates for commercial applications, more so for CO selective oxidation than for WGS, based on the lower operation temperature in the case of the former. Results reported by several groups on the performance of Au formulations for PROX agree on the very high selectivities obtained at low temperature [94–96]. Bethke and Kung [96] found improved selectivity of Au/γ-Al_2O_3 when Mg citrate is added to the preparation procedure, and rationalized the results based on increased dispersion. Grisel et al. [94,95] further rationalized the effect of MgO or MnOx through the stabilization of Au particles.

Improved CO conversions at low temperatures ($<100°C$) and stoichiometric ratio O_2/CO were reported for Ce-promoted Pt/γ-alumina [97]; no improvement in conversion has been observed in the presence of Ce at higher temperature.

Oh and Sinkevitch [98] studied a variety of materials for the preferential oxidation, among which Pt, Pd, Rh, and Ru at low loading on γ-alumina, at 20 000 h^{-1} in very dilute CO and H_2 feed. Improved selectivities were found for Ru and Rh only at very low H_2 concentrations compared to Pt/γ-alumina. Comparing activities and selectivities of Pt/alumina and Pt/zeolites at ~6% Pt loading, Igarachi et al. found improved selectivity for zeolite-supported Pt over Pt-alumina [99]. Depending on temperature and water content in the feed, water adsorption may, however, be an issue for these materials [99, 100].

CuO—CeO_2 mixed oxides have been reported as promising candidate systems for PROX [101]. In a parallel study of Pt/γ-Al_2O_3, Au/αFe_2O_3, and CuO—CeO_2, the Au formulation was the most active at low temperature, while the selectivity of CuO—CeO_2 was remarkably higher than that of both Au and Pt formulations. Pt/γ-Al_2O_3 was the most resistant toward H_2O and CO_2 in the feed.

Other formulations studied for PROX include Pd [102,103], Ru [104] and Pt loaded on zeolite membranes [105].

Research carried out in Johnson Matthey laboratories in the development of PROX catalysts have resulted in catalysts specifically tailored for high temperature applications (200–250°C), with minimal or no reverse WGS and no methanation activity, as well as in selective catalysts for lower temperature applications. With the new PGM-promoted formulations coated on monolithic supports, CO conversions of 88–90% have been obtained at high space velocities ($\sim 2 \times 10^6$ cc/g cat/h) for inlet CO concentrations of 0.75–1%. The new formulations led to reducing the complexity and size of the CO clean-up system in Johnson Matthey.

5. CONCLUSIONS

Over the past several years, the challenge for small and medium scale, dynamic fuel processors to produce H_2 for PEM fuel cells has led to the development of active and selective catalysts, particularly containing precious metals, that result in small, reliable ATR, WGS, or PROX units. Efforts in the research community – academic, industrial or government – are continuing the development of more active and cost-efficient catalytic systems. Studies are directed toward understanding such catalysts under real, steady state and transient conditions, and improving their durability under the stringent requirements of fuel processing.

ACKNOWLEDGEMENTS

The author thanks Suzanne Ellis, Jessica Reinkingh and Dr. Mike Petch from Johnson Matthey Fuel Cells for insightful discussions and valuable contributions to this review and for sharing technical results on catalyst performance and testing.

REFERENCES

1. P.G. Gray and M.I. Petch. Advances with HotSpot™ fuel processing-efficient hydrogen production for use with solid polymer fuel cells. *Platinum Metals Rev.* **44(3)** (2000) 108–111.
2. S. Golunski. HotSpot™ fuel processor. Advancing the case for fuel cell powered cars. *Platinum Metals Rev.* **42(1)** (1998) 2–7.
3. N. Edwards, S.R. Ellis, J.C. Frost, S.E. Golunski, A.N.J. van Keulen, N.G. Lindewald *et al.* On-board hydrogen generation for transport applications: the HotSpot™ methanol processor. *J. Power Sources* **71** (1998) 123–128.
4. Carpenter, N. Edwards, S. Ellis, J. Frost, S. Golunski, N. van Keulen *et al.* On-board hydrogen generation for PEM fuel cells in automotive applications. *SAE Tech. Paper Ser.* **1** (1999) 1320.
5. P.G. Gray and J.C. Frost. Impact on clean energy in road transportation. *Energ. Fuel.* **12** (1998) 1121–1129.
6. A. Docter and A. Lamm. Gasoline fuel cell systems. *J. Power Sources* **84** (1999) 194–200.
7. D.R. Brown. PEM fuel cells for commercial building. Office of Building Technology, State and Community Programs, Document No. PNNL-12–51, prepared for the US Department of Energy by the Pacific Northwest National Laboratory, November 1998.
8. EG&G Services, Parson, *Fuel Cells Handbook*, 5th edn. Science Applications International Corporation–US Department of Energy, Office of Fossil Energy, National Energy Technology Laboratory (2000).
9. G.J.K. Acres, J.C. Frost, G.A. Hards, R.J. Potter, T.R. Ralph, D. Thompsett *et al.* Electrocatalysts for fuel cells. *Catal. Today* **38** (1997) 393–400.
10. T.R. Ralph and G. Hards. Fuel cells: clean energy production for the new millennium. *Chem. Ind. Lond.* **8** (1998) 334–335; T.R. Ralph and G. Hards. Powering the cars and homes for tomorrow. *Chem. Ind. Lond.* **8** (1998) 337–342.
11. S.H. Clarke, A.L. Dicks, K. Pointon, T.A. Smith and A. Swann. Catalytic aspects of the steam reforming of hydrocarbons in internal reforming fuel cells. *Catal. Today* **38** (1997) 411–423.
12. S.S. Bharadwaj and L.D. Schmidt. Catalytic partial oxidation of natural gas to syngas. *Fuel Process Technol.* **42** (1995) 109–127 and references therein.
13. S. Ahmed and M. Krumpelt. Hydrogen from hydrocarbon fuels for fuel cells. *Int. J. Hydrogen Energ.* **26(4)** (2001) 291–301.
14. A. Heinzel, B. Vogel and P. Hubner. Reforming of natural gas–hydrogen generation for small scale stationary fuel cell systems. *J. Power Sources* **105** (2002) 202–207.
15. C. Palm, P. Ciemer, R. Peters and D. Stolten. Small-scale testing of a precious metal catalyst in the autothermal reforming of various hydrocarbon feeds. *J. Power Sources* **106** (2002) 231–237.
16. K. Geissler, E. Newson, F. Vogel, T.B. Truong, P. Hottinger and A. Wokaun. Autothermal methanol reforming for hydrogen production in fuel cell applications. *Phys. Chem. Chem. Phys.* **3(3)** (2001) 289–293.
17. P. Miszey, E. Newson, T.B. Truong and P. Hottinger. The kinetics of methanol decomposition: a part of autothermal partial oxidation to produce hydrogen for fuel cells. *Appl. Catal. A Gen.* **213(2)** (2001) 233–237.
18. S.H. Chan and H.M. Wang. Methanol autothermal reforming for fuel cell applications. *Fuel. Int.* **1(1)** (2000) 17–29.
19. S. Velu, K. Suzuki, M.P. Kapoor, F. Ohashi and T. Osaki. Selective production of hydrogen for fuel cells via oxidative steam reforming of methanol over CuZnAl(Zr)-oxide catalysts. *Appl. Catal. A Gen.* **213** (2001) 47–63.

20. J.R. Rostrup-Nielsen. *Steam Reforming Catalysts – An Investigation of Catalysts for Tubular Steam Reforming of Hydrocarbons*. Copenhagen Teknisk Forlag A/S, Danish Technical Press, 1975.
21. J.R. Rostrup-Nielsen. Production of synthesis gas. *Catal. Today* **18** (1993) 305–324.
22. M.V. Twigg. *Catalyst Handbook*, 2nd edn. Wolfe Press, London, 1989. (a) Chapter 5, Steam reforming; (b) Chapter 6, Water-gas shift; (c) Chapter 7, Methanation.
23. R. Craciun, W. Daniell and H. Knozinger. The effect of CeO_2 structure on the activity of supported Pd catalysts used for methane steam reforming. *Appl. Catal. A Gen.* **230** (2002) 153–168; R. Craciun, B. Shereck and R.J. Gorte. Kinetic studies of methane steam reforming on ceria-supported Pd. *Catal. Lett.* **51** (1998) 149–153.
24. W. Chu, W. Yang and L. Lin. The partial oxidation of methane to syngas over the nickel-modified hexaaluminate catalysts $NaNi_yAl_{12-y}O_{19-\delta}$. *Appl. Catal. A Gen.* **235(1)** 39–45.
25. A.T. Ashcroft, A.K. Cheetham, J.S. Foord, M.L.H. Green, C.P. Grey, A.J. Murrell *et al.* Selective oxidation of methane to synthesis gas using transition metal catalysts. *Nature* **344** (1990) 319–321.
26. A. Anumakonda, I. Yamanis and J. Ferrall. *Catalytic Partial Oxidation of Hydrocarbon Fuels to Hydrogen and Carbon Monoxide*. US Patent 6,221,280, April 24, 2001.
27. C. Elmasides, T. Ioannides and X.E. Verykios. Kinetic model of the partial oxidation of methane to synthesis gas over Ru/TiO_2 catalyst. *AIChE J.* **46(6)** (2000) 1260–1270.
28. J.-D. Grunwaldt, P. Kappen, L. Basini and B.S. Clausen. Iridium clusters for catalytic partial oxidation of methane – an in situ transmission and fluorescence XAFS study. *Catal. Lett.* **78(1–4)** (2002) 13–21.
29. W.-S. Dong, K.-W. Jun, H.-S. Roh, Z.-W. Liu and S.-E. Park. Comparative study on partial oxidation of methane over Ni/ZrO_2, Ni/CeO_2 and $Ni/Ce—ZrO_2$ catalysts. *Catal. Lett.* **78(1–4)** (2002) 215–222.
30. P. Pantu, K. Kim and G.R. Gavalas. Methane partial oxidation on $Pt/CeO_2—ZrO_2$ in the absence of gaseous oxygen. *Appl. Catal. A Gen.* **193** (2000) 203–214.
31. V.A. Sadykov, T.G. Kuznetsova, S.A. Veniaminov, D.I. Kochubey, B.M. Novgorodov, E.B. Burgina *et al.* Cation/anion modified ceria–zirconia solid solutions promoted by Pt as catalyst of methane oxidation into syngas by water in reversible redox cycles. *React. Kinet. Catal. Lett.* **76(1)** (2002) 83–92.
32. A. Trovarelli. Catalytic properties of ceria and CeO_2-containing materials. *Catal. Rev. Sci. Eng.* **38(4)** (1996) 439–520 and references therein.
33. S. Bedrane, C. Descorme and D. Duprez. Investigation of the oxygen storage process on ceria and ceria–zirconia-supported catalysts. *Catal. Today* **75** (2002) 401–405; S. Bedrane, C. Descorme and D. Duprez. Towards the comprehension of oxygen storage processes on model three-way catalysts. *Catal. Today* **73** (2002) 233–238 and references therein.
34. I.W. Carpenter and J.W. Hayes. *Catalytic Generation of Hydrogen*. CA 2325506, filed March 23, 1999.
35. M. Krumpelt, R. Wilkenhoener, D.J. Carter, J.-M. Bae, J.P. Kopasz and T. Krause. Catalytic autothermal reforming. In: *Annual Progress Report*, US DOE, Energy Efficiency and Renewable Energy, Office of Transportation Technologies (2000) 66–70.
36. T. Krause, J. Mawdsley, C. Rossignol, J. Kopasz, D. Applegate, M. Ferrandon *et al.* Catalytic autothermal reforming. In: *Electrochemical Technology Program–Argonne National Laboratory, 2002 National Laboratory R&D Meeting, DOE Fuel Cells for Transportation Program*, Denver, CO, May 9–10, 2002.
37. M. Krumpelt, A. Shabbir, R. Kumar and D. Rajiv. *Partial Oxidation Catalyst*. US 6,110,861, August 29, 2000.
38. M. Krumpelt, A. Shabbir, R. Kumar and D. Rajiv. *Method for Making Hydrogen Rich Gas from Hydrocarbon Fuel*. US 5,929,286, July 27, 1999.
39. K.W. Kramarz, I.D. Bloom, R. Kumar, S. Ahmed, R. Wilkenhoener and M. Krumpelt. *Steam Reforming Catalyst*. US Patent 6,303,098, October 16, 2001.
40. B. Lindström, U. Pettersson and P.G. Menon. Activity and characterization of Cu/Zn, Cu/Cr and Cu/Zr on γ-alumina for methanol reforming for fuel cell vehicles. *Appl. Catal. A Gen.* **234** (2002) 111–125; B. Lindström and U. Pettersson. Steam reforming of methanol over copper-based monoliths: the effects of zirconia doping. *J. Power Sources* **106** (2002) 264–273.
41. I.P. Breen and J.R.H. Ross. Methanol reforming for fuel-cell applications: development of zirconia-containing Cu—Zn—Al catalysts. *Catal. Today* **51** (1999) 521–533.
42. X.D. Hu and J.P. Wagner. *Promoted and Stabilized Copper Oxide and Zinc Oxide Catalyst and Preparation.* US Patent 5,990,040, November 23, 1999 and references therein.
43. B.A. Peppley, J.C. Amphlett, L.M. Kearns and R.F. Mann. Methanol-steam reforming on $Cu/ZnO/Al_2O_3$ catalysts. Part 2. A comprehensive kinetic model. *Appl. Catal. A Gen.* **179** (1999) 31–49.

44. J. Sun, I.S. Metcalfe and M. Sahibzada. Deactivation of $Cu/ZnO/Al_2O_3$ methanol synthesis catalyst by sintering. *Ind. Eng. Chem. Res.* **38** (1999) 3868–3872.
45. E. Xue, M. O'Keeffe and J.R.H. Ross. Water-gas shift conversion using a feed with a low steam to carbon monoxide ratio and containing sulfur. *Catal. Today* **30** (1996) 107–118.
46. M. Schneider, I. Pohl, K. Kochloefl and O. Boch. *Iron Oxide–Chromium Oxide Catalyst and Process for High Temperature Water-Gas Shift Reaction*. US 4598062, July 1, 1986.
47. C.V. Ovesen, B.S. Clausen, B.S. Hammershøi, G. Steffensen, T. Askgaard, I. Chorkendorff *et al.* A microkinetic analysis of the water-gas shift reaction under industrial conditions. *J. Catal.* **158** (1996) 170–180.
48. C. Rhodes, G.J. Hutchings and A.M. Ward. Water-gas shift reaction: finding the mechanistic boundary. *Catal. Today* **23** (1995) 43–58.
49. M.S. Wainwright and D.L. Trimm. Methanol and water-gas shift reaction on Ranney copper catalysts. *Catal. Today* **23** (1995) 29–42.
50. K. Klier, R.G. Herman and G.A. Vedage. *Water Gas Shift Reaction With Alkali-Doped Catalyst.* US 5,021,233, June 4, 1991.
51. C.T. Campbell. Promoters and poisons in the water-gas shift reaction. In: *Chemistry and Physics of Solid Surfaces*, Vol. 6. 1993, pp. 287–310.
52. M.V. Twigg and M.S. Spencer. Deactivation of supported copper metal catalysts for hydrogenation reactions. *Appl. Catal. A Gen.* **212(1–2)** (2001) 161–174.
53. T.P. Yu. *Down-Sized Water-Gas Shift Reactor*. CA 2317992, filed September 12, 2000.
54. T. Utaka, K. Sekizawa and K. Eguchi. CO removal by oxygen-assisted water gas shift reaction over supported Cu catalysts. *Appl. Catal. A Gen.* **194–195** (2000) 21–26; Y. Tanaka, T. Utaka, R. Kikuchi, K. Sasaki, K. Eguchi. CO removal from reformed fuel over $Cu/ZnO/Al_2O_3$ catalysts prepared by impregnation and coprecipitation methods. *Appl. Catal.* **238** (2002) 11–18.
55. N.C. Schiodt, P.E.H. Nielsen, P. Lehrmann and K. Aasberg-Petersen. *Process for the Production of a Hydrogen Rich Gas*. CA 2,345,515, application filed April 26, 2001.
56. G. Sauvignon and J. Caillod. *Process for the Conversion of Carbon Monoxide by Steam Using a Thioresistant Catalyst*. EP 0189701, March 6, 1990.
57. D.S. Newsome. *Catal. Rev. Sci. Eng.* **21** (1980) 275.
58. A.A. Andreev, V.J. Kafedjiyski and R.M. Edreva-Kardjieva. Active forms for water-gas shift reaction on NiMo-sulfide catalysts. *Appl. Catal. A Gen.* **179(1–2)** (1999) 223–228.
59. D.J. Meyers, J.F. Krebs, T.R. Krause and M. Krumpelt. In: *Annual Progress Report*, US DOE, Energy Efficiency and Renewable Energy, Office of Transportation Technologies (2000) 70–74; D.J. Meyers, D.J. Krebs, T.R. Krause and J.D. Carter. *Water-Gas Shift Catalysts with Improved Durability for Automotive Fuel Cell Applications*. ABSTR PAP AM CHEM S 222: 110-FUEL Part 1, August 2002.
60. J. Patt, D.J. Moon, C. Phillips and L. Thompson. Transition metal carbide water gas shift catalysts. *Catal. Lett.* **65** (2000) 193.
61. W. Ruettinger, J. Lampert, O. Korotkikh and R.J. Farrauto. *Non-Pyrophoric Water-Gas Shift Catalysts for Hydrogen Generation in Fuel Cell Applications*. ABSTR AP AM CHEM S 221: 11-Fuel Part 1, April 1, 2001.
62. Y. Li, Q. Fu and M.F. Stephanopoulos. Low-temperature water-gas shift reaction over Cu- and Ni-loaded cerium oxide catalysts. *Appl. Catal. B Environ.* **27** (2000) 179–191.
63. L. Mendelovici and M. Steinberg. Reaction of ethylene with oxygen on a Pt/CeO_2 catalysts. *J. Catal.* **93** (1985) 353–359.
64. L. Mendelovici and M. Steinberg. Methanation and water-gas shift reaction over Pt/CeO_2. *J. Catal.* **96** (1985) 285–287.
65. S.L. Swartz, M.M. Seabaugh, C.T. Holt and W.J. Dawson. Fuel processing catalysts based on nanoscale ceria, *Fuel. Cell. Bull.* **4(30)** (2001) 7–10.
66. T. Bunluesin, R.J. Gorte and G.W. Graham. Studies of the water-gas shift reaction on ceria-supported Pt, Pd, and Rh: implication for oxygen-storage properties. *Appl. Catal. B Environ.* **15** (1998) 107–114.
67. S. Hilaire, X. Wang, T. Luo, R.J. Gorte and J. Wagner. A comparative study of the water-gas shift reaction over ceria supported metallic catalysts. *Appl. Catal. A Gen* **215** (2001) 271–278.
68. Y. Iwasawa. The effects of coadsorbate on the behaviour of surface species and sites in catalysis by means of EXAFS and FTIR. In: R.W. Joyner and R.A. Van Santen (Eds.), *Elementary Reaction Steps in Heterogeneous Catalysis*. Kluwer Academic, 1993, pp. 287–301.

69. S. Golunski, R. Rajaram, N. Hodge, G.J. Hutchings and C.J. Kiely. Low-temperature redox activity in co-precipitated catalysts: a comparison between gold and platinum-group metals. *Catal. Today* **72** (2002) 107–113.
70. S.E. Golunski, J.M. Gascoyne, A. Fulford and J.W. Jenkins. *Metal Oxide Catalyst and Use Thereof in Chemical Reactions*. US 5,877,377, March 2, 1999.
71. A. Basinska, L. Kepinski and F. Domka. The effect of support on WGSR activity of ruthenium catalysts. *Appl. Catal. A Gen.* **183** (1999) 143–153.
72. J.M. Zalc, V. Sokolovskii and D.G. Loffler. Are noble metal-based water-gas shift catalysts practical for automotive fuel processing. *J. Catal.* **206** (2002) 169–171.
73. M. Haruta and M. Date. Advances in the catalysis of Au nanoparticles. *Appl. Catal. A Gen.* **222** (2001) 427–437.
74. G.C. Bond. Gold: a relatively new catalyst. *Catal. Today.* **72** (2002) 5–9.
75. Q. Fu, A. Weber and M.F. Stephanopoulos. Nanostructured Au—CeO_2 catalysts for low-temperature water-gas shift. *Catal. Lett.* **77** (2001) 87–95.
76. D. Andreeva, V. Idakiev, T. Tabakova, L. Ilieva, P. Falaras and A. Bourlinos *et al.* Low-temperature water-gas shift reaction over Au/CeO_2 catalysts. *Catal. Today* **72** (2002) 51–57.
77. F. Boccuzzi, A. Chiorino, M. Manzoli, D. Andreeva, T. Tabakova, L. Ilieva *et al.* Gold, silver and copper catalysts supported on TiO_2 for pure hydrogen production. *Catal. Today* **75** (2002) 169–175.
78. D. Andreeva, T. Tabakova, V. Idakiev, P. Christov and R. Giovanoli. Au/α-Fe_2O_3 catalyst for water-gas shift reaction prepared by deposition–precipitation. *Appl. Catal. A Gen.* **169** (1998) 9–14.
79. A.G. Gunner, I. Hyde, R.I. Potter and Thompsett D. *Catalyst.* US 5,939,220, August 17, 1999.
80. B. Rohland and V. Plazak. The PEMFC-integrated CO oxidation – a novel method of simplifying the fuel cell plant. *J. Power Sources* **84** (1999) 183–186.
81. A.N.J. Van Keulen and J.G. Reinkingh. *Hydrogen Purification*. US patent 6,403,049, June 11, 2002.
82. G.W. Skala, M.A Brundage, R.L. Borup, W.H. Pettit, K. Stukey, D.J. Hart-Predmore and J. Fairchok. US 6,132,689, October 17, 2000.
83. C.D. Dudfield, R. Chen and P.L. Adcock. A compact selective oxidation reactor for solid polymer fuel cell powered vehicle application. *J. Power Sources* **86** (2000) 214–222.
84. J.G.E. Cohn. *Process for Selectively Removing Carbon Monoxide from Hydrogen-Containing Gases*. US patent 3,216,783, November 9, 1965.
85. X. Liu, O. Korotkikh and R. Farrauto. Selective catalytic oxidation of CO in H_2: structural study of Fe oxide-promoted Pt/alumina catalyst. *Appl. Catal. A Gen.* **226** (2002) 293–303.
86. O. Korotkikh and R. Farrauto. Selective catalytic oxidation of CO in H_2: fuel cell applications. *Catal. Today* **62** (2000) 249–254.
87. M.J. Kahlich, H.A. Gasteiger and R.J. Behm. Kinetics of the selective CO oxidation in H_2-rich gas on Pt/Al_2O_3. *J. Catal.* **171** (1997) 93–105 and see also references therein.
88. M.M. Shubert, M.J. Kahlich, H.A. Gasteiger and R.J. Behm. Correlation between CO surface coverage and selectivity/kinetics for the preferential oxidation over Pt/γ-Al_2O_3 and Au/α-Fe_2O_3: an in-situ DRIFTS study. *J. Power Sources* **84** (1999) 175–182.
89. M.M. Shubert, H.A. Gasteiger and R.J. Behm. Surface formates as side products in the selective CO oxidation on Pt/γ-Al_2O_3. *J. Catal.* **172** (1997) 256–258.
90. A. Manasilp and E. Gulari. Selective CO oxidation over Pt/alumina catalysts for fuel cell applications. *Appl. Catal. B Environ.* **37** (2002) 17–25.
91. G.C. Bond and D.T. Thompson. Catalysis by gold. *Catal. Rev. Sci. Eng.* **41(3&4)** (1999) 319–388.
92. G.J. Hutchings. Gold catalysis in chemical processing. *Catal. Today* **72** (2002) 11–17.
93. N.A. Hodge, C.J. Kiely, R. Whyman, M.R.H. Siddiqui, G.J. Hutchings, Q.A. Pankhurst *et al.* Microstructural comparison of calcined and uncalcined gold/iron-oxide catalysts for low-temperature CO oxidation. *Catal. Today* **72** (2002) 133–144.
94. R.J.H. Griesel, C.J. Weststrate, A. Goossens, M.W.J. Crajé, A.M. van der Kraan and B.E. Nieuwenhuys. Oxidation of CO over Au/MOx/Al_2O_3 multi-component catalysts in a hydrogen-rich environment. *Catal. Today* **72** (2002) 123–132.
95. R.J. Griesel and B.E. Nieuwenhuys. Selective oxidation of CO over supported Au catalysts. *J. Catal.* **199** (2001) 48–50.
96. G.K. Bethke and H.H. Kung. Selective CO oxidation in a hydrogen-rich stream over Au/γ-Al_2O_3 catalysts. *Appl. Catal. A Gen.* **194–195** (2000) pp. 43–53.

97. H. Son and A.M. Lane. Promotion of $Pt/\gamma\text{-}Al_2O_3$ by Ce for preferential oxidation of CO in H_2. *Catal. Lett.* **76(3–4)** (2001) 151–154.
98. S.H. Oh and R.M. Sinkevitch. Carbon monoxide removal from hydrogen-rich fuel cell feedstreams by selective catalytic oxidation. *J. Catal.* **142** (1993) 254–262.
99. H. Igarashi, H. Uchida, M. Suzuki, Y. Sasaki and M. Watanabe. Removal of carbon monoxide from hydrogen-rich fuels by selective oxidation over platinum catalyst supported on zeolite. *Appl. Catal. A Gen.* **159** (1997) 159–169.
100. G.G. Xia, Y.G. Yin, W.S. Willis, J.Y. Wang and S.L. Suib. Efficient stable catalysts for low temperature carbon monoxide oxidation. *J. Catal.* **185** (1999) 91–105.
101. G. Avgouropoulos, T. Joannides, Ch. Papadopoulou, J. Batista, S. Hicevar and H.K. Matralis. A comparative study of $Pt/\gamma\text{-}Al_2O_3$, $Au/\alpha\text{-}Fe_2O_3$ and $CuO[SB]CeO_2$ catalysts for the selective oxidation of carbon monoxide in excess hydrogen. *Catal. Today* **75** (2002) 157–167.
102. W.H. Cheng. *React. Kinet. Catal. Lett.* **58(2)** (1996) 329.
103. K. Sekizawa, S. Yano, K. Eguchi and H. Arai. *Appl. Catal. A Gen.* **169** (1998) 191.
104. S.F. Abdo, C.A. DeBoy and G.F. Schroeder. *Process for Carbon Monoxide Preferential Oxidation for Use in Fuel Cells*. US 6,299,995, October 9, 2001.
105. Y. Hasegawa, K. Kusakabe and S. Morooka. Selective oxidation of carbon monoxide in hydrogen-rich mixtures by permeation through a platinum-loaded Y-type zeolite membrane. *J. Membr. Sci.* **190(1)** (2001) 1–8.

Chapter 6

Conversion of hydrocarbons and alcohols for fuel cells

Finn Joensen and Jens R. Rostrup-Nielsen

Abstract

The growing demand for clean and efficient energy systems is the driving force in the development of fuel processing technology for providing hydrogen or hydrogen-containing gaseous fuels for power generation in fuel cells. Successful development of low cost, efficient fuel processing systems will be critical to the commercialisation of this technology. This article reviews various reforming technologies available for the generation of such fuels from hydrocarbons and alcohols. It also briefly addresses the issue of carbon monoxide clean-up and the question of selecting the appropriate fuel(s) for small-medium-scale fuel processors for stationary and automotive applications.

Keywords: Fuel processor; Steam reforming; Partial oxidation; Fuel cell; Hydrocarbon; Alcohol

Article Outline

1. Introduction 107
2. Steam reforming 108
3. Partial oxidation 111
4. Carbon monoxide clean-up 112
5. The fuel choice 114
6. Conclusions 115
References 115

1. INTRODUCTION

Catalysis is likely to become a key element in the conversion of liquid or gaseous fuels into hydrogen for fuel cells. This conversion, commonly referred to as fuel processing, most often involves either *hydrocarbons*, like methane, propane/LPG, and higher liquid hydrocarbons or *alcohols*, e.g. methanol and ethanol, although, in principle, any hydrogen-containing compound may be applied, such as dimethyl ether and ammonia. It is even possible to convert other fuels than hydrogen – directly or indirectly – in the fuel cell [1,2]. Known examples comprise methanol, methane and carbon monoxide. However, most fuel cells are based on the electrochemical oxidation of hydrogen, although higher electrical efficiencies may be achieved when applying other fuels, for instance, methane in the solid oxide fuel cell and indeed, in some cases, ideal electrical efficiencies in excess of 100% may, in principle, be obtained [3].

Fuel Cells Compendium

The generation of hydrogen, or hydrogen-rich product streams, by reforming of hydrocarbons or alcohols, may from a thermodynamic point of view, be categorised in two basically different types of processes. One is – endothermic – steam reforming in which the hydrocarbon or alcohol feed is reacted with steam. The heat required for the reaction is supplied from external sources – either by combustion of part of the feed, by burning combustible off gases or by a combination of both. The other is – exothermic – partial oxidation, where the feed reacts directly with air, enriched air or (in large plants) pure oxygen at a carefully balanced oxygen to fuel ratio. In this case, the overall process becomes net heat producing. In either of the processes heat management and, as part thereof, thermal integration of the fuel processor and the fuel cell becomes key to achieve high overall plant efficiencies.

2. STEAM REFORMING

The most important route to hydrogen is steam reforming of either natural gas (Eq. (1)) or liquid hydrocarbons as exemplified in (2) by pure component *n*-heptane. For small scale hydrogen plants and for automotive applications, steam reforming of methanol (3) may be an attractive alternative.

$$CH_4 + H_2O = CO + 3H_2, \quad -\Delta H^\circ_{298} = -198\,\text{kJ mol}^{-1} \tag{1}$$

$$C_7H_{16} + 7H_2O = 7CO + 15H_2, \quad -\Delta H^\circ_{298} = -1175\,\text{kJ mol}^{-1} \tag{2}$$

$$CH_3OH + H_2O = CO_2 + 3H_2, \quad -\Delta H^\circ_{298} = -49\,\text{kJ mol}^{-1} \tag{3}$$

In all of reactions (1)–(3), the water-gas shift (WGS) reaction participates independently:

$$CO + H_2O = CO_2 + H_2, \quad -\Delta H^\circ_{298} = 41\,\text{kJ mol}^{-1} \tag{4}$$

All reactions are reversible. Under the preferred reforming conditions, however, the position of the thermodynamic equilibrium makes reactions (2) and (3) essentially irreversible, but due to the WGS reaction, the equilibrated product inevitably contains steam and carbon oxides along with the desired hydrogen product. Steam reforming of hydrocarbons is catalysed by group VIII metals, with Ni being the most cost-effective.

Steam reforming of *methane* is strongly endothermic – more so per carbon atom than any of the higher hydrocarbon homologues. Methane steam reforming is, also in practice, a reversible reaction. In order to ensure a high methane conversion, therefore, it is necessary to operate at high temperature, low pressure and relatively high steam to carbon ratios (Fig. 1).

In conventional tubular steam reforming, the heat is transferred to the process by placing the reformer tubes in a fired furnace. One constraint imposed by this layout is that only about 50% of the furnace heat is transferred to the reforming process. The remainder is recovered in a waste heat section typically serving to produce steam and to preheat feed streams. This figure may, however, be significantly increased when applying heat exchange reforming in which both the flue gas and the high-temperature product gas is cooled by heat exchange with the process gas within the reformer itself [4–6].

Light distillate naphtha is an attractive feedstock in areas where natural gas is not readily available. The conversion of *higher hydrocarbons* takes place by irreversible adsorption to the nickel surface, subsequent breakage of terminal C—C bonds one by one until, eventually, the hydrocarbon is converted into C_1 components. However, reaction rates of individual hydrocarbons over a given catalyst are often quite different from one particular component to the other. And, even though most higher hydrocarbons react faster than methane, they are at the same time also susceptible to non-catalytic thermal cracking [7]. At temperatures above 600–650°C, i.e.

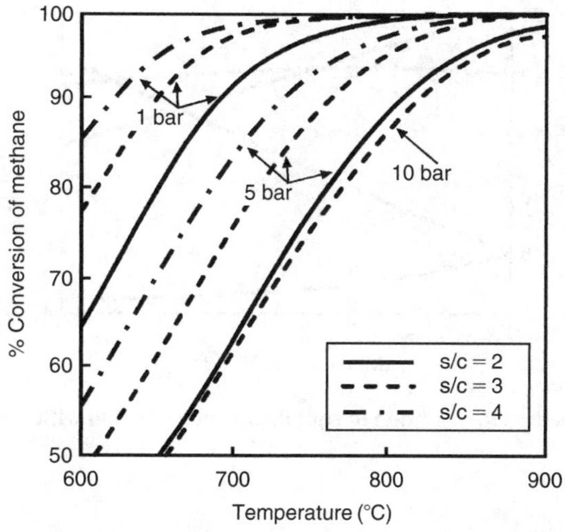

Fig. 1. Steam reforming of methane. Equilibrium conversion against temperature, pressure and steam/carbon ratio

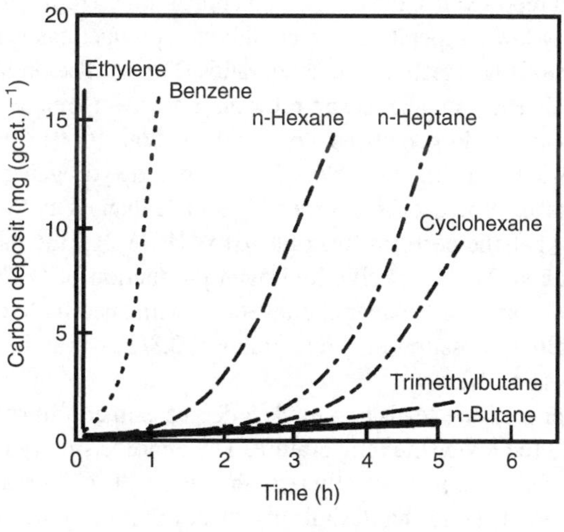

Fig. 2. Rate of carbon formation for selected different hydrocarbons [7]

at the temperatures characteristic to steam reforming, the thermal reactions begin to compete with the catalytic processes and increasingly so as the catalyst activity decreases, e.g. due to sulphur poisoning. The thermal cracking – or pyrolysis – of the higher hydrocarbons produces olefins which are precursors for coke formation. In particular, ethylene leads to rapid carbon formation (Fig. 2).

In general, the heavier the hydrocarbon feedstock, the slower the reaction rate [8] and the higher the risk of pyrolysis. This problem of carbon formation may be solved effectively by the insertion of a low temperature, fixed bed adiabatic *prereformer* [7] prior to the primary steam reformer. In the adiabatic prereformer, the higher hydrocarbons are completely converted into C_1 fragments (CH_4, CO and CO_2). This is quite similar to the process taking place in conventional reforming, but the relatively low temperatures (350–550°C) in the prereformer eliminate the potential for carbon formation. Moreover, the prereformer allows for higher inlet temperatures in the primary reformer, thereby reducing its size.

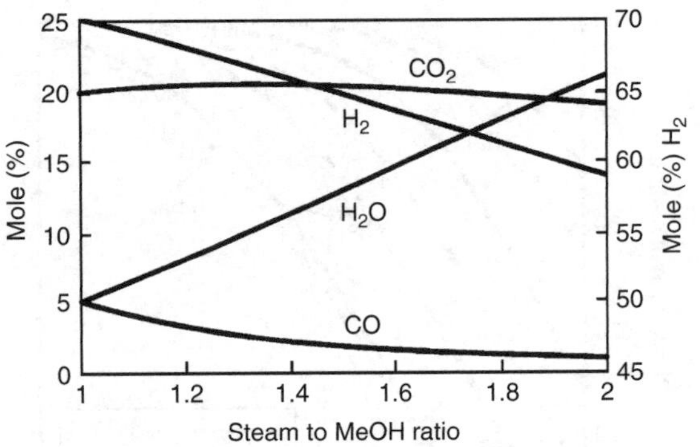

Fig. 3. Steam reforming of methanol. Variation in equilibrium composition with steam to methanol ratio (280°C, 5 bar)

The *steam reforming of methanol* is much less endothermic (Eq. (3)) than that of hydrocarbons. Thus, more than 99% conversion is readily achieved at low temperatures, 200–300°C, using copper-based catalysts. No methanation occurs and conversion and product distribution is not very pressure-sensitive. Although highly active Pd catalysts have been reported [9], those based on copper are preferred for economical reasons.

As a result of the relatively low temperatures the equilibrated product gas is rich in hydrogen and, consequently, low in CO even at moderate steam to methanol ratios (Fig. 3). The optimum steam to carbon ratio is typically between 1.2 and 1.5. The methanol steam reforming process is relatively well-understood [10–13]. The main application is in small hydrogen plants, i.e. less than about 1000 $N\,m^3h^{-1}$. The technology is also considered by several groups to be an attractive solution for on-board hydrogen generation for automotive purposes, because the methanol fuel processor is considerably simple than its hydrocarbon-based counterpart.

As part of a recent study [14], the performance of a 50 kW (LHV H_2) methanol reformer was evaluated. Stationary performance tests showed a specific hydrogen production of 6.7 $N\,m^3/(kg_{cat}\,h)$ at a methanol conversion of 95%. Figure 4 correlates methanol conversion with specific hydrogen productivity in terms of $N\,m^3/(kg_{cat}\,h)$ at two different pressures and temperatures, 3.8/21 bar and 260–280°C, respectively, and a molar steam to methanol ratio of 1.5.

As can be seen, the higher pressure leads to a modest decrease in equilibrium conversion, this decrease becoming more pronounced the lower the temperature. The endothermicity of the steam reforming reaction is reflected by the large increase in productivity upon a mere 20°C increase in operating temperature. This effect of temperature emphasises the desirability of developing more temperature tolerant copper-based reforming catalysts.

Ethanol and higher alcohols may also be converted. However, as this involves breaking of one or more C—C bonds, the process is far from as facile as it is for methanol and higher temperatures are generally required. Ethanol easily becomes dehydrated, forming ethylene which leads to carbon formation. At modest temperatures, below approximately 450°C, efficient bond cleavage is obtained over a Ni/Al_2O_3 catalyst to yield a mixture of essentially methane and carbon dioxide [15]. However, nickel is a very effective catalyst for carbon formation from ethylene, if formed (cf. Fig. 2). Cu/Ni-based catalysts active below 300°C have been reported [16]. Other catalyst systems studied include those based on Co, Cu/Zn, Cu/Zn/Cr and noble metals supported on different carriers [17–19]. These studies indicate that steam reforming of ethanol proceeds via an acetaldehyde intermediate [18–21]. Acetic acid is also commonly observed and a reaction mechanism involving two parallel pathways from common intermediate acetaldehyde has been proposed [20]: one involves the direct decarbonylation of acetaldehyde forming CO and CH_4; the other goes from acetaldehyde via acetic acid to CO, CO_2 and H_2.

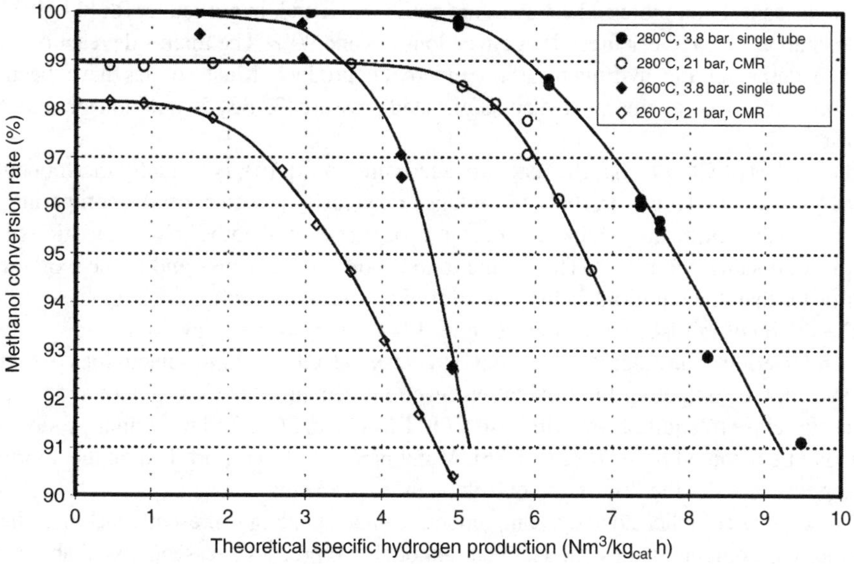

Fig. 4. Methanol reforming (steam/methanol molar ratio = 1.5): conversion against specific hydrogen production at different pressures and temperatures [14]

3. PARTIAL OXIDATION

Partial oxidation reactions may be carried out either by catalytic partial oxidation (CPO), by non-catalytic partial oxidation or by autothermal reforming (ATR), the latter being a combination of non-catalytic oxidation and steam reforming.

The *autothermal reformer* [22] consists of a thermal and a catalytic zone. The feed is introduced to a burner and mixed intensively with steam and a substoichiometric amount of oxygen or air. In the combustion (thermal) zone, part of the feed reacts essentially according to

$$CH_4 + \tfrac{3}{2}O_2 = CO + 2H_2O, \quad -\Delta H^\circ_{298} = 519\ \text{kJ mol}^{-1} \tag{5}$$

By proper adjustment of oxygen to carbon and steam to carbon ratios, the partial combustion in the thermal zone (Eq. (5)) supplies the heat for the subsequent endothermic steam reforming (Eq. (1)) and shift (Eq. (4)) reactions taking place in the catalytic zone in which soot precursors are effectively broken down. Thus, the product gas composition is fixed thermodynamically through the pressure, exit temperature, steam to carbon and oxygen to carbon ratios [23].

The *non-catalytic partial oxidation* [24] needs high temperature to ensure complete conversion of methane and to reduce soot formation. Some soot is normally formed and is removed in a separate scrubber system downstream of the partial oxidation reactor. The thermal processes typically result in a product gas with H_2/CO = 1.7–1.8.

Catalytic partial oxidation has been subject to intensified research efforts in recent years. In CPO, the reaction is initiated catalytically (flameless) as opposed to ATR and POX. It has been shown [25–29] that under extremely short residence times, in the order of milliseconds, methane may be partially oxidised forming H_2 and CO as the main products:

$$CH_4 + \tfrac{1}{2}O_2 = CO + 2H_2, \quad -\Delta H^\circ_{298} = 38\ \text{kJ mol}^{-1} \tag{6}$$

For natural gas conversion preferred catalysts are based on Ni and, in particular, Rh [27,30,31] and selectivities higher than 90% may be achieved at conversions beyond 90%. The main side reactions are competitive, further oxidation of the hydrogen/carbon monoxide product. Most studies have been made near atmospheric pressure. Experiments carried out at elevated pressure [32,33] do not indicate dramatic changes in product distribution.

Among the virtues of CPO is that the reaction, according to Eq. (6), is virtually thermoneutral and has a low net energy demand. However, these advantages may easily be offset by the competing total oxidation reactions which significantly enhance process exothermicity. Another characteristic is that, ideally, the reaction is kinetically controlled. This is due to the short contact times and to the oxidation reactions being much faster than the equilibrating steam reforming and shift reactions. Thus, by selecting a proper interval in the region of millisecond residence times [34], it is possible to avoid the slower steam reforming reactions interfering to any significant extent. However, despite the high selectivities to carbon monoxide and hydrogen, the competing total oxidation reactions remain a major problem and, in practice, gas compositions are close to equilibrium with respect to Eqs (1) and (4). Today, similar product selectivities as for CPO may be obtained by ATR [23,35,36]. Many papers fail to report data on the approach to equilibrium. Often, this renders the discussion on selectivities confusing.

The CPO reaction is complicated and a comprehensive understanding is presently lacking. The complexity of the technology is compounded by the fact that, although the process is essentially adiabatic, it is characterised by high catalyst surface temperatures [28,34] which leads to thermal non-equilibrium between the solid and gaseous phases. Given the process being conducted at high temperatures and extremely short contact times, i.e. within the domain of kinetic control, it is evident that heat and mass transfer play a decisive role in determining process characteristics, temperature and concentration profiles [34,37], which eventually may change the entire product spectrum. As addressed in several papers [29,34,38] careful examination of factors such as gas mixing and flow patterns, radiation, reactor and catalyst geometry, etc. are of utmost importance to further uncover the fundamentals of CPO.

For gasoline fuel processing for automotive applications, small air blown reformers are the preferred choice and CPO reformer prototypes have been developed [39].

Also, *methanol* can be converted as proposed by the so-called HotSpot™ fuel processor concept [40–42]. The first step in this process is believed to be total oxidation of part of the methanol, supplying the heat for subsequent steam reforming. Thus, in the current terminology, we are dealing here with *flameless* ATR rather than methanol-based CPO fuel processing. The catalyst is a combined noble metal/base metal catalyst with the noble metal acting as a process initiator by catalysing the total oxidation reaction. The fuel processor features a modular design to enable different maximum power output at similar response times. Ethanol has also been proposed as a feed for partial oxidation [43].

4. CARBON MONOXIDE CLEAN-UP

Whereas high temperature fuel cells (MCFC and SOFC) are capable of converting methane, CO and alcohols, etc. in the anode chamber by internal reforming, the PAFC and the PEM cells do not tolerate excessive amounts of CO as indicated in Table 1.

Table 1. Fuel cell characteristics

	Cell temperature (°C)	Maximum CO content (ppm)	Primary fuel
PEMFC	70–80	50	H_2, MeOH
PAFC	200	500	H_2
MCFC	600–650	No limit	H_2, CH_4, CO, MeOH
SOFC	700–1000	No limit	H_2, CH_4, CO, MeOH

For the PAFC, it is possible to reach the level of <500 ppm (0.05 vol.%) CO by means of high- and low-temperature WGS reactors as shown in Fig. 5. The fuel processor system (FPS) consists of a hydrodesulphurisation unit, a heat exchange reformer (to minimise waste heat production) and two WGS reactors bringing the CO content down to <0.05%. The anode off-gas is used as fuel for the reformer and the flue gas is used to preheat the feed and the cathode air.

The PEMFC does not tolerate more than in the order of 50 ppm CO; the lower the CO concentration, the higher the efficiency of the cell. Therefore, further purification is required, which makes hydrocarbon-powered fuel processors for PEMFC's so complicated relative to the methanol-based ones which do not require shift of the reformate gas prior to the final CO clean-up. Currently, among the processes for CO removal, selective oxidation appears to be the preferred solution [44–47]. Alternatives include CO-selective methanation [40,48] or the use of hydrogen-selective Pd-alloy membranes [49–51]. Figure 6 shows a principal flow diagram of fuel processing system for automotive use based on methanol steam reforming.

One inherent problem to the PEMFC's is that they require low operating temperatures in order to avoid deterioration of the Nafion®-type polymer membranes. Enabling a higher working temperature of the fuel cell

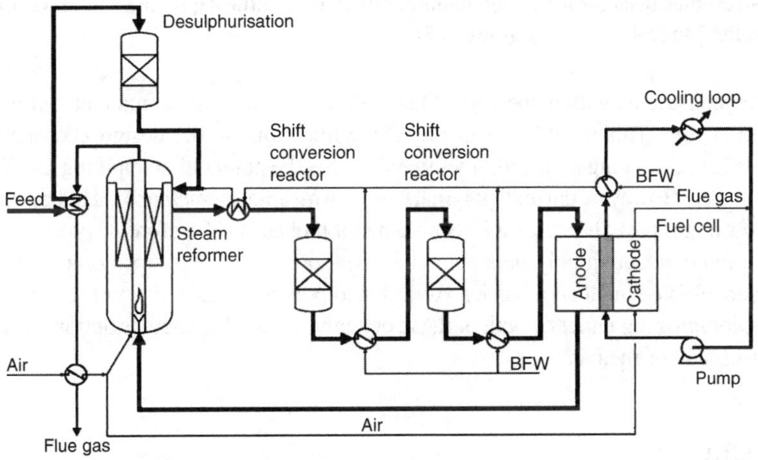

Fig. 5. Schematic diagram of FPS for phosphoric acid fuel cell

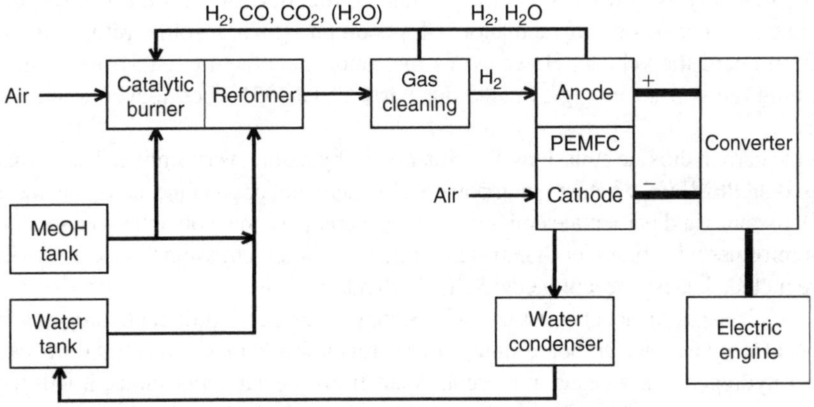

Fig. 6. Principal flow diagram of methanol-powered FPS for automotive application [52]

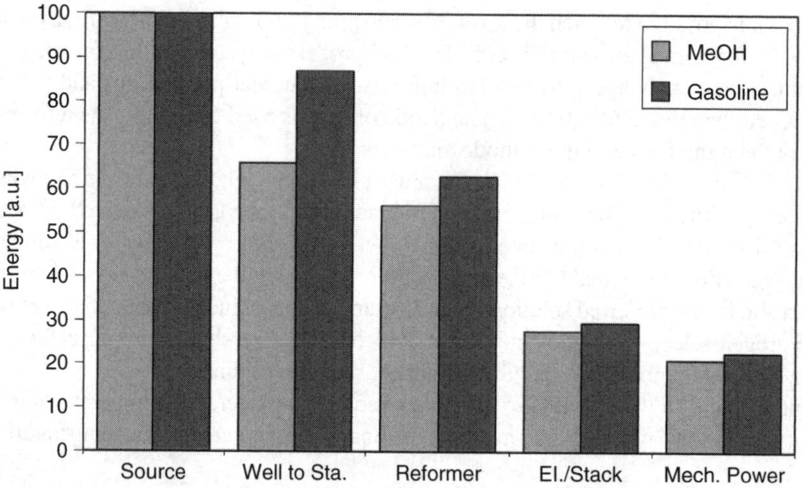

Fig. 7. Well-to-wheel efficiencies for methanol (manufactured from natural gas) and "gasoline" (no penalty applied for polishing "gasoline" to fuel processing grade) [55]

would alleviate the constraints with respect to CO content and might even completely eliminate the need for the final purification step. This would lead to great simplifications in the design and operation of the fuel processor. Another aspect relating to the polymer membrane is the potential of applying the PEMFC as a direct methanol fuel cell [2]. However, current polymers are permeable to methanol. Higher resistance towards methanol crossover might pave the way for direct methanol fuel cells. Therefore, very interesting perspectives arise if alternative ion conducting polymers could be developed, capable of operating at higher temperatures and/or providing an effective methanol barrier. Acid-doped polybenzimidazole membranes appear to possess the potential of accomplishing either or both of these objectives [53,54]. Another option might involve the use of dimethyl ether instead of methanol.

5. THE FUEL CHOICE

The PEMFC is the preferred solution for automotive application. However, while there is little doubt about hydrocarbons – natural gas, LPG or liquids – being the preferred fuels for *stationary* plants, there is considerable more ambiguity as to what is currently being the fuel of choice for *automotive* on-board hydrogen manufacture. The conversion of methanol is less complex than gasoline reforming and, moreover, more efficient on-board the vehicle. However, as methanol manufacture, e.g. from natural gas is more energy consuming the two technologies come out fairly equal [55] when looked at on a well-to-wheel basis (Fig. 7).

With respect to carbon dioxide emissions, the situation is, by nature, very similar. The primary advantages of vehicles based on PEMFCs, whether being powered by methanol or gasoline, consist in the extremely low local NO_x, CO, organics and particulate emissions. Table 2 compares data obtained in a study [14] of a compact methanol reformer with future emission standards. Even when emissions associated with the well-to-station step are included, these are still below SULEV standards.

Obviously, such "near-zero" local emissions will become "true-zero" if direct hydrogen vehicles become available on a commercial scale. On-board storage of hydrogen would make drive trains far less complicated. However, direct hydrogen is not viable at present. Apart from special applications, it will require a breakthrough in materials development [56] as well as establishment of a hydrogen infrastructure.

Table 2. Comparison of specific passenger car emissions (mg/km)

	CO	NO_x	Organic gases
Euro 2005 (gasoline)	1000	50	100
Euro 2005 (diesel)	500	250	50
Calif. ULEV	1312	43	34
Calif. SULEV	625	12	6
PEMFC (MeOH)	0.3	<0.01	0.9

6. CONCLUSIONS

Catalysis offers a variety of options for the conversion of hydrocarbons and alcohols into hydrogen for fuel cells.

Natural gas, whenever available, is the obvious choice for large scale hydrogen manufacture. This will also be true for smaller scale stationary units, e.g. for future on-site hydrogen manufacture at gas stations and for an emerging market for decentralised combined heat and power generation in residential areas, companies and institutions. In remote areas or in other locations where a natural gas infrastructure is not established, alternative fuels such as propane/LPG and higher, liquid hydrocarbons represent competitive alternatives, thus emphasising the aspect of fuel flexibility.

For automotive purposes, pure hydrogen is the preferred fuel due to simplicity in design, low cost and high efficiency. However, on a medium term time horizon, it is difficult to envisage hydrogen penetrating the market beyond centrally fuelled fleets. In the meantime, fuel processor development, integration with the fuel cell stack and auxiliaries remains a difficult technical challenge. Furthermore, as to fuel choice and specifications, there still appears to be a profound lack of consensus among world class organisations for what may be a transitional solution bridging the gap to direct hydrogen. This is as other, in defining the basis for a rational fuel and system selection. These uncertainties must be resolved in order for fuel cell power trains to constitute a serious alternative to conventional – and currently improving – engine technology.

The call for clean energy is the main driver in the development of fuel processing technology for both stationary and mobile applications. Several fuel processor designs are likely to emerge, for different applications and fuel cell types and for different fuels. The technology is available, but further progress in catalyst and component development, system design and integration, cost reduction and not least successful field test programmes is required for fuel processors to eventually gain general customer acceptance.

REFERENCES

1. S. Park, J.M. Voks and R.J. Gorte. *Nature* **404** (2000) 265.
2. S. Wasmus and A. Kuver. *J. Electroanal. Chem.* **461** (1999) 14.
3. J.R. Rostrup-Nielsen and L.J. Christiansen. *Appl. Catal. A Gen.* **126** (1995) 381.
4. J.R. Rostrup-Nielsen, I. Dybkjær and L.J. Christiansen. In: H. Lasa (Ed.), *Chemical Reactor Technology for Environmentally Safe Reactors and Products.* Kluwer Academic Publishers, Dordrecht, 1992, p. 249.
5. N.R. Udengaard, L.J. Christiansen and D.M. Rastler. Abstracts, Fuel Cell Seminar 1988, Long Beach, CA, 1988, p. 47.
6. H. Stahl and C.L. Laursen. Abstracts, Fuel Cell Seminar, Tuscon, AZ, 1992, p. 465.
7. J.R. Rostrup-Nielsen, I. Dybkjaer and T.S. Christensen. *Stud. Surf. Sci. Catal.* **113** (1998) 81.
8. J.R. Rostrup-Nielsen. In: J.R. Anderson and M. Boudart (Eds.), *Catalytic Steam Reforming in Catalysis, Science and Technology*, Vol. 5. Springer, Berlin, 1983.
9. N. Iwasa, S. Masuda, N. Ogawa and N. Takezawa. *Appl. Catal. A* **125** (1995) 145.
10. H.G. Düsterwald, B. Höhlein, H. Kraut, J. Meusinger, R. Peters and U. Stimming. *Chem. Eng. Technol.* **20** (1997) 617.

11. B. Höhlein, M. Boe, J. Bøgild Hansen, P. Bröckerhoff, G. Colsman, B. Emonts, R. Menzer and E. Riedel. *J. Power Sources* **61** (1996) 143.
12. C.J. Jiang, D.L. Trimm, M.S. Wainwright and N.W. Cant. *Appl. Catal. A* **97** (1993) 145.
13. B.A. Peppley, J.C. Amphlett, L.M. Kaerns and R.F. Mann. *Appl. Catal. A* **179** (1999) 31.
14. B. Emonts, J. Bøgild Hansen, H. Schmidt, T. Grube, B. Höhlein, R. Peters and A. Tschauder. *J. Power Sources* **86** (2000) 228.
15. A. Williams. Eur. Patent No. 34407 (1981).
16. F.J. Marino, E.G. Cerrella, S. Duhalde, M. Jobbagy and M.A. Laborde. *Int. J. Hydrogen Energ.* **23** (1998) 1095.
17. F. Haga, T. Nakajima, H. Miya and S. Mishima. *Catal. Lett.* **48** (1997) 223.
18. S. Cavallaro and S. Freni. *Int. J. Hydrogen Energ.* **21** (1996) 465.
19. F. Haga, T. Nakajima, K. Yamashita and S. Mishima. *React. Kinet. Catal. Lett.* **63** (1998) 253.
20. J.G. Highfield, F. Geiger, E. Uenala and Th.H. Schucan. *Hydrogen Energy Progress X*, **2** (1994) p. 1039.
21. N. Iwasa and N. Takezawa. *Bull. Chem. Soc. Jpn.* **64** (1991) 2619.
22. J.R. Rostrup-Nielsen. *Phys. Chem. Chem. Phys.* **3** (2001) 283.
23. T.S. Christensen, P.S. Christensen, I. Dybkjær, J.-H. Bak Hansen and I.I. Primdahl. *Stud. Surf. Sci. Catal.* **119** (1998) 883.
24. A. Docter and A. Lamm. *J. Power Sources* **84** (1999) 194.
25. V.R. Choudary, A.S. Mamman and S.D. Sansare. *Angew. Chem. Int. Ed. Engl.* **31** (1992) 1189.
26. H. Hickman and L.D. Schmidt. *J. Catal.* **138** (1992) 267.
27. H. Hickman, E.A. Haupfear and L.D. Schmidt. *Catal. Lett.* **17** (1993) 223.
28. L. Basini, A. Guarinoni and K. Aasberg-Petersen. *Stud. Surf. Sci. Catal.* **119** (1998) 699.
29. A.S. Bodke, S.S. Bharadwaj and L.D. Schmidt. *J. Catal.* **179** (1998) 138.
30. S.S. Bharadwaj and L.D. Schmidt. *J. Catal.* **146** (1994) 11.
31. P.M. Torniainen, X. Chu and L.D. Schmidt. *J. Catal.* **146** (1994) 1.
32. A.G. Dietz and L.D. Schmidt. *Catal. Lett.* **33** (1995) 15.
33. L. Basini, K. Aasberg-Petersen, A. Guarinoni and M. Østberg. *Catal. Today* **64** (2001) 9.
34. P.M. Witt and L.D. Schmidt. *J. Catal.* **163** (1996) 465.
35. K. Aasberg-Petersen, J.-H. Bak Hansen, T.S. Christensen, I. Dybkjaer, P. Seier Christensen, C. Stub Nielsen, S.E.L. Winter Madsen, J.R. Rostrup-Nielsen. *Appl. Catal.* (submitted for publication).
36. J.R. Rostrup-Nielsen, I. Dybkjaer and K. Aasberg-Petersen. *Symp., Prepr. Am. Chem. Soc. Div. Pet. Chem.* **45** (2000) 186.
37. F. Basile, L. Basini, M. D'Amore, G. Fornasari, A. Guarinoni, D. Matteuzzi, G. Del Piero, F. Trifirò and A. Vaccari. *J. Catal.* **173** (1998) 247.
38. L.D. Schmidt. *Stud. Surf. Sci. Catal.* **130** (2000) 61.
39. Hart Eur. *Fuels News* **4 (4)** (2000) 3.
40. S. Golunski. *Platinum Met. Rev.* **42** (1998) 2.
41. N. Edwards, S.R. Ellis, J.C. Frost, S.E. Golunski, A.N.J. van Keulen, N.G. Lindewald and J.G. Reinkingh. *J. Power Sources* **71** (1998) 123.
42. P.G. Gray and M.I. Petch. *Platinum Met. Rev.* **44** (2000) 108.
43. W.L. Mitchell. *SAE Paper* 952761 (1995).
44. M.J. Kahlich, H.A. Gasteiger and R.J. Behm. *J. Catal.* **182** (1999) 430.
45. A.I. Kozlov, A.P. Kozlova, H. Liu and Y. Iwasawa. *Appl. Catal. A* **182** (1999) 9.
46. S.H. Oh and R.M. Sinkevitch. *J. Catal.* **142** (1993) 254.
47. R.M. Torres-Sanchez, A. Ueda, K. Tanaka and M. Haruta. *J. Catal.* **125** (1997) 125.
48. S.-I. Fujita and N. Takezawa. *Chem. Eng. J.* **68** (1997) 63.
49. A. Li, W. Liang and R. Hughes. *Catal. Today* **56** (2000) 45.
50. M. Kajiwara, S. Uemiya, T. Kojima and E. Kikuchi. *Catal. Today* **56** (2000) 65.
51. S.L. Jørgensen, P.E.H. Nielsen and P. Lehrmann. *Catal. Today* **25** (1995) 303.
52. W. Wiese, B. Emonts and R. Peters. *J. Power Sources* **84** (1999) 187.
53. J.T. Wang, R.F. Savinell, J. Wainwright, M. Litt and H. Yu. *Electrochim. Acta* **41** (1996) 193.
54. B.S. Pivovar, Y. Wang and E.L. Cussler. *J. Membrane Sci.* **154** (1999) 155.
55. J. Bøgild Hansen. Unpublished data.
56. M.S. Dresselhaus, K.A. Williams and P.C. Eklund. *MRS Bull.* **24** (1999) 45.

Chapter 7
An assessment of alkaline fuel cell technology

G.F. McLean, T. Niet, S. Prince-Richard and N. Djilali

Abstract

This paper provides a review of the state of the art of alkaline fuel cell (AFC) technology based on publications during the past 25 years. Although popular in the 1970s and 1980s, the AFC has fallen out of favor with the technical community in the light of the rapid development of Proton Exchange Membrane Fuel Cells (PEMFCs). AFCs have been shown to provide high-power densities and achieve long lifetimes in certain applications, and appear to compete favorably with ambient air PEMFCs. In this report we examine the overall technology of AFCs, and review published claims about power density and lifetime performance. Issues surrounding the sensitivity of the AFC to CO_2 in the oxidant stream are reviewed and potential solutions discussed. A rough cost comparison between ambient air AFCs and PEMFCs is presented. Overall, it appears the AFCs continues to have potential to succeed in certain market niche applications, but tends to lack the R&D support required to refine the technology into successful market offerings.

Article Outline

1. Introduction . . . 118
2. Alkaline fuel cell background and development status . . . 119
   2.1. Principle of operation . . . 119
   2.2. Research activity level . . . 121
   2.3. Corporate activities . . . 121
3. Technical review . . . 122
   3.1. Power density . . . 122
      3.1.1. Space applications . . . 122
      3.1.2. Atmospheric pressure cells . . . 123
      3.1.3. Performance of components . . . 124
         3.1.3.1. Three-dimensional electrodes . . . 124
         3.1.3.2. Electrode materials . . . 125
         3.1.3.3. Electrode fabrication methods . . . 125
      3.1.4. Comparison to PEM . . . 126
   3.2. Poisoning and contamination issues . . . 126
      3.2.1. Effect of carbon dioxide on the cathode . . . 127
      3.2.2. Carbon dioxide strategies for the cathode . . . 128
      3.2.3. Effect of impurities on the anode . . . 129
      3.2.4. Strategies for the anode . . . 129
      3.2.5. Summary of contamination effects . . . 129

Fuel Cells Compendium

3.3. System issues 130
3.3.1. Electrolyte circulation 130
3.3.2. Salvage 131
3.4. Lifetime and duty cycle information 131
3.4.1. Zevco long-term tests 131
3.4.2. Other long-term tests 132
3.4.3. Summary of AFC lifetimes 132
4. Cost analysis 133
4.1. Gross costs and commercial estimates 134
4.2. Materials and manufacturing 134
4.2.1. AFC stack materials 134
4.2.2. PEMFC stack materials 135
4.2.3. AFC system costs 136
4.2.4. PEMFC stack costs 136
4.3. Impact of production volume 136
4.4. Extrapolation to ambient air PEM 137
4.5. Balance of plant 138
4.5.1. AFC peripherals 138
4.5.1.1. Air blower 138
4.5.1.2. CO_2-scrubber 138
4.5.1.3. Electrolyte recirculation loop 138
4.5.1.4. Water management 138
4.5.1.5. Nitrogen purge 139
4.5.2. Alkaline peripheral costs 139
4.5.3. Compressed PEMFC peripherals 139
4.5.4. Ambient air PEM peripherals 139
4.6. Cost of consumables 140
4.6.1. Soda lime 140
4.6.2. KOH 140
4.7. System cost estimates 140
5. Conclusions 141
Appendix A. 7 kW PEMFC stack cost development 142
References 143

1. INTRODUCTION

The alkaline fuel cell (AFC) was the first fuel cell technology to be put into practical service and make the generation of electricity from hydrogen feasible. Starting with applications in space the alkaline cell provided high-energy conversion efficiency with no moving parts and high reliability. AFCs were used as the basis for the first experiments with vehicular applications of fuel cells, starting with a farm tractor in the late 1950s equipped with an Allis Chalmers AFC (Kordesch and Simader, 1996). This was followed by the now famous Austin A40 operated by Karl Kordesch in the early 1970s [1] and continuing today with the commercialization activities of the ZEVCO company [2,3]. However, despite its early success and leadership role in fuel cell technology, AFCs have fallen out of favor with the research community and have been eclipsed by the rapid development of the Proton Exchange Membrane fuel cell (PEMFC) as the technology of choice for vehicular applications.

This paper provides a critical overview of the state of the art of AFC technology and attempts to synthesize the published information on AFCs to provide a unified view of the technology. A re-examination of the economics of AFC technology is also presented. The issues generally assumed to have caused the demise of interest in AFCs, namely low-power density and electrolyte poisoning are addressed in detail to provide as complete a picture as possible, based primarily on published and publicly available information.

The PEMFC has recently emerged as the technology of choice for low temperature, moderate power applications and has largely displaced the AFC in this application. Because of this, we have provided a comparison between alkaline and PEM technology wherever possible. In particular, a detailed cost comparison between PEM and AFCs is included.

The public domain literature has been reviewed including the most recently published results on alkaline electrode materials and manufacture as well as older publications describing the state of the art around 1980. Earlier publications, which largely describe the now defunct space applications of AFC technology, have not been reviewed. The overall purpose of this review has been to establish a technical opinion about the viability of AFCs and to identify key areas for research.

The report is structured as follows. In Section 2 we provide a general orientation to AFC technology and review the nature of the published research and recent corporate activities. In Section 3 we discuss the major technical issues confronting AFCs, including the reported power densities, poisoning issues, lifetime, duty cycles and systems considerations. This section also provides a hint at some new AFC technologies that may be of interest. Section 4 provides a detailed cost analysis and includes a comparison to published PEM cost projections. In Section 5 we provide conclusions and state our general technical position.

2. ALKALINE FUEL CELL BACKGROUND AND DEVELOPMENT STATUS

2.1. Principle of operation

AFCs use an aqueous solution of potassium hydroxide as the electrolyte, with typical concentrations of about 30%. The overall chemical reactions are given by

Anode reaction $$2H_2 + 4OH^- \rightarrow 4H_2O + 4e^-$$

Cathode reaction $$O_2 + 2H_2O + 4e^- \rightarrow 4OH^-$$

Overall cell reaction $$2H_2 + O_2 \rightarrow 2H_2O^+ \text{ electric energy} + \text{heat}$$

By-product water and heat have to be removed. This is usually achieved by recirculating the electrolyte and using it as the coolant liquid, while water is removed by evaporation. A schematic of the recirculating electrolyte AFC is shown in Fig. 1 (after De Geeter [4]).

The electrodes consist of a double-layer structure: an active electrocatalyst layer, and a hydrophobic layer. According to the dry manufacturing method described by Kivisaari et al. [5] and De Geeter [4], the active layer consists of an organic mixture (carbon black, catalyst and PTFE) which is ground, and then rolled at room temperature to cross-link the powder to form a self-supporting sheet. The hydrophobic layer, which prevents the electrolyte from leaking into the reactant gas flow channels and ensures diffusion of the gases to the reaction site, is made by rolling a porous organic layer, again to cross-link the layer and form a self-supporting sheet. The two layers are then pressed onto a conducting metal mesh. The process is eventually completed by sintering. The total electrode thickness is of the order of 0.2–0.5 mm. A major operating constraint is the requirement for low carbon dioxide concentrations in the feed oxidant stream. In the presence of CO_2, carbonates form and precipitate.

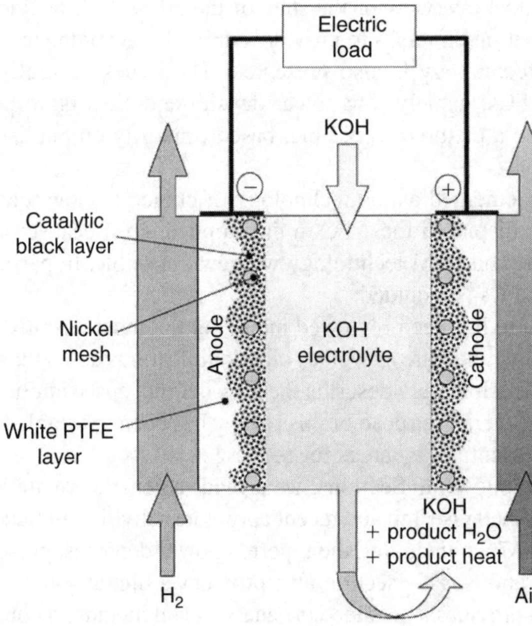

Fig. 1. Alkaline fuel cell composition

$$CO_2 + 2OH^- \rightarrow (CO_3)^{2-} + H_2O.$$

The carbonates can lead to potential blockage of the electrolyte pathways and/or electrode pores. This issue is discussed in detail in Section 3.2.1.

The inherently faster kinetics of the oxygen reduction reaction in an alkaline cell allows the use of non-noble metal electrocatalysts. It is useful to compare the eletrochemical performance of AFCs and PEMFCs in terms of the relationship between cell potential, E, and current density, i. When mass transport limitations are negligible (low to intermediate current density), E and i are approximately related by Blomen and Mugerwa [6]:

$$E = E_0 - \beta \log i - Ri \tag{1}$$

with

$$E_0 = E_r + \beta \log i_0, \tag{2}$$

where E_r is the reversible thermodynamic potential, β and i_0 are the Tafel slope and the exchange current density for the oxygen reaction, and R is the differential resistance of the cell.

Differentiating Eq. (1) provides further insight into the relative importance of losses associated with electrode kinetics and electric resistance:

$$\frac{\partial E}{\partial i} = -\frac{\beta}{i} - R. \tag{3}$$

At low current densities, the first term on the RHS is dominant and corresponds to the typical steep fall of the cell potential with increasing current. At higher current densities, $\beta \ll R$ and the second term becomes dominant, resulting in a quasi-linear drop of cell potential with current, until mass transport limitations become important. Optimal performance is obtained for low Tafel slopes (β) and cell resistance (R), and high exchange current density (i_0). The better electrode kinetics of AFCs results in Tafel slopes lower by about 30% than for PEMFC, when Pt is used as a catalyst in both [7].

The main contribution to cell resistance is due to the ionic/protonic resistivity of the electrolyte. Again, AFCs appear to have lower electrolyte resistivities (0.05 vs. 0.08 Ω/cm^2 for PEMFC). It should be pointed out that new generation ultra-thin acidic polymer membranes [8] achieve low resistance. Nonetheless AFCs have an intrinsic advantage over PEMFC on both cathode kinetics and ohmic polarization. A puzzling aspect of all published AFC data is that polarization curves are invariably presented for maximum current densities of about 400 mA/cm^2, with no indication that mass transport limitations have been reached. A possible explanation is that, for cost reasons, the catalysts of choice are nickel alloys. Nickel is, however, susceptible to oxidation, leading to high-performance degradation over time. This problem would presumably be exacerbated at higher current densities.

2.2. *Research activity level*

The strongest and most consistent advocate for AFC research has been Professor Kordesch, who has been researching, developing and promoting this technology for over 30 years and who continues to be involved in its commercialization. Kordesch gained notoriety for his development of a fuel cell powered Austin A40 car in the early 1970s [1] and his accounts of the experience gained with that test platform and fuel cell stack have formed the core of his publications ever since [1,9–13].

The work of Kordesch links directly into the commercially driven research of the Elenco/Zevco companies. The vast majority of the publications presenting system design or performance information is produced either directly by ZEVCO researchers, or is based on ZEVCO AFC modules [3,4,14–17]. This makes it difficult to separate potential AFC system performance from Zevco system performance.

During the early 1990s Olle Lindstrom published several excellent reviews of fuel cell technology in general, and AFC in particular [18–20]. However, his recent death prevents the possibility of using him as an arms length expert for the purpose of qualifying the ZEVCO claims.

The remainder of the published research on AFC technology consists largely of detailed studies of component parts of AFCs, conducted by individual researchers. The work is dominated by detailed characterization, analyses and evaluations of components, mostly electrodes. Information on AFC systems is scarce. Research publications of this type continue to be published at a roughly constant rate. Lindstrom's 1993 review of the state of the art of fuel cell technology revealed that 10% of publications over the 10-year period from 1983 to 1993 were associated with AFCs.

2.3. *Corporate activities*

Corporate development activities have been shifting away from AFCs and more toward PEM for low-temperature and mobile fuel cell applications since the mid 1980s. Major European projects conducted by Siemens, Hoechts and DLR were all cancelled prior to 1996 [21]. North American development of AFCs for space applications is continued by United Technologies/International fuel Cells. However, this work appears to be limited to providing fuel cells to the space shuttle program and appears to have no aspirations for entering other markets [22].

The remaining developers of AFC technology are almost exclusively related to Zetek Corporation. Zetek is the parent organization of three companies involved in developing products for transportation

(Zevco plc), marine (ZeMar Ltd) and stationary power (ZeGen Ltd) applications. Recent developments from Zetek include the announcement of a new 5 MW automated production line in Germany that will see Zetek manufacture more fuel cells than the combined production capacity of the rest of the world [23].

Astris Energi [24] recently announced a 4-kW prototype systems and offer the only off the shelf commercial source of low power AFCs. The Electric Auto Corporation is working with the Technical University of Graz (with Kordesch) to develop a lead-cobalt battery/AFC hybrid vehicle [25]. These two initiatives are small compared to the magnitude of Zetek's activities.

3. TECHNICAL REVIEW

3.1. Power density

Fuel cell systems have typically been evaluated on the basis of their volumetric and gravimetric power density, probably due to the historical challenge posed by developing a fuel cell system of adequate power within the volume and weight constraints imposed by equivalent power internal combustion engines. Such measures of evaluation must be based on overall system volume or weight, thus making it difficult to assess power density on the basis of the narrowly defined performance of the fuel cell electrochemical reaction. In the absence of absolute volumetric or gravimetric system power densities, polarization information is often used to assess the merits of particular fuel cell designs. This approach is reasonable in providing a figure of relative merit as a cell with higher current density, at an equivalent voltage, will provide higher overall power density so long as the stack geometry and ancillary systems remain constant, which is a reasonable assumption for the majority of PEM and AFCs. Therefore, in either case it is possible to judge the relative merits of fuel cells.

In half-cell testing particular components of the fuel cell are evaluated with potentials measured against some reference electrode. Direct comparison between different experimental results becomes difficult to assess in these situations because the full details of the experiments are not provided, different reference electrodes are used, or different test conditions employed.

The foregoing discussion serves to point out the difficulty encountered in trying to assess the power densities reported by AFC researchers. Results are often incomplete, and very few reported results discuss *system* performance. Nonetheless, we can review the partial results that have been reported and infer from them a reasonable picture of the power density achieved by AFCs.

3.1.1. Space applications

The AFC was initially used in space applications to produce electrical power for mission critical services. As such, these fuel cells were designed to provide reliable power, with low volume and weight, at virtually unconstrained cost.

A matrix type alkaline H_2–O_2 fuel cell is discussed by Matryonin et al. [26]. The cell is indicated as operating at 100°C and pressures of 4–4.5 $\times 10^5$ Pa. The presentation discusses the effects of varying these operating parameters on system performance. The performance is impressive, showing 3.2 A/cm^2 at 600 mV. The reported results show very good current density at high voltages with pressures between 30 and 60 psig.

Martin and Manzo [27] present performance data from the Orbiter fuel cell which is even more impressive. Operating with gas pressures of 200 psi and temperatures of 300 F and a 50% weight KOH electrolyte solution they report current densities up to 9 A/cm^2 at just over 0.7 V. Further information on the Orbiter fuel cell and the Siemens BZA4 is provided by Jo and Yi [28]. Their Orbiter data reports 1000 mA/cm^2 at 900 mV. The performance of the Siemens fuel cell indicates 1 A/cm^2 at 0.74 V. The orbiter cell is reported at an operating pressure of 60 psig compared to 30 psig for the Siemens cell.

Table 1. Summary of space application AFC performance

Operating point mV	Operating point mA/cm^2	Power (W/cm^2)	Pressure (psig)	Temperature (°C)	Source
950	140	0.133	29	98	[26]
950	220	0.209	58	98	[26]
950	310	0.2945	116	98	[26]
950	150	0.1425	58	65	[26]
950	280	0.266	58	96	[26]
950	440	0.418	58	130	[26]
600	3200	1.92	58	98	[26]
600	4200	2.52	58	121	[26]
800	6730	5.384	299	149	[27]
740	1000	0.74	29	80	[28]
900	320	0.288	29	80	[28]
900	1000	0.9	60	80	[28]

Although these reported results for space-based AFCs do not typically provide complete polarization curves, they nonetheless indicate very high current densities. It is important to remember that most of these results were obtained in the early 1970s or with 1970s technologies, well before today's highly optimized PEM systems had even been thought of. The available performance information from these space-based approaches is shown in Table 1.

3.1.2. Atmospheric pressure cells

The second distinct class of AFCs is based on operation at atmospheric pressure. This is the type of cell being developed commercially today, so these data are perhaps most relevant in validating any claims made by contemporary fuel cell manufacturers.

The oldest published results can be found in Binder et al. [29] which provides a summary of the state of the art of the AFC at that time. Current densities of 100 mA/cm^2 are reported there.

The most recently published ZEVCO performance [4] indicates normal operation at 100 mA/cm^2 range at 0.67 V/cell. Operation at current densities between 200 and 400 mA/cm^2 is discussed briefly, but no voltage information is provided. Kordesch et al. [12] discuss system performance, but like De Geeter do not provide a complete description of the operating parameters used in the system. Even so, current densities of 250 mA/cm^2 are reported. Similar performance is apparently achieved with Ag or a low cost Spinel catalyst. Weight and volume information is provided for the original Kordesch Austin fuel cell vehicle [10], but these data are now over 30 years old. Zevco's more recent designs easily supersede the original Kordesch system. Summary results with an Elenco stack are reported in Vegas et al. [14]. The reported data indicate very low current densities of only 50 mA/cm^2.

The systems discussed in the preceding paragraph are based on unipolar cell construction. Performance of a bipolar plate AFC are reported in Tomantschger et al. [30] where current density similar to De Geeter is presented, i.e. 100 mA/cm^2 at 0.85 V running on air. The cell voltage increased to 0.9 V when pure oxygen was used.

Performance of an AFC using a solid ionomer alkaline membrane is reported in Swette et al. [31]. The system was operated at 44 psi gas pressures at 40°C, which is a unique operating point. Using a Platinum–Irridium catalyst produced the best results, but still only 100 mA/cm^2 was produced at 800 mV. The solid ionomer alkaline membrane is intriguing because it suggests one possible path for developing

Table 2. Summary of terrestrial application AFC performance

Current density 0.7 V (mA/cm^2)	Power at 0.7 V (W/cm^2)	Point 2 mV	Point 2 MA/cm^2	Power at Point 2 (W/cm^2)	Pressure (psig) and gases	Temperature (°C)	Source
290	0.203	800	260	0.208	atm H_2–air	75	[12]
450	0.315	800	280	0.224	atm H_2–air	75	[12]
N/A	N/A	670	100	0.067	atm		[4]
90	0.063	800	35	0.028	44 H_2–air	40	[31]
108	0.076	800	102	0.082	44 H_2–air	40	[31]
115	0.081	570	225	0.128	atm H_2–air	40	[32]
125	0.088	700	125	0.88	atm H_2–O_2	40	[32]
88	0.062	700	88	0.062	atm H_2–air	40	[32]
N/A	N/A	750	106	0.140	atm H_2–O_2	40	[32]
157	0.110	700	157	0.110	atm H_2–air	40	[32]
N/A	N/A	850	100	0.085	atm H_2–air	65	[30]
N/A	N/A	900	100	0.090	atm H_2–O_2	65	[30]
87	0.061	670	100	0.067	atm H_2–air	70	[2]
40	0.028	N/A	N/A	N/A	atm H_2–air	60	[14]

AFC systems combining the desirable properties of a solid electrolyte with the fast anode reaction kinetics of an alkaline cell. Unfortunately, no further developments achieved with this technology have been published, leading us to conclude that the work has been discontinued.

An AFC stack developed specifically for operation with Biomass produced hydrogen is reported by Kiros et al. [32]. Both H_2–air and H_2–O_2 performance were reported, but the operating data were incomplete with no gas pressure information being provided. Performance of two generations of the design indicates 88/125 and 157/186 mA/cm^2 for air and O_2, respectively. Given the range of these performance values, we assume that atmospheric pressures are used.

Taken together, the atmospheric AFC performance results reported in the literature suggest that power densities between 100 and 200 mA/cm^2 have been achieved for several decades. The performance results are summarized in Table 2.

3.1.3. Performance of components

There have been numerous publications describing the performance of specific components of AFC systems, notably electrodes, where the object of the research has been to develop some incremental improvement over the state of the art. The results reported for such investigations tend to consist of half-cell reactions reporting current densities at different cell overpotentials with respect to a mercury or silver reference electrode. Generally these investigations do not provide useful performance data, but rather indicate the extent of work being undertaken to refine and optimize the performance of AFC systems.

3.1.3.1. Three-dimensional electrodes

Several reports have presented the idea of using a fluidized bed electrode structure in which a bed of electrode particles mixed with liquid electrolyte is subject to reactant gas flow through the bed. The fluidized bed is thus formed from the electrode particle, electrolyte, gas mix. A coarse membrane separates the anode and cathode reactions and electrodes are inserted into the fluidized beds to gather current.

A co-axial cylindrical single cell described in Nakagawa et al. [33] produced >1 A at 0.8 V. The volume of this single cell was not reported. In Matsuno et al. [34,35] the design and performance of an alternative structure for fluidized-bed electrodes is described. Operation of an AFC using a flooded gas diffusion electrode is reported in Holeschovsky et al. [36]. Only half-cell results were reported here, with the promise of a forthcoming article describing the performance of a practical system based on this idea. No such paper has been found.

Most conventional AFC designs devise methods to contain the liquid electrolyte, creating a system with operational features resembling those of solid membrane cells. By contrast, this approach uses the fluid properties of the electrolyte to eliminate the need for construction of gas diffusion electrodes completely. Although performance is still correlated to the surface area of the triple interface, this area is now contained within the volume of the fluidized bed. There is thus potential for very high power densities in a very low cost package. We are also tempted to speculate that issues surrounding electrode poisoning due to formation of carbonates may be far less serious in this design due to the elimination of the porous gas diffusion electrode.

The fluidized bed approach to AFC design represents a completely new direction for fuel cell technology development that is a direct consequence of the liquid electrolyte utilized in AFCs. Although the preliminary results we have seen with fluidized-bed AFCs are not impressive, this technology is worth watching.

3.1.3.2. Electrode materials

Research continues in the development of electrode materials to improve alkaline cell performance. Baseline performance of Ni electrodes is described in Al-Saleh et al. [37,38] which is then improved upon through the impregnation of copper into the Ni electrode [39] with the improvement being attributed to reduced contact resistance due to the copper. Introduction of Sn into a Ni cathode is shown to reduce H overpotential [40]. Performance of Ag catalyst is compared to Pl in Lee et al. [41] with the conclusion that the two perform equivalently producing up to 200 mA/cm^2 at 0.8 V. Electrodes based on Raney Silver are reported in Gultekin et al. [42] but with no useful power or current density information reported. Commercial electrodes are reported to provide similar current densities [43]. These reports show the steady performance improvement in electrode performance but do not introduce any radically new insights or technologies.

3.1.3.3. Electrode fabrication methods

The manner of production of electrodes for AFCs has significant impact on the performance of the overall cell. In general, the electrodes are manufactured by a method of wet fabrication followed by sintering or by a method of dry fabrication through rolling and pressing components into the electrode structure. In all cases the resulting electrode consists of a hydrophobic catalyzed layer on top of a gas porous conductive layer which is in turn bonded to a porous backing material that is usually metallic. The best results appear to be achieved when the electrode structure is built up from several layers and most of the current literature describes two-layer fabrication techniques.

A good overall description of the alternative methods of electrode fabrication is provided in Kivisaari et al. [5]. In seeking an optimal air electrode structure some 30 or so different electrodes were developed and tested. The best results showed half-cell results with current densities of 500 mA/cm^2 and no tendency to reaching current limits.

The effect of platinum loading on a multi-layer rolled electrode is reported in Han et al. [44]. 125 mA/cm^2 is reported with 0.3 mg/cm^2 of platinum catalyst. The current density increases to 225 mA/cm^2 when the platinum loading was increased to 2.0 mg/cm^2.

Alternatives to the basic fabrication techniques have been reported, but the results do not seem to suggest any great improvements in performance. Composite electrodes with carbon fibers pressed into a metal backing are reported in Ahn [45]. Use of an oxygen plasma treatment to increase the surface area of carbon black on a metallic substrate is reported in Li and Horita [46]. No clear performance improvement is reported in either case. A filtration method combining the best features of wet and dry fabrication is presented in Sleem et al. [47] with performance roughly equivalent to other AFCs (current densities approximately 180 mA/cm^2).

The volumetric and gravimetric power density of the ZEVCO module is reported to be on the order of 0.09 kW/kg and 0.06 kW/l, respectively [3]. It is conceivable that these densities could be doubled if current densities can be increased. However, there do not appear to be any huge breakthroughs on the horizon (either at ZEVCO or elsewhere) that would vastly improve the power density of the systems. The most promising area, perhaps, is in the fluidized-bed electrode structures. This research, however, is a long way from producing commercial products.

3.1.4. Comparison to PEM

Typical PEM fuel cell performance describes a system in which current densities are greater than 1 A/cm^2 at 0.6 V or higher, volumetric densities exceed 1 kW/l and gravimetric densities exceed 1 kW/kg. However, caution must be used when applying these rough numbers as a measure of fuel cell system performance.

There is no doubt that PEMFC technology is now producing power densities well in excess of the performance reported for ambient AFC technology. However, the PEM systems are typically pressurized to at least 30 psig. In pressurized AFC systems similar or even higher power densities were reported many years ago. Rather than optimizing these high power technologies and driving price down through volume manufacturing the AFC community has evolved toward lower power ambient air systems. Therefore PEM and AFC systems are fundamentally different.

Published results for ambient air operated PEM systems suggest performance that is on the same order of magnitude as ambient air alkaline systems. Kordesch and Simader [48] provide comparative information between ambient air and pressurized operation of an undisclosed PEM FC using a Dow membrane. Sugawara et al. [49] provide polarization information for a 40 cell ambient air PEM system and Koschany [50] presents polarization information for a small air-breathing cell designed to power cellular phones. The current density achieved at 0.7 and 0.6 V for each case is presented in Table 3. Note that these data are gathered from available literature and do not necessarily reflect state of the art performance of ambient air PEM systems.

Comparing the results presented in Table 2 and Table 3, it is apparent that available alkaline and PEM technologies achieve roughly equivalent current densities when operated on ambient air oxidant streams. This means that in applications where ambient air alkaline technology is proposed (as in Zevco's planned hybrid vehicle system) there is no reason to think that the alkaline technology will be easily displaced by a better, more efficient PEM system.

3.2. Poisoning and contamination issues

AFCs, like all fuel cells, have limits to the amount of impurities they can tolerate in their feed gas streams. The "poisoning" of the fuel cell by impurities can be caused by any number of different gases. In

Table 3. Summary of ambient air PEMFC performance

Current density 0.7 V (mA/cm^2)	Power at 0.7 V (W/cm^2)	Current density 0.6 V (mA/cm^2)	Power at 0.6 V (W/cm^2)	Source
200	0.140	425	0.255	[50]
250	0.175	500	0.300	[48]
125	0.088	450	0.270	[49]

published reports, carbon dioxide contained in the oxidant stream has received the most attention, since it is perceived as the only major issue preventing the commercialization of terrestrial AFCs running with air. No information on other impurities in the oxidant stream has been mentioned. The effect of carbon dioxide, as well as carbon monoxide and oxygen, on the anode side of an AFC has also been studied.

3.2.1. Effect of carbon dioxide on the cathode

The common perception of the AFC is that they cannot operate if there is any carbon dioxide in the cathode feed gas streams. Since terrestrial application AFCs will in all likelihood operate on ambient air, this is a significant issue.

It is suspected [37,38,48] that the poisoning reaction involves the alkaline electrolyte directly by the following reaction(s):

$$CO_2 + 2OH^- \rightarrow CO_3^{2-} + H_2O$$

and/or

$$CO_2 + 2KOH \rightarrow K_2CO_3 + H_2O.$$

This has the effect of reducing the number of hydroxyl ions available and therefore reducing the ionic conductivity of the electrolyte solution. It may also have the effect of blocking the pores in the electrodes. The carbonate "may precipitate and block the micro pores of the Raney catalyst or may stay as a liquid and reduce the ionic conductivity of the electrolyte" [37 and 38]. Kordesch states that the carbon dioxide reduces electrode "breathing" [9]. As well as these bulk effects, the effect on water management due to a change in vapor pressure and/or a change in electrolyte volume can be detrimental [51].

A number of papers present a point blank dismissal of this problem, as illustrated by the following quote: "it is often reported that the AFC … must be fed with pure oxygen because it is poisoned by CO_2 in the atmosphere … None of these myths can be substantiated" [4,15]. However, these papers present no data to substantiate their claim.

Al-Saleh et al. [37] showed that concentrations of up to 1% CO_2 in the oxidant stream of Ag/PTFE electrode at 72°C did not significantly affect the cell performance over a period of 200 h. However, at 25°C the CO_2 did adversely affect the performance. It is thought that the solubility of K_2CO_3 is lower at 25°C and therefore precipitates out and blocks the electrode pores. Al-Saleh et al. verified the presence of precipitated K_2CO_3 in the electrodes for the 25°C run using X-ray diffraction, thereby proving that the electrodes were directly affected in this experiment.

The presence of K_2CO_3 in the electrolyte by itself does not appear to produce any degradation in performance over the course of a 48 h period. Al-Saleh et al. showed this by mixing K_2CO_3 into the electrolyte and observing the current supplied at a specific over potential for 48 hours. In this test, the K_2CO_3 did not penetrate the pores of the electrodes or degrade the performance.

Appleby and Foulkes [51] discuss the fact that improved electrode formulations and structures can give dramatically varied results. Gulzow [21], in investigations with DLR, developed this concept and found that the effect of CO_2 on the electrodes was minimal if the electrodes were properly constructed. Gulzow used silver electrodes, which do not show the same fine pore structure as the standard Raney-nickel electrodes used in most AFC systems. This ensured that the carbonates did not block the pore structure and allowed the cells to work much more effectively. Gulzow found a 17 μV/h degradation for all cases, with and without carbon dioxide in the feed gas streams.

Gulzow [21] also discussed the changing of the electrolyte every 800 h. This ensured that the carbonates did not precipitate out of the electrolyte solution and damage the electrodes while also maintaining the electrolyte concentrations. In another paper, Gulzow [66] discusses the addition of water to the electrolyte to maintain the hydroxyl ion concentration.

Van den Broek et al. [16] states that "feeding a module over 6000 operating hours with CO_2-free air and with air containing 50 ppm CO_2, respectively, did not show any difference in performance or endurance". This may imply that the poisoning mechanism is not entirely CO_2 based and that other effects may be taking place. This is supported by Kinoshita [52] who discusses other effects including the oxidation of the carbon electrode to produce carbon dioxide and, consequently, carbonates. Kinoshita suggests that a highly active catalyst, although allowing for higher potentials, will cause a more rapid oxidation of the carbon electrodes.

Michael [2] reported that at 670 mV with 50 ppm CO_2 in the air stream over 6000 h the power output was reduced from 70 to 50 mW/cm^2 (approximately a 30% reduction) for a 500 W stack. The paper stated that this was a non-continuous test but did not provide information on electrolyte replenishment or replacement.

A test with intermittent operation was also performed by Zevco [3]. They found that the decrease in performance over time was greater with intermittent operation than with continuous operation. However, the draining of the electrolyte when the cells were shutdown seemed to prevent a large part of this deterioration. This seems to imply corrosion of the electrode materials by the electrolyte and not necessarily a CO_2 poisoning effect. Kordesch [9] has also discussed this effect and discusses that the cells do not seem to degrade at all if draining and purging are employed.

3.2.2. Carbon dioxide strategies for the cathode

From the above discussion, it is evident that there is a carbon dioxide poisoning issue. A number of papers mention methods for dealing with this problem.

Kordesch and Simader [48] mention that the "Removal of the 0.03% carbon dioxide from the air can be accomplished by chemical absorption in a tower filled with, e.g., soda lime". One kilogram of soda lime has the ability to clean 1000 m^3 of air from 0.03 to 0.001% CO_2 [51]. Zevco, who use soda lime, typically use 1 kg of limestone per 8 kWh of operation in present testing [3]. However, this corresponds to only 7% of the limestone being reacted and Zevco believes that efficiencies up to 80% may be achievable by proper selection of column conditions and grain size [2]. If this efficiency is achieved, over 90 kWh of cleaned air could be produced from 1 kg of limestone. The numbers given by Appleby and Foulkes [51] indicate 135–250 kWh/kg of limestone at 20–30% oxygen utilization.

Regenerable absorbers using molecular sieves can easily achieve the reduction of atmospheric carbon dioxide to acceptable levels [51]. The requirement for dry air for these processes, since water is preferentially absorbed, and the cost of regeneration of these systems increases both their capital and operating costs. However, they can be used for both reformed gases on the anode side and for air on the cathode side.

The concept of electrochemical removal of the carbonates from the electrolyte is very interesting. By drawing a large current out of an AFC, the concentration of hydroxyl ions is reduced at the anode. At the same time, the carbonate ions migrate toward the anode. An acidic solution at the anode is produced with hydrogen carbonate being the major component. When the current density is increased further, a number of cascading effects occur with the end result being the electrolyzing of the carbonates out of the solution at the anode by the following reaction [51]:

$$H_2CO_3 \rightarrow H_2O + CO_2.$$

With a regeneration period where the cells are run at a higher current density performed at 7000 and 15 000 h, the lifetime of a cell was doubled to 20 000 h [51]. No reference is given to substantiate this claim. Pratt and Whitney developed a system, which incorporated special regeneration cells into a regular fuel cell stack, regenerating the electrolyte continuously for the stack running on air. They found that the loss of efficiency was less than 1% from the incorporation of these cells and that the cells could run with 3000–4000 ppm carbon dioxide without a serious effect [51].

One alternate strategy for CO_2 management involves the synergistic possibility of using liquid hydrogen to condense the carbon dioxide out of the air. Ahuja and Green [53 and 54] discuss this at length and develop a model for the heat exchanger required for this. Liquid hydrogen is a strong fuel candidate for fuel cells, especially in Germany, where there is a large amount of research being performed, and in situations where there are captive fleets of vehicles. This system would enable the recovery of the energy of cold, which is around 30% of the total energy available from liquid hydrogen.

There are two technologies that alleviate carbon dioxide poisoning. Fyke [55] discusses the possibility of a solid ionomer alkaline membrane that would enable a cell to run without the possibility of carbon dioxide poisoning, as there would be no free potassium cations to which the carbonate anions could attach. This is an intriguing concept but no progress in solid ionomer alkaline membranes has been reported since Swette et al. [31] discussed this possibility for regenerative fuel cells.

There are a number of possibilities involving the modification of the fuel cell operating parameters. Operating the electrodes at higher temperatures would increase the solubility of the K_2CO_3 in the electrolyte and prevent it from precipitating out [52]. The circulating of the electrolyte improves the AFC tolerance to carbon dioxide significantly [12]. In general, modification of the operating conditions can prolong electrode life, but it is clear that the life expectancy of air cathodes is lower when CO_2 is present in the fuel cell [52].

3.2.3. Effect of impurities on the anode

Published information discusses the effect of carbon dioxide, carbon monoxide (CO) and oxygen on the anode. Al-Saleh et al. [37] tested the effect of CO_2 in the anode gas stream. They found that, although the presence of the CO_2 adversely affected the performance, the effect was entirely reversible under all experimental conditions. No details were given regarding the process used to determine the reversibility, but it seems that the cell was tested with and without CO_2 a number of times in a cyclic manner. At 40 mA/cm^2 they found a 75 mV polarization effect between 0 and 4% CO_2 in the hydrogen stream.

The effect of CO in the anode gases of an AFC is, intriguingly enough, often reversible. At temperatures above 72°C, this effect was completely reversible and below this temperature it was at least partially reversible [56]. This effect was also found to be a specific polarization loss and did not seem to cause a continual loss over time.

Kiros [57] found that both CO and O in the anode gases significantly affected the polarization of an AFC at 55°C in 6 M KOH due to a change in the surface properties of the electrode. No information on the reversibility of the effect was given in the conference abstract.

3.2.4. Strategies for the anode

The use of reformed fuels as the anode gases for AFCs has mostly been discounted. Kordesch suggests that the removal of the CO_2 from the feed gases would be very expensive and impractical, particularly for small systems [10,11]. It is therefore usually assumed that high-purity hydrogen, either liquefied or compressed, will be used with AFC systems. Michael [2] also mentions this and states that the successful development of economic palladium membranes or molecular sieves might allow the AFC to use reformed hydrocarbon fuels.

3.2.5. Summary of contamination effects

CO_2 in the oxidant stream has a distinct effect on the performance of AFC systems even though questions remain about the exact cause. There is strong evidence that a large amount of this poisoning is reversible and that effective electrolyte management will mitigate a large part of the problem. This could be done in a similar manner as an oil change is performed on vehicles today.

The only method currently employed to alleviate the oxidant side carbon dioxide poisoning is CO_2 scrubbing using soda lime. Technically, this system works, but is not a strong option for commercial systems. This suggests that significant benefits could be obtained from the use of other scrubbing techniques.

3.3. System issues

The majority of published descriptions of AFC systems are based on the early work of Kordesch, followed by descriptions of Elenco [16 and 58] and then ZEVCO [4,12–14,30] systems. Complete lab scale systems are described in Khalidi et al. [59] and Ergul [60]. These do not provide descriptions of practical fuel cell stacks, but do provide alternative descriptions of means of electrolyte circulation, heat and water management. Some discussion of stationary systems is provided in Kiros et al. [32] and Lindstrom [61], however, no specific information concerning system configuration or operating conditions is provided. Consideration of systems issues must therefore be considered on the basis of the published ZEVCO experience alone.

The alkaline system requires the control of three fluid loops including the reactant fuel and oxidant and the recirculating electrolyte. The fuel and oxidant loops are operated at marginally higher than ambient pressures and are thus very simple. The fuel loop contains a simple water knockout and re-injection into the input stream via a venturi pump. The air loop contains no recirculation. No details of the connection of the air loop to air scrubbing apparatus are provided in any of the published reports. This connection is, however, quite important for successful system operation.

Although no specific systems descriptions of entrapped electrolyte AFCs are provided, they have been discussed by Kordesch and Simader [48]. Entrapment of the electrolyte by suspension in an asbestos matrix forces the system design to rely on fuel or oxidant flow to pick up product water and heat. This significantly complicates the design of these gas flow loops, forcing similar considerations as are applied in the design of PEM systems. Cell cooling via the air loop would result in the flow of large amounts of scrubbed air through the cell, a practice that would be wasteful of the soda lime scrubbing in conventional alkaline cell operation. Availability of low cost or more effective means of purifying the air stream in an alkaline cell may therefore improve the feasibility of this different mode of operation.

There is no discussion of sealing in any of the published reports. This is not unusual as the problem of sealing is seldom discussed in open literature for any type of fuel cell (but is nonetheless a critical component for successful operation). The dominant design continues to be based on edge-collected cells assembled in "modules" that have a stacked arrangement (cathode–anode–cathode). This is strange, as we would have expected a bipolar stacking arrangement to replace this edge-collected structure by now. A bipolar stacking arrangement is called for by Kordesch and Simader [48] and was developed by Tomantschger et al. [30]. Gas manifolding, electrolyte recirculation and current collection all require different approaches in the bipolar stack arrangement and effective sealing technology is key to the success. Since Tomantschger et al. published their paper; there has been no other reported progress on bipolar AFC technology. We remain curious about the demise of the bipolar alkaline stack.

3.3.1. Electrolyte circulation

The liquid electrolyte is circulated, allowing the possibility of removing product water and heat from the cell and also allowing the possibility of removing carbonates from the electrolyte to maintain cell performance. The circulation of the electrolyte within the alkaline cell is analogous to the circulation of cooling within PEM cells with roughly equivalent complexity for both. The major difference between the two is that the alkaline cell must deal with a highly caustic electrolyte, requiring more care than the simple deionized water used in PEM technology.

Given the opportunity to develop alkaline cell electrodes that will not clog with precipitated carbonates, the circulation may provide an opportunity to clean or replace the electrolyte much as engine oil is replaced in conventional internal combustion engines. None of the reports discusses this as a normal operating strategy, however.

Kordesch and Simader [48] point out that the circulation of the electrolyte can introduce parasitic current loops within the stack but they do not indicate any serious negative effects, which may result from these. Tomantschger et al. [30] include an explicit electrolyte heating loop in his designs, suggesting the use of the electrolyte to bring the cell up to operating temperature for low temperature startup.

3.3.2. Salvage

There are no published reports comparing the lifecycle environmental impact of AFCs with PEMFCs or internal combustion engines. We would expect the alkaline cell to fare well in such a comparison due to the material composition of the cell. The alkaline cell can be manufactured without the use of a noble metal catalyst that while contributing directly to lower short-term costs also has environmental benefits [62].

The simplicity of the electrolyte used in the alkaline cells provides a distinct advantage compared to PEM. There are no major supply security issues associated with AFCs, and while familiar proton exchange membrane manufacturers presently make their products widely available it is possible that large market players or government regulations could limit the distribution of these materials in future. Further, the disposal of current PEM membranes presents an environmental hazard due to the reliance on fluoropolymers, which are not recyclable [63]. These considerations are important for the environmentally sensitive European market or for global markets in developing countries where material supply security is a major concern.

3.4. Lifetime and duty cycle information

A number of long-term tests have been performed with AFC systems and cells. Most of these long-term tests were performed in an attempt to gain an understanding of carbon dioxide poisoning with few test reporting the result of operation to ultimate failure. Nonetheless it is possible to obtain a good sense of AFC life from the published reports. As with other technical aspects of AFC systems, the literature divides naturally into two groups consisting of reported performance for the Kordesch/Elenco/ZEVCO technology, which tends to provide the most complete system performance information, and reports from other research groups which tends to be more oriented toward single cell and component testing. Results reported from these two groups will be discussed separately.

3.4.1. Zevco long-term tests

The earliest mention found for a long-term test with a Zevco (or Elenco, as it was then called) stack stated that the degradation of the stacks over the course of 5000 operating hours was 12–14 mV/1000 h [16]. De Geeter et al. [4] re-iterate this 5000 h figure as the minimum operational life of a standard Zevco module. Operational parameters are missing from these reports, except the statement that some of the modules tested were operated at full power.

Vegas et al. [14] tested an Elenco module over the course of 1000 operational hours with a varied and largely unstructured operating duty cycle over the duration of the test. Highlights of this test include over 100 startup and shutdown procedures, repeated continuous operation for more than 100 h at a time and at least one 6 month period where the system was not operated at all. The results they obtained show that performance of the module degraded significantly throughout the duration of the test. Even so, the system remained operational and reasonably functional after 1000 h of operation.

The information presented above is supported by Michael [2] who states that a 50–70 mW/cm^2 power reduction (nearly 30%) is found for the 6000 h test of a 500 W stack. This paper stated that the test was non-continuous but did not give any information on electrolyte replenishment or replacement. The tests were performed at the standard operating conditions for the Zevco modules of 100 mA/cm^2.

Zevco's standard for determining if their cells are operating appropriately consists of measuring the degradation after 3000 h of operation. The cells are considered to be working effectively if the drop over these 3000 h of operation is less than 10% [3].

3.4.2. Other long-term tests

A number of other long-term tests on half cells and space system fuel cells have been reported as well as some data on the performance of other fuel cell systems.

Tomantschger et al. [30] performed electrode tests under continuous operation for over 3500 h. These tests, which were performed at 100 mA/cm^2, 65°C with 12 N KOH, showed a significant decrease in the hydrogen electrode voltage of 50–100 mV over the course of 3000 h. The oxidant electrode, which was operated on air, showed a slightly lowered potential from 1500–3000 h but this drop was reversed after 3000 h. No further discussion is given regarding this effect.

Tomantschger et al. [64] also reported on 1000 h tests at 100 mA/cm^2 with varied temperature. From these results, they determined that the internal resistance of the cell increased with time and a change to silver electrodes from the nickel that was originally used gave significantly improved performance. The magnitude of the improvement was not quantified.

Strasser [58] tested four Siemens BZA4 modules and found that each module showed a similar performance drop over the course of the test. A drop of approximately 50 mV was observed over the course of this 700 h test performed at 80°C with pure hydrogen and oxygen at 2.3 and 2.1 bar, respectively. Tests of several thousand hours are briefly mentioned but no details are provided.

Lamminen et al. [65] developed electrodes that were constructed with different catalysts. At 100 mA/cm^2 they ran one sample intermittently for 424 h while another sample was tested for over 600 h. They found that the performance degradation was as great or greater when no power was drawn from the cell.

Kordesch, who learned from his early Austin A40 experiment that draining the electrolyte from the cells when the system was not operational greatly enhanced the lifetime, has made this observation repeatedly.

Gulzow et al. [66] found that they could run a cell for 1000 h at 100 mA/cm^2 with a 17 μV/h degradation of cell voltage with or without CO_2 in the air stream. Al-Saleh et al. [37,38] performed an excellent 200 h test of an AFC system in relation to carbon dioxide performance, which was discussed in Section 3.2.1.

Kordesch and Simader [48] and Khalidi et al. [59] both refer to a 15 000 λh lifetime for AFCs. Kordesch gives no information regarding the source of this figure while Khalidi et al. mention that this is for small installations of the Orbiter fuel cell with no other information. Kordesch et al. [13] quote a figure of 4000 h from JPL with no other data given.

Kiros et al. [32] showed the performance of an anode electrode over the course of 110 000 h. At 100 mA/cm^2 they found a decay rate of 3–4 μV/h. They do mention changing their electrolyte at regular intervals to avoid carbonate buildup, which could change the performance characteristics of the test.

3.4.3. Summary of AFC lifetimes

The lifetime of an AFC can, in general, be well over 5000 h for inexpensive terrestrial AFCs and has been shown to be significantly over 10 000 h for space application AFCs. It would not be unreasonable to assume, given a significant development effort that the lifetime of AFC cells could be well over 15 000 h. Table 4 summarizes the different operational lifetime figures discussed above.

The only discussion of operating duty cycles and procedures for alkaline cells is provided by Kordesch [9] in the context of the Austin A40 fuel-cell-powered car. The electrolyte was drained nightly and a

Table 4. Summary of reported AFC lifetimes

Date	Hours	Current density (mA/cm^2)	System information	Source
1986	3500	100	Electrodes	[30]
1987	5000	100	Elenco module	[16]
1990	>2000	Unknown	Siemens BZA4	[58]
1991	424 & 600	100	Electrode tests	[65]
1994	3500	100	Electrode tests	[66]
1995	200	160 mV Overpotential	Electrode tests	[37,38]
1996	15 000	Unknown	Not stated	[48]
1996	15 000	Unknown	Orbiter fuel cell	[59]
1998	1000	Varied	Elenco module	[14]
1999	>5000	100	Zevco module	[4]
1999	4000	Unknown	From JPL	[13]
1999	11 000	100	Anode electrode	[32]
2000	6000	100	Zevco module	[2]

nitrogen purge was used to neutralize the cells during the shutdown and into inactive operation. Despite these operational issues, Kordesch claims that the system could be returned to operational condition within a few minutes.

4. COST ANALYSIS

AFC technology, evaluated from a purely technical perspective, has the potential to compete with other low temperature fuel cell technologies. Although the alkaline technology has been largely neglected in the last 10 years, mostly due to the apparent CO_2 poisoning issue, the foregoing section has shown that there are no obvious technical reasons to discount its potential for useful applications. This general conclusion raises the questions of relative costs between AFC technology and its competitors in order to better understand the potential economic competitiveness of the AFC technology.

In this section, we present a review of the cost information available for AFC technology and where possible compare these cost estimates to equivalent PEM costs. The reader is cautioned that while the cost comparisons included in this section are based on the best available information at the time of writing, we have been faced with extrapolating costs in some cases, estimating costs in others and comparing cost information provided for different purposes. For example, most of the AFC cost information presented is based on ZEVCO provided data from recent conference publications. Such data, though likely optimistic, is nonetheless rooted in a hard estimate of short-term costs and must be believable by investors with a 3–5 year time horizon for return on investment. Similar cost estimates for PEM systems are not available due to their proprietary nature. The PEM cost estimates we show are derived from sources, which apply standard industrial forecasting methods assuming full-scale automotive production, is achieved [67] as a means of forecasting long-term trends in technology development. As a result we tend to be more confidant in the accuracy of the alkaline cost data than in the potentially over-optimistic PEM cost projections.

Beyond costs of components and stacks, overall fuel cell power plant costs must be considered. The overall cost of a 50 kW FC power plant is estimated at US$ 2103 [67]. This is the ultimate cost against which any engine replacement system will have to compete. The ZEVCO strategy is to develop a hybrid fuel cell/battery system for vehicle applications. This goal may be achieved using an AFC, an ambient air

PEMFC or a compressed air PEMFC. In the following analysis, we will derive production volume cost estimates for these three options and compare them to the overall fuel cell engine costs.

4.1. Gross costs and commercial estimates

There is general agreement in the literature that AFC costs are lower or at least equivalent to other fuel cell technologies, both in terms of material and production costs. Citing calculations of the costs of fuel cell systems by DLR (Germany), ZSW (Germany), Hoechst (Germany) and the Royal Institute of Technology (Sweden), Gulzow [21] states: "All calculations show that the stack costs are similar to all other low temperature systems [and] the production cost for the AFC systems seem to be the lowest". Their estimates give a conservative price of **US$ 400–500/kW** using 1996 technologies and knowledge in large-scale production. They also mention a 5–10 times higher production cost for small-scale production.

Other projected general estimates for AFC material or stack costs range from US$ 80/kW to US$ 265/kW (figures adjusted to US$ 2000 [18,30,13].

Although lacking any hard numerical figures, a recent report from ETSU (UK) makes some interesting remarks regarding the cost of AFCs [2]. One underlines the fact that "current AFC stack designs have a modest performance compared with [PEMFC], but the AFC stack is relatively cheap even at low manufacturing volumes". The author then emphasizes a key difference at this point between PEMFC and AFC system: "Whilst high manufacturing volumes will be essential for [PEMFCs] to beat the target costs for cars, the AFC could become competitive as a battery charger for electric vehicles". Regarding the battery-charger approach taken by Zevco to commercialize its AFC systems, he states: "There seems to be no evidence to suggest that the AFC could not be manufactured in small volumes at prices necessary to be competitive as a battery range extender…".

The cost of commercially available low-power AFCs and PEMFCs is shown in Table 5. Although not representative of higher power fuel cell costs and based on a very small sample, these figures seem to support the cost advantage of AFCs over PEMFCs (2–4 times cheaper in this case).

4.2. Materials and manufacturing

Stack cost has a large influence upon the system cost of any fuel cell system. As such, a detailed breakdown of the stack materials and manufacturing costs is needed to determine the competitive position of alkaline and PEMFC technologies. This section first presents an overview of the different material components of both AFC and PEM stacks and then compiles the available cost estimates to provide the basis for the remainder of the analysis.

4.2.1. AFC stack materials

Table 6 lists the materials currently used in the various AFC cells and systems discussed in the literature.

Table 5. Summary of low power ambient air fuel cell prices [71–73]

Company (fuel cell product)	Nominal power	Type of fuel cell	Price (US$)
Astris (LC200-16)	240 W	AFC	2400
H-Power (PowerPEM-PS250)	250 W	PEMFC	5700
DAIS-Analytic (DAC-200)	200 W	PEMFC	8500

Table 6. Materials and manufacturing processes for AFC stacks [4,5,21]

Component	Materials	Manufacturing processes
Electrodes		
Anode	PTFE powder graphite powder catalyst: (Pt or Pd 0.12–0.5 mg/cm^2) Ni–Al, Ag	Mechanical process involving grinding, dispersion, filtering, rolling and drying
Cathode	PTFE powder graphite powder catalyst: Pt	Mechanical process involving grinding, dispersion, filtering, rolling and drying
White layer (for both anode and cathode)	PTFE powder	Pre-forming and rolling
Module current collectors	Nickel mesh	Pressed to black and white layers (as above)
Plastic frames	ABS plastic	Injection molding and manual assembly with electrodes
Spacers	Unknown	
Stack assembly		Plastic frames are friction-welded to module casing for sealing

Table 7. Materials and manufacturing processes for PEMFC stacks [6,51]

Component	Materials	Manufacturing processes
MEA Membrane	Polymer matrix with attached sulfonic acid groups e.g., Nafion, BAM 3G, etc.	Complex chemical process
Electrode Substate	Carbon Paper, PTFE	Attached to membrane through hot pressing
Catalyst	Pt (0.4–4 mg/cm^2)	Deposited between the electrode substrate and the membrane
Other stack components		
Flow field plates (including cooling plates)	Graphite, Stainless steel, carbon polymers, etc.	Machined out of bulk material, stamped, injection molded
Non-repeating components	Off the shelf components	Simple machining

Potential improvements in the AFC stack materials include the reduction of the catalyst loadings, as well as development of cobalt oxide-based catalysts and replacement of the nickel mesh current collectors with a cheaper metal mesh [3].

4.2.2. PEMFC stack materials

Table 7 gives a summary of PEMFC stack materials and manufacturing processes.

A number of potential improvements are foreseen for both manufacturing and materials in PEMFC stacks. These include the reduction of the catalyst loading down to 0.04 mg/cm^2 through improved deposition techniques, different nanostructure catalyst supports, the use of carbon composite materials and stamped metal sheets for the flow field plates and the reduction of the MEA thickness [68].

Table 8. Costs of AFC stack components [8][a]

Component	Current (US$/kW)	Projected (US$/kW)
Total stack costs	1750	205

[a]Note: Converted from ECU/kW to US$/kW on a 1:0.925 basis.

Table 9. Costs of PEMFC components [67,70]

Component	1998 PEMFC materials (US$/kW)	500 000 unit per year production (US$/kW)	% (500 000 units per year)
Membrane	120	0.40	2
Catalyst	243	8.20	41
Gas diffusion electrode	31	3.00	15
Flow field plates (including cooling plates)	825	6.00	30
Non-repeating components	1	1.00	5
Assembly		1.40	7
Total	1220	20	100

4.2.3. AFC system costs

Table 8 lists the projected costs of a Zevco stack module. This table is based on the data provided by Zevco in recent conferences and public presentations but has not been published in a citable reference.

No information on the assembly or manufacturing costs has been specifically stated for AFCs. However, it is reasonable to assume that the manufacturing costs are included in the component costs. The cost of final assembly, especially for larger volume manufacturing, is assumed to be minimal.

4.2.4. PEMFC stack costs

PEMFC stack costs have been reported in a number of papers and reports, with current stack cost estimates ranging from $500/kW [69] to $5000/kW [70]. Optimistic cost projections for a 70 kW stack, for a typical automotive production volume of 50 000 units per year, produce a lower bound cost estimate of $20/kW [67].

Table 9 summarizes the two extreme cost estimates available in the literature. The data of Ekdunge and Raberg [70] are summarized in the second column and covers material costs only for small-scale laboratory production of a 75 kW unit using "conventional" materials. The data of James et al. [67], summarized in column 3, are on the other hand an estimate that includes material, manufacturing and assembly costs for large-scale production of 30–90 kW stacks using "advanced materials".

Ekdunge assumed a catalyst loading of 16 g/kW, which corresponds to around 8 mg/cm^2. With today's catalyst loadings, this could be decreased by nearly an order of magnitude, reducing the cost estimate by around $200. Also, the cost of the flow field plates as stated by Ekdunge and Raberg [70] may currently be significantly lower with the use of different materials.

4.3. Impact of production volume

Fuel cell technology presently has no established market and will thus inevitably go through several phases of niche market penetration before widespread deployment of the technology occurs. This means that while ultimate high volume production costs may favor a particular technological option there may be short-term

Table 10. Effect of production volumes on fuel cell costs

	AFC stack (US$/kW)	PEMFC stack (US$/kW)	$ PEMFC ($ AFC)
Small batch fabrication	1750	2000–5000 [70,74]	1.2–2.9
Small-scale manufacturing (100s, 1000s?)	205	500–1500* [69,75]	2.5–7
Improved AFC performance			
High volume production (Unknown volume for AFC) (500 000 units/yr for PEMFC)	155	20 [67] (50-kW unit) 60 extrapolated from [67] (7-kW unit)	0.13–0.4

* $1500/kW includes ancillaries.

cost advantages for other technologies. In particular, we are interested to determine if AFCs possess any inherent cost advantage in small volume production that is more indicative of early fuel cell markets.

No study presenting cost estimates of AFC stacks at very high volumes has been found. To estimate low volume production costs we have used the lowest power density ZEVCO Mark II costs described in the previous section. Conversely, our high volume cost estimate is derived on the highest performance projections provided by ZEVCO.

There are many sources providing high-volume mass production cost estimates of PEMFC systems. All these results are within the **US$ 20–50/kW range**. Directed Technologies completed one particularly thorough report for the Ford Motor Company [67]. Most cost estimates for PEMFC system components at high volumes, used in the present analysis, are taken from this report. The Directed Technologies report provided cost estimates for PEMFC systems in the 30–90 kW range. These cost estimates have been extrapolated to estimate the costs in the 7 kW range, details of this extrapolation are included in Appendix A.

The estimated stack costs for both alkaline and PEMFC technology are compared to each other at different production volumes in Table 10. The high volume cost estimate for AFC was deduced by extrapolating the data of DeGeeter [4] and Michael [2]. At low volumes anticipated in early markets the PEM technology is between 3 and 9 times more costly than the alkaline technology. However, this trend reverses at very high volume production rates. The lower costs are attributable to lower material costs and simple manufacturing technologies. The PEM cost reductions are achieved through the anticipated gains that will be made by exploiting economies of scale in manufacturing materials and stack components.

4.4. Extrapolation to ambient air PEM

The possibility of an ambient air PEMFC being used as a "battery range extender" constitutes a major threat to low power density applications of AFC system technology. Therefore, we are interested in estimating the cost of such a system. To produce this estimate, the Directed Technologies calculations for PEMFC stack costs [67] have been extrapolated to an ambient air PEMFC.

Current ambient air PEMFC systems, as mentioned in Section 2.1.5, operate at roughly 200 mA/cm^2 compared to compressed air PEMFCs that produce over 1 A/cm^2. Therefore, for the same power output an air breathing PEM stack will require at least 5 times the active area of a compressed air stack. While ambient air operation may imply a simpler stack design overall, the amount of membrane material and catalyst required will be several times higher than for the compressed air counterpart.

Upper and lower bound cost estimates for ambient air PEM stacks can be constructed from this information as shown in Table 11. The high cost estimate is formed by multiplying the high compressed stack cost estimate by 5 to estimate a "worst case" scenario. The low cost, best-case estimate is formed by multiplying the low compressed stack costs for a 7 kW stack by 3.

Table 11. Ambient air PEMFC cost estimate table

	High cost estimate	Low cost estimate
Compressed PEMFC cost (US$/kW)	1220	60
Multiplication factor for running at ambient	5	3
Ambient PEMFC cost (US$/kW)	6100	180

4.5. Balance of plant

Balance of plant components that need to be considered for AFCs include the air blower, CO_2 scrubber, electrolyte circulation and nitrogen purging. PEM balance of plant requirements differ because of the need for air compression at significantly higher pressures than alkaline, humidification of reactant gases and cooling systems. Cost estimates for PEM balance of plant have been previously published [67].

Although the overall control system required for operation an alkaline hybrid system should be significantly less complex than for a PEMFC engine, this saving may be more than offset by cost of power electronics involved in managing the battery system. Furthermore, if an ambient air PEM system were used to replace the alkaline system in vehicle applications the controller costs would be equivalent.

4.5.1. AFC peripherals

Most of the components used in the balance of plant of the AFC system are relatively standard equipment, with the notable exception of the CO_2-scrubber. Their aggregate cost does not seem to be a major obstacle to AFC system commercialization.

4.5.1.1. Air blower
For a 5-kW system, Zevco uses a 350-W pump in the air circuit, but no optimization appears to have gone into the selection of this component. De Geeter [4] suggests that reduction of blower power by a factor of five is anticipated through easily achievable redesign of the airflow path.

4.5.1.2. CO_2-scrubber
No cost data has been found for the soda lime scrubber proposed, but some numbers are available for the required quantity and cost of soda lime. We anticipate that the cost of the soda lime will become the significant component of CO_2 scrubbing because the reactor vessel containing the soda lime is composed of a passive container operating without any high pressures or temperatures. The cost of soda lime used over the lifetime of the cell is provided in Section 4.6.1.

4.5.1.3. Electrolyte recirculation loop
The main components of the alkaline electrolyte loop are the heater, a 50-W electrolyte pump, a small heat exchanger and a ventilator. The cost of this subsystem are assumed to be roughly equivalent to the cost of the coolant loop in conventional PEM technology, which we estimate to be US$ 100 in mass production [67]. Unlike PEM fuel cells, the electrolyte in AFCs requires maintenance and incurs an operational cost over the lifetime of the cell. This is discussed in Section 3.3.1.

4.5.1.4. Water management
Water management is relatively straightforward in the Zevco AFC system and incurs a minimal cost. The associated components are mainly a small water tank and a water condenser.

Table 12. AFC peripheral costs

	Cost (US$)	%
Air blower	14	5.5
CO_2 scrubber	14	5.5
H_2 recirculation ejector	22	8.6
Electrolyte recirculation	100	39.2
Nitrogen purge	15	5.9
Electronic engine control(EEC)	50	19.6
Piping, valving, misc.	40	15.7
Total peripheral cost (inclusive mark-up contingency)	255	100

Table 13. PEMFC peripheral costs

	Cost (US$)	%
Air compression subsystem (Compr./Expander/Motor Unit-CMEU)	330	41.4
Air humidifier subsystem	65	8.1
H_2 recirculation ejector	22	2.8
Radiator subsystem	92	11.5
DI filter	14	1.8
Electronic engine control (EEC)	220	27.5
Piping, valving, misc.	55	6.9
Total peripheral system cost (incl mark-up and cost contingency)	798	100

4.5.1.5. Nitrogen purge

The use of a nitrogen purge to remove reactant gases from AFCs is shown on most system diagrams from the ZEVCO system [4,15]. However, no details of the nitrogen purge system are available, and some designs claim not to require this component.

4.5.2. Alkaline peripheral costs

Based on the foregoing we are able to construct an estimate of the cost for balance of plant components required in AFC systems, as shown in Table 12. Most estimates are adapted from those of a PEMFC component with equivalent function.

4.5.3. Compressed PEMFC peripherals

In a PEMFC system the oxidant compression system, (EECs), radiator system and humidifier system contribute 89% of the cost of peripheral components. An extrapolation to a 7 kW system of the peripheral costs given in the study by Directed Technologies/Ford (for systems in the range 30–90 kW) [67] is presented in Table 13.

4.5.4. Ambient air PEM peripherals

For an ambient air PEMFC system, the peripheral costs are somewhat different than the compressed PEMFC system. The air compressor is replaced with a simpler blower unit and the EEC is vastly reduced in complexity due to the simpler operation of the fuel cell as a battery charger (Table 14).

Table 14. Estimated peripheral component costs for ambient air PEM systems

	Cost (US$)	%
Air blower	14	5.5
Air humidifier subsystem	65	25.8
H_2 recirculation ejector	22	8.6
Radiator subsystem	50	19.5
DI filter	14	5.5
Electronic engine control (EEC)	50	19.5
Piping, valving, misc.	40	15.6
Total peripheral system cost (inclusive mark-up and cost contingency)	255	100

Table 15. Soda lime cost estimate based on bulk cost of $0.2/kg 5000 h cell lifetime at 7 kW

% utilisation	Consumption rate (kWh/kg)	Lifetime mass (kg)	Cost ($)
7	8	4500	900
80	92	394	80

4.6. Cost of consumables

AFCs consume electrolyte and soda lime. In this section we estimate these costs.

4.6.1. Soda lime

Soda lime is consumed in significant quantities in AFCs. In fact it appears that the mass of soda lime used is approximately equal to the mass of hydrogen used in normal cell operation. Therefore, regardless of the simple costs associated with maintaining the soda lime scrubbing unit there is a potentially large intangible cost associated with the regular maintenance required.

Presently, scrubbing technology is able to make use of only 7% of the limestone contained in the scrubber unit, but utilization up to 80% is achievable [2]. Using the rate of 8 kWh/kg for the present technology, we produce the cost estimates for CO_2 scrubbing over the 5000 h, lifetime of the system as shown in Table 15.

4.6.2. KOH

A 7 kW AFC system requires 13 kg of 6–9 N KOH solution for the electrolyte, representing a mass of roughly 3 kg of KOH. Although KOH is considered to be a cheap bulk material its cost must be factored into the overall system cost for an AFC.

In small lab scale quantities KOH is available at US$ 9.00/kg, leading to an estimated cost of US$ 3.00/kg in bulk. Based on this information and the frequency of electrolyte replenishment over the lifetime of the cell the total electrolyte costs are estimated as shown in Table 16.

4.7. System cost estimates

By combining the information presented in the previous sections upper and lower bounds for the cost of three competing 7 kW fuel cell systems are produced, as shown in Table 17. Apart from being deduced from commercial data, the alkaline estimates have a total cost range that is a factor of 6, compared to a

Table 16. Cost of KOH electrolyte

Electrolyte life (h)	Lifetime # changes	Lifetime cost ($US)
300	17	153
500	10	90
1000	5	45
5000	1	9

Table 17. Total system costs comparison

	Compressed PEMFC		Ambient air PEMFC		Ambient air alkaline	
Component	**Upper bound**	**Lower bound**	**Upper bound**	**Lower bound**	**Upper bound**	**Lower bound**
Stack cost($/kW)	1220	60	6100	180	643	155
Stack cost	8540	420	42 700	1260	10 942	1084
Balance of plant	798	798	256	256	255	255
Consumables	N/A	N/A	N/A	N/A	N/A	N/A
Total	9338	1218	42 956	1516	12 250	1428
Total per kW	1334	174	6136	217	1750	204

range of 28 for the ambient air PEM estimate, reflecting our uncertainty here. Gulzow [21] quotes a figure of $400–$500/kW for an AFC system using the technology of the time for high volume production. This number falls in between the two figures obtained for an AFC system.

This analysis indicates AFC systems are cost competitive with comparably sized PEMFC systems, at least for low power. This advantage remains for all production volumes, but is most significant at low and medium production volumes.

However, it should be noted that the 7 kW alkaline system would be competing with a 50 kW PEMFC system. Directed Technologies has estimated that the total system cost for a 50 kW PEMFC system would be about $2100 for production volumes of 500 000 [67]. This means that extra components required to complete the alkaline hybrid power system (namely batteries) must cost no more than about $670.

Of interest with this report is the opportunity available for CO_2 scrubbing with novel technologies. The total cost of CO_2 scrubbing in the above system is $94 ($14 for the Canister and $80 for the Soda Lime). Therefore, if we consider the cost of the system without batteries or CO_2 scrubbing we have a cost of $1334. This implies that, to be competitive with a 50 kW PEMFC system, the cost of batteries and CO_2 scrubbing must be less than about $750.

5. CONCLUSIONS

In this report we have presented a review of AFC technology to assess its potential from both technical and economic perspectives. Research and development in AFC technology has become largely stagnant during the past decade although we can find no obvious technical or economic reasons for the relative neglect it has received.

AFCs can theoretically outperform PEMFCs and some of the earliest pressurized AFC systems showed current densities much higher than those achieved today with current PEM technology. Concerns about the low power density achieved by current AFC technology are misplaced, as the current AFC designs are directed at low power applications. Ambient air operated AFCs produce current densities comparable to ambient air operated PEMFCs.

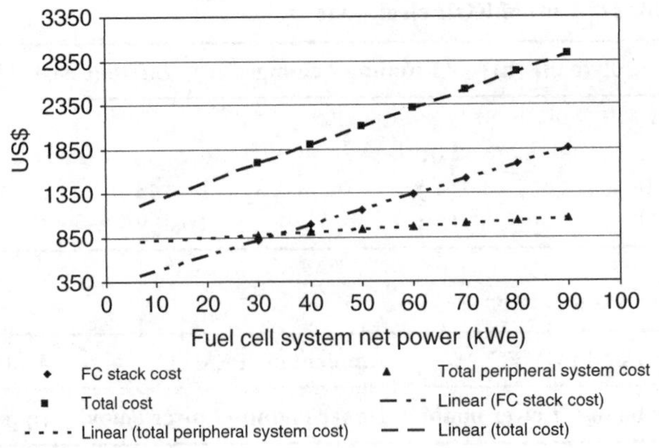

Fig. 2. PEMFC power system costs

Only a single design paradigm has been explored in commercial AFC systems. There is considerable scope for improvement of AFC technology through further research, in particular for the development of new architectures for AFC operation. There is no strong IP position to prohibit the further development of AFCs, nor are there any material supply issues to potentially impede AFC development. AFC technology has the potential to yield major improvements for modest R&D investments.

Contamination of AFCs due to the presence of CO_2 is an issue for sustained system operation. CO_2 in the Cathode air stream definitely poisons the electrolyte and in turn can cause some designs of electrodes to become clogged with carbonate. The use of high current draw from a cell to "electrolyze" the carbonates should be investigated further. The only practical solution to the CO_2 problem currently employed is the use of soda lime for scrubbing CO_2 from the air-stream. This is cumbersome, comparatively costly (estimated as US$ 94 per system) and has not been optimized as yet. Development of new means of CO_2 removal from the oxidant stream for an AFC system would address many operational issues associated with AFC stacks. Contamination due to impure hydrogen is another problem that may prohibit the use of AFCs with reformed hydrogen streams, though this contamination seems to be totally reversible.

Current AFC systems have been demonstrated to easily meet the 5000 h lifetime required for traction applications. However, electrolyte management issues in AFCs imply a degree of ongoing maintenance not necessary with PEMFC technology. Periodic maintenance is also required for the existing soda lime CO_2 scrubbing. Minimizing maintenance in AFCs is an important topic for development.

Our analysis of costs shows that AFC systems for low power applications including hybrid vehicles are at least competitive with the cost of any equivalent system constructed using PEMFC technology. The AFC system has a low cost stack and low cost peripheral components. An ambient air operated PEM system, while having low cost peripheral components has prohibitively high stack costs. A high pressure PEM system, while enjoying low stack costs, requires expensive peripheral components. Further improvements in AFC technology will only strengthen this competitive position.

APPENDIX A. 7 KW PEMFC STACK COST DEVELOPMENT

James et al. [67] develop stack and system costs for 30–90 kW PEMFC systems. The data given by James allows for the extrapolation of these costs down to 7 kW. Figure 2 illustrates this extrapolation.

It can be seen that the cost of the stack is much more sensitive to scale than the cost of peripherals. The cost figures developed using this linear extrapolation for a 7 kW stack are given in Table 18.

Table 18. Cost summary for kW PEMFC stack

	Total cost (US$)	Cost per kW (US$/kW)
Stack	419	60
System peripherals	798	114
Total system cost	1217	174

REFERENCES

1. K. Kordesch. *Power Sources for Electric Vehicles. Modern Aspects of Electrochemistry*, Vol. 10. Plenum Press, New York, 1975, pp. 339–443.
2. P.D. Michael. *An Assessment of the Prospects for Fuel Cell-Powered Cars.* ETSU, United Kingdom, 2000.
3. Zevco Website: www.zevco.co.uk, last accessed April 5, 2000.
4. E. De Geeter, M. Mangan, S. Spaegen, W. Stinissen and G. Vennekens. Alkaline fuel cells for road traction. *J. Power Sources* **80** (1999) 207–212.
5. J. Kivisaari, J. Lamminen, M.J. Lampinen and M. Viitanen. Preparation and measurement of air electrodes for alkaline fuel cells. *J. Power Sources* **32** (1990) 233–241.
6. L.J.M.J. Blomen and M.N. Mugerwa. *Fuel Cell Systems.* Plenum Press, New York, 1993.
7. S. Srinivasan, M.A. Enayetullah, S. Somasundaram, D.H. Swan, D. Manko, H. Koch and A. John Appleby. Recent advancements in solid polymer electrolyte fuel cell technology with low platinum loading electrodes. *Proceedings of the 24th Intersociety Energy Conversion Engineering Conference, IEEE*, 1989, pp. 1623–1629.
8. B. Bahar, C. Cavalca, S. Cleghorn and J. Kolde. Effective selection and use of advanced membrane electrode assemblies. *Fuel Cell Seminar 1998*, Palm Springs, 1998, pp. 531–534.
9. K.V. Kordesch. *Outlook for Alkaline Fuel Cell Batteries. From Electrocatalysis to Fuel Cells*, Seattle, WA, 1972, pp. 157–164.
10. K. Kordesch. Electrochemical Power Generation. *International Symposium on Electrochemistry in Industry: New Directions.* Plenum Press, New York, 1982.
11. K. Kordesch. Fuel cell R&D – toward a hydrogen economy. *Electric. Vehicle Develop.* **8** (1989) 25–26.
12. K. Kordesch, J. Gsellmann, M. Cifrain. Revival of alkaline fuel cell hybrid systems for electric vehicles. *Proceedings of Fuel Cell Seminar*, Palm Springs, 1998.
13. K. Kordesch. Intermittent use of low-cost alkaline fuel cell-hybrid system for electric vehicles. *J. Power Sources* **80** (1999) 190–197.
14. A. Vegas, C. Garcia and A. Gonzalez. Experience with an alkaline fuel cell stack: test plant, measurements and analysis of the results. *Hydrogen Energy Progress XII*, Buenos Aires, 1998, pp. 1621–1628.
15. M. Mangan and E. De Geeter. Gaseous hydrogen/alkaline fuel cell car. In: P.F. Howard (ed.), *Ninth Canadian Hydrogen Conference*, Canadian Hydrogen Association, Vancouver, BC, 1999.
16. H. Van den Broeck *et al.* Status of Elenco's alkaline fuel cell technology. *Intersociety Energy Conversion Engineering Conference Proceedings*, Philadelphia, 1987.
17. H. Van den Broek *et al.* Hybrid fuel cell battery city bus eureka project. *Fuel Cell Seminar 1994*, San Diego, CA, 1994.
18. O. Lindstrom. *A Critical Assessment of Fuel Cell Technology.* Department of Chemical Engineering and Technology, Royal Institute of Technology, Stockholm, Sweden, 1993.
19. O. Lindstrom. A critical assessment of fuel cell technology. *Fuel Cell Seminar 1994*, San Diego, CA, 1994.
20. O. Lindstrom *et al.* Small scale AFC stack technology. *Fuel Cell Seminar 1994*, San Diego, CA, 1994.
21. E. Gulzow. Alkaline fuel cells: a critical view. *J. Power Sources* **61** (1996) 99–104.
22. International Fuel Cells Website: www.internationalfuelcells.com, last accessed April 25, 2000.
23. HyWeb-Gazette: www.hydrogen.org/news/gazette.html, last accessed April 27, 2000.
24. J. Nor. *Personal Communication.* Astris Energi, Ontario, Canada, April 6, 2000.
25. Electric Auto Website: www.electricauto.com, last accessed April 25, 2000.

26. V.I. Matryonin, A.T. Ovchinkov and A.P. Tzedilkin. Investigation of the operating parameters influence on H_2–O_2 alkaline fuel cell performance. *Int. J. Hydrogen Energ.* **22** (1997) 1047–1052.
27. R.E. Martin and M.A. Manzo. Alkaline fuel cell performance investigation. In: D.Y. Goswami (ed.), *Proceedings of the 23rd Intersociety Energy Conversion Engineering Conference*, Vol. 2. Denver, Colorado, 1988, pp. 301–304.
28. J.-H. Jo and S.-C. Yi. A computational simulation of an alkaline fuel cell. *J. Power Sources* **84** (1999) 87–106.
29. H. Binder, A. Kohling, W.H. Kuhn, W. Lindner and G. Sandstede. *Hydrogen and Methanol Fuel Cells with Air Electrodes in Alkaline Electrolyte. From Electrocatalysis to Fuel Cells*, Seattle, WA, 1972, pp. 157–164.
30. K. Tomantschger, F. McClusky, L. Oporto, A. Reid and K. Kordesch. Development of low cost alkaline fuel cells. *J. Power Sources* **18** (1986) 317–335.
31. L. Swette, J.A. Kopek, C.C. Copley and A.B. LaConti. Development of single unit acid and alkaline regenerative solid ionomer fuel cells. In: *Intersociety Energy Conversion Engineering Conference Proceedings*, Atlanta, GA, 1993, pp. 1227–1232.
32. Y. Kiros, C. Myren, S. Schwartz, A. Sampathrajan and M. Ramanathan. Electrode R&D, stack design and performance of biomass-based alkaline fuel cell module. *Int. J. Hydrogen Energ.* **24** (1999) 549–564.
33. T. Nakagawa, A. Tsutsumi and K. Yoshida. Performance of an alkaline fuel cell system using three-dimensional electrodes. *Buenos Aires: Hydrogen Energy Progress XII*, 1998, pp. 1333–1338.
34. Y. Matsuno, K. Suzawa, A. Tsutsumi and K. Yoshida. Characteristics of three-phase fluidized-bed electrodes for an alkaline fuel cell cathode. *Int. J. Hydrogen Energ.* **21** (1996) 195–199.
35. Y. Matsuno, A. Tsutsumi and K. Yoshida. Improvement in electrode performance of three-phase fluidized bed electrodes for an alkaline fuel cell cathode. *Int. J. Hydrogen Energ.* **22** (1997) 615–620.
36. U.B. Holeschovsky, J.W. Tester and W.M. Den. Flooded flow fuel cells: a different approach to fuel cell design. *J. Power Sources* **63** (1996) 63–69.
37. M.A. Al-Saleh, S. Gultekin, A.S. Al-Zakri and H. Celiker. Effect of carbon dioxide on the performance of Ni/PTFE and Ag/PTFE electrodes in an alkaline fuel cell. *J. Appl. Electrochem.* **24** (1994) 575–580.
38. M.A. Al-Saleh, S. Gultekin, A.S. Al-Zakri and H. Celiker. Performance of porous nickel electrode for alkaline H_2/O_2 fuel cell. *Int. J. Hydrogen Energ.* **19** (1994) 713–718.
39. M.A. Al-Saleh, S. Gultekin, A.S. Al-Zakri and A.A.A. Khan. Steady state performance of copper impregnated Ni/PTFE gas diffusion electrode in alkaline fuel cell. *Int. J. Hydrogen Energ.* **21** (1996) 657–661.
40. S. Tanaka, N. Hirose and T. Tanaki. Evaluation of Raney-nickel cathodes prepared with aluminum powder and tin powder. *Int. J. Hydrogen Energ.* **25** (2000) 481–485.
41. H.-K. Lee, J.-P. Shim, M.-J. Shim, S.-W. Kim and J.-S. Lee. Oxygen reduction behavior with silver alloy catalyst in alkaline media. *Mater. Chem. Phys.* **45** (1996) 238–242.
42. S. Gultekin, A.S. Al-Zakri, M.A. Al-Saleh and H. Celiker. Experimental studies and electrochemical kinetic parameters of Raney-silver gas diffusion electrode in hydrogen–oxygen fuel cell. *Arab. J. Sci. Eng.* **20** (1995) 635–647.
43. FCT gas diffusion electrodes: www.fuelcell.kosone.com/elec.html, last accessed February 23, 2000.
44. E. Han, I. Eroglu and L. Turker. Performance of an alkaline fuel cell with single or double layer electrodes. *Int. J. Hydrogen Energ.* **25** (2000) 157–165.
45. S. Ahn and B.J. Tatarchuk. Fibrous metal–carbon composite structures as gas diffusion electrodes for use in alkaline electrolyte. *J. Appl. Electrochem.* **27** (1997) 9–17.
46. X. Li and K. Horita. Electrochemical characterization of carbon black subjected to RF oxygen plasma. *Carbon* **38** (2000) 133–138.
47. U.R. Sleem, M.A. Al-Saleh, A.S. Al-Zakri and S. Gultekin. Preparation of Raney-Ni gas diffusion electrode by filtration method for alkaline fuel cells. *J. Appl. Electrochem.* **27** (1997) 215–220.
48. K. Kordesch and S. Gunter. *Fuel Cells and Their Applications.* Wiley-VCH, Berlin, Germany, 1996.
49. Y. Sugawara *et al.* High performance PEFC in dry condition at atmospheric pressure. *Fuel Cell Seminar 1998*, Palm Springs, 1998.
50. P. Koschany. Small multi-purpose fuel cell systems. *Portable Fuel Cells Proceedings*, Lucerne, Switzerland, 1999.
51. A.J. Appleby and F.R. Foulkes. *Fuel Cell Handbook.* Krieger Publishing Company, Malabar, Florida, 1993.
52. K. Kinoshita. *Electrochemical Oxygen Technology*, Wiley, New York, 1992.

53. V. Ahuja and R.K. Green. CO_2 removal from air for alkaline fuel cells operating with liquid hydrogen – heat exchanger development. *Int. J. Hydrogen Energ.* **21** (1996) 415–421.
54. V. Ahuja and R. Green. Carbon dioxide removal from air for alkaline fuel cells operating with liquid hydrogen – a synergistic advantage. *Int. J. Hydrogen Energ.* **23** (1998) 131–137.
55. A. Fyke. An investigation of alkaline and PEM fuel cells. *Institute for Integrated Energy Systems*, University of Victoria, Canada, 1995.
56. S. Gultekin, M.A. Al-Saleh and A.S. Al-Zakri. Effect of CO impurity in H_2 on the performance of Ni/PTFE diffusion electrodes in alkaline fuel cells. *Int. J. Hydrogen Energ.* **19** (1994) 181–185.
57. Y. Kiros. Tolerance of gases by the anode for the alkaline fuel cell. *Fuel Cell Seminar 1994*, San Diego, CA, 1994.
58. K. Strasser. The design of alkaline fuel cells. *J. Power Sources* **29** (1990) 149–166.
59. A. Khalidi, B. Lafage, G. Gave, M.J. Clifton and P. Cezac. Electrolyte and water transfer through the porous electrodes of an immobilized-alkali hydrogen–oxygen fuel cell. *Int. J. Hydrogen Energ.* **21** (1996) 25–31.
60. M.T. Ergul, L. Turker and I. Ergolu. An investigation on the performance optimization of an alkaline fuel cell. *Int. J. Hydrogen Energ.* **22** (1997) 1039–1045.
61. O. Lindstrom and W. Lavers. Cost engineering of power plants with alkaline fuel cells. *Int. J. Hydrogen Energ.* **22** (1997) 815–823.
62. G. Gerhart. A comparative life-cycle analysis of proton exchange membrane fuel cells and internal combustion engines. M.A.Sc. Thesis. University of Victoria, Victoria, BC, 1996.
63. J. Ferguson. The Future is Green for BWT: dailynews.yahoo.com/h/nm/20000313/tc/austria_water_1.html, last accessed March 14, 2000.
64. K. Tomantschger, R. Findlay, M. Hanson, K. Kordesch and S. Srinivasan. Degradation modes of alkaline fuel cells and their components. *J. Power Sources* **39** (1992) 21–41.
65. J. Lamminen, J. Kivisaari, M.J. Lampinen, M. Viitanen and J. Vuorisalo. Preparation of air electrodes and long run tests. *J. Electrochem. Soc.* **138** (1991) 905–908.
66. E. Gulzow *et al.* Carbon dioxide tolerance of gas diffusion electrodes for alkaline fuel cells. *Fuel Cell Seminar 1994*, San Diego, CA, 1994.
67. B.D. James, F.K. Lomax Jr., C.I. Thomas and W.G. Colella. PEM fuel cell power system cost estimates: sulfur-free gasoline partial oxidation and compressed direct hydrogen. Directed Technologies, Inc. for the Ford Motor Co., 1997.
68. DOE Contractor's Report. Fuel Cells for Transportation 1998, DOE, Washington, DC, 1998.
69. M. Nundin. World fuel cell council. *Keynote Lecture Presented at the Fuel Cells 2000 Conference*, Lucerne, Switzerland, July 10–14, 2000.
70. P. Ekdunge and M. Raberg. The fuel cell vehicle analysis of energy use, emissions and cost. *Int. J. Hydrogen Energ.* **23** (1998) 381–385.
71. Astris Energi Website: www.astrisfuelcell.com, last accessed April 5, 2000.
72. Dais Analytical Website: www.daisanalytic.com, last accessed April 27, 2000.
73. H-Power Promotional Material. H-Power, New Jersey, 2000.
74. F. Barbin. Technical challenges in PEM fuel cell development. In: *Proceedings of the 12th World Hydrogen Energy Conference*. Buenos Aires, Argentina, June 1999, pp. 1589–1597.
75. C.E. Thomas, B.D. James and F.D. Lomax. Market penetration scenarios for fuel cell vehicles. *Int. J. Hydrogen Energ.* **23(10)** (1998) 949–966.

Chapter 8
Molten carbonate fuel cells

Andrew L. Dicks

Abstract

Lithium-sodium carbonate is emerging as the preferred electrolyte for molten carbonate fuel cells and has been tested at the 10 kW scale. Nickel oxide cathodes can be made more resistant to dissolution by coating with nanomaterial. Conventional NiO anodes continue to develop, but recent tests using ceramic oxide anodes suggest that future cells could be run on dry methane.

Article Outline

1. Introduction 147
2. Cell construction 148
3. Electrolyte 148
4. Cathode materials 150
5. Anode materials 150
6. Corrosion protection 151
7. Conclusion 151
Acknowledgement 151
References 151

1. INTRODUCTION

In 1960, G.H.J. Broers and J.A.A. Ketelaar reported a high-temperature fuel cell, that had run for 6 months, employing an electrolyte comprising a mixture of alkali metal carbonates constrained within a disc of magnesium oxide. The cell operated well above the melting point of the carbonates, and the carbonate ion (CO_3^{2-}) was found to be the means of charge transport within the molten electrolyte. Over the past 40 years many changes have been made in the materials of construction of the molten carbonate fuel cell (MCFC) but the operating principle remains the same. This is illustrated schematically in Fig. 1 [1], which also shows one of the characteristics that distinguish this type of fuel cell from all others, i.e. there is a net transfer of CO_2 from the cathode side of the cell to the anode through the electrolyte.

Following the early work in the Netherlands, development of the MCFC was taken up by several university groups and organizations such as the Gas Technology Institute in the USA, and ECN in the Netherlands. By the 1990s, the materials of construction had become well established and their limitations understood. Although the materials costs were relatively low compared with other fuel cell types, there were major problems associated with degradation of materials and the poor lifetime of cells. At this stage, research was focused in three main areas: selection of materials to increase the operating life of the cell (mainly through improvement of electrolyte and its support, and reduction of electrolyte losses), scale up of stacks and system design to maximize the benefit of internal reforming. R&D in all these areas continues, with commercialization of the technology being pursued in the USA, Japan and Europe. This review

Fuel Cells Compendium

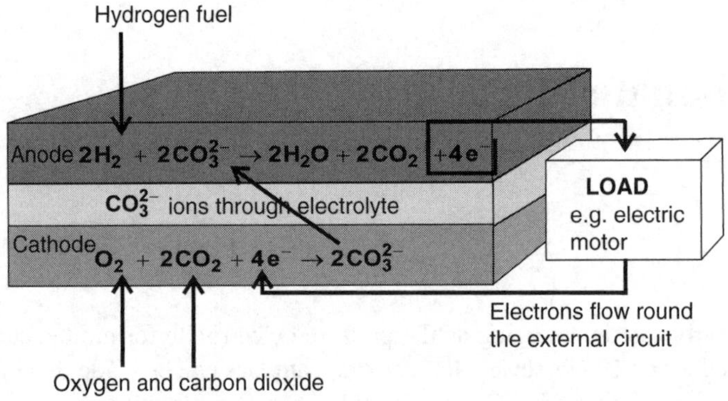

Fig. 1. Operating principle of the molten carbonate fuel cell, showing the anode and cathode reactions when hydrogen is used as fuel

focuses on R&D carried out over the past 12 months associated with materials for the electrolyte and electrodes of the MCFC.

2. CELL CONSTRUCTION

In the early days of the MCFC, the electrodes more often than not employed precious metals. As the technology developed, nickel was found to be adequate both as a metal for the anode and as oxide (NiO) for the cathode. For the electrolyte, most developers have adopted a eutectic mixture of lithium and potassium carbonates (62 wt.% Li and 38 wt.% K), which has a melting point around 550°C. This mixture is usually impregnated into a porous solid support matrix made of lithium aluminate ($LiAlO_2$). However, since both anode and cathode also need to be porous to allow the reacting gases to reach the electrode/electrolyte interfaces, the pore structure of the cell components needs to be carefully controlled so that electrolyte loss is minimized. Unlike all other fuel cell types, the MCFC relies on a balance in capillary pressures within the pores of the anode and cathode to establish the interfacial electrode/electrolyte boundaries.

The MCFC is invariably of planar construction and each of the porous components is normally made by tape-casting. This has enabled significant scale-up to be achieved and cells of 1 m^2 are now routinely manufactured. The normal operating temperature of between 600 and 700°C means that the cells can be held together in a housing made of stainless steel. A particular issue is that of ensuring gas tight seals around each cell. This is achieved by using the liquid molten carbonate to form a wet seal against the metal cell housing. Where the carbonate is in contact with the metal, corrosion is likely unless the metal is protected, and to achieve this a coating of alumina is normally employed.

3. ELECTROLYTE

In the early 1990s it was found that, using a eutectic mixture of Li_2CO_3/K_2CO_3, electrolyte segregation (i.e. separation of the Li and K carbonates) can occur within both the cell and the stack. In the cell, the segregation increases the potassium concentration near the cathode and this leads to increased cathode solubility and a decline in performance. In the stack, it was found that electrolyte segregation can lead to a severe decline in performance of the end cells within the stack. These observations, particularly the issue of cathode dissolution in the electrolyte, caused several developers in the late 1990s to investigate alternative electrolyte materials. An early study by Ang and Sammells [2] had suggested that a Li/Na should exhibit a high ionic conductivity compared with Li/Na/K, Li/K and Li/Na carbonates at the same temperatures. This has been verified by several workers, for example by Morita et al. [3]. Tanimoto et al. [4] also carried out a series of laboratory-scale cell tests with different Li, Na and K carbonate electrolytes, and investigated the effect of adding Ca, Sr and Ba to the materials.

NiO is less soluble in Na_2CO_3 than K_2CO_3 and therefore Li/Na carbonate should be better as an electrolyte than Li/K in terms of cathode dissolution. This has been tested recently by workers at Mitsubishi [5], who examined the cathode, cathode current collector and electrolyte of cells that had run for extended periods with Li/Na electrolytes. They found that the amount of Ni deposited within the electrolyte (commonly occurring via reduction of dissolved Ni^{2+} ions originating from the cathode by H_2 originating from the anode side) was reduced, and that particle growth of NiO in the cathodes was suppressed by the use of Li/Na. They also found little effect on degradation of the $LiAlO_2$ electrolyte matrix.

To understand the mechanism of Ni dissolution within Li/Na electrolytes, Belhomme et al. [6] have carried out cyclic voltammetry of Ni in molten Li/Na carbonates at 650°C. By performing measurements on different electrodes (gold, nickel and aged nickel), they have been able to identify the main reactions occurring at the MCFC cathode. In addition to a reduction peak corresponding to the Ni/Ni^{2+} transition, other reduction peaks, occurring at lower potentials, were observed and their presence attributed to phase transitions. Evidence was given for the existence of a Ni/Ni_2 system and the formation of Ni(III), most probably $NaNiO_2$, by the oxidation of NiO at relatively high potentials.

Recent refinement of the Li/Na electrolyte composition by Tanimoto et al. [7] showed that the addition of 9 mol% $CaCO_3$ or 9 mol% $BaCO_3$ to an electrolyte comprising 52 mol% Li_2CO_3/Na_2CO_3 significantly reduced the solubility of NiO in the electrolyte and resulted in a 15–20% longer lifetime compared with the undoped material. The test basis was a single 81 cm^2 fuel cell operating at 0.85 MPa with a cathode feed comprising 85% CO_2/O_2. $SrCO_3$ addition has also been shown to have a beneficial effect on the electrolyte performance [8], and the latest results of a 10 kW scale MCFC stack [9] verify that Li/Na electrolyte performs well for prolonged periods.

The advantage of using Li/Na in terms of reduced cathode dissolution is offset slightly by the increased solubility of oxygen in Li/Na compared with Li/K, which results in a higher polarization of the cathode. The effect of this is reduced somewhat if the cell operating pressure is increased. It is also worth remarking that at low temperatures (<600°C) Li/Na gives an inferior performance to Li/K, which may be due to differences in wetting behavior of the electrodes [10].

It has been known for some time that the matrix of lithium aluminate used to support the molten electrolyte degrades over time. In Li/Na carbonates, the particle size of γ-$LiAlO_2$ has been shown to increase along with a phase change from γ- to α-$LiAlO_2$. The phase transition appears to be dependent on factors such as the particle size of the starting material, the operating temperature and gas atmosphere, particularly CO_2 partial pressure and the composition of the molten electrolyte [11]. Under MCFC conditions with Li/K carbonates, the α-$LiAlO_2$ undergoes a phase change to the γ form; this is the reason why γ-$LiAlO_2$ has been preferred as the matrix material in the past. Danek et al. [12] suggest that the α- to γ-$LiAlO_2$ transition may occur via a chemical step such as the decomposition of the double lithium-aluminium oxide at the surface of α-$LiAlO_2$ particles via a surface nucleation mechanism. With Li/Na electrolyte however, γ appears to be less stable than α-$LiAlO_2$ [13]. Vidya et al [14] have recently shown that α-$LiAlO_2$ can be fabricated as thin layer by tape-casting from the raw powder using a binder of polyvinyl butyral and a plasticizer of polyethylene glycol in a non-aqueous solvent (butanol and *iso*-propanol). The resulting matrix had pores in the range 0.2–0.9 μm, and a porosity of 70 vol.%. While a test cell with this material gave a good performance on hydrogen over 10 h, it is too early to say whether this material will fare better in the long-term with α-Li/Na electrolyte than the conventionally prepared γ-$LiAlO_2$ material.

Table 1. New MCFC cathode materials

Composition	Preparation method	Comments	Reference
CoO/NiO	Mechanical coating on Ni filaments	25% of the solubility of NiO	[15–17]
Co_3O_4	Electrochemical deposition	Cubic Co_3O_4 structure 40% less soluble than NiO	[18,19]
	Co_3O_4 coated on Ni powder from polymer precursor – Pechini method	Stable $LiCo_{1-y}Ni_yO_2$ formed on NiO surface	[20,21]
$MgFe_2O_4$/NiO	Mechanical coating on Ni filaments	50% less soluble than NiO	[15,28]
$LiCoO_2$/NiO	Solgel 1.5 μm thick	40% less soluble than NiO	[22,23]
	Acetate-ascorbate sols dip-coated 1 μm thick	No measureable solubility	[24]
	Electrophoretic deposition		[31]
ZnO/NiO	Impregnation of NiO	With 2 mol% 10% ZnO, dissolution in Li/K carbonate was 10% that of undoped NiO	[25,26]
LiO/NiO		10% of the solubility of NiO	[30]
$LiFeO_2$-$LiCoO_2$/NiO	Pechini method	Electronic conductivity of 50:50 wt. $LiFeO_2$:NiO increases significantly with increasing $LiCoO_2$ content up to ca. 25%	[27]
$LiMgO_{0.05}Co_{0.95}O_2$	Acetate-ascorbate sols dip-coated ~1 μm thick		[24]
$La_{0.8}Sr_{0.2}CoO_3$/NiO	Sol gel, then sintered	3 cm pot tests showed good short-term stability	[29]

4. CATHODE MATERIALS

Significant advances in MCFC cathode materials have been made in recent months, brought about through the need to reduce NiO dissolution, which is exacerbated if the MCFC is operated at elevated pressures. While other metallic or ceramic oxides are available, which do not dissolve in the electrolyte (e.g. $LiCoO_2$), most have an inferior electronic and/or ionic conductivity. NiO therefore continues to be the basic cathode material of choice. The various NiO composite materials reported over the past 18 months are summarized in Table 1. While many of these are prepared by conventional coating methods such as dip-coating, increasingly more sophisticated techniques such as the sol-gel Pechini method [21,27] and electrophoretic deposition [31] are being employed. It is to be expected that these will produce far more ordered structures at the nanoscale, and thereby improve long-term stability of the cathode material. Already cathode dissolution rates of an order of magnitude or more lower than normal are being reported for some preparations [24,29].

5. ANODE MATERIALS

The main challenge with existing MCFC anode material is the susceptibility to creepage. Traditionally, Cr or Cu is added prior to tape-casting the nickel as oxide or metal powder to help stabilize the nickel. There have been some improvements reported for creep resistance in which Cr-doped Ni is subjected to oxidation or partial oxidation, followed by reduction. These processes affect the dispersion of the metal and it is found that partial oxidation followed by reduction produces a more creep-resistant material than full oxidation and reduction [32,33]. Alumina can also be added to Ni to improve creep resistance, and improved understanding of the use of alumina has been discussed recently by Lee et al. [34].

By far, the most significant development in anode materials recently, however, is the use of ceramic oxides. As with solid oxide fuel cells, these offer the prospects of being able to be fuelled by dry methane. Tagawa et al. [35] fed dry methane to an MCFC employing a composite anode made of La_2O_3/Sm_2O_3 (incorporating titanium powder to provide electronic conductivity). The cell performed well over a period of 144 h following an initial decrease of open-circuit voltage. Long-term testing of such materials is now required.

6. CORROSION PROTECTION

Stainless-steel bipolar plates are usually coated with a layer of alumina to provide protection against corrosion from the electrolyte in the areas providing the gas seals. Alternative corrosion protection materials investigated recently include borosilicate glass [36], a Ti/Al/N/O composite material [37] and co-deposited chromium and aluminium [38]. The latter deposited onto austenitic stainless steel (310S) demonstrated complete corrosion protection for the cathode side of the cell for 480 h.

7. CONCLUSION

Li/Na carbonate is emerging as a preferred electrolyte for MCFCs, although further optimization is likely. There are several means of improving the NiO cathode and future application of nanotechnology in this area could prove beneficial. While stabilized nickel anodes will continue to be used for some time, the ability to run on dry methane makes the use of ceramic materials particularly attractive. Although little has been reported in the past 12 months on improvement of corrosion protection, this is an area that could also yield benefits in terms of extending cell lifetime.

ACKNOWLEDGEMENT

This work was produced with the assistance of the Australian Research Council under the ARC Centres of Excellence Program.

REFERENCES

1. J. Larminie and A. Dicks. *Fuel Cell Systems Explained*, 2nd edn. Wiley, New York, 2003.
2. P.G.P. Ang and A.F. Sammells. *J. Electrochem. Soc.* **127**(6) (1980) 1287–1294.
3. H. Morita, M. Komoda, Y. Mugikura, Y. Izaki, T. Watanabe, Y. Masuda and T. Matsuyama. Performance analysis of molten carbonate fuel cell using a Li/Na electrolyte. *J. Power Sources* **112** (2002) 509–518.
4. K. Tanimoto, Y. Miyazaki, M. Yanagida, S. Tanase, T. Kojima, N. Ohtori, H. Okuyama and T. Kodama. *J. Power Sources* **39** (1992) 285–297.
5. Y. Fujita, T. Nishimura, T. Yagi and M. Matsumura. Degradation of the components in molten carbonate fuel cells with Li/Na electrolyte. *Electrochemistry* (Japan) **71(1)** (2003) 7–13.
6. C. Belhomme, J. Devynck and M. Cassir. *J. Electroanal. Chem.* **545** (2003) 7–17.
7. K. Tanimoto, T. Kojima, M. Yanagida, K. Nomura and Y. Miyazaki. Optimization of the electrolyte composition in a $(Li_{0.52}Na_{0.48})$2-2xAExCO$_3$ (AE = Ca and Ba) molten carbonate fuel cell. *J. Power Sources* **131** (2004) 256–260.

8. S. Scaccia. Investigation on NiO solubility in binary and ternary molten alkali metal carbonates containing additives. *J. Mol. Liquids* **116** (2004) 67.
9. F. Yoshiba, H. Morita, M. Yoshikawa, Y. Mugikura, Y. Izaki, T. Watanabe, M. Komoda, Y. Masuda and M. Zaima. Improvement of electricity generating performance and life expectancy of MCFC stack by applying Li/Na carbonate electrolyte: test results and analysis of 0.44 m^2/10 kW- and 1.03 m^2/10 kW-class stack. *J. Power Sources* **128(2)** (2004) 152–164.
10. S.-G. Hong and J.R. Selman. Wetting characteristics of carbonate melts under MCFC operating conditions. *J. Electrochem. Soc.* **151(1)** (2004) A77–A84.
11. K. Takizawa and A. Hagiwara. The transformation of $LiAlO_2$ crystal structure in molten Li/K carbonate. *J. Power Sources* **109** (2002) 127–135.
12. V. Danek, M. Tarniowy and L. Suski. Kinetics of the phase transformation in $LiAlO_2$ under various atmospheres within the 1073–1173 K temperatures range. *J. Mater. Sci.* **39(7)** (2004) 2429–2435.
13. S. Terada, K. Higaki, I. Nagashima and Y. Ito. Stability and solubility of electrolyte matrix support material for molten carbonate fuel cells. *J. Power Sources* **83(1–2)** (1999) 227–230.
14. V.S. Batra, S. Maudgal, S. Bali and P.K. Tewari. Development of alpha lithium aluminate matrix for molten carbonate fuel cell. *J. Power Sources* **112** (2002) 322–325.
15. T. Fukui, S. Ohara, H. Okawa, M. Naito and K. Nogi. Synthesis of metal and ceramic composite particles for fuel cell electrodes. *J. Eur. Ceram. Soc.* **23** (2003) 2835–2840.
16. M.Z. Hong, S.C. Bae, H.S. Lee, H.C. Lee, Y.-M. Kim and K. Kim. A study of the Co-coated Ni cathode prepared by electroless deposition for MCFCs. *Electrochim. Acta* **48** (2003) 4213–4221.
17. S.-G. Kim, S.P. Yoon, J. Han, S.W. Nam, T.-H. Lim, S.-A. Hong and H.C Lim. A stabilized NiO cathode prepared by sol-impregnation of $LiCoO_2$ precursors for molten carbonate fuel cells. *J. Power Sources* **112** (2002) 109–115.
18. L. Mendoza, V. Albin, M. Cassir and A.J. Galtayries. *Electroanal. Chem.* **548** (2003) 95–107.
19. L. Mendoza, R. Baddour-Hadjean, M. Cassir and J.P. Pereira-Ramos. Raman evidence of the formation of LT-$LiCoO_2$ thin layers on NiO in molten carbonate at 650°C. *Appl. Surface Sci.* **225** (2004) 356–361.
20. H. Lee, M. Hong, S. Bae, H. Lee, E. Park and K. Kim. A novel approach to preparing nano-size Co_3O_4 coated Ni powder by the Pechini method for MCFC cathodes. *J. Mater. Chem.* **13(10)** (2003) 2626–2632.
21. H.S. Lee, H.C. Lee, Y.-M. Kim and K. Kim. A study of the Co-coated Ni cathode prepared by electroless deposition for MCFCs. *Electrochim. Acta* **48(28)** (2003) 4213–4221.
22. S.S.-G. Kim, S.P. Yoon, J. Han, S.W. Nam, T.H. Lim, I.-H. Oh and S.-A. Hong. A study on the chemical stability and electrode performance of modified NiO cathodes for molten carbonate fuel cells. *Electrochim. Acta* **49(19)** (2003) 3081–3089.
23. C. Huang and J. Li. Fuel Cell Energy Inc., USA. U.S. Patent Appl. Publ., 2002.
24. W. Lada, A. Deptula, B. Sartowska, T. Olczak, A.G. Chmielewski, M. Carewska, S. Scaccia, E. Simonetti, L. Giorgi and A. Moreno. Synthesis of $LiCoO_2$ and $LiMgO_{0.05}Co_{0.95}O_2$ thin films on porous Ni/NiO cathodes for MCFC by complex sol–gel process (CSGP). *J New Mater. Electrochem. Syst.* **6(1)** (2003) 33–37.
25. B. Huang, F. Li, Q.-C. Yu, G. Chen, B.-Y. Zhao and K.-A. Hu. Study of NiO cathode modified by ZnO additive for MCFC. *J. Power Sources* **128** (2004) 135–144.
26. B. Huang, F. Li, G. Chen, B.-Y. Zhao and K.-A. Hu. Stability of NiO cathode modified by ZnO additive for molten carbonate fuel cell. *Mater. Res. Bull.* **39** (2004) 1359–1366.
27. A. Wijayasinghe, B. Bergman and C. Lagergren. $LiFeO_2$-$LiCoO_2$-NiO cathodes for molten carbonate fuel cells. *J. Electrochem. Soc.* **150(5)** (2003) A558–A564.
28. H. Okawa, J.-H. Lee, T. Hotta, S. Ohara, S. Takahashi, T. Shibahashi and Y. Yamamasu. Performance of NiO/$MgFe_2O_4$ composite cathode for a molten carbonate fuel cell. *J. Power Sources* **131(1–2)** (2004) 251–255.
29. P. Ganesan, H. Colon, B. Haran and B.N. Popov. Performance of La0.8Sr0.2CoO_3 coated NiO as cathodes for molten carbonate fuel cells. *J. Power Sources* **115** (2003) 12–18.
30. M.J. Escudero, T. Rodrigo, J. Soler and L. Daza. Electrochemical behaviour of lithium–nickel oxides in molten carbonate. *J. Power Sources* **118** (2003) 23–34.
31. L.-K. Chen, J. Zuo and C.-J. Lin. A novel MCFC cathode material modified by the EPD technique. *Fuel Cells* **3(40)** (2003) 220–223.
32. D. Jung, I. Lee, H. Lim and D. Lee. On the high creep resistant morphology and its formation mechanism in Ni-10 wt.% Cr anodes for molten carbonate fuel cells. *J. Mater. Chem.* **13(7)** (2003) 1717–1722.

33. C. Lee, D. Jung, I. Lee, K. Byun and H. Lim. Simplified and cost-effective sintering processes for creep resistant Ni-10wt.% Cr MCFC anodes. *Metals Mater. Int.* **9(6)** (2003) 605–611.
34. C.-G. Lee, K.-S. Ahn, H.-C. Lim and J.-M. Oh. Effect of carbon monoxide addition to the anode of a molten carbonate fuel cell. *J. Power Sources* **125** (2004) 166–171.
35. T. Tagawa, A. Yanase, S. Goto, M. Yamaguchi and M. Kondo. Ceramic anode catalyst for dry methane type molten carbonate fuel cell. *J. Power Sources* **126** (2004) 1–7.
36. M.J. Pascual, F.G. Valle, A. Duran and R. Berjoan. *J. Am. Chem. Soc.* **83(11)** (2003) 1918–1926.
37. H.C. Kim and T.L. Alford. Fuel cell having TiAlxNyOz deposited as a protective layer on metallic surface. *PCT Int. Appl.* 2004; WO 2004038842.
38. H.H. Park, M.H. Lee, J.S. Yoon, I.S. Bae and B.I. Kim. Corrosion resistance of austenitic stainless steel separator plate for molten carbonate fuel cell. *Metals Mater. Int.* **9(3)** (2003) 311–317.

Chapter 9
Phosphoric acid fuel cells: fundamentals and applications

Nigel Sammes, Roberto Bove and Knut Stahl

Abstract

Phosphoric acid fuel cells (PAFC) currently represent one of the fuel cell technologies that have been demonstrated in many countries around the world and for many applications. PAFCs can be purchased, complete with a warranty, maintenance and spare parts service. The first PAFC power plants were installed in the 1970s, and now more than 500 units have been installed all around the world. In the present paper, the principles of PAFC are presented, together with a state of the art. Finally, operational experiences are presented.

Article Outline

1. Introduction 155
2. Fundamentals of the PAFC 156
3. PAFC components: state of the art 158
   3.1. Electrolyte and matrix 158
   3.2. Electrodes 158
   3.3. Bipolar plates 159
4. PAFC in operation: the German case 160
   4.1. First PC25C fuel cell installations in Germany 160
   4.2. Hydrogen operation 160
   4.3. Utilization of anaerobic digester gas 161
   4.4. Energy supply for hospitals 162
   4.5. Fuel cell power plant overhaul 163
   4.6. Service and maintenance 163
5. Conclusions 164
References 164

1. INTRODUCTION

The phosphoric acid fuel cell (PAFC) is the most widely used and best-documented type of fuel cell. Since the 1970s, more than 500 PAFC power plants have been installed and tested around the world. With every new product release, the number of units sold became larger as well as the power rating per unit. The largest fuel cell ever built to date is an 11 MW PAFC power plant for the Tokyo Electric Power Co. (TEPCO) in Japan, which was operated for more than 23 000 h between 1991 and 1997 [1]. The most important PAFC developers are UTC Fuel Cells (formerly ONSI/International Fuel Cells), Toshiba and Fuji Electric. All the installations have been used for stationary applications, with the exception of the Georgetown University Fuel Cell Transit Program, in which a 100 kW UTC fuel cell was deployed [2]. PAFCs have shown a

Fuel Cells Compendium

Table 1. Example of hourly cost of electricity outage in US [4]

Premium power user	Typical cost for 1 h interruption (US$)
Cellular communication	41 000
Telephone ticket sale	72 000
Air reservation system	90 000
Semiconductor manufacturer	2 000 000
Credit card operation	2 580 000
Brokerage firm	6 480 000

remarkable reliability. UTC PAFC systems are characterized by a mean time before failure (MTBF) that ranges between 2500 h for the PC25 and 6750 h for the 400 kW advanced PAFC [3]. Owing to the high reliability, that is well above those of traditional systems, PAFC represents, for some applications, the answer to electricity quality and availability needs. For some applications, in fact, the interruption of the electricity availability, even for a short period of time, represents a relevant economical loss. Some examples of these applications and the relative costs of 1 h of electricity interruption are reported in Table 1. [4].

The efficiency of PAFC is also relatively high, the UTC PC25, for example, has a proven net electrical efficiency of 37%, and a total efficiency of 87% in combined heat and power (CHP) applications [5], when running on natural gas. The consequence is a low fuel consumption, and a reduction of pollution emissions, compared to traditional power systems.

Despite the excellent technical characteristics, the large diffusion of PAFC in the market has slowed down because of economical issues. A recent study survey shows that the number of power plant installations grew from 1990 to 2000, while, from 2000 to 2003, more attention has been given to other fuel cell technologies, such as molten carbonate fuel cells (MCFC), solid oxide fuel cells (SOFC) and proton-exchange membrane (PEM) fuel cells [6]. In particular, in 2003, the number of MCFC units installed exceeded that of PAFC. For this reason, some authors have referred to a drop in the interest toward PAFC [7]. However, in 2004, the number of commissioned PAFC units for large stationary applications represented about 40% of the total fuel cell units manufactured, despite the 23% of MCFC and less than 20% of PEM.

2. FUNDAMENTALS OF THE PAFC

A schematic representation of the PAFC operating configuration is depicted in Fig. 1. Hydrogen, or a hydrogen-rich gas mixture, is provided to the anode side, where the following reaction takes place:

$$H_2 \rightarrow 2H^+ + 2e^- \tag{1}$$

The electrolyte, primarily composed of phosphoric acid (H_3PO_4), is a proton conductor, thus the protons migrate from the anode to the cathode, while the electrons migrate through an external circuit. At the cathode side, air is provided, where oxygen reacts with the protons and the electrons, coming from the electrolyte and the external load, respectively,

$$\tfrac{1}{2}O_2 + 2H^+ + 2e^- \rightarrow H_2O \tag{2}$$

The overall reaction is

$$H_2 + \tfrac{1}{2}O_2 \rightarrow H_2O \tag{3}$$

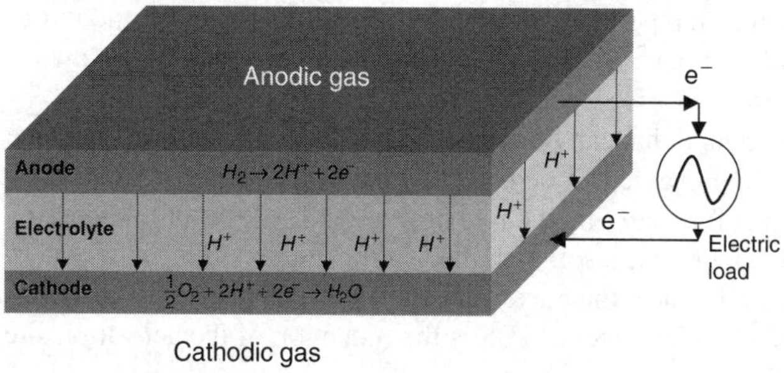

Fig. 1. Schematic representation of a PAFC

The operating temperature of the PAFC is typically between 150 and 200°C [8]. The operating temperature is found to be a compromise between the electrolyte conductivity (that increases with temperature) and cell life (that decreases when the temperature is increased).

Although most of the installed systems operate at atmospheric pressure, experience with operating pressure exceeding 8 atm has been reported [9].

The ideal voltage of a PAFC is given by the Nernst equation

$$E = -\frac{\Delta G}{2F} = E^0 + \frac{RT}{2F}\ln\left(\frac{P_{H_2}P_{O_2}^{\ 0.5}}{P_{H_2O}}\right) = E^0 + \frac{RT}{2F}\left[\ln\left(\frac{X_{H_2}X_{O_2}^{\ 0.5}}{X_{H_2O}}\right) + \ln\left(\frac{P_{tot}}{P_{ref}}\right)^{0.5}\right] \quad (4)$$

where E is the ideal voltage, E^0 the ideal voltage at standard pressure, F the Faraday constant, P_{ref} the reference pressure, and P_i and X_i the partial pressure and the molar fraction of the ith species, respectively.

When the total pressure is varied from P_1 to P_2, the related increased Nernst voltage is

$$\Delta E = \frac{RT}{4F}\ln\frac{P_2}{P_1} \quad (5)$$

Experimental data confirm the improved performance under pressurized conditions, and the following expression is reported, when the cell operates at 190°C and 323 mA/cm^2 [10]:

$$\Delta V_p(\text{mV}) = 146\log\frac{P_2}{P_1} \quad (6)$$

The improved performance under pressurized conditions is not only due to an increase in the reversible potential, but also due to the reduced diffusion polarization of the cathode, and the reduction of the ohmic losses.

However, a pressurized system presents a more complex balance of plant (BoP), thus, the fuel consumption reduction is counterbalanced by a capital cost increase. For this reason, most of the PAFC systems operate at atmospheric pressure, while keeping the BoP complexity as low as possible.

As is found for all the other low-temperature fuel cells, the presence of carbon monoxide in the anodic gas affects the performance of the cell itself. The main reason for this reduction is the poisoning effect of CO on the Pt electrode catalyst. Benjamin et al. [10] derived the following equation for evaluating the voltage reduction:

$$\Delta V_{CO} = k(T)([\text{CO}]_2 - [\text{CO}]_1) \quad (7)$$

where $k(T)$ is a function of temperature, and $[CO]_1$ and $[CO]_2$ represent the initial and final CO concentration. In the later work of Song and Shin [11], the voltage reduction was shown to depend upon the current density, thus showing that k is also a function of the current density.

Sulfur is another compound that can be present in the anodic gas, however, since the fuel, before entering the anode, is usually processed in dedicated reactors (typically a reformer and two shift reactors), sulfur must be removed before entering the fuel processor. The level of tolerance of the fuel cell is usually higher than that of the fuel processor, thus the sulfur content is usually safe for the cell operation. Nevertheless, it is prudent to know the maximum allowable concentration of sulfur in the anodic gas. Chin and Haward [12] report that the effect of H_2S is the reduction of the activation sites of Pt. The following reactions are presumed:

$$Pt + HS^- \rightarrow Pt - HS_{ads} + e^- \tag{8}$$

$$Pt - H_2S_{ads} \rightarrow Pt - HS_{ads} + H^+ + e^- \tag{9}$$

$$Pt - HS_{ads} \rightarrow Pt - S_{ads} + H^+ + e^- \tag{10}$$

In [9], the effect of NH_3 in the cathodic or anodic gas is also assessed. The following reaction is postulated as taking place:

$$H_3PO_4 + NH_3 \rightarrow (NH_4)H_2PO_4 \tag{11}$$

The reduction in H_3PO_4 content reduces the O_2 reduction; thus the molecular nitrogen concentration must be maintained below 4%.

3. PAFC COMPONENTS: STATE OF THE ART

3.1. *Electrolyte and matrix*

The choice of phosphoric acid is dictated by good thermal, chemical and electrochemical stability that this inorganic acid represents. At the same time, H_3PO_4 is tolerant to CO_2, which is always present in a reformate gas mixture. In the first PAFC systems, H_3PO_4 was diluted to avoid material corrosion, while 100% H_3PO_4 is currently used [13]. Due to possible liquid loss, it is necessary to refill the electrolyte or to provide the cell with an excess, before it is operated. The current solution is to create an electrolyte reservoir plate (ERP) that provides enough electrolyte to allow the cell to operate for more than 40000 h [9].

H_3PO_4 is retained in a 0.1–0.2 mm thick SiC matrix. The ohmic resistance of the matrix is very low, due to the small thickness, while the mechanical properties are somewhat limited. The maximum pressure difference between anode and cathode, in fact, cannot exceed 200 mbar [13].

3.2. *Electrodes*

As for the other fuel cell technologies, the anode and cathode have the function of allowing the gas to diffuse from the gas channel to the electrolyte. Every electrode faces the gas channel on one side and the electrolyte on the other. On the electrolyte side, an electrocatalyst, whose function is primarily to favor the gas reaction, is placed. Since the electrolyte is in a liquid form, and to expel the produced water, electrodes need to be hydrophobic. This is generally achieved, by immersing the backing layer into a polytetrafluoroethylene (PTFE) solution. PTFE is also used as a binder, in order to prevent pore flooding. Chan and Wan [14]

Table 2. Evolution of electrodes materials [9]

Component	ca. 1965	ca. 1975	Current status[a]
Anode	PTFE-bonded Pt black	PTFE-bonded Pt/C Vulcan XC-72[a]	PTFE-bonded Pt/C
	9 mg Pt/cm^2	0.25 mg Pt/cm^2	0.25 mg Pt/cm^2
Cathode	PTFE-bonded Pt black	PTFE-bonded Pt/C Vulcan XC-72[a]	
	9 mg Pt/cm^2	0.5 mg Pt/cm^2	0.5 mg Pt/cm^2
Electrode support	Ta mesh screen	Graphite structure	Graphite structure

[a] Over 40000 h component life demonstrated in commercial power plant.

monitored the dispersion of PTFE in the catalysts by means of an X-ray diffraction technique, and pointed out that the dispersion efficiency increases when the bank temperature and the PTFE content are increased. In addition, electrodes need a good electrical conductivity to enable electrons to flow through it, without significant resistance, from the catalyst layer to the current collector (anode) or vice versa (cathode).

The evolution of PAFC electrode materials is summarized in Table 2 [9]. In the mid-1960s, both the anode and the cathode were made of PTFE-bound Pt black and the Pt load was 9 mg/cm^2. A major breakthrough occurred at the end of the 1960s, with the deployment of Pt supported on carbon and graphite [15], that allowed a significant reduction in Pt loading. Yang et al. [16] reported a Pt loading of 0.6 mg/cm^2 for the anode, whereas for the alloy cathodic catalyst they considered Pt, Fe and Co. The use of carbon and graphite, however, imposes some limitations on the FC operation. In particular, the FC should be run at potentials of less than 0.8 V, otherwise there is a possibility of corrosion occurring. Passalacqua et al. [17] investigated the influence of Pt content on the corrosion phenomena. Their results indicate that at high potentials, anodic dissolution of Pt takes place, thus no metal is available to catalyze the corrosion of carbon. Another limitation related to the use of carbon is the tendency of Pt to migrate to the surface of the carbon to agglomerate in large areas, thus reducing the active surface [13].

Although experimental tests are the only way to choose among different options for material selection as well as geometry configurations, mathematical models allow the developer to predict the performance related to specific solutions. Mathematical models have the advantage of being inexpensive and relatively quick, compared to experimental tests. Mathematical models, however, need to be validated by experiments. Examples of PAFC simulations are given in [18–21].

3.3. Bipolar plates

As for the other fuel cell technologies, the open-circuit voltage of a single cell is slightly over 1 V, thus, more single cells are connected in series for achieving a reasonable operating voltage. The interconnection is made via bipolar plates (BP) that connect the anode of one cell with the cathode of the next one, forming a stack. Together with the electrical connection, the BPs are usually machined so that they can act as gas channels. A full description of the BP for PAFC is given in [13].

A new design for PAFC is based on manufacturing the BP in several layers. Two external layers made of porous materials are interposed on a conductive, impermeable material, such as carbon. Through this configuration the separation of the anodic and cathodic gases is ensured by the carbon layer, while, at the same time, porous layers hold phosphoric acid, i.e. a refill for the electrolyte losses, which is the ERP solution, previously mentioned.

4. PAFC IN OPERATION: THE GERMAN CASE

In Germany, PAFC power plants were first tested in the early 1990s, when several large German power and gas companies purchased four PC25A fuel cells from UTC Fuel Cells, for field-testing purposes. In 1995, the PC25A was replaced by the PC25C, and in the following years 12 power plants of this type were installed in Germany. Today, some of these power plants have been in operation in excess of 40000 h, verifying the expected lifetime of the fuel cell stack.

4.1. First PC25C fuel cell installations in Germany

The first PC25C fuel cell power plant in Germany was installed and commissioned during the turn of the year 1996/1997 by the Erdgas-Energie-Systeme GmbH (today ABB New Ventures GmbH). This fuel cell power plant is combined with a new condensing boiler and it serves as a local district heating system in a residential area in Saarbrücken-Nachtweide. Although the fuel cell power plant has surpassed 40000 load hours in 2004, it is still capable of running at rated power and remains in service.

Only a few months later, the Energieversorgung Halle installed the first PC25C fuel cell in the new federal states that were formed after the reunification with Eastern Germany (Fig. 2). The PC25C power plant provides power and heat for a public bath in Halle (Saale). As expected, public baths have a continuous demand for power and heat almost throughout the entire year, and therefore this power plant has been in operation for more than 46500 h, most likely a European record.

4.2. Hydrogen operation

Although PAFC systems usually present a steam reformer, in the eventuality of a "hydrogen economy", it is useful to test fuel cells running on pure hydrogen. The Hamburg-based utility companies HEW and HGW demonstrated the utilization of hydrogen in a fuel cell to generate power and heat for the residential area Lyser Straße in Hamburg between 1997 and 2000. After this successful demonstration, in 2001 the PC25C power plant was relocated into the industrial area of Höchst in Frankfurt. At this chemical

Fig. 2. Fuel cell installed at the public bath of Halle

Fig. 3. Installation at Celanese Ventures, in Frankfurt

industry site, hydrogen is available as a by-product of chlor-alkali-electrolysis, which now serves as fuel for the PC25C power plant. In September 2002, the small power plant again was relocated within the Höchst industrial area, and now it provides power and heat for the new Celanese Ventures factory, a manufacturer of fuel cell components (Fig. 3). The power plant has been in operation without major disruption.

4.3. Utilization of anaerobic digester gas

The use of biogas as an energy source for the PAFC has already been demonstrated in previous applications in the USA. In particular, operation and design of a UTC PC 25, operating on landfill gas is extensively reported in the literature [22–25], while in [26] the application for an anaerobic digester is described. However, the composition of biogas depends on various factors, such as the application, location, climate, etc.; thus the American experience is not directly exportable to European needs, although this represents a useful source of information. In 2000, the GEW RheinEnergie AG together with the city of Cologne has installed the first fuel cell power plant in Europe that operates on anaerobic digester gas (ADG) in the wastewater treatment plant in Köln-Rodenkirchen [27]. The fuel cell installed was a PC25C (Fig. 4). Operating a PC25C fuel cell on ADG has several challenges. A methane-enrichment system was considered first to upgrade the ADG to natural gas quality, but it was quickly rejected due to high cost and increased system complexity. Instead, the PC25C fuel cell was modified to operate on low-BTU ADG by installation of a second fuel gas train with increased pipe diameters and larger control valves to accept higher gas flows. The fuel cell has dual-fuel capability and can run either on natural gas or on ADG, although the natural gas option was never used on this site.

Of particular interest was the design of a gas-processing unit (GPU) that removes contaminants from the ADG before it enters the fuel cell. The GPU design includes a gas dryer stage, a gas chiller stage and an adsorption stage. The gas dryer stage removes condensate, and is cooled by the cold gas coming from the chiller stage. The chiller stage further cools the gas to below −25°C, which is lower than the dew point of

Fig. 4. Fuel cell-anaerobic digester integrated system in Cologne

many higher organic compounds, particularly siloxanes. The adsorption stage removes organic sulfur and halides, if present. The gas-processing unit design was verified using a continuous run in excess of 5000 h. Unfortunately, the gas-processing unit was too maintenance-intensive and was, thus, redesigned after 2 years to simplify it. The fuel cell power plant has been in operation for more than 4 years and has proven its durability on ADG. The fuel cell covers approximately 50% of the power demand of the wastewater treatment plant. More than 1.3 million m^3 of ADG were converted into clean power, and when compared to the German power plant average, more than 1700 tons of carbon dioxide were abated.

4.4. Energy supply for hospitals

Since early 2001, a PC25C fuel cell provides power and heat for the St.AgnesHospital in Bocholt, Germany. The fuel cell operates in combination with two existing gas engine cogeneration units, and the total power of all three cogeneration units is above 600 kW. The fuel cell provides the electrical base load of the hospital. In summer, the high-grade heat from the fuel cell is used to drive an absorption chiller, to provide some base load for the hospital's air-conditioning system. The fuel cell has been in continuous operation in excess of 8000 h per year, and has surpassed 30000 load hours, generating more than 6 million kWh of electrical power. The PC25C at the St. Agnes Hospital in Bocholt is the most recent phosphoric acid fuel cell installation in Germany. The project was funded by the Thyssengas GmbH and the Bocholter Energie- und Wasser GmbH, and was granted an amount of US$200000 US through the US DOE's Climate Change Fuel Cell Program.

Due to its high efficiency, and its excellent operational results, the fuel cell at the St. Agnes Hospital in Bocholt received a German Gas Industry Award in 2002. Figure 5 depicts the installation at the St. Agnes Hospital. Beside the St. Agnes Hospital in Bocholt, a second PC25C fuel cell power plant is installed at the Malteser-Krankenhaus in Kamenz, Germany.

Fig. 5. Fuel cell for trigeneration at the St. Agnes Hospital in Bocholt

4.5. Fuel cell power plant overhaul

In 2003, the Fernwärmeversorgung Niederrhein GmbH in Dinslaken, a large provider of district heating, took an old PC25C power plant from the Studiengesellschaft Brennstoff-zelle in Nuremberg, for testing. This unit had been in operation in Nuremberg since 1997, but had experienced some damage to its cell stack, and its previous owner had planned to terminate the fuel cell project.

The relocation of the power plant from Nuremberg to Dinslaken and the recommissioning procedure were performed without major problems. However, due to the damage in the cell stack, a continuous operation was questionable. Since the balance of plant had been tested and verified, and since the overall condition of the power plant was satisfactory, it was decided to perform an overhaul of this PC25C unit.

The damaged cell stack was conditioned for transportation and was shipped back to the manufacturer, UTC Fuel Cells. In return, UTC Fuel Cells supplied an overhauled cell stack, for replacement (Fig. 6). During the downtime of the fuel cell power plant, the internal coolant loop was cleaned, repaired, and hydro-tested by a pressure vessel inspector. The internal water treatment system was rebuilt, and several components were replaced or repaired. Finally, the thermal insulation was renewed, and the unit was painted. The fuel cell was successfully restarted and shown to the public in September 2003. Since then, the fuel cell power plant is back in operation, with only minor problems.

4.6. Service and maintenance

Fuel cells are often said to be maintenance-free, since they differ from gas engines and gas turbines in that they have no mechanical wear and no thermal stresses. The PC25C power plant requires less maintenance than competing conventional power generators of similar size. However, all PC25C power plant owners agree that this product is still too maintenance-intensive, and needs to be redesigned to further reduce this effort. The PC25C power plant contains consumables such as filters, water-treatment resins and activated charcoal. The units are typically installed outdoors and are exposed to rain water, dust and dirt, making it necessary to clean the interior and to inspect the unit for water leakages. Further, the manufacturer, UTC Fuel Cells, recommends that certain components are periodically replaced, for preventive maintenance reasons. Over the years, the list of such components has grown significantly. Although service and maintenance can

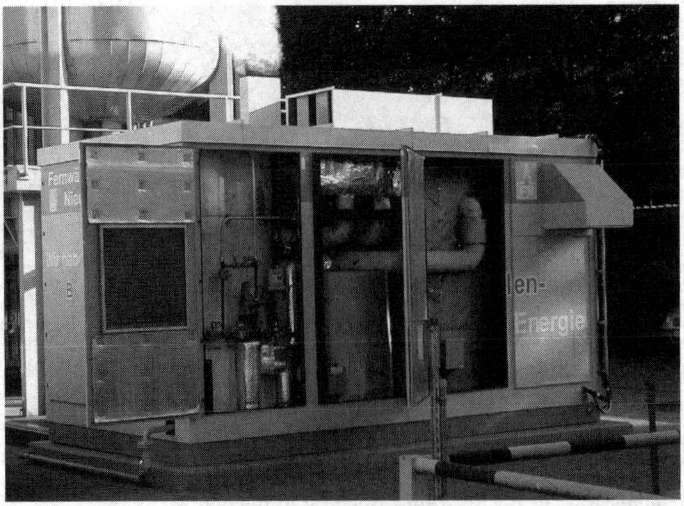

Fig. 6. PC25C fuel cell in Dinslaken, after overhaul

be carried out by power plant operators, after some training, in case of troubleshooting issues and major repairs, engineering support is required. A design flaw in early PC25C models has led to reformer thermocouple malfunctions, requiring a replacement of the entire reformer. Although UTC Fuel Cells offers a new reformer free of charge, such a reformer replacement requires heavy machinery and several man-days of work, and in some cases the power plant owners decided to terminate operation, especially on older units.

The PC25C fuel cell power plants in Germany are approaching the expected cell stack life of 40000 h. To continue operation, an overhaul of the cell stack is required. Besides new cell stacks, UTC Fuel Cells also offers repaired or refurbished components. The initial operational results using these components are good, but there are still very few power plants that have been operated with a refurbished stack in excess of 10000 h. New procedures are currently under investigation to replenish evaporated electrolyte inside the cells.

5. CONCLUSIONS

The efforts in phosphoric acid fuel cell development of the last three decades enabled PAFC to be available on the market. Although UTC Fuel Cells is the most important PAFC manufacturer, other companies, such as Toshiba and Fuji Electric have also shown excellent results for PAFC operations.

A large number of installations all around the world demonstrated that phosphoric acid fuel cells have high reliability, high efficiency and flexibility for a variety of applications. In particular, PAFC demonstrated excellent performance for most of the distributed power generation applications, in terms of power, efficiency and low emissions. However, the challenge that this technology needs to overcome for becoming a mass product is the cost. Without the help of incentives, in fact, the economical revenue is guaranteed only for niche markets, that is in most of the cases a premium power application.

REFERENCES

1. Fuel Cell Today Bank Knowledge. *PAFC System Survey* (2001). Report downloadable from www.fuelcelltoday.com
2. S. Romano and J.T. Larkins. Georgetown University fuel cell transit bus program. *Fuel Cells Fundam. Syst.* **3(3)** (2003) 128–132.

3. F. Preli. Trends in research – What are we aiming for? *Fuel Cell Science & Technology Conference*, October 6–7, Munich, Germany, 2004.
4. M. Pehnt and S. Ramesohl. Fuel cells for distributed power: benefits, barriers and prospective. *Final Report.* IFEU, Wuppertal Institut (2003).
5. UTC Power Corporation, PC25 data sheet.
6. A. Baker and D. Jollie. Fuel call market survey: large stationary applications. *Fuel Cell Today Bank Knowledge* (2004). www.fuelcelltoday.com
7. M. Cropper. Why is interest in phosphoric acid fuel cells falling? *Fuel Cell Today Bank Knowledge* (2003). www.fuelcelltoday.com
8. L. Carrette, K.A. Friederich and U. Stimming. Fuel cells – fundamentals and applications. *Fuel Cells Fundam. Syst.* **1**(**1**) (2001) 5–39.
9. EG&GT Technical Services Inc. *Fuel Cell Handbook*, 6th edn, US Department of Energy publication (2002).
10. T.G. Benjamin, E.H. Camara and L.G. Marianowski. *Handbook of Fuel Cell Performance.* Prepared by the Institute of Gas Technology for the US Department of Energy, under the contract number EC-77-C-03-1545 (1980).
11. R. Song and D.R. Shin. Influence of CO concentration and reactant gas pressure on cell performance in PAFC. *Int. J. Hydrogen Energ.* **26** (2001) 1259–1262.
12. D.T. Chin and P.D. Haward. Hydrogen sulfide poisoning of platinum anode in phosphoric acid fuel cell electrolyte. *J. Electrochem. Soc.* **133** (1986) 2447–2450.
13. J. Larmine and A. Dicks. *Fuel Cell Systems Explained.* Wiley, UK, 2000.
14. D.S. Chan and C.C. Wan. Influence of PTFE dispersion in the catalyst layer of porous gas-diffusion electrodes for phosphoric acid fuel cells. *J. Power Sources* **50**(**1–2**) (1994) 163176.
15. P. Appleby. In: S. Sarangapani, J.R. Akridge and B. Shumm. *Proceedings of the Workshop of the Electrochemistry of Carbon*, The Electrochemical Society, Inc., Pennington, NJ, 1984, p. 251.
16. J.C. Yang, Y.S. Park, S.H. Seo, H.J. Lee and J.S. Noh. Development of a 50 kW PAFC power generation system *J. Power Sources* **106** (2002) 68–75.
17. E. Passalacqua, P.L. Antonucci, M. Vivaldi, A. Patti, V. Antonucci, N. Giordano and K. Kinoshita. The influence of Pt on the electrooxidation behavior of carbon in phosphoric acid. *Electrochim. Acta* **37**(**15**) (1992) 2725–2730.
18. L. Qingfeng, X. Gang, H.A. Hjuler, R.W. Berg and N.J. Bjerrum. Limiting current of oxygen reduction on gas-diffusion electrodes for phosphoric acid fuel cells. *J. Electrochem. Soc.* **141**(**11**) (1994) 3114–3118.
19. L. Qingfeng, X. Gang, H.A. Hjuler, R.W. Berg and N.J. Bjerrum. Oxygen reduction on gas-diffusion electrodes for phosphoric acid fuel cells by a potential decay method. *J. Electrochem. Soc.* **142**(**10**) (1994) 3250–3255.
20. S.C. Yang, M.B. Cutlip and P. Stonehart. Simulation and optimization of porous gas-diffusion electrodes used in hydrogen/oxygen phosphoric acid fuel cells. I. Application of cathode model simulation and optimization to PAFC cathode development. *Electrochim. Acta* **35**(**5**) (1990) 869–878.
21. M.B. Cutlip, S.C. Yang and Stonehart. Simulation and optimization of porous gas-diffusion electrodes used in hydrogen oxygen phosphoric acid fuel cells. II. Development of a detailed anode model. *Electrochim. Acta* **36**(**3–4**) (1991) 547–553.
22. R.J. Spiegel, J.C. Trocciola and J.L. Preston. Test results for fuel cell operation on landfill gas. *Energy* **22**(**8**) (1997) 777–786.
23. R.J. Spiegel, J.L. Preston and J.C. Trocciola. Fuel cell operation on landfill gas at Penrose power station. *Energy* **24** (1999) 723–742.
24. R.J. Spiegel and J.L. Preston. Technical assessment of fuel cell operation on landfill gas at Groton, CT, landfill. *Energy* **28** (1999) 397–409.
25. SCS Engineering. Comparative analysis of landfill gas utilization technologies. Final report prepared for Northeast Regional Biomass Program CONEG Policy Research Center, Inc., 1997.
26. R.J. Speigel, S.A. Thorneloe, J.C. Trocciola and J.L. Preston. Fuel cell operation on anaerobic digester gas: conceptual design and assessment. *Waste Manage.* **19** (1999) 389–399.
27. U. Langnickel. First European fuel cell application using digester gas. *Fuel Cells Bull.* **3**(**22**) (2000) 10–12.

Chapter 10

International activities in DMFC R&D: status of technologies and potential applications

R. Dillon, S. Srinivasan, A.S. Aricò and V. Antonucci

Abstract

Technological improvements in direct methanol fuel cells (DMFCs) are fuelled by their exciting possibilities in portable, transportation and stationary applications. In this paper, a synopsis of the worldwide efforts resulting in inventions of a plethora of DMFC prototypes with low, medium and high power capacities by a number of Companies, Research Institutions and Universities is presented. The most promising short-term application of DMFC appears to involve the field of portable power sources. Recent advances in the miniaturization technology of DMFCs devices make these systems attractive to replace the current Li-ion batteries. In the field of electrotraction recent demonstration of DMFC stacks with specific power densities and efficiencies approaching those of the combined system methanol reformer-polymer electrolyte fuel cell (PEMFC) have stimulated further investigation on the development of materials with higher performance and lower cost. The most appropriate range of operation temperatures for applications in transportation appears to lie between 100 and 150°C. These operating conditions may be sustained by using new high temperature electrolyte membranes or composite perfluorosulfonic membranes containing inorganic materials with water retention properties at high temperature. The most challenging problem for the development of DMFCs is the enhancement of methanol oxidation kinetics. At present, there are no practical alternatives to Pt-based catalysts. High noble metal loading on the electrodes and the use of perfluorosulfonic membranes significantly contribute to the cost of these devices. Critical areas include the design of appropriate membrane electrode assemblies for specific DMFC applications and the reduction of methanol crossover. This latter aspect is strictly related to the use of membrane alternatives to Nafion, but it may also be conveniently addressed by the development of methanol-tolerant oxygen reduction catalysts.

Keywords: Direct methanol fuel cells; Polymer electrolyte membrane; Electrocatalysis; Electro-traction; Portable power sources

Article Outline

1. Introduction . . . 168
2. Current status of technology and potential applications . . . 168
   2.1. Portable power . . . 168
   2.2. Transportation . . . 175
3. Status of knowledge in basic research areas and needed breakthroughs . . . 179
   3.1. Electrode kinetics and electrocatalysis of methanol oxidation . . . 179

Fuel Cells Compendium

3.1.1. Overall reaction, intermediate steps and rate determining steps 179
3.1.2. Electrocatalysis 180
3.2. Methanol crossover 182
3.2.1. Mechanism and its effects on DMFC performance 182
3.2.2. Methods for inhibition of methanol crossover 182
3.3. Electrode kinetics and electrocatalysis of oxygen reduction 184
4. Conclusions 185
Acknowledgements 185
References 185

1. INTRODUCTION

Most of the world energy requirements are presently addressed by burning fossil fuels in low-efficiency thermal processes. Related consequences in terms of atmospheric pollution, global warming, green house effect, etc. are the objects of many debates between developed countries that are searching for a common legislation to properly restrict the polluting emissions and protect the environment. Transportation represents a significant portion of world energy consumption and contributes considerably to the atmospheric pollution. Although modern cars emit a lower amount of toxic gases and particulate than their older predecessors, their increasing number result in growing levels of pollution from transportation sources.

Reduced levels of transportation-related pollution may be achieved by replacing a significant number of internal combustion engine vehicles with electric cars in the near future. In this regard, polymer electrolyte fuel cells (PEMFCs) and direct methanol fuel cells (DMFCs) have been envisaged as suitable power sources for electric cars. DMFCs which directly employ methanol as fuel have good potentialities since they eliminate the need for a complex reformer unit in the system [1]. Furthermore, since methanol is fed with large amount of water to the anode it also avoids complex humidification and thermal management problems associated to PEMFCs. DMFCs provide the advantage of smaller system sizes and weight in relation to other fuel systems and the concept of the DMFC device may be extended to alternative fuels obtained from natural gas (e.g., dimethylether) or from biomasses as well as fermentation of agricultural products such as ethanol reducing the dependence on insecure energy resources. DMFC devices presently suffer from methanol crossover across polymer electrolyte membranes (crossover affects the performance of the cathode as well as fuel efficiency) and poor methanol electrooxidation kinetics [1]. Other relevant aspects are the cost of materials (noble metal catalysts, perfluorosulfonic membranes) and the cost of production of the various components of the device which are presently higher than conventional energy conversion systems. Automation and large-scale production, however, may significantly reduce the latter. Although DMFC systems have been primarily investigated for their potential use in portable power and electro-traction applications, fundamental research into distributed power sources for residential applications have shown exciting progress. Different applications imply different system design characteristics, operation parameters as well as materials employed in the device. The aim of the present DMFC review is to provide an overview on the international state of the art, recent progress, R&D focal areas and current problems to be solved for the different applications.

2. CURRENT STATUS OF TECHNOLOGY AND POTENTIAL APPLICATIONS

2.1. *Portable power*

Several organizations are actively engaged in the development of low-power DMFCs for cellular phone, laptop computer, portable camera and electronic game applications [2–6]. The initial goal of this research is to develop proof of concept DMFCs capable of replacing high-performance rechargeable batteries in the US$ 6-billion portable electronic devices market. Theoretically, methanol has a superior specific energy density (6000 Wh/kg) in comparison with the best rechargeable battery, lithium polymer and lithium ion polymer (theoretical, 600 Wh/kg) systems. This performance advantage translates into longer conversation times using cell phones, longer times for use of laptop computers between replacement of fuel cartridges and more power available on these devices to support consumer demand. In relation to consumer convenience, another significant advantage of the DMFC over the rechargeable battery is its potential for instantaneous refueling. Unlike rechargeable batteries that require hours for charging a depleted power pack, a DMFC can have its fuel replaced in minutes. These significant advantages make DMFCs an exciting development in the portable electronic devices market. Noteworthy accomplishments in these areas are reported below and in Table 1.

Motorola Labs – Solid State Research Center (US), in collaboration with Los Alamos National Laboratory ((LANL), US), is actively engaged in the development of low-power DMFCs (greater than 300 mW) for cellular phone applications [7]. Motorola has recently demonstrated a prototype of a miniature DMFC (Fig. 1) based on a membrane electrode assembly (MEA) set between ceramic fuel delivery substrates. Motorola utilized their proprietary low-temperature co-fired ceramic (LTCC) technology to create a ceramic structure with embedded microchannels for methanol/water mixing and delivery to the MEA and, exhausting by-product CO_2. In addition, processing of the ceramic material into a grid screen design facilitated the delivery of ambient air to the MEA. Substrates are processed in multiple layers after aligning, tacking and laminating at approximately 3.45×10^6 Pa. The final monolithic integrated ceramic substrate is formed after sintering at 850°C.

In the current design as represented in Fig. 1, the MEA is mounted between two porous ceramic plates. Thin films of electrocatalysts were applied in a proprietary process using carbon cloth gas diffusion layers. For the anode, an unsupported Pt/Ru (1:1) alloy at a high loading of 6–10 mg/cm^2, and for the cathode Pt black were used as electrocatalysts. Nafion 117 membranes were used as the electrolyte and were hydrated by running deionized (DI) water through the cell for 18 h. The active electrode area for a single cell is approximately 3.5–3.6 cm^2. In the stack assembly, four cells are connected in series in a planar configuration with an MEA area of 13–14 cm^2, the cells exhibited average power densities between 15 and 22 mW/cm^2. Four cells (each cell operating at 0.3 V) are required for portable power applications because DC–DC converters typically require 1 V to efficiently step up to the operating voltage for electronic devices. The fuel cell consumed oxygen from ambient air (21°C and 30% RH) and the fuel from 1.0 M methanol pumped at a rate of 0.45 ml/min using a peristaltic pump. Variations in time of operation, temperature, fuel mixing, flow rate and humidity gradually led to improved performance characteristics of the system. In addition, improved assembly and fabrication methods have led to peak power densities greater than 27 mW/cm^2. Motorola is currently improving their ceramic substrate design to include micro-pumps, methanol concentration sensors and supporting circuitry for second generation systems.

Energy Related Devices Inc. (ERD, USA) is working in alliance with Manhattan Scientific Inc., (USA) to develop miniature fuel cells for portable electronic applications [2,8]. A relatively low-cost sputtering method, similar to the one used by the semiconductor industry for production of microchips, is being used for deposition of electrodes (anode and cathode) on either side of a microporous plastic substrate; the micropores (15 nm–20 μm) are etched into the substrate using nuclear particle bombardment. Micro-fuel arrays, with external connections in series, are precisely fabricated and have a thickness of about a millimeter. The principal advantages of the cell include the high utilization of catalyst, controlled pore geometry, low-cost materials and minimum cell thickness and weight. A MicroFuel Cell™ was reported to have achieved a specific energy density of 300 Wh/kg using methanol and water and air as the anodic and cathodic fuels, respectively.

Figure 2 describes the schematic cross section of the fuel cell. The anode design is a critical new advance in the development of a cost-effective pore-free electrode that is only permeable to hydrogen ions.

Table 1. Portable power

Single cell/ stack developer	Power density	Temperature	Oxidant	Methanol concentration (M)	Anode catalyst	Membrane electrolyte	Cathode catalyst	Number of cells/ surface area
Motorola Labs	12–27 mW/cm^2	21	Ambient air	1 (0.45 ml/min)	Pt/Ru alloy, 6–10 mg/cm^2	Nafion 117	6–10 mg/cm^2	4/13–15 cm^2 planar stack
Energy Related Devices	3–5 mW/cm^2	25	Ambient air	1 (pure)	Pt/Ru alloy	Nafion	Pt	Planar stack
Jet Propulsion Labs	6–10 mW/cm^2	20–25	Ambient air	1	Pt/Ru alloy, 4–6 mg/cm^2	Nafion 117	Pt, 4–6 mg/cm^2	6/6–8 cm^2 flat-pack
Los Alamos National Labs	300 W/l	60	Air flowed at 3–5 times stoichiometry	0.5	Pt/Ru, 0.8–16.6 mg/cm^2	Nafion	Pt, 0.8–16.6 mg/cm^2	5/45 cm^2
Forschungszentrum Julich GmbH	45–55 mW/cm^2	50–70	3 atm O_2	1	Pt/Ru, 2 mg/cm^2	Nafion 115	Pt, 2 mg/cm^2	40/100 cm^2 bipolar plate
Samsung Advanced Institute of Technology	10–50 mW/cm^2 (single cell)	25	Ambient air	2–5	Pt/Ru	Hybrid membrane	Pt	12/24 cm^2 monopolar
Korea Institute of Energy Research	121–207 mW/cm^2	25–50	Ambient pressure, O_2 (300 cc/min)	2.5	Pt/Ru/C metal powder	Nafion 115 & 117	Pt-black	6/52 cm^2 bipolar
Korea Institute of Science and Technology	3–9 mW/cm^2	25	Ambient air		Pt/Ru, 8 mg/cm^2	Nafion 117	Pt, 8 mg/cm^2	15/90 cm^2 monopolar
More Energy Ltd	60–100 mW/cm^2	25	Ambient air	30–5% methanol	Pt/Ru	Liquid Electrolyte	Pt	/20 cm^2

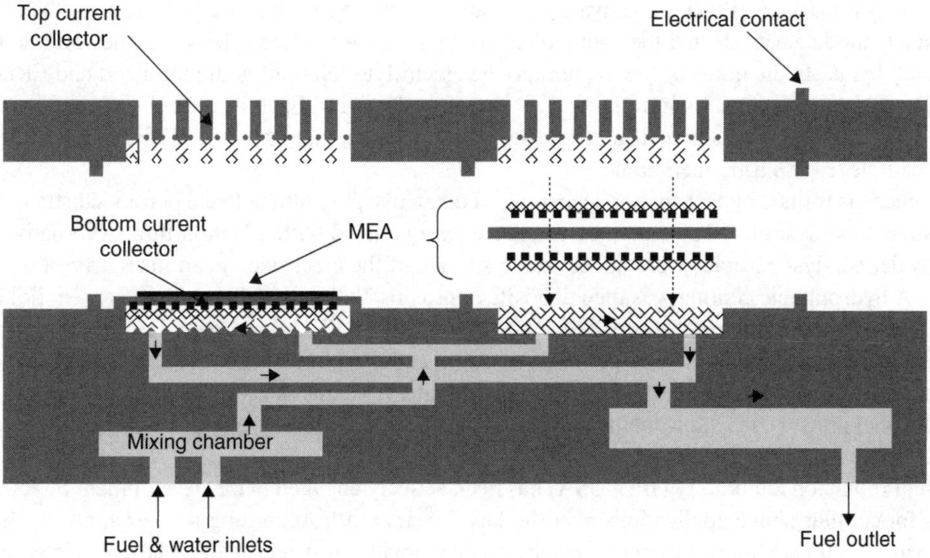

Fig. 1. Schematic of Motorola's miniature DMFC prototype [7]

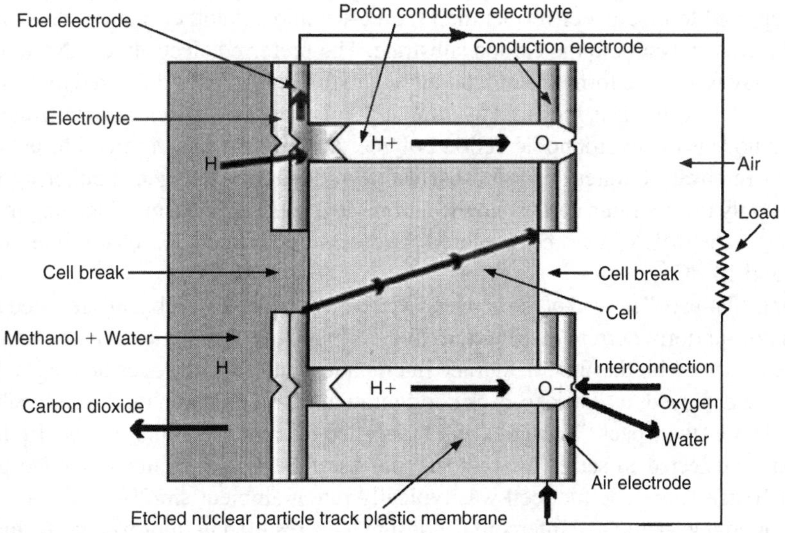

Fig. 2. MicroFuel Cell™ schematic cross-sectional view [2,8]

This increases the efficiency of a methanol fuel cell because it blocks the deleterious effect of methanol crossover across the membrane. The first layer of the anode electrode forms a plug in the pore of the porous membrane; an example is a 20 nm-thick palladium metal film on a Nuclepore filter membrane with 15 nm diameter pores. The second layer (platinum) is deposited to mitigate the hydration-induced cracking that occurs in many of these films. The third layer is deposited over the structural metal film and is the most significant layer because it needs to be catalytically active to methanol and capable of accepting hydrogen ions. An alternate method of forming the electrode is to include on the surface of the metal films powder catalyst particles (Pt/Ru on activated carbon) to enhance the catalytic properties of the electrode. Between the anode electrode and the cathode electrode is the electrolyte-filled pore, the cell interconnect and the cell break. In the pores of the membrane the electrolyte (Nafion) is immobilized and ERD claims this collimated structure results in improved protonic conductivity. Each of the cells is electrically separated from the adjacent cells by cell breaks, useless space occupying the central thickness of the etched nuclear particle track plastic membrane.

The cathode is formed by first sputter depositing a conductive gold film onto the porous substrate followed by a platinum catalyst film. The electrode is subsequently coated with a Nafion film. Alternatively, platinum powder catalyst particles were added to the surface of the electrode via an ink slurry of 5% Nafion solution. A hydrophobic coating was then deposited onto this Nafion layer in order to prevent liquid product water from condensing on the surface of the air electrodes. ERD developed a novel configuration to utilize their fuel cell as a simple charger in powering a cellular phone. The fuel cell is configured into a plastic case that is in close proximity to a rechargeable battery. Methanol is delivered to the fuel cell via fuel needle and fuel ports, which allow methanol to wick or evaporate out into the fuel manifold, and is delivered to the fuel electrodes.

The Jet Propulsion Laboratory (JPL, USA) has been actively engaged in the development of "miniature" DMFCs for cellular phone applications over the last 2 years [4,9]. According to their analysis, the power requirement of cellular phones during the standby mode is small and steady at 100–150 mW. However, under operating conditions the power requirements fluctuates between 800–1800 mW. In the JPL DMFC the anode is formed from Pt–Ru alloy particles, either as fine metal powders (unsupported) or dispersed on high surface area carbon. Alternatively, a bimetallic powder made up of submicron platinum and ruthenium particles was reported to give better results than the Pt–Ru alloy. Another method describes the sputter-deposition of Pt–Ru catalyst onto the carbon substrate. The preferred electrolyte is Nafion 117; however, other materials may be used to form proton-conducting membranes. Air is delivered to the cathode by natural convection and the cathode is prepared by applying a platinum ink to a carbon substrate. Another component of the cathode is the hydrophobic Teflon polymer utilized to create a three-phase boundary and to achieve efficient removal of water produced by electro-reduction of oxygen. Sputtering techniques can also be used to apply the platinum catalyst to the carbon support. The noble metal loading in both electrodes was 4–6 mg/cm^2. The MEA may be prepared by pressing the anode, electrolyte and cathode at 8.62×10^6 Pa and 146°C.

JPL opted for a "flat-pack" instead of the conventional bipolar plate design, but this resulted in higher ohmic resistances and non-uniform current distribution. In this design the cells are externally connected in series on the same membrane, with through membrane interconnect and air electrodes on the stack exterior. Two "flat packs" can be deployed in a back-to-back configuration with a common methanol feed to form a "twin pack" (Fig. 3). Three "twin-packs" in series will be needed to power a cellular phone. In the stack assembly, six cells are connected in series in a planar configuration, which exhibits average power densities between 6 and 10 mW/cm^2. The fuel cell was typically run at ambient air, 20–25°C with 1 M methanol. Improvements of configuration and interconnect design have resulted in improved performance characteristics of the six-cell "flat-pack" DMFC. Based on the results of the current technology, the JPL researchers predict that a 1 W DMFC power source, with the desired specifications for weight and volume and having an efficiency of 20% for fuel consumption, can be developed for a 10-h operating time, prior to replacement of methanol cartridges.

As stated earlier LANL has been in collaboration with Motorola Labs – Solid State Research Center to produce a ceramic-based DMFC, which provides better than 10 mW/cm^2 power density. LANL researchers have also been engaged in a project to develop a portable DMFC power source, capable of replacing the "BA 5590" primary lithium battery, used by the US Army in communication systems [10]. A 30-cell DMFC stack, with electrodes having an active area of 45 cm^2, was constructed, an important feature of this stack being a narrow width (i.e., 2 mm) of each cell. MEAs are made by the decal method; that is, thin film

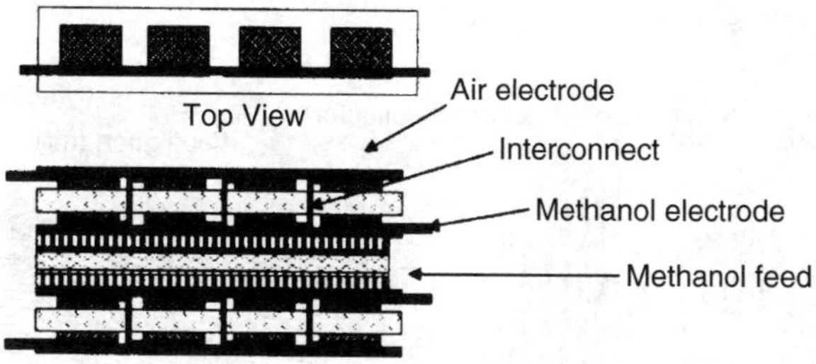

Fig. 3. Schematic of JPL's interconnected cells in a flat-pack DMFC design [4,9]

catalysts bonded to the membrane resulting in superior catalyst utilization and overall cell performance. Anode catalyst loading of Pt between 0.8 and 16.6 mg/cm^2 in unsupported PtRu and carbon-supported PtRu are used. A highly effective flow field for air made it possible to use a dry air blower for operation of the cathode at three to five times stoichiometry. The stack temperature was limited to 60°C and the air pressure was 0.76 atm, which is the atmospheric pressure at Los Alamos (altitude of 2500 m). To reduce the crossover rate, methanol was fed into the anode chamber at a concentration of 0.5 M. Since water management becomes more difficult at such low methanol concentrations, a proposed solution was to return water from the cathode exhaust to the anode inlet, while using a pure methanol source and a methanol concentration sensor to maintain the low methanol concentration feed to the anode. The peak power attained in the stack near ambient conditions was 80 W at a stack potential of 14 V and approximately 200 W near 90°C. From this result, it was predicted that this tight-packed stack could have a power density of 300 W/l. An estimate of an energy density of 200 Wh/kg was made for a 10 h operation, assuming that the weight of the auxiliaries is twice the weight of the stack.

Forschungszentrum Julich GmbH (FJG, Germany) has developed and successfully tested a 40-cell 50 W DMFC stack [11]. The FJG system consists of the cell stack, a water/methanol tank, a pump and ventilators as auxiliaries. The stack is designed in the traditional bipolar plate configuration, which results in lower ohmic resistances but heavier material requirements. To circumvent the weight limitations current collectors are manufactured from stainless steel (MEAs are mounted between current collectors) and are inserted into plastic frames to reduce stack weight. The 6 mm distance between MEAs (cell pitch) reveals very tight packaging of the stack design. Each frame carries two DMFC single cells that are connected in series by external wiring (Fig. 4). MEAs are fabricated in-house with anode loading of 2 mg/cm^2 PtRu black, catalyst loading of 2 mg/cm^2 Pt black and cell area of 100 cm^2 for each of the 40 cells. At the anode a novel construction allows the removal of CO_2 by convection forces at individual cell anodes. The conditions for running the stack were 1 M methanol, 60°C and 3 bar O_2 which led to peak energy densities of 45–55 mW/cm^2. The cathode uses air at ambient or elevated pressures, when the stack operates at temperatures above 60°C the air is fed into the cathode by convection forces. Further evaluation of the system revealed that current collectors made of stainless steel showed an inhomogeneous distribution of contact resistance and as a result single cells displayed fluctuating power densities. It was postulated that the pressure of the current collectors on the MEAs is not high enough to prevent delamination of the electrocatalyst layer. Recent developments include a three-cell short stack design which has reduced the cell pitch to only 2 mm. The individual cell area of this design is larger, 145 cm^2, than the previous prototype and although it is not air-breathing it works with low air stoichiometric rates (more efficient cathodic flow distribution structure). The short stack was tested under ambient pressure (a low-power-consuming compressor provided air to the cathode) and operated at 45°C.

Samsung Advanced Institute of Technology (SAIT, South Korea) has developed a small monopolar DMFC cell pack (2 cm^2, 12 cells, CO_2 removal path, 5–10 M methanol, air-breathing and room temperature) of 600 mW for mobile phone applications [12]. Single cell are constructed with a PtRu black anode

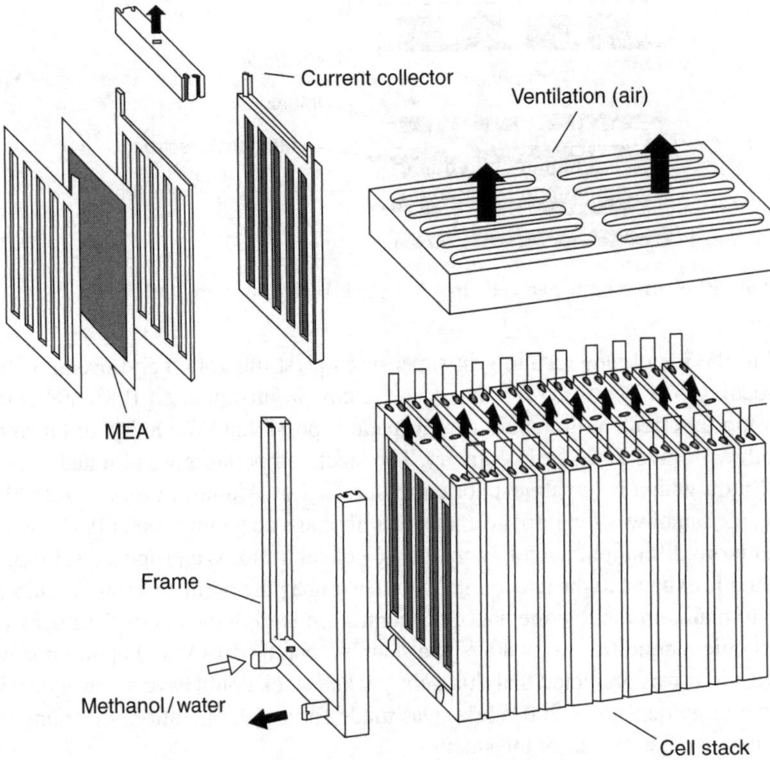

Fig. 4. Schematic of the Forscungszentrum Julich GmbH 50 W DMFC stack [11]

electrode (a methanol flow field and a capillary wicking structure are part of the anode structure), a hybrid membrane material (ionomer/$ZrO(HPO_4)_2$, ionomer/$BaTiO_3$ or ionomer-laminated material) and a Pt black cathode electrode with an air flow field (breathing structure). The catalyst layers have been fabricated by magnetron sputtering methods on Nafion membrane surfaces. Single cell experiments at room temperature utilizing 2–5 M methanol and ambient air result in power densities on the order of 10–50 mW/cm^2 (air-breathing and air blowing). The cellular phone used to test the DMFC prototype is reported to be functional for up to 40 days on standby and 20 h of talk time.

The Korea Institute of Energy Research (KIER, South Korea) has developed a 10 W DMFC stack (bipolar plate, graphite construction) fabricated with six single cells of 52 cm^2 total electrode area [13]. The stack was tested at 25–50°C using 2.5 M methanol supplied without a pumping system and at ambient pressure O_2 at a flow rate of 300 cc/min. The maximum power densities obtained in this system were 6.3 W (121 mW/ cm^2) at 87 mA/cm^2 at 25°C and 10.8 W (207 mW/cm^2) at 99 mA/cm^2 at 50°C. MEAs using Nafion 115 and 117 were formed by hot pressing and the electrodes were produced from carbon-supported Pt–Ru/C metal powders and Pt-black for anode and cathode electrodes, respectively.

Korea Institute of Science and Technology (KIST, South Korea) has developed a 15-cell monopolar stack of 90 cm^2 total electrode area and a maximum power density of 3.2 mW/cm^2 [14]. The MEAs were fabricated by hot pressing catalyst layered carbon paper and Nafion 117 membrane, catalyst layers were formed by spray-coating catalyst ink. Anode electrocatalyst loading of 8 mg PtRu/cm^2 and cathode electrocatalyst loading of 8 mg Pt/cm^2 are utilized in this system. The performance tests were conducted at room temperature under static feed conditions: air was fed only through natural convection and the methanol solution was stored in the engraved plate that contacted the MEA on the anode side. Single cell tests at various methanol concentrations resulted in maximum power density of 9 mW/cm^2. The poor performance of the stack power density was attributed to the poor air diffusion into the cathode with a resulting mass transfer limitation at that electrode.

More Energy Ltd (MEL, ISRAEL), a subsidiary of Medis Technologies Ltd (MDTL, US), is developing a direct liquid methanol (DLM) fuel cells (a hybrid PEM/DMFC system) for portable electronic devices [15]. The key features of the DLM fuel cell are as follows: (i) the anode catalyst extracts hydrogen from methanol directly, (ii) the DLM fuel cell uses a proprietary liquid electrolyte that acts as the membrane in place of a solid polymer electrolyte (Nafion) and (iii) novel polymer and electrocatalyst enable the fabrication of more effective electrodes. The company's fuel cell module delivers approximately 0.9 V and 0.24 W at 60% of its nominal capacity for 8 h. This translates into energy densities of approximately 60 mW/cm^2 with efforts underway to improve that result to 100 mW/cm^2. The high power capacity of the cell is attributed to the proprietary electrode ability to efficiently oxidize methanol. In addition, Medis claims the use of high concentrations of methanol (30%) in its fuel stream with plans for increasing that concentration to 45% methanol. The increased concentration of methanol in the feed stock results in concentration gradients that should lead to higher methanol crossover rates. However, this technical concern is not mentioned in the company's literature.

2.2. Transportation

DMFC technology offers a solution for transportation applications in the transition toward a zero emission future. Using methanol as a fuel circumvents one of the major hurdles plaguing PEMFC technology, that is, the development of an inexpensive and safe hydrogen infrastructure to replace the gasoline/diesel fuel distribution network. It has been well established that the infrastructure for methanol distribution and storage can be easily modified from the current gasoline intensive infrastructure. Another drawback in using PEMFC technology is the need to store hydrogen (at very high pressures) or carry a bulky fuel processor to convert the liquid fuel into hydrogen on board the vehicle. Methanol is an attractive fuel because it is a liquid under atmospheric conditions and its energy density is about half of that of gasoline.

Despite the compelling advantages of using DMFCs in transportation applications, major obstacles to their introduction remain. These barriers include the high costs of materials used in fabricating DMFCs (especially the high cost of platinum electrocatalysts), the crossover of methanol through the electrolyte membrane from the anode to the cathode and, the lower efficiency and power density performance of DMFCs in comparison to PEMFCs. Despite these obstacles a number of institutions (particularly in the last 5 years) have become actively engaged in the development of DMFCs for transport applications. The most remarkable results achieved in this field are summarized in Table 2. These institutions have directed their resources toward improving every facet of the DMFC in the quest for competitive balance with PEMFCs, as stated below.

Ballard Power Systems Inc (BPSI, Canada) in collaboration with Daimler-Chrysler (Germany) recently reported the development of a 3 kW DMFC system that is at a very preliminary stage in comparison to Ballard's PEMFC products [16]. Daimler-Chrysler (Germany) demonstrated this system for the transportation application in a small one-person vehicle at its Stuttgart Innovation Symposium in November 2000. The DMFC go-cart weighed approximately 100 kg, required an 18 V/1 Ah battery system for starting the electric motor on its rear wheels, and had a range of 15 km and a top speed of 35 km/h. The stack used 0.5 l methanol (the concentration of methanol was unclear) as fuel and operated at approximately 100°C. In January 2001, our private communication with Ballard revealed that they have built and operated a 6 kW stack (60 V) based on the same stack design as the prototype shown in Stuttgart. No details are available at this time with respect to the stack design and performance of the DMFC power source. However, the patent literature indicates fabrication techniques for producing DMFC electrodes [17].

Table 2. Transportation

Single cell/ stack developer	Power/power density	Temperature (°C)	Oxidant	Methanol concentration (M)	Anode catalyst	Membrane electrolyte	Cathode catalyst	Number of cells/surface area (cm^2)
Ballard Power Systems Inc	3 Kw	100	Air	1 (pure)	Pt/Ru	Nafion	Pt	—
IRD Fuel Cell A/S	100 mW/cm^2	90–110	1.5 atm air	—	Pt/Ru	Nafion	Pt	4/154 cm^2 bipolar
Thales CNR-ITAE Nuvera Fuel Cells	140 mW/cm^2	110	3 atm air	1	Pt/Ru	Nafion	Pt	5/225 cm^2 bipolar
Siemens AG	250 mW/cm^2 (90)	110 (80)	3 atm O_2 (1.5 atm air)	0.5 (0.5)	Pt/Ru	Nafion 117	Pt-black 4 mg/cm^2	3 cm^2 per cell
Los Alamos National Labs	1 kW/l	100	3 atm air	0.75	Pt–Ru	Nafion 117	Pt	30/45 cm^2 bipolar

The anode was prepared by first oxidizing the carbon substrate (carbon fiber paper or carbon fiber non-woven) via electrochemical methods in acidic aqueous solution (0.5 M sulfuric acid) prior to incorporation of the proton-conducting ionomer. Oxidation results in the formation of various acidic surface oxide groups on the carbonaceous substrate and can be achieved by constructing a simple electrochemical cell comprising the carbonaceous electrode substrate as the working electrode. During the treatment of the carbon substrate a voltage of greater than 1.2 V and more than 20 coulombs/cm^2 was used in the process. The second step involves the impregnation of a proton-conducting ionomer such as a poly(perfluorosulfonic acid) into the carbon substrate and then drying off the carrier solvent; the amount impregnated into the substrate was usually greater than 0.2 mg/cm^2. The anode preparation is completed by applying aqueous electrocatalyst ink to the carbon substrate without extensive penetration in the substrate. This method ensures that less electrocatalyst is used and, the catalyst is applied to the periphery of the electrode where it will be utilized more efficiently. The performance enhancements associated with the treatment of the carbonaceous substrate may be related to the increase in the wettability of the carbonaceous substrate. This may result in the more intimate contact of an ionomer coating with the electrocatalyst thereby improving proton access to the catalyst. Another theory concludes that the presence of the acidic groups on the carbon substrate itself may improve proton conductivity or, the surface active acidic groups may affect the reaction kinetics at the electrocatalyst sites. The assembly of the MEA and single cell occurred via conventional methods, that is, hot pressing the anode and cathode to a solid polymer membrane electrolyte. Oxygen and methanol flow fields are subsequently pressed against cathode and anode substrates, respectively but details of this assembly have not been forthcoming.

IRD Fuel Cell A/S (Denmark) has developed DMFCs primarily for transportation applications (0.7 kW) [18]. The stack was constructed with separate water and fuel circuits and the bipolar flow plates are made of a special graphite/carbon polymer material for corrosion reasons. The MEAs have an active cell area of 154 cm^2 with cell dimensions of 125 mm^2. The air pressure was 1.5 bar at the cathode. A nominal cell voltage of 0.5 V was observed for IRDs stack at a current density at 0.2 A/cm^2 and electric power was generated at 15 W per cell.

A consortium composed of Thales-Thompson (France), Nuvera Fuel Cells (Italy), LCR (France) and Institute CNR–ITAE (Italy) has developed a five-cell 150 W stainless steel-based air-fed DMFC stack with financial support of the European Union Joule Program [19]. Bipolar plates were utilized in the stack design and MEAs were fabricated using Nafion as the solid polymer electrolyte and high surface area carbon-supported Pt–Ru and Pt electrocatalyst for methanol oxidation and oxygen reduction, respectively. The electrode area was 225 cm^2 and stack was designed to operate at 110°C, using 1 M methanol and 3 atm air achieving an average power density of 140 mW/cm^2. Figure 5 shows the overall stack performance. A comparison of the polarization curves for single cells in the stack and a prototypal cell is shown in Figure 6. The different diffusion characteristics of the cells in the stack indicate that the stack fluidodynamics should be enhanced in terms of homogeneity of distribution of reactant over the electrodes.

Siemens Ag (Germany) optimized its DMFC system (high oxygen pressure operation) for a niche market and, examined DMFCs in the low temperature, low pressure air operation for more general purposes [20]. MEAs in single cells experiments are constructed using a Nafion 117 membrane, Pt-black with a catalyst loading of 4 mg/cm^2 for the cathode and a high surface area Pt–Ru alloy (either unsupported or carbon supported) for the anode (2 mg/cm^2). A maximum power density of ~250 mW/cm^2 is achieved for operating conditions of 110°C, 3 bar O_2, 0.5 M methanol and an electrode surface area of 3 cm^2. Single cell experiments exploring operating conditions at lower temperatures, lower pressures and air being supplied to the cathode electrode utilize similar MEA components as described previously. A maximum power density of ~90 mW/cm^2 is achieved for operating conditions of 80°C, 1.5 bar air and 0.5 M methanol. These conditions result in a maximum power density that is significantly lower than results obtained for previous experiments using O_2 as the cathodic fuel. We should also note that there is a positive correlation between the air flow rate (25–100 standard cubic centimeter per minute (sccm)) and the cell performance.

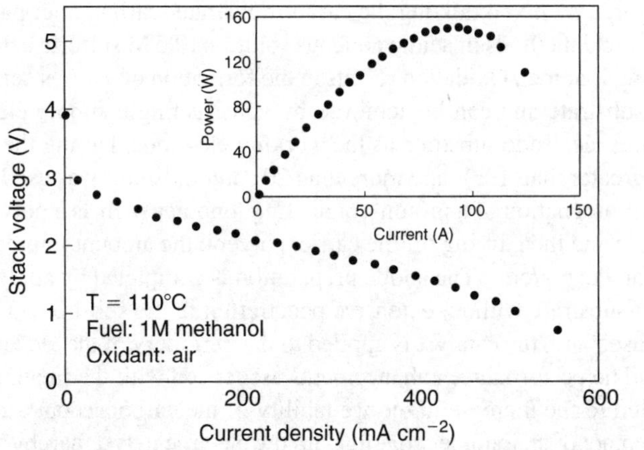

Fig. 5. Galvanostatic polarization and power density for a five-cell air-feed DMFC stack at 110°C. Electrolyte: Nafion 117. Catalysts: 85% Pt–Ru/C and 85% Pt/C; 2 mg Pt cm^{-2}; methanol 1 M, electrode surface 225 cm^2 [19]

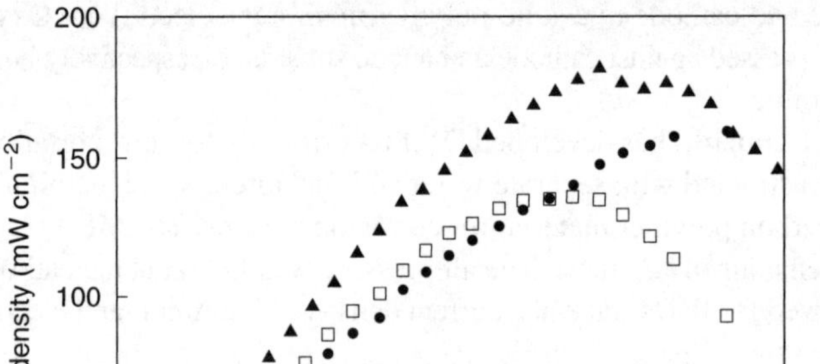

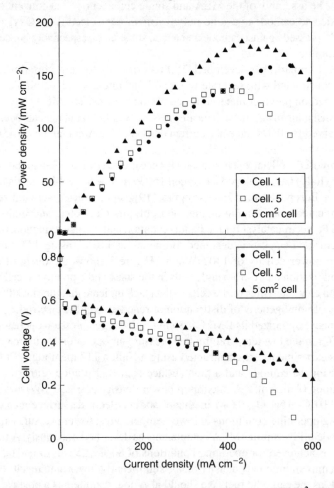

Fig. 6. Galvanostatic polarization data and power densities at 110°C for two 225 cm^2 cells along a 150 W air-fed DMFC stack section (from the reactant inlet) and comparison with a 5 cm^2 graphite single cell operating under same conditions and equipped with the same M&E assembly. Electrolyte: Nafion 117. Catalysts: 85% Pt–Ru/C; 2 mg Pt cm^{-2}; methanol 1 M [19]

Siemens AG in Germany, in conjunction with IRF A/S in Denmark and Johnson Matthey Technology Center in the United Kingdom has developed a DMFC stack with an electrode area of 550 cm^2 under the auspices of the European Union Joule Program [21–23]. The projected cell performance is a potential of 0.5 V at a current density of 100 mA/cm^2, with air pressure at 1.5 atm and the desirable stoichiometric flow rate. A three-cell stack has been demonstrated by operating at a temperature of 110°C and a pressure of 1.5 atm and using 0.75 M methanol, this stack exhibited a performance level of 175 mA/cm^2 at 0.5 V per cell; at 200 mA/cm^2 the cell potential was 0.48 V. These performances were obtained at a high stoichiometric air flow rate (factor of 10) but in order to reduce auxiliary power requirements, one of the goals at Siemens is to improve the design to lower the air stoichiometric flow to the desired value of about a factor of 2. A 0.85 kW air-fed stack composed of 16 cells and operating at 105°C was successively demonstrated with maximum power density of 100 mW/cm^2.

LANL is also actively pursuing the design and development of DMFC cell stacks for electric vehicle applications. According to the latest available information, a five-cell short stack with an active electrode area of 45 cm^2 per cell has been demonstrated [24]. The cells were operated at 100°C, an air pressure of 3 atm and a methanol concentration of 0.75 M. The maximum power of this stack was 50 W, which corresponds to a power density of 1 kW/l. At about 80% of the peak power, the efficiency of the cell stack with respect to the consumption of methanol was 37%.

3. STATUS OF KNOWLEDGE IN BASIC RESEARCH AREAS AND NEEDED BREAKTHROUGHS

3.1. Electrode kinetics and electrocatalysis of methanol oxidation

3.1.1. Overall reaction, intermediate steps and rate determining steps

Since methanol is the most electroactive organic fuel for fuel cells, extensive, fundamental studies have been carried out to elucidate the reaction mechanism in a multitude of laboratories in the USA, England, Russia, France, Germany and Japan starting as early as the 1960s. Two recent reviews cover in detail the analyses of the reaction mechanisms as well as the conflicting views that still exist on reaction pathways and rate determining steps [25,26]. Thus, only a summary of the present status of knowledge and a survey of the current international activities is presented in this section. The electrooxidation of methanol to carbon dioxide is a six-electron transfer reaction; due to the slow kinetics of this reaction (even on the best possible electrocatalysts) and poisoning by adsorbed intermediates, partial oxidation to products such as formaldehyde, formic acid and methyl formate occurs. A generally accepted schematic for the reaction pathways leading to the partial or complete oxidation of methanol on Pt–Ru catalysts is reported below:

$$CH_3OH + Pt \rightarrow Pt\text{ - - - }CH_2OH + H^+ + le^- \qquad (1)$$

$$Pt\text{ - - - }CH_2OH + Pt \rightarrow Pt\text{ - - - }CHOH + H^+ + le^- \qquad (2)$$

$$Pt\text{ - - - }CHOH + Pt \rightarrow PtCHO + H^+ + le^- \qquad (3)$$

A surface rearrangement of the methanol oxidation intermediates gives carbon monoxide, linearly bonded to Pt sites, as following:

$$PtCHO \rightarrow Pt\text{ - - - }CO + H^+ + le^- \qquad (4)$$

In the presence of Ru as promoter, water discharging occurs at low anodic overpotentials on Ru with the formation of Ru-OH species at the catalyst surface.

$$Ru + H_2O \rightarrow RuOH + H^+ + 1e^- \quad (5)$$

The final step is the reaction of Ru-OH groups with neighboring methanolic residues to give carbon dioxide:

$$RuOH + PtCO \rightarrow Pt\text{ - - - }Ru + CO_2 + H^+ + 1e^- \quad (6)$$

On the better electrocatalysts, such as a Pt–Ru alloy, CO_2 is the main product, while on an inferior electrocatalyst such as Pt small amounts of formic acid and formaldehyde have been detected. In situ product analyses have been carried out by on-line gas or liquid chromatography by Lamy and co-workers [27] and differential electrochemical mass spectrometry by Vielstich et al., Bonn University [28]. Another method used by Lamy and co-workers is in situ Fourier transform infrared reflectance spectroscopy, used as single potential alteration infrared spectroscopy [29]. A similar technique, electrochemically modulated infrared spectroscopy, was used by this group to identify intermediate species strongly adsorbed on the electrode [29]. The strongly adsorbed CO species was identified as the main poisoning species, blocking the electrode sites for further intermediate formation during methanol oxidation. The vital step appears to be the formation of the $(^{\bullet}CHO)_{ads}$ species, which facilitates the overall reaction; it subsequently forms the strongly adsorbed CO species. The rate determining step is the oxidation of adsorbed CO with adsorbed OH species, according to the publications of Swathirajan and Mikhail at General Motors [30], Bockris and Kahn, Texas A&M University [31] and Kauranen et al. [32].

Increase in operation temperature produces a significant increase in DMFCs performance. This effect is nearly related to the faster CO removal from electrocatalyst surface as revealed by CO stripping analysis [33]. Figure 7 shows DMFC polarization curves in the presence of a Pt–Ru anode catalyst and Nafion 112 membrane at various operating temperatures.

3.1.2. Electrocatalysis

The interest in the electrocatalytic oxidation of methanol is only surpassed by investigations of the hydrogen and oxygen electrode reactions. Platinum and platinum–ruthenium alloys have been the most investigated electrocatalysts. For a more complete review of electrocatalyst studies and the conclusions reached, the reader is referred to the aforementioned review articles. Highlights of recent and ongoing investigations may be summarized as follows: (i) With a Pt–Ru alloy electrocatalyst, the water discharge occurs at low potentials on Ru sites while methanol chemisorption requires three neighboring Pt sites. The removal of carbon monoxide needs the presence of OH species on adsorbed Ru sites. According to work at LANL by Dinh et al. [33], the above processes are accelerated by the presence of low index planes. (ii) A ruthenium content of 50% is optimal for methanol oxidation. (iii) X-ray absorption spectroscopic studies by McBreen and Mukerjee at Brookhaven National Laboratory [34] have shown that an increase in d band vacancy is produced by alloying Pt with Ru and this modifies the adsorption energy of methanol residues on Pt. Thus, the reaction rate is not only influenced by the bifunctional mechanism but also by electronic effects. (iv) Promotional effects of Ru and Sn with Pt have also been extensively analyzed – Aricò et al. [35], at the CNR–TAE Laboratory in Italy observed a shift by 1.1 eV in the X-ray near edge spectrum (XANES) of the Pt–Sn/C electrocatalyst, which suggests that Sn atoms in Pt–Sn donate electrons to Pt atoms and are thereby oxidized. A charge transfer from Sn to Pt was also shown using XPS analysis by Shukla and co-workers in an International Collaboration with CNR–ITAE in Italy, Indian Institute of Science and Seoul National University Korea [36]. (v) Iwasita et al., Bonn University [37], have attributed the shift toward higher frequencies for CO stretching in their FTIR experiments to a lower chemisorption energy for CO on the Pt–Ru alloy. (vi) In a related study at Eindhoven University, The Netherlands, Frelink et al. [38] have observed a shift to higher frequencies at various coverage's due to changes in binding energies to the alloy surface, induced by Ru through a ligand effect on Pt. (vii) A combinatorial catalytic approach

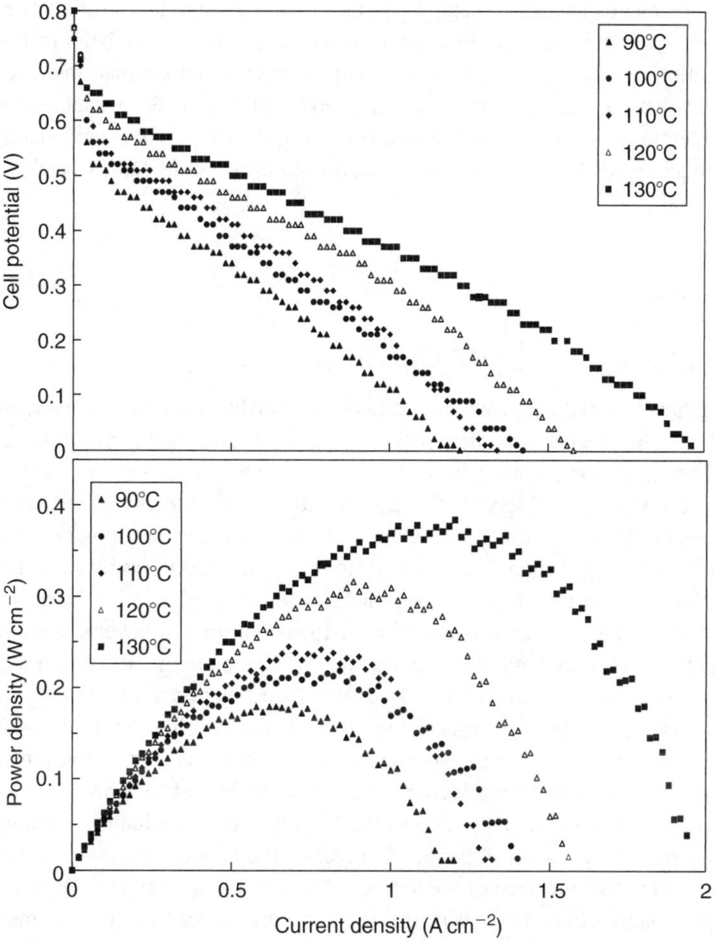

Fig. 7. Galvanostatic polarization data for a DMFC equipped with CNR–ITAE Pt–Ru (anode) and E-TEK 30% Pt/C (cathode) catalysts; 2 M CH_3OH, oxygen feed

was used at Pennsylvania State University and the Illinois Institute of Technology [39] and (viii) Several studies have been carried out and are ongoing as well to elucidate the morphologic aspects of electrocatalysts. Significant information has also been obtained on the behavior of ternary as well as multifunctional catalysts which aided the interpretation of the role of the various promoting elements [40].

Watanabe et al. [41], Yamanashi University, reported that the electrocatalytic activity for methanol oxidation does not increase with particle size above 20 Å. This means that the mass activity increases with the increasing dispersion of the electrocatalyst. Wieckowski and co-workers [42] at the University of Illinois found that the (1 1 1) Pt crystallographic plane, partially covered with Ru ad-atoms performs better than when Ru is adsorbed on any other plane. (ix) Studies to investigate the role of carbon black have shown, in the work of Kaurenan and Skou [43], Ravikumar and Shukla [44] from the Indian Institute of Science, and Goodenough et al. [45] New South Wales, that (a) a low surface area carbon black (e.g., acetylene black) does not yield a high dispersion of metal phase, (b) a high surface area carbon black (e.g., Vulcan XC-72, Ketjen Black) accommodates a high amount of metal phase with a high degree of dispersion due to the significant amount of micropores. However, there will be no homogeneous distribution of the electrocatalyst – this leads to mass transport limitations of reactant and limited access to inner electrocatalytic sites. (x) Several alternatives to Pt or Pt alloy electrocatalysts have been investigated worldwide. These include transition metal alloys and transition metal oxide/metal combinations. The latter are attractive because they could assist the decomposition of water and facilitate a redox route for electrooxidation of metal. To date, problems encountered with the stabilities of these materials in acid electrolytes have arisen. However, this approach needs further examination in the state-of the-art DMFC with a proton-conductive membrane electrolyte.

3.2. Methanol crossover

3.2.1. Mechanism and its effects on DMFC performance

The crossover of methanol from the anode to the cathode in a DMFC has serious consequences of reducing its coulombic and voltage efficiencies. The main reason for the crossover is that the methanol fuel is soluble in water over the full range of composition from 0% to 100%. This is unlike the case of gaseous hydrogen and oxygen fuels oxidized and reduced at the anode and cathode electrodes, respectively. As a consequence, the diffusion rate of methanol from the anode to the cathode is extremely high (corresponding to an equivalent current loss larger than 100 mA/cm^2 under open circuit conditions). This should be compared with a diffusion rate that results in an equivalent current loss of a few mA/cm^2 or less for hydrogen or oxygen in a proton exchange membrane fuel cell. Investigations in several laboratories in the USA, Japan, Canada, Korea and Germany have examined the extent of methanol crossover in DMFC's as a function of operating temperatures and current density using electrochemical on-line gas chromatographic and mass spectrometry techniques [46–50]. Apart from the high rate of diffusion transport from anode to cathode, one encounters the electroosmotic transport whereby methanol is carried with the proton (ion–dipole interaction) as in the case of a water molecule being strongly bound to a proton. The rate of crossover decreases with increasing current density, due to the higher rate of methanol consumption at the anode. This induces a concentration gradient in the active layer of the anode electrode and a considerably lower methanol concentration at the interface of the active layer with the membrane. Higher operating temperatures and a lower methanol concentration in the feed stream reduces the rate of methanol crossover in DMFCs. As a compromise for optimizing performance and reducing crossover, most researchers are using a concentration of 1 M methanol. Apart from the crossover problem reducing the coulombic efficiency of a DMFC, its voltage efficiency is also decreased because of a lowering of the open circuit potential (OCV) (caused by the depolarizing of the oxygen electrode under open circuit conditions) and the poisoning effects of the Pt electrocatalyst by methanol-derived species at this electrode.

3.2.2. Methods for inhibition of methanol crossover

The crossover of methanol is due to its high rate of permeability through the membrane, caused by the high concentration gradient of methanol from the anode to the cathode. Several projects have been and are being carried out to minimize the permeation rate. Researchers at LANL [51] in the USA have shown that the permeation rate is markedly reduced at current densities above 300 mA/cm^2. Researchers at CNR-ITAE [52] have also shown reduced methanol crossover rates with thinner membrane electrolytes and higher temperature operations when the DMFC cell is working at high current density. Modified composite membranes (Nafion–silica), with SiO_2 particles entrapped in the polymeric structure [53], serves as a physical barrier for methanol crossover; even though the ohmic resistance is increased (depends on the concentration of silica). Low crossover rate (equivalent to 40 mA/cm^2, at a DMFC operating current density of 500 mA/cm^2) have been demonstrated in the presence of Nafion–Silica composite membranes [54]. At Pennsylvania State University in the USA, Allcock and co-workers [55] are investigating phosphazene membranes, prepared by thin film casting of a poly(aryloxy) phosphazene from a solution of tetrahydrofuran, phosphorus oxychloride and water. The microporous membrane contained either phosphoric acid entrapped in the membrane or the acid coordinately bound in the polymer backbone. Another type of membrane was prepared by treating benzenoid side groups (the side groups are attached to a polyphosphazene mainchain) with sulfonating agents such as concentrated sulfuric acid, fuming sulfuric acid and sulfur trioxide. Although lower crossover rates (by a factor of 5), as compared with Nafion membranes, were reported the ohmic resistances were higher. Savinell and coworkers [56,57] at Case Western Reserve University imbibed polybenzimidazole membranes with a large amount of phosphoric acid (about 400% content) and found low crossover rates at a temperature of 150–200°C. Such a membrane in a DMFC environment exhibited a crossover rate of 5–10%. This membrane is akin to a silicon carbide matrix impregnated with phosphoric acid in a phosphoric acid fuel cell International Fuel Cells, USA, had evaluated a fuel cell with the latter type of membrane and found that some methanol was consumed to form methyl phosphate and dimethyl ether. Researchers at Samsung Advanced Institute of Technology, Korea [58], are evaluating inorganic–organic hybrid polymer membranes – mainly composites of Nafion with silica, TiO_2 and zirconyl phosphate, prepared by hydrolysis or sol–gel reactions.

In an international cooperation between Princeton University and CNR-ITAE [59], a composite Nafion 115 zirconium hydrogen phosphate (23%, w/w) membrane was investigated for application in DMFC at high temperatures (150°C). This membrane shows lower methanol crossover with respect to recast Nafion–SiO_2 (3%, w/w) membrane due to the higher content of inorganic compound inside the polymer electrolyte channels acting as diffusion barrier for methanol (Fig. 8). Yet, larger ohmic resistances were observed up to 150°C due to the reduced proton/water mobility inside Nafion channels (Fig. 9). The resulting effect is a better performance for zirconium hydrogen phosphate based membrane in the activation controlled region and a lower performance in the ohmic and diffusion controlled region of the polarization curve with respect to the SiO_2 based membrane (Fig. 8). Further, increase in performance for the Nafion–SiO_2 membrane is achieved by adsorbing a strong acid on the surface of colloidal SiO_2 particles indicating that the surface acidity of the inorganic oxide is probably governing the conductivity and water retention properties at high temperature (Fig. 9). The highest conductivity for the Nafion–SiO_2 membrane is achieved

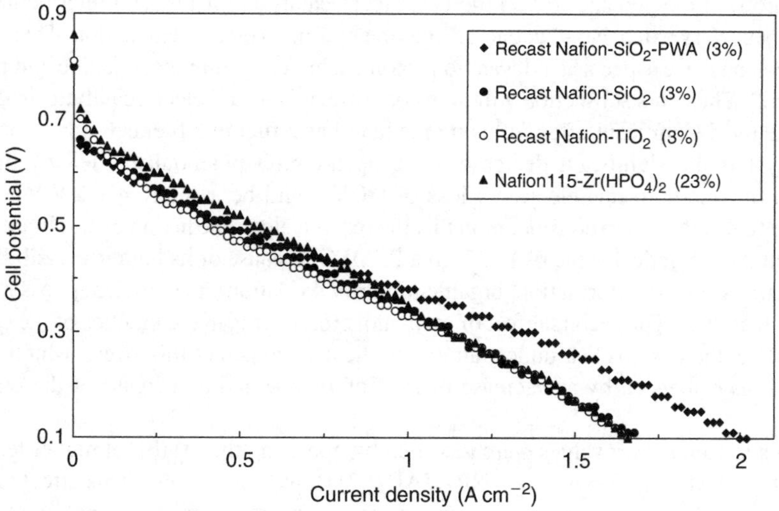

Fig. 8. Polarization curves for DMFC equipped with different composite Nafion membranes and operating at 145°C. Catalysts: 60% Pt–Ru/C and 30% Pt/C (E-TEK); oxygen feed, methanol 2 M

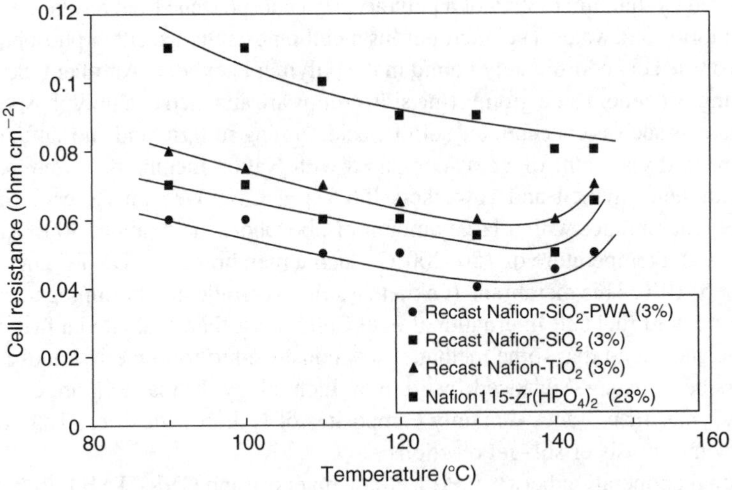

Fig. 9. Variation of cell resistance with temperature during DMFC operation with different composite Nafion membranes

at 145°C; at higher temperatures water losses determine an increase in the cell resistance. As opposite, the Nafion 115-zirconium hydrogen phosphate membrane shows a progressive decrease of cell resistance (Fig. 9). In fact, in the presence of $ZrH(PO_4)_3$ which is a protonic conductor at intermediate temperatures, the proton mobility inside the system is strongly activated by temperature in presence of reduced water content inside the membrane.

3.3. Electrode kinetics and electrocatalysis of oxygen reduction

Worldwide research activities to elucidate the mechanism of the complex electrode reaction (both evaluation and reduction) of oxygen and to find the best electrocatalysts for low and intermediate temperature fuel cells have been very extensive – second most to the hydrogen electrode reaction. There have also been excellent reviews on these aspects as relevant to proton exchange membrane, alkaline and phosphoric acid fuel cells [60–62]. Thus, this subsection will focus on mechanistic and electrocatalytic aspects, as relevant to DMFCs. The most significant effect of using methanol as a fuel in a fuel cell on the oxygen electrode Pt electrocatalyst is the significant decrease in its open-circuit potential, caused by the crossover of methanol from the anode to this electrode; loss of OCV could be as high as 0.2 V when using 1–2 M methanol as the fuel. Taking into consideration that the oxygen electrode has an open-circuit potential about 0.2 V lower than the theoretical value of 1.23 V in a PEMFC, because of its high irreversibility and competing anodic reactions (Pt oxide formation, organic impurity oxidation), the efficiency loss under these conditions is as high as 30%. The second effect of methanol crossover is in the kinetics of the electroreduction of oxygen. To date, there is no clear understanding of the mechanism of this effect, which slows down the rate of reaction, as evidenced by an increase of the Tafel slope and a decrease of the exchange current density.

In some cases low crossover values were recorded by operating the DMFC at higher temperatures – as illustrated by the LANL [51] (USA) and CNR-ITAE [52] (Italy) researchers. This effect is mainly related to the higher achievable current densities which produces a fast methanol consumption at the anode/electrolyte interface and thus a lower methanol concentration gradient across the electrolyte. As stated in the preceding subsection, the crossover can be reduced by use of alternate or composite membranes. Apart from the as-mentioned effects of these methods to increase the coulombic efficiencies, other benefits are the increase of the open-circuit potential of the oxygen electrode and the improvement of the kinetics of oxygen reduction. The latter is probably due to the decreased concentration of organic species (derived from methanol) adsorbed on the electrode. Research studies have revealed that Co and Fe porphyrins and phthalocyanins electrocatalysts are insensitive to the presence of methanol, when functioning as oxygen electrodes [63–66]. These electrocatalysts were supported on a high surface area carbon (Vulcan XC 72R) and thermally treated at 800°C. An alternative approach, used by these workers was to disperse these metal-organic macrocyclics in a film of a conducting polymer (polyaniline or polypyrrole) [67].

4. CONCLUSIONS

An analysis of the international activities carried out over the last 2 years in the field of DMFC stack and system development technology has been made. It is widely recognized that to reduce greenhouse gases and obey recent environmental laws it is necessary to develop highly efficient and low-cost energy conversion systems. Direct methanol fuel cells possess good potentialities in this regard due to intrinsically low polluting emissions and system simplicity. Recent results on DMFC stacks in terms of power density output ($\approx$1 kW/l) and overall conversion efficiency (37% at the design point of 0.5 V/cell) indicate that these systems are quite competitive with respect to the reformer-H_2/air PEMFC units for application in electrotraction as well as in distributed power generation. Yet, significant progress is necessary to further decrease the gap that still exists with respect to conventional power generation systems in terms of power density and costs. The major hurdles concern the reduction of noble metal loading, methanol crossover drawbacks and fabrication costs. At present, the most appealing application for DMFCs is in the field of portable power sources where device costs are less critical and power densities are close to those of Li-batteries. The present analysis indicates that the targets for each application may be achieved through a thoughtful development of materials device design as well as through an appropriate choice of operating conditions.

ACKNOWLEDGEMENTS

We would like to thank V. Baglio, P. Cretì, A. Di Blasi, E. Modica, G. Monforte, F. Urbani of CNR-ITAE, P.L. Antonucci (University of Reggio Calabria, Italy) and C. Yang (Princeton University, US) for their cooperation. We would also like to thank Mai Pham-Thi (Thomson-LCR France), E. Ramunni (De Nora, Italy), R. Ornelas (Nuvera Fuel cells-Europe), R. Gille, D. Buttin, M. Straumann, M. Dupont and E. Rousseau (Thales, France) and all other members of the Nemecel EU Joule project.

REFERENCES

1. K. Kordesch and G. Simader. *Fuel Cells and their Applications.* Wiley-VCH, Weinheim, 1996.
2. R.G. Hockaday, M. DeJohn, C. Navas, P.S. Turner, H.L. Vaz and L.L. Vazul. In: *Proceedings of the Fuel Cell Seminar*, Portland, OR, USA, October 30– November 2, 2000, p. 791.
3. S.C. Kelley, G.A. Deluga and W.H. Smyrl. *Electrochem. Solid-State Lett.* **3** (2000) 407.
4. S.R. Narayanan, T.I. Valdez and F. Clara. In: *Proceedings of the Fuel Cell Seminar*, Portland, OR, USA, October 30–November 2, 2000, p. 795.
5. X. Ren, P. Zelenay, S. Thomas, J. Davey and S. Gottesfeld. *J. Power Sources* **86** (2000) 111.
6. D.-H. Jung, Y.-K. Jo, J.-H. Jung, S.-Y. Cho, C.-S. Kim and D.-R. Shin. In: *Proceedings of the Fuel Cell Seminar*, Portland, OR, USA, October 30–November 2, 2000, p. 420.
7. J. Bostaph, R. Koripella, A. Fisher, D. Zindel and J. Hallmark. In: *Proceedings of the 199th Meeting on Direct Methanol Fuel Cell*, Electrochemical Society, Washington, DC, USA, March 25–29, 2001.
8. R.G. Hockaday. US Patent No. 5 759 712 (1998).

9. C.K. Witham, W. Chun, T.I. Valdez and S.R. Narayanan. *Electrochem. Solid-State Lett.* **3** (2000) 497.
10. S. Gottesfeld, X. Ren, P. Zelenay, H. Dinh, F. Guyon and J. Davey. In: *Proceedings of the Fuel Cell Seminar*, Portland, OR, USA, October 30–November 2, 2000 799.
11. H. Dohle, J. Mergel, H. Scharmaan and H. Schmitz. In: *Proceedings of the 199th Meeting Direct Methanol Fuel Cell Symposium*, Electrochemical Society, Washington, DC, USA, March 25–29, 2001.
12. H. Chang. In: *The Knowledge Foundation's Third Annual International Symposium on Small Fuel Cells and Battery Technologies for Portable Power Applications*, Washington, DC, USA, April 22–24, 2001.
13. D.-H. Jung, Y.-K. Jo, J.-H. Jung, S.-H. Cho, C.-S. Kim and D.-R. Shin. *Proceedings Fuel Cell Seminar*, Portland, OR, USA, October 30–November 2, 2000, p. 420.
14. H.Y. Ha, S.Y. Cha, I.-H. Oh and S.-A. Hong. In: *Proceedings Fuel Cell Seminar*, Portland, OR, USA, October 30–November 2, 2000, p. 175.
15. R.F. Lifton. In: *The Knowledge Foundation's Third Annual International Symposium on Small Fuel Cells and Battery Technologies for Portable Power Applications*, Washington, DC, USA, April 22–24, 2001.
16. D. Harris. Ballard Power Systems Inc. *News Release*, November 9, 2000.
17. J. Zhang, K.M. Colbow and D.P. Wilkinson. US Patent No. 6 187 467 (2001).
18. http://www.ird.dk/product.htm
19. D. Buttin, M. Dupont, M. Straumann, R. Gille, J.C. Dubois, R. Ornelas, G.P. Fleba, E. Ramunni, V. Antonucci, V. Arico, P. Creti, E. Modica, M. Pham-Thi and J.P. Ganne. *J. Appl. Electrochem.* **31** (2001) 275.
20. M. Baldauf and W. Preidel. *J. Power Sources* **84** (1999) 161.
21. M. Baldauf, M. Frank, R. Kaltschmidt, W. Lager, G. Luft, M. Poppinger, W. Preidel, H. Seeg, V. Tegeder, M. Odgaard, U. Ohlenschloeger, S. Yde-Andersen, M.H. Lindic, T. Drews, C.V. Nguyen, I. Tachet, E. Skou, J. Engell, J. Lundsgaard, M.P. Hogarth, G.A. Hards, I. Kelsall, B.R.C. Theobald, E. Smith, D. Thompsett, A. Gunner and N. Walsby. *Publishable Final Report of The European Commission Joule III Programme*, 1996–1999.
22. M. Baldauf and W. Preidel. Book of abstracts. In: *Proceedings of the Third International Symposium on Electrocatalysis: Workshop, Electrocatalysis in direct and indirect methanol PEM fuel cells*, Portoroz, Slovenia, September 12–14, 1999.
23. M. Baldauf and W. Preidel. *J. Appl. Electrochem.* **31** (2001) 781.
24. X. Ren, P. Zelenay, S. Thomas, J. Davey and S. Gottesfeld. *J. Power Sources* **86** (2000) 111.
25. C. Lamy, J.-M. Leger and S. Srinivasan. In: J.O'M. Bockris, B.E. Conway and R.E. White (Eds.), *Modern Aspects of Electrochemistry*, Vol. 34, 2001, p. 53.
26. A.S. Arico, S. Srinivasan and V. Antonucci. *Fuel Cells* **1** (2001) 133.
27. B. Beden, C. Lamy and J.-M. Léger. Electrocatalytic oxidation of oxygenated aliphatic organic compounds at noble metal electrodes. In: J.O'M. Bockris, B.E. Conway, R.E. White (Eds.), *Modern Aspects of Electrochemistry*, Vol. 22. Plenum Press, New York, 1992, p. 97.
28. T. Iwasita-Vielstich. In: H. Gerischer and C.W. Tobias (Eds.), *Advances in Electrochemical Science and Engineering*, Vol. 1. VCH Verlag (Weinheim), 1990, p. 127.
29. B. Beden and C. Lamy. In: R.J. Gale (Ed.), *Spectroelectrochemistry, Theory and Practice.* Plenum Press, New York, 1998, p. 189.
30. S. Swathirajan and Y.M. Mikhail. *J. Electrochem. Soc.* **138** (1991) 1321.
31. J.O'M. Bockris and S.U.M. Khan. *Surface Electrochemistry, A Molecular Approach.* Plenum Press, New York, 1993.
32. P.S. Kauranen, E. Skou and J. Munk. *J. Electroanal. Chem.* **404** (1996) 1.
33. H.N. Dinh, X. Ren, F.H. Garzon, P. Zelenay and S. Gottesfeld. *J. Electroanal. Chem.* **491** (2000) 222.
34. J. McBreen and S. Mukerjee. *J. Electrochem. Soc.* **142** (1995) 3399.
35. A.S. Aricò, V. Antonucci, N. Giordano, A.K. Shukla, M.K. Ravikumar, A. Roy, S.R. Barman and D.D. Sarma. *J. Power Sources* **50** (1994) 295.
36. A.K. Shukla, A.S. Aricò, K.M. el-Khatib, H. Kim, P.L. Antonucci and V. Antonucci. *Appl. Surf. Sci.* **137** (1999) 20.
37. T. Iwasita, F.C. Nart and W. Vielstich. *Ber. Bunsenges Phys. Chem.* **94** (1990) 1030.
38. T. Frelink, W. Visscher and J.A.R. vanVeen. *Langmuir* **12** (1996) 3702.
39. E. Reddington, A. Sapienza, B. Gurau, R. Viswanathan, S. Sarangapani, E.S. Smotkin and T.E. Mallouk. *Science* **280** (1998) 1735.
40. K. Lasch, L. Jorissen and J. Garche. In: C. Lamy and H. Wendt (Eds.), *Proceedings of the Workshop on Electrocatalysis in Indirect and Direct Methanol PEM Fuel Cells*, Portoroz, Slovenia, September 12–14, 1999, p. 104.

41. M. Watanabe, S. Saegusa and P. Stonehart. *J. Electroanal. Chem.* **271** (1989) 213.
42. W. Chrzanowski and A. Wieckowski. *Langmuir* **14** (1998) 1967.
43. P.S. Kaurenan and E. Skou. *J. Electroanal. Chem.* **408** (1996) 189.
44. M.K. Ravikumar and A.K. Shukla. *J. Electrochem. Soc.* **143** (1996) 2601.
45. J.B. Goodenough, A. Hamnett, B.J. Kennedy and S.A. Weeks. *Electrochim. Acta* **32** (1987) 1233.
46. H. Dohle, J. Divisek, J. Mergel, H.F. Oetjen, C. Zingler and D. Stolten. In: *Proceedings Fuel Cell Seminar*, Portland, OR, USA, October 30–November 2, 2000, p. 126.
47. A. Heinzel and V.M. Barragan. *J. Power Sources* **84** (1999) 70.
48. M. Walker, K.M. Baumgartner, M. Kaiser, J. Kerres, A. Ullrich and E. Rauchle. *J. Appl. Polym. Sci.* **74** (1999) 67.
49. A. Kuver and K. Potje-Kamloth. *Electrochim. Acta* **43** (1998) 2527.
50. J. Cruickshank and K. Scott. *J. Power Sources* **70** (1998) 40.
51. X. Ren, P. Zelenay, S. Thomas, J. Davey and S. Gottesfeld. *J. Power Sources* **86** (2000) 111.
52. A.S. Arico, P. Creti, V. Baglio, E. Modica and V. Antonucci. *J. Power Sources* **91** (2000) 202.
53. A.S. Arico, P. Creti, P.L. Antonucci and V. Antonucci. *Electrochem. Solid-State Lett.* **1** (1998) 66.
54. P.L. Antonucci, A.S. Arico, P. Creti, E. Ramunni and V. Antonucci. *Solid State Ionics* **125** (1999) 431.
55. S.N. Lvov, X.Y. Zhou, E. Chakova, M.V. Fedkin, H.R. Allcock and M.A. Hofmann. In: *Proceedings of the 199th Meeting Direct Methanol Fuel Cell Symposium*, Electrochemical Society, Washington, DC, USA, March 25–29, 2001 (Abstracts 69).
56. J. Wang, S. Wasmus and R.F. Savinell. *J. Electrochem. Soc.* **142** (1995) 4218.
57. D. Weng, J.S. Wainright, U. Landau and R.F. Savinell. *J. Electrochem. Soc.* **143** (1996) 1260.
58. H. Kim, C. Lim and H. Chang. In: *Proceedings of 199th Meeting Direct Methanol Fuel Cell Symposium*, Electrochemical Society, Washington, DC, USA, March 25–29, 2001 (Abstract 66).
59. C. Yang, S. Srinivasan, A.S. Aricò, P. Cretì, V. Baglio and V. Antonucci. *Electrochem. Solid-State Lett.* **4** (2001) 31.
60. J. Maruyama, M. Inaba, T. Morita and Z. Ogumi. *J. Electroanal. Chem.* **504** (2001) 208.
61. D. Chu. *Electrochim. Acta* **43** (1998) 3711.
62. S.B. Lee and S.I. Pyun. *J. Appl. Electrochem.* **30** (2000) 795.
63. G. Faubert, G. Lalande, R. Cote, D. Guay, J.P. Dodelet, L.T. Weng, P. Bertrand and G. Denes. *Electrochim. Acta* **41** (1996) 1689.
64. S. Gupta, D. Tryk, S.K. Zecevic, W. Aldred, D. Guo and R.F. Savinell. *J. Appl. Electrochem.* **28** (1998).
65. R.Z. Jiang and D. Chu. *J. Electrochem. Soc.* **147** (2000) 4605.
66. O. ElMouahid, C. Coutanceau, E.M. Belgsir, P. Crouigneau, J.M. Leger and C. Lamy. *J. Electroanal. Chem.* **426** (1997) 117.
67. C. Coutanceau, P. Crouigneau, J.M. Leger and C. Lamy. *J. Electroanal. Chem.* **379** (1994) 389.

Chapter 11

Transport properties of solid oxide electrolyte ceramics: a brief review

V.V. Kharton, F.M.B. Marques and A. Atkinson

Abstract

This work is centered on the comparative analysis of oxygen ionic conductivity, electronic transport properties and thermal expansion of solid electrolyte ceramics, providing a brief overview of the materials having maximum potential performance in various high-temperature electrochemical devices, such as solid oxide fuel cells (SOFCs). Particular emphasis is focused on the oxygen ionic conductors reported during the last 10–15 years, including derivatives of γ-$Bi_4V_2O_{11}$ (BIMEVOX), $La_2Mo_2O_9$ (LAMOX), $Ln_{10-x}Si_6O_{26}$-based apatites, $(Gd,Ca)_2Ti_2O_{7-\delta}$ pyrochlores and perovskite-related phases based on $LaGaO_3$ and $Ba_2In_2O_5$, in order to identify their specific features determining possible applications. The properties of the new ion-conducting phases are compared to data on well-known solid electrolytes, such as stabilized zirconia, δ-Bi_2O_3-based ceramics, doped ceria and $LaAlO_3$. The compositions exhibiting highest ionic conductivity are briefly discussed.

Keywords: Solid oxide electrolyte; Ceramics; Oxygen ionic conductivity; Electronic conduction; Thermal expansion
PACS: 66.10.Ed; 67.55.Hz; 81.05.Je

Article Outline

1. Introduction . . . 190
2. Zirconia-based solid electrolytes . . . 190
3. $LaGaO_3$-based electrolytes . . . 194
4. Doped ceria electrolytes . . . 196
5. δ-Bi_2O_3- and $Bi_4V_2O_{11}$-based ceramics . . . 200
6. Materials based on $La_2Mo_2O_9$ (LAMOX) . . . 203
7. Perovskite- and brownmillerite-like phases derived from $Ba_2In_2O_5$ . . . 204
8. Perovskites based on $LnBO_3$ (B = Al, In, Sc, Y) . . . 205
9. Solid electrolytes with apatite structure . . . 207
10. Pyrochlores and fluorite-type $(Y,Nb,Zr)O_{2-\delta}$ . . . 208
Acknowledgements . . . 209
References . . . 209

Fuel Cells Compendium

1. INTRODUCTION

Technologies based on the use of high-temperature electrochemical cells with oxygen ion-conducting solid electrolytes, provide important advantages with respect to the conventional industrial processes. In particular, solid oxide fuel cells (SOFCs) are considered as alternative electric power generation systems due to high energy conversion efficiency, fuel flexibility including the prospects to operate directly on natural gas and environmental safety [1–5]. Practical application of SOFCs is, however, still limited for economical reasons, particularly as result of the high costs of component materials. The SOFC-based generators can be commercially viable only if their production costs drop significantly [2,3]. This can be addressed in part through development of new materials, including solid oxide electrolytes as they are key components of electrochemical cells [5]. Solid electrolytes should satisfy numerous requirements, including: fast ionic transport, negligible electronic conduction and thermodynamic stability over a wide range of temperature and oxygen partial pressure. In addition, they must have thermal expansion compatible with that of electrodes and other construction materials, negligible volatilization of components, suitable mechanical properties and negligible interaction with electrode materials under operation conditions.

Due to growing interest in environmentally benign and energy-saving technologies, developments in the field of SOFCs and other high-temperature electrochemical devices have been considerably intensified during the last 10–15 years. In particular, some novel oxygen ionic conductors have been reported in the literature, including $LaGaO_3$-based perovskites, derivatives of $Bi_4V_2O_{11}$ (BIMEVOX) and $La_2Mo_2O_9$ (LAMOX), perovskite- and brownmillerite-like phases (e.g., derived from $Ba_2In_2O_5$), several new pyrochlores with relatively high ionic transport such as $(Gd,Ca)_2Ti_2O_{7-\delta}$, and apatite materials derived from $Ln_{10-x}Si_6O_{26\pm\delta}$ where Ln is a rare earth cation. In many aspects such materials exhibit an improved performance with respect to ZrO_2-, ThO_2-, HfO_2-, CeO_2- and δ-Bi_2O_3-based solid electrolytes known and used since the early 1960–1970s [6–10]. On the other hand, the rapid developments in this field have resulted in a significant lack of knowledge regarding key properties of these potential new electrolytes, even regarding important characteristics such as their electronic conductivity and stability.

The aim of this review is to compare available data on the newly reported materials with conventional ionic conductors, placing special emphasize on their transport properties. Some data on thermal expansion coefficients (TECs), and compatibility with electrode materials, are also briefly covered. Readers interested further in the well-established materials are referred to relevant reviews and monographs [11–15]. In this brief review, it has been impossible to cover all promising materials, important theoretical aspects and numerous technological approaches developed in the field of solid oxide electrolytes. Priority has been given mainly to comparing transport properties and thermal expansion of single-phase ceramics exhibiting the highest ionic conductivity. Selection of the references for this review is focused on the last 10–15 years, with the main emphasis on the newly reported materials. For earlier work the review by Etsell and Flengas [6], published in 1970 and covering almost 700 literature sources, is still an invaluable source. Furthermore, the references included in this article were selected in order to show typical relationships between properties of different materials (rather than to provide a comprehensive literature survey), with particular attention on recent publications. It should be noted that some of the materials considered here are strictly speaking mixed ionic–electronic conductors rather than perfect solid electrolytes and, in many cases, the level of electronic conductivity is still unknown. In such cases, we have assumed dominant oxygen-ionic conductivity in common with the original references, although this assumption might not be valid.

2. ZIRCONIA-BASED SOLID ELECTROLYTES

The maximum ionic conductivity in ZrO_2-based systems is observed when the concentration of acceptor-type dopant(s) is close to the minimum necessary to completely stabilize the cubic fluorite-type phase

[6,10,16,17]. This concentration (so-called low stabilization limit) and the corresponding conductivity are, to a degree, dependent on the processing history and microstructural features, such as: dopant segregation, impurities, kinetically limited phase transitions and formation of ordered microdomains. Nevertheless, despite minor contradictions still existing in the literature, for most important systems, the dopant concentration ranges providing maximum ionic transport are well established. For example, the highest conductivity levels in $Zr_{1-x}Y_xO_{2-x/2}$ and $Zr_{1-x}Sc_xO_{2-x/2}$ ceramics are observed at $x = 0.08$–0.11 and 0.09–0.11, respectively. Further additions decrease the ionic conductivity due to increasing association of the oxygen vacancies and dopant cations into complex defects of low mobility. It is commonly accepted that this tendency increases with increasing difference between the host and dopant cation radii [11–14,16]. Similar phenomena explain the conductivity variations in numerous fluorite, perovskite and pyrochlore systems. This trend is illustrated in Fig. 1 which shows the highest conductivity in ZrO_2–Ln_2O_3 systems, and the association and oxygen-ion migration enthalpies as a function of the Ln^{3+} radius [16]. Because the size of Zr^{4+} is smaller than that of the trivalent rare earth cations, the maximum ionic transport is observed for Ln = Sc^{3+}. Taking into account the costs of different dopants and also the known ageing of Sc-containing materials (Table 1; Refs. [18–20]) at moderate temperatures, yttria-stabilized zirconia is used for most practical applications up to now. However, the data in Fig. 1 appear to indicate that the association enthalpy decreases as the dopant ionic radius increases. This is contrary to the generally accepted view of dopant–vacancy interactions. These figures probably are erroneous due to an over-simplified analysis of the Arrhenius conductivity plots and demonstrate that even in zirconia, there are significant basic issues still remaining to be understood.

Doping of ZrO_2 with alkaline earth metal cations (A^{2+}) is much less effective compared to rare earth dopants. This is due to a higher tendency to defect association and to a lower thermodynamic stability of the cubic fluorite-type solid solutions in ZrO_2–AO systems [6,10]. Some attempts have been made to search for new solid–electrolyte compositions in ternary systems. These aim to increase the conductivity by optimizing the average size of dopant cations, to increase the stability of Sc-containing materials by codoping, or to decrease the cost of Ln^{3+}-stabilized phases by mixing rare-earth with cheaper alkaline earth dopants [10]. However, no worthwhile improvement has been observed. Another approach to improving zirconia solid electrolytes has been to add small amounts of highly dispersed alumina [21–23].

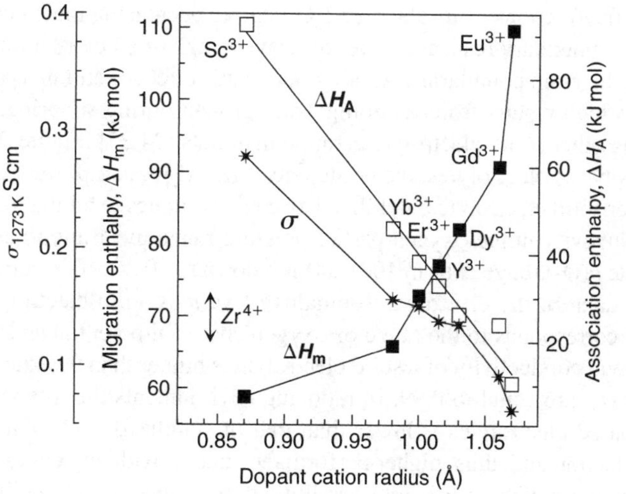

Fig. 1. Maximum conductivity in the binary ZrO_2–Ln_2O_3 systems at 1273 K, and the oxygen-ion migration and association enthalpies vs. radius of Ln^{3+} cations [16]

Table 1. Selected examples of the conductivity degradation of zirconia-based solid electrolytes under isothermal conditions in air

Dopant content (mol%)		T	Ageing time	Conductivity drop	Reference
Sc_2O_3	Y_2O_3	(K)	(h)	($\Delta\sigma/\sigma$, %)	
–	3.0	1273	83	13	[17]
–	8.0		83	17	[17]
–	10.0		83	0	[17]
	10.0		100	1	[18]
–	12.0		83	0	[17]
	13.3		100	4	[18]
	15.0		100	10	[18]
4.0	4.0		83	27	[17]
7.0	–		83	28	[19]
7.8	–		83	34	[17]
9.0	–		83	10	[19]
9.5	–		83	7	[19]
10.0	–		83	2	[19]
11.0	–		83	4	[19]
6.0	4.0	1173	300	0	[20]
7.0	3.0		300	3	[20]
3.0	5.0		300	2	[20]
7.0	–	1123	83	13	[19]
9.0	–		83	7	[19]
9.0	–		1330	35	[19]
9.3	–		83	3	[19]
9.5	–		83	7	[19]
10.0	–		83	7	[19]
11.0	–		83	5	[19]

While significant volume fractions of insulating Al_2O_3 lead to decreasing ionic conductivity, minor additions decrease the grain-boundary resistance due to "scavenging" of silica-rich impurity phases into new phases that do not wet the grain boundaries. Another noticeable effect of alumina additions is an improved mechanical strength, which results from retarding grain growth during sintering. This approach is also effective for modifying other solid–electrolyte ceramic materials [24,25]. Figure 2 compares the conductivity of selected ZrO_2-based electrolytes, including one YSZ–Al_2O_3 composite [26–29].

With respect to other solid electrolytes, stabilized zirconia ceramics exhibit a minimum electronic contribution to total conductivity in the oxygen partial pressure range most important for practical applications. This approximate $p(O_2)$ range is from 100–200 atm down to 10^{-25}–10^{-20} atm. As an example, Fig. 3 presents comparative data on the electrolytic domains of various solid electrolytes [30–33]. The term "electrolytic domain" corresponds to the range of oxygen chemical potential and temperature, where the ionic contribution to total conductivity of a solid electrolyte is higher than 99%; at the boundaries, the ion transference numbers (t_o) are equal to 0.99. In reducing environments the n-type electronic transport in ThO_2- and $LaGaO_3$-based electrolytes is lower than that of stabilized ZrO_2, but the latter shows lower p-type electronic conduction and, thus, higher performance under oxidizing conditions. Note also that the performance of lanthanum gallate at low $p(O_2)$ is limited by reduction and gallium oxide volatilization rather than the n-type electronic conductivity. The oxygen pressure dependence of total conductivity of $La(Sr)Ga(Mg)O_{3-\delta}$ (LSGM) at reduced $p(O_2)$ is quite complex, suggesting possible decomposition

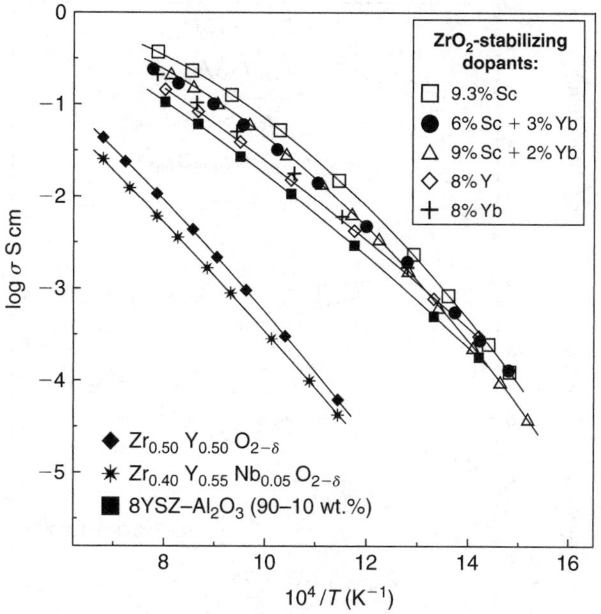

Fig. 2. Total conductivity of ZrO_2-based solid electrolytes in air [19,26–29]

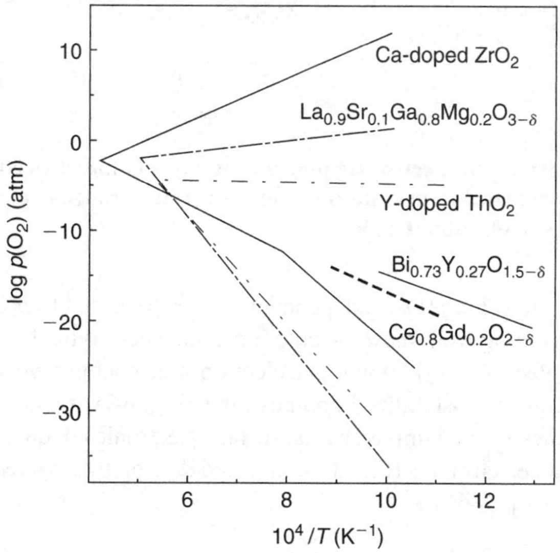

Fig. 3. Boundaries of the electrolytic domain ($t_o \geqslant 0.99$) for several ionic conductors [30–33]

or a decrease in the ionic conductivity due to association of oxygen vacancies [34], whereas the electrolytic domain boundary given in Fig. 3 was estimated assuming $p(O_2)$-independent ionic transport and the n-type conductivity proportional to $p(O_2)^{-1/4}$ [33].

Figure 4A compares the electrolytic domains of several ZrO_2-based compositions, evaluated from oxygen permeation data [35]. Again, these estimates are based the commonly used assumptions that the ionic

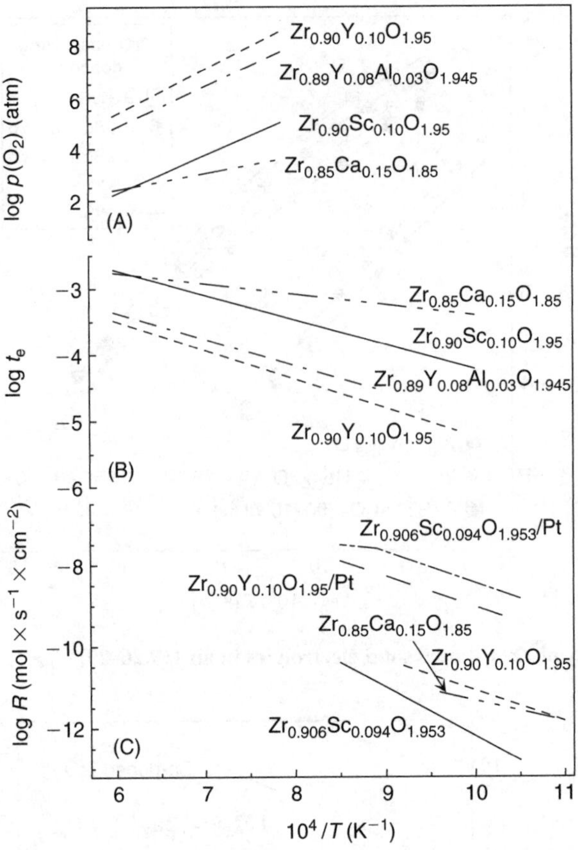

Fig. 4. High-$p(O_2)$ boundary of the electrolytic domains of zirconia-based electrolytes (A) [10,35], electron transference numbers approximated to the intermediate temperature range in air (B), and interfacial oxygen exchange rate at $p(O_2) = 1.3 \times 10^{-2}$ atm (C) [36]

and electron hole conductivities are $p(O_2)$-independent and proportional to $p(O_2)^{1/4}$, respectively. Within the limits of uncertainty resulting from these assumptions, the electrolytic domain boundaries seem quite similar for all ZrO_2-based electrolytes (although the electron-hole conductivity of Ca-stabilized zirconia is definitely higher than that of yttria-stabilized compositions, Fig. 4B). Analogously, the oxygen isotopic exchange rate, which is believed to be limited by the surface electronic conductivity [36], has similar magnitude for all zirconia-based electrolytes (Fig. 4C). Selected data on thermal expansion of stabilized ZrO_2 ceramics are listed in Table 2 [37–56].

3. $LaGaO_3$-BASED ELECTROLYTES

Perovskite-type ABO_3 phases derived from lanthanum gallate, $LaGaO_3$, possess a higher ionic conductivity than that of stabilized zirconia and are thus promising materials for electrochemical cells operating in the intermediate temperature range, 770–1100 K [24,33,34,57–72]. Compared with CeO_2-based electrolytes (which are also of interest for lowering the operating temperature of SOFCs), the electrolytic domain of doped $LaGaO_3$ extends to substantially lower oxygen chemical potentials (Fig. 3). Materials derived from lanthanum gallate have relatively low thermal expansion, similar to that of stabilized zirconia

Table 2. Thermal expansion coefficients of selected solid oxide electrolytes

Composition		T (K)	$\alpha \times 10^6$ (K^{-1})	Reference
Zirconia-based electrolytes				
Stabilizing dopant (mol%)	Other additives (wt.%)			
8 (Y_2O_3)	–	300–1273	10.0	[27]
		300–1073	10.5	[37]
		300–1273	10.8	[41,42]
9 (Y_2O_3)	–	1273	9.5	[28,43]
10 (Y_2O_3)	–	30–1073	10.6	[38,39]
8 (Y_2O_3)	10 (Al_2O_3)	300–1273	9.7	[27]]
8 (Sc_2O_3)	–	300–1273	10.4	[44]
10 (Sc_2O_3)	–	300–1273	10.9	[38,39]
Ceria-based electrolytes				
$Ce_{0.9}Gd_{0.1}O_{2-\delta}$		773	12.4	[45]
$Ce_{0.8}Gd_{0.2}O_{2-\delta}$		773	12.5	[45]
		300–1100	11.8	[46]
		323–1273	12.5	[47]
$LaGaO_3$- and $LaAlO_3$-based electrolytes				
$La_{0.9}Sr_{0.1}Ga_{0.8}Mg_{0.2}O_{3-\delta}$		300–1073	10.4	[24]
		300–1473	11.9	[34]
$La_{0.9}Sr_{0.1}Ga_{0.8}Mg_{0.2}O_{3-\delta}$: 2 wt.% Al_2O_3		300–1073	10.6	[24]
$La_{0.8}Sr_{0.2}Ga_{0.8}Mg_{0.2}O_{3-\delta}$		300–1473	12.4	[34]
$La_{0.90}Sr_{0.10}Ga_{0.76}Mg_{0.19}Co_{0.05}O_{3-\delta}$		300–1473	12.7	[48]
$La_{0.90}Sr_{0.10}Ga_{0.76}Mg_{0.19}Fe_{0.05}O_{3-\delta}$		300–1473	11.6	[48]
$La_{0.90}Sr_{0.10}Ga_{0.45}Al_{0.45}Mg_{0.10}O_{3-\delta}$		300–1223	10.9	[49]
$La_{0.9}Sr_{0.1}Al_{0.9}Mg_{0.1}O_{3-\delta}$		300–1223	11.2	[49]
Electrolytes with pyrochlore and apatite structure				
$Gd_2Ti_2O_{7\pm\delta}$		323–1273	10.8	[50]
$Gd_{1.86}Ca_{0.14}Ti_2O_{7-\delta}$		400–1300	10.4	[51]
$La_{9.83}Si_{4.5}AlFe_{0.5}O_{26-\delta}$		373–1273	8.9	[52]
$La_7Sr_3Si_6O_{24}$		373–1273	8.9	[53]
$La_2Mo_2O_9$- and Bi_2O_3-based electrolytes				
$La_{1.7}Bi_{0.3}Mo_2O_9$		373–1073	16.0	[54]
$La_2Mo_{1.7}W_{0.3}O_9$		373–623	14.4	[54]
		623–1073	19.8	
$Bi_2V_{0.9}Cu_{0.1}O_{5.5-\delta}$		300–730	15.3	[55]
		730–1030	18.0	
$(Bi_{0.95}Zr_{0.05})_{0.85}Y_{0.15}O_{1.5+\delta}$		320–710	13.8	[56]
		710–1120	16.6	

(Table 2). High oxygen ionic conductivity in $LaGaO_3$ can be achieved by substituting lanthanum with alkaline earth elements and/or incorporating divalent metal cations, such as Mg^{2+}, into the gallium sublattice in order to increase the oxygen vacancy concentration. Following the principle of minimum lattice distortion giving maximum oxygen ion mobility, doping with Sr leads to a higher ionic conductivity in comparison with calcium or barium [57,60,73]. For the $La_{1-x}Sr_xGa_{1-y}Mg_yO_{3-\delta}$ (LSGM) series, the maximum ionic transport is achieved at $x = 0.10-0.20$ and $y = 0.15-0.20$; further acceptor-type doping leads to progressive vacancy association processes. A decrease in conductivity is also observed on doping

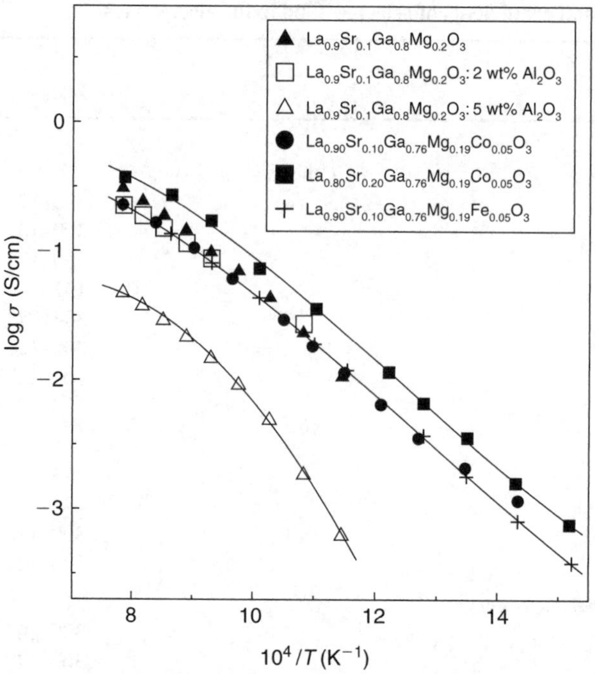

Fig. 5. Total conductivity of $LaGaO_3$-based solid electrolytes in air [24,34,48]

with a smaller radius A-site cation, or creating A-site deficiency. Rather surprisingly, introduction of small amounts of cations with variable valence, such as cobalt, onto the gallium sites increases the ionic conduction in LSGM, and produces only a small increase in the electronic conductivity [68–70]. However, the concentration of transition metal dopants should be limited to below 3–7% as further additions lead to increasing electronic and decreasing ionic conductivities. Disadvantages of $LaGaO_3$-based materials include possible reduction and volatilization of gallium oxide, formation of stable secondary phases in the course of processing, the relatively high cost of gallium and significant reactivity with perovskite electrodes under oxidizing conditions as well as with metal anodes in reducing conditions [34,74,75]. These problems may be, to some extent, suppressed by the optimization of processing techniques and additional B-site substitution. Data on total conductivity, predominantly oxygen ionic, of selected $LaGaO_3$-based compositions are presented in Fig. 5. Figure 6 compares the levels of partial ionic conductivity in various solid-electrolyte ceramics [76–78].

4. DOPED CERIA ELECTROLYTES

The properties of solid electrolytes based on doped cerium dioxide, $Ce(Ln)O_{2-\delta}$, have been considered in numerous reviews and survey papers [3,11,14,32,46,79–84]. The main advantages of this group of oxygen ion conductors include a higher ionic conductivity with respect to stabilized ZrO_2 (particularly at lower temperatures) and a lower cost in comparison with LSGM and its derivatives.

Among ceria-based phases, the highest level of oxygen ionic transport is characteristic of the solid solutions $Ce_{1-x}M_xO_{2-\delta}$, where M = Gd or Sm, $x = 0.10-0.20$. Selected data on their total conductivity, which is predominantly ionic under oxidizing conditions, are shown in Fig. 7. The composition of doped ceria is often abbreviated, e.g., $Ce_{0.9}Gd_{0.1}O_{1.95}$, is abbreviated to CGO10. CGO10 has a lattice ionic

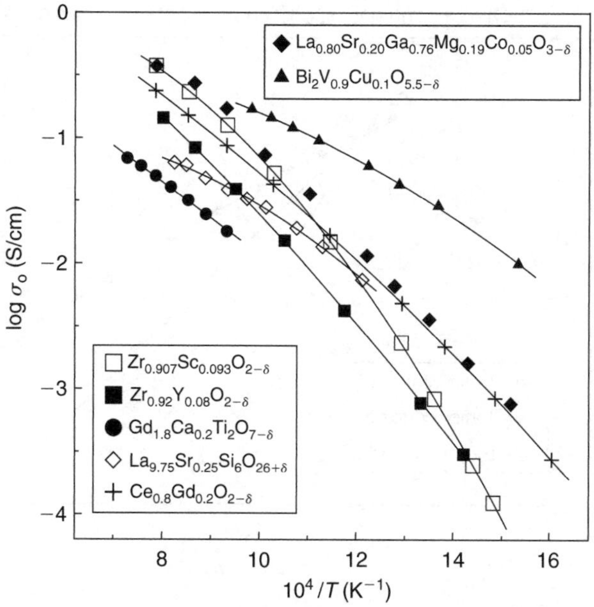

Fig. 6. Oxygen ionic conductivity of various solid–electrolyte materials [19,27,46,48,55,76–78]

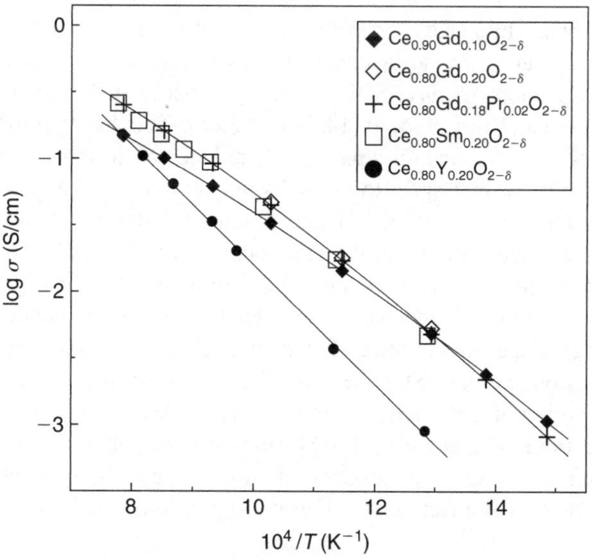

Fig. 7. Selected data on the total conductivity of ceria-based solid electrolytes in air [46,83]

conductivity of 0.01 S cm^{-1} at 500°C. As with all ceramic electrolytes, the grain boundaries are partially blocking to ionic transport across them. This is an extra contribution to the total resistance that is dependent on impurities that segregate to the boundaries and is therefore highly variable from one source to another. This has been a source of considerable confusion and misleading conclusions in measurement of

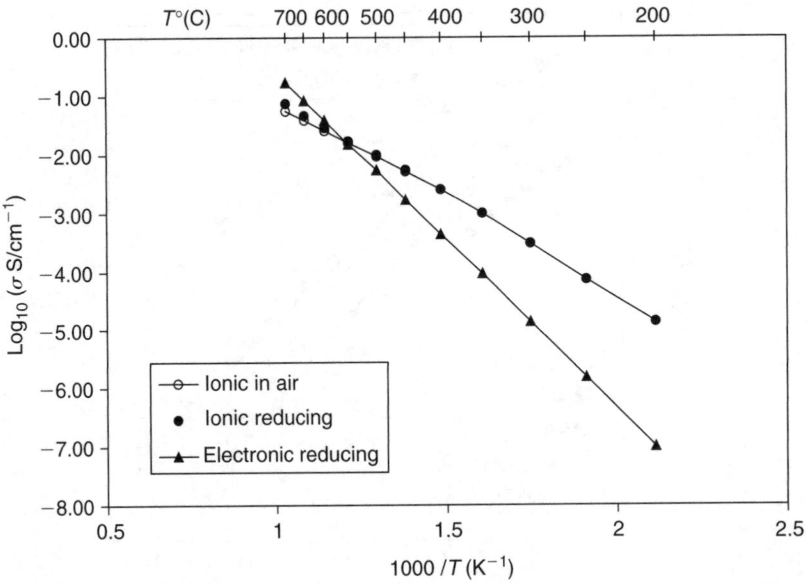

Fig. 8. Ionic conductivity of CGO10 in air and ionic and electronic conductivities in reducing atmosphere (10% H_2, 2.3% H_2O)

total conductivity data. Thus, while CGO10 has the highest lattice conductivity, CGO20 often has higher total conductivity because its grain boundary contribution seems to be more tolerant of impurities [84].

The main problems in using doped ceria as SOFC electrolyte arise from the partial reduction of Ce^{4+}– Ce^{3+} under the reducing conditions of the anode [9,14,79,80,85,86]. This has two main effects: first, it gives n-type electronic conductivity which causes a partial internal electronic short circuit in a cell, and second, it generates nonstoichiometry (with respect to normal valency in air) and expansion of the lattice which can lead to mechanical failure. The reduction phenomena are described in detail in the review by Mogensen et al. [14]. Parameters describing the electronic conductivity of CGO10 have been given by Steele [84], and the resulting comparison with ionic conductivity is shown in Fig. 8. This indicates that the electronic conductivity at the anode side will be greater than the ionic conductivity for temperatures greater than about 550°C. Experiments have shown that CGO10 is more resistant to reduction than CGO20, but that the inclusion of other codopants (e.g., Pr) has little effect [46]. Such problems can be partially solved by a combination of $Ce(M)O_{2-\delta}$ with other solid electrolytes such as stabilized zirconia or doped lanthanum gallate, in multilayer cells [80–88]. However, the performance of multilayer cells is relatively poor due to formation of reaction products with low conductivity at the interface between the solid–electrolyte phases, as well as differences in thermal expansion of the electrolytes, resulting in microcracks [87–89].

The effect of lattice expansion on mechanical integrity depends on the geometry of the cell and the way the ceria is supported. Simulations of various configurations indicate that ceria electrolytes will be unstable mechanically in SOFCs at temperatures above about 700°C [90]. The mechanical properties of bulk-doped ceria ceramic seem to be inferior to those of YSZ. However, the reason for this is not understood. If ceria is to be used in supported form, then the demands on mechanical properties are in any case not as severe as in a self-supported membrane.

The high ionic conductivity (compared with YSZ) at lower temperatures, combined with the problems arising from reduction at high temperatures make ceria electrolytes viable only for low temperature operation. The

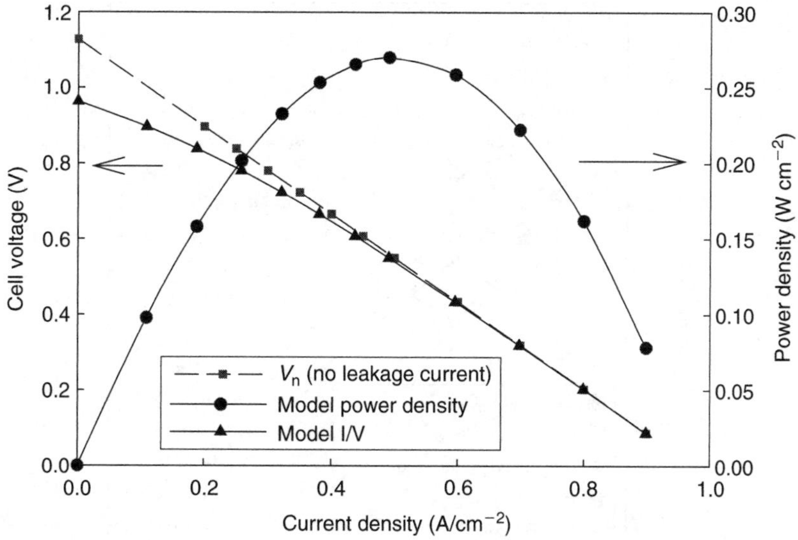

Fig. 9. Simulated SOFC characteristic with CGO10 electrolyte at 500°C in humidified H_2 (3% H_2O)

average TEC of gadolinia-doped ceria is slightly high if compared to stabilized zirconia and LSGM, but similar to those of lanthanum gallate codoped with alkaline earth and transition metal cations (Table 2; Refs. [91,92]). The thermal expansion coefficient is compatible with ferritic stainless steels, which makes ceria attractive for metal-supported SOFCs. The potential problems arising from electronic conductivity have been assessed by parametric simulations of the operating characteristics of ceria SOFCs. These show that temperature, electrolyte thickness, electrode polarization resistances and fuel utilization, all influence the effect of electronic conductivity. An example is shown in Fig. 9 in comparison with a simulation assuming no electronic conductivity. The electronic conductivity reduces the OCV, but as current is drawn from the cell, the I–V curve for the ceria electrolyte soon approaches that of a pure ionic conductor. This is because polarization at the anode suppresses the reduction of the ceria. The efficiency of the ceria SOFC compared with that using a pure ionic conductor of the same ionic conductivity is given by the ratio of the terminal voltages at a given current density. This shows that the electronic conductivity only causes an efficiency penalty at low current densities and that at a terminal voltage of 0.6–0.7 V, the penalty is small. This conclusion is valid if the operating temperature is kept below about 600°C.

Doped ceria is relatively unreactive towards potential electrode materials. At the present time reasonable performance has been reported with Ni/ceria anodes and LSCF cathodes. There are now many studies reported in the literature that have demonstrated that high-power densities (of the order of 400 mW cm^{-2}) can be achieved with doped ceria electrolytes in this temperature range, e.g., Refs. [93–95].

Because doped ceria is only viable for operating temperatures below 600°C, it must be used in a supported thick-film form. In order to preserve high activity electrodes (and be compatible with metal supports if these are chosen), then the ceria must be processed at modest temperatures (e.g., 1000°C). Small additions (e.g., 1 mol%) of divalent transition metals (e.g., Co) have been found to be effective sintering aids for CGO, and processing to full density can be achieved at these modest temperatures. Furthermore, the transition metal can be at such a low concentration that it has negligible effect on either the ionic or the electronic performance of the CGO [96–98].

Nanocrystalline-undoped ceria has increased electronic conductivity due to electronic conduction along the grain boundaries [99,100]. Some experiments with nanocrystalline-doped ceria have been reported to

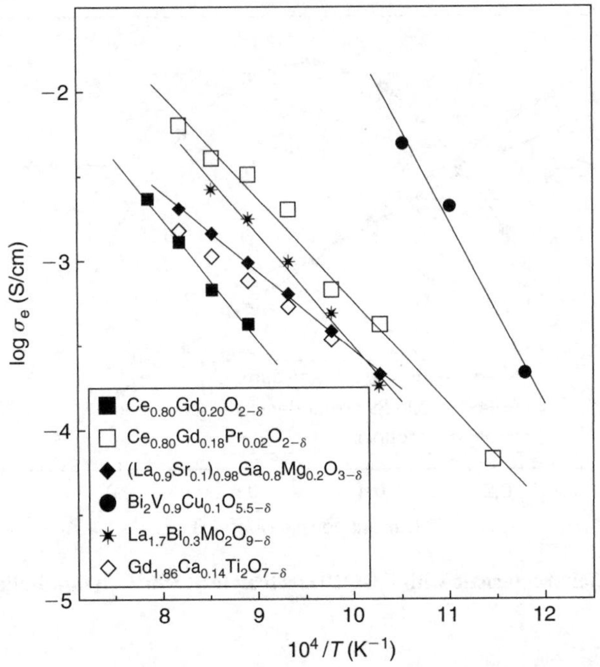

Fig. 10. Total electronic (p- and n-type) conductivity of various solid electrolytes in oxidizing conditions. Data for LSGM [91], CGO [92] and $Gd_{1.86}Ca_{0.14}Ti_2O_{7-\delta}$ [51] correspond to $p(O_2) = 0.21$ atm. Data on other solid electrolytes [54,78,92] corresponds to the oxygen partial pressure range of 1.0/0.21 atm

show increased ionic conductivity [101]. The possible benefits of nanocrystalline ceria electrolytes are thus not yet clear.

Zhu et al. [102,103] have reported high ionic conductivity at low temperatures for composite electrolytes comprising doped ceria and a molten salt. These materials look interesting, but as yet, their properties have not been confirmed by other groups. Moreover, practical use of such materials seems problematic due to possible decomposition of molten salts, and their high corrosion activity under the SOFC operation conditions.

Finally, the p-type electronic conductivity of gadolinia-doped ceria (CGO) in air is 0.5–3.0 times lower than that of LSGM, and this difference increases with reducing temperature (Fig. 10). Similar tendency is observed for the electron transference numbers (t_e), as illustrated by Fig. 11. This makes CGO ceramics promising candidates for electrochemical oxygen pumps operating at intermediate temperatures [104,105].

5. δ-Bi_2O_3- AND $Bi_4V_2O_{11}$-BASED CERAMICS

Among oxygen ion-conducting materials, oxide phases derived from Bi_2O_3 are particularly interesting due to their high ionic conductivity with respect to other solid electrolytes [12,13,106,107–108]. The fast ionic transport is characteristic of stabilized δ-Bi_2O_3, which has a fluorite-type structure with a highly deficient oxygen sublattice, and γ-$Bi_4V_2O_{11}$ (parent of the so-called BIMEVOX materials), which belongs to the Aurivillius series. Unfortunately, Bi_2O_3-based materials possess a number of disadvantages, including thermodynamic instability in reducing atmospheres (Fig. 3), volatilization of bismuth oxide at moderate temperatures, a high corrosion activity and low mechanical strength. Hence, the applicability of these oxides in electrochemical cells is considerably limited.

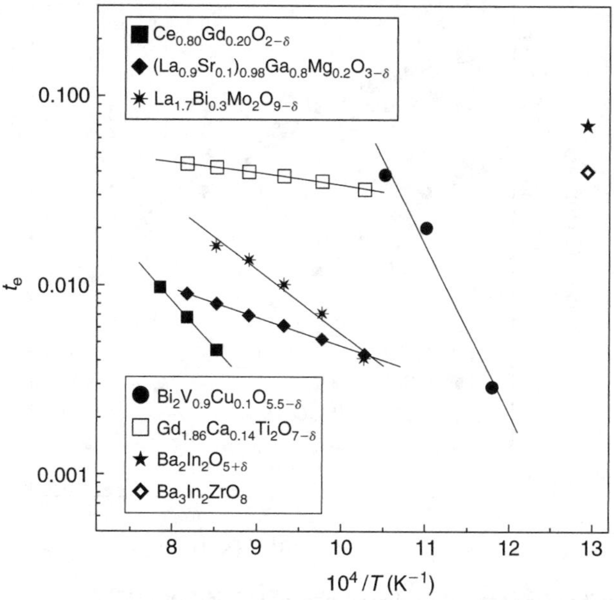

Fig. 11. Electron transference numbers of various solid electrolytes under oxidizing conditions. Data for LSGM [91], CGO [92], $Gd_{1.86}Ca_{0.14}Ti_2O_{7-\delta}$ [51], $Ba_2In_2O_5$ [106] and $Ba_3In_2ZrO_8$ [106] correspond to $p(O_2) = 0.21$ atm. Data on $Bi_2V_{0.9}Cu_{0.1}O_{5.5-\delta}$ [78] and $La_{1.7}Bi_{0.3}Mo_2O_{9-\delta}$ [54] correspond to the range of 1.0/0.21 atm

The stabilization of the high-diffusivity δ-Bi_2O_3 phase down to temperatures significantly lower than the $\alpha \rightarrow \delta$ transition temperature (978–1013 K) can be achieved by the substitution of bismuth with rare-earth dopants (such as Y, Dy or Er) and their combinations with higher-valency cations, such as W or Nb [9,13,56,107–110]. Figure 12 compares the total conductivity, mainly ionic, of several Bi_2O_3-based phases [13,55,110,111]. As for ZrO_2-based electrolytes, the highest ionic transport is found for materials containing the minimum addition necessary to stabilize the cubic fluorite phase. Further doping decreases oxygen ion mobility due to decreasing unit cell volume and increasing average strength of the cation–anion bonds. Because Bi^{3+} cations are relatively large, the oxygen ionic conduction (at a given doping level) increases with increasing rare earth dopant radius. However, the minimum stabilization limit also increases with Ln^{3+} size and this leads to a decrease in the maximum ionic conductivity with increasing radius of the stabilizing cation. Therefore, the maximum conductivity in the binary systems is observed for Er- and Y-containing phases, namely $Bi_{1-x}Er_xO_{1.5}$ ($x \approx 0.20$) and $Bi_{1-x}Y_xO_{1.5}$ ($x = 0.23-0.25$). However, both binary and ternary solid solutions with the disordered fluorite structure are metastable at temperatures below 770–870 K, and they undergo a slow phase transformation and a decrease in conductivity with time [56,112–116]. Although such degradation can be partly suppressed by the incorporation of higher-valence dopants including Zr^{4+}, Ce^{4+}, Nb^{5+} or W^{6+} [44,110,113,117], so far it has not proved possible completely to avoid phase transformation.

Solid solutions based on γ-$Bi_4V_2O_{11}$ [25,55,78,118–130], stabilized by partial substitution of vanadium with transition metal cations such as Cu, Ni or Co, exhibit high ionic conductivity and oxygen ion transference numbers close to unity at temperatures below 900 K (Figs 11 and 12). Compared to the fluorite-like Bi_2O_3-based oxides, the phase stability of doped γ-bismuth vanadate (BIMEVOX) at moderate temperatures is substantially better, though still questionable [127]. The crystal lattice of the BIMEVOX family is based on the Aurvillius series and consists of alternating $Bi_2O_2^{2+}$ and perovskite-like $VO_{3.5}^{2-}$ layers, with

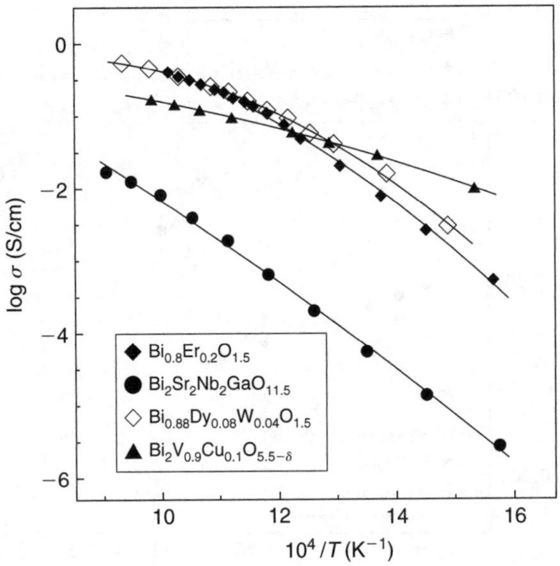

Fig. 12. Total conductivity of several Bi_2O_3-based phases [13,55,110,111]

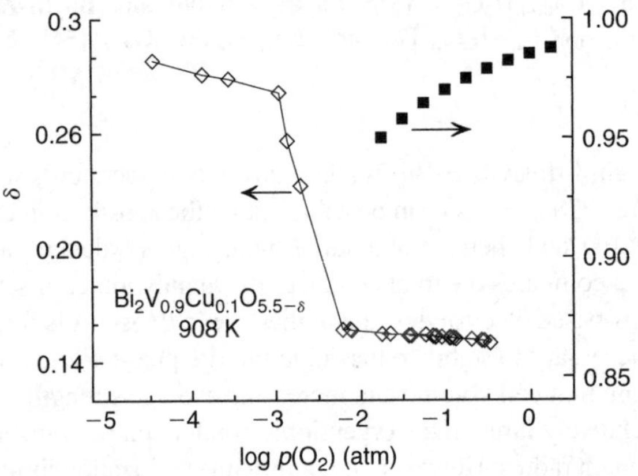

Fig. 13. Oxygen partial pressure dependence of the oxygen non-stoichiometry and ion transference numbers of $Bi_2V_{0.9}Cu_{0.1}O_{5.5-\delta}$ at 908 K [78,129]

oxygen vacancies in the perovskite layers providing ion migration. Solid solutions of the type $Bi_2V_{1-x}Me_xO_{5.5-\delta}$ (Me = Cu, Ni and $0.07 \leq x \leq 0.12$) have some of the highest oxygen ion conductivity levels reported [118–120]. Despite the high ionic conduction, practical use of bismuth vanadate-based ceramics for electrochemical applications is complicated due to their extremely high chemical reactivity (Fig. 13) and low mechanical strength [56,121]. As for other solid electrolytes, one of the most promising approaches to solving these problems is the addition of highly dispersed alumina or zirconia [25,130]. Rapid reaction of BIMEVOX ceramics with electrode materials can be reduced by the development of multilayer electrode structures [121].

An alternative approach to solving these problems is the use of multilayer cells having a layer of Bi_2O_3-based ionic conductor applied on another material, which acts as a mechanical support and protects it

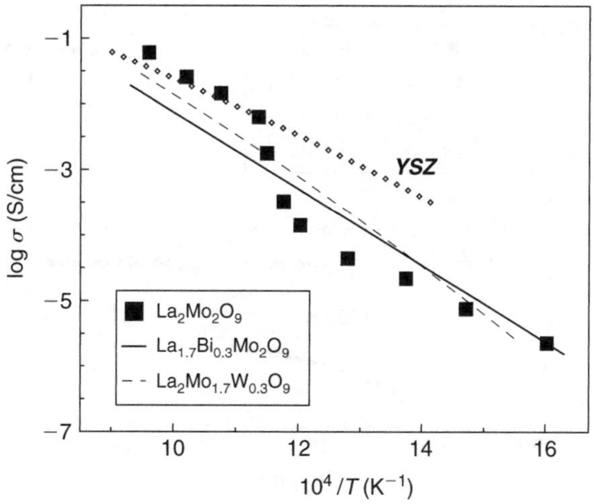

Fig. 14. Total conductivity of $La_2Mo_2O_9$-based materials in air [132,134], compared to the ionic transport in 8 mol% yttria-stabilized zirconia [27]

against reduction (e.g., Ref. [131]). Unfortunately, the high thermal expansion coefficients of bismuth oxide-based ceramics (Table 2) complicate this solution. At temperatures above about 900 K, decreasing oxygen partial pressure down to 10^{-2}atm results in electron transference numbers of $Bi_2V_{0.9}Cu_{0.1}O_{5.5-\delta}$ increasing up to 0.05; further reduction causes phase decomposition [129]. Similar decomposition is observed under applied DC voltage [128].

6. MATERIALS BASED ON $La_2Mo_2O_9$ (LAMOX)

The parent compound of the so-called LAMOX series [132–136] is $La_2Mo_2O_9$, the high-temperature polymorph of which (β-phase) has a cubic lattice isostructural with β-$SnWO_4$. In order to explain the ionic transport mechanism, Lacorre et al. [132–134] proposed the lone-pair substitution (LPS) concept by which lone electron pairs of cations could stabilize oxygen vacancies. In β-$SnWO_4$, the outer electrons of Sn^{2+} (lone electron pairs) project into a vacant oxygen site, whereas the La^{3+} cations occupying Sn^{2+} sites in $La_2Mo_2O_9$ have no lone-pair electrons. Thus, half of the cation sites occupied by lone pairs in β-$SnWO_4$ are not occupied by lone pair cations in $La_2Mo_2O_9$; consequently, more anion sites are occupied by oxygen ions. Minor doping of $La_2Mo_2O_9$ can suppress the $\beta \leftrightarrow \alpha$ vacancy ordering transition which normally occurs at approximately 853 K. The maximum conductivity was found for the compositions $La_{1.7}Bi_{0.3}Mo_2O_{9-\delta}$, $La_2Mo_{1.7}W_{0.3}O_{9-\delta}$ and $La_2Mo_{1.95}V_{0.05}O_{9-\delta}$. Figure 14 compares the total conductivity of these materials with the conductivity of 8% yttria-stabilized zirconia electrolyte [27].

When discussing $La_2Mo_2O_9$-based materials, one should note that in air their electronic conductivity, mainly n-type, is comparable to the p-type transport in LSGM only at temperatures below 1000–1100 K (Fig. 10). In other words, LAMOX is indeed a solid electrolyte with electron transference numbers lower than 0.01 (Fig. 11), but only under oxidizing conditions and only in the intermediate temperature range. The electronic contribution to total conductivity of LAMOX increases with temperature and with reducing $p(O_2)$. Similar behavior is characteristic of numerous Mo- and W-containing phases, which are well known as mixed oxygen-ionic and n-type electronic conductors [107,137,138]. The TEC values of LAMOX ceramics are high, close to those of Bi_2O_3-based solid electrolytes (Table 2). Moreover, some $La_2Mo_2O_9$-based materials exhibit degradation at moderate oxygen pressures (Fig. 15). The properties of

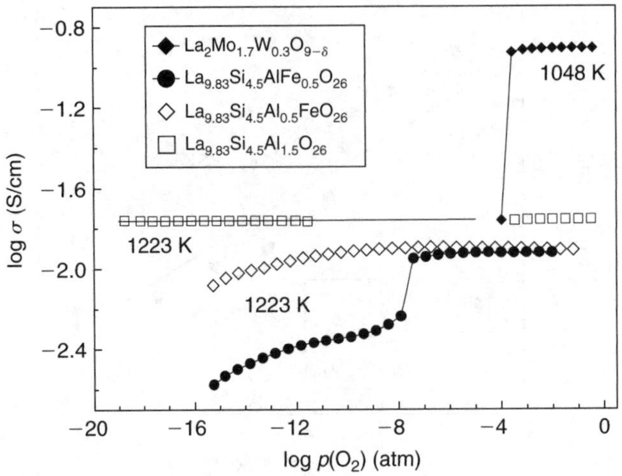

Fig. 15. Examples of the oxygen partial pressure dependencies of total conductivity of $La_{9.83}Si_{5.4}Al_{1-x}Fe_xO_{26-\delta}$ apatites and $La_2Mo_{1.7}W_{0.3}O_{9-\delta}$ [52,53,54]

LAMOX seem, therefore, rather inappropriate for practical applications. Nonetheless, the lone-pair substitution concept developed by Lacorre et al. can be used as a promising strategy in the search for new solid electrolytes.

7. PEROVSKITE- AND BROWNMILLERITE-LIKE PHASES DERIVED FROM $Ba_2In_2O_5$

Materials with substantially high conductivity can be derived by partial substitution of brownmillerite-type $Ba_2In_2O_5$ [106,139–151]. The structure of brownmillerite, $A_2B_2O_5$, consists of alternating perovskite layers of corner-sharing BO_6 octahedra and layers of BO_4 tetrahedra and can be considered as an oxygen-deficient perovskite where the oxygen vacancies are ordered along (010) planes. These vacancies may contribute to ionic transport, forming one-dimensional diffusion pathways for oxygen ion migration in the tetrahedral layers, and facilitate water absorption. Electrical conductivity is typically oxygen-ionic in dry atmospheres with moderate $p(O_2)$, mixed ionic and p-type electronic under oxidizing conditions, and protonic in H_2O-containing gas mixtures. The parent compound, $Ba_2In_2O_5$, possesses mixed conductivity with dominant oxygen ionic transport in dry air; the ion transference number at 773 K is about 0.93 (Fig. 11). Heating up to 1140–1230 K causes a transition into the disordered perovskite phase, which leads to a drastic increase of the ionic conduction. Substitution of indium with higher-valence cations, such as Zr, Ce, Sn or Hf, makes it possible to stabilize the disordered cubic perovskite structure and thus to increase the ionic transport in the intermediate temperature range. A similar effect is provided by the incorporation of lanthanum in the A sublattice. Figure 16 presents conductivity data for $BaIn_{0.7}Zr_{0.3}O_{3-\delta}$ perovskite. This material exhibits a p-type electronic contribution to total conductivity lower than 5% at 773 K [106,104]. Compared to stabilized zirconia electrolytes, the use of doped $Ba_2In_2O_5$ might be advantageous at moderate temperatures, where the oxygen ionic conductivity of the latter is higher. A significant benefit might also be expected from the high protonic conductivity under SOFC operation conditions [141,142, 144–147]. However, due to the instability of $Ba_2In_2O_5$-based ceramics in humid atmospheres, high reactivity with CO_2 and easy reducibility [140,144,147,148], it is difficult to imagine practical applications taking into account that both the stability and ionic conductivity of LSGM are higher [139].

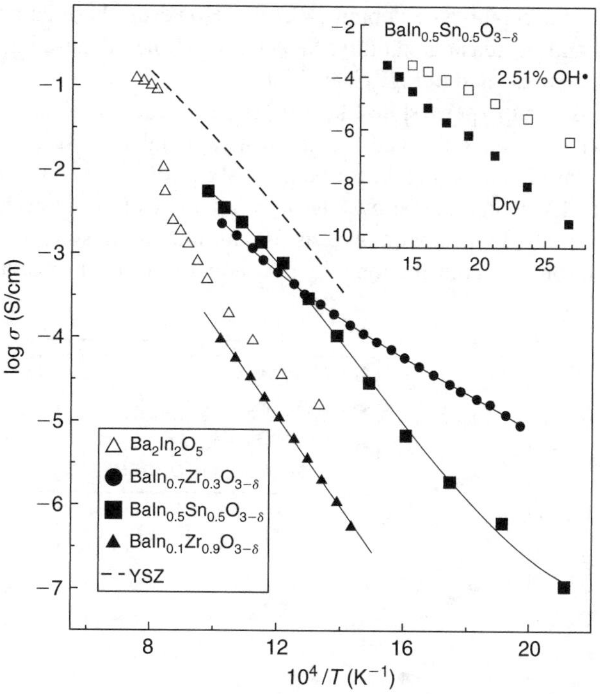

Fig. 16. Total conductivity of $Ba_2In_2O_5$-based ceramics [106,139,140,144] in comparison with 8 mol% yttria-stabilized zirconia electrolyte [27]. Data for $Ba_2In_2O_5$ [139] correspond to $p(O_2) = 10^{-6}$atm, where the conductivity is predominantly oxygen ionic [106]. Data on other materials corresponds to dry air. Inset compares conductivity values of dried $BaIn_{0.5}Sn_{0.5}O_{3-\delta}$ and the same material containing 2.51% intercalated $OH^{\bullet}$ ions [144]

8. PEROVSKITES BASED ON $LnBO_3$ (B = Al, In, Sc, Y)

Oxygen ion-conducting oxide materials based on acceptor doping of perovskite aluminates such as $LnAlO_3$ [49,73,153–161] have attracted significant interest since the 1970s. Their advantages include relatively low cost, moderate thermal expansion and higher stability with respect to reduction and volatilization compared to doped $CeO_{2-\delta}$ and $LaGaO_{3-\delta}$. The major disadvantages relate to a rather low level of the oxygen ionic conduction, relatively high p-type electronic transport under oxidizing conditions, and, in many cases, poor sinterability. Therefore, promising applications of perovskite-type aluminates may be restricted to protective layers for the anode side of $LaGaO_3$-based materials or isomorphic additives to composite solid electrolytes [24,49]. The values of TEC are similar to that of LSGM, thus giving mechanical compatibility in high-temperature electrochemical cells (Table 2). In common with other perovskite-type oxides, the oxygen ion conductivity in aluminates decreases with decreasing A-site cation radius. Among alkaline earth dopants which can be used to increase oxygen deficiency in $LaAlO_3$, the maximum conductivity and minimum vacancy-association enthalpy are provided by Sr^{2+}. The literature contains numerous contradictions resulting from poor sintering, porosity and high grain-boundary resistivity of aluminate ceramics. Nevertheless, the highest level of the ionic conductivity seems to be found for $La_{0.9}Sr_{0.1}AlO_{3-\delta}$, possibly substituted with less than 10% Mg^{2+} on the B sublattice, [153,154,161]. Further improvement in the ionic conduction can be achieved by the substitution of aluminum with gallium, accompanied by additional acceptor-type doping [49,159]. However, these compositional modifications lead to increasing costs and decreasing stability. Selected examples illustrating transport properties of $LaAlO_3$-based ceramics are presented in Fig. 17 and Table 3.

Other materials of this group (perovskite-like $LnBO_3$, B = In, Sc, Y and their derivatives; Refs. [73,98,153,154,162–164], represent only academic interest, although a considerable level of protonic conduction in some $LaYO_3$-based phases might also be interesting for some applications. The oxygen ionic conductivity of In-, Sc- and Y-containing perovskites is comparable or even lower than that of aluminates, while the costs are substantially higher. Moreover, as for aluminates, most doped $LnBO_3$ (B = In, Sc, Y) phases exhibit a high electron-hole contribution to total conductivity under oxidizing conditions (Table 3).

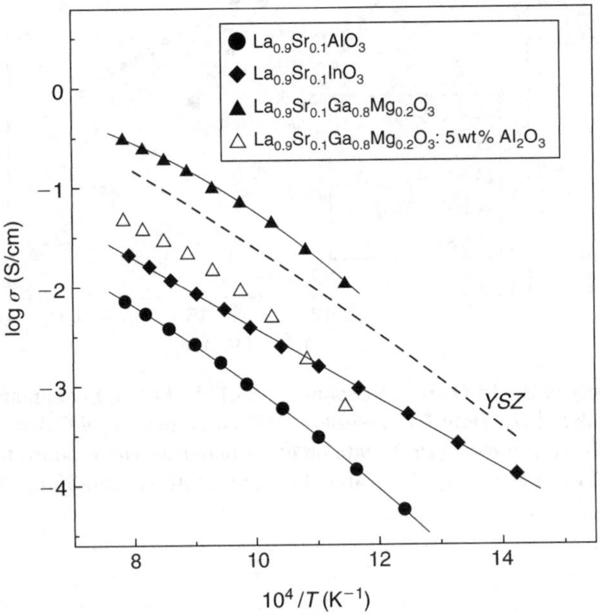

Fig. 17. Total conductivity of $La_{0.9}Sr_{0.1}AlO_{3-\delta}$ and $La_{0.9}Sr_{0.1}InO_{3-\delta}$ in nitrogen atmosphere where the conduction is predominantly ionic [153], compared to 8% YSZ [27], LSGM [24] and one LSGM-Al_2O_3 composite [24]

Table 3. Transport properties of perovskite phases, derived from $LaBO_3$ (B = Al, In, Sc), at 1073 K in air

Composition	Total conductivity $\sigma \times 10^2$ (S/cm)	Ionic transport parameters* t_o	σ_o (S/cm)	E_a (kJ/mol)	Reference
$La_{0.9}Sr_{0.1}AlO_{3-\delta}$	0.6	0.35	2×10^{-3}	92	[153]
$La_{0.9}Sr_{0.1}Al_{0.9}Mg_{0.1}O_{3-\delta}$	0.7	2.9×10^{-2}	2×10^{-4}	87	[154]
$La_{0.9}Sr_{0.1}ScO_{3-\delta}$	4.6	0.15	7×10^{-3}	68	[153]
$LaSc_{0.9}Mg_{0.1}O_{3-\delta}$	2.0	1.0×10^{-2}	2×10^{-4}	92	[155]
$La_{0.9}Sr_{0.1}Sc_{0.9}Mg_{0.1}O_{3-\delta}$	1.9	2.6×10^{-2}	5×10^{-4}	48	[154]
$La_{0.9}Sr_{0.1}InO_{3-\delta}$	2.3	0.29	7×10^{-3}	81	[153]
$La_{0.9}Sr_{0.1}In_{0.9}Mg_{0.1}O_{3-\delta}$	0.2	0.45	9×10^{-4}	–	[154]

*t_o, σ_o and E_a are the ion transference number, partial ionic conductivity and activation energy for ionic transport, respectively.

9. SOLID ELECTROLYTES WITH APATITE STRUCTURE

Extensive studies of another group of oxygen ionic conductors, the apatite-type phases $A_{10-x}(MO_4)_6O_{2\pm\delta}$ where M = Si or Ge, and A corresponds to rare earth and alkaline earth cations, began in the 1990s [165–175]. While germanate apatites seem unlikely to find practical applications due to high volatilization, tendency to glass formation and high costs of GeO_2; the silicate-based systems are of practical interest. These low-cost oxides possess relatively low TECs (Table 2) and a significant level of oxygen ion conductivity (Fig. 18). In the apatite lattice, A-site cations are located in the cavities created by MO_4 tetrahedra with four distinct oxygen positions; additional oxygen sites (O_5) form channels through the structure. As for perovskites, the oxygen ionic transport in $Ln_{10}Si_6O_{27}$ (Ln = La, Pr, Nd, Sm, Gd, Dy) increases with increasing radius of Ln^{3+} cations, with maximum conductivity for the La-containing phase [165,166]. Due to relatively poor sintering, different processing techniques and substantial anisotropy of ionic transport in the apatite lattice, the conductivity values reported in the literature vary in a very broad range even for similar compositions, for example, from 8.4×10^{-5} to 4.3×10^{-3} S/cm at 773 K for $La_{10}Si_6O_{27}$ ceramics [165,166,169]. The highest ionic transport is observed when apatite phases contain more than 26 oxygen ions per unit formula, suggesting a significant role of the interstitial migration mechanism. Decreasing oxygen concentration below the stoichiometric value leads to the vacancy mechanism becoming dominant and there is a considerable drop in ionic conductivity. Similar effects are observed for transition metal-containing apatites in reducing atmospheres (Fig. 15). Another important factor influencing oxygen diffusion is A-site deficiency, which affects the unit cell volume and may cause O5 ion displacement into interstitial sites, thus creating vacancies in the O5 sites at fixed total oxygen content (oxygen Frenkel disorder; Refs. [168,171,175,176]). In particular, an enhanced ionic conduction was found in the system $La_{9.33+x/3}Si_{6-x}Al_xO_{26}$, where Al doping is compensated by the A-site vacancy concentration without oxygen content variations [168]. As for other solid oxide electrolytes [10,15,177], an improvement in the sinterability can be achieved by doping with transition metal cations. Fe in particular has significant solubility in the Si sublattice [52,174]. Incorporation of iron in the lattice of $La_{10-x}(Si,Al)_6O_{26\pm\delta}$ results also in increasing oxygen ionic and electronic transport; the total

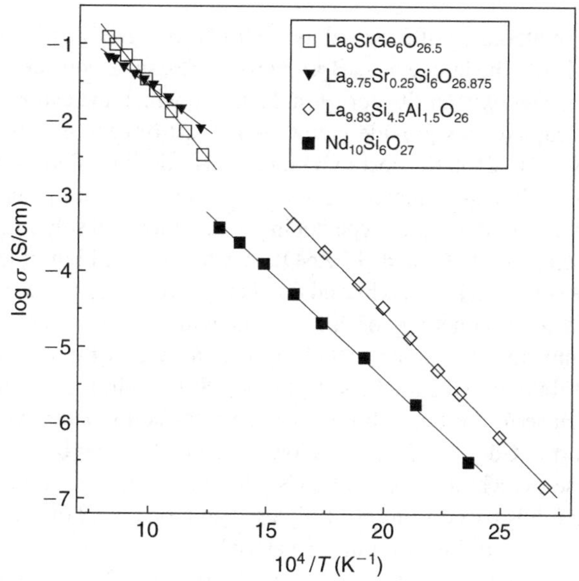

Fig. 18. Total conductivity of several apatite phases in air [168,170,176]

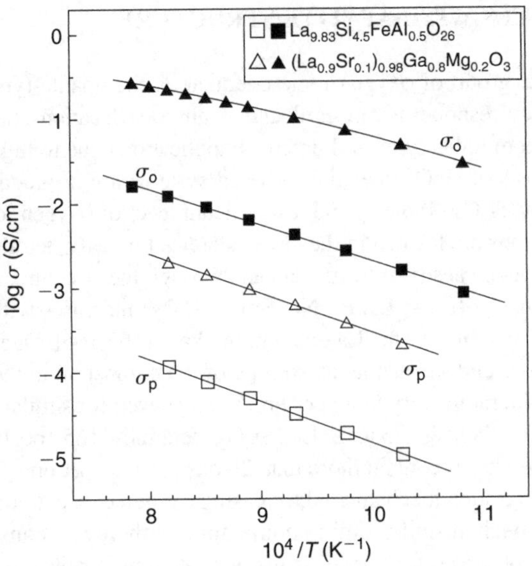

Fig. 19. Comparison of the partial oxygen ionic and p-type electronic conductivities of $La_{9.83}Si_{4.5}FeAl_{0.5}O_{26\pm\delta}$ apatite [52] and $(La_{0.9}Sr_{0.1})_{0.98}Ga_{0.8}Mg_{0.2}O_{3-\delta}$ perovskite [91] in air

conductivity of apatite silicates, where up to 30% of Si sites is occupied with Fe, remains dominantly ionic [52,174]. As an example, Fig. 19 presents data on the ionic and electron-hole conductivities in one Fe-substituted apatite phase.

10. PYROCHLORES AND FLUORITE-TYPE $(Y,Nb,Zr)O_{2-\delta}$

Oxygen ion-conducting materials with pyrochlore structure have been studied since the 1960s [5,6,10,50,51,76,81,178–186]. The lattice of $A_2B_2O_7$ pyrochlore can be considered as an $(A,B)O_2$ fluorite-based structure with one vacant oxygen site per formula unit in which the cations and anion vacancies are both ordered. These unoccupied sites provide pathways for fast oxygen transport. At elevated temperatures, typically as high as 1650–2500 K, most pyrochlores disorder into fluorite polymorphs. Decreasing A-site cation radius favours the "pyrochlore → fluorite" transition. Usually, the maximum conductivity (which can be further enhanced by acceptor-type doping within the solid solution formation limits) occurs for cation stoichiometry (e.g., $Gd_2Ti_2O_7$ and $Gd_2Zr_2O_7$). Up to now, the highest level of oxygen ionic conduction in pyrochlore-type compounds was achieved for $Gd_{2-x}Ca_xTi_2O_{7-\delta}$ with $x \approx 0.20$ [76,180]. Figure 20 compares partial oxygen ionic conductivity of this material with data on other pyrochlores.

Incorporation of calcium into the lattice of $Gd_2Ti_2O_{7-\delta}$ increases p-type electronic conduction and decreases the n-type contribution to the total conductivity [180]. In air, the electron transference numbers of $(Gd,Ca)_2Ti_2O_{7-\delta}$ are close to the upper limit acceptable for solid electrolytes (Fig. 11). Taking into account that ionic transport in $Gd_{2-x}Ca_xTi_2O_{7-\delta}$ is lower than that in stabilized zirconia (Fig. 6), the most likely applications of these pyrochlores are in SOFCs with thick-film electrolytes, or as protective layers applied onto $LaGaO_3$- or CeO_2-based solid electrolytes. The moderate TEC values of $Gd_{2-x}Ca_xTi_2O_{7-\delta}$ ceramics (Table 2) enable compatibility with these materials.

Among other interesting materials, one should note fluorite-type $Y_4NbO_{8.5}$ and $(Y,Nb,Zr)O_{2-\delta}$ solid solutions [29,187,188,189]. The total conductivity of $Y_4NbO_{8.5-\delta}$ was found essentially independent of the

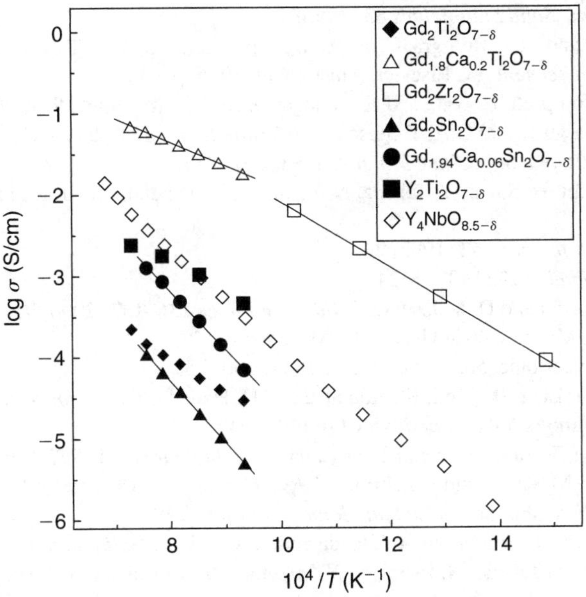

Fig. 20. Oxygen ionic conductivity of selected pyrochlores [76,179,181,182]. Data on total conductivity of fluorite-type $Y_4NbO_{8.5}$ [187], presumably oxygen ionic [188], are shown for comparison

oxygen partial pressure, which may suggest dominant ionic transport [188]. However, the ionic conductivity in this system is rather low (Figs 2 and 20), although further improvement can be achieved by zirconia additions [29].

ACKNOWLEDGEMENTS

Helpful discussions, experimental contributions and the assistance in preparation of this review, made by A. Shaula, E. Tsipis, N. Vyshatko, O. Smirnova, A. Yaremchenko, I. Marozau, F. Figueiredo, A. Viskup and E. Naumovich, are gratefully acknowledged. The authors would like to thank the FCT (Portugal), the SOFCNET contract (CEC, Brussels) and the NATO Science for Peace program for partial financial support.

REFERENCES

1. O. Yamamoto. *Electrochim. Acta* **45** (2000) 2423.
2. T.P. Chen, J.D. Wright and K. Krist. In: U. Stimming, S.C. Singhal, H. Tagawa and W. Lehnert (Eds.), *SOFC V*. The Electrochemical Society, Pennington, NJ, 1997, p. 69 PV 97-40.
3. B.C.H. Steele. *J. Mater. Sci.* **36** (2001) 1053.
4. P. Holtappels, F.W. Poulsen and M. Mogensen. *Solid State Ionics* **135** (2000) 675.
5. H.L. Tuller. In: H. Tuller, J. Schoonman and I. Riess (Eds.), *Oxygen Ion and Mixed Conductors and their Technological Applications*. Kluwer (NATO ASI series), Dordrecht, 2000, p. 245.
6. T.H. Etsell and S.N. Flengas. *Chem. Rev.* **70** (1970) 339.
7. H. Rickert. *Electrochemistry of Solids. An Introduction*. Springer-Verlag, Berlin, 1982.
8. V.N. Chebotin and M.V. Perfilyev. *Electrochemistry of Solid Electrolytes*. Khimiya, Moscow, 1978.
9. M.V. Perfilyev, A.K. Demin, B.L. Kuzin and A.S. Lipilin. *High-Temperature Electrolysis of Gases*. Nauka, Moscow, 1988.
10. V.V. Kharton, E.N. Naumovich and A.A. Vecher. *J. Solid State Electrochem.* **3** (1999) 61.

11. H. Inaba and H. Tagawa. *Solid State Ionics* **83** (1996) 1.
12. H.J.M. Bouwmeester and A.J. Burggraaf. In: A. Burggraaf and L. Cot (Eds.), *Fundamentals of Inorganic Membrane Science and Technology*. Elsevier, Amsterdam, 1996, p. 435.
13. N.M. Sammes, G.A. Tompsett, H. Nafe and F. Aldinger. *J. Eur. Ceram. Soc.* **19** (1999) 1801.
14. M. Mogensen, N.M. Sammes and G.A. Tompsett. *Solid State Ionics* **129** (2000) 63.
15. J.P.P. Huijsmans. *Curr. Opin. Solid State Mater. Sci.* **5** (2001) 317.
16. O. Yamamoto, Y. Arachi, H. Sakai, Y. Takeda, N. Imanishi, Y. Mizutani, M. Kawai and Y. Nakamura. *Ionics* **4** (1998) 403.
17. S.P.S. Badwal. *Solid State Ionics* **52** (1992) 23.
18. A.N. Vlasov. *Elektrokhimiya* **19** (1983) 1624.
19. S.P.S. Badwal, F.T. Ciacchi and D. Milosevic. *Solid State Ionics* **136–137** (2000) 91.
20. A.I. Ioffe. PhD Thesis. Moscow State University, Moscow, 1977.
21. J. Drennan and G. Auchterlonie. *Solid State Ionics* **134** (2000) 75.
22. J.-H. Lee, T. Mori, J.-G. Li, T. Ikegami, M. Komatsu and H. Haneda. *J. Electrochem. Soc.* **147** (2000) 2822.
23. A. Yuzaki and A. Kishimoto. *Solid State Ionics* **116** (1999) 47.
24. I. Yasuda, Y. Matsuzaki, T. Yamakawa and T. Koyama. *Solid State Ionics* **135** (2000) 381.
25. M.C. Steil, J. Fouletier, M. Kleitz and P. Labrune. *J. Eur. Ceram. Soc.* **19** (1999) 815.
26. R. Chiba, T. Ishii and F. Yoshimura. *Solid State Ionics* **91** (1996) 249.
27. M. Mori, T. Abe, H. Itoh, O. Yamomoto, Y. Takeda and T. Kawahara. *Solid State Ionics* **74** (1994) 157.
28. O. Yamomoto, Y. Arati, Y. Takeda, N. Imanishi, Y. Mizutani, M. Kawai and Y. Nakamura. *Solid State Ionics* **79** (1995) 137.
29. J.-H. Lee and M. Yoshimura. *Solid State Ionics* **124** (1999) 185.
30. J.W. Patterson. *J. Electrochem. Soc.* **118** (1971) 1033.
31. T. Takahashi, T. Esaka and H. Iwahara. *J. Appl. Electrochem.* **7** (1977) 303.
32. T. Kudo and H. Obayashi. *J. Electrochem. Soc.* **123** (1976) 415.
33. J.-H. Kim and H.-I. Yoo. *Solid State Ionics* **140** (2001) 105.
34. J.W. Stevenson, T.R. Armstrong, L.R. Pederson, J. Li, C.A. Levinsohn and S. Baskaran. *Solid State Ionics* **113–115** (1998) 571.
35. V.K. Gilderman, A.D. Neuimin and S.F. Palguev. *Dokl. Akad. Nauk Ukr. SSR* **218** (1974) 133.
36. E.Kh. Kurumchin. DSc Thesis. Institute of High-Temperature Electrochemistry RAS, Ekaterinburg, Russia, 1997.
37. R. Maenner, E. Ivers-Tiffée, W. Wersing and W. Kleinlien. In: G. Ziegler and H. Hausner (Eds.), *Proceedings of the 2nd European Ceramic Society Conference (Euro-Ceramics II)*. Deutshe Keramische Gesellschaft, 1991, p. 2085.
38. F. Tietz, G. Stochniol and A. Naoumidis. In: L. Sarton and H. Zeedijk (Eds.), *Proceedings of the 5th European Conference on Advanced Materials, Processes and Applications (Euromat 97)*, Vol. 2. Netherlands Society for Materials Science, 1997, p. 271.
39. F. Tietz. *Ionics* **5** (1999) 129.
40. E. Eldre. *Science and Technology of Zirconia II*. American Ceramic Society, Columbus, OH, 1984, p. 685.
41. E. Ivers-Tiffée, W. Wersing, M. Schießl and H. Greiner. *Berich. Bunsen Gesell. Phys. Chem.* **94** (1990) 978.
42. P. Larsen, C. Bagger, M. Mogensen and J. Larsen. In: M. Dokiya, O. Yamamoto, H. Tagawa and S.C. Singhal (Eds.), *SOFC-IV*. The Electrochemical Society, Pennington, NJ, 1995, p. 69 PV95-1.
43. O. Yamomoto, Y. Takeda, R. Kanno, K. Kohno and T. Kamiharai. *J. Mater. Sci. Lett.* **8** (1989) 198.
44. Y. Mizutani, M. Tamura, M. Kawai and O. Yamamoto. *Solid State Ionics* **72** (1994) 271.
45. H. Hayashi, M. Kanoh, C. Quan, H. Inaba, S. Wang, M. Dokiya and H. Tagawa. *Solid State Ionics* **132** (2000) 227.
46. V.V. Kharton, F.M. Figueiredo, L. Navarro, E.N. Naumovich, A.V. Kovalevsky, A.A. Yaremchenko, A.P. Viskup, A. Carneiro, F.M.B. Marques and J.R. Frade. *J. Mater. Sci.* **36** (2001) 1105.
47. M. Mogensen, T. Lindegaard, U.R. Hansen and G. Mogensen. *J. Electrochem. Soc.* **141** (1994) 2122.
48. J.W. Stevenson, K. Hasinska, N.L. Canfield and T.R. Armstrong. *J. Electrochem. Soc.* **147** (2000) 3213.
49. T.L. Nguen and M. Dokiya. *Solid State Ionics* **132** (2000) 217.
50. M. Mori, G.M. Tompsett, N.M. Sammes, E. Suda and Y. Takeda. *Solid State Ionics* **158** (2003) 79.
51. V.V. Kharton, E.V. Tsipis, A.A. Yaremchenko, N.P. Vyshatko, A.L. Shaula, E.N. Naumovich and J.R. Frade. *J. Solid State Electrochem.* **7** (2003) 468.
52. A.L. Shaula, V.V. Kharton, J.C. Waerenborgh, D.P. Rojas, E.V. Tsipis, N.P. Vyshatko, M.V. Patrakeev and F.M.B. Marques. *Mater. Res. Bull.* **39** (2004) 763.

53. V.V. Kharton, A.L. Shaula, M.V. Patrakeev, J.C. Waerenborgh, D.P. Rojas, N.P. Vyshatko, E.V. Tsipis, A.A. Yaremchenko and F.M.B. Marques. *J. Electrochem. Soc.* **151** (2004) A1236.
54. I.P. Marozau, A.L. Shaula, V.V. Kharton, N.P. Vyshatko, A.P. Viskup, J.R. Frade and F.M.B. Marques. *Mater. Res. Bull.* (2005) (in press).
55. A.A. Yaremchenko, V.V. Kharton, E.N. Naumovich and F.M.B. Marques. *J. Electroceram.* **4** (2000) 235.
56. A.A. Yaremchenko, V.V. Kharton, E.N. Naumovich, A.A. Tonoyan and V.V. Samokhval. *J. Solid State Electrochem.* **2** (1998) 308.
57. T. Ishihara, H. Matsuda and Y. Takita. *J. Am. Chem. Soc.* **116** (1994) 3801.
58. M. Feng and J.B. Goodenough. *Eur. J. Solid State Inorg. Chem.* **31** (1994) 663.
59. P. Huang and A. Petric. *J. Electrochem. Soc.* **143** (1996) 1644.
60. J.W. Stevenson, T.R. Armstrong, D.E. McGready, L.R. Pederson and W.J. Weber. *J. Electrochem. Soc.* **144** (1997) 3613.
61. J. Drennan, V. Zelizko, D. Hay, F.T. Ciacci, S. Rajendran and S.P. Badwal. *J. Mater. Chem.* **7** (1997) 79.
62. R.T. Baker, B. Gharbage and F.M.B. Marques. *J. Electrochem. Soc.* **144** (1997) 3130.
63. V.V. Kharton, A.P. Viskup, E.N. Naumovich and N.M. Lapchuk. *Solid State Ionics* **104** (1997) 67.
64. T. Ishihara, M. Higuchi, H. Furutani, T. Fukushima, H. Nishiguchi and Y. Takita. *J. Electrochem. Soc.* **144** (1997) L122.
65. M.S. Khan, M.S. Islam and D.R. Bates. *J. Phys. Chem. B* **102** (1998) 3099.
66. K. Huang, R.S. Tichy and J.B. Goodenough. *J. Am. Ceram. Soc.* **81** (1998) 2565.
67. T. Ishihara, J.A. Kilner, M. Honda, N. Sakai, H. Yokokawa and Y. Takita. *Solid State Ionics* **113–115** (1998) 593.
68. T. Ishihara, T. Akbay, H. Furutani and Y. Takita. *Solid State Ionics* **113–115** (1998) 585.
69. N. Trofimenko and H. Ullmann. *Solid State Ionics* **118** (1999) 215.
70. V.V. Kharton, A.P. Viskup, A.A. Yaremchenko, R.T. Baker, B. Gharbage, G.C. Mather, F.M. Figueiredo, E.N. Naumovich and F.M.B. Marques. *Solid State Ionics* **132** (2000) 119.
71. P.R. Slater, J.T.S. Irvine, T. Ishihara and Y. Takita. *Solid State Ionics* **107** (1998) 319.
72. K. Huang, M. Feng, J.B. Goodenough and C. Milliken. *J. Electrochem. Soc.* **144** (1997) 3620.
73. H. Hayashi, H. Inaba, M. Matsuyama, N.G. Lan, M. Dokiya and H. Tagawa. *Solid State Ionics* **122** (1999) 1.
74. E. Djurado and M. Labeau. *J. Eur. Ceram. Soc.* **18** (1998) 1397.
75. K. Yamaji, T. Horita, M. Ishikawa, N. Sakai and H. Yokokawa. *Solid State Ionics* **121** (1999) 217.
76. S.A. Kramer and H.L. Tuller. *Solid State Ionics* **82** (1995) 15.
77. H. Arikawa, H. Nishiguchi, T. Ishihara and Y. Takita. *Solid State Ionics* **136–137** (2000) 31.
78. A.A. Yaremchenko, M. Avdeev, V.V. Kharton, A.V. Kovalevsky, E.N. Naumovich and F.M.B. Marques. *Mater. Chem. Phys.* **77** (2002) 552.
79. B.C.H. Steele. *Solid State Ionics* **129** (2000) 95.
80. M. Goedickemeier and L.J. Gauckler. *J. Electrochem. Soc.* **145** (1998) 414.
81. V.V. Kharton, A.A. Yaremchenko, E.N. Naumovich and F.M.B. Marques. *J. Solid State Electrochem.* **4** (2000) 243.
82. S.J. Hong, K. Mehta and A.V. Virkar. *J. Electrochem. Soc.* **145** (1998) 638.
83. H. Yahiro, K. Eguchi and H. Arai. *Solid State Ionics* **36** (1989) 71.
84. B.C.H. Steele. *Solid State Ionics* **129** (2000) 95–110.
85. A. Atkinson. *Solid State Ionics* **95** (1997) 249.
86. I. Yasuda and M. Hishinuma. In: T.A. Ramanarayanan (Ed.), *Ionic and Mixed Conducting Ceramics III*. The Electrochemical Society, Pennington, NJ, 1998, p. 178 PV97-24.
87. F.M.B. Marques and L.M. Navarro. *Solid State Ionics* **100** (1997) 29.
88. A. Tsoda, A. Gupta, A. Naoumoidis, D. Skarmoutsos and P. Nikolopoulos. *Ionics* **4** (1998) 234.
89. M. Hrovat, A. Ahmad-Khanlou, Z. Samardzija and J. Holz. *Mater. Res. Bull.* **34** (1999) 2027.
90. A. Atkinson and T. Ramos. *Solid State Ionics* **129** (2000) 259.
91. V.V. Kharton, A.L. Shaula, N.P. Vyshatko and F.M.B. Marques. *Electrochim. Acta* **48** (2003) 1817.
92. V.V. Kharton, A.P. Viskup, F.M. Figueiredo, E.N. Naumovich, A.A. Yaremchenko and F.M.B. Marques. *Electrochim. Acta* **46** (2001) 2879.
93. R. Doshi, V.L. Richards, J.D. Carter, X.P. Wang and M. Krumpelt. *J. Electrochem. Soc.* **146** (1999) 1273.
94. J. Liu, B.D. Madsen, Z.Q. Ji and S.A. Barnett. *Electrochem. Solid State Lett.* **5** (2002) A122.
95. C.R. Xia and M.L. Liu. *J. Am. Ceram. Soc.* **84** (2001) 1903.
96. C. Kleinlogel and L.J. Gauckler. *Solid State Ionics* **135** (2000) 567.

97. G.S. Lewis, A. Atkinson, B.C.H. Steele and J. Drennan. *Solid State Ionics* **152–153** (2002) 567.
98. D.P. Fagg, V.V. Kharton and J.R. Frade. *J. Electroceram.* **9** (2002) 199.
99. H.L. Tuller. *Solid State Ionics* **131** (2000) 143.
100. S. Kim and J. Maier. *J. Electrochem. Soc.* **149** (2002) J73.
101. T. Suzuki, I. Kosacki and H.U. Anderson. *Solid State Ionics* **151** (2002) 111.
102. B. Zhu. *Solid State Ionics* **119** (1999) 305.
103. B. Zhu. *J. Power Sources* **114** (2003) 1.
104. P.N. Dyer, R.E. Richards, S.L. Russek and D.M. Taylor. *Solid State Ionics* **134** (2000) 21.
105. D. Waller, J.A. Kilner and B.C.H. Steele. In: H.U. Anderson, A.C. Khandkar and M. Liu (Eds.), *Ceramic Membranes I*. The Electrochemical Society, Pennington, NJ, 1997, p. 48 PV95-24.
106. J.B. Goodenough, J.E. Ruiz-Dias and Y.S. Zhen. *Solid State Ionics* **44** (1990) 21.
107. V.V. Kharton, E.N. Naumovich, A.A. Yaremchenko and F.M.B. Marques. *J. Solid State Electrochem.* **5** (2001) 160.
108. P. Shuk, H.-D. Wiemhofer, U. Guth, W. Gopel and M. Greenblatt. *Solid State Ionics* **89** (1996) 179.
109. T. Takahashi, H. Iwahara and T. Arao. *J. Appl. Electrochem.* **5** (1975) 187.
110. N. Jiang, E.D. Wachsman and S. Jung. *Solid State Ionics* **150** (2002) 347.
111. K.R. Kendall, J.K. Thomas and H.-C. zur Loye. *Solid State Ionics* **70/71** (1994) 221.
112. H. Kruidhof, H.J.M. Bouwmeester, K.J. de Vries, P.J. Gellings and A.J. Burggraaf. *Solid State Ionics* **50** (1992) 181.
113. K.-Z. Fung, J. Chen and A.V. Virkar. *J. Am. Ceram. Soc.* **76** (1993) 2403.
114. A. Watanabe. *Solid State Ionics* **86–88** (1996) 1427.
115. A.A. Yaremchenko, V.V. Kharton, E.N. Naumovich and A.A. Tonoyan. *Mater. Res. Bull.* **35** (2000) 515.
116. S. Boypati, E.D. Wachsman and N. Jiang. *Solid State Ionics* **140** (2001) 149.
117. K. Huang, M. Feng and J.B. Goodenough. *Solid State Ionics* **89** (1996) 17.
118. F. Abraham, J.C. Boivin, G. Mairesse and G. Nowogrocki. *Solid State Ionics* **40/41** (1990) 934.
119. R.N. Vanier, G. Mairesse, F. Abraham and G. Nowogrocki. *Solid State Ionics* **70/71** (1994) 248.
120. E. Pernot, M. Anne, M. Bacmann, P. Strobel, J. Fouletier, R.N. Vannier, G. Mairesse, F. Abraham and G. Nowogrocki. *Solid State Ionics* **70/71** (1994) 259.
121. G. Mairesse, J.C. Boivin, G. Lagrange and P. Cocolios. International Patent Application PCT WO 94/06544 (1994).
122. F. Krok, W. Bogusz, W. Jakubowski, J.R. Dygas and D. Bangobango. *Solid State Ionics* **70/71** (1994) 211.
123. Y.L. Yang, L. Qiu, W.T.A. Harrison, R. Christoffersen and A.J. Jacobson. *J. Mater. Chem.* **7** (1997) 243.
124. A.J. Francklin, A.V. Chadwick and J.W. Couves. *Solid State Ionics* **70/71** (1994) 215.
125. C.K. Lee, D.C. Sinclair and A.R. West. *Solid State Ionics* **62** (1993) 193.
126. J.C. Boivin, C. Pirovano, G. Nowogrocki, G. Mairesse, Ph. Labrune and G. Lagrange. *Solid State Ionics* **113–115** (1998) 639.
127. A. Watanabe and K. Das. *J. Solid State Chem.* **163** (2002) 224.
128. R.N. Vannier, R.J. Chater, S.J. Skinner, J.A. Kilner and G. Mairesse. *Solid State Ionics* **160** (2003) 327.
129. V.N. Tikhonovich, E.N. Naumovich, V.V. Kharton, A.A. Yaremchenko, A.V. Kovalevsky and A.A. Vecher. *Electrochim. Acta* **47** (2002) 3957.
130. C.M. Steil, J. Fouletier, M. Kleitz, G. Lagrange, P. Del Gallo, G. Mairesse and J.-C. Boivin. US Patent 6207038 (2001).
131. E.D. Wachsman, P. Jayaweera, N. Jiang, D.M. Lowe and B.G. Pound. *J. Electrochem. Soc.* **144** (1997) 233.
132. P. Lacorre, F. Goutenoire, O. Bohnke, R. Retoux and Y. Laligant. *Nature* **404** (2000) 856.
133. F. Goutenoire, O. Isnard and P. Lacorre. *Chem. Mater.* **12** (2000) 2575.
134. F. Goutenoire, O. Isnard, E. Suard, O. Bohnke, Y. Laligant, R. Retoux and P. Lacorre. *J. Mater. Chem.* **10** (2000) 1.
135. X.P. Wang and Q.F. Fang. *Solid State Ionics* **146** (2002) 185.
136. J.B. Goodenough. *Nature* **404** (2000) 821.
137. A.V. Kovalevsky, V.V. Kharton and E.N. Naumovich. *Mater. Lett.* **38** (1999) 300.
138. A.Y. Neiman. *Solid State Ionics* **83** (1996) 263.
139. J.B. Goodenough. *Solid State Ionics* **94** (1997) 17.
140. A. Manthiram, J.F. Kuo and J.B. Goodenough. *Solid State Ionics* **62** (1993) 225.
141. M. Schwartz, B.F. Link and A.F. Sammells. *J. Electrochem. Soc.* **140** (1993) L62.
142. G.B. Zhang and D.M. Smyth. *Solid State Ionics* **82** (1995) 153.
143. C.A.J. Fisher, B. Derby and R.J. Brook. *Br. Ceram. Proc.* **56** (1996) 25.

144. T. Schober. *Solid State Ionics* **109** (1998) 1.
145. T. Schober and J. Friedrich. *Solid State Ionics* **113–115** (1998) 369.
146. C.A.J. Fisher and M.S. Islam. *Solid State Ionics* **118** (1999) 355.
147. T. Schober, J. Friedrich and F. Krug. *Solid State Ionics* **99** (1997) 9.
148. T. Hashimoto, Y. Inagaki, A. Kishi and M. Dokiya. *Solid State Ionics* **128** (2000) 227.
149. H. Yamamura, Y. Yamada, T. Mori and T. Atake. *Solid State Ionics* **108** (1998) 377.
150. T. Yao, Y. Uchimoto, M. Kinuhata, T. Inagaki and H. Yoshida. *Solid State Ionics* **132** (2000) 189.
151. P. Berastegui, S. Hull, F.J. Garcia-Garcia and S.-G. Eriksson. *J. Solid State Chem.* **164** (2002) 119.
152. T. Takahashi and H. Iwahara. *Energ. Convers.* **11** (1971) 105.
153. K. Nomura and S. Tanase. *Solid State Ionics* **98** (1997) 229.
154. D. Lybye, F.W. Poulsen and M. Mogensen. *Solid State Ionics* **128** (2000) 91.
155. H. Fujii, Y. Katayama, T. Shimura and H. Iwahara. *J. Electroceram.* **2** (1998) 119.
156. J. Mizusaki, I. Yasuda, J. Shimoyama, S. Yamauchi and K. Fueki. *J. Electrochem. Soc.* **140** (1993) 467.
157. H. Matsuda, T. Ishihara, Y. Mizuhara and Y. Takita. In: S.C. Singhal and H. Iwahara (Eds.), *SOFC III*. The Electrochemical Society, Pennington, NJ, 1993, p. 129 PV93-4.
158. J. Ranlov, M. Mogensen and F.W. Poulsen. In: F.W. Poulsen, J.J. Bentzen, T. Jacobsen, E. Skou and M.J.L. Ostergard (Eds.). *Proceedings of the 14th Riso International Symposium on Materials Science*. Riso National Laboratory, Roskilde, Denmark, 1993, p. 389.
159. T. Ishihara, H. Matsuda and Y. Takita. *J. Electrochem. Soc.* **141** (1994) 3444.
160. P.S. Anderson, F.M.B. Marques, D.C. Sinclair and A.R. West. *Solid State Ionics* **118** (1999) 229.
161. T.L. Nguen, M. Dokiya, S. Wang, H. Tagawa and T. Hashimoto. *Solid State Ionics* **130** (2000) 229.
162. H. He, X. Huang and L. Chen. *Solid State Ionics* **130** (2000) 183.
163. H. Yamamura, K. Yamazaki, K. Kakinuma and K. Nomura. *Solid State Ionics* **150** (2002) 255.
164. V. Thangadurai and W. Weppner. *J. Electrochem. Soc.* **148** (2001) A1294.
165. S. Nakayama, T. Kageyama, H. Aono and Y. Sadaoka. *J. Mater. Chem.* **5** (1995) 1801.
166. S. Nakayama and M. Sakamoto. *J. Eur. Ceram. Soc.* **18** (1998) 1413.
167. C. Barthet, B. Pintault and J.-Y. Poinso. In: S.C. Singhal and H. Yokokawa (Eds.), *SOFC-VII*. The Electrochemical Society, Pennington, NJ, 2001, p. 431 PV2001-16.
168. E.J. Abram, D.C. Sinclair and A.R. West. *J. Mater. Chem.* **11** (2001) 1978.
169. S. Tao and J.T.S. Irvine. *Mater. Res. Bull.* **36** (2001) 1245.
170. H. Arikawa, H. Nishiguchi, T. Ishihara and Y. Takita. *Solid State Ionics* **136–137** (2000) 31.
171. J.E.H. Sansom, L. Hildebrandt and P.R. Slater. *Ionics* **8** (2002) 155.
172. P. Berastegui, S. Hull, F.J. Garcia-Garcia and J. Grins. *J. Solid State Chem.* **168** (2002) 294.
173. P.R. Slater, J.E.H. Sansom, J.R. Tolchard and M.S. Islam. *Mater. Res. Soc. Symp. Proc.* **756** (2003) 467.
174. J. McFarlane, S. Barth, M. Swaffer, J.E.H. Sansom and P.R. Slater. *Ionics* **8** (2002) 149.
175. J.R. Tolchard, M.S. Islam and P.R. Slater. *J. Mater. Chem.* **13** (2003) 1956.
176. S. Nakayama, M. Sakamoto, M. Higuchi, K. Kodaira, M. Sato, S. Kakita, T. Suzuki and K. Itoh. *J. Eur. Ceram. Soc.* **19** (1999) 507.
177. D.P. Fagg, V.V. Kharton and J.R. Frade. *J. Electroceram.* **9** (2002) 199.
178. M.P. van Dijk, K.J. de Vries and A.J. Burggraaf. *Solid State Ionics* **9–10** (1983) 913.
179. M.P. van Dijk, K.J. de Vries and A.J. Burggraaf. *Solid State Ionics* **16** (1985) 211.
180. S. Kramer, M. Spears and H.L. Tuller. *Solid State Ionics* **72** (1994) 59.
181. T.-H. Yu and H.L. Tuller. *J. Electroceram.* **2** (1998) 49.
182. S. Yamaguchi, K. Kobayashi, K. Abe, S. Yamazaki and Y. Iguchi. *Solid State Ionics* **113–115** (1998) 393.
183. R.E. Williford and W.J. Weber. *J. Am. Ceram. Soc.* **82** (1999) 3266.
184. L. Minervini and R.W. Grimes. *J. Am. Ceram. Soc.* **83** (2000) 1873.
185. B.J. Wuensch, K.W. Eberman, C. Heremans, E.M. Ku, P. Onnerud, E.M.E. Yeo, S.M. Haile, J.K. Stalick and J.D. Jorgensen. *Solid State Ionics* **129** (2000) 111.
186. M. Pirzada, R.W. Grimes, L. Minervini, J.F. Maguire and K.E. Sickafus. *Solid State Ionics* **140** (2001) 201.
187. J.-H. Lee, M. Kakihana and M. Yoshimura. *J. Am. Ceram. Soc.* **81** (1998) 894.
188. J.-H. Lee, M. Yashima and M. Yoshimura. *Solid State Ionics* **107** (1998) 47.
189. H. Yamamura, K. Matsui, K. Kakinuma and T. Mori. *Solid State Ionics* **123** (1999) 279.

Chapter 12

A review on the status of anode materials for solid oxide fuel cells

W.Z. Zhu and S.C. Deevi

Abstract

Present review is aimed at providing a state-of-the art development of anode for solid oxide fuel cell (SOFC) with principal emphasis on the materials aspect. The criteria for the anode of SOFC are first presented. The prospects and problems of the currently developed anode materials are elucidated. In particular, the electrochemical properties of the Ni/YSZ cermet anode that is the most commonly employed in the establishment of SOFC stack is described along with various approaches attempted for their improvements. The advantages and disadvantages of other anode materials are compared to offer some insights for the research and development of new generation of anode materials for SOFC.

Keywords: Solid oxide fuel cell; Anode; Mixed ionic-electronic conductor

Article Outline

1. Introduction 215
2. Development of anode materials 217
   2.1. Ni–$ZrO_2(Y_2O_3)$ cermet 217
   2.2. CeO_2 (rare-earth doped) anode 226
   2.3. Other anode materials 229
3. Conclusions 230
Acknowledgements 230
References 231

1. INTRODUCTION

Solid oxide fuel cell (SOFC) is an electrochemical device that converts the energy of a chemical reaction directly into electrical energy. Owing to the utilization of solid electrolyte and highest operating temperatures (typically at 700–900°C), it offers many advantages over conventional power-generating systems in terms of efficiency, reliability, modularity, fuel flexibility, and environmental friendliness [1,2]. In addition, SOFCs offer the possibility of co-generation with gas turbine power systems to enable full exploitation of both electricity and heat, thereby enhancing the efficiency up to approximately 70% [3–7]. Single SOFC cell essentially consists of two porous electrodes separated by a dense, oxygen-conducting electrolyte. The operating principle of such a cell is schematically illustrated in Fig. 1. On the cathode (air electrode)

Fuel Cells Compendium

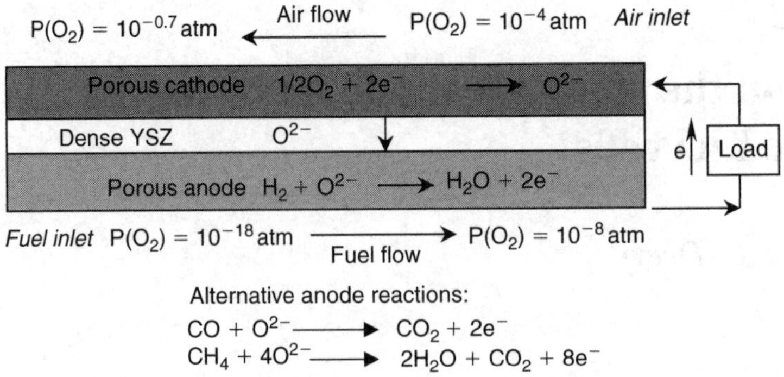

Fig. 1. Schematic drawing showing the working principle of a solid oxide fuel cell operating on hydrogen

side, oxygen reacts with incoming electrons from external load to become oxygen ions that migrate through the electrolyte. On the anode (fuel electrode) side, fuel is oxidized by incoming oxygen ions to liberate the electrons that flow through the external electrical circuit. The charge flow in the external circuit is balanced by ionic current flow within the electrolyte. More specifically, the oxygen is disassociated and converted to oxygen ions at cathode/electrolyte interface, whereas the electrochemical oxidation of fuel takes place at anode/electrolyte interface. The ideal voltage ($E°$) from a single cell under open circuit conditions is close to 1.0 V dc as calculated from the Nernst equation. However, the useful voltage output (V) under load conditions, that is, when a current passes through the cell, is given by [8]

$$V = E° - IR - \eta_c - \eta_a \tag{1}$$

where I is the current passing through the cell, R the electrical resistance of the cell, and η_c and η_a the polarization losses associated with the cathode and anode, respectively. The voltage loss due to internal electrical resistance encompasses the contributions from electrodes and electrolyte with overwhelming contribution coming from electrolyte on account of its ionic conduction in nature. To minimize the IR loss, the increasingly preferred practice is to fabricate dense, gas-tight electrolyte membrane as thin as possible [9,10].

Among the SOFC components, the porous anode serves to provide electrochemical reaction sites for oxidation of the fuel, allow the fuel and byproducts to be delivered and removed from the surface sites, and to provide a path for electrons to be transported from the electrolyte/anode reaction sites to the interconnect in SOFC stacks. Interconnect is a component to connect anode of one cell with cathode of another so that voltage output can be enhanced for practical application. The characteristics of high operating temperature of SOFC present special challenges related to materials degradation, in particular, constitute one of the toughest criteria for the dimensional and chemical stability of anode material itself in reducing atmospheres. Besides, desired anode is supposed to be mixed conductor with predominant electronic conductivity to permit the passage of electrons. It should display no reaction with neighboring electrolyte and interconnect (chemical compatibility with adjacent components). It should also have thermal expansion coefficient close to those of adjoining components. It must show high electrocatalytic activity toward oxidation of fuel gases, and preferably, desired catalytic activity toward the hydrocarbon reforming. High wettability with respect to the electrolyte substrate is highly advantageous for competitive anode. It is evident that anode should have continuous channels made of pores to allow rapid transport of fuel and reactant gases. Anode is supposed to exhibit excellent carburization and sulfidation resistance. Most advantageously, the anode ought to be fuel flexible, ease of fabrication, and low cost are of tremendous importance for a wide range of commercial applications. Unfortunately, no current working anodes can fulfil all of the above requirements.

The electrical resistance of anode is essentially comprised of internal resistance, contact resistance, concentration polarization resistance, and activation polarization resistance. The internal resistance refers to the resistance to the transport of electrons within the anode, and therefore, is determined by the magnitude of electronic resistivity and thickness of anode. The contact resistance is caused by poor adherence between anode and electrolyte. Concentration polarization is related to the transport of gaseous species through the porous electrodes and, thus, its magnitude is dictated by the microstructure of the electrode, specifically, the volume percent porosity, the pore size, and the tortuosity factor. Activation polarization is related to the charge transfer processes and depends on the area of electrode/electrolyte/gas triple-phase boundaries (TPB) and the electrocatalytic activity of the electrode itself. It is worthwhile to stress that a significant ionic contribution to the overall conductivity, a property related to catalytic activity, seems to be a critical requirement for anode design. The effective electrochemical reaction zone (ERZ) at anode of SOFC is mainly limited to the physical TPB (electrolyte/anode/fuel), if the anode exhibits solely electronic conduction. The double-phase boundary (contact plane) between electrolyte and anode cannot work effectively in the anodic reaction. In contrast, the use of mixed-conducting anode is expected to drastically enlarge the ERZ over the entire anode-gas interfacial area. This can lead to a significant drop in the activation polarization and yield remarkable improvement in electrical efficiency. Present overview is aimed at providing a state-of-art development of anode for SOFCs with principal emphasis on the materials aspects. The advantages and disadvantages of various recently developed anodes are described and compared to provide some guidelines in search for new generation of anode materials for SOFCs.

2. DEVELOPMENT OF ANODE MATERIALS

2.1. *Ni–ZrO_2(Y_2O_3) cermet*

Porous Ni/YSZ cermet (YSZ, yttria stabilized zirconia) is currently the most common anode material for SOFC applications because of its low cost. It is also chemically stable in reducing atmospheres at high temperatures and its thermal expansion coefficient is close to that of YSZ-electrolyte. More importantly, the intrinsic charge transfer resistance that is associated with the electrocatalytic activity at Ni/YSZ boundary is low. More than 30%v/v, of continuous porosity is required to facilitate the transport of reactant and product gases. Nickel serves as an excellent reforming catalyst and electrocatalyst for electrochemical oxidation of hydrogen. It also provides predominant electronic conductivity for anode. The YSZ constitutes a framework for the dispersion of Ni particles and acts as an inhibitor for the coarsening of Ni powders during both consolidation and operation. Additionally, it offers a significant part of ionic contribution to the overall conductivity, thus effectively broadening the three-phase boundaries. Finally, the thermal expansion coefficient of anode can be managed to match with those of other SOFC components as YSZ is mixed with Ni in an arbitrary ratio [11].

Ni/YSZ cermet is currently a preferred anode since, Ni and YSZ are essentially immiscible in each other and non-reactive over a very wide temperature. This enables the preparation of a NiO + YSZ composite via conventional sintering followed by reduction upon exposure to fuel gases. The subsequent development of a very fine microstructure can be maintained during service for relatively a long period of time.

The Ni/YSZ anode was previously applied onto the electrolyte by a two-step process as exemplified in the fabrication of tubular-configured SOFCs. Nickel powder slurry was deposited over the electrolyte followed by the growth of YSZ around the nickel particles induced by the electrochemical vapor deposition (EVD) approach. However, a more cost-effective fabrication method involving the simultaneous

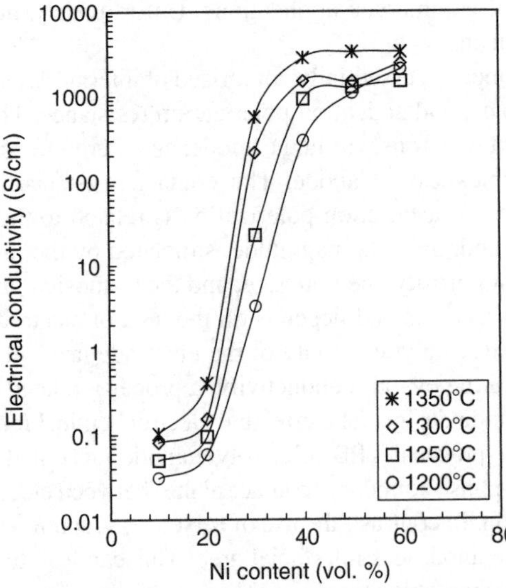

Fig. 2. Variation of electrical conductivity measured at 1000°C as a function of nickel concentration of $Ni/ZrO_2(Y_2O_3)$ cermet fired at different temperatures [17]

deposition of Ni/YSZ slurry over the electrolyte coupled with adequate final sintering has yielded anodes that are equivalent in performance to those fabricated by the EVD process. The use of this non-EVD process results in a substantial reduction in the manufacture cost of SOFCs.

A two-layered anode structure for the anode-supported solid oxide fuel cell has been proposed by Virkar et al. [12] in an attempt to minimize both the concentration and activation polarizations. The anode interlayer with a thickness of 10–50 μm is prepared by spray coating the slurry of NiO + YSZ mixture of one composition. The outer anode support layer is fabricated by tape casting the slurry of NiO + YSZ mixture of another composition. By controlling the slurry formulations, the microstructure of the interlayer is made finer than that of outer support layer in terms of volume percent porosity, pore size, and its distribution. The coarse outer layer facilitates rapid transport of fuel gases into and removal of reactant gases out of anode so that the concentration polarization can be retarded. The fine interlayer is intended to maximize the Ni/YSZ/gas triple boundary area so that the number of electrochemical reaction sites can be multiplied and activation polarization can be lowered. Another promising approach of preparing interlayer anode, as demonstrated by Muller et al. [13] in their effort to develop multilayer anode, is directly screen printing anode paste onto electrolyte green tape followed by co-sintering. A low anode polarization resistance is recorded after relatively long-term exposure to anode-like environment, which is largely attributed to the desirable anode/electrolyte adherence. However, the rapid mechanical degradation of bilayer anode under compression mode is brought to attention by a recent study [14] which indicates that compressive creep behavior of the composite at elevated temperatures is governed by the deformation of Ni rather than that of the ceramic phase. This suggests that stress endured by the anode be kept very low so that mechanical integrity of entire stack can be preserved throughout the operation.

The electrical conductivity of Ni/YSZ cermet is strongly dependent on its Ni content. The conductivity of the cermet as a function of Ni content shows the S-shaped curve predicted by percolation theory (Fig. 2) [15–17]. The percolation threshold for the conductivity is at about 30 vol.% nickel. Below this threshold, the cermet exhibits predominantly ionic conducting behavior. Above 30 vol.% nickel, the conductivity is about three-orders of magnitude higher, corresponding to a change in mechanism to electronic

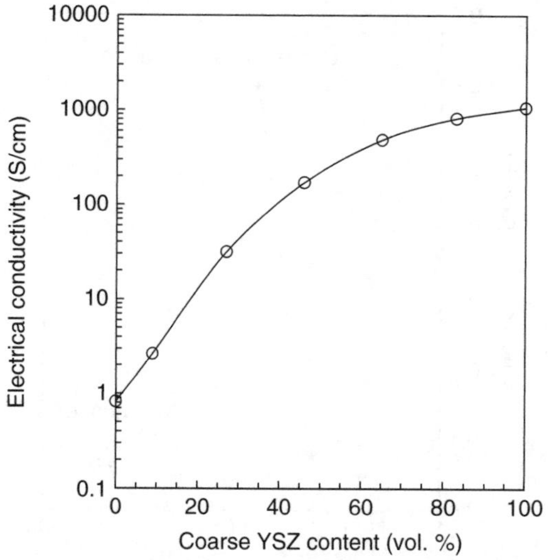

Fig. 3. Electrical conductivity measured at 1000°C of Ni/YSZ cermet containing both coarse and fine YSZ particles as a function of coarse YSZ content of total YSZ [25]

conduction through metallic phase. Alternatively, a switch in controlling mechanism from ionic conduction to electronic conduction as Ni concentration is raised over 30 vol.% is responsible for the steep rise in overall electrical conductivity. The percolation threshold is revealed to be influenced by many variables such as the porosity, pore size, size distribution, and size of raw powders as well as contiguity of each constituent component. The electrical behavior of Ni/YSZ cermets is, therefore, a strong function of these factors [18–20]. For instance, it can be seen from Fig. 2 that increasing sintering temperature results in lower percolation threshold, which might be due to the decreased porosity as well as narrowed pore size distribution. It is also noted that larger Ni particles give rise to larger threshold [21]. Besides, electrical conductivity is also affected by size ratio between YSZ and Ni (d_{YSZ}/d_{NiO}). For a fixed Ni content and porosity of anode with an overwhelming electronic conductor, the higher the ratio, the larger the electrical conductivity tends to be [22]. Additionally, a broader size spectrum of YSZ powders shows improved packing efficiency, and hence enhanced electrical conductivity. However, the coarse YSZ particles are more likely to show large shrinkage during firing and reducing of NiO. If the shrinkage becomes too high, the force due to adhesive constraint and structural movement could cause some macrocracks and rapid degradation of a cell [20]. Moreover, from an electrocatalyst point of view, another likely consequence of use of coarse YSZ powder in formulating anode cermet, is the decline in the TPB area where electrochemical oxidation of fuels takes place [23,24], thus noticeably driving up the activation polarization. As a compromise, in order to achieve a high conductivity and hence an excellent cell performance along with a low overall shrinkage, novel conceptual microstructure that is composed of coarse YSZ, fine YSZ, and NiO, which is subsequently reduced to Ni, has been proposed. Even with the fixed overall YSZ concentration (60 vol.%), the electrical conductivity is found to be profoundly altered as a function of coarse YSZ content as illustrated in Fig. 3 [25]. It can be seen that the overall electrical conductivity differs by several orders of magnitude as the fraction of coarse YSZ is raised from 0 to 100 vol.%. This implies that the predominant conduction mechanism shifts from ionic to electronic as contagious phase changes gradually from YSZ to Ni. In this microstructure, roles of coarse YSZ particles are: (1) adjust thermal expansion to an acceptable level with other components, especially YSZ electrolyte; (2) prevent the anode layer from significant shrinkage during the fabrication and operation; (3) inhibit the Ni agglomeration; and last but not the least

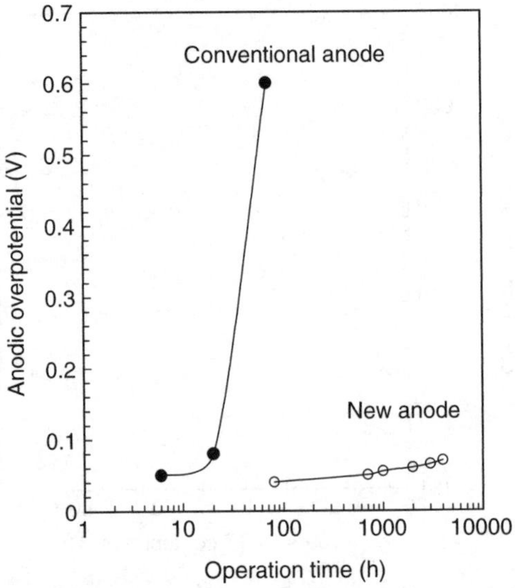

Fig. 4. Comparison of polarization resistances at 0.2 A cm^{-2} for the conventional and new anodes; the conventional anode is a cermet of Ni with one type of YSZ ceramic, while the new anode consists of Ni, fine and coarse YSZ powders [25]

(4) provide and retain micropores among these particles. The function of fine YSZ powder is to provide strong adhesion with coarse YSZ. The major advantage of this new anode over conventional one is long-term stability during SOFC generation as shown in Fig. 4 [25]. The incorporation of fine YSZ leads to the development of a more stable anode without sacrificing the electrical conductivity. Instead, more electrochemically active sites are introduced, contributing to a considerably suppressed polarization resistance [26]. Fig. 4 indicates that the overpotential (proportional to polarization resistance) of conventional anode rises sharply after 20 h service, while the new anode is capable of maintaining reasonably low resistive loss even after 2500 h of operation.

In the development of SOFCs, optimization of anode microstructure to achieve low overpotential has been the subject of intensive investigation [27–32]. The overpotential of Ni/YSZ cermet is found to be enormously dependent upon the content of Ni powder, the characteristics of starting raw material as well as the processing approach. The effect of Ni content on anodic overpotential is well documented [33,34]. It is revealed that as long as the Ni concentration falls into the range of 40–45 vol.% in this particular experimental condition, a minimum overpotential is achieved over entire current density range studied. This presumably corresponds to the enlargement of three-phase boundaries where anode reaction takes place. Comparison study of anodes fabricated using two different processing methods [35], as shown in Fig. 5, clearly illustrates that undesirable microstructure such as anisotropic particle packing, insufficient porosity, poor connectivity of conducting or pore phases contributes to the drastically reduced cell performance. To prepare an anode with high electrical conductivity, stable microstructure, as well as sufficient porosity, all processing variables should be optimized and meticulously controlled. Optimization of prior treatment parameters and the sintering temperature is also necessary to minimize the anode overpotential [32]. Besides, the overpotential of such anode is also influenced by such extrinsic factors as fuel composition and gas flow rate. For instance, it is reported that the overpotential is relatively independent of H_2 concentration under dry condition [36,37] and can be significantly reduced in the presence of steam (Fig. 6) [37]. This indicates that the electrochemical reaction process dominating the overpotential is related to the reaction involving H_2O. It turns out, as validated by the electrochemical impedance

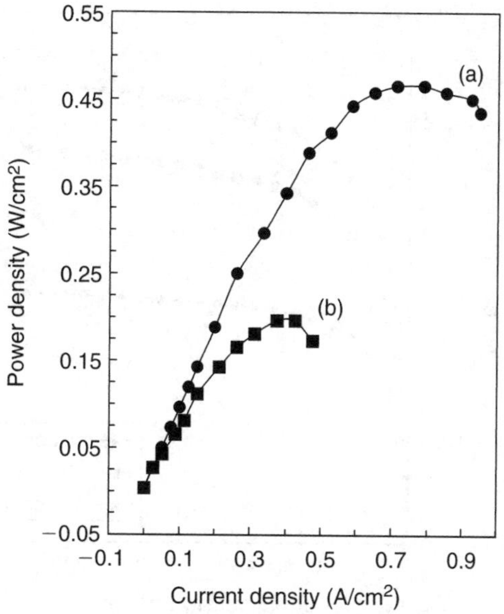

Fig. 5. Comparison of the power-generating characteristics of the unit cell fabricated using two different approaches [35]: (a) Liquid condensation process, and (b) spray drying method

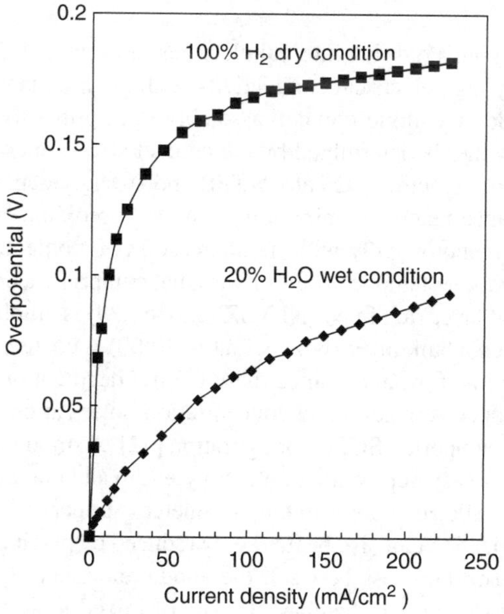

Fig. 6. Dependence of overpotential on the current density for a Ni/YSZ cermet with 40 vol.% Ni measured in the dry and wet conditions at 850°C [37]

spectroscopy investigation [38,39] that the anode reaction is a complex process involving activation and diffusion of hydrogen, oxygen, and steam, where presence of steam likely accelerates the adsorption/desorption of hydrogen. Besides, the overpotential can equally be suppressed by decreasing the gas flow rate of the humid fuel as shown in Fig. 7 [37].

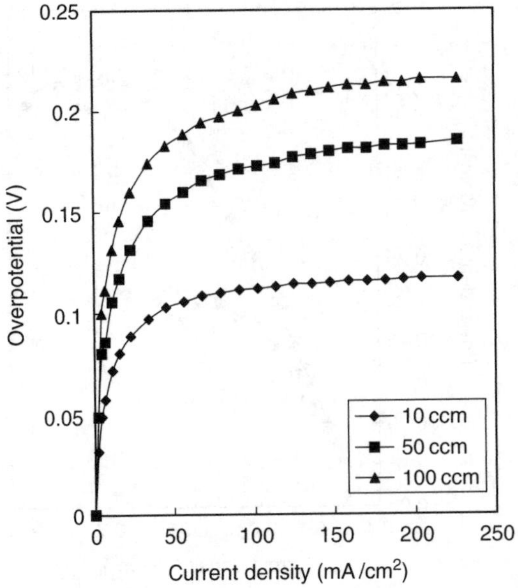

Fig. 7. Dependence of overpotential on the current density for a Ni/YSZ cermet with 40 vol.% Ni measured at various gas flow rates [37]

Study by Koide et al. [40] points out that internal electrical resistance of Ni/YSZ anode itself is negligibly small owing to its fairly high electrical conductivity with predominant electronic contribution, and electrical losses of such anode in a single fuel cell assembly stem primarily from contact resistance and polarization resistance, the former is determined by the contact area of nickel on the electrolyte, and the latter is related to the number of reaction sites along TPB and strongly dependent upon the Ni:YSZ ratio [41]. The polarization resistance reaches a minimum value at approximately 40 vol.% Ni, whereas the contact resistance decreases monotonically with the increase in Ni content, as presented in Fig. 8 [40]. Consequently, a new configuration of anode is put forward that comprises a double cermet layer with varying Ni contents. An interfacial layer rich in Ni (Ni:YSZ ratio, 61:39) is employed in intimate contact with the electrolyte, on top of which a bulk layer (Ni:YSZ ratio, 40:60) is coated. It has been tested that anode like this exhibits the optimum performance, since there is little degradation in regards to voltage output after 8000 h operation. To reduce activation and concentration polarization, anode substrate with graded porosity is proposed in anode-supported SOFC configuration [42]. A smooth anode surface with fine pores on the micrometer scale allows easy deposition of electrolyte film and maximizes the area of TPB. On the lower substrate side, presence of coarse pores in the millimeter is imperative to guarantee the undisturbed gas flow. The activation polarization can also be markedly reduced by placing Sm_2O_3 or Y_2O_3 doped CeO_2 porous layer of around 0.5 μm thickness between the anode and electrolyte [43–45]. The three-phase boundary can be vastly increased due to the characteristic of mixed ionic-electronic conductivity of the coating.

Thermal expansion mismatch between anode and electrolyte is another concern that has attracted widespread attention. An unacceptably high degree of mismatch in thermal expansion coefficients of SOFC components can generate stresses large enough to result in cracking or delamination [46]. To alleviate this concern, various options have been assessed which include improving the fracture toughness of electrolyte to boost stress tolerance [47], controlling critical processing flaws and fine-tuning anode formulation via addition of trace constituent [48 and 49]. It is also reported that [50] a graded structure along the thickness

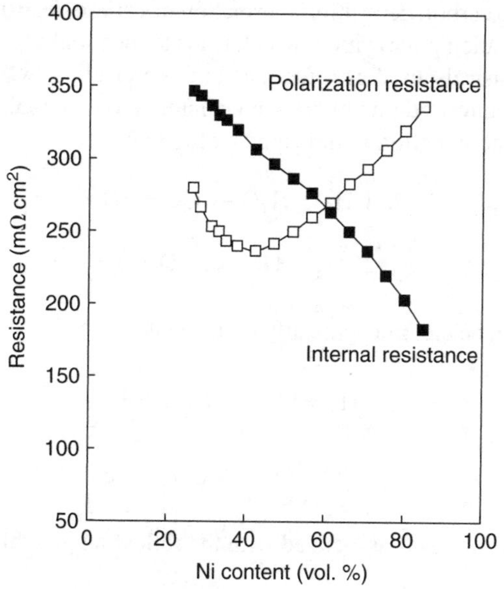

Fig. 8. Effect of Ni content on the polarization and internal resistances of Ni/YSZ anode, 200 ml min^{-1} of hydrogen as anode gas, 1000 ml min^{-1} of oxygen as cathode gas [40]

direction in terms of both Ni content and Ni:YSZ particle size ratio is quite effective in tailoring the thermal expansion coefficient. In other words, the thermal expansion coefficient of Ni:YSZ composite is not only controlled by the content of constituent phases, but also affected by Ni/YSZ particle size ratio [13]. This arises from the fact that a well-defined Ni/YSZ particle size ratio leads to an optimal package configuration of the cermet. Since Ni has a low melting temperature, it is liable to agglomerate and change shape during consolidation and subsequent operation. A marked drop in the electrical conductivity of anode cermet is observed after being exposed to fuel gases for prolonged period of time, which is associated with the growth of Ni-particle size [51]. It is now well established that the major mechanism of anode performance degradation is either Ni coarsening or agglomeration after long-term operation, leading to a reduction in both the TPB and electrical conductivity. More importantly, the rate of Ni-agglomeration is strongly enhanced in a water vapor containing reducing gas that is conventionally employed, in comparison with dry fuels. Once the water vapor is included, the extent of the agglomeration appears to be independent of the water content within the investigated range. To prevent the Ni coarsening (or agglomeration), some low surface energy oxides such as MgO, TiO_2, Mn_3O_4, and Cr_2O_3 are added to the anode cermet [52]. The roles of these additives are threefold: retard the coarsening of Ni particles during the high-temperature operation of cell, improve the mechanical properties of anode by assisting in the sintering of YSZ, and enhance the wettability of Ni particles by acting as anchoring sites at the anode/electrolyte interface. Suppression of Ni grain growth can also be achieved through microstructural modification by appropriate synthesis approach [53]. It is worth noting that Ni/YSZ cermet anode consisting of fine YSZ particles dispersed on the surface of Ni particles by the spray pyrolysis method exhibits no conceivable performance deterioration even after 1000 h service at 1000°C.

Aside from being prone to coarsening at operating temperatures, poor carbon and sulfur resistances are other two prominent drawbacks of Ni/ZrO_2 anode cermet [54–56]. Direct oxidation of dry CH_4 and CO results in rapid anode failure due to carbon deposition unless an appropriate steam to carbon ration of the fuel gas or sufficiently high current density is employed [57]. Nickel is an excellent electrocatalyst for electrochemical oxidation of hydrogen. However, this activity is impaired when the natural gas or methane is directly used as fuel due to carbon deposition. Formation of carbon deposits on Ni particles is responsible for excessively high activation polarization, which leads to the rapid deterioration of cell performance. Therefore, hydrogen is exclusively used as a fuel gas in the prevailing SOFCs. If methane is to be used with this anode, it has to be converted into hydrogen via internal or external steam reforming as shown by Eq. (2), or via catalytic partial oxidation as indicated by Eq. (3):

$$CH_4 + H_2O \rightarrow CO + 3H_2 \tag{2}$$

$$2CH_4 + O_2 \rightarrow 2CO + 4H_2 \tag{3}$$

The gas produced can then be electrochemically oxidized as

$$H_2 + O^{2-} \rightarrow H_2O + 2e^- \tag{4}$$

$$CO + O^{2-} \rightarrow CO_2 + 2e^- \tag{5}$$

The steam reforming reaction (2) is associated with the following gas shift reaction (6):

$$CO + H_2O \rightarrow CO_2 + H_2 \tag{6}$$

If the water content in the feed gas is insufficient for reaction (2) to occur, carbon will be deposited on the anode surface by methane cracking (7) or by the Boudouard reaction (8):

$$CH_4 \rightarrow C + 2H_2 \tag{7}$$

$$2CO \rightarrow C + CO_2 \tag{8}$$

In addition to the accelerated coarsening of Ni particles, the immediate disadvantage of steam addition, however, is the reduction of anode electrical potential due to the presence of oxygen atoms in fuels according to Nernst equation. The feasibility of steam reforming and partial oxidation of methane over Ni/YSZ anode has been thoroughly evaluated by Cunningham and Ormerod [58]. It is beyond doubt that Ni is a fairly good catalyst for both reactions (2) and (3), though the effect may vary somewhat. In the case of steam reforming, the product selectivity depends upon both the temperature and CH_4/H_2O ratio. When it comes to the methane partial oxidation, the selectivity for CO and H_2 can be increased by increasing temperature, while increasing CH_4/O_2 ratio only promotes H_2 selectivity. By highlighting the importance of reaction (3), Cunningham et al. [59,60] further concluded that carbon deposition can be adequately controlled by choosing the correct operating temperature and fuel composition in both cases. However, it must be reiterated that the additional reforming step, either internal or external, reduces the overall efficiency and adds to the cost. Moreover, methane steam reforming is strongly endothermic and when it takes place internally, it creates temperature gradients across the cell thus reducing its performance. In the case of internal reforming, although addition of water into methane does suppress carbon deposition on anode, an operating temperature in excess of 800°C is required in order to obtain high equilibrium conversions for the steaming reforming reaction [61]. Other concerns related to the internal fuel reforming have been revealed to be cooking of anode particles at high operating temperatures as well as leaching and delamination of the anode materials. To mitigate these problems, alternative anodes are being actively pursued for use in dry or low-humidity methane fuels. Doping Ni/YSZ anode with molybdenum and gold has been proved to be a promising option to enhance carbon deactivation resistance. The doped cermets have shown

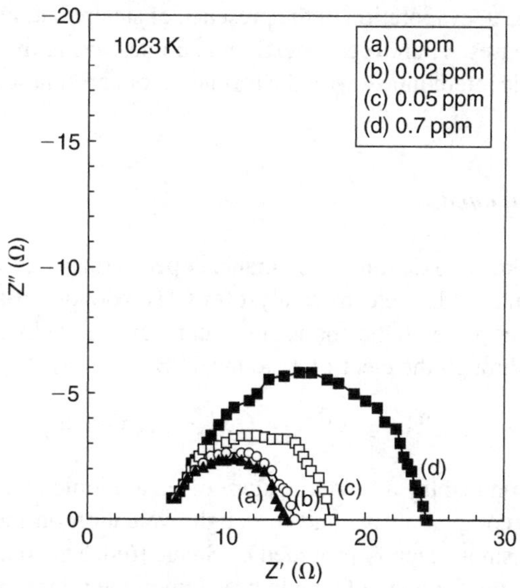

Fig. 9. The ac impedance spectra of SOFC with Ni/YSZ cermet anode measured after equilibration in humidified H_2 mixed with several concentrations of H_2S [62]

increased resistance to carbon deposition, increased methane conversion, and increased durability, particularly, at lower reforming temperatures and lower methane/steam ratios.

Low sulfur tolerance of Ni/YSZ anode constitutes another hurdle to the use of natural gas as fuel. It has been demonstrated that addition of H_2S enlarges the semicircle associated with activation process on the impedance plots shown in Fig. 9 [62]. The deactivation from sulfur poisoning occurs because H_2S strongly absorbs on active sites of nickel, leading to the substantial reduction in the rate of electrochemical reaction occurring at TPBs [40]. Matsuzaki and Yasuda [62,63] performed systematic research on the influence of sulfur impurity on Ni/YSZ anode performance. Their results distinctly show that within certain H_2S concentration level, the performance loss caused by sulfur-poisoning is reversible upon removal of sulfur source. The critical H_2S concentration above which a pronounced rise in polarization resistance of anode is triggered displays strong temperature dependence. This might be viewed as compelling evidence that adsorption of sulfur on Ni particles is physical in nature.

One issue that received relatively little attention is the long-term degradation of Ni/YSZ anode performance resulting from the loss of metallic conductivity caused by the oxidation of nickel. It has been indicated [64] that SOFC anode is only operative in the oxygen potential range between −0.8 and −1.0 V. When the oxygen potential rises above −0.68 V, hydroxide $Ni(OH)_2$ becomes the dominating volatile compound whose amount increases with the increase in either oxygen potential or operating temperature. It is further noticed that Ni loss is even higher at fuel entrance where a mixture of methane and water vapor is supplied.

Another problem associated with the Ni/YSZ anode arises in the utilization of chromia-forming metallic interconnects for SOFC stack. NiO particles, which are subsequently reduced to Ni once the reducing fuel is incorporated along the anode side of the interconnect, react much more readily with Cr_2O_3 than with ceramic $LaCrO_3$-based perovskites to yield Ni–Cr spinels. This phenomenon might cause some problems for the metallic/Ni-anode interface. Although the spinel is reduced during anode reduction, this process seems to result in the formation of large Cr_2O_3 crystals, which prevents alignment between adjacent surfaces in the stack during operation and hence generates the contact problem. Moreover, the large

Cr_2O_3 crystals are more prone to evaporation in the presence of steam that, in many cases, is an indispensable component of the fuel gases. Therefore, special precautions during the start up of the stacks should be exercised when the metallic interconnect with the formation of chromia scale is employed together with Ni anodes.

2.2. CeO_2 (rare-earth doped) anode

To allow for direct electrochemical oxidation of methane, doped ceria anode was developed since ceria has long been confirmed to be an excellent electrocatalyst for CH_4 oxidation [65–67]. Accordingly, the electrochemical reaction that takes place on the anode side can be expressed by the following equation where oxygen ions are transferred through the electrolyte to the TPB:

$$CH_4 + 4O^{2-} \rightarrow CO_2 + 2H_2O + 8e^- \quad (9)$$

Both doped and undoped ceria display a unique feature of mixed ionic and electronic conduction at low oxygen partial pressures. A comprehensive and thorough evaluation on the mixed conductivity and its underlying physical mechanism is recently provided by Steele [68]. The ionic contribution to the conductivity can be described using the following formula based upon the migration of oxygen ions via oxygen vacancies:

$$\sigma T = \sigma_0 \exp\left(-\frac{\Delta E}{kT}\right) \quad (10)$$

where σ_0 is a pre-exponential constant and the term ΔE, activation energy for conduction, involves both the enthalpy for the migration of oxygen (ΔH_m) and association enthalpy of defect complexes (ΔH_a) at low temperatures, while at elevated temperature, it is solely determined by the magnitude of ΔH_m. It is the minimization of association enthalpy when the trivalent rare-earth oxides such as Gd_2O_3, Sm_2O_3, and Y_2O_3 are in solid solution with CeO_2 that these dopants are invariably favored. This has been correlated with the close proximity between dopant and host ions in terms of ionic radius. The ionic conductivity data for some doped ceria is given in Table 1. It is particularly noteworthy that even at identical temperatures, the ionic conductivity of optimally doped ceria is about one order of magnitude larger than that of yttria-doped zirconia, making it a viable candidate for the electrolyte used in the intermediate temperature SOFCs. The electronic contribution to the conductivity is usually interpreted in terms of a small polaron activated hopping process and described by the following formula:

$$\sigma_n T = \sigma_n^0 \exp\left(-\frac{\Delta H_n}{kT}\right) P_{O_2}^{-1/4}$$

Table 1. Ionic conductivity data for some rare earth-doped ceria ceramics [68]

Composition	Dopant	r_d (Å)	ΔE (eV)	σ_0	σ(S cm^{-1}) 500°C	600°C	700°C
$Ce_{0.9}Gd_{0.1}O_{1.95}$	Gd^{3+}	1.053	0.64	1.09×10^5	0.0095	0.0253	0.0544
$Ce_{0.9}Sm_{0.1}O_{1.95}$	Sm^{3+}	1.079	0.66	5.08×10^4	0.0033	0.0090	0.0200
$Ce_{0.887}Y_{0.113}O_{1.9435}$	Y^{3+}	1.019	0.87	3.16×10^6	0.0087	0.0344	0.1015
$Ce_{0.8}Gd_{0.2}O_{1.9}$	Gd^{3+}	1.053	0.78	5×10^5	0.0053	0.018	0.047

The oxygen partial pressure dependence indicates that doped ceria has a significant contribution from electronic conductivity when it is used as anode in the fuel atmosphere. Furthermore, the temperature imposes a profound effect on the electronic conductivity of doped ceria. Increase in temperature leads to a dramatic rise in σ_n, which incidentally explains the reason why the doped ceria is only competent as electrolyte for intermediate temperature SOFCs (below 800°C).

The prominent drawback of ceria is the degradation of mechanical integrity due to the lattice expansion in the low oxygen partial pressure environment arising from the transition of Ce^{4+} to Ce^{3+} [69]. This may result in the formation of cracks at the electrode–electrolyte interface and subsequent delamination of the electrode from the electrolyte, especially in the case of YSZ electrolyte. Doping with lower valent cations such as Gd^{3+}, Sm^{3+}, or Y^{3+} can significantly decrease the dimensional contraction during reduction. A composition where 40–50% of the Ce^{4+} ions are substituted with Gd^{3+} ($Ce_{0.6}Gd_{0.4}O_{1.8}$, CG4, for example) or a similar rare-earth cation has been shown to be a reasonable compromise between conductivity and dimensional stability. As a further improvement, the concern of thermodynamic stability can be alleviated by reducing the thickness of the anode as well as placing an intermediate metal or oxide layer, or alternatively, by anchoring a thin anode layer with YSZ particles and sintering at relatively low temperatures [70,71]. Both thermal and redox cycling tests indicate sufficient adhesion between the thin CG4 layer and the YSZ electrolyte obtained by introducing the anchoring YSZ layer to withstand both the reduction expansion and the thermal expansion coefficient mismatch. Cells with such ceria anodes sustained several rapid thermal cycles and full redox cycles without measurable performance degradation. The reaction between anode and YSZ electrolyte can be avoided by sintering at low temperatures. Another alternative to surmount the reaction issue during co-firing is to replace the YSZ with doped ceria as electrolyte for intermediate temperature SOFC applications [72]. However, the electronic conductivity part of the electrolyte contributes adversely to the overall operating efficiency since it virtually increases the internal electrical resistance of ionic-conducting electrolyte. Additionally, the issue of mechanical integrity of the electrolyte in both oxidizing (near cathode) and reducing (near anode) atmospheres deserves special attention.

Ceria-based anode such as CG4 is widely recognized to be effective in suppressing carbon deposition. This benefits the utilization of methane-rich fuels with a low steam-to-carbon ratio. However, the electrocatalytic activity of CG4 toward methane oxidation is recently challenged by Marina et al. [73], who attributed the previously observed high electrocatalytic activity of CG4 to the presence of platinum as a current collector. This argument seems to be indirectly supported by recent finding that addition of small amount of Ni to various mixed ionic-electronic conductors (MIEC) including CG4 is very effective in improving the anode performance [74]. Figure 10 illustrates that the low-frequency impedance arc of CG4 is substantially suppressed as a result of minor Ni addition. This indicates that electrode reaction on MIEC anode is altered because the rate-limiting step, specifically, the adsorption and/or dissociation of hydrogen, is bypassed. It appears, however, that a bond breaking catalyst is called for as a new component of the anode. Further investigation is warranted to gain an unambiguous understanding of this point.

Uchida et al. [75] examined the applicability of yttria-doped ceria (YDC) and samaria-doped ceria (SDC) as potential anodes for intermediate temperature SOFCs. The common feature of these two candidate materials is, in reducing atmosphere typical of anode environment, they exhibit characteristic of mixed ionic-electronic conductivity. Since YDC shows greater electronic conductivity than SDC and comparable ionic conductivity with YSZ, the anode reaction can be appreciably enhanced. It is, therefore, concluded that high electronic conductivity in the MIEC anode is beneficial in reducing anodic polarization, in particular, activation polarization. Performance of anode based on either SDC or YDC can be further improved by uniformly dispersing trace amounts of noble metal catalysts like Ru, especially at lower operating temperatures.

Similar to Ni/YSZ cermet anode, Ni/doped ceria cermet has also been formulated and tested for use as potential anode in intermediate temperature SOFCs [72,76]. Experimental results by Livermore et al. [77] indicate that Ni/CGO (ceria–gadolinia) cermet exhibits high activity toward methane steam reforming with

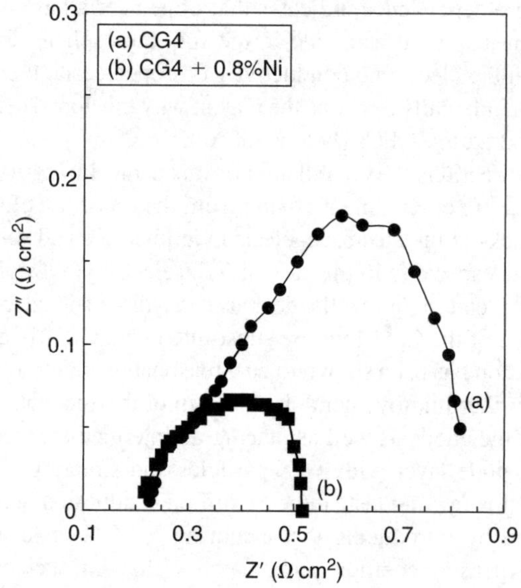

Fig. 10. The ac impedance spectra of CG4 electrode fed with humidified H_2 showing dramatic suppression of the semicircle at the low-frequency portion as a result of minor Ni addition [74]

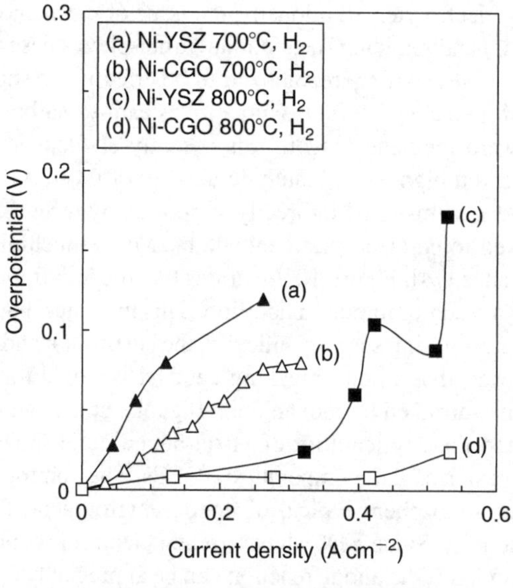

Fig. 11. Comparison of anodic overpotential of two different cermet anodes operating on dry hydrogen at two temperatures, indicating that Ni/CGO anode greatly outperforms Ni/YSZ anode [78]

the onset of methane activation occurring at temperatures as low as 482 K, and no appreciable carbon deposition was observed. Comparison of anodic overpotential of Ni/YSZ and Ni/CGO even fueled with hydrogen [78], as presented in Fig. 11, indicates that Ni/CGO anode outperforms Ni/YSZ anode. Doped ceria has also been utilized as a functional layer between anode and electrolyte, where the anode itself is a composite with

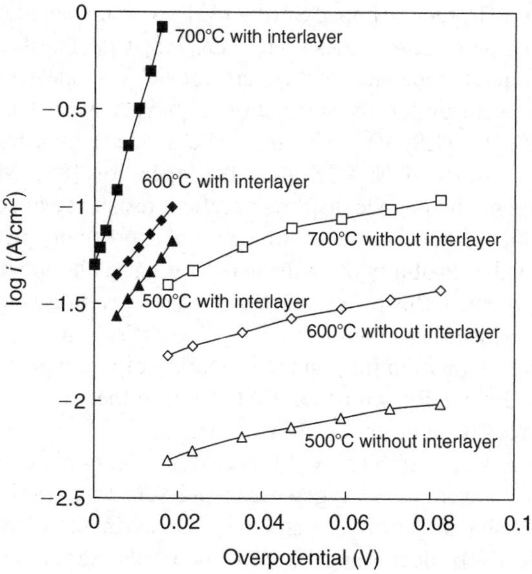

Fig. 12. Polarization responses of Ni/CSO (ceria-samaria) anodes with and without interlayer of the same composition at different temperatures [80]

mixed ionic-electronic conductivity whose composition is different from the functional layer [79]. Since doped ceria is considered to be more electrocatalytically active than doped yttria, the electrochemical reaction kinetics at anode can be favorably enhanced by introducing a porous interlayer with MICE such as SDC. As a result, the ERZ is effectively extended into the entire functional interlayer. Alternatively, by placing a dense interlayer of anode composition underneath the porous Ni/CSO (ceria–samaria) cermet, the polarization resistance can be remarkably reduced as compared to the situation without the interlayer (Fig. 12) [80]. Apparently, the improved contact between the anode and CSO electrolyte is believed to be responsible for the observed reduction in overpotential.

The competitiveness of ternary component anode consisting of Cu, CeO_2, and YSZ on direct activation of dry hydrocarbon gases is evaluated by Park et al. [81] by performing a short-term single cell test at 700°C. Their results are encouraging since Cu/CeO_2/YSZ anode demonstrates an excellent activity toward direct electrochemical oxidation of a variety of hydrocarbon gases. Anode performance does not suffer measurable degradation resulting from carbon deposition. Given that long-term stable anode performance operating on hydrocarbon fuels can be sustained, this development shows the potential of significantly reducing the cost and complexity of SOFC power generation system.

2.3. Other anode materials

In the search for alternative anode materials that are capable of withstanding sulfur contamination and carbon deposition, oxides with perovskite structure have drawn considerable attention. Lanthanum chromite itself is not qualified as anode material due to the poor mechanical response and lattice expansion in reducing atmosphere. However, its desired thermal and chemical stability is highly attractive, and consequently many efforts have been directed towards the exploration of this type of anode. A substitution of La with Sr and Cr with Ti leads to $SrTiO_3$-based materials with n-type conductivity rather than p-type as well as reduced size changes in reducing atmospheres [82]. Unfortunately, the electrochemical performance of $La_{0.7}Sr_{0.3}Cr_{0.8}Ti_{0.2}O_3$ and doped $SrTiO_3$ [83] perovskite anode is very much inferior to that of Ni/YSZ cermet. Studies on perovskite oxides such as $La_{0.6}Sr_{0.4}Co_{0.2}Fe_{0.8}O_3$ [84] and $LaNi_{1-x}M_xO_3$ (M: Ti, V, Nb, Mo, W) [83] also clearly indicate that these are not viable anode materials, because they are very unstable in reducing atmosphere due to the generation of oxygen vacancies. In contrast, experimental results on $La_{0.8}Sr_{0.2}Cr_{0.97}V_{0.03}O_3$ (LSCV)–YSZ composite anode demonstrate that the electrochemical performance is comparable to that of Ni/YSZ after 48 h operation [85]. Most advantageously, unlike Ni/YSZ anode, the above-mentioned anode displays excellent resistance to carbon deposition despite the fact that its activity toward the methane steam reforming is not appreciably promoted. By impregnating Ru into LSCV–YSZ composite, the capability of methane reforming can be substantially improved. Given its microstructure is further optimized, this material is expected to be a promising anode. Considerable electronic conductivity can be induced by introducing TiO_2 into $Zr(Y, Sc)O_{(2-x)}$ to form $Zr(Ti, Y, Sc)O_{(2-x)}$ solid solution. Preliminary investigation [86] on the suitability of this type of MIEC conductor as potential anode suggests that, in combination with Ni or Cu to form cermet, it may offer promise as an alternate anode operating on direct electrochemical oxidation of methane.

Niobate and titanate perovskites exhibit mixed ionic/n-type electronic conducting behavior with conductivity proportional to $P_{O_2}^{-1/4}$ over entire oxygen partial pressure range [87]. Preliminary study [88] confirms that Mn-doping at B-site of perovskite structure assists in the electrochemical reaction at the electrode surface, thus effectively decreasing the activation polarization. However, the electrocatalytic response of these newly developed anodes towards methane steam reforming as well as their performance at extended times still remain to be evaluated.

Recently, materials with the composition $(Ba/Sr/Ca/La)_{0.6}M_xNb_{1-x}O_{3-\delta}$ (M: Mg, Ni, Mn, Cr, Fe, In, Sn) of the tetragonal tungsten bronze structure have been considered as anodes for SOFCs because they exhibit considerable electronic conductivity and might additionally show some ionic contribution [89]. This is also true of niobate-based tungsten bronze materials [90,91]. Moreover, these materials are also reasonably stable in both oxidizing and reducing atmospheres. The unique feature of mixed conductivity imparts these materials the ability to accelerate the electrode gas reaction. The charge-transfer reaction between lattice oxygen and fuel may occur over the entire electrode area, whereas in Ni/YSZ cermet this reaction is only limited to the three-phase contacts between fuel, electrode, and electrolyte. It should be admitted that plenty of work including the measurements of thermal expansion coefficient, sulfidation and carburization tolerance, chemical compatibility with electrolyte as well as catalytic behavior with regards to the oxidization or reforming of methane has to be carried out before a full assessment of this type of potential anode is achieved.

3. CONCLUSIONS

In spite of its inherent drawbacks, Ni/YSZ cermet is the most preferred anode material for SOFCs operating on hydrogen because of the acceptable thermodynamic stability and desirable electrochemical properties. The microstructure of this anode can be optimized via compositional modification and fabrication techniques to minimize the polarization resistances. Rare earth doped CeO_2 is a viable anode material for low temperature SOFC due to its advantageously high electrocatalytic activity that enables the direct oxidation of low hydrocarbon gases, unfortunately, the occurrence of pronounced ionic conductivity along with mechanical degradation in reducing atmospheres pose a great concern for its practical applications.

ACKNOWLEDGEMENTS

The authors gratefully appreciate the reviewer's valuable comments and suggestions on the manuscript.

REFERENCES

1. N.Q. Minh. *J. Am. Ceram. Soc.* **76** (1993) 563.
2. S.C. Singhal. *MRS Bull.* **25** (2000) 16.
3. S.D. Vora. In: A.J. McEvoy (Ed.), *Proceedings of the Fourth European Solid Oxide Fuel Cell Forum*, Vol. 2, Lucerne, Switzerland, 10–14 July, 2000, p. 175.
4. J. Sukkel. In: A.J. McEvoy (Ed.), *Proceedings of the Fourth European Solid Oxide Fuel Cell Forum*, Vol. 2, Lucerne, Switzerland, July 10–14, 2000, p. 159.
5. H. Yokoyama, A. Miyahara and S.E. Veyo. In: U. Stimming, S.C. Singhal, H. Tagawa and W. Lehnert (Eds.), *Proceedings of the Fifth International Symposium on Solid Oxide Fuel Cells (SOFC-V)*, Aachen, Germany, June 2–5, 1997, p. 94.
6. H. Mori, H. Omura, N. Hisatome, K. Ikeda and K. Tomida. In: S.C. Singhal, M. Dokiya (Eds.), *Proceedings of the Sixth International Symposium on Solid Oxide Fuel Cells (SOFC-VI)*, Honolulu, HI, October 17–22, 1999, p. 52.
7. K. Krist, K.J. Gleason and J.D. Wright. In: S.C. Singhal and M. Dokiya (Eds.), *Proceedings of the Sixth International Symposium on Solid Oxide Fuel Cells (SOFC-VI)*, Honolulu, HI, October 17–22, 1999, p. 107.
8. J.H. Hirschenhofer, D.B. Sauffer, R.R. Engleman and M.G. Klett (Eds.), *Fuel Cell Handbook*, US Department of Energy, Morgantown, WV, 1998, p. 3.
9. S.L. Swartz, M.M. Seabaugh and W.J. Dawson. In: S.C. Singhal and M. Dokiya (Eds.), *Proceedings of the Sixth International Symposium on Solid Oxide Fuel Cells (SOFC-VI)*, Honolulu, HI, October 17–22, 1999, p. 135.
10. H. Nagamoto and Z.H. Cai. In: S.C. Singhal and M. Dokiya (Eds.), *Proceedings of the Sixth International Symposium on Solid Oxide Fuel Cells (SOFC-VI)*, Honolulu, HI, October 17–22, 1999, p. 163.
11. M. Mori, T. Yamamoto, H. Itoh, H. Inaba and H. Tagawa. In: U. Stimming, S.C. Singhal, H. Tagawa and W. Lehnert (Eds.), *Proceedings of the 5th International Symposium on Solid Oxide Fuel Cells (SOFC-V)*, Aachen, Germany, June 2–5, 1997, p. 869.
12. A.V. Virkar, J. Chen, C.W. Tanner and J.W. Kim. *Solid State Ionics* **131** (2000) 189.
13. A.C. Muller, D. Herbstritt and E.I. Tifffee. *Solid State Ionics* **152–153** (2002) 537.
14. F.G. Mora, J.M. Ralph and J.L. Routbort. *Solid State Ionics* **149** (2002) 177.
15. D.W. Dees, T.D. Claar, T.E. Easler, D.C. Fee and F.C. Mrazek. *J. Electrochem. Soc.* **134** (1987) 2141.
16. E.I. Tiffee, W. Wersing, M. Schiebi and H. Greiner. *Ber. Bunsen-Ges. Phys. Chem.* **94** (1990) 978.
17. S.K. Pratihar, R.N. Basu, S. Mazumder and H.S. Maiti. In: S.C. Singhal and M. Dokiya (Eds.), *Proceedings of the Sixth International Symposium on Solid Oxide Fuel Cells (SOFC-VI)*, Honolulu, HI, October 17–22, 1999, p. 513.
18. H. Itoh, T. Yamamoto, M. Mori, T. Horita, N. Sakai, H. Yokokawa and M. Dokiya. *J. Electrochem. Soc.* **144** (1997) 641.
19. T. Iwata. *J. Electrochem. Soc.* **143** (1996) 1521.
20. A. Tintinelli, C. Rizzo, G. Giunta and A. Selvaggi. In: U. Bossel (Ed.), *Proceedings of the First European Solid Oxide Fuel Cells Forum*, Vol. 1, Lucerne, Switzerland, October 3–7, 1994, p. 455.
21. W. Huebner, H.U. Anderson, D.M. Reed, S.R. Sehlin and X. Deng. In: M. Dokiya, O. Yamamoto, H. Tagawa and S.C. Singhal (Eds.), *Proceedings of the Fourth International Symposium on Solid Oxide Fuel Cells (SOFC-IV)*, Yokohama, Japan, June 6–9, 1995, p. 159.
22. B.C.H. Steele. In: U. Bossel (Ed.), *Proceedings of the First European Solid Oxide Fuel Cell Forum*, Vol. 1, Lucerne, Switzerland, October 3–7, 1994, p. 375.
23. T. Norby, O.J. Velle, H.L. Oslen and R. Tunold. In: S.C. Singhal, H. Iwahara (Eds.), *Proceedings of the Third International Symposium on Solid Oxide Fuel Cells (SOFC-III)*, Honolulu, HI, September 22–25, 1993, p. 473.
24. J. Muzusaki, H. Tagawa, T. Saito and T. Tamamura. *Solid State Ionics* **70–71** (1990) 52.
25. H. Itoh, T. Yamamoto, M. Mori, N. Mori and T. Watanabe. In: B. Thorstensen (Ed.), *Proceedings of the Second European Solid Oxide Fuel Cell Forum*, Vol. 1, Oslo, Norway, May 6–10, 1996, p. 453.
26. H. Itoh, Y. Hiei, T. Yamamoto, M. Mori and T. Watanabe. In: H. Yokokawa and S.C. Singhal (Eds.), *Proceedings of the Seventh International Symposium on Solid Oxide Fuel Cells (SOFC VII)*, Tsukuba, Ibaraki, Japan, June 3–8, 2001, p. 750.
27. T. Inoue, K. Eguchi, Y. Setoguchi and H. Arai. *Solid State Ionics* **40–41** (1990) 407.
28. S. Primdahl and M. Morgensen. *J. Electrochem. Soc.* **144** (1997) 3409.
29. S. Primdahl and M. Morgensen. *J. Electrochem. Soc.* **145** (1998) 2431.

30. N. Nakagawa, H. Sakurai, K. Kondo, T. Morimoto, K. Hatanaka and K. Kato. *J. Electrochem. Soc.* **142** (1995) 3474.
31. R. Wilkenhoner, T. Kloidt and W. Mallener. In: U. Stimming, S.C. Singhal, H. Tagawa and W. Lehnert (Eds.), *Proceedings of the Fifth International Symposium on Solid Oxide Fuel Cells (SOFC-V)*, Aachen, Germany, June 2–5, 1997, p. 851.
32. M.S. Brown, N.M. Sammes and M. Mogensen. In: U. Stimming, S.C. Singhal, H. Tagawa and W. Lehnert (Eds.), *Proceedings of the Fifth International Symposium on Solid Oxide Fuel Cells (SOFC-V)*, Aachen, Germany, June 2–5, 1997, p. 861.
33. W. Huebner, D.M. Reed and H.U. Anderson. In: S.C. Singhal and M. Dokiya (Eds.), *Proceedings of the Sixth International Symposium on Solid Oxide Fuel Cells (SOFC-VI)*, Honolulu, HI, October 17–22, 1999, p. 503.
34. J.H. Lee, H. Moon, H.W. Lee, J. Kim, J.D. Kim and K.H. Yoon. *Solid State Ionics* **148** (2002) 15.
35. J.H. Lee, J.W. Heo, D.S. Lee, J. Kim, G.H. Kim, H.W. Lee, H.S. Song and J.H. Moon. *Solid State Ionics* **158** (2003) 225.
36. C.J. Wen, R. Kato, H. Fukunaga, H. Takahashi and K. Yamada. In: A.J. McEvoy (Ed.), *Proceedings of the Fourth European Solid Oxide Fuel Cell Forum*, Vol. 2, Lucerne, Switzerland, July 10–14, 2000, p. 497.
37. C.J. Wen, R. Kato, H. Fukunago, H. Ishitani and K. Yamada. *J. Electrochem. Soc.* **147** (2000) 2076.
38. A. Bieberle and L.J. Gauckler. In: H. Yokokawa and S.C. Singhal (Eds.), *Proceedings of the Seventh International Symposium on Solid Oxide Fuel Cells (SOFC VII)*, Tsukuba, Ibaraki, Japan, June 3–8, 2001, p. 728.
39. M. Guillodo, P. Vernoux and J. Fouletier. *Solid State Ionics* **127** (2000) 99.
40. H. Koide, Y. Someya, T. Yoshida and T. Maruyama. *Solid State Ionics* **132** (2000) 253.
41. S.P. Jing and S.P.S. Badwal. *Solid State Ionics* **123** (1999) 209.
42. B. Gut. In: A.J. McEvoy (Ed.), *Proceedings of the Fourth European Solid Oxide Fuel Cell Forum*, Vol. 2, Lucerne, Switzerland, July 10–14, 2000, p. 561.
43. N.M. Sammes and Z.H. Cai. *Solid State Ionics* **121** (1999) 121.
44. J.J. Bentzen and H. Schwartzbach. *Solid State Ionics* **40–41** (1990) 942.
45. P. Ihringer, V.V. Herle and A.J. McEvoy. In: P. Stevens (Ed.), *Proceedings of the Third European Solid Oxide Fuel Cell Forum*, Nantes, France, June 2–5, 1998, p. 407.
46. S. Majumdar, T. Claar and B. Flandermeyer. *J. Am. Ceram. Soc.* **69** (1986) 628.
47. J.P. Singh, A.L. Bosak, D.W. Dees and C.C. McPheeters. In: *1988 Fuel Cell Seminar Abstracts*, Long Beach, CA, Oct. 23–26, 1988, Courtesy Associates, Washington, DC, 1988, p. 145.
48. F.P.F. van Berkel, B.D. Boer, G.S. Schipper and G.M. Christie. In: P. Stevens (Ed.), *Proceedings of the Third European Solid Oxide Fuel Cell Forum*, Nantes, France, June 2–5, 1998, p. 279.
49. C.M. Finnerty, R.H. Unningham and R.M. Ormerod. In: S.C. Singhal and M. Dokiya (Eds.), *Proceedings of the Sixth International Symposium on Solid Oxide Fuel Cells (SOFC-VI)*, Honolulu, HI, October 17–22, 1999, p. 568.
50. A.C. Muller, D. Herbstritt, A. Weber and E.I. Tiffee. In: A.J. McEvoy (Ed.), *Proceedings of the Fourth European Solid Oxide Fuel Cell Forum*, Vol. 2, Lucerne, Switzerland, July 10–14, 2000, p. 557.
51. D. Simwonis, F. Tietz and D. Stover. *Solid State Ionics* **132** (2000) 241.
52. A. Tsoga, P. Nikolopoulos, A. Kontogeorgakos, F. Tietz and A. Naoumidis. In: U. Stimming, S.C. Singhal, H. Tagawa and W. Lehnert (Eds.), *Proceedings of the Fifth International Symposium on Solid Oxide Fuel Cells (SOFC-V)*, Aachen, Germany, June 2–5, 1997, p. 823.
53. T. Fukui, T. Oobuchi, Y. Ikuhara, S. Ohara and K. Kodera. *J. Am. Ceram. Soc.* **80** (1997) 261.
54. I. Alstrup, J.R. Rostrup-Nielsen and S. Roen. *Appl. Catal.* **1** (1981) 303.
55. S. Morita and T. Inoue. *Int. Chem. Eng.* **5** (1965) 180.
56. N. Nakagawa, H. Sakurai, K. Kondo, T. Morimoto, K. Hatanaka and K. Kato. *J. Electrochem. Soc.* **142** (1995) 3474.
57. A. Weber, B. Sauer, A.C. Muller, D. Herbstritt and E.I. Tiffee. *Solid State Ionics* **152–153** (2002) 543.
58. R.H. Cunningham and R.M. Ormerod. In: A.J. McEvoy (Ed.), *Proceedings of the Fourth European Solid Oxide Fuel Cells Forum*, Vol. 2, Lucerne, Switzerland, July 10–14, 2000, p. 507.
59. R.H. Cunningham, C.M. Finnerty, K. Kendall and R.M. Ormerod. In: U. Stimming, S.C. Singhal, H. Tagawa and W. Lehnert (Eds.), *Proceedings of the Fifth International Symposium on Solid Oxide Fuel Cells (SOFC-V)*, Aachen, Germany, June 2–5, 1997, p. 965.
60. R.H. Cunningham, C.M. Finnerty, K. Kendall and R.M. Ormerod. In: U. Stimming, S.C. Singhal, H. Tagawa and W. Lehnert (Eds.), *Proceedings of the Fifth International Symposium on Solid Oxide Fuel Cells (SOFC-V)*, Aachen, Germany, June 2–5, 1997, p. 973.

61. M. Ihara, T. Kawai, K. Matsuda and C. Yokoyama. In: H. Yokokawa and S.C. Singhal (Eds.), *Proceedings of the Seventh International Symposium on Solid Oxide Fuel Cells (SOFC VII)*, Tsukuba, Ibaraki, Japan, June 3–8, 2001, p. 684.
62. Y. Matsuzaki and I. Yasuda. *Solid State Ionics* **132** (2000) 261.
63. Y. Matsuzaki and I. Yasuda. In: H. Yokokawa and S.C. Singhal (Eds.), *Proceedings of the Seventh International Symposium on Solid Oxide Fuel Cells (SOFC VII)*, Tsukuba, Ibaraki, Japan, June 3–8, 2001, p. 769.
64. A. Gubner, H. Landes, J. Metzger, H. Seeg and R. Stubner. In: U. Stimming, S.C. Singhal, H. Tagawa and W. Lehnert (Eds.), *Proceedings of the Fifth International Symposium on Solid Oxide Fuel Cells (SOFC-V)*, Aachen, Germany, June 2–5, 1997, p. 844.
65. T. Takahashi, H. Iwahara and Y. Suzuki. In: *Proceedings of the Third International Symposium on Fuel Cells*, Presses Academiques Europeennes, Bruxelles, June 16–20, 1969, p. 113.
66. B.C.H. Steele, I. Kelly, H. Middleton and R. Rudkin. *Solid State Ionics* **28–30** (1988) p. 1547.
67. M. Mogensen, B. Kindl and B.M. Hansen. In: *Proceedings of the Program and Abstracts of Fuel Cell Seminar*, Phoenix, AZ, Courtesy Associate Inc., Washington, DC, 1990, p. 195.
68. B.C.H. Steele. *Solid State Ionics* **129** (2000) 95.
69. M. Mogensen, T. Lindegaard, U.R. Hansen and G. Mogensen. *J. Electrochem. Soc.* **141** (1994) 2122.
70. E.P. Murray, T. Tsai and S.A. Barnett. *Nature* **400** (1999) 649.
71. T. Tsai and S.A. Barnett. *J. Electrochem. Soc.* **145** (1998) 1696.
72. M. Godickemeier, K. Sasaki and L.J. Gauckler. In: M. Dokiya, O. Yamamoto, H. Tagawa and S.C. Singhal (Eds.), *Proceedings of the Fourth International Symposium on Solid Oxide Fuel Cells (SOFC-IV)*, Yokohama, Japan, June 6–9, 1995, p. 1072.
73. O.A. Marina, C. Bagger, S. Primdahl and M. Mogensen. *Solid State Ionics* **123** (1999) 199.
74. S. Primdahl and M. Mogensen. *Solid State Ionics* **152–153** (2002) 597.
75. H. Uchida, M. Sugimoto and M. Watanabe. In: H. Yokokawa and S.C. Singhal (Eds.), *Proceedings of the Seventh International Symposium on Solid Oxide Fuel Cells (SOFC VII)*, Tsukuba, Ibaraki, Japan, June 3–8, 2001, p. 653.
76. R.M. Ormerod. *Stud. Surf. Sci. Catal.* **122** (1999) 35.
77. S.J.A. Livermore, J.W. Cotton and R.M. Ormerod. In: S.C. Singhal and M. Dokiya (Eds.), *Proceedings of the Sixth International Symposium on Solid Oxide Fuel Cells (SOFC-VI)*, Honolulu, HI, October 17–22, 1999, p. 593.
78. M.B. Joerger and L.J. Gauckler. In: H. Yokokawa and S.C. Singhal (Eds.), *Proceedings of the Seventh International Symposium on Solid Oxide Fuel Cells (SOFC VII)*, Tsukuba, Ibaraki, Japan, June 3–8, 2001, p. 662.
79. Y. Matsuzaki and I. Yasuda. *Solid State Ionics* **152–153** (2002) 463.
80. S. Wang, T. Kato, S. Nagata, T. Honda, T. Kaneko, N. Iwashita and M. Dokiya. In: A.J. McEvoy (Ed.), *Proceedings of the Fourth European Solid Oxide Fuel Cell Forum*, Vol. 2, Lucerne, Switzerland, July 10–14, 2000, p. 479.
81. S. Park, H. Kim, S. Mcintosh, W. Worrell, R.J. Gorte and J.M. Vohs. In: H. Yokokawa and S.C. Singhal (Eds.), *Proceedings of the Seventh International Symposium on Solid Oxide Fuel Cells (SOFC VII)*, Tsukuba, Ibaraki, Japan, June 3–8, 2001, p. 712.
82. G. Pudmich. *Solid State Ionics* **135** (2000) 433.
83. S.Q. Hui and A. Petric. *Solid State Ionics* **143** (2001) 275.
84. M. Weston and I.S. Metcalfe. *Solid State Ionics* **113–115** (1998) 247.
85. P. Vernoux, M. Guillodo, J. Fouletier and A. Hammon. *Solid State Ionics* **135** (2000) 425.
86. J.T.S. Irvine, S. Tao, A.J. Feighery and T.D. Mcolm. In: H. Yokokawa and S.C. Singhal (Eds.), *Proceedings of the Seventh International Symposium on Solid Oxide Fuel Cells (SOFC VII)*, Tsukuba, Ibaraki, Japan, June 3–8, 2001, p. 738.
87. J.T.S. Irvine, P.R. Later, A. Kaiser, J.L. Bradley, P. Holtappels and M. Mogensen. In: A.J. McEvoy (Ed.), *Proceedings of Fourth European Solid Oxide Fuel Cell Forum*, Vol. 2, Lucerne, Switzerland, July 10–14, 2000, p. 471.
88. S.Q. Hui, A. Petric and W. Gong. In: S.C. Singhal and M. Dokiya (Eds.), *Proceedings of the Sixth International Symposium on Solid Oxide Fuel Cells (SOFC-VI)*, Honolulu, HI, October 17–22, 1999, p. 632.
89. P.R. Slater and J.T.S. Irvine. *Solid State Ionics* **124** (1999) 61.
90. P.R. Slater and J.T.S. Irvine. *Solid State Ionics* **120** (1999) 125.
91. C.M. Reich, A. Kaiser and J.T.S. Irvine. In: A.J. McEvoy (Ed.), *Proceedings of Fourth European Solid Oxide Fuel Cell Forum*, Vol. 2, Lucerne, Switzerland, July 10–14, 2000, p. 517.

Chapter 13

Advances, aging mechanisms and lifetime in solid-oxide fuel cells

Hengyong Tu and Ulrich Stimming

Abstract

Solid-oxide fuel cells (SOFCs) are an energy conversion device that theoretically has the capability of producing electrical energy for as long as the fuel and oxidant are supplied to the electrodes but performance is expected for, at least, 40 000 h. In reality, performance degradation is observed in planar SOFC with metallic bipolar plate under steady and repeated thermal cycling conditions, which limits the practical operating life of these SOFCs. In this paper, the advances in SOFC are briefly summarized and the aging mechanisms of some components (anode, cathode and interconnect) in SOFC are discussed. The emphasis is given to aging mechanisms due to instability of materials and microstructures under real operation conditions. Identification of aging kinetics contributes to improvement in the stability of SOFC. It is indicated that development of new materials, optimization of microstructures and lower operating temperatures are desirable for the long-term stability of SOFC. Beneficial operation condition of SOFC is also proposed.

Keywords: Solid-oxide fuel cell (SOFC); Reduced temperature; Hydrocarbon fuels; Aging mechanism of components; Operation condition

Article Outline

1. Introduction 236
2. Advances in SOFC research and development 236
   2.1. SOFC operating at temperature <800°C 237
   2.2. Direct supply of hydrocarbon fuels 238
      2.2.1. Reforming 239
      2.2.2. Direct oxidation 239
3. Aging mechanism of components in SOFC 241
   3.1. Aging mechanism of anode 242
   3.2. Aging mechanism of cathode 243
   3.3. Aging mechanism of interconnect 245
4. Benefit of SOFC operation at higher cell voltage 247
5. Conclusions 248
References 248

Fuel Cells Compendium

1. INTRODUCTION

A fuel cell is an energy conversion device that generates electricity and heat by electrochemically combining a gaseous fuel and an oxidizing gas via an ion-conducting electrolyte. The chief characteristic of a fuel cell is its ability to convert chemical energy directly into electrical energy without the need for combustion, giving much higher conversion efficiencies than most conventional thermo-mechanical methods (e.g., steam turbines). Consequently fuel cells have much lower carbon dioxide emissions than fossil fuel-based technologies for the same power output. They also produce negligible emissions of NO_x, the main constituents of acid rain and photochemical smog. As with other technologies, it is very important to achieve a service life of several thousand hours without impacting performance levels. Several types of fuel cells are now being developed around the world, the chief difference between them being the material used for the electrolyte (and thus also their operating temperature).

Solid-oxide fuel cells (SOFCs) consist entirely of solid-state materials; they utilize a fast oxygen ion conducting ceramic as the electrolyte, and operate in the temperature range of 700–1000°C.

SOFCs have several features that make them more attractive than most other types of fuel cells:

- The highest efficiencies of all fuel cells (50–60%).
- A potential long-life expectancy of more than 40 000–80 000 h.
- Constructed from readily obtainable ceramic materials, not precious metals like platinum.
- Few problems with electrolyte management (cf. liquid electrolytes, which are typically corrosive and difficult to handle).
- High-grade waste heat is produced, for combined heat and power (CHP) applications increasing overall efficiencies to over 80%.
- Internal reforming of hydrocarbon fuels is possible.

SOFC systems are being advanced by a number of companies and organizations with three major fuel cell stack designs emerging. The major design types are tubular, planar, and monolithic. Only the first two are currently being developed. Tubular SOFC designs are closer to commercialization and are being mainly produced by Siemens Westinghouse Power Corporation (SWPC). The tubular SOFC design constructs the stack as a bundle of tubular electrode – electrolyte assemblies. Air is typically introduced to the inside of each tube while fuel bathes the outside of the tubes to produce electricity. One distinct feature of this design is that it has no seals. Demonstrations of tubular SOFC technology have produced as much as 220 kW. Compared to tubular designs, planar SOFC designs consist of flat plates bonded together to form the electrode – electrolyte assemblies. This is a compact stack design. Individual cells can be electrolyte self-supporting or anode-supported. The planar SOFC requires high-temperature gas seals at the edges of the plates. The planar designs have been demonstrated in single cell and smaller stack sizes in the single to multiple kilowatt range. In comparison with tubular SOFC, the planar SOFC could have the following advantages:

- Higher power densities, and better overall performance.
- Manufacturing process relatively simple and high potential for low cost manufacturing.
- Operating temperature below 800°C possible.
- High efficiency.

2. ADVANCES IN SOFC RESEARCH AND DEVELOPMENT

SOFC is emerging as a potential breakthrough technology for the low cost production of electricity. The low-product cost is a prerequisite for the successful market introduction and penetration of SOFC power plants. SOFC has the advantage of fuel flexibility. Hydrogen is currently expensive to produce and deliver.

Therefore, hydrocarbon fuels are a realistic choice in the immediate future for SOFC. In recent years, significant advances have been achieved in SOFC research and development, although a number of problems still have to be solved. The research and development is necessary to decrease capital costs, as well as to avoid the use of hydrogen as the primary fuel. Main recent advances in SOFC include realization of SOFC operating at temperature $<800°C$ and direct supply of hydrocarbon fuels.

2.1. SOFC operating at temperature <800°C

Conventional high-temperature SOFC operates at around 950°C. This temperature implies high material costs, particularly for interconnect and construction materials. In high-temperature SOFC, the interconnect may be a ceramic such as lanthanum chromite, or a sophisticated refractory alloy, e.g., based on mechanically alloyed Y/Cr. In either case, the interconnect represents a major proportion of the cost of the stack. Operation of the SOFC at a reduced temperature can overcome some of these problems and bring additional benefits.

Advantages of SOFC at $T < 800°C$:

- Low-cost metallic materials, e.g., ferritic stainless steels can be used as interconnect and construction materials. This makes both the stack and balance of plant cheaper and more robust (balance of plant is widely assumed to constitute 50% of the cost of the SOFC system).
- Potential for rapid startup and high capability of thermal cycle.
- Simplification of the design and materials requirements of the balance of plant.
- Significant reduction of corrosion rates, higher long-term stability of the system.

As the operating temperature of the SOFC is reduced, the ionic conductivity of the electrolyte decreases, which results in a rapid deterioration of the performance of the SOFC. There are two ways of minimizing ohmic losses across the electrolyte at temperature $<800°C$. One way is to use higher conductivity materials (doped ceria and lanthanum gallate). However, this will result in uncertainties of long-term stability and material compatibility. Another way is to use thin YSZ electrolyte membranes (5–50 μm). For SOFC application, it is desirable that the oxygen fluxes through YSZ electrolyte membranes attain values around $1\,A\,cm^{-2}$, and so with typical values of the electric potentials in the range 0.5–1.0 V. It follows that the area-specific resistance (ASR) term should be as low as possible. Adopting a typical target value of $0.15\,\Omega cm^2$ enables the membrane thickness at different temperatures to be determined from Fig. 1 [1]. As shown in Fig. 1, the thin YSZ electrolyte membranes can work at temperatures 600–800°C. When the electrolyte thickness is reduced from ~150 to ~5 μm, the resistance is decreased by more than one order of magnitude across the electrolyte.

Conventionally, planar cells use the electrolyte to support thin electrodes on either face. This limits minimum electrolyte thickness to around 150 μm. By using a thick, structural anode as the substrate, the mechanical stability of the cells is transferred from the electrolyte to the anode. Reduction of the YSZ electrolyte thickness can be achieved in a number of ways. If electrodes with optimized microstructure are used, electrochemical experiments show that at temperatures 600–800°C it is easy to meet a minimum requirement of $0.2\,W/cm^2$ at an operating voltage of 0.7 V with anode as support. Figure 2 shows the two kinds of cell concept.

A concerted effort is also being made by researchers around the globe to develop this substrate technology [2]. The most advanced development in this field is currently located at the Research Center Jülich in Germany and at Global Thermoelectric, Canada, whose technique is based on the Jülich concept. The Research Center Jülich works on the construction of SOFC stacks using the substrate cell technique. The design of these stacks is optimized with the aid of model calculations of temperature distribution, gas stream and mechanical load. The thermo-mechanical data required for these calculations are determined experimentally. Since a stack does not only consist of single cells but also of other components such as interconnect foils, contact layers and glass ceramic sealings, materials research plays an essential role.

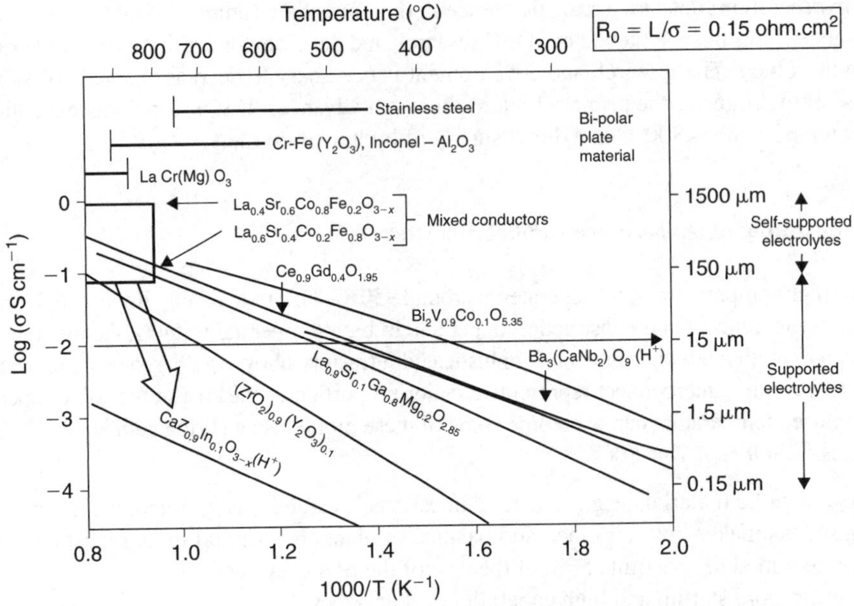

Fig. 1. Specific ionic conductivity values for selected ceramic oxide ion conducting membranes as a function of reciprocal temperature [1]

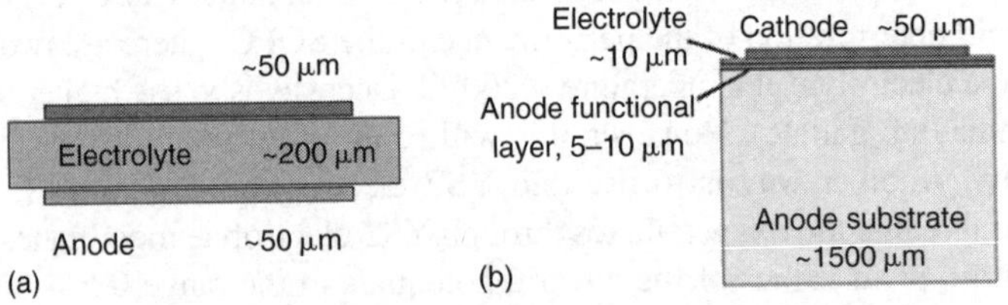

Fig. 2. The two kinds of cell concept: Conventional planar cell concept with self-supporting electrolyte (a) and Jülich membrane–electrode assembly (MEA) concept with the anode as the cell support (b)

Another priority is stack construction. Within the technology development program on anode supported SOFC in Jülich short-stack (two cells 10 cm × 10 cm; 81 cm^2) tests were performed in order to improve, among others, the durability [3]. First results with interconnect plates machined from a newly developed ferritic steel JS-3 show degradation rates less than 1%/1000 h during the first thousand hours of continuous operation. Several kW-class stacks (10–40 cells 20 cm × 20 cm; 361 cm^2) were assembled and operated. The 40-cell stack produced a maximum power of 9.2 kW (300 A at 30.2 V) operating on hydrogen (10% humidified, 76% fuel utilization). Operating on simulated partially pre-reformed methane the stack produced a maximum power of 5.4 kW (182 A at 30.2 V). A medium-term goal at Research Center Jülich is the design and construction of a 25 kW system for combined heat and power generation with natural gas in order to demonstrate the feasibility of an anode – substrate-based SOFC.

2.2. Direct supply of hydrocarbon fuels

One of the main attractions of SOFC over other types of fuel cells is their ability to handle more convenient hydrocarbon fuels – other types of fuel cell have to rely on a clean supply of hydrogen for their operation.

Hydrogen derived from renewable resources would be an appropriate fuel. However, its availability and distribution are very limited. In the short-to medium-term future, the only realistic fuels are hydrocarbon-based, especially natural gas. Because SOFCs operate at high temperature there is the opportunity to reform hydrocarbons within the system either indirectly in a discrete reformer or directly on the anode of the cell. A further process, direct oxidation can also occur at low steam partial pressures. This process offers the ultimate in thermodynamic efficiency, almost 100% theoretically.

2.2.1. Reforming

Reforming enables SOFC to use hydrocarbon directly as fuel. Reaction (1), for steam reforming can be formulated for methane as the main component of natural gas, which is generally associated with a water gas shift equilibrium reaction (2). Additionally, methane may be reformed with CO_2 (Eq. (3)):

$$CH_4 + H_2O \rightleftarrows CO + 3H_2, \ \Delta H^0 = 206 \text{ kJ mol}^{-1} \tag{1}$$

$$CH_4 + H_2O \rightleftarrows CO_2 + H_2, \ \Delta H^0 = 41 \text{ kJ mol}^{-1} \tag{2}$$

$$CH_4 + H_2O \rightleftarrows 2CO + 2H_2, \ \Delta H^0 = 247 \text{ kJ mol}^{-1} \tag{3}$$

Two approaches have been developed, external and internal reforming. In the first case, the reaction occurs in a separate reactor, which consists of heated tubes filled with nickel or noble metal. In internal reforming, natural gas reforming can be directly carried out at the anode of the fuel cell. Advantages of internal reforming over external reforming are the good and direct heat transfer between fuel cell stack and reforming zone and the high degree of chemical integration. Steam as a product of the electrochemical reaction in the cell can be directly coupled in as a base product of the reforming reaction, so that less steam for reforming must be produced in comparison with external reforming, improving the electrical efficiency. Other advantages include reduced system costs, more homogeneous hydrogen formation and higher methane conversion rate. However, as to the state-of-the-art anode Ni/YSZ, two major problems occur: the risk of carbon deposition and the creation of temperature gradients. Firstly, when insufficient steam is present, carbon may be deposited according to the reaction

$$CH_4 \rightleftarrows C + 2H_2, \ \Delta H^0 = 75 \text{ kJ mol}^{-1} \tag{4}$$

which deactivates the anode. This cracking reaction is even more problematic with higher hydrocarbons, which also are present in natural gas. Therefore, a steam/methane ratio >2 is required to avoid carbon deposition. Secondly, high-temperature gradients arise in the region of the fuel inlet due to the significant cooling effect of the reforming process, which is on account of its very fast reaction rate in comparison to electrochemical reaction rate. This leads to strong thermal stressing of the materials of the fuel cell. Research efforts were made toward overcoming these problems. Some new materials were proposed to decrease the catalytic activity of Ni for reforming, in order to avoid carbon deposition without using excess steam. For example, the influence of adding iron to the Ni/YSZ cermet has been studied [4]. The iron achieves a decrease in the catalytic activity of nickel and adjusts the thermal expansion coefficient closer to YSZ. Furthermore, a new concept "gradual internal methane reforming" has been proposed [5]. It is based on a local coupling between the steam reforming and the electrochemical oxidation of hydrogen. Thus, the reaction is distributed over the entire anode surface.

2.2.2. Direct oxidation

Direct oxidation of methane, reaction (5), has the thermodynamic possibility of 99.2% conversion efficiency. Operating fuel cells directly on hydrocarbons would obviously eliminate the need for reformer and

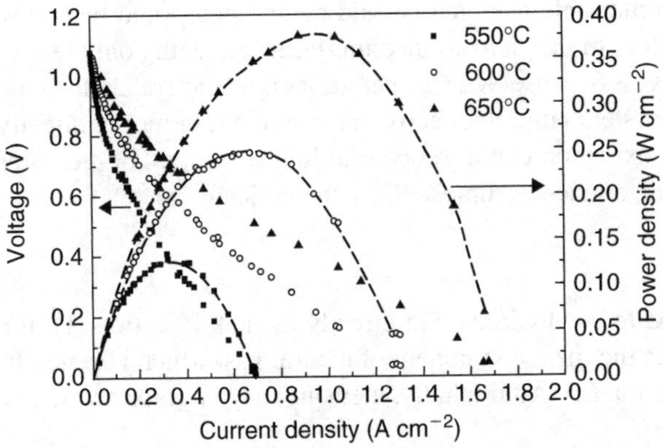

Fig. 3. Cell voltage and power density versus current density for an SOFC operated on air and wet methane. The measurements were collected in atmospheric-pressure air, and the methane fuel was supplied at ~50 cm^3 STP min^{-1} [6] (Reprinted with permission from Nature, 12 August, vol. 400, p. 649. Copyright 1999, Macmillan.)

improve efficiency. If this reaction is to be achieved, it is necessary to avoid or inhibit methane cracking. There is also considerable controversy as to whether the reaction is actually a direct anode oxidation of methane (Eq. (6)) or a reaction by a steam reforming process (Eq. (1)) including some intermediate reactions. The steam-reforming process is associated with a water gas shift reaction (2). The produced H_2 and CO are then electrochemically converted into H_2O (Eq. (7)) and CO_2 (Eq. (8)), respectively. Therefore, the anode reaction of methane by a steam-reforming process also leads to the net result shown in reaction (6):

$$CH_4 + 2O_2 \rightleftarrows CO_2 + 2H_2O,\ \Delta H^0 = -802\ \text{kJ mol}^{-1} \quad (5)$$

$$CH_4 + 4O^{2-} \rightarrow CO_2 + 2H_2O + 8e^-,\ \text{electrochemical} \quad (6)$$

$$H_2 + O^{2-} \rightarrow H_2O + 2e^-,\ \text{electrochemical} \quad (7)$$

$$CO + O^{2-} \rightarrow CO_2 + 2e^-,\ \text{electrochemical} \quad (8)$$

There have been some successes reported for direct electrochemical oxidation with ceria-based materials. Firstly, Perry Murray et al. [6] reported the direct electrochemical oxidation of methane in SOFCs by using ceria/nickel composite electrodes at lower temperatures, <700°C, i.e., at temperatures below the cracking of the hydrocarbon. The SOFCs were fabricated on porous $La_{0.8}Sr_{0.2}MnO_3$ (LSM) cathodes. The LSM pellets were ~2 cm in diameter and 1 mm thick, and were produced using standard ceramic processing techniques. All SOFC layers, starting with a 0.5-mm thick $(Y_2O_3)_{0.15}(CeO_2)_{0.85}$ (YDC) porous film, were deposited on the LSM pellet using d.c. reactive magnetron sputtering. The electrolyte, 8 mol% Y_2O_3-stabilized ZrO_2 (YSZ), was then deposited under conditions yielding a dense, 8-μm-thick film. To complete the cell, another 0.5-μm-thick YDC film was deposited, followed by a porous, 2-μm-thick Ni-YSZ anode. Figure 3 shows measurements of current density and power density versus voltage, performed on a typical cell using air and methane. The results for dry and wet (with 3% H_2O) methane were nearly identical. Cell performance was stable in preliminary 100 h life tests, except for wet methane at low voltages where the anode Ni gradually oxidized. The results in Fig. 3 are similar to those obtained for the

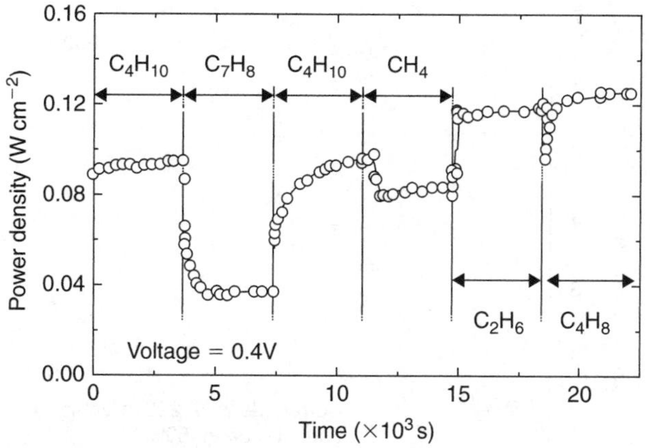

Fig. 4. Effect of switching fuel type on the cell with the Cu-(doped ceria) anode at 700°C. The power density is shown as a function of time. The fuels were: *n*-butane (C_4H_{10}), toluene (C_7H_8), *n*-butane, methane (CH_4), ethane (C_2H_6), and 1-butene (C_4H_8) [7] (Reprinted by permission from *Nature*, 16 March, vol. 404, p. 265. Copyright 2000, Macmillan.)

same cells operated with humidified hydrogen fuel, except that the power densities are ~20% lower. Examination of the anodes after the cell tests (by visual observation, energy-dispersive X-ray, and scanning electron microscopy) showed no evidence of carbon deposition after ~100 h of operation.

Secondly, Gorte and co-workers [7] report the direct, electrochemical oxidation of various hydrocarbons (methane, ethane, 1-butene, *n*-butane, and toluene) using a SOFC at 700 and 800°C with a composite anode of copper and ceria (or samaria-doped ceria). The SOFCs used here were prepared with a 60-μm-thick, YSZ electrolyte, 12.5 mm in diameter, and a cathode formed from a 50:50 mixture of YSZ and $La_{0.8}Sr_{0.2}MnO_3$ powders. The anodes were 40 wt.% Cu and 20 wt.% CeO_2, held in place by a YSZ matrix formed from zircon fibers. In a second cell, a 20% Sm_2O_3–80% CeO_2 mixed oxide replaced CeO_2. The first cell was operated at the maximum power density of 0.12 W cm^{-2} in dry butane at 700°C for a period of 48 h with no observable change in performance. Visual inspection of a cell after 2 days in *n*-butane at 800°C showed that the anode itself remained free of the carbon deposits. The data in Fig. 4 show further improvements in cell performance of the second cell. For these experiments, the current densities were measured at a potential of 0.4 V at 700°C. The power densities for H_2 and *n*-butane in this particular cell were approximately 20% lower than for the first cell, which is within the range of the ability to reproduce cells. However, the power densities achieved for some other fuels were significantly higher. In particular, stable power generation was now observed for toluene. Similarly, Fig. 4 shows that methane, ethane and 1-butene could be used as fuels to produce electrical energy. The data show transients for some of the fuels, which are at least partially due to switching.

3. AGING MECHANISM OF COMPONENTS IN SOFC

The commercial requirements for SOFC systems include a cell and stack life in the order of 40–50 000 h with very small degradation rates. All resistive limitations in an SOFC may suffer from degradation. Demonstration of total stack durability is of primary interest, but an analysis of aging mechanism in each component is desirable. It is therefore essential to increase the understanding of the degradation of each component in SOFC.

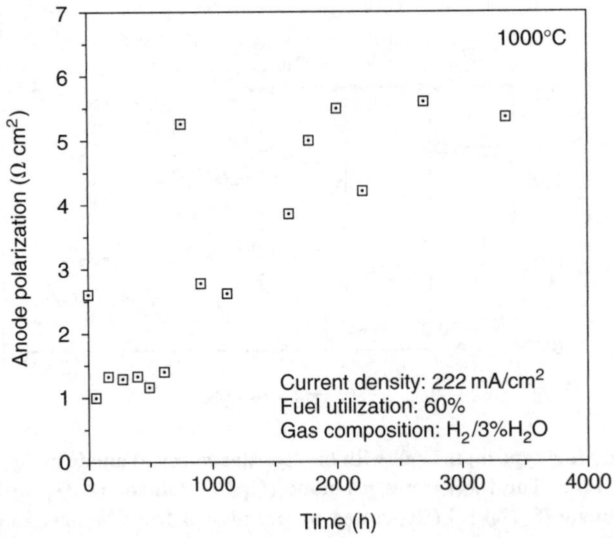

Fig. 5. Effect of nickel sintering on cermet anode polarization [11] (Reproduced with permission from The Electrochemical Society.)

3.1. Aging mechanism of anode

State-of-the-art SOFC anodes consist of a Ni-YSZ cermet. While Ni plays the role of the catalytically active as well as of the electronically conducting phase, the YSZ is added in order to support the Ni particles, to inhibit coarsening of the Ni, and to provide a thermal expansion coefficient which is similar to that of the zirconia-based electrolyte. A homogeneous or graded structure consisting of three phases, Ni, yttria-stabilized zirconia (YSZ), and porosity, should be obtained, providing percolation paths for electrons, oxide ions, and gaseous hydrogen and water, respectively. This requirement originates in the poor conductivity of electrons in YSZ and oxide ions in Ni. The line where the three phases meet is referred to as the triple-phase boundary (TPB) and is considered to be electrochemically active only if percolation paths for the active species are provided. The reaction rate for electrochemical oxidation of hydrogen has been demonstrated to correlate with the length of the TPB. In the research and development of SOFCs, serious reductions in cell performance have been observed due to degradation of the anode. The polarization characteristics of the anode are highly dependent on its morphology. The apparent activation energy has a distinct tendency toward lower value for fine cermets than that for coarse cermets [8]. It is widely accepted that sintering of nickel plays an important role [9]. The SOFC anode cermet is commonly made from YSZ and NiO powders. The NiO is then reduced in situ to nickel metal when exposed to the fuel in the fuel cell. Since nickel particles are high-surface-area solids, there will always be a thermodynamically driving force to decrease free energy, i.e., to minimize surface area. Thus, the sintering behavior of the nickel/YSZ anode is strongly dependent on the wetting properties of the nickel on the YSZ. The agglomeration of Ni particles under SOFC operating conditions leads a reduction in electrochemical reaction sites, namely TPB, and the cutting-off on current paths [10]. Figure 5 shows, as an example, the effect of nickel sintering on the polarization of the anode [11]. Furthermore, shrinkage of the electrode upon firing also leads to decrease in the gas permeability of the electrode layer, the occurrence of shear stresses at the electrode – electrolyte interfaces, and increases in contact resistance between the electrode and the current collector when assembled as a stack.

There are a number of programs dedicated to long-term stability studies and the prevention of Ni agglomeration at the operating conditions of the fuel cell. In this regard, a novel anode microstructure has been

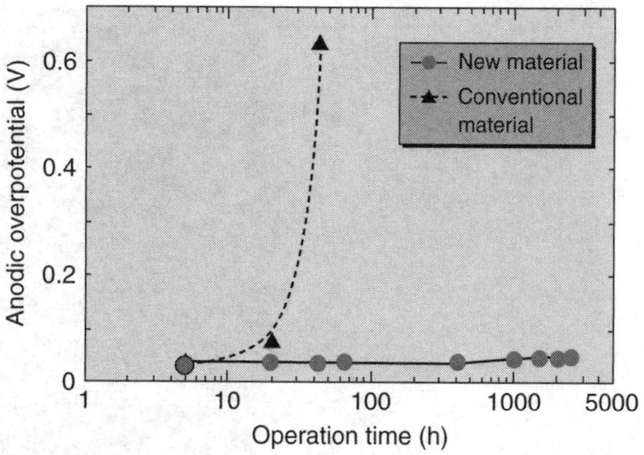

Fig. 6. Change in anodic polarizations between the previous and the new anodes at 0.2 A/cm^2 as a function of cell operating time [12] (Reproduced with permission from the Central Research Institute of Electric Power Industry, Japan.)

proposed [12]. A feature of this new anode is that, unlike other materials, the YSZ powders are divided into coarse and fine particles. Upon sintering, the coarse YSZ particles are connected by a network of fine YSZ particles. This forms a strong framework, which prevents the agglomeration and coarsening of Ni particles. The shrinkage and porosity behavior for both sintering under reducing atmosphere at 1000°C (as operating condition) and in air at 1400°C (as producing condition) were measured and compared. The previous material showed considerable shrinkage and a large decline in porosity. On the other hand, the new material showed little shrinkage under producing condition and virtually no change in volume and porosity under operating condition. Figure 6 shows the overpotentials for the two anodes during a long-term operation. The overpotential of previous anode rapidly rises after a few tens of hours, whereas the new one keeps the overpotential constant, allowing more 2500 h of continuous generation, which indicates superior long-term stability. Therefore, the durability of the anodes can be expected with optimization of the cermet structure.

3.2. Aging mechanism of cathode

A composite of A-site-deficient strontium-doped lanthanum manganite (LSM) and the electrolyte material, yttria-stabilized zirconia [13,14] is presently used as cathode in SOFC. The formation of low-conductive reaction products such as lanthanum zirconate at the cathode – electrolyte interface can be avoided during cell fabrication. Furthermore, the electrochemically active reaction zone may be extended from the interface between the electrode and the electrolyte to the bulk of the electrode. The cathode reaction depends on the catalytic activity and microstructure of composite cathodes. Cathode overpotential is often the main factor limiting SOFC performance. An attempt has been made to investigate whether temperature, current load or kinetic processes are responsible for the degradation [15]. The cathodes were kept at constant, realistic operating conditions ($-300\,\mathrm{mA\,cm^{-2}}$ at 1000°C in air) for up to 2000 h. Nominally identical cathodes are kept for 2000 h at 1000°C in air without current load for comparison. After 2000 h test, the increase in electrode overvoltage exceeded 100% of the initial value for the electrodes with the galvanostatic load. However, the electrode without load showed little or no degradation. The pore formation, observed after the galvanostatic durability test, was a common feature for all the cathodes investigated. Figure 7 shows the pore formation at the composite electrode layer (C-layer) interfaces, which might have caused a decrease in the length of active triple phase boundary between electrode, electrolyte and gas phase, as the contact area between composite electrode layer and electrolyte was decreased considerably.

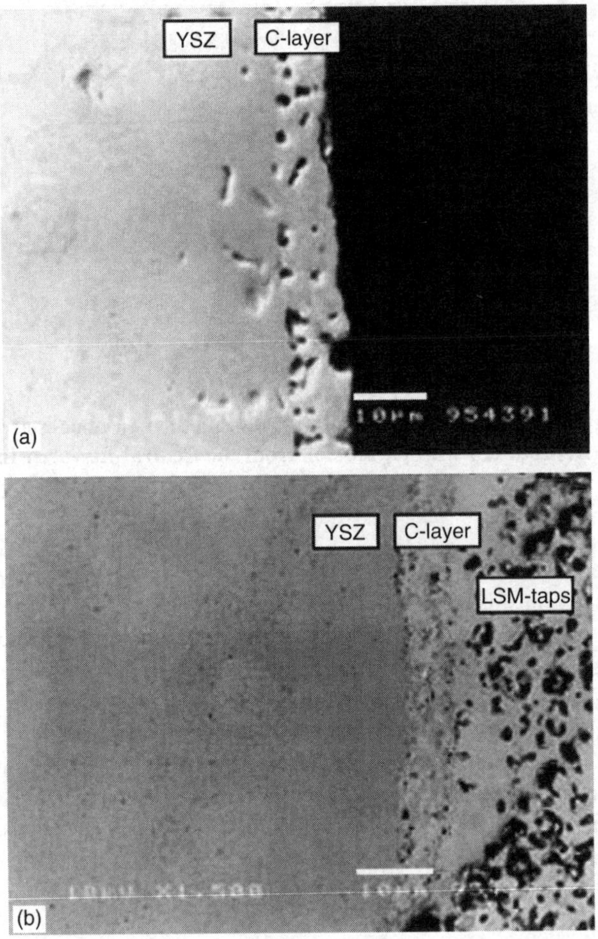

Fig. 7. Structural change in the composite cathode during operation at 1000°C, 300 mA cm^{-2} in air: reference sample (a) and after 2000 h with current load (b) [15] (Reproduced with kind permission from Kluwer Academic Publishers.)

As the polarization resistance scales with the inverse of TPB length, the microstructural changes might be responsible for the observed increase in polarization resistance during galvanostatic tests. When exposing a cation-deficient oxide material to an oxygen potential gradient cations can migrate toward the higher oxygen potential interface, while vacancies move in the opposite direction. This may lead to pore formation and pore movement [16]. The pores will preferably form at the interface with the lowest oxygen potential in places, where there is an indentation or a notch, as this part of the interface is unstable. The pores are transported toward the interface with the highest oxygen potential. The applied current creates an oxygen potential difference across the composite electrode layer and this difference may be the driving force of the pore formation.

The decrease in operation temperature of SOFC seems to assure the stability of cathodes [17]. The planar SOFC in Siemens design is made of cells or membrane – electrode assemblies (MEAs) connected in parallel or/and in series to a stack. MEA consists of planar solid electrolyte YSZ with a thickness of 150 m and of two porous electrodes: cathode and anode. The cathode consists of two layers: electrochemically active layer from $La_{0.75}Sr_{0.2}MnO_3$ and YSZ composite and electronic conductive $La_{0.75}Sr_{0.2}MnO_3$ layer. After 5000 h of stack operation at 850°C, 300 mA cm^{-2} in air/H_2, no changes in the porosity of both layers

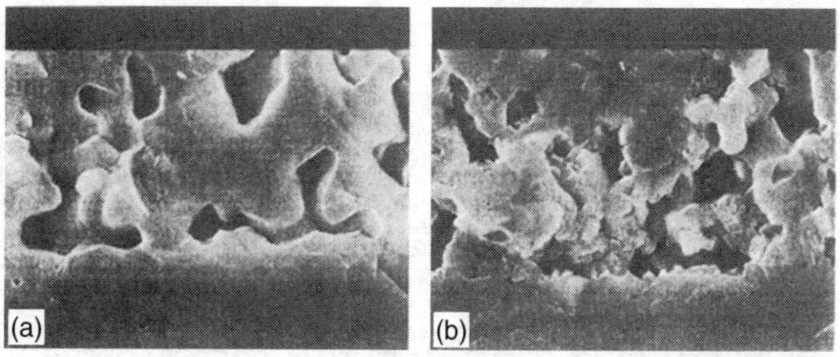

Fig. 8. Structural stability in the composite cathode during operation at 850°C, 300 mA cm^{-2} in air: initial state (a) and after operation of 5000 h (b) [17] (Reproduced with permission from European Fuel Cell Forum.)

were found. It means that no post-sintering takes place at 850°C. The cross section of the cathode at initial state and after 5000 h of operation is shown in Fig. 8.

3.3. Aging mechanism of interconnect

Interconnect is one of the key components in planar SOFC, which provide the electrical connection between the individual cells in a series to make SOFC stacks and separate the anode and the cathode gases. There are two types of interconnect materials commonly used in SOFC, doped $LaCrO_3$-based ceramic materials and high-temperature oxidation resistant alloy. The latter is more attractive because doped $LaCrO_3$-based ceramic materials have low mechanical strength and high manufacturing costs. The alloys used as metallic interconnects contain Cr in order to render possible formation of a chromia scale for corrosion protection. Novel Cr-based alloys are, for example, the oxide dispersion strengthened (ODS) alloy $Cr5Fe1Y_2O_3$. However, the use of chromium containing alloys can lead to a rapid degradation of the electrical properties of an SOFC due to chromium evaporation at the cathode side of the fuel cell [18]. The alloys form Cr_2O_3 under cathodic conditions. A thin layer of this oxide grows on the surface of the alloy. The oxide completely covers the alloy surface and determines the chromium vaporization. Depending on temperature and the partial pressures of H_2O and O_2, volatile chromium species are formed by the Cr_2O_3 layer:

$$Cr_2O_3(s) + 1.5O_2(g) = 2CrO_3(g) \tag{9}$$

$$Cr_2O_3(s) + 1.5O_2(g) + 2H_2O(g) = 2CrO_2(OH)_2(g) \tag{10}$$

$$Cr_2O_3(s) + O_2(g) + H_2O(g) = 2CrO_2OH(g) \tag{11}$$

The high valent oxides and oxyhydroxides of chromium in the vapor phase can also undergo cathodic reduction. Then oxide ions are formed react subsequently with the fuel at the anode side as is the case in the reduction of oxygen. The electrochemical reactions involved in the reduction of $CrO_2(OH)_2(g)$ are given as an example in detail.

$$\text{At cathode: } CrO_2(OH)_2(g) + 3e^- = \tfrac{1}{2}Cr_2O_3(s) + H_2O(g) + \tfrac{3}{2}O^{2-} \tag{12}$$

$$\text{At electrolyte: } \tfrac{3}{2}O^{2-}\text{, cathode} = \tfrac{3}{2}O^{2-}\text{, anode} \tag{13}$$

$$\text{At anode: } \tfrac{3}{2}O^{2-} + \tfrac{3}{2}H_2(g) = \tfrac{3}{2}H_2O(g) + 3e^- \tag{14}$$

$$\text{Total reaction: } CrO_2(OH)_2(g) + \tfrac{3}{2}H_2(g) = \tfrac{1}{2}Cr_2O_3(s) + H_2O(g\text{, cathode}) + \tfrac{3}{2}H_2O(g\text{, anode}) \tag{15}$$

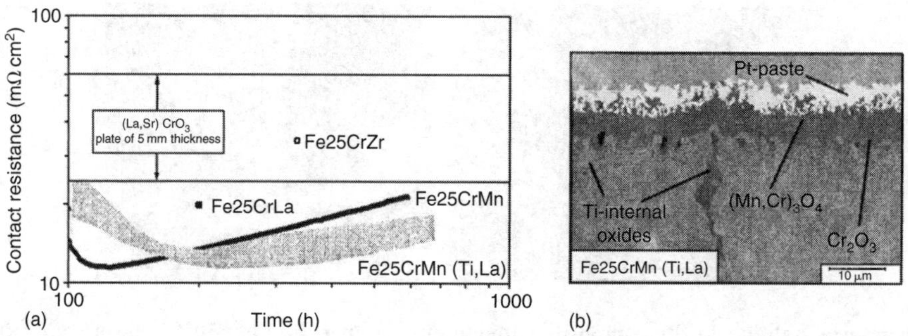

Fig. 9. (a) Electrical contact resistance of selected model alloys at 800°C in air compared with ceramic interconnect of 5 mm thickness. (b) Microstructure of the scale formed on model alloy Fe25CrMn (Ti,La) after 600 h conductivity testing in air [19] (Reproduced with permission from The Electrochemical Society.)

The following reactions for the overall cell result in the case of the involvement of CrO_3(g) and $CrO_2(OH)$(g) in the cathodic reduction:

$$CrO_3(g) + \tfrac{3}{2}H_2(g) = \tfrac{1}{2}Cr_2O_3(s) + \tfrac{3}{2}H_2O(g) \tag{16}$$

and

$$CrO_2(OH)(g) + H_2(g) = \tfrac{1}{2}Cr_2O_3(s) + \tfrac{3}{2}H_2O(g) \tag{17}$$

The result of such processes is the precipitation of Cr_2O_3(s) phase at the cathode–electrolyte phase boundary. The formation of Cr_2O_3(s) by the electrochemical reduction of high valent chromium oxide and oxyhydroxide species, in particular CrO_3(g) and $CrO_2(OH)_2$(g), is most important for the observed degradation of the electrochemical properties of the cell since Cr_2O_3(s) is formed at those sites, which are particularly well suited for the cathodic oxygen reduction. The formation of Cr_2O_3(s) can thus inhibit the oxygen reduction necessary for the operation of an SOFC and may lead to polarization losses.

Furthermore, the formation of Cr_2O_3(s) can be coupled to other chemical reaction with perovskites $La_{1-x}Sr_xMnO_{3-\delta}$, as well as with both perovskite $La_{1-x}Sr_xMnO_{3-\delta}$ and YSZ. YSZ is treated as ZrO_2

$$La_{1-x}Sr_xMnO_3(s) + \tfrac{1}{2}Cr_2O_3(s) = La_{1-x}Sr_xMn_{1-y}Cr_yO_3(s) + (Cr_{1-y}Mn_y)O_{1.5-\delta}(s) + \tfrac{1}{2}\delta O_2(g) \tag{18}$$

$$La_{1-x}Sr_xMnO_3(s) + yZrO_2(s) + \tfrac{1}{2}(1-y)Cr_2O_3(s) = La_{1-x}Sr_xMn_{1-y}Zr_yO_3(s) + (Cr_{1-y}Mn_y)O_{1.5-\delta}(s) + \tfrac{1}{2}(\delta + 0.5y)O_2(g) \tag{19}$$

The driving force for the reaction (18) is essentially the formation of the $(Cr_{1-y}Mn_y)O_{1.5-\delta}$ oxide solution or spinel. Reaction (19) occurs only if the negative Gibbs energy exceeds the Gibbs energy change of $SrZrO_3$ formation and its subsequent dissolution in a perovskite compatible with ZrO_2. The dissolution of the chromia and precipitation of the spinel phase may deteriorate the electrochemical properties of the perovskite.

A reduction in the evaporation rate of volatile chromium oxides and hydroxides can be achieved by using a steel that forms an oxide layer on top of the chromia scale. Fe—Cr model alloys with variations in chromium content and additions of Ti and/or Mn as well as additions of La, Ce, Zr and Y were studied for potential application in interconnect [19]. Figure 9 shows the electrical resistances of the oxide scales during exposure up to 500 h in air at 800°C. These results are compared with those for a typical ceramic $(La,Sr)CrO_3$ interconnect of 5 mm thickness and chromia-forming alloys Fe25CrLa and Fe25CrZr. After reaching stable conditions, all tested alloys showed lower electrical resistances than the ceramic interconnect.

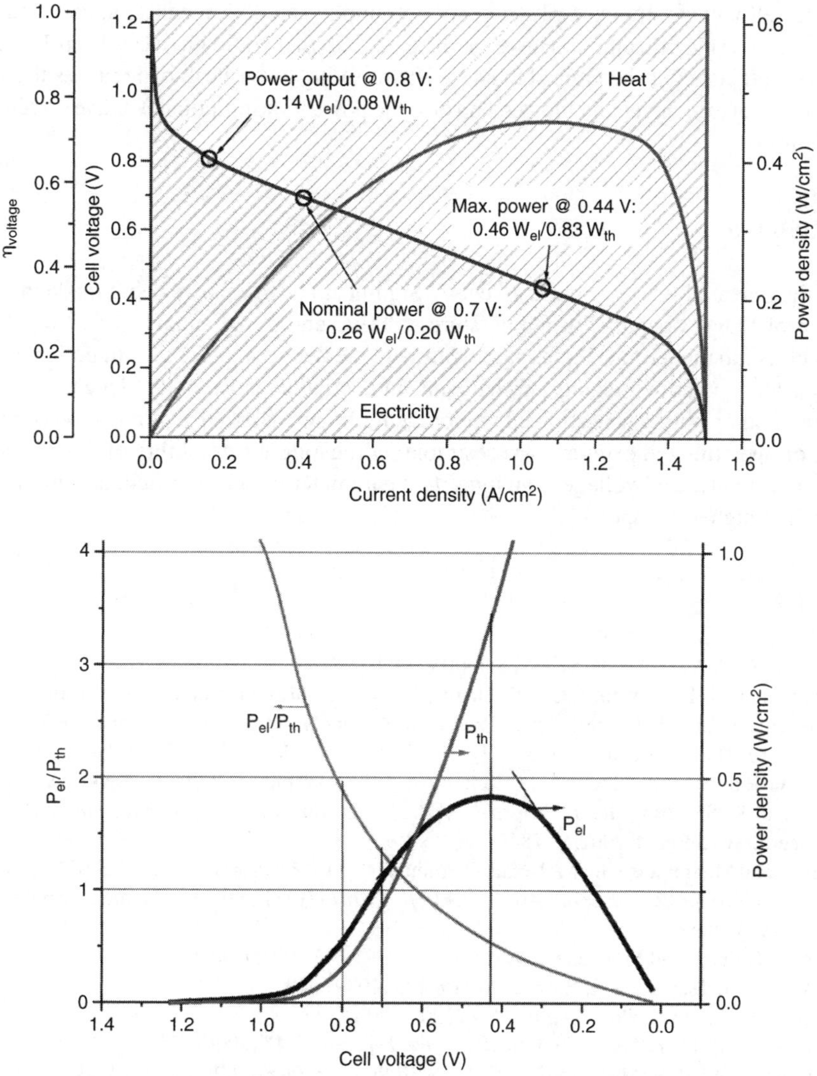

Fig. 10. Comparison of specific electricity and heat generation in an SOFC

The alloys containing Ti and La exhibited relatively low electrical resistance. The $(Mn,Cr)_3O_4$ spinel-forming steels showed lower chromium evaporation rates than pure chromia-forming alloys.

4. BENEFIT OF SOFC OPERATION AT HIGHER CELL VOLTAGE

SOFC systems are presently rated at a cell voltage of 0.7 V/cell. While this is a reasonable approach for performance comparisons and exhibits a high electric power output (>50% of the maximum), there are also some disadvantages. As seen in Fig. 10, the cell efficiency is restricted to 56% by the electricity/heat ratio of 1.3. The cell generates approximately 0.75 W heat/W electricity. Removal of the heat needs large amounts of air, which affect stack, heat exchanger and blower design.

A rise in the cell voltage, for example, to 0.8 V/cell implies an 80% increase in cell area and therefore stack size for the same electric output. However, the electricity/heat ratio increases from 1.3 at 0.7 V to 1.75 at 0.8 V. The cell efficiency is then raised to 64%. The resulting lower heat load reduces the size of peripherals like tubing, heat exchanger and blower, and also the power consumption of active cooling systems like blowers and pumps.

5. CONCLUSIONS

SOFCs have the potential market competitiveness. Significant advances are being made in SOFCs including realization of SOFC operating at temperature <800°C and direct supply of hydrocarbon fuels. The aging of the cell components contributes significantly to the decrease of long-term performance of SOFCs under steady-state operation. Aging mechanisms in components strongly depend on the materials, microstructures and operating conditions. Development of new materials, optimization of microstructures and lowering of operating temperatures are desirable for the long-term stability of SOFC and the possibility for direct fuel supply. Cell voltage is an important parameter in SOFC system design. It is beneficial to operate the cell at higher voltage.

REFERENCES

1. B.C.H. Steele. *C. R. Acad. Sci.* Paris, t. 1, Séris II c, (1998) 533.
2. H.P. Buchkremer, U. Diekmann, L.G.J. de Haart, H. Kabs, U. Stimming and D. Stöver. In: U. Stimming, S.C. Singhal, H. Tagawa and W. Lehnert (Eds.), *Proceedings of the 5th International Symposium on Solid-Oxide Fuel Cells (SOFC-V)*, The Electrochemical Society, Pennington, NJ, PV 97, 40, 1997, p. 160.
3. L. Blum, L.G.J. de Haart, I.C. Vinke, D. Stolten, H.-P. Buchkremer, F. Tietz, G. Blass, D. Stöver, J. Remmel, A. Cramer and R. Sievering. In: J. Huijsmans (Ed.), *Proceedings of the 5th European Solid-Oxide Fuel Cell Forum*, Lucern, Switzerland, 2002, p. 784.
4. K. Morimoto and M. Shimotsu. In: M. Dokia, O. Yamamoto, H. Tgawa and S.C. Singhal (Eds.), *Proceedings of the 4th International Symposium on Solid-Oxide Fuel Cells (SOFC-IV)*, The Electrochemical Society, Pennington, NJ, PV 95-1, 1995, p. 269.
5. P. Vernoux, J. Guindet and M. Kleitz. *J. Electrochem. Soc.* **145** (1998) 3487.
6. E. Perry Murray, T. Tsai and S.A. Barnett. *Nature* **400** (1999) 649.
7. S. Park, J.M. Vohs and R.J. Gorte. *Nature* **404** (2000) 265.
8. M. Brown, S. Primdahl and M. Mogensen. *J. Electrochem. Soc.* **147** (2000) 475.
9. L.J.M.J. Blomen and M.N. Mugerva (Eds.), *Fuel Cell Systems*. Plenum Press, New York, 1993.
10. D.W. Dees, T.D. Claar, T.E. Elser and F.C. Marzek. *J. Electrochem. Soc.* **134** (1987) 2141.
11. S. Elangovan and A. Khandkar. In: T.A. Ramanarayanan and H.L. Tuller (Eds.), *Proceedings of the 1st International Symposium on Ionic and Mixed Conducting Ceramics*, 16–17 October 1991, Phoenix, AZ, The Electrochemical Society, Pennington, NJ, 1991, p. 122.
12. H. Itoh. Report: research and development on high performance anode for solid-oxide fuel cells – improvement of the microstructure for new long life anode, *CRIEPI Report* W94016, 1995.
13. T. Kenjo and M. Nishiya. *Solid State Ionics* **57** (1992) 295.
14. M.J.L. Østergård, C. Clausen, C. Bagger and M. Mogensen. *Electrochim. Acta* **40** (1995) 1971.
15. M.J. Jørgensen, P. Holtappels and C.C. Appel. *J. Appl. Electrochem.* **30** (2000) 411.
16. G.J. Yurek and H. Schmalzried. *Ber. Bunsenges. Phys. Chem.* **79** (1975) 255.
17. M. Kuznecov, H. Greiner, M. Wohlfart, Klaus Eichler and P. Otschik. In: A.J. McEvoy (Ed.), *Proceedings of the 4th European Solid-Oxide Fuel Cell Forum*, Lucerne, Switzerland, 2000, p. 261.
18. K. Hilpert, D. Das, M. Miller, D.H. Peck and R. Weiß. *J. Electrochem. Soc.* **143** (1996) 3642.
19. J.P. Abellán, V. Shemet, F. Tietz, L. Singheiser, W.J. Quadakkers and A. Gil. In: H.Yokokawa and S.C. Singhal (Eds.), *Proceedings of the 7th International Symposium on Solid-Oxide Fuel Cells (SOFC VII)*, The Electrochemical Society, Pennington, NJ, PV 2001–16, 2001, p. 811.

Chapter 14
Components manufacturing for solid oxide fuel cells

F. Tietz, H.-P. Buchkremer and D. Stöver

Abstract

A worldwide overview of processing technology of solid oxide fuel cell (SOFC) components is given and the fabrication techniques of ceramic components are summarized for the different types of SOFCs. Generally, a tendency towards up-scalable and automatizable processes is observed. In addition, critical points of interconnect materials and interconnect fabrication are stressed. Especially for planar cell designs, the chromium contamination of the cathode and interfacial corrosion is regarded as the weak points to be solved for demonstration of planar SOFC units.

Keywords: Solid oxide fuel cell (SOFC); Components manufacturing; Processing; Interconnect materials

Article Outline

1. Introduction 249
2. Cell design 250
3. Ceramic components 250
4. Interconnect materials 254
5. Summary 256
References 256

1. INTRODUCTION

Worldwide, several developers of solid oxide fuel cell (SOFC) technology have made significant progress in recent years and are scheduling SOFC systems in the 1–1000 kW range for concept studies, field tests or even as commercial products between 2001 and 2005. This shift of development from laboratory testing to near-market products implies not only a change in operating conditions from "artificial" comparative tests with hydrogen to real conditions using steam-reformed methane, high fuel utilization and long operating periods. It also implies consistently defined material combinations and especially the choice of up-scalable ceramic processing technologies.

Two main issues in SOFC development can be identified as driving forces during recent years: cost reduction with respect to low-cost materials and simpler processing techniques, and the improvement of durability in long-term operation. In this context, also the aim of decreasing the operating temperature can be understood in terms of longer SOFC operation. Previously, the planar electrolyte-supported SOFCs operated at temperatures around 1000°C [1,2]. Therefore, either ceramic interconnect materials or rather expensive chromium alloys had to be used. It was not possible to take advantage of much cheaper ferritic

Fuel Cells Compendium

steels due to the too high operating temperatures. Reducing the electrolyte thickness to 10–20 μm [3,4], corresponding to a decrease in electrical resistance, enabled the same power output at 150–200°C lower operating temperatures than for electrolyte-supported cells with an electrolyte thickness of about 200 μm [5]. Also ferritic steels could be applied [6] because operating temperatures remained below the temperature limit of these materials. However, reducing the electrolyte thickness further does not lead to decreased operating temperatures or increased power densities, because the contribution of the ohmic resistance to the overall electrical losses is already rather small. Any further decrease in the operating temperature can only be achieved by improving the electrode performance, i.e. reducing the electrode overvoltages especially at the cathode. Hence, this corresponds to a longer component life with respect to lower corrosion of the metallic interconnect. The introduction of fine-grained composite layers of the electrolyte material (yttria-stabilized zirconia, YSZ) and the electrocatalyst (lanthanum manganite for the cathode and nickel metal for the anode) as an interlayer between the electrolyte and the original electrode has led to remarkable improvements in cell performance. Today power densities of 0.35 W/cm^2 at 0.7 V and 800°C can easily be achieved [6], a value which was previously obtained under the same experimental conditions at 900–1000°C.

The benefit of such improved electrode performance can be used in three ways: (a) directly as enhanced power output, (b) by lower operating temperature in terms of longer component life and (c) as a buffer for peak electricity demands during operation. This last possibility can be regarded as an option for increased long-term stability because long performance at high loads can lead to deterioration of the fuel cell [7] due to cationic de-mixing [8,9] and subsequent decomposition. Such a phenomenon has already been observed during operation of oxygen permeation membranes [10]. Improved cell performance can also be achieved by optimizing the microstructure of the electrochemically active layers [6,11]. Therefore, improved materials as well as improved processing of the components are the key to obtaining SOFCs with sufficient long-term stability at an acceptable cost level.

As SOFC materials have already been reviewed in the past [12,13], this paper gives a detailed overview of the current SOFC manufacturing techniques applied by actual SOFC developers worldwide.

2. CELL DESIGN

SOFCs can be grouped into tubular and planar designs. Both types may consist of one or several single cells per stacking unit, i.e. on a single tube or in a single layer (Table 1). Furthermore, the planar designs can be divided into stack systems with metallic or ceramic interconnect material as well as into cells with thick (electrolyte-supported) or thin (electrode-supported) membranes with thicknesses usually of 100–250 and 5–20 μm, respectively. Table 1 gives an overview of the diversification of SOFCs together with the main developers involved.

3. CERAMIC COMPONENTS

Different processes are used for the low-cost fabrication of electrodes and electrolyte for the SOFC. The components produced must display specific properties to ensure cyclable and long-term stable operation in the fuel cell stack.

The fabrication routes for the individual cell components of the different SOFC designs (planar or tubular) differ greatly depending on which cell component is to perform the supporting function in the cell. Whereas for the planar concept in most cases the electrolyte or the anode ensures the mechanical stability of the cells (Tables 2 and 3), in the tubular concepts the cathode or an inert tube is the supporting component (Table 4). The SOFC developers of tubular cells have been looking for alternative, cost-effective coating techniques for electrolyte membrane and anode deposition for more than 5 years to replace the expensive

Table 1. SOFC typology and currently leading SOFC developers

Solid oxide fuel cells						
Tubular		**Planar**				
One cell per tube	**Several cells per tube**	**One cell per layer**			**Several cells per layer**	
Ceramic interconnect	**Ceramic interconnect**	**Metallic interconnect**		**Ceramic interconnect**	**Metallic interconnect**	
		Thick electrolyte	Thin electrolyte	Thick electrolyte	Thick electrolyte	Thin electrolyte
SWPC (D/USA)	MHI + EPDC (Nagasaki) (JP)	Sulzer (CH)	FZJ (D)	SOFCo(USA)	CFCL (AUS)	Rolls-royce (GB)
Toto(JP)		ECN (NL)	ECN (NL)	Tokyo Gas(JP)	Sanyo (JP) (terminated)	
		TMI (USA)	Risø (DK)	MHI + CEPC (Himeji) (JP)	Siemens (D) (terminated)	
		Ztek (USA)	Global			
		Fuji Electric (JP) (terminated)	Thermo electric (CAN)	Mitsui (JP)		
		Murata + Osaka Gas (JP) (terminated)	Allied Signal (USA)	Risø (DK) (terminated)		
			CFCL (AUS)	Toho Gas (JP)		
				Donier (D) (terminated)		

Table 2. Developers of SOFC in electrolyte-supported planar cell design and corresponding fabrication and design details

Company	Country	Component	Material	Production process	Thickness	Reference
Sulzer Hexis	CH	Electrolyte	YSZ	Tape casting	ns	[14]
		Cathode	$(La,Sr)MnO_3$	Screen printing	ns	
		Anode	Ni/YSZ	Screen printing	ns	
ECN/InDec	NL	Electrolyte	YSZ	Tape casting	ns	[15,16]
		Cathode	$(La,Sr)MnO_3$	Screen printing	50 μm (two layers)	[15,16]
		Anode	Ni/YSZ	Screen printing	graded composite	[15]
Fraunhofer Ges., IKTS	D	Electrolyte	YSZ	Tape casting	150 μm	[17]
		Cathode	$(La,Sr)MnO_3$	Screen printing	ns, two layers	[17]
		Anode	Ni/YSZ	Screen printing	ns	
CFCL	AUS	Electrolyte	3YSZ, 8YSZ	Tape casting	100 μm	[18,19]
		Cathode	$(La,Sr)MnO_3$	Screen printing	50–60 μm	[18,19]
		Anode	Ni/YSZ	Screen printing	50 μm	[18,19]
SOFCo	USA	Electrolyte	YSZ, $(Ce,Sm)O_2$	Pressing and sintering	180 μm, 300 μm	[20,21]
		Cathode	$(La,Sr)CoO_3$	Screen printing	ns	[21]
		Anode	Ni/YSZ	Screen printing	ns	[21]
Tokyo Gas	JP	Electrolyte	3YSZ	Tape casting	50–100 μm	[22]
		Cathode	$(La,Sr)MnO_3$	Screen printing	150 μm	[22]
		Anode	Ni/(Ce,Y)SZ	Screen printing	30 μm	[22]
Mitsui Eng. & Shipbuilding	JP	Electrolyte	8YSZ	Tape casting	300 μm	[23,24]
		Anode	Ni/YSZ	Painting	150 μm	[23,24]
		Cathode	$(La,Sr)(Mn,Cr)O_3$	Painting	150 μm	[23,24]

ns = not specified.

Table 3. Developers of SOFC in anode-supported planar cell design and corresponding fabrication and design details

Company	Country	Component	Material	Production process	Thickness	References
Sulzer Hexis	CH	Anode substrate	Ni/YSZ	Tape casting	250–500 μm	[25]
		Electrolyte	YSZ/(Ce,Y)O_2	Reactive magnetron sputtering	5/1 μm	[26]
		Cathode	$La_{0.6}Sr_{0.4}Co_{0.2}Fe_{0.8}O_3$	Screen printing	ns	[26]
ECN/InDec	NL	Anode substrate	Ni/YSZ	Tape casting	500–800 μm	[15,16,27]
		Anode	Ni/YSZ	Screen printing	3–7 μm	[27]
		Electrolyte	YSZ	Screen printing	7–10 μm	[28]
		Cathode	(La,Sr)MnO_3+YSZ	Screen printing	ns	[27]
FZJ	D	Anode substrate	Ni/YSZ	Tape casting	200–500 μm	[29]
		Anode substrate	Ni/YSZ	Warm pressing	1500 μm	[6,30]
		Anode	Ni/YSZ	Vacuum slip casting	5–15 μm	[6,30]
		Electrolyte	YSZ	Vacuum slip casting	5–30 μm	[6]
		Electrolyte	YSZ	Reactive magnetron sputtering	2–10 μm	[31]
		Cathode	(La,Sr)MnO_3 + YSZ	Wet powder spraying	50 μm	[6,32]
Risø	DK	Anode substrate	Ni/YSZ	Tape casting	200–300 μm	[33]
		Electrolyte	YSZ	Wet powder spraying	10–25 μm	[34]
		Cathode	(La,Sr)MnO_3 + YSZ	Screen printing	50 μm	[34]
Global Thermoelectric	CAN	Anode substrate	Ni/YSZ	Tape casting	1000 μm	[35]
		Electrolyte	YSZ	Vacuum slip casting	10 μm	[35]
		Electrolyte	YSZ	Screen printing	ns	[36]
		Cathode	(La,Sr)MnO_3	Screen printing	40 μm	[35]
Allied Signal	USA	Anode	Ni/YSZ	Tape casting and calendaring	100 μm	[37,38]
		Electrolyte	YSZ	Tape calendering	5–10 μm	[37,39]
		Cathode	Doped $LaMnO_3$	Tape calendering	ns	[39]
CFCL	AUS	Anode substrate	Ni/YSZ	Tape casting	500–700 μm	[18,19,40]
		Electrolyte	YSZ	Lamination and sintering	10–30 μm	[19,40]
		Electrolyte	YSZ	Reactive magnetron sputtering	<16 μm	[41]
		Cathode	(La,Sr)MnO_3	Screen printing	ns	[19,40]
Mitsui Eng. and Shipbuilding	JP	Anode substrate	Ni/YSZ	ns	1000 μm	[23]
		Electrolyte	8YSZ	ns	30 μm	[23]
		Cathode	(La,Sr)(Mn,Cr)O_3	ns	150 μm	[23]

ns = not specified.

and complicated electrochemical vapor deposition. Wet ceramic techniques are favored due to cost aspects and such work is under way.

In the case of electrolyte-supported cells (Table 2), the fabrication of the electrolyte and of the electrodes is dominated by tape casting and screen printing, respectively. Both fabrication processes are well-established methods in the electroceramics industry and a scale-up is easily feasible. The thickness of the components varies only little: the membrane foils have a thickness between 100 and 200 μm, and both the cathode and the anode are screen printed onto the electrolyte sheets with thicknesses of 40–60 μm. The tape cast electrolyte foils usually have a size of up to 10 × 10 cm, because larger tapes are difficult to handle after sintering.

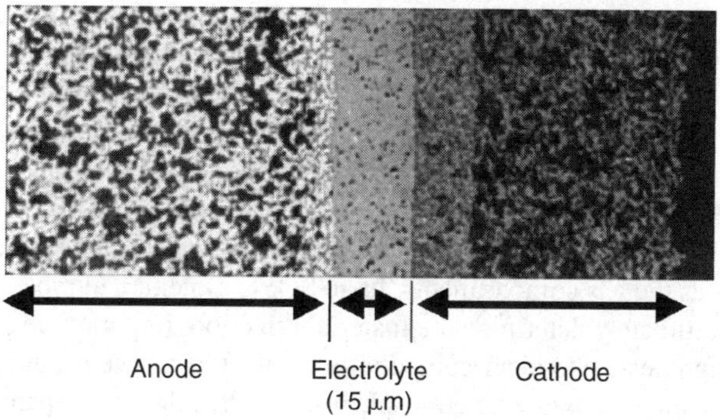

Fig. 1. Microstructure of an anode-supported SOFC with a 15-μm-thick solid electrolyte. Note that the anode substrate is about 30–100 times thicker than the electrolyte layer (see Table 3)

Other developers aiming at the kW range during the next few years prefer planar anode-supported SOFCs due to their potential of lowering the operating temperature. Here several companies and research organizations in Australia, America and Europe have concentrated on cells with a thick, porous anode substrate and a 5–20 μm-thin electrolyte membrane. Besides the frequently used ceramic processing techniques, i.e. tape casting and screen printing, also alternative methods like warm pressing, tape calendaring and wet powder spraying are under investigation. The main selection criteria for the future fabrication route are the cost aspects, the potential for automation, reproducibility and precision of the different techniques.

In the anode substrate concept, the anode is the supporting component of the cell and must therefore display sufficient mechanical stability. The substrates are predominantly produced by tape casting. Pressing processes are very rarely applied and extrusion molding as for the supporting tubes not at all (Table 4). The reason is presumably that the substrates in such processing routes can hardly be fabricated thinner than 1 or 1.5 mm and most of the developers aim at substrates with thicknesses of around 0.5 mm. Usually an anode functional layer of a few micrometers in thickness is then deposited onto the substrate to enhance the electrochemical performance [6]. A widely used deposition technique for the thin anode, electrolyte and cathode layers is screen printing (Table 3). A cross-section through the electrochemical active layers of an anode-supported cell is shown in Fig. 1. In few cases slip casting and wet powder spraying is applied, whereas magnetron sputtering is a curiosity for achieving very thin layers rather than a realistic approach for cost-effective products with high throughput. Also rather seldom is plasma spraying either as vacuum plasma spraying (VPS) for anodes and electrolytes [50] or as atmospheric (APS) or flame spraying (FS) for electrolytes and cathodes [51]. However, the cost targets of commercial SOFC are difficult to achieve with plasma spraying techniques. Therefore these techniques, commonly applied for tubular systems in the past, are being increasingly replaced by slurry processing (see Table 4). In the tubular fuel cell system of Siemens Westinghouse or Toto the cathode is the supporting cell component. The cathode tubes are produced using an extrusion process with subsequent sintering and the other components are fabricated by slurry coating (Toto) or by plasma spraying and electrochemical vapor deposition (Siemens Westinghouse). Similarly, Mitsubishi Heavy Industries (MHI) also now applies slurry coatings but the supporting tubes are made of stabilized zirconia. Finally, Rolls-Royce's concept is also based on an inert spinel-type rectangular tube on which the fuel cell components are deposited by screen printing (Fig. 1).

Apart from tape casting, calendering, screen printing, slip casting, plasma spraying, wet powder spraying, electrochemical vapor deposition for the manufacturing of electrolytes or the coating of substrates with electrolyte layers, other methods have been applied and tested such as laser ablation, multiple spin coating, colloidal deposition, reactive magnetron sputtering, chemical vapor deposition, spray pyrolysis and electrophoresis. All these alternative methods are of scientific interest rather than having the potential to be commercially relevant. A brief overview is also provided by Minh [52].

4. INTERCONNECT MATERIALS

The interconnect in SOFCs is the component which electrically connects the single cells and in planar systems additionally separates the gas compartments. In an SOFC system, a number of demands are made on the interconnect, which ultimately determine the material selection. Important requirements are good electrical conductivity, gastightness, chemical compatibility with the adjacent components of the fuel cell, chemical stability in reducing and oxidizing atmospheres, matched thermal expansion and last but not least reasonable costs. In order to meet these requirements, two classes of materials are commonly used for the interconnect, namely, ceramic and metallic materials. Whereas ceramic interconnects played a dominant role in the early SOFC developments and are still essential in tubular designs, metallic interconnects have been frequently used in recent developments. Both variants have benefits and disadvantages and the final choice is therefore always a compromise depending, among other aspects, on the design, the operating temperature, the required service life as well as on the material and production costs of these components.

Practically all the ceramic interconnects of present SOFC systems are based on the perovskite structure of the $LaCrO_3$ type. By modifying the stoichiometry with other elements it is possible to adapt this interconnect material with respect to thermal expansion and behavior in the presence of reaction gases [48,53]. However, the material costs of perovskites are rather high and their application as ceramic interconnect is only meaningful as long as the stack design requires only small amounts of the material as in the case of the tubular system of MHI (Table 4). Furthermore, lanthanum chromites are often plasma-sprayed (Table 4) although this technique is expensive. In this case, however, one has to consider the frequently observed low sinterability of the interconnect material, which prevents gastightness and low-cost production by sintering since it requires sintering temperatures between 1450 and 1600°C.

Whereas practically all activities are related to $LaCrO_3$ modifications for the ceramic interconnects, clearly more material systems are under development for the metallic interconnects. In general, advantages for metallic arrangements are considered to be high electrical conductivity, good processability and the lower costs to be expected, whereas especially long-term resistance, corrosion behavior, chromium evaporation and high expansion coefficients are disadvantageous. The success of metallic interconnects for use in the SOFC system will decisively depend on solving of these problems.

The long-term stability of the metallic interconnect is essentially governed by its corrosion characteristics. The materials used for interconnects are chromia forming alloys which ensure sufficiently high conductivity for thin oxide scales.

With respect to thermal expansion, electrical conductivity and corrosion behavior, the $Cr5Fe1Y_2O_3$ ODS alloy developed by Plansee in cooperation with Siemens shows excellent behavior at temperatures up to 950°C [54]. A disadvantage of the alloy produced by powder metallurgy is the currently high price which could be drastically reduced by suitable production techniques. A relevant approach is near-net-shape processing producing the interconnects practically without a finishing operation [55].

From the aspect of costs, ferritic chromium steels are attractive candidates for metallic bipolar plates. On the one hand, they form chromium oxides, have a lower thermal expansion compared to austenitic alloys and can be mechanically easily deformed and machined. On the other hand, they have a number of properties limiting their application such as lower high-temperature strength, insufficient corrosion protection at high temperatures and brittle phase formation. R&D work on ferritic steels therefore concentrates on application temperatures $<800°C$. The application range of interest for this material class coincides with the development goals for planar anode-supported fuel cells. For this reason, such materials are being used or developed by all companies and research institutions working on this concept (e.g. Sulzer, CFCL, Plansee, Sanyo, FZJ)

Table 4. Developers of SOFC in tubular cell design and corresponding fabrication and design details

Company	Country	Component	Material	Production process	Thickness	References
SWPC	USA	Cathode tube	Doped $LaMnO_3$	Extrusion and sintering	2200 μm	[42]
		Electrolyte	YSZ	Electrochemical vapor deposition (EVD)	40 μm	[42]
		Anode	Ni/YSZ	Slurry coating or EVD	100 μm	[42]
		Interconnect	Doped $LaCrO_3$	Plasma spraying	85 μm	[42]
Toto-KEPC	JP	Cathode tube	$(La,Sr)MnO_3$	Extrusion and sintering	ns	[43,44]
		Electrolyte	YSZ	Slurry coating	40 μm	[44]
		Anode	Ni/YSZ	Slurry coating	thick film	[43,44]
		Interconnect	$(La,Ca)CrO_3$	Slurry coating	ns	[43]
MHI + EPDC	JP	Substrate tube	Ca–SZ	Extrusion	(out) 21	[45–47]
		Cathode	$LaCoO_3$	APS	150–200 μm	[46]
			new: $(La,Sr)MnO_3$	New: Slurry coating		[48]
		Electrolyte	YSZ	Low pressure plasma spraying	100–150 μm	[47]
				New: Slurry coating		[48]
		Anode	Ni/YSZ	APS	80–100 μm	[46,47]
				new: Slurry coating		[48]
		Interconnect	$NiAl/Al_2O_3$	APS	80–100 μm	[46,47]
			new: $(Ln,AE)TiO_3^*$	New: Slurry coating	ns	[48]
Rolls-Royce	GB	Substrate tube	ns	ns	ns	
		Anode	ns	Printing	ns	[49]
		Electrolyte	ns	Wet slurry printing	<20 μm	[49]
		Cathode	ns	Printing	ns	[49]

* Ln = lanthanide element, AE = alkaline earth element, ns = not specified.

[56–58]. In general, it can be stated that the long-term corrosion behavior of commercially available materials is not yet sufficient. Recent developments [59] have attempted to achieve acceptable long-term behavior by selected doping with reactive elements (Ti, Y, La) and spinel formers.

In operating high-temperature fuel cells with metallic interconnects, a time-dependent degradation is observed, which is attributable to poisoning of the active centers of the cathode by chromium evaporating from the interconnect. This familiar process, already described in detail [60–62], is caused by highly volatile chromium species which form when chromium oxide is in contact with oxidizing gas atmospheres. $CrO_2(OH)_2$ and CrO_3 are particularly critical here [60] and can react with the perovskite of the cathode to more stable but less catalytically active Cr–Mn spinels [62].

Attempts are being made to reduce the damaging effect of chromium vapors on the cathode side by suitable protective layers of $LaCrO_3$ or by neutralizing the chromium atoms in the applied layers by gettering. This development is most advanced for the $Cr5Fe1Y_2O_3$ chromium-base interconnect, on which La–Sr–Cr oxides have yielded the best results [63]. Vacuum plasma spraying and physical vapor deposition have proved to be effective application processes [64,65]. The best result so far has been obtained with a $La_{0.9}Sr_{0.1}CrO_3$ coating for operation temperatures of about 950°C. At a temperature of 850°C, the chromium vaporization rate was only half as great [66]. Future developments and tests must above all demonstrate the long-term stability and thermal cycling resistance of such coatings.

Except for the Cr-base material, the metallic interconnects are manufactured using the casting-rolling-(forging) route with mechanical machining of the semi-finished products. The solutions have not yet reached the cost goal. Intensive work is being performed to find cheaper solutions by simplifying and adapting the design and using other modern manufacturing techniques (e.g. punching, laser cutting, brazing, etc.).

5. SUMMARY

The overview of this report shows the wide diversification in SOFC technology. The manufacturing and processing of materials and components has reached a converging stage where technologies with a known potential for mass production and cost reduction are preferred. For systems close to market launch, electrode and electrolyte materials are favored which have been well known and investigated for several decades and which have demonstrated the best long-term stability and reliability. Other materials and fabrication processes needed for stacking and construction have to be tested in more detail to ascertain whether they can fulfil all the requirements for such a complicated joined system.

REFERENCES

1. L. Blum, W. Drenckhahn, H. Greiner and E. Ivers-Tiffée. In: M. Dokiya, O. Yamamoto, H. Tagawa and S.C. Singhal (Eds.), *Proceedings of the 4th International Symposium on Solid Oxide Fuel Cells (SOFC-IV)*, The Electrochemical Society, Pennington, NJ, 1995p. 163.
2. Stolten, In: G. Ziegler (Ed.), *Verbundwerkstoffe und Werkstoffverbunde*, DGM Informationsgesellschaft mbH, Oberursel, Germany, 1996p. 283.
3. Arai, K. Eguchi, T. Setoguchi, R. Yamaguchi, K. Hashimoto and H. Yoshimura. In: F. Grosz, P. Zegers, S.C. Singhal and O. Yamamoto (Eds.), *Proceedings of the 2nd International Symposium on Solid Oxide Fuel Cells (SOFC-II)*, Commission of the European Communities, Brussels, 1991 p. 167.
4. P. Buchkremer, U. Diekmann and D. Stöver. In: B. Thorstensen (Ed.), *Proceedings of the 2nd European SOFC Forum*, Vol. 1, The European Fuel Cell Forum, Oberrohrdorf, Switzerland, 1996 p. 221.
5. Stolten, R. Späh and R. Schamm. In: U. Stimming, S.C. Singhal, H. Tagawa and W. Lehnert (Eds.), *Proceedings of the 5th International Symposium on Solid Oxide Fuel Cells (SOFC-V)*, The Electrochemical Society, Pennington, NJ, 1997 p. 88.
6. P. Buchkremer, U. Diekmann, L.G.J. de Haart, H. Kabs, U. Stimming and D. Stöver. In: U. Stimming, S.C. Singhal, H. Tagawa and W. Lehnert (Eds.), *Proceedings of the 5th International Symposium on Solid Oxide Fuel Cells (SOFC-V)*, The Electrochemical Society, Pennington, NJ, 1997 p. 160.
7. M.J. Jørgensen, P. Holtappels and C.C. Appel. *J. Appl. Electrochem.* **30** (2000) 411.
8. M. Martin. In: S.C. Singhal and M. Dokiya (Eds.), *Proceedings of the 6th International Symposium on Solid Oxide Fuel Cells (SOFC-VI)*, The Electrochemical Society, Pennington, NJ, 1999 p. 308.
9. O. Teller and M. Martin. *Electrochemistry* **68** (2000) 478.
10. R.H.E. van Doorn, H.J.M. Bouwmeester and A.J. Burggraaf. *Solid State Ionics* **111** (1998) 263.
11. R. Wilkenhöner, W. Malléner, H.-P. Buchkremer, Th. Hauber and U. Stimming. In: B. Thorstensen (Ed.), *Proceedings of the 2nd European SOFC Forum*, Vol. 1, The European Fuel Cell Forum, Oberrohrdorf, Switzerland, 1996 p. 279.
12. N.Q. Minh. *J. Am. Ceram. Soc.* **76** (1993) 563.
13. S.P.S. Badwal and K. Föger. *Mater. Forum* **21** (1997) 187.
14. R. Diethelm and E. Batawi. In: A.J. McEvoy (Ed.), *Proceedings of the 4th European SOFC Forum*, The European Fuel Cell Forum, Oberrohrdorf, Switzerland, 2000 p. 183.
15. G.M. Christie and J.P.P. Huijsmans. In: U. Stimming, S.C. Singhal, H. Tagawa and W. Lehnert (Eds.), *Proceedings of the 5th International Symposium on Solid Oxide Fuel Cells (SOFC-V)*, The Electrochemical Society, Pennington, NJ, 1997 p. 718.

16. G.M. Christie, J.P. Ouweltjes, R.C. Huiberts, E.J. Siewers, F.P.F. van Berkel and J.P.P. Huijsmans. In: *Proceedings of the 3rd International Fuel Cell Conference,* Nagoya, Japan, 1999 p. 361.
17. M. Kuznecov, H. Greiner, M. Wohlfahrt, K. Eichler and P. Otschik. In: A.J. McEvoy (Ed.), *Proceedings of the 4th European SOFC Forum,* The European Fuel Cell Forum, Oberrohrdorf, Switzerland, 2000 p. 261.
18. K. Föger and B. Godfrey. In: A.J. McEvoy (Ed.), *Proceedings of the 4th European SOFC Forum,* The European Fuel Cell Forum, Oberrohrdorf, Switzerland, 2000 p. 167.
19. R. Bolden, K. Föger and T. Pham. In: S.C. Singhal and M. Dokiya (Eds.), *Proceedings of the 6th International Symposium on Solid Oxide Fuel Cells (SOFC-VI),* The Electrochemical Society, Pennington, NJ, 1999 p. 80.
20. A. Khandkar, S. Elangovan, J. Hartvigsen, D. Rowley, R. Privette and M. Tharp. In: S.C. Singhal and M. Dokiya (Eds.), *Proceedings of the 6th International Symposium on Solid Oxide Fuel Cells (SOFC-VI),* The Electrochemical Society, Pennington, NJ, 1999 p. 88.
21. W. Bakker, C. Milliken, J. Hartvigsen, S. Elangovan and A. Khandkar. In: U. Stimming, S.C. Singhal, H. Tagawa and W. Lehnert (Eds.), *Proceedings of the 5th International Symposium on Solid Oxide Fuel Cells (SOFC-V),* The Electrochemical Society, Pennington, NJ, 1997 p. 254.
22. K. Ogasawara, I. Yasuda, Y. Matsuzaki, T. Ogiwara and M. Hishinuma. In: U. Stimming, S.C. Singhal, H. Tagawa and W. Lehnert (Eds.), *Proceedings of the 5th International Symposium on Solid Oxide Fuel Cells (SOFC-V),* The Electrochemical Society, Pennington, NJ, 1997 p. 143.
23. M. Izumi, T. Makino, N. Nishimura, K. Murata and M. Shimotsu. In: *Proceedings of the 3rd International Fuel Cell Conference,* Nagoya, Japan, 1999 p. 379.
24. M. Shimotsu, M. Izumi and K. Murata. In: S.C. Singhal and H. Iwahara (Eds.), *Proceedings of the 3rd International Symposium on Solid Oxide Fuel Cells (SOFC-III),* The Electrochemical Society, Pennington, NJ, 1993 p. 732.
25. K. Honegger, E. Batawi, Ch. Sprecher and R. Diethelm. In: U. Stimming, S.C. Singhal, H. Tagawa and W. Lehnert (Eds.), *Proceedings of the 5th International Symposium on Solid Oxide Fuel Cells (SOFC-V),* The Electrochemical Society, Pennington, NJ, 1997 p. 321.
26. K. Honegger, J. Krumeich and R. Diethelm. In: A.J. McEvoy (Ed.), *Proceedings of the 4th European SOFC Forum,* The European Fuel Cell Forum, Oberrohrdorf, Switzerland, 2000 p. 29.
27. J.P. Ouweltjes, F.P.F. van Berkel, P. Nammensma and G.M. Christie. In: S.C. Singhal and M. Dokiya (Eds.), *Proceedings of the 6th International Symposium on Solid Oxide Fuel Cells (SOFC-VI),* The Electrochemical Society, Pennington, NJ, 1999 p. 803.
28. G.M. Christie, P. Nammensma and J.P.P. Huijsmans. In: A.J. McEvoy (Ed.), *Proceedings of the 4th European. SOFC Forum,* The European Fuel Cell Forum, Oberrohrdorf, Switzerland, 2000 p. 3.
29. D. Simwonis, H. Thülen, F.J. Dias, A. Naoumidis and D. Stöver. *J. Mater. Process. Technol.* **92–93** (1999) 107.
30. D. Stöver, U. Diekmann, U. Flesch, H. Kabs, W.J. Quadakkers, F. Tietz and I.C. Vinke. In: S.C. Singhal and M. Dokiya (Eds.), *Proceedings of the 6th International Symposium Solid Oxide Fuel Cells (SOFC-VI),* The Electrochemical Society, Pennington, NJ, 1999 p. 813.
31. B. Hobein, F. Tietz, D. Stöver, M. Čekada and P. Panjan. *J. Eur. Ceram. Soc.* **21** (2001) 1843.
32. R. Wilkenhöner, W. Malléner, H.P. Buchkremer, T. Hauber and U. Stimming. In: B. Thorstensen (Ed.), *Proceedings of the 2nd European SOFC Forum,* Vol. 1, The European Fuel Cell Forum, Oberrohrdorf, Switzerland, 1996 p. 279.
33. S. Primdahl, M.J. Jørgensen, C. Bagger and B. Kindl. In: S.C. Singhal and M. Dokiya (Eds.), *Proceedings of the 6th International Symposium on Solid Oxide Fuel Cells (SOFC-VI),* The Electrochemical Society, Pennington, NJ, 1999 p. 793.
34. M.J. Jørgensen, P.H. Larsen, S. Primdahl and C. Bagger. In: A.J. McEvoy (Ed.), *Proceedings of the 4th European SOFC Forum,* The European Fuel Cell Forum, Oberrohrdorf, Switzerland, 2000 p. 203.
35. D. Gosh, G. Wang, R. Brule, E. Tang and P. Huang. In: S.C. Singhal and M. Dokiya (Eds.), *Proceedings of the 6th International Symposium on Solid Oxide Fuel Cells (SOFC-VI),* The Electrochemical Society, Pennington, NJ, 1999 p. 822.
36. M. Pastula, R. Boersma, D. Prediger, M. Perry, A. Horvath, J. Devitt and D. Gosh. In: A.J. McEvoy (Ed.), *Proceedings of the 4th European SOFC Forum,* The European Fuel Cell Forum, Oberrohrdorf, Switzerland, 2000 p. 123.

37. N.Q. Minh and K. Montgomery. In: U. Stimming, S.C. Singhal, H. Tagawa and W. Lehnert (Eds.), *Proceedings of the 5th International Symposium on Solid Oxide Fuel Cells (SOFC-V)*, The Electrochemical Society, Pennington, NJ, 1997 p. 153.
38. N.Q. Minh, T.R. Amstrong, J.R. Esopa, J.V. Guiheen, C.R. Horne and J. Van Ackeren. In: S.C. Singhal and H. Iwahara (Eds.), *Proceedings of the 3rd International Symposium on Solid Oxide Fuel Cells (SOFC-III)*, The Electrochemical Society, Pennington, NJ, 1993 p. 801.
39. N.Q. Minh. In: M. Dokiya, O. Yamamoto, H. Tagawa and S.C. Singhal (Eds.), *Proceedings of the 4th International Symposium on Solid Oxide Fuel Cells (SOFC-IV)*, The Electrochemical Society, Pennington, NJ, 1995 p. 138.
40. K. Föger, R. Donelson and R. Ratnaraj. In: S.C. Singhal and M. Dokiya (Eds.), *Proceedings of the 6th International Symposium on Solid Oxide Fuel Cells (SOFC-VI)*, The Electrochemical Society, Pennington, NJ, 1999 p. 95.
41. P.K. Srivastava, T. Quach, Y.Y. Duan, R. Donelson, S.P. Jiang, F.T. Ciacchi and S.P.S. Badwal. *Solid State Ionics* **99** (1997) 311.
42. S.C. Singhal. In: S.C. Singhal and M. Dokiya (Eds.), *Proceedings of the 6th International Symposium on Solid Oxide Fuel Cells (SOFC-VI)*, The Electrochemical Society, Pennington, NJ, 1999 p. 39.
43. H. Takeuchi, H. Nishiyama, A. Ueno, S. Aikawa, A. Aizawa, H. Taijiri, T. Nakayama, S. Suehiro and K. Shukuri. In: S.C. Singhal and M. Dokiya (Eds.), *Proceedings of the 6th International Symposium on Solid Oxide Fuel Cells (SOFC-VI)*, The Electrochemical Society, Pennington, NJ, 1999 p. 879.
44. M. Aizawa, M. Kuroishi, A. Ueno, H. Tajiri, T. Nakayama, K. Eguchi and H. Arai. In: U. Stimming, S.C. Singhal, H. Tagawa and W. Lehnert (Eds.), *Proceedings of the 5th International Symposium on Solid Oxide Fuel Cells (SOFC-V)*, The Electrochemical Society, Pennington, NJ, 1997 p. 330.
45. S. Kaneko, T. Gengo, S. Uchida and Y. Yamauchi. In: S.C. Singhal and H. Iwahara (Eds.), *Proceedings of the 2nd International Symposium on Solid Oxide Fuel Cells (SOFC-II)*, The European Commission, Brussels, Belgium, 1991 p. 35.
46. H. Mori, H. Omura, N. Hisatome, K. Ikeda and K. Tomida. In: S.C. Singhal and M. Dokiya (Eds.), *Proceedings of the 6th International Symposium on Solid Oxide Fuel Cells (SOFC-VI)*, The Electrochemical Society, Pennington, NJ, 1999 p. 52.
47. S. Kakigani, T. Kurihara, N. Hisatome and K. Nagata. In: U. Stimming, S.C. Singhal, H. Tagawa and W. Lehnert (Eds.), *Proceedings of the 5th International Symposium on Solid Oxide Fuel Cells (SOFC-V)*, The Electrochemical Society, Pennington, NJ, 1997 p. 180.
48. T. Nishi, N. Hisatome, H. Yamamoto and N. Murakami. In: P. Vincenzini (Ed.), *Proceedings of the 9th Cimtec – World Forum on New MaterialsInnovative Materials in Advanced Energy Technologies*, Vol. 24. Techna Publishers S.r.l, Faenza, Italy, 1999 p. 41.
49. F.J. Gardner, M.J. Day, N.P. Brandon, M.N. Pashley and M. Cassidy. *J. Power Sources* **86** (2000) 122.
50. M. Lang, R. Henne, G. Schiller and N. Wagner. In: U. Stimming, S.C. Singhal, H. Tagawa and W. Lehnert (Eds.), *Proceedings of the 5th International Symposium on Solid Oxide Fuel Cells (SOFC-V)*, The Electrochemical Society, Pennington, NJ, 1997 p. 461.
51. T. Iwata, N. Kadokawa and S. Takenoiri. In: M. Dokyia, O. Yamamoto, H. Tagawa and S.C. Singhal (Eds.), *Proceedings of the 4th International Symposium on Solid Oxide Fuel Cells (SOFC-IV)*, The Electrochemical Society, Pennington, NJ, 1997 p. 110.
52. N.Q. Minh. In: S.C. Singhal and M. Dokiya (Eds.), *Proceedings of the 6th International Symposium on Solid Oxide Fuel Cells (SOFC-VI)*, The Electrochemical Society, Pennington, NJ, 1999 p. 127.
53. G. Pudmich, B.A. Boukamp, M. Gonzalez-Cuenza, W. Jungen, W. Zipprich and F. Tietz. *Solid State Ionics* **135** (2000) 433.
54. W. Thierfelder, H. Greiner and W. Köck. In: U. Stimming, S.C. Singhal, H. Tagawa and W. Lehnert (Eds.), *Proceedings of the 5th International Symposium on Solid Oxide Fuel Cells (SOFC-V)*, The Electrochemical Society, Pennington, NJ, 1997 p. 1306.
55. W. Glatz, E. Batawi, M. Janousek, W. Kraussler, R. Zech and G. Zobl. In: S.C. Singhal and M. Dokiya (Eds.), *Proceedings of the 6th International Symposium on Solid Oxide Fuel Cells (SOFC-VI)*, The Electrochemical Society, Pennington, NJ, 1999 p. 783.
56. P.H. Hou, K. Huang and W.T. Bakkerm. In: S.C. Singhal and M. Dokiya (Eds.), *Proceedings of the 6th International Symposium Solid Oxide Fuel Cells (SOFC-VI)*, The Electrochemical Society, Pennington, NJ, 1999 p. 737.

57. Th. Malkow, U. von der Crone, A.M. Laptev, T. Koppitz, U. Breuer and W.J. Quadakkers. In: U. Stimming, S.C. Singhal, H. Tagawa and W. Lehnert (Eds.), *Proceedings of the 5th International Symposium on Solid Oxide Fuel Cells (SOFC-V)*, The Electrochemical Society, Pennington, NJ, 1997 p. 1244.
58. Y. Miyake, Y. Akiyama, T. Yasuo, S. Taniguchi, M. Kadowaki and K. Nishio. In: *Proceedings on Fuel Cell Seminar,* November 17–20, Orlando, 1996 p. 36.
59. W.J. Quadakkers, T. Malkow, J. Pirón-Abellán, U. Flesch, V. Shemet and L. Singheiser. In: A.J. McEvoy (Ed.), *Proceedings of the 4th European SOFC Forum*, The European Fuel Cell Forum, Oberrohrdorf, Switzerland, 2000 p. 827.
60. K. Hilpert, D. Das, M. Miller, D.H. Peck and R. Weiß. *J. Electrochem. Soc.* **143** (1996) 3642.
61. S.P. Jiang, J.P. Zhang and K. Föger. *J. Electrochem. Soc.* **147** (2000) 3195.
62. S.P. Jiang, J.P. Zhang, L. Apateanu and K. Föger. *J. Electrochem. Soc.* **147** (2000) 4013.
63. R. Ruckdäschel, R. Henne, G. Schiller and H. Greiner. In: U. Stimming, S.C. Singhal, H. Tagawa and W. Lehnert (Eds.), *Proceedings of the 5th International Symposium Solid Oxide Fuel Cells (SOFC-V)*, The Electrochemical Society, Pennington, NJ, 1997 p. 1273.
64. N. Oishi and Y. Yamazaki. In: S.C. Singhal and M. Dokiya (Eds.), *Proceedings of the 6th International Symposium on Solid Oxide Fuel Cells (SOFC-VI)*, The Electrochemical Society, Pennington, NJ, 1999 p. 759.
65. E. Batawi, A. Plas, W. Straub, K. Honegger and R. Diethelm. In: S.C. Singhal and M. Dokiya (Eds.), *Proceedings of the 6th International Symposium on Solid Oxide Fuel Cells (SOFC-VI)*, The Electrochemical Society, Pennington, NJ, 1999 p. 767.
66. C. Gindorf, K. Hilpert, H. Nabielek, L. Singheiser, R. Ruckdäschel and G. Schiller. In: A.J. McEvoy (Ed.), *Proceedings of the 4th European SOFC Forum*, Vol. 2, The European Fuel Cell Forum, Oberrohrdorf, Switzerland, 2000 p. 845.

Chapter 15

Engineered cathodes for high performance SOFCs

R.E. Williford and P. Singh

Abstract

Computational design analysis of a high performance cathode is a cost-effective means of exploring new microstructure and material options for solid oxide fuel cells. A two-layered porous cathode design has been developed that includes a thinner layer with smaller grain diameters at the cathode/electrolyte interface followed by a relatively thicker outer layer with larger grains at the electrode/oxidant interface. Results are presented for the determination of spatially dependent current generation distributions, assessment of the importance of concentration polarization, and sensitivity to measurable microstructural variables. Estimates of the electrode performance in air at 700°C indicate that performance approaching 3.1 A/cm^2 at 0.078 V is *theoretically* possible. The limitations of the model are described, along with efforts needed to verify and refine the predictions. The feasibility of fabricating the electrode configuration is also discussed.

Keywords: Solid oxide fuel cells; Cathodes; Microstructure

Article Outline

1. Introduction . . . 262
2. Engineered cathode development . . . 263
 2.1. Estimation of current distribution . . . 263
 2.2. Material properties and microstructural sensitivities of the model . . . 264
3. Design optimization for a two-layer cathode . . . 267
 3.1. Outer layer . . . 267
 3.2. Inner layer . . . 268
  3.2.1. Small grain diameter (0.25 μm) . . . 268
  3.2.2. Larger grain diameter (0.5 μm) . . . 268
4. Conclusions . . . 269
Acknowledgements . . . 269
Appendix A. Model parameters . . . 269
Appendix B . . . 270
Appendix C . . . 272
 C.1. Microstructural design including concentration polarization . . . 272
References . . . 274

Fuel Cells Compendium

1. INTRODUCTION

A solid oxide fuel cell (SOFC) is composed of a dense electrolyte sandwiched between porous electrodes. The key electrochemical reactions occur mostly on the *surfaces* at the electrolyte/electrode interfaces, thus enabling the harvesting of electrons in an electrical circuit to produce useful power. The dense electrolyte conducts oxygen ions from the cathode to the anode, but prevents mixing of the fuel and oxidant in the *gas* phase, where electrons cannot be harvested. The porous electrodes permit passage of the gases to the electrolyte/electrode interfaces where the reactions occur. A relatively thick porous anode is often used to provide structural support for the assembly. The cathode is often relatively thin to minimize its contribution to the overpotential by polarization losses. It is generally recognized that a significant portion of these polarization losses originate in the cathode. Consequently, much attention has recently been focused on improving cathode performance through two means: improved materials and improved microstructural designs. Improved materials include mixed ionic electronic conductors (MIECs), which essentially increase the surface area available for conversion of gaseous oxygen molecules into oxygen ions. Such a material has been developed in our laboratory [1], and was employed in this work. The present paper focuses on methods to improve the microstructural design of the cathode.

An objective of this work is to design a SOFC cathode exhibiting a low area-specific resistance ($ASR = 0.1\,\Omega cm^2$) and capable of high current output. Such an objective can be attained best by a coupled experimental-modeling approach: the modeling helps to guide the experiments and the experiments provide data for fitting the parameters of the model. Since modeling is often cheaper than a long series of Edisonian experiments, costs are often minimized with this approach. This paper describes modeling efforts and includes selected experimental data, which are reported in more detail elsewhere [1].

An initial step was to review the existing models in the literature, and select an approach that was both pragmatic and thorough, in addition to exhibiting direct linkages with the experimental data. We found four categories of models, which are described below in terms of complexity and input data requirements. The following paragraphs are not an exhaustive review, and contain only representative examples of each model type.

In the first category are detailed models by Svensson et al. [2]. This model treats the classic three phase boundary (TPB) problem explicitly by addressing the individual mechanisms involved (surface adsorption, dissociation, electronic exchange, surface diffusion). Non-linear second order differential equations are derived and solved numerically, with the proper boundary conditions (six are required). Although the model certainly contains enough technical depth, it also requires the estimation of about a dozen parameters (such as surface diffusion coefficients) that generally have uncertainties of two to four orders of magnitude. Estimation of these parameters would require an extensive series of well-controlled experiments, at appreciable cost. This model was not suitable for the present program. Similar conclusions have been reached elsewhere [3].

In the second category are models based on a homogenized, effective medium concept for the porous cathode material [4,5]. The TPB mechanisms described above are lumped into a single parameter called the surface exchange coefficient (K_s, cm/s), which is experimentally measurable on a routine basis, e.g. [6]. A significant database exists in the literature for a range of MIEC that are important in the present effort. In this respect, the models in [4, 5] are suitable because of the direct connection to experimental tasks. However, two factors present problems. The first was the mathematical complexity resulting from treatment of four polarization terms (Butler–Volmer, chemical resistance caused by the exchange process, concentration polarization, and gas capacitance at the cathode/gas interface). Second, communication with the author [4,5] indicated that several typographical errors existed in the literature, thus requiring re-derivation of the model to ensure correctness. The investment required for this type of model also rendered it unsuitable for the present work.

In the third category are models by Deng et al. (DZA) [7] (without concentration polarization), and [8] (with concentration polarization). These models use the same surface exchange concept as in [4,5], but treat only the chemical resistance term [7] and the concentration polarization term [7,8]. Numerical input requirements are comparatively modest, and the models are mathematically tractable. Although solutions are available in closed-form, these solutions are for a homogeneous material, so that spatially variant material properties are not treated. Consequently, numerical solutions of the differential equations will ultimately be required for multilayered cathode concepts. The models are also well suited for the high conductivity materials used in this investigation, and were selected for the calculations herein.

In the fourth category are the more classical models, e.g. [9], which generally treat all polarization terms in a semi-empirical manner. They require input of a modest amount of experimental data, most of which are available, and are mathematically tractable. However, linearizations in the model have apparently caused the loss of important effects, particularly for gas diffusion through the porous material. The manifestation of this problem was the rather high values derived for the tortuosity, indicating that several unknowns and uncertainties had numerically accumulated in this parameter. This type of model was thus not suitable for the present effort.

We selected the models by Deng et al. (DZA) [7,8] for this work because of their direct relationship to experimental measurables, treatment of the two most important polarization terms, mathematical tractability, and the availability of closed-form solutions for initial estimates. The following paragraphs describe preliminary results using these models.

2. ENGINEERED CATHODE DEVELOPMENT

2.1. Estimation of current distribution

In the early stages of this investigation, the lanthanum strontium copper ferrite (LSCuF) cathode material being tested exhibited very low ASR and high current. It was therefore thought that concentration polarization was negligible up to about 0.1 V. The DZA [7] model was extended to permit estimation of possible oxygen potential gradients, and used to estimate the current distribution throughout the cathode material. Model parameters are defined in Appendix A, and the model details are given in Appendix B, along with a benchmark case for LSCoF.

Preliminary results for the very efficient MIECs under consideration indicated that most of the ionic current was generated near the cathode/electrolyte interface (at $x/L = 1$), as shown in Fig. 1 for a single layer of LSCuF. The approximately 5–8 μm region of high current generation is larger than the usual three phase boundary width discussed in the literature, probably because of errors in the estimated material parameters at

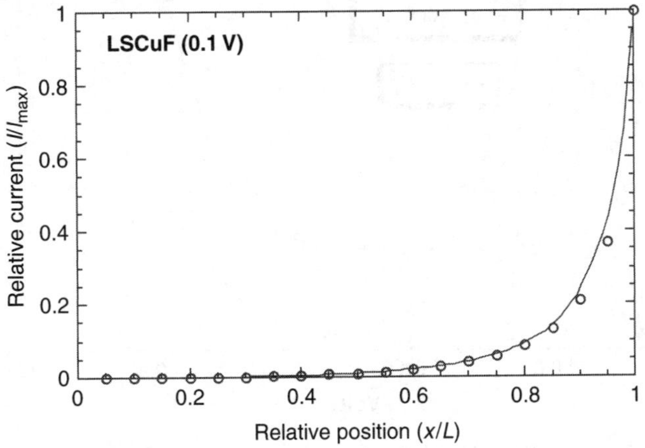

Fig. 1. Current distribution in a homogeneous LSCuF cathode

this early stage in the analysis. The results are very sensitive to the pore surface area/volume ratio (S) defined in Appendix A. Although the results are consistent with LSCuF performance at low overpotentials, a detectable amount of concentration polarization was indicated, which motivated the next stage of the analysis.

2.2. *Material properties and microstructural sensitivities of the model*

The second DZA model [8] also treated concentration polarization along with the polarization due to chemical resistance (R_{ch}), and is briefly summarized in Appendix C. A second computer simulation code was constructed using this model. The code has three main functions: (a) fitting to experimental V–I data to extract the value of the surface exchange coefficient directly from the performance data, (b) variation of model parameters to reveal the sensitivities to microstructural variables, and (c) computation of the current and voltage as a function of position in the cathode material.

Although the closed-form solution provided by this model is very useful, its limitation is that it treats only homogeneous materials. That is, the microstructural parameters may not depend on spatial position. It will be seen that a piecewise numerical solution of the model is required for the final design of the cathode.

Analysis using the extraction mode was applied to one of the best performing materials developed in this laboratory: LSCuF. The operating parameters were $T = 700°C$ and $P_{O_2} = 2.12 \times 10^4$ Pa (air) [1]. The ionic and electronic conductivities were taken as 0.01 and 100 S/cm, respectively. The thickness was 25 μm, porosity 50%, grain size 1 μm, pore size 1 μm, with solid and gas tortuosities of 1.89 and 2.5 (see Appendix C). Several dozen runs revealed that the best ionic diffusivity value was 5×10^{-8} cm²/s, which is slightly higher than that for LSC and LSCoF (1×10^{-8} to 2×10^{-8}) [6], but still quite reasonable in light of the chemically active role copper plays in many technologies (e.g., getters, superconductivity, etc.). This is probably due to the electronic structure of copper, analysis of which is outside the scope of this work.

The solutions for K_s for the various current and voltage data pairs are shown in Fig. 2. The curve labeled curs_p (with concentration polarization) reproduces the data at an ASR of about 0.06 Ω cm² in the ohmic regime, and shows the apparent concentration polarization becoming important above about 0.1 V as the non-polarized (curs_np) curve departs noticeably from curs_p. When compared to the solid material

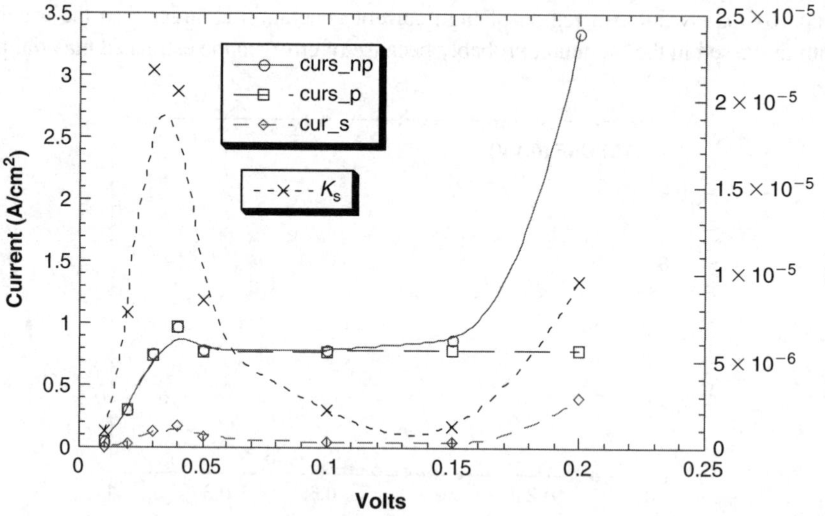

Fig. 2. Analysis of LSCuF data in the extraction mode. The squares are the experimental data points. The long-dashed curve is the model fit to the data. Other symbols are explained in the text

(cur_s), both curves show the significant enhancement provided by the porous material. L_p (the characteristic length of the three phase region) at 0.1 V was 1.7 μm and L_g (the characteristic length for concentration polarization) was 113 μm, indicating minimal concentration polarization at this voltage (see Appendix A for definitions of parameters). The P_{O_2} at the cathode/electrolyte interface had fallen to 1.52 Pa at 0.1 V. The value of K_s at 0.1 V was 2.17×10^{-6} cm/s, which is quite reasonable compared to that for LSC and LSCoF at 700°C (3×10^{-6} to 4×10^{-6} cm/s).

The above results were then used in the model's sensitivity mode to study the impact of varying microstructural parameters as follows. Figure 3 shows the effect of varying the cathode thickness. It appears that the present 25 μm thickness is just sufficient to establish a stable enhancement of current due to the ion collection activity of the outer regions of the MIEC cathode material. At thicknesses greater than 25 μm, concentration polarization effects become noticeable.

Variations in porosity cause variations in the pore surface area/volume ratio (S) and the tortuosities, and are shown in Fig. 4. Although the enhancement remains significant, it is interesting that a lower porosity

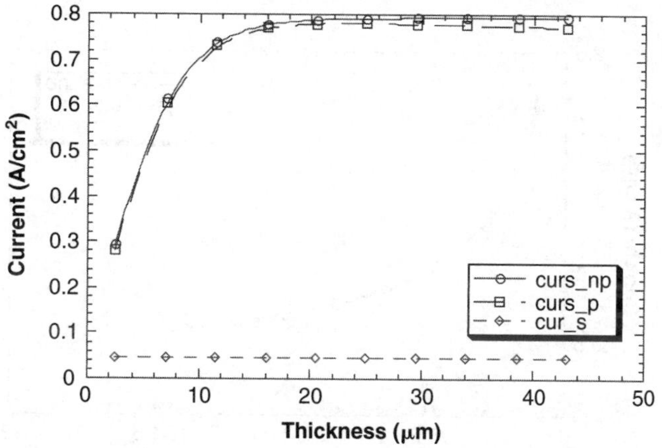

Fig. 3. Effect of varying cathode thickness with all other parameters held constant: overvoltage = 0.1 V

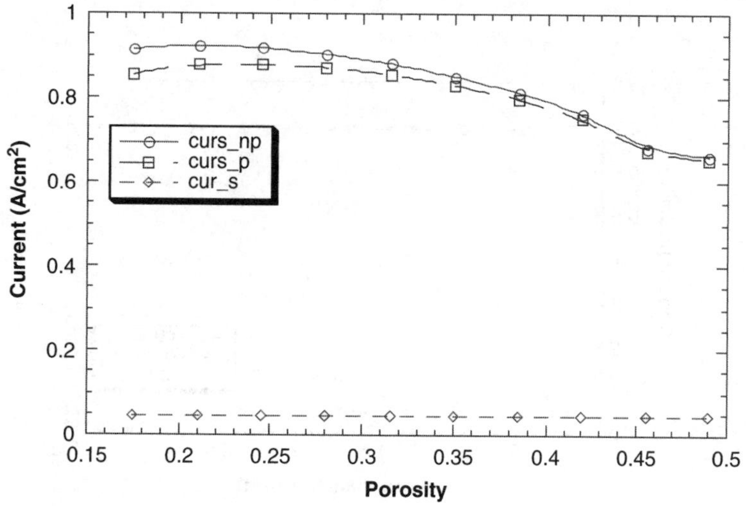

Fig. 4. Effect of varying porosity at 0.1 V overvoltage

results in a higher current. This may be due to part of the particular relationship (see Appendix C) between the ionic diffusivity and the surface exchange coefficient. The characteristic distance for the transition between reactivity and diffusion dominance was $L_d = 230\,\mu m$. Another data set and associated benchmarking exercise may indeed give different results. For the present combination of parameters, at 0.1 V overvoltage, it appears that lower porosity may give a slightly higher current. However, this enhancement is smaller than that provided by other microstructural parameters, as shown below.

The pore surface area decreases when grain diameter increases, but the tortuosities are not greatly affected in the present formulations. The decreasing pore surface area reduces the total ionic current, as shown in Fig. 5.

Figure 6 shows that variations in pore diameter have negligible effect in this regime because the micron-sized pores are greater than the mean free path of the molecules, thus minimizing the Knudsen effect.

The conclusion was that smaller grain diameters would provide the greatest increase in currents, e.g., reducing the grain size from 1.0 to 0.25 μm could double the current.

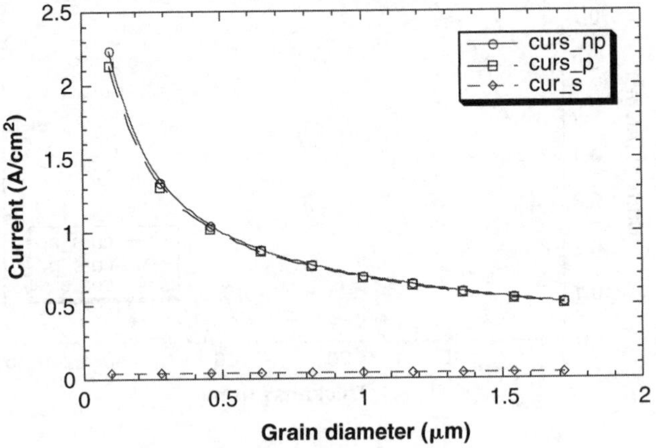

Fig. 5. Effect of grain size on currents at 0.1 V overvoltage

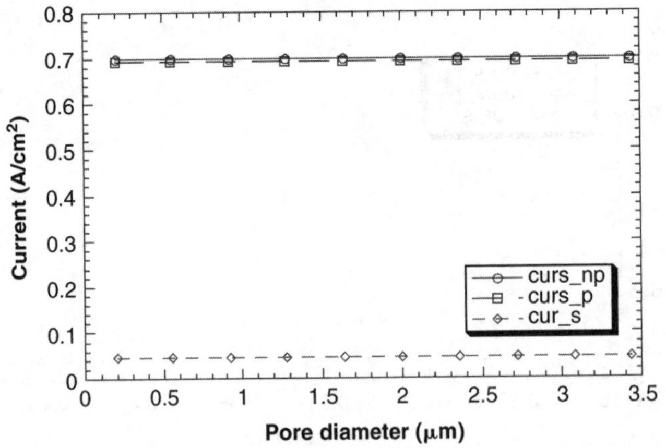

Fig. 6. Effect of varying pore diameters at 0.1 V overvoltage

3. DESIGN OPTIMIZATION FOR A TWO-LAYER CATHODE

The above results indicated that the most effective way to increase the current output of the cathode was to decrease the grain size, while maintaining the porosity at a high level. This means that the cathode internal pore surface area to cathode volume ratio will rise appreciably. It also means that the cathode will be a very fragile structure, and difficulties may be expected in attaching current collectors to its outer surface. One solution to this problem is to employ a two-layer cathode system: a more structurally robust outer layer with larger grains and an inner layer (next to the electrolyte) with smaller grains. The outer layer should have enough porosity to easily transport a steady supply of air/O_2 to the more chemically active inner layer. The following design is based on the above cited material properties extracted from experimental data (0.78 A/cm^2 at 0.1 V). The only change is that the base porosity was reduced to 45% to take advantage of the small enhancement in Fig. 4, and to ensure numerical stability in the calculations.

The design of each layer is described below. It is important to note how the model was applied in these cases. Each layer was treated as an individual cathode, i.e., the model was applied twice, once for the outer layer and once for the inner layer. This was necessary because of the single-material limitations of the model. Thus, the results should be considered as approximate until they can be confirmed by more detailed analysis with a model that treats the spatial variation of material properties explicitly.

3.1. Outer layer

The following microstructural parameters were defined: thickness = 25 μm, porosity 45%, pore diameter = 1 μm, grain diameter = 1 μm, surface/volume ratio S = 50 000 cm^{-1}, gas tortuosity = 2.22, solid tortuosity = 1.83 (effective path for ionic and electronic conduction). Results are shown in Fig. 7.

The 5 μm active region adjacent to the electrolyte (at $x = 0$) is evident. The P_{O_2} at the 5 μm distance was estimated to be 1.62×10^4 Pa, the overvoltage (η) 4.69×10^{-3} V, and the ionic current 0.0106 A/cm^2. These values define the boundary conditions for the inner layer, at the inner/outer layer interface. While the overvoltage and current at the 5 μm position are good numbers, it is important to note that the P_{O_2} value was very approximate. This is because of limitations in the closed-form solution of the model,

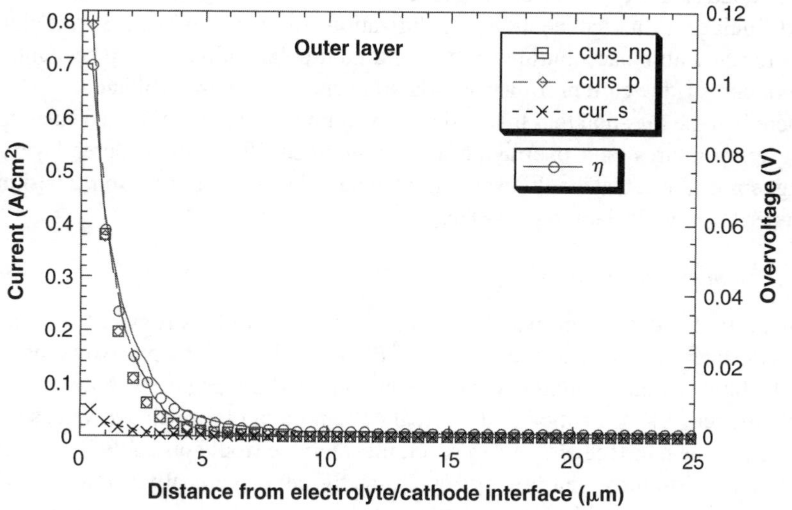

Fig. 7. Ionic current and overvoltage for the cathode outer layer

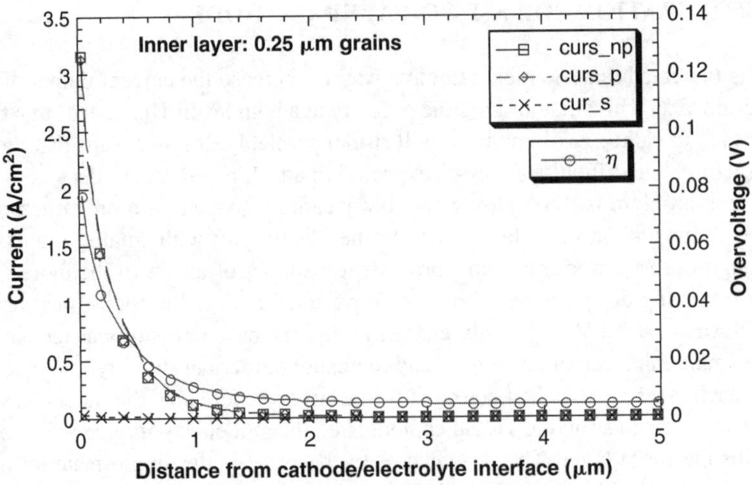

Fig. 8. Ionic current and overvoltage (η) for the small grain inner layer

which did not supply a solution for the oxygen chemical potential for this particular geometry. It was thus necessary to estimate the P_{O_2} using a Nernstian approximation, which will be improved upon in later work.

Two designs for the inner layer are discussed next, the objective of which is to increase the current output above that seen for the single layer in Fig. 7.

3.2. Inner layer

3.2.1. *Small grain diameter (0.25 μm)*

Because of the smaller grains, the surface/volume ratio is increased to $S = 1.22 \times 10^6 \text{cm}^{-1}$. The pore diameter and gas tortuosity were unchanged, but the solid tortuosity is reduced to 1.67. Figure 8 shows that it is theoretically possible to increase the output of the two-layered cathode to about 3.1 A/cm^2 at 0.078 V. The P_{O_2} at the cathode/electrolyte interface is about 5.17×10^2 Pa using the above mentioned Nernst estimation method, indicating that concentration polarization was not dominant. In the figure, recall that curs_np is the current with no accounting for concentration polarization, curs_p accounts for concentration polarization, cur_s is the current for a solid material, and η is the overvoltage.

However, there is some question that this design may not be manufacturable: the smaller grains require lower sintering temperatures, and overlaying a larger-grained 20 μm thick outer layer would require higher sintering temperatures that would over-sinter the inner layer. For this reason, a design was also generated for an inner layer with a larger grain size.

3.2.2. *Larger grain diameter (0.5 μm)*

The surface/area ratio for this design was 62 777 cm^{-1}, the solid tortuosity was 1.65, and the P_{O_2} at the cathode/electrolyte interface was less than 1×10^{-5} Pa by the Nernstian approximation. Because of the low P_{O_2}, the calculation actually failed at 0.4 μm from the interface. Figure 9 shows that the overvoltage was appreciably higher (0.12 V), although the actual current output may achieve values near those of the 0.25 μm grain case at the interface proper. However, the interface would probably be oxygen-starved, thus eliminating any benefit from oxygen impingement directly on a chemically stable thin layer (i.e., 1 μm ceria) between the electrolyte and LSCuF cathode material.

It appears that the best design is the 0.25 μm grain, 5 μm thick inner layer. However, the manufacturability of this design must be proven experimentally.

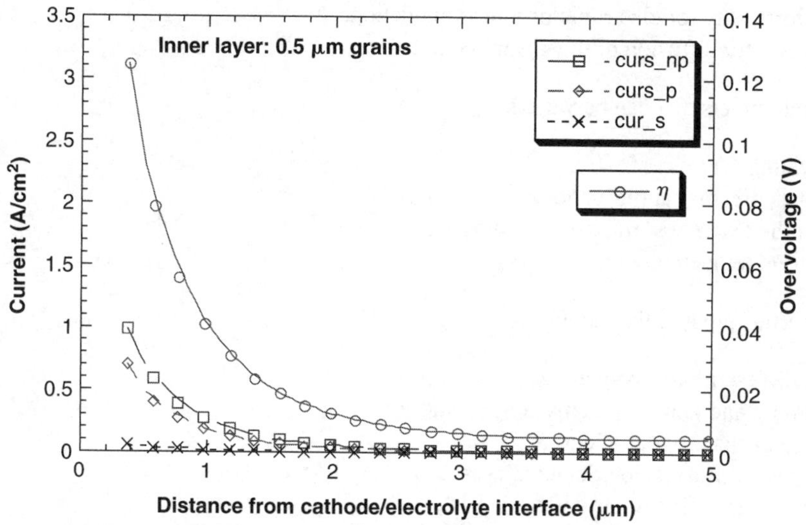

Fig. 9. Ionic current and overvoltage (η) for the cathode inner layer with larger grain diameter

4. CONCLUSIONS

The models employed to this point are suitable for a homogeneous material where the properties and microstructures do not vary spatially. However, the microstructural design of the two-layered cathode described above has extended these closed-form models appreciably beyond their intended range of application, and should be considered as approximate.

The candidate material LSCuF investigated in this analysis has exhibited a high affinity for oxygen, and may be suitable for a cathode material if structural and chemical stability can be maintained during long-term high power density operations at elevated operating temperatures. Experimental efforts are underway to investigate the stability of this material, and will be reported in a separate paper.

ACKNOWLEDGEMENTS

The work described in this report was performed for DARPA under contract MDA972-01-C-0067. Work performed by Pacific Northwest National Laboratory which is operated by Battelle Memorial Institute for the US Department of Energy under contract DE-AC06-76RLO1830. The authors wish to thank Drs. Nguyen Q. Minh and Rajiv Doshi (GE Power Systems) for technical discussions, Dr. V. Browning (DARPA) for support and encouragement, and Mr. G. Coffey and Mr. E. Thompson for their experimental contributions.

APPENDIX A. MODEL PARAMETERS

The microstructural parameters are:

- the porosity, ϕ;
- the grain diameter, d_g;
- the pore diameter, d_p;
- the internal pore surface area per unit volume, S;
- the thickness of the cathode, L;

- the solid tortuosity for electronic or ionic conduction, t_s;
- the tortuosity for diffusion of the gas in the pores, t_g.

Operational parameters that can be varied:

- the overvoltage, V;
- the total gas pressure at the cathode surface, P_{tot};
- the P_{O_2} at the external surface of the cathode;
- the operating temperature, T.

Material properties input to the model:

- the ionic diffusivity for oxygen, D_{ion};
- the electronic and ionic conductivities, σ_e and σ_I;
- the surface exchange coefficient, K_s;
- the ambipolar diffusion coefficient, $D_{IE} = D_{ion}/(1 + \sigma_I/\sigma_e)$;
- the ionic concentration in the solid material, C_i;
- the characteristic distance for the transition between reactivity and diffusion domination, $L_d = D_{IE}/K_s$;
- the characteristic length of the three phase boundary region, or region of high reactivity near the cathode/electrolyte interface, L_p;
- the characteristic length for concentration polarization effects, L_g;
- the chemical resistance due to ionic conduction and surface exchange, R_{ch}.

Calculated output from the model:

- the ionic current for a solid material, I_I^s;
- the ionic current for a porous material with no concentration polarization, I_{np};
- the ionic current for a porous material including concentration polarization effects, I_p;
- the ASR of the cathode material, given by the current divided by the overvoltage.

APPENDIX B

The DZA model [7] for small concentration polarization is summarized as follows.
The ionic current for a solid material is defined as

$$I_i^s = \frac{K_s C_i \Delta\mu_g}{RT}, \tag{B.1}$$

where

$$\Delta\mu_g = \frac{RT}{4} \ln\left(\frac{P_{O_2\,\mathrm{external}}}{P_{O_2\,\mathrm{interface}}}\right). \tag{B.2}$$

The corresponding current for a porous material is given by

$$I_i(x) = I_I^s M G(x), \tag{B.3}$$

where the material factor is

$$M = \frac{L_d S(1-\phi)}{t_s}, \tag{B.4}$$

and G is a geometric factor given by

$$G(x) = \frac{a\exp[(L-x)/L_{\mathrm{p}}] + \exp[x/L_{\mathrm{p}}]}{a\exp[L/L_{\mathrm{p}}] + 1}, \tag{B.5}$$

where x is the distance from the cathode/electrolyte interface, and

$$a = \frac{1+u}{1-u}, \tag{B.6}$$

$$u = \left[\frac{t_{\mathrm{s}}(1-\phi)}{SL_{\mathrm{d}}}\right]^{1/2}, \tag{B.7}$$

and

$$L_{\mathrm{p}} = \left[\frac{L_{\mathrm{d}}(1-\phi)}{St_{\mathrm{s}}}\right]^{1/2}. \tag{B.8}$$

Eq. (B.3) describes the ionic current in the porous solid in the Gerisher limit, where the oxygen potential is constant. The source of this ionic current is the oxygen molecular flux through the pores: the molecular current is converted into ionic current through the surface exchange mechanism. The molecular flux through the pores can be expressed as:

$$I_{\mathrm{g}} = -\frac{\phi}{t_{\mathrm{g}}D_{\mathrm{g}}}\frac{\mathrm{d}C_{\mathrm{g}}}{\mathrm{d}x}, \tag{B.9}$$

where D_{g} is the diffusivity of the O_2 in the pores (corrected for Knudsen effects) and C_{g} is the concentration of the O_2 in the pores, which depends on position x in the porous material. When all the molecular O_2 is converted to ionic current, we have

$$I_{\mathrm{g}} = \frac{1}{2}I_{\mathrm{I(max)}}. \tag{B.10}$$

But this does not admit spatial dependence of P_{O_2}.

To obtain that spatial dependence, we depart from the original developments by DZA [7], as follows. Note that at any position x, the time rate of change of the molecular current is proportional to the spatial gradient in the ionic current:

$$\frac{\mathrm{d}I_{\mathrm{g}}}{\mathrm{d}t} = -\frac{1}{2}\frac{\mathrm{d}I_{\mathrm{I}}}{\mathrm{d}x}. \tag{B.11}$$

From the continuity equation, we also have

$$\frac{\mathrm{d}I_{\mathrm{g}}}{\mathrm{d}t} = -\frac{\mathrm{d}I_{\mathrm{g}}}{\mathrm{d}x}. \tag{B.12}$$

Substituting Eqs (B.3), (B.9) and (B.12) into Eq. (B.11) results in a second order differential equation in the oxygen partial pressure as a function of spatial position x. This is a two-point boundary value problem, and is integrated numerically by a relaxation method. The numerical solution is then substituted into the

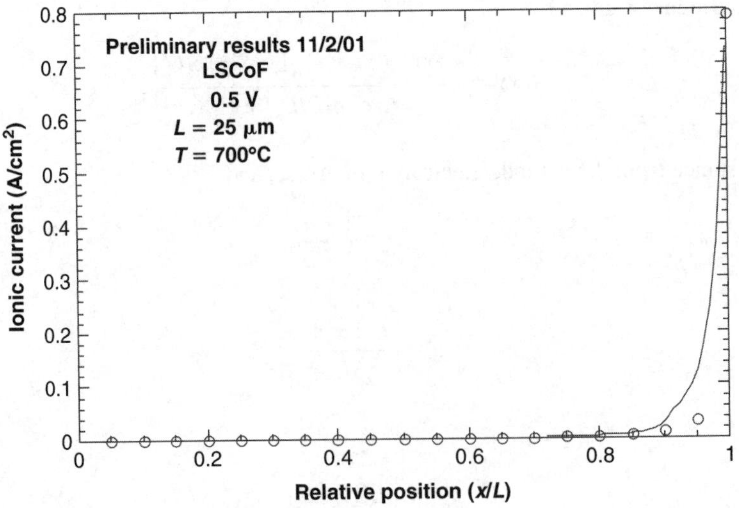

Fig. 10. Benchmark case for LSCoF at 0.5 V, $L = 25$ μm, $T = 700$°C

Nernst equation (since concentration polarization is expected to be negligible) to obtain an estimate of the local overpotential between two points $x(i)$ and $x(i-1)$:

$$\eta = \frac{RT}{4F}\ln\left[\frac{P_{O_2}(x(i))}{P_{O_2}(x(i-1))}\right]. \tag{B.13}$$

The chemical resistance R_{ch} (or ASR for no concentration polarization) is calculated from

$$R_{ch} = \frac{RT}{2F_2K_sC_{ion}}\left(\frac{t_s}{L_dS(1-\phi)}\right)^{1/2}, \tag{B.14}$$

and the current increment between $x(i)$ and $x(i-1)$ is calculated from Ohm's law, including the geometric factor G for the ionic flux, by

$$\Delta I(x(i)) = \frac{\Delta\eta(x(i))}{R_{ch}\times G(x(i))}. \tag{B.15}$$

The current increments are then added to find the local current as a function of position. The method was benchmarked using a case for LSCoF discussed in detail in Ref. [5]. Results (shown in Fig. 10) indicate that the majority of the current is generated within about 2.5 μm of the cathode/electrolyte interface (at $x/L = 1$), in reasonably good agreement with the previous calculation of $L_d = 3$ μm, shown in Fig. 8a of [5]. Results for LSCuF are shown in the main text.

APPENDIX C

C.1. Microstructural design including concentration polarization

The second DZA model [8] solved two coupled differential equations for the chemical potential μ of the ambipolar (ion + electron) species and for the chemical potential μ of the oxygen molecules, as a function of position x in the porous cathode material. The ionic current is given by

$$I_{\mathrm{P}} = -\frac{1-\phi}{t_{\mathrm{S}}}\frac{D_{\mathrm{IE}}C_{\mathrm{i}}}{RT}\frac{\mathrm{d}\mu}{\mathrm{d}x}, \tag{C.1}$$

where

$$\mu - \mu_0 = \left[g_{11}\exp\left(-\frac{x}{L_{\mathrm{m}}}\right) + g_{12}\exp\left(\frac{x}{L_{\mathrm{m}}}\right)\right]\Psi_{11} + [g_{21}x + g_{22}]\Psi_{12} - \Delta\mu_{\mathrm{g}}, \tag{C.2}$$

where μ_0 is the ambipolar chemical potential at the cathode/electrolyte interface ($x = 0$), and the g_{ii} are constants determined from the boundary conditions The g_{ii} equations are several pages long and are not repeated here. Several typographical errors in the original publication have been corrected for the present work. The Ψ_{ii} are components of the eigenvector matrix, and L_{m} is given by

$$\frac{1}{L_{\mathrm{m}}^2} = \frac{1}{L_{\mathrm{p}}^2} + \frac{1}{L_{\mathrm{g}}^2}, \tag{C.3}$$

where

$$L_{\mathrm{g}} = 2L_{\mathrm{P}}\left[\frac{t_{\mathrm{s}}}{t_{\mathrm{g}}}\frac{\phi}{1-\phi}\frac{D_{\mathrm{g}}}{D_{\mathrm{IE}}}\frac{C_{\mathrm{g}}}{C_{\mathrm{i}}}\right]^{1/2} \tag{C.4}$$

is the characteristic length for concentration polarization to occur, i.e., if this length is smaller than the cathode thickness concentration polarization is important. $\Delta\mu_{\mathrm{g}}$ is the molecular chemical potential difference between the two surfaces of the cathode.

It is important to note that mathematical relationships between the microstructural parameters ϕ, S, t_{s}, and t_{g} have been derived for cubic systems in [10], and were also employed in this analysis:

$$S = \frac{3b(1-\phi)}{r_{\mathrm{g}}}, \tag{C.5}$$

where

$$r_{\mathrm{g}} = \frac{1}{2}d_{\mathrm{g}}, \tag{C.6}$$

$$b = \frac{1-3q}{1-\frac{9q^2}{2}\left(1-\frac{q}{3}\right)}, \tag{C.7}$$

$$1-\phi = \frac{3\pi/4[2/9 - q^2(1-q/3)]}{(1-q)^3}, \tag{C.8}$$

$$t_{\mathrm{s}} = \frac{2}{\pi}(1-\phi)(1-q)\ln\left(\frac{2}{q}-1\right), \tag{C.9}$$

$$t_g \simeq \frac{1}{\phi}. \tag{C.10}$$

and q is the fractional consolidation of the grains, a sintering parameter.

This model thus provides a closed-form, simultaneous solution for the two polarization terms of interest in this work, chemical and concentration polarization. Although very useful (see main text), its limitation is that it treats only a homogeneous material. That is, the microstructural parameters are not dependent on spatial position.

REFERENCES

1. G.W. Coffey, J. Hardy, O. Marina, L.R. Pederson, P.C. Rieke and E. Thompson. *Solid State Ionics* **175** (2004) 73–78.
2. A.M. Svensson, S. Sunde and K. Nisancioglu. *J. Electrochem. Soc.* **145** (1998) 1390.
3. G.W. Coffey, L.R. Pederson and P.C. Rieke. *J. Electrochem. Soc.* **150** (2003) A1139.
4. S.B. Adler, J.A. Lane and B.C.H. Steele. *J. Electrochem. Soc.* **143** (1996) 3554.
5. S.B. Adler. *Solid State Ionics* **111** (1998) 125.
6. H.J.M. Bouwmeester and A.J. Burgraaf. In: P.J. Gellings and H.J.M. Bouwmeester (Eds.), *CRC Handbook of Solid State Electrochemistry*, CRC Press, Boca Raton, 1996, pp. 482–553 (Table 14.2).
7. H.M. Deng, M.-Y. Zhou and B. Abeles. *Solid State Ionics* **74** (1994) 75–84.
8. H.M. Deng, M.-Y. Zhou and B. Abeles. *Solid State Ionics* **80** (1995) 213–222.
9. J.W. Kim, A.V. Virkar, K.Z. Fung, K. Mehta and S.C. Singhal. *J. Electrochem. Soc.* **146** (1999) 69.
10. M.-Y. Zhou and P. Sheng. *Phys. Rev. B* **39(16)** (1989) 12027.

Chapter 16

Surface science studies of model fuel cell electrocatalysts

N.M. Marković and P.N. Ross, Jr.

Abstract

The purpose of this review is to discuss the progress in the understanding of electrocatalytic reactions through the study of model systems with surface spectroscopies. Pure metal single crystals and well-characterized bulk alloys have been used quite successfully as models for real (commercial) electrocatalysts. Given the sheer volume of all work in electrocatalysis that is on fuel cell reactions, we will focus on electrocatalysts for fuel cells. Since Pt is the model fuel cell electrocatalyst, we will focus entirely on studies of pure Pt and Pt bimetallic alloys. The electrode reactions discussed include hydrogen oxidation/evolution, oxygen reduction, and the electrooxidation of carbon monoxide, formic acid, and methanol. Surface spectroscopies emphasized are FTIR, STM/AFM, and surface X-ray scattering. The discussion focuses on the relation between the energetics of adsorption of intermediates and the reaction pathway and kinetics, and how the energetics and kinetics relate to the extrinsic properties of the model system, e.g. surface structure and/or composition. Finally, we conclude by discussing the limitations that are reached by using pure metal single crystals and well-characterized bulk alloys as models for real catalysts, and suggest some directions for developing more realistic systems.

CTR	crystal truncation rods
E_{ad}	adsorption energy
E_{diss}	dissociation energy
$E_{Pt—H}$	Pt—hydrogen bond energy
$E_{Pt—O}$	Pt—oxygen bond energy
E_{RHE}	potential with respect to the reversible hydrogen electrode
E_{SCE}	potential with respect to the standard calomel electrode
f	Frumkin interaction parameter ($=r/RT$)
F	Faraday constant
$\Delta G_{H_{upd}}$	apparent Gibbs energy of H_{upd} at $\Theta = 0$
$\Delta G^{\Theta=0}_{H_{upd}}$	apparent Gibbs energy of H_{upd}
$\Delta G_{OH_{ad}}$	apparent Gibbs energy of adsorption for OH_{ad}
$\Delta H_{H_{upd}}$	apparent enthalpy of adsorption for H_{upd}
HER/HOR	hydrogen evolution/oxidation reaction
IRRAS	infrared reflection/absorption spectroscopy
LEED	low-energy electron diffraction
LEIS	low energy ion scattering
ORR	oxygen reduction reaction

Fuel Cells Compendium

q_{st}^{H}	isosteric heat of adsorption for H_{upd}
Q	partial coverage by adsorbates
r	rate of change of ΔG with Θ
R	gas constant
RRDE	rotating ring disk electrode
$\Delta S_{H_{upd}}$	apparent entropy of adsorption for H_{upd}
STM	scanning tunneling microscopy
SXS	surface X-ray scattering
T	temperature of the electrolyte
UHV	ultrahigh vacuum
UPD	underpotential deposition
XPS	X-ray photoelectron spectroscopy

Article Outline

1. Introduction . . . 277
2. Surface structures and energetics of Pt($h\,k\,l$) in UHV . . . 278
   2.1. Clean surfaces . . . 278
   2.2. Adsorbate-induced changes in surface structure . . . 279
      2.2.1. Hydrogen adsorption . . . 280
      2.2.2. Oxygen adsorption . . . 281
      2.2.3. Carbon monoxide adsorption . . . 283
3. Structures and chemistry of Pt bimetallic surfaces in UHV . . . 285
4. Surface structures and energetics of Pt($h\,k\,l$) surfaces in electrolyte . . . 287
   4.1. Reconstruction of Pt($h\,k\,l$) surfaces . . . 288
      4.1.1. Pt(1 1 1) . . . 288
      4.1.2. Pt(1 0 0) . . . 290
      4.1.3. Pt(1 1 0) . . . 291
   4.2. Energetics of Pt(1 1 1)—H_{upd} and Pt(1 1 1)—OH_{ad} systems . . . 294
      4.2.1. Pt(111)—H_{upd} system . . . 294
      4.2.2. Pt(1 1 1)—OH_{ad} system . . . 297
   4.3. Relaxation of Pt($h\,k\,l$) surfaces induced by H_{upd} and OH_{ad} . . . 298
   4.4. Surface structures and energetics of anion adsorption on Pt($h\,k\,l$) surfaces . . . 299
      4.4.1. (Bi)sulfate adsorption . . . 300
      4.4.2. Halide adsorption . . . 302
      4.4.3. Summary of anion adsorption on Pt($h\,k\,l$) . . . 307
   4.5. Surface structures of UPD and irreversibly adsorbed metals on Pt($h\,k\,l$) surfaces . . . 307
      4.5.1. Cu UPD . . . 308
      4.5.2. Pb UPD . . . 311
      4.5.3. Irreversibly adsorbed Bi . . . 313
   4.6. Surface structure of thin metal films . . . 316
5. Electrocatalysis at well-defined surfaces . . . 317
   5.1. HER/HOR . . . 319
      5.1.1. Structure sensitivity on Pt($h\,k\,l$) surfaces . . . 319
      5.1.2. HOR on Pt($h\,k\,l$) surfaces modified with Cu_{upd}, Pb_{upd}, and Bi_{ir} . . . 324
      5.1.3. HER/HOR on Pt(1 1 1)-modified with a pseudomorphic Pd film . . . 325

5.2. ORR 327
5.2.1. Reaction pathway 327
5.2.2. Structure sensitivity on Pt(*h k l*) surfaces 328
5.2.3. ORR on Pt(*h k l*) surfaces modified with UPD metals 331
5.2.4. ORR on the Pt(1 1 1)-modified with a pseudomorphic Pd film 332
5.2.5. ORR on Pt alloy surfaces 332
5.3. Electrooxidation of CO 335
5.3.1. Surface structures of CO_{ad} on Pt(*h k l*) surfaces 335
5.3.2. Energetics and kinetics of CO electrooxidation on Pt(*h k l*) surfaces 342
5.3.3. Surface chemistry of CO on Cu_{upd}, Pb_{upd}, Sn_{upd} and Bi_{ir}-modified Pt(*h k l*) surfaces 346
5.3.4. Surface chemistry of CO on Pt bimetallic alloy surfaces 349
5.4. Oxidation of formic acid on Pt(*h k l*) and bimetallic surfaces 353
5.5. Oxidation of methanol on Pt(*h k l*) and bimetallic surfaces 358
6. Future developments 365
Acknowledgments 366
References 366

1. INTRODUCTION

The role of pure metal single crystals and well-characterized bulk alloys as models for real (commercial) electrocatalysts parallels the role these same materials have played in gas-phase heterogeneous catalysis. Consequently, in studying such model systems, ultrahigh vacuum (UHV) surface analytical tools such as low energy electron diffraction (LEED), Auger electron spectroscopy (AES), X-ray photoelectron spectroscopy (XPS), and low-energy ion scattering (LEIS) have been critical to the development of the surface science of electrocatalytic reactions. The low-index faces of the pure metals provide information about the structure sensitivity of the reaction, while the higher index or stepped surfaces are used to test the "active site" hypothesis, i.e. whether low coordination sites are particularly active. For understanding binary alloy electrocatalysis, and identifying mechanisms of action of alloying constitutents, analysis of the true surface (first-layer) composition of model surfaces by LEIS has been essential. As for gas-phase catalysis, these UHV surface analytical tools can only be applied ex situ, and not under reacting conditions. One must infer from before and after measurements the stability of the model surface under reaction conditions. Because of the intrinsically corrosive nature of the environment in which electrocatalytic reactions take place, strong acids or bases, in situ surface analytical tools have been pursued perhaps even more vigorously by the electrocatalysis community than by our colleagues in gas-phase catalysis. One particular example is surface X-ray scattering (SXS), where the number of papers on studies at the solid–liquid interface is in the dozens versus a handful at the gas–solid interface. In situ SXS has been a critical tool for determining the stability of specific surface structures in electrolyte under reaction conditions. In the present volume, we review the progress in the surface science of electrocatalytic reactions. Given the sheer volume of all work in electrocatalysis that is on fuel cell reactions, we will focus on electrocatalysts for fuel cells. Since Pt is the model fuel cell electrocatalyst, we will focus entirely on studies of pure Pt and Pt bimetallic alloys. The initial sections cover the characterization of the structure and composition of the surface under both vacuum and reaction (in electrolyte) conditions by various techniques. The later sections discuss reaction kinetics on the model surfaces and the relation of catalytic activity to structure and composition. Finally, we conclude with some suggestions for future developments and the need to develop and study more complex model systems.

2. SURFACE STRUCTURES AND ENERGETICS OF Pt(*hkl*) IN UHV

The last two decades have witnessed substantial advances in our understanding of the atomic scale structures at the solid–electrolyte interface. These ongoing developments have been driven by the concurrent emergence of new in situ structural probes like scanning tunneling microscopy (STM) and SXS, to complement the earlier ex situ application of the many tools available in UHV surface science. Today, UHV surface science and electrochemical surface science are so interrelated that any review of the surface processes at metal surfaces in aqueous electrolytes should begin with a review of the relevant advances from purely UHV studies.

2.1. Clean surfaces

Modern surface crystallographic studies have shown that on the atomic scale most clean metals tend to minimize their surface energy by two kinds of surface atom rearrangements, *relaxation* and *reconstruction*. Relaxation of metal surfaces is usually defined as small interlayer spacing changes (usually in the direction perpendicular to the surface plane) relative to the ideal bulk lattice [1–5]. Typically (but not always) in the near-surface region the top layer of metal atoms relaxes inward toward the bulk while one or more of the lower layers of atoms relaxes *outward* slightly. The low coordination number of surface atoms is the main driving force for the relaxation of the clean metal surfaces. In some metals, the surface relaxation can result in a change in the equilibrium position and bonding of surface atoms (displacements both perpendicular to and parallel to the surface plane) and can give rise to reconstruction of the outermost layers. Reconstruction is usually observed straightforwardly in measurements by LEED, reflection-high energy electron diffraction (REED), LEIS, field ion microscopy (FIM), SXS, and STM. Relaxation, on the other hand, can be determined by a more limited number of surface sensitive probes, such as analysis of LEED intensity data or modeling the so-called crystal truncation rod (CTR) data in SXS experiments.

The hexagonal face centered cubic (fcc) (1 1 1) surface of Pt (surface density of 1.53×10^{15} atoms/cm^2), (Fig. 1a), has no tendency to reconstruct [5]. Although there is a general trend toward contraction of the topmost interlayer spacing in many metal surfaces (see below), on the clean unreconstructed Pt(1 1 1) surface the topmost interlayer spacing (Δd_{12}) is expanded by ca. $2.5 \pm 1.3\%$, see the side view of Fig. 1a. The fcc Pt(1 0 0) surface, (Fig. 1b), is atomically less dense than the Pt(1 1 1) surface. As a result, one finds a stronger tendency for both reconstruction and relaxation. A metastable surface with the (1×1) bulk truncation structure can be prepared with the hydrogen–oxygen titration method of Griffiths et al. [6]. The resulting (1×1) phase transforms *irreversibly* into a reconstructed phase above 390 K. Reconstruction of the Pt(1 0 0) surface is characterized by a contraction of the top layer and the formation of a quasi-hexagonal coincidence lattice, which is corrugated due to its varying registry with the substrate. This structure has been labeled the Pt(1 0 0)-"hex" structure, and has a surface Pt density of 1.55×10^{15} atoms/cm^2, as compared to 1.28×10^{15} atoms/cm^2 in the (1×1) bulk truncation structure [7]. Above a surface temperature of 1070 K, the "hex" surface is modified by a slight rotation of the top layer, of ca. 0.7°; this structure has been labeled Pt(1 0 0)-"hex-rot" [8] and is described in detail in the literature. For the "almost clean" Pt(1 0 0)-(1×1) surface containing some adsorbed hydrogen (from the hydrogen–oxygen titration procedure), the interlayer spacing was found to be within $\Delta d_{12} \approx 0.2\%$ of the bulk spacing, e.g. the surface is unrelaxed [9].

Pt(1 1 0) exhibits $(1 \times n)$ reconstructions, where the higher periodicity *n* arises along the [0 0 1] direction which is perpendicular to the close-packed atomic [1 1 0] chains. It is now well-established that the clean fcc Pt(1 1 0) face has at room temperature a (1×2) periodicity [3] that is called the "missing row" structure, since every other row is lost in going from the (1×1) phase (surface Pt density of 0.94×10^{15} atoms/cm^2, Fig. 1c) to the (1×2) phase (Fig. 1d). As a result, the atom surface density of

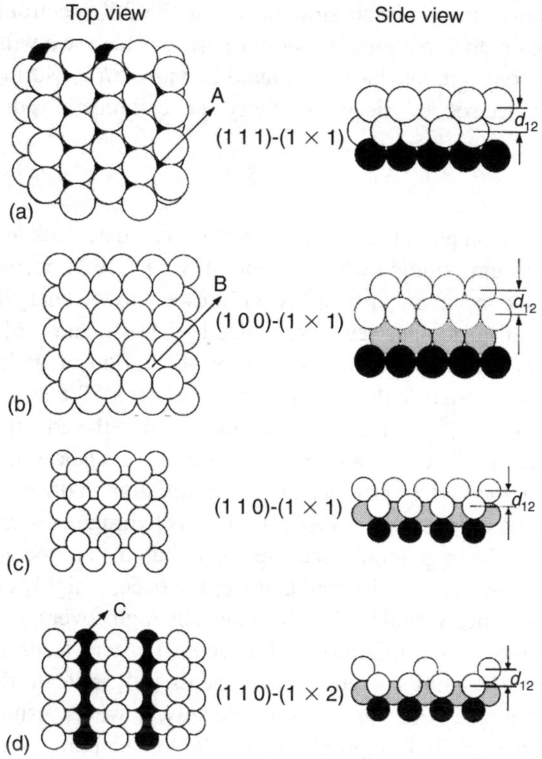

Fig. 1. Top views and side views of the face-centered cubic (fcc) crystal surfaces: (a) Pt(111)-(1 × 1), (b) Pt(100)-(1 × 1), (c) Pt(110)-(1 × 1), and (d) Pt(110)-(1 × 2). Top views: sites A, B, and C represent the most probable centers for adsorption of H_{upd}. Side views: a schematic of platinum atoms surface relaxation, d_{12}

(1 × 2) decreases by 50% from the (1 × 1) surface. The effect of removing every other row is to make the surface even more corrugated than the (1 × 1) surface. The higher ($n > 2$) reconstructions appear to be stabilized by the smallest amounts of impurities [3]. At a critical temperature, $T_c \approx 855 - 1050$ K, the clean surface undergoes a (1 × 2) ↔ (1 × 1) structure transition [10–13].

A contraction of the interlayer spacing on fcc (1 1 0) metal surfaces is more prevalent than expansion. This behavior is also followed for clean Pt(1 1 0)-(1 × 1) and Pt(1 1 0)-(1 × 2) surfaces, which show an alternating pattern of contractions and expansions of the first four to five layers of atoms, e.g. $\Delta d_{12} = -18.4\%$, $\Delta d_{23} = -12.6\%$, $\Delta d_{34} = -8.7 \times 5$, etc. [14,15]. More details about the relaxation and reconstruction of Pt($h\,k\,l$), including theories developed to rationalize or predict surface reconstructions of clean metal surfaces, may be found in reviews published by Thiel and Estrup [3] and Van Hove [5].

2.2. *Adsorbate-induced changes in surface structure*

In general, adsorption of atoms and/or molecules on clean metal surfaces has a dramatic effect on surface structure. The presence of the adsorbed layer usually alters the interlayer spacing and can cause clean metal surfaces to reconstruct or to deconstruct ("lift") back to the (1 × 1) phase. The thermodynamic driving force for adsorbate-induced restructuring is the formation of strong adsorbate–substrate bonds that are comparable to or stronger than the bonds between the substrate atoms in the clean surfaces. Within the framework of this paper, however, it is impossible to review the full spectrum of the adsorbate-induced changes in surface structure on different metals. For our purposes here, we will focus on UHV results for the interaction of hydrogen, oxygen, and carbon monoxide with Pt(*h k l*) surfaces, since these molecules are of most relevance to the electrocatalysis of the related fuel cell reactions discussed in Section 5.

2.2.1. Hydrogen adsorption

Studies of hydrogen adsorption on platinum single crystal surfaces date back to the early 1970s, and it was one of the first adsorption systems studied with modern UHV surface science tools. However, the results from that period were very inconsistent, particularly with respect to the energetics of H adsorption on the (1 1 1) surface. An excellent historic overview is provided by Christmann [16]. Whereas Christmann and Ertl [17] found a rather low value for the initial heat of adsorption of hydrogen on Pt(1 1 1) (around 50–60 kJ/mol), other groups reported higher values, lying between 70 and 90 kJ/mol [18,19]. It is now well-established that these higher heats of adsorption correspond to the adsorption of hydrogen on defect sites, e.g. step sites [16]. As Fig. 2 shows, a linear decrease in the adsorption energy with the hydrogen coverage, Θ_H, is observed on the Pt(1 1 1) surface [17]. The variation in the heat of adsorption with coverage arises mainly from H–H repulsion that comes into play as hydrogen coverage reaches a certain critical level. Until recently, it has been generally accepted that H atoms are located in the three-fold hollow sites, presumably due to the tendency of adsorbed hydrogen to occupy highly coordinated sites. The occupation of all three-fold hollow sites would lead to the unusually high coverage of 2 H/Pt. Occupation of all three-fold *next-nearest*-neighbor sites minimizes H–H repulsion and leads to a coverage of 1 H/Pt in agreement with experiment. He atom scattering experiments [20] indicate these are the hcp-type sites above a second layer Pt atom, site A in Fig. 1a. Very recently, however, using density functional theory (DFT) within the generalized gradient approximation (GGA), including scalar relativistic effects and modeling the Pt(1 1 1) surface as a slab, Olsen et al. [21] found that hydrogen preferentially occupies top sites. Nevertheless, regardless of the preference for the adsorption site, the difference in the adsorption energy among various sites is very small [21]. It should also be noted that the formation of a hydrogen monolayer (ML = 1 H/Pt) on Pt(1 1 1) does not induce surface reconstruction, and produces no measurable change in the Δd_{12} interlayer spacing [5]. This is common for systems with a relatively low heat of adsorption on the close-packed (1 1 1) surface of a metal with an appreciable cohesive energy [16].

Hydrogen adsorption on Pt(1 0 0) is more complex than the adsorption on Pt(1 1 1), complicated by the fact that the H adsorption deconstructs or "lifts" the "hex" or "hex-rot" clean surface reconstruction. In an early work, Norton et al. [7] demonstrated that at low temperature upon adsorption of H_2 the reconstruction is only partially lifted by adsorbed hydrogen, site B in Fig. 1b. More recently, Wandelt and co-workers [22] showed that the "hex" reconstruction is lifted by the adsorption of H_2 but not above a threshold temperature, e.g. at 35 K hydrogen adsorption does not bring about lifting of the reconstruction, but after subsequent heating to 100 K the reconstruction is partially lifted. Hydrogen adsorption induced lifting of Pt(1 0 0)-"hex" to (1 × 1) was also observed by Hu and Lin [23]. On the (1 × 1) surface stabilized with a monolayer of adsorbed hydrogen, a surface expansion of $\Delta d_{12} = 3.1\%$ was calculated from the modeling of LEED intensity data. Pasteur et al. [24] made dynamic measurements of the (1 × 1) island growth. At very low temperatures, the deconstruction is a very slow process which can be accelerated during a typical thermal programmed desorption (TPD) experiment. As a consequence, unreconstructed (1 × 1) Pt islands are formed resulting in a large variety of H binding sites and possibly a strong energetic heterogeneity which must be accounted for in analyzing the TPD data. However, the initial heat of adsorption ($\Theta_H < 10\%$), as calculated from TPD, differed only slightly between the two surface structures, e.g. from ≈90 kJ/mol on Pt(1 0 0)-(1 × 1) to ≈98 kJ/mol on Pt(1 0 0)-"hex" [25]. This unexpected result was attributed to the small effect of steps on the average *electronic* coordination for the two (1 0 0) surface structures. The Pt—H bond energy is, however, significantly higher on either of the (1 0 0) surface structures versus the (1 1 1) [16].

In contrast to the Pt(1 0 0) surface, the Pt(1 1 0)-(1 × 2) phase is not affected by the adsorption of hydrogen, and so no (1 × 2) ≈ (1 × 1) transition is observed for this surface. This is perhaps not too surprising result. There is a very large difference in the atomic density between the unreconstructed and reconstructed surfaces. As a result, a much larger thermal energy will be required to displace the respective Pt atoms and to make them diffuse over greater distances on a more corrugated surface. This larger energy cannot be provided by the relatively low heat of adsorption of hydrogen on Pt. Although one does not have a structural phase change to deal with in studying hydrogen on (1 1 1)-(1 × 2), the system is complicated by a large variety of possible adsorption sites that need to be considered, see Fig. 1d. According to Weinberg and co-workers [26], H atoms may initially occupy the deep trough sites in the third Pt layer, and that after further H uptake less deep sites would be populated. Kirsten et al. [27], however, suggested that initial adsorption of hydrogen on the missing row (1 1 0) structure takes place on the three-fold hollow sites below the topmost Pt rows (sites C in Fig. 1c), whereby the initial ca. 15% inward relaxation (Δd_{12}) changes into ca. 20% outward relaxation. With a 20% expansion of the topmost Pt atoms, the average bonding distance of H atoms in the subsurface three-fold hollow sites becomes ≈2.1 Å, in very good agreement with the value of 2.08 Å observed in Pt hydride [27]. Further adsorption of hydrogen induced a pronounced decrease in expansion, attributed to adsorption of hydrogen within the troughs of the missing row structure. The energetics of the Pt(1 1 0)–H interaction was examined initially by Engstrom et al. [26] and more recently by Lee et al. [25]. The former authors found a complex coverage dependence. Due to an attractive H–H interaction for $0 < \Theta_H < 0.15$, the bond energy increases from 74 to 110 kJ/mol. At higher coverages, however, the H–H interaction becomes repulsive, so the heat of adsorption decreases continuously and at saturation reaches a value of only ≈55 kJ/mol. The latter authors found that the initial heat of adsorption of hydrogen on Pt(1 1)-(1 × 2), ca. 73 kJ/mol, is significantly smaller than on an unreconstructed Pt(1 1 0)-(1 × 1) surface, ca. 115 kJ/mol. This difference was not fully explained, but it would appear from the results of Kirsten et al. [27] that the "trough sites", which are unique to the (1 × 2) structure and the last to be populated, are the sites of the weakly adsorbed state with a heat of adsorption of only ≈55 kJ/mol.

The absolute value of the Pt—H bond energy on Pt surfaces can be obtained from the heat of adsorption ($\Delta H_{H_{ad}}$) through the relation

$$2E_{Pt-H} = |\Delta H_{H_{ad}}| + E_{diss} \tag{1}$$

where E_{diss} is the dissociation energy of H_2, 432 kJ/mol. It can be seen that the Pt—H bond energy (E_{Pt-H}) varies from about 240 to 270 kJ/mol depending on the specific adsorption site and coverage. This is a relatively weak chemical bond. Nonetheless, the dissociation of H_2 on Pt surfaces is an activationless process, i.e. the activation energy is of the order of kT. The relatively weak Pt—H bond makes H_{ad} on Pt a very reactive intermediate in catalytic hydrogenation reactions, but also makes the catalyst very susceptible to poisoning by impurities in the hydrogen that are more strongly adsorbed. As we shall see later, this is a very important fundamental problem for low temperature fuel cells.

2.2.2. *Oxygen adsorption*

The adsorption of oxygen is more complex than hydrogen adsorption, since molecular adsorption, dissociative chemisorption, and oxide formation are all possible even in relatively mild conditions of temperature and pressure (like those employed in UHV studies). Oxygen adsorption on Pt(1 1 1) have yielded consistent results. Gland et al. [28] characterized three states of oxygen as a function of temperature (dosing with O_2 at $<10^{-6}$ Torr): molecular adsorption predominates below 120 K, adsorbed atomic oxygen predominates in the 150–500 K temperature range, while subsurface or "oxide" formation may occur in the range 1000–1200 K. Adsorption into the molecular state does not have substantial activation energy, and, therefore, the heat of adsorption is approximately equal to the heat of desorption, ca. 37 kJ/mol.

Molecular oxygen forms a dioxygen species of the peroxo-type (O_2^{2-}) with a single O—O bond. Heating the overlayer of molecular oxygen resulted in the formation of atomic oxygen, which interacts more strongly with the Pt(1 1 1) surface. Gland et al. [28] have suggested that dissociation of O_2 molecules proceeds by sequential population of chemisorbed precursor states, i.e. the so-called molecular precursor route. Shortly after Gland's report, Campbell et al. [29] suggested that part of O_2 molecules may also adsorb directly into the atomic state, without previous accommodation in the molecular state. In a later study, ab initio local-spin-density calculations for the adsorption of O_2 on Pt(1 1 1) identified two distinct, but energetically almost degenerate chemisorbed precursors [30]. A superoxo-like paramagnetic precursor is formed at the bridge site, with molecular parallel to the surface. A second peoxo-like non-magnetic precursor is formed in the three-fold hollow, with the atom slightly canted in a top-hollow-bridge geometry. As for the Pt–H system, an appreciable decrease in the heat of adsorption was observed with increasing coverage by oxygen, indicative of repulsive long-range interactions between adsorbed oxygen atoms. The values ranged from about 160 kJ/mol at $\Theta_O = 0.8\Theta_{max}$ to about 250 kJ/mol at $\Theta_O = 0.2\Theta_{max}$, where Θ_{max} is the absolute oxygen coverage at saturation [28] (Fig. 2). Under relatively low pressures and temperatures (300–500 K), the coverage by atomic oxygen saturates at 0.25 ML (1 ML = 1 O/Pt) in a well-ordered $p(2 \times 2)$ overlayer with the O atoms in three-fold hollow sites [31]. If the surface temperature exceeds 800 K during O_2 dosing, a stable "oxide" (subsurface) state is formed which desorbs at a temperature around 1200 K. At elevated pressures and temperatures, however, coverages higher than 0.25 ML can be achieved at Pt(1 1 1) [29,32]. According to [29,32], the population of the (1 1 1) surface above 0.25 ML requires a different (presumably activated) pathway for dissociation than the molecular precursor route.

Oxygen adsorption on Pt(1 0 0) again exhibits more complex behavior than Pt(1 1 1) and previous studies are in lesser accord. One complicating factor is the reconstruction of the clean Pt(1 0 0), as was the case for the adsorption of hydrogen. Norton et al. [33] showed that the Pt(1 0 0)-(1 × 1) surface is more reactive to oxygen than the Pt(1 0 0)-"hex" surface. On the (1 × 1) surface oxygen dissociates even at 123 K. On warming to 240 K, a (2 × 1) LEED pattern is observed that is believed to originate from an oxygen overlayer on an unreconstructed substrate. On the other hand, because Pt(1 0 0)-"hex" is an energetically stable and a more chemically inert surface, oxygen is adsorbed entirely in the molecular state with an activation energy for desorption ca. 37 kJ/mol [33]. Accordingly, O_2 dissociative adsorption would be an activated process on the "hex" surface, but non-activated on the (1 × 1) surface. Furthermore, while on (1 × 1) the molecular state serves as a precursor to O_2 dissociative adsorption, on the "hex" surface O_2 dissociation occurs through a direct mechanism, for details see [34]. However, on either surface the coverage from low temperature/low pressure dosing saturates at ca. 0.2–0.25 ML, with the underlying Pt structure being (1 × 1). As with Pt(1 1 1), dosing with oxygen at both higher partial pressures (ca. 10^{-3} Pa) and surface temperatures (570 K) leads to much higher coverages, close to 1 ML. Considerable controversy raged on the nature of this high coverage state (see [32] for an overview). However, angle-resolved XPS revealed that there was no significant amount of subsurface oxygen in this state [32], i.e. it is not an "oxide", as some had suggested.

Adsorption of O_2 on the Pt(1 1 0)-(1 × 2) surface has also been the subject of considerable attention. As for other two Pt surfaces, the kinetic and dynamic aspects of the adsorption and desorption of O_2 on Pt(1 1 0) surface is dependent on the temperature of the surface and the exposure of oxygen. In general, the adsorption of oxygen takes place in stages. At very low temperatures, oxygen molecules are first adsorbed in the valley of the surface, then at higher coverages adsorption on the (1 1 1) microfacets occurs [35]. Schmidt et al. [36] have proposed that at low temperatures oxygen molecules are adsorbed as two different peroxo-like molecular species in on-top and two-fold coordinated sites. When the oxygen exposure is increased, oxygen is additionally adsorbed in superoxo form. Heating the surface above 35 K leads to partial dissociation and partial desorption of the oxygen molecules. At about 100 K the superoxo species desorb completely: desorption of the peroxo species begins at 125 K and is completed at 200 K. At 300 K oxygen is adsorbed entirely in atomic form. As with Pt(1 0 0), the reconstruction is lifted by higher coverages of atomic oxygen, with low pressure–low temperature dosing producing the highest coverages of the three low-index surfaces, ca. 0.45–0.5 ML. Very recently, oxygen dissociation on stepped Pt(1 1 1) surfaces has been studied by using a combined experimental and theoretical approach. Since the Pt(1 1 0)-(1 × 2) surface consists of three atom wide Pt(1 1 1) terraces separated by monoatomic Pt(1 1 1) steps, O binding sites on this surface are most likely identical with the one proposed for Pt $\{n(1\,1\,1) \times (1\,1\,1)\}$ stepped surfaces. In particular, first principle calculations showed that on Pt surfaces vicinal to (1 1 1) O adatoms are attracted to step edges, gaining 0.2–0.3 eV per nearest-neighbor step-edge Pt atoms, and that they favor "fcc-like" over "hcp-like" sites by 0.4 eV [37]. STM and thermal energy atom scattering in combination with DFT supported this view [38] by showing that on the surfaces vicinal to (1 1 1) O_2 molecules dissociate from a molecular precursor state (MPS) on the upper site of the Pt step edges, occupying "fcc-like" sites. The authors also demonstrated that the transition state (TS) has the key role for the local reactivity, e.g. the enhanced reactivity at the Pt step sites is not caused by a decrease of the local dissociation barrier from the MPS but is related to a stabilization of both the MPS and TS [38].

Interestingly, the initial heats of adsorption of (atomic) oxygen do not vary significantly between the three low-index surfaces. Since oxygen always desorbs in the molecular form, the strength of the Pt—O bond, $E_{\text{Pt—O}}$, is simply related with the adsorption energy, E_{ad}, and the dissociation energy of O_2, $E_{\text{diss}} = 497$ kJ/mol, through

$$2E_{\text{Pt—O}} = |E_{\text{ad}}| + E_{\text{diss}} \tag{2}$$

For the initial heat of adsorption, ca. $E_{\text{ad}} = 250$ kJ/mol [39], the strength of the Pt(1 1 1)—O bond is close to $E_{\text{Pt—O}} = 350$ kJ/mol.

2.2.3. *Carbon monoxide adsorption*

Carbon monoxide is perhaps the molecule whose adsorption properties have been investigated in greatest detail with modern UHV systems. Blyholder [40] proposed a simple model to explain how CO can bind to a transition metal surface. It is commonly accepted that chemisorption of CO on transition metals occurs in molecular form through electron transfer from the 5σ orbital of CO to the metal, and back-donation of metallic d electrons to the unfilled anti-bonding $2\pi^*$ orbital of CO. The back-donation weakens the C—O bond, and the amount of back-donation increases as one moves to the left in the periodic table from (Ni, Pd, Pt). In particular, this model correctly predicts the dissociation of CO on transition metals depending on their electronic configuration, i.e. their position in the periodic table. CO on Pt(1 1 1) is adsorbed non-dissociatively with carbon bonded to the surface in both linear and bridged configurations [39]. At ca. 170 K and at a fractional coverage $\Theta_{\text{CO}} = 0.33$, a $(\sqrt{3} \times \sqrt{3})\,R30°$ overlayer structure was reported. At $\Theta_{\text{CO}} = 0.5$, a sharp $c(4 \times 2)$ structure is observed which with increasing coverage transforms into a compressed hexagonal close-packed complex domain wall structure with $\Theta_{\text{CO}} = 0.68$ [41]. The heat of adsorption of CO on Pt(1 1 1) at low coverages, i.e. at surface concentrations that are small enough to rule out energetics owing to mutual interactions between the adsorbates, is ca. 140 kJ/mol [42]. The change of heat of adsorption for CO on Pt(1 1 1) with surface coverage from the classical work by Ertl et al. [42] is shown in Fig. 2. On the single crystal surface, the isosteric heat of adsorption varies with coverage primarily from lateral repulsive interactions, and to a much lesser extent the occupation of different adsorption sites, e.g. defects or steps. The heat of adsorption shows a sharp decrease as the coverage increases above ca. 0.5 CO/Pt. Between 0.55 and the saturation coverage of 0.68 CO/Pt, the heat of adsorption falls from 80 to 40 kJ/mol, presumably due to strong repulsive interaction between adsorbate molecules at high coverage. A heat of adsorption of only 40 kJ/mol lies at the transition between "chemisorption" and "physisorption", hence this high coverage state of CO_{ad} may be termed as the "weakly adsorbed" state of CO. As we shall discuss later in this review, we have, in fact, adopted the terminology "weakly" and "strongly" adsorbed states of CO in discussing the electrocatalytic oxidation of CO at the Pt($h\,k\,l$)–electrolyte interface.

Numerous studies for the Pt(1 0 0)—CO_{ad} system have shown that the adsorption of CO on the Pt(1 0 0)-"hex" surface leads to lifting of the reconstruction, resulting in a (1 × 1) phase of Pt covered by CO. On the basis of Thiel et al. [43] CO removes the Pt(1 0 0) reconstruction by a mechanism in which CO adsorption on the "hex" phase is followed by migration, cluster formation, rapid "hex" ↔ (1 × 1) conversion of the local substrate area, and trapping of the CO molecules on the resultant (1 × 1) phase. The difference in the low coverage heat of adsorption of CO on the "hex" and (1 × 1) phases, 106 kJ/mol versus 156 kJ/mol, respectively, has been proposed as the driving force for the Pt phase transformation during adsorption. The CO itself forms a $c(2 \times 2)$ overlayer with ideal coverage of 0.5 ML. Beyond $\Theta_{CO} = 0.5$, the unit cell of the adsorbed layer is continuously compressed until the $c(4 \times 2)$ adlayer with saturation coverage of $\Theta_{CO} \approx 0.8$ is reached [44]. The infrared spectra for this system show a dominant atop band together with a weaker bridging feature, indicating that the $c(4 \times 2)$ adlayer contains predominantly CO in atop sites [45].

The adsorption of CO on Pt(1 1 0) has been studied extensively using a variety of techniques. Many of these studies focused on the effect of the CO on the (1 × 2) ↔ (1 × 1) surface phase transition. It is now well-established that the exposure of the Pt(1 1 0)-(1 × 2) structure to CO above 250 K lifts the reconstruction, forming either a (1 × 1) or (2 × 1)-$p1g1$-CO phase at saturation depending on the adsorption temperature [46–50]. Direct observation by STM [48] of the removal of the "missing row" reconstruction under the influence of CO_{ad} revealed that at 300 K this process is initiated by homogenous nucleation of small characteristic (1 × 1) patches. Their further growth is limited by thermal activation, and only at higher temperatures the enhanced surface mobility leads to the formation of larger, strongly anisotropic islands. Photoemission studies have shown that the molecular axis of the CO molecules in these islands is tilted, even at low coverages ($\Theta_{CO} < 0.5$) [47]. Sharma et al. [50] suggested that the driving force for the (1 × 2) ↔ (1 × 1) surface phase transition is the increase in the heat of adsorption for on-top CO on top row atoms compared with second layer atoms, which must exceed the energy difference between two clean surface phases. Below 250 K, the (1 × 2) reconstruction remains, due to immobility of the Pt atoms at these temperatures. In the ordered (1 × 2)-plgl-CO phase, the molecules occupy on-top sites on each top layer Pt, i.e. on the atomic ridges separated by troughs, consistent with a saturation coverage value of $\Theta_{CO} = 1$. The strong repulsive interaction between neighboring CO molecules that result from this close-packed arrangement are overcome by the tilting of alternate molecules by 25° to the surface normal in

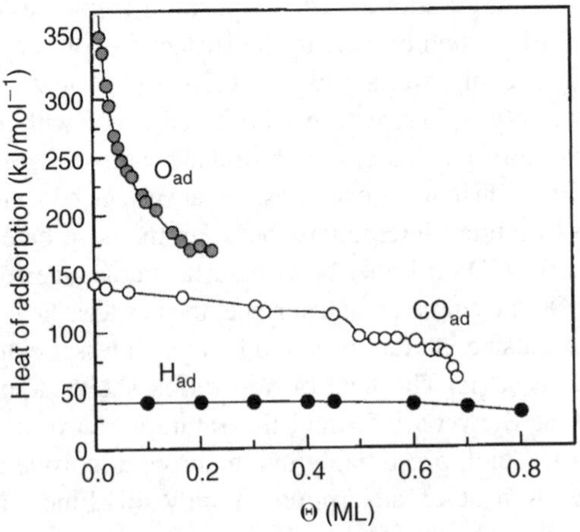

Fig. 2. The heat of adsorption of H_2, O, and CO on Pt(1 1 1) [17,28,42]

opposite directions; for details see recent paper by King and co-workers [50]. For coverage in the range $0.2 < \Theta_{CO} < 0.5$, the isosteric heat of adsorption is 160 ± 15 kJ/mol [46].

As a useful summary for later discussions, the isosteric heats of adsorption of oxygen, carbon monoxide, and hydrogen on Pt(1 1 1) are all shown as function of coverage in Fig. 2.

3. STRUCTURES AND CHEMISTRY OF Pt BIMETALLIC SURFACES IN UHV

In this section, we review the UHV characterization of the surface structures and compositions of four Pt alloy systems that have been used as electrodes in electrocatalytic reactions: Pt—Ru, Pt—Sn, Pt—Ni and Pt—Co. The results of the electrocatalytic studies are discussed in 5.2 and 5.3.

The difference between the surface and the bulk composition of alloys has been the subject of intensive research, in both theory and experiment, in the last two decades, and a review of the subject is beyond the scope of this chapter. Excellent reviews of both theory and experiment have been presented by Campbell [51], and Dowben and Miller [52], and earlier by Sachtler and van Santen [53] and Chelikowski [54]. The encyclopedia of surface structures by Watson et al. [55] is also an excellent place to find references to a specific alloy or metal-on-metal system. It is now widely recognized that surface segregation, i.e. the enrichment of one element at the surface relative to the bulk, is a ubiquitous phenomenon in bimetallic alloys, and the theory for accounting for and predicting this segregation is well developed. The most reliable method for determining the composition of the outermost layer of atoms of a polycrystalline bulk alloy is by LEIS using inert gas ions like helium and neon. An excellent review of the physics of LEIS is provided by Heiland and Taglauer [56]. An example from our laboratory of an LEIS spectrum from a Pt—Ru alloy electrocatalyst is shown in Fig. 3a. Details of the analysis are given in [57]. The bulk alloy composition is 70.2% Pt, but the annealed surface, which is the spectrum shown, has a composition of

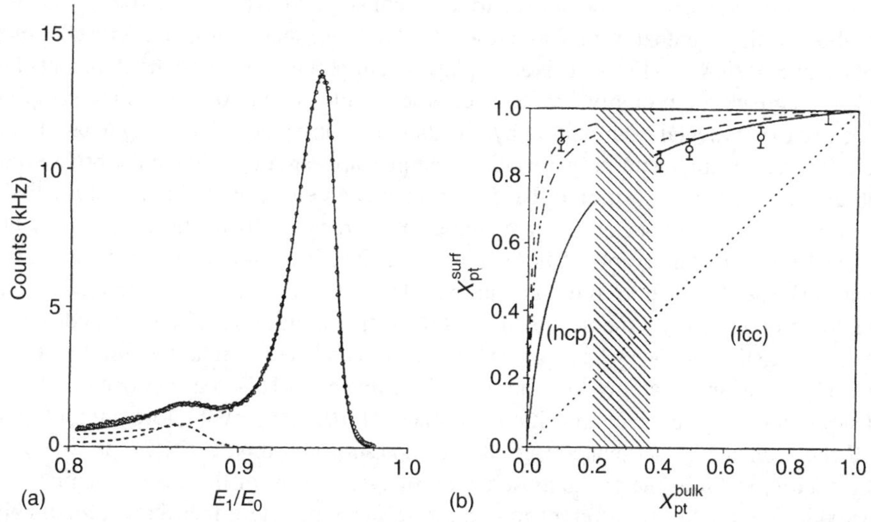

Fig. 3. (a) LEIS spectrum from an annealed $Pt_{70}Ru_{30}$ alloy surface using a 2 keV He^+ ion beam (18 nÅ/cm^2) rastered over an area of 3 mm × 3 mm. Calculated values of E'/E_0 for Pt and Ru atoms are 0.936 and 0.881, respectively. Circles are experimental data, and solid lines are fitted using Gaussian line shape and calculated values of E'. (b) Experimental surface composition of annealed Pt—Ru alloys versus predictions from a thermodynamic model for different low-index planes: (- - -) fcc(1 1 1) and hcp (0 0 0 1); (· — ·) fcc (1 1 0) and hcp (1 1 $\bar{2}$ 0); (- - -) fcc (1 0 0) and hcp (1 0 $\bar{1}$ 0); (\\\\) indicates two-phase region of the bulk alloy [57]

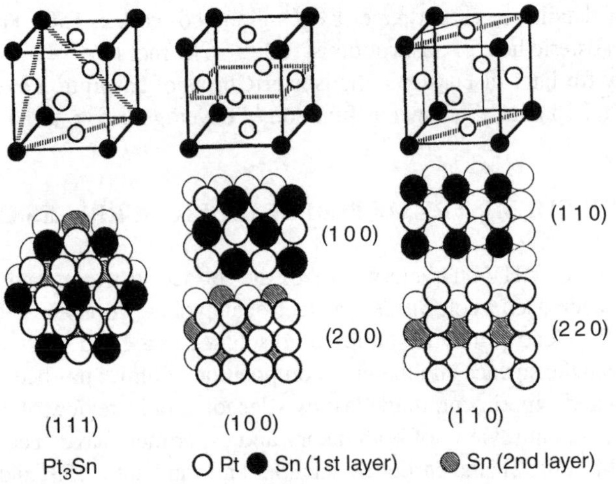

Fig. 4. Models of the Ll_2 fcc structure of Pt_3Sn showing the bulk termination planes of the three low-index faces [59]

92.1% Pt. The Ru peak, with a ΔM of nearly 100, is easily resolved with He^+ ions even at the low surface concentration of ca. 8%. The bulk concentrations can be produced on the surface as well by sputtering off the Pt-enriched layer(s) with argon ions, e.g. The Pt—Ru system is a classic example of surface segregation of the element having the lower heat of sublimation. The equilibrium surface composition of the Pt—Ru system as determined by LEIS [57] is shown in Fig. 3b. A compilation of other surfaces analyzed by LEIS has been made by Watson et al. [55].

Pt_3Sn occupies a special place in both alloy surface chemistry and in alloy electrocatalysis. It is a highly exothermic alloy, with an enthalpy of formation of −50.2 kJ/g atom [58], that crystallizes in the Ll_2 (Cu_3Au-type) lattice, as shown in Fig. 4. This alloy played an important role in the development of the broken-bond model for segregation in highly exothermic, ordered alloys and polycrystalline samples were the subject of intense experimental examination by Sachtler and van Santen [53] using a variety of methods, including LEIS. These results with polycrystalline samples appeared to validate the broken-bond model, since the observed surface composition of ca. 50% Sn is in agreement with the model prediction. In collaboration, Bardi and Ross conducted the first studies of segregation with single crystals of Pt_3Sn, using both LEIS and LEED crystallography. These studies are described in a series of papers [59,60], and a review of the work has already been presented in [61]. The single crystal results revealed that the broken-bond model does not, in fact, get the details of the segregation quite right. The model correctly predicted an absence of segregation on the (1 1 1) surface, but incorrectly predicts that the pure Pt(2 0 0) planes are enriched in Sn by exchange of atoms with the second atomic layer. This does not occur, rather the crystal is terminated preferentially in the compositionally mixed (1 0 0) plane by the formation of double-height steps. A similar preferential termination in the compositionally mixed (2 2 0) plane occurs for the <1 1 0> orientation. Pt_3Sn is one of the most active catalyst known for the electrochemical oxidation of carbon monoxide (CO), and the different low-index surfaces just described have remarkably different activity for this reaction, as discussed in detail in a later section.

Pt—Ni alloys are very interesting from a theoretical perspective because Pt and Ni have essentially identical surface energies, yet strong enrichment of the surface in Pt was observed by LEIS [62]. By LEED crystallography, Gauthier et al. [63] found that the near-surface region exhibits a highly structured compositional oscillation in the first three atomic layers of the (1 1 1) crystal, as shown in Fig. 5. Such composition oscillations were previously thought to occur only in highly exothermic alloys, i.e. alloys with a high

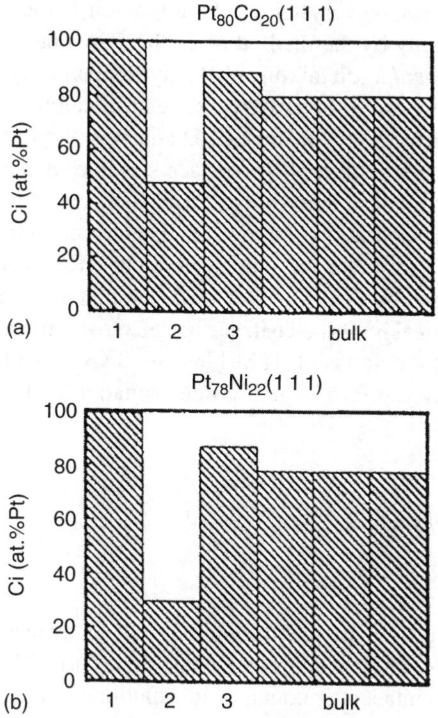

Fig. 5. Compositional oscillation at the (1 1 1) surface of (a) Pt—Co and (b) Pt—Ni alloys inferred by LEED crystallography [63]

enthalpy of mixing, which are usually ordered in the bulk as well, Pt_3Sn being the prime example [53]. The Pt—Ni alloys are disordered in the bulk and have only small enthalpies of mixing. The so-called size effect, i.e. the tendency for the larger atom to be at the surface [64], must be treated rigorously to account for the compositional oscillation in this system, as done by Treglia and Legrand [65].

The (1 1 1) and (1 0 0) surfaces of Pt_3Co were first studied by Bardi and co-workers [66,67], and later by Gauthier [68]. Interestingly, the LEED studies showed that the (1 0 0) surface forms an hexagonal reconstruction quite similar to that on the pure Pt(1 0 0) surface [67,68]. The LEIS studies revealed that for both the (1 1 1) and (1 0 0) crystals the outermost layer of the clean annealed surfaces are pure Pt, with Pt depletion in the second layer. The compositional oscillation in $Pt_{20}Co_{80}(1\,1\,1)$ and $Pt_{78}Ni_{22}(1\,1\,1)$ as determined by LEED crystallography is shown in Fig. 5a and b, respectively. Bardi et al. [66] also studied adsorption of these surfaces and found a significant reduction (ca. 10%) in the CO binding energy versus the pure Pt surface of the same orientation. This was attributed to an electronic effect of intermetallic bonding of the Ni/Co-rich second layer with the topmost Pt atoms. We shall demonstrate in a later section that this electronic effect plays an important role in the electrocatalytic properties of Pt—Co alloys.

4. SURFACE STRUCTURES AND ENERGETICS OF Pt(*h k l*) SURFACES IN ELECTROLYTE

A well-defined and clean Pt(*h k l*) single crystal electrode for the study of the electrochemical surface processes can be prepared either by conventional UHV sample preparation, e.g. sputtering and annealing in UHV [69], or by annealing the single crystal in a hydrogen-air flame. There are several variants to the latter method, distinguished mainly by the method of cooling: by quenching in water [70,71], or slower cooling in hydrogen [72], hydrogen/argon mixtures [71], hydrogen/nitrogen mixtures [73], vapor phase of I_2 [74], or in CO/nitrogen mixtures [73]. In early work, determination of the stability of the well-defined structures in electrolyte was derived from ex situ LEED analysis of emerged surfaces [75–77]. Although LEED is a well-developed and powerful tool for surface structure determination in UHV, the emersion process itself may disrupt the structural integrity of Pt(hkl)–electrolyte interface, and may also leave a salt deposit on the surface after evaporation of the water. Therefore, establishing the relationship between the structure of the interface in electrolyte and that observed in UHV was always problematic, and had to be carefully examined on a case-by-case basis. The development of the flame-annealing method by Clavilier [70] has become of critical importance in surface electrochemistry. Clavilier's method along with the recent advances in the in situ methods of STM [78] and SXS [79–83] have alleviated this "emersion gap", and has provided definitive surface structure determination of Pt(hkl) while in electrolyte under potential control.

4.1. *Reconstruction of Pt(***h k l***) surfaces*

4.1.1. *Pt(1 1 1)*

Examination of the stability of a UHV prepared Pt(1 1 1)-(1 × 1) surface in solution has been the subject of numerous ex situ experiments. These studies date back to the early 1970s, when it was reported that the (1 1 1)-(1 × 1) structure remains intact after contact with solution [84,85]. More recently, systematic in situ SXS and STM studies have demonstrated that well-ordered Pt(1 1 1)-(1 × 1) structures are observed in solution following preparation by flame-annealing and cooling in different atmospheres. Direct evidence to support the existence of well-characterized Pt(1 1 1)-(1 × 1) in solution was demonstrated by Itaya's group using STM [78,86], Fig. 6. By utilizing SXS, Tidswell et al. [87] showed that cooling in H_2 leads to a well-ordered Pt(1 1 1)-(1 × 1) surface in solution. Kibler et al. [73] confirmed that cooling in H_2 indeed produces flat terraces with straight triangular step edges, indicative of a high surface quality. The latter authors have also found that when cooling in the presence CO + N_2 mixtures, the resulting adlayer of CO provides a protection against surface contamination and produces a defect-free (1 × 1) surface. Besides probing the ability of the flame-annealing method to produce well-ordered Pt(1 1 1)-(1 × 1)structure, SXS results unambiguously showed that in aqueous solutions this structure is stable between the hydrogen evolution and the "oxide" formation potential region [87]. The important consequence of the structural stability of the Pt(1 1 1)-(1 × 1) surface is that electrochemical surface processes can be examined on the geometrically simple surface having well-known adsorption sites.

The current–potential curve of Pt(1 1 1)-(1 × 1) gives a distinctive voltammogram with a broad nearly flat hydrogen desorption/adsorption peaks between $\approx 0.05 < E < 0.375$ V in both 0.05 M H_2SO_4 (Fig. 7a) and 0.1 M KOH (Fig. 7c), which is indicative of hydrogen adsorption which is *not accompanied* with concomitant anion adsorption. Depending on the pH of solution, this state is generated either from protons or water molecules

$$Pt + H_3O^+ + e^- \rightarrow Pt\text{—}H_{upd} + H_2O \qquad (pH \leq 7) \tag{3}$$

$$Pt + H_2O + e^- \rightarrow Pt\text{—}H_{upd} + OH^- \qquad (pH \geq 7) \tag{4}$$

For consistency with the current nomenclature used in electrochemistry, the state of hydrogen (atomic), which is adsorbed onto a metal substrate at a potential that is positive of the Nernst potential for the

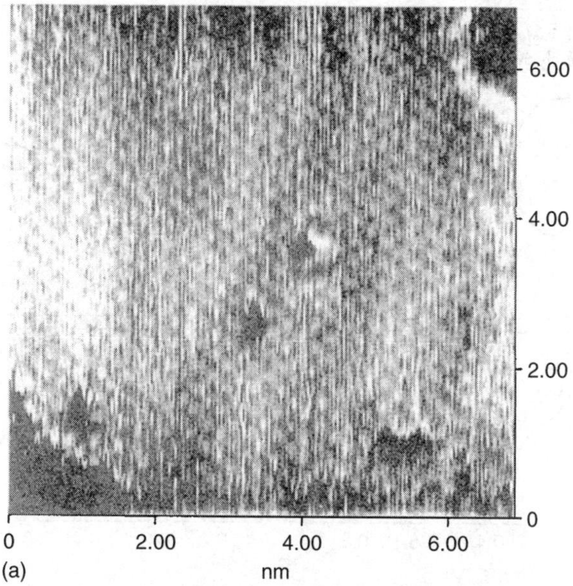

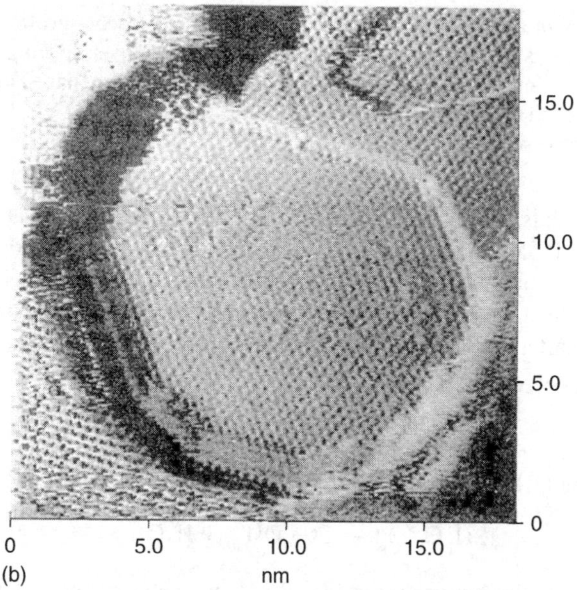

Fig. 6. Atomically resolved in situ STM of the Pt(1 1 1) substrate obtained at 0.1 V: (a) a higher resolution STM scan with some point defects; (b) a larger scan showing the long-range ordered Pt(1 1 1) substrate and well-defined step ledges [86]

hydrogen electrode reaction (HER), which has the overall stoichiometry

$$2H_3O^+ + 2e^- \rightarrow H_2 + 2H_2O \qquad (pH \leq 7) \tag{5}$$

$$2H_2O + 2e^- \rightarrow H_2 + 2OH^- \qquad (pH \geq 7) \tag{6}$$

will be referred to as underpotentially deposited hydrogen (H_{upd}).

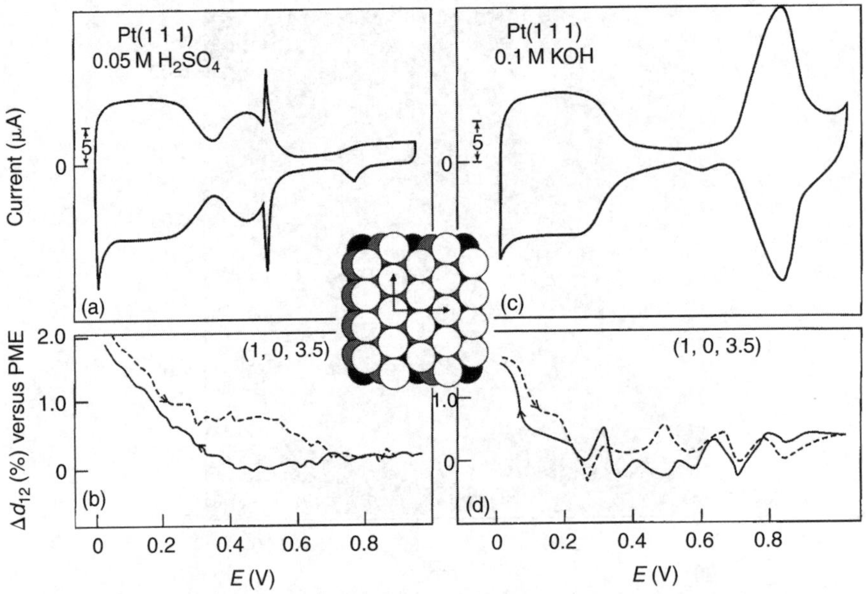

Fig. 7. Cyclic voltammetry of the Pt(1 1 1)-(1 × 1) surface in an electrochemical cell: (a) in H_2SO_4 and (c) in 0.1 M KOH. The potential was scanned at 50 mV/s. Changes in inter-layer spacing (Δd_{12}) measured from the potential of minimum expansion (PME) (e.g. the least coverage by any adsorbates) on scanning the potential at 2 mV/s (b) in H_2SO_4 and (d) in 0.1 M KOH. Inset: ideal model for the Pt(1 1)-(1 × 1) surface. Electrode potential *E* is given versus the reversible hydrogen electrode (RHE) [126]

At more positive potentials, the so-called "anomalous" peaks are observed at ≈0.4 − 0.6 V in sulfuric acid solution and ≈0.6 − 0.85 V in alkaline (Fig. 7c) and perchloric acid solutions. These reversible peaks are now understood to be anion adsorption pseudo-capacitance: in 0.05 M H_2SO_4 adsorption/desorption of (bi)sulfate anions, see Section 4.4.1; in 0.1 M KOH and 0.1 M $HClO_4$ reversible adsorption/desorption of hydroxyl species, hereafter denoted as OH_{ad}. The adsorption of OH_{ad} in alkaline solution is just OH^- adsorption with charge transfer

$$OH^- + Pt \rightarrow Pt\text{—}OH_{ad} + e^- \tag{7}$$

while in acid solution, the OH_{ad} adsorption proceeds according to

$$2H_2O + Pt \rightarrow Pt\text{—}OH_{ad} + H_3O^+ + e^- \tag{8}$$

As long as the potential remains between 0 and ca. 0.8 V, the well-ordered (1 1 1) structure remains well-ordered and (1 × 1), i.e. as long as there is only reversible adsorption of anions, H_{upd} or reversibly adsorbed OH_{ad}. When the potential is increased to above ca. 1.0 V so as to form "oxide" by place exchange between OH_{ad} and Pt atoms in the first layer [88], the surface becomes irreversibly roughened [89].

4.1.2. Pt(1 0 0)

Considerable effort has been devoted to the question of whether a reconstructed UHV prepared Pt(1 0 0) surface is stable in contact with aqueous electrolytes. A review of the stability of Pt(*h k l*) surfaces at solid–liquid interfaces by Kolb [90] is an excellent place to find references to a specific Pt(*h k l*)–electrolyte system. Briefly summarizing, the first ex situ experiments indicated that the Pt(1 0 0)-"hex" structure transforms into the (1 × 1) phase upon contact with aqueous electrolytes, or even when in contact with water vapor or liquid water [77]. On the other hand, using STM Baltruschat et al. [91] reported that when the irreversibly adsorbed iodine layer (using vapor-phase I_2 after the flame-annealing) was desorbed from the surface, the Pt(1 0 0) surface reconstructs. Zei et al. [92] also found experimental evidence that the UHV prepared "hex" structure remains stable in contact with electrolytes, but no evidence of potential induced reconstruction was ever detected. An important observation in [92] was the stability of the "hex"-phase within the H_{upd} potential region. In contrast, using a similar combination of voltammetric and UHV-based surface sensitive probes, Al-Akl et al. [93] reported that Pt(1 0 0)-"hex-rot" phase is unstable when a monolayer of H_{upd} is formed. Apart from controversy about the potential stability of the hex structure, there is consensus that an initially reconstructed surface is irreversibly transformed into (1 × 1) phase upon the adsorption [92] or desorption [93] of ions from the supporting electrolytes.

Besides the UHV preparation and characterization of Pt(1 0 0), the structure of this surface prepared by the flame-annealing method was also studied. For example, closely following the UHV methodology, Tidswell et al. [82] reported from in situ SXS measurements that a Pt(1 0 0)-(1 × 1) surface is observed in solution following preparation by flame-annealing a Pt(1 0 0) crystal with cooling in a stream of pure H_2 or H_2/argon mixtures. In recent STM measurements, Sashikata et al. [94] and Kibler et al. [73] have confirmed that hydrogen cooling of a flame-annealed Pt(1 0 0) electrode produces a (1 × 1) phase, (Fig. 8). Kibler et al. [73] examined the influence of a CO-cooling atmosphere on the final morphology of the flame-annealed Pt(1 0 0) surface. In accord with CO-induced deconstruction of the "hex" surface in UHV, the STM results revealed that cooling in CO led to a flat (1 × 1) surface with large terraces of islands or holes. In situ SXS results also showed that once formed the Pt(1 0 0)-(1 × 1) surface remains unreconstructed throughout the hydrogen evolution and "oxide" formation potential regions [82]. Unlike the case of Au(1 0 0) [80,95], there is no evidence for a potential-induced reconstruction of the Pt(1 0 0) surface from (1 × 1) to "hex". The weight of evidence suggests that electrode processes on Pt(1 0 0) electrodes prepared using most flame-annealing methods should be considered to take place on a Pt(1 0 0)-(1 × 1) surface.

The main characteristic of well-ordered Pt(1 0 0) voltammetry in sulfuric acid, Fig. 9a, is that two well-delineated peaks at 0.4 and 0.25 V corresponds to the *coupled processes* of hydrogen adsorption and bisulfate anion desorption on (1 0 0) terrace sites at 0.4 and 0.25 V at the step sites produced by lifting the reconstruction. The relative amounts of charge under the two peaks is extremely sensitive to the quality of the Pt crystal being used, reflecting the variations in step densities that can be experienced. The main characteristic of well-ordered Pt(1 0 0) in alkaline solution, Fig. 9c, is that the hydrogen adsorption/desorption and hydroxyl anion adsorption/desorption are not completely decoupled processes, and the desorption of hydrogen $0.05 < E 0.25$ V is immediately followed by the reversible adsorption of hydroxyl anions, even in the H_{upd} potential region. In the potential range between $0.75 < E < 0.95$ V, however, the observed asymmetry in a current–voltage curve traces between positive and negative sweep directions clearly represents an irreversible state of adsorption of OH_{ad} species, which we shall call "oxide", hereafter.

4.1.3. Pt(1 1 0)

As was already mentioned in Section 2, clean Pt(1 1 0) undergoes reconstruction in UHV. In the early ex situ LEED analyses of UHV prepared Pt(1 1 0)-(1 × 2) electrode, it was found that upon contact with several solutions the reconstructed (1 × 2) surface remained stable if potential cycling was restricted to certain potential regions, with the window of stability differing in different reports [76,96]. A voltammetric study concerning the nature of the Pt(1 1 0) surface structure, prepared by flame-annealing, followed by quenching in water, has been reported by Armand and Clavilier [97]. They suggested that, as for clean annealed crystals in UHV, the Pt(1 1 0) surface has predominantly a (1 × 2) structure. On the basis of infrared spectra, and using the same preparation method, Kinomoto et al. [98] proposed that the adsorption of CO on clean reconstructed Pt(1 1 0)-(1 × 2) induces a (1 × 2) ↔ (1 × 1) surface phase transition.

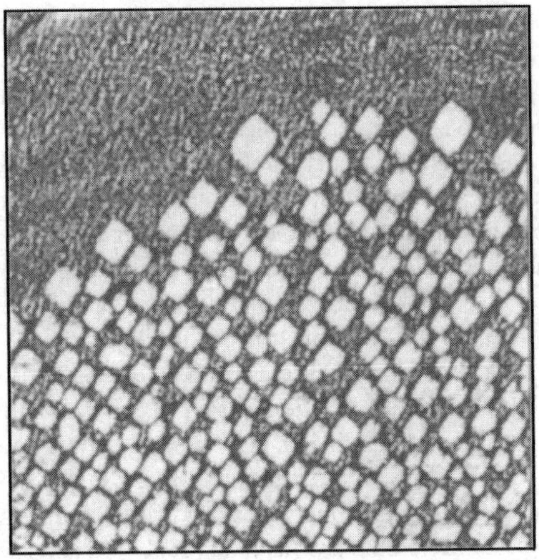

(a)

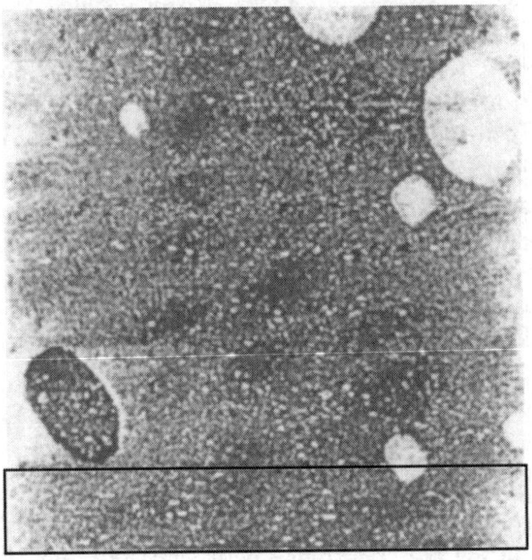

(b)

Fig. 8. (a) STM image (225 nm × 225 nm) of Pt(1 0 0), prepared by flame-annealing and cooling in H_2 + N_2 mixtures, obtained at 0.55 V (versus. the standard calomel electrode, SCE) in 0.1 M H_2SO_4 with Pt/Ir tip. I_T = 2 nA; U_T = 0.10 V (tip negative). (b) STM image (730 nm × 730 nm) in 0.1 M H_2SO_4 of a Pt(1 0 0) electrode, prepared by flame-annealing and cooling under a strong H_2 stream, W-tip; I_T = 10 nA; U_T = 0.1 V (tip positive) [73]

By utilizing the Clavilier method, voltammetric results for the adsorption of CO on Pt(1 1 0), reported by Morallon et al. [99], suggested that the irreversible adsorption of CO might produce some surface modifications of Pt(1 1 0), but these modifications were different from the (1 × 2) ↔ (1 × 1) surface phase transition observed in vacuum. The preparation method of Wieckowski and co-workers [74] involving flame-annealing followed by cooling in an inert atmosphere over iodine and subsequent replacement of iodine by CO, seemed to produce the Pt(1 1 0)-(1 × 1) structure. A recent study using STM of the Pt(1 1 0) surface prepared by this technique confirmed that, after desorption of iodine at negative potentials, the

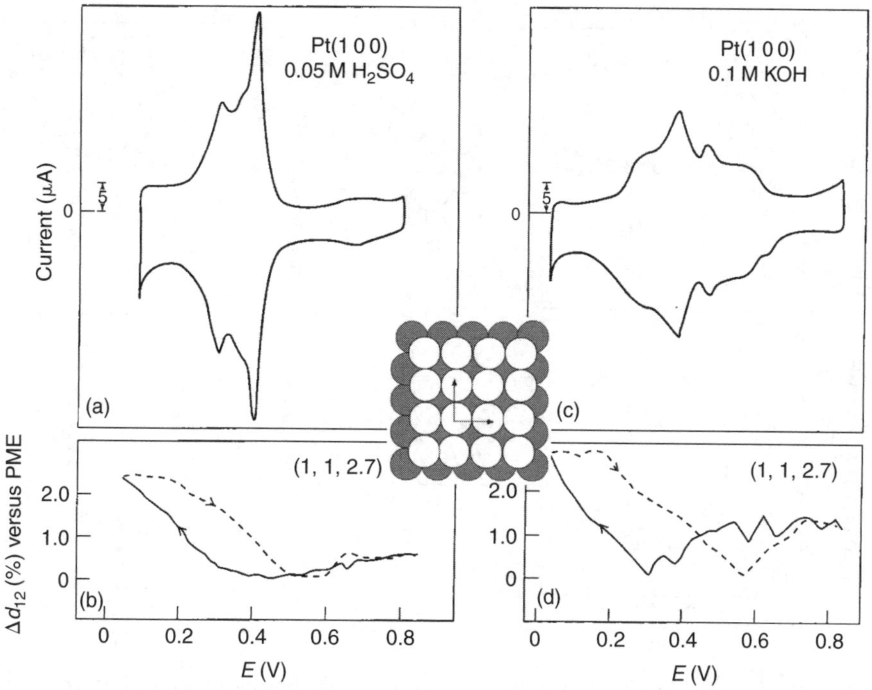

Fig. 9. Cyclic voltammetry of the Pt(1 0 0)-(1 × 1) surface in an electrochemical cell: (a) in H_2SO_4 and (c) in 0.1 M KOH. The potential was scanned at 50 mV/s. Changes in inter-layer spacing (Δd_{12}) measured on scanning the potential at 2 mV/s (b) in H_2SO_4 and (d) in 0.1 M KOH. Inset: ideal model for the Pt(1 0 0)-(1 × 1) surface. *E* versus RHE [126]

Pt(1 1 0) surface does indeed have the (1 × 1) geometry [100]. Most recently, the stability and structure of the Pt(1 1 0) surface in several electrolytes was examined by in situ SXS. The experimental details of both methods of preparation and the electrocatalytic properties of these two surfaces have been discussed at length in recent publications [101,102]. It was found that depending on the particular flame-annealing procedure, a reconstructed Pt(1 1 0)-(1 × 2) or an unreconstructed Pt(1 1 0)-(1 × 1) surface can be created. The inset of Fig. 10 shows a rocking scan (approximately along [*h* 0 0]) through the (0, 1.5, 0.1) reciprocal lattice position. The solid line is a fitted Lorentzian lineshape to the data, which allows the correlation length of the (1 × 2) unit cell to be derived. From these measurements, and longitudinal scans along [0 *k* 0] (not shown), a correlation length of ≈350 Å was derived along the [1 0 0] direction and ≈250 Å along the [0 1 0] direction, where these correspond to the parallel and perpendicular directions to the rows in the missing row model of the (1 × 2) reconstruction, see Fig. 1c. The (1 × 2) surface reconstruction was stable over a wide potential range, e.g. even at 1.2 V there was still a strong (1 1 0)-(1 × 2) diffraction pattern. Upon sweeping the potential negatively, however, the (1 × 2) reconstruction is finally lifted as the surface atoms move to accommodate the oxide reduction. It should also be noted that the Pt(1 1 0)-(1 × 2) structure was so stable in aqueous electrolytes that adsorption of CO on this surface did not induce the (1 × 2) ↔ (1 × 1) transition that is observed in UHV upon adsorption of CO [101,102].

Figure 10 contains representative cyclic voltammograms for the Pt(1 1 0)-(1 × 2) surface recorded in both alkaline and acid solution. In 0.05 M H_2SO_4, the voltammetric features are the coupled processes of hydrogen adsorption/desorption with the anion deposition/adsorption at 0.05–0.35 V, followed by a double-layer charging potential region, and the hysteretic peaks associated with oxide formation and reduction at more positive potentials. As with the (1 0 0) surface, a distinguishing characteristic of Pt(1 1 0) in 0.1 M KOH is

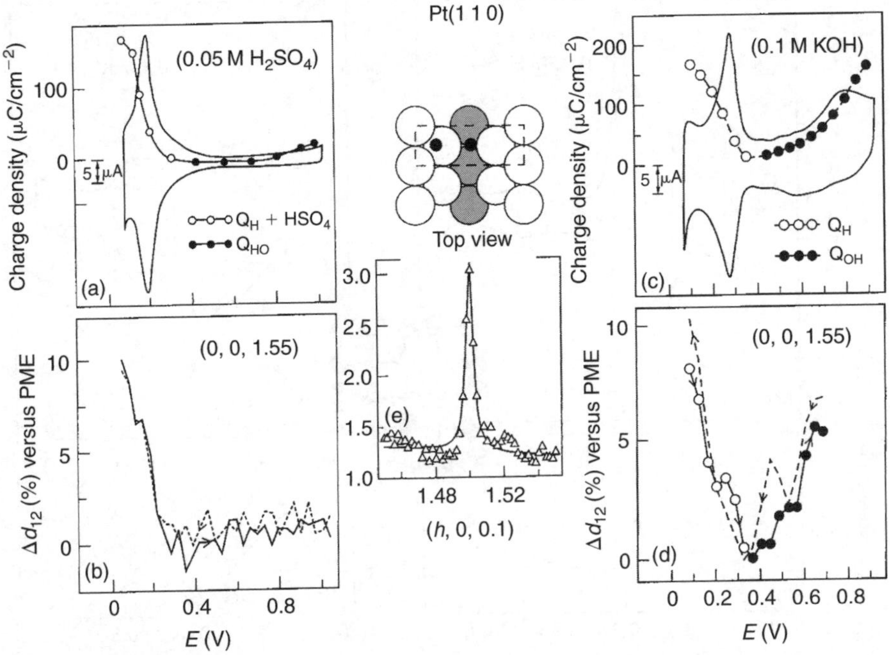

Fig. 10. Cyclic voltammetry of the Pt(1 1 0)-(1 × 2) surface in electrochemical cell: (a) in H_2SO_4 and (c) in 0.1 M KOH. The potential was scanned at 50 mV/s. Changes in inter-layer spacing (Δd_{12}) measured on scanning the potential at 2 mV/s (b) in H_2SO_4 and (d) in 0.1 KOH. (e) The measured X-ray intensity at (0, 1.5, 0.1) along the [0 1 0] direction along with an ideal model for the (1 × 2) structure: solid dots represent H_{upd} and OH_{rv}. *E* versus RHE [126]

that on the positive going sweep the desorption of hydrogen, $0.05 < E < 0.375$ V is associated with simultaneous adsorption of OH_{ad} at the step sites. The fact that OH_{ad} can be adsorbed even in the H_{upd} potential region is a new observation which, as will be discussed in Section 5, has a considerable effect on the kinetics of the hydrogen oxidation and CO oxidation reactions. Fig. 10c shows that the "H_{upd}" potential region is followed first by the additional reversible adsorption of OH, and then by the irreversible formation of "oxide".

4.2. Energetics of Pt(1 1 1)—H_{upd} and Pt(1 1 1)—OH_{ad} systems

As with adsorption systems studied in UHV, the energetics of adsorption on Pt surfaces in electrolyte is derived from measurements of the equilibrium coverage by adsorbates as a function of temperature. The coverages are usually obtained by measuring the charge transferred through the interface as a function of potential at different temperatures. For recent overviews see [103,104]. Unfortunately, because of close coupling of specific anion adsorption with the hydrogen adsorption on Pt(1 0 0) and Pt(1 1 0), an unambiguous measurement of the energetics of hydrogen on these two surfaces cannot be obtained by the same thermodynamic analysis. Furthermore, due to the narrow temperature range available in aqueous solution, the thermodynamic functions for irreversibly adsorbed CO also cannot be determined at solid–electrolyte interfaces, as we discuss in Section 5.3.

4.2.1. Pt(1 1 1)—H_{upd} system

Thermodynamic state functions on Pt(1 1 1) can be obtained in an unambiguous way from the experimentally measured temperature dependence of H_{upd} adsorption isotherms (see inset of Fig. 11) when the

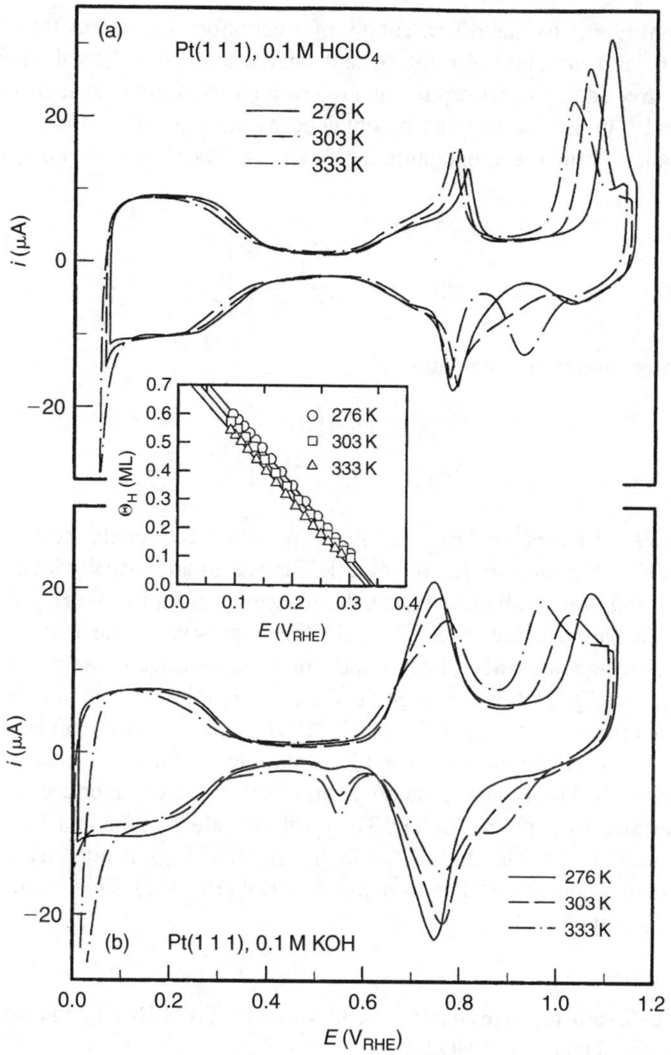

Fig. 11. Cyclic voltammograms of Pt(1 1 1) in (a) 0.1 M $HClO_4$ and (b) 0.1 M NaOH at various temperatures. Sweep rate 50 mV/s. Inset: potential dependent surface coverage by H_{upd} on Pt(1 1 1) in 0.1 M $HClO_4$ at 276, 303, and 333 K [114]

electrode potential is referenced to the potential of the hydrogen electrode, E_{RHE}, in the same electrolyte at the same temperature [103,105]. This methodology was adopted from original works by Breiter [106,107] and Conway et al. [108–110] for the measurements of the thermodynamic state functions on polycrystalline platinum electrode. For the Pt(1 1 1)–H_{upd} system, the apparent Gibbs free energy of adsorption as a function of coverage, $\Delta G_{H_{upd}}(\Theta)$, is well represented by the general (non-ideal) form of the Langmuir isotherm [103,109,111–113], where the Gibbs energy of adsorption is assumed to vary linearly with coverage,

$$\left(\frac{\Theta}{1-\Theta}\right)\exp\left(\frac{r\Theta}{RT}\right)=\exp\left(\frac{-E_{RHE}F}{RT}\right)\exp\left(\frac{-\Delta G^{0}_{H_{upd}}}{RT}\right) \quad (9)$$

and $\Delta G^0_{H_{upd}}$ is the initial (zero coverage) free energy of adsorption. The interaction parameter, $f = r/RT$, is repulsive for $f > 0$, and attractive for $f < 0$, and characterizes the lateral interaction in the H_{upd} adlayer. The apparent free energy of adsorption at any given temperature is then characterized by just two parameters, $\Delta G^0_{H_{upd}}$ and f, both of which can be obtained by simple curve fitting. The isosteric heats of adsorption can be obtained from the temperature dependence of the Gibbs free energy of adsorption from the relation

$$q^{H}_{st} = -\frac{\partial\left(\Delta G_{H_{upd}}/T\right)_{\theta}}{\partial T^{-1}} \tag{10}$$

and the entropy of adsorption from the relation

$$\Delta S_{H_{upd}} = -\frac{\partial\left(\Delta G_{H_{upd}}\right)_{\Theta}}{\partial T^{-1}} \tag{11}$$

The resulting values ($\Theta = 0$) of $q^{H}_{st} - \Delta H_{H_{upd}}$, $\Delta S_{H_{upd}}$, $\Delta G^0_{H_{upd}}$ and f, and finally the $E_{M\text{—}H}$ bond energies for three different electrolytes are summarized in Table 1. The values are virtually identical *independently of the anion or the* pH *of solution*. In all electrolytes, the energetics of Pt(1 1 1)—H_{upd} follows the UHV-type of linear decrease with the coverage (Fig. 12), reflecting repulsive lateral interaction in the hydrogen adlayer. Although the change of isosteric heat of adsorption of hydrogen does not show a sharp decrease as the coverage increases (Fig. 2), this coverage-dependent lateral interaction may be one of factors causing that the saturation coverage of H_{upd} to be much <1 H_{upd} per Pt, only 0.66 H_{upd} per Pt, at chemical potentials equivalent to an H_2 partial pressure of $\leq$1 atm. A fundamental measure of interest for the Pt(1 1 1)—H_{upd} system is the Pt—H_{ad} bond energy at the electrochemical interface, which may be obtained from the heat of adsorption by applying Eq. (1). The resulting values of the Pt(1 1 1)—H_{upd} bond energies in alkaline and acid solutions are 240–250 kJ/mol, indicating that $E_{M\text{-}H}$ is not only independent of the pH of solution, but also close to the Pt—H_{ad} bond energy in UHV [103,105]. This result is consistent with the weak interaction of water with Pt[1 1 1].

Table 1. Thermodynamic state functions for H_{upd} in 0.1 M NaOH, 0.1 M $HClO_4$, and 0.05 M H_2SO_4, evaluated using Eq. (3) (see also Figs 3 and 4); r = 40 kJ/mol

T (K)	$\Delta G^{\Theta=0}_{H_{upd}}$ (kJ/mol)	$f = r/RT$	$\Delta H_{H_{upd}}$ (kJ/mol)	$\Delta S_{H_{upd}}$ (kJ/mol)	$E_{Pt\text{—}H}$ (kJ/mol)
0.1 M NaOH					
276	−24	16			
303	−22	15	−41	≈−63	≈240
333	−20	13			
0.1 M $HClO_4$					
276	−28	13			
303	−27	12	−42	≈−45	≈240
333	−25	10			
0.05 M H_2SO_4					
276	−29	14			
303	−28	13	−42	≈−48	≈240
333	−27	11			

4.2.2. Pt(1 1 1)—OH_{ad} system

The thermodynamic analysis of the temperature dependence of the current–potential curves for the OH_{ad} adsorption process is not as straightforward as for the H adsorption process. The interested reader should consult the original reference for details [114]. For coverages to about 0.5 OH/Pt, the isosteric heat of adsorption for OH_{ad} on Pt(1 1 1) is ca. 200 kJ/mol and is *independent* of pH of solution, as it was the case for the isosteric heat of adsorption for H_{upd} [114]. From the measured isosteric heat of adsorption of

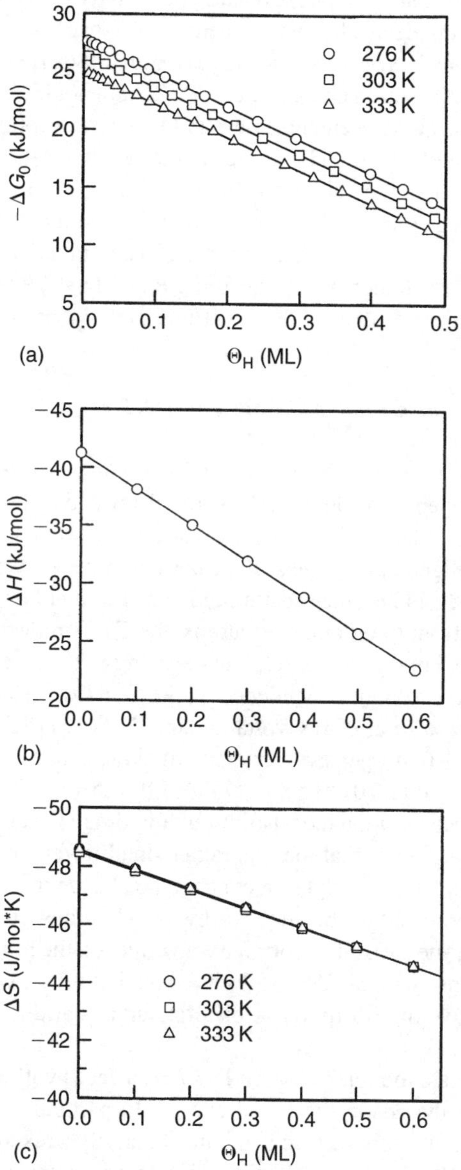

Fig. 12. Change of thermodynamic functions for H_{upd} on Pt(111) in 0.1 M $HClO_4$ with surface coverage of H_{upd} (a) apparent free energy of adsorption with (b) apparent enthalpy of adsorption, and (c) apparent entropy of adsorption [114]

OH_{ad} in alkaline solution (ca. ≈ 200 kJ/mol) and the tabulated enthalpy of formation of OH radical (ca. ≈ +39 kJ/mol), the Pt(1 1 1)—OH_{ad} bond energy was estimated to be ≈136 kJ/mol [114], which is much less than the Pt—O_{ad} bond energy at a gas–solid interface (≈350 kJ/mol) [115,116]. Wagner and Ross [89] studied the thermal decomposition in UHV of anodically formed $Pt(OH)_4$, and found that it evolved water ($2OH_{ad} \rightarrow H_2O + O_{ad}$) and molecular oxygen in two distinct steps, the latter at a temperature about 300 K lower than for O_2. This result is at least qualitatively consistent with the difference in bond energy between Pt—OH_{ad} and Pt—O_{ad}. As we discuss in Section 5.3, a relatively weak interaction of OH_{ad} with the platinum surface supports a high catalytic activity of the OH_{ad} state in the surface electrochemistry of CO_{ad} on platinum electrodes. Very recently, DFT has been used to study the adsorption of hydroxyl on the Pt(1 1 1) surface [128]. At low coverages (from 1/9 to 1/3 ML) OH binds preferentially at bridge and top sites with a chemisorption energy of ca. 225 kJ/mol. At high coverages (1/2–1 ML) the formation of an OH network causes a 15% enhancement in OH chemisorption energy, and a strong preference for OH adsorption at top sites. It would be interesting to see that if the DFT calculations included the effect of solvating water molecules the calculated Pt—OH_{ad} bond strength would approach the experimental value determined in electrolyte. Finally, we mention that although there is no true thermodynamic information about the Pt—OH_{ad} bond strength at the Pt(1 1 0) and Pt(1 0 0) interfaces, the voltammetry in KOH suggests that the bond strength increases in the order Pt(1 1 1) < Pt(1 0 0) < Pt(1 1 0). The latter two surfaces appear to have bond energies that are 50–70 kJ/mol higher than on Pt(1 1 1).

4.3. Relaxation of Pt(h k l) surfaces induced by H_{upd} and OH_{ad}

The concept of structural relaxation of surface atoms in UHV can be extended to the Pt(*h k l*)–electrolyte interface. The effects of hydrogen adsorption and hydroxyl formation on the Pt(1 1 1) surface structure were monitored by in situ SXS [82,87,95]. Surface relaxation was probed by measuring the CTRs, which are rods of scattering and are aligned along the surface normal in the Pt reciprocal space and pass through bulk Pt Bragg reflections [79–81,117]. Since scattering from H_{upd} and OH_{ad} makes a negligible contribution relative to the diffraction from the topmost Pt atoms, the CTRs reflect adsorbate-induced changes in the Pt surface normal structure, i.e. the Δd_{12} interplanar spacing in the insets of Fig. 7, Fig. 9 and Fig. 10. This method of measuring the potential dependence of the Pt(*h k l*) surface relaxation in aqueous electrolytes will hereafter be referred to as X-ray voltammetry (XRV) [118,119]. Fig. 7, Fig. 9 and Fig. 10 clearly revealed that close to the hydrogen evolution potential the expansion of surface atoms increases in the sequence Pt(1 1 1) (1.5%) < Pt(1 0 0) (2.5%) ≪ Pt(1 1 0) (25%) in both H_2SO_4 and KOH solutions. The change in the interlayer spacing upon the adsorption of hydrogen is again very similar to the situation in UHV. This is a further confirmation that there is rather small difference in the energetics between the Pt—H bond in the gas-phase and in electrolyte. In contrast, no change in the interlayer spacing is observed upon the adsorption/desorption of either bisulfate or hydroxyl anions on the Pt(1 1 1) and Pt(1 0 0) surfaces. On the other hand, while the (bi)sulfate ions have no effect on the relaxation of Pt(1 1 0)-(1 × 2) surface atoms, an additional expansion of ca. 25% is observed on the surface covered by a monolayer of OH_{ad} species. This could be indicative of adsorption of bisulfate anions at different specific surface sites than either H_{upd} or OH_{ad}.

It is tempting to try to correlate the relaxation on Pt(*h k l*) surfaces with other independent observations of the adsorption sites and/or the adsorption bond energy. In particular, the smallest expansion on the Pt(1 1 1) surface may arise because this surface is the most densely packed, and thus the H_{upd} is adsorbed *onto* the Pt(1 1 1) surface. To describe the interaction of hydrogen with Pt(1 1 1) surface, Olsen et al. [21] have performed extensive computational study. Density functional calculations favored the on-top site, although the differences in adsorption energies are very small. On the (1 0 0) surface, however, some of H_{upd} adatoms might sit in deeper wells, e.g. the four-fold hollow sites, and thus be more in the surface than on the surface. In the case of Pt(1 1 0)-(1 × 2), the large relaxation possibly indicates that the sites for H_{upd} are the three-fold coordinated sites *below* the topmost (1 1 0) rows of Pt atoms [102], as proposed for the adsorption of hydrogen on Pt(1 1 0) in UHV, sites C in Fig. 1.

Recent studies using vibrational spectroscopy provide a mixed picture about the adsorption sites of the H_{upd} state. On the basis of IR reflectivity spectra, Nichols reported that hydrogen adsorption on Pt(111) (Fig. 11) is quite different from Pt(1 0 0) [120], suggesting different sites for H_{upd} on the two surfaces. Conventional infrared spectra by Ogasawara and Ito [121] found Pt—H bands indicative of on-top adsorption sites on both Pt(1 0 0) and Pt(1 1 0) surfaces, but only at potentials very close to the Nernst potential. It is not clear whether Ogasawara and Ito studied the (1 1 0)-(1 × 2) structure or the (1 × 1) structure, which can also result from the flame-annealing procedure. The results of Ogasawara and Ito were similar to an earlier report by Bewick and Russell [122] for polycrystalline Pt. Using visible–infrared sum frequency generation (SFG), Tadjeddine and co-workers [123–125] reported that the frequencies of the bands in the 0–0.4 V region corresponded only to terminally bonded hydrogen on all three low-index Pt surfaces. The weight of observations favors only terminally bonded hydrogen, considering that the spectra from conventional IR reflection experiments have relatively poor signal-to-noise and are not compelling.

Detailed knowledge about the adsorption/formation of hydroxyl layer on the platinum surface is in a similarly mixed circumstance. The simplest case to understand, in principle, is the reversible adsorption of the OH^- anion in alkaline solution, which presumably is just adsorption with charge transfer (Eq. (7)), like H_{upd} or halide ion adsorption, but this has not been proven. Unfortunately there are no available data on the binding geometry of OH and O at the Pt(hkl)–solution interfaces. On the basis of the fact that the relaxation of the Pt (1 1 0)-(1 × 2) surface is the same in both the H_{upd} and OH_{ad} potential regions it was proposed that the adsorption sites for the reversible adsorption of OH_{ad} and H_{upd} are the same [102], e.g. the three-fold hollow sites in the troughs (site C in Fig. 1c). Nevertheless, regardless of the true binding sites, it is now well-documented that the transition from reversible to irreversible adsorption of oxygenated species on this surface also produces a surface roughening, presumably by the place-exchange mechanism. It is important to note that a transition from the reversible to irreversible adsorption occurs at a potential at which almost all three-fold hollow sites are occupied by OH_{ad}. Therefore, "oxide" starts to form on the Pt(1 1 0)-(1 × 2) surface by insertion of OH_{ad} under the top row of Pt atoms. As for Pt(1 1 0), there is very little direct evidence for adsorption geometry of OH_{ad} on either the (1 0 0) or (1 1 1) surfaces. On Pt(1 0 0), in the OH_{ad} potential region the relaxation of platinum surface atoms is smaller than in the H_{upd} region, and on Pt(1 1 1) there is no relaxation at all. On the basis of this fact, we suggested that reversibly adsorbed hydroxyl is on these surfaces rather than in or below the surface (see [126] for an overview). Using UHV STM and high-resolution electron energy loss spectroscopy (HREELS), Bedurftig et al. [127] proposed that on Pt(1 1 1) the hydroxyl molecules occupy on-top sites. However, the interaction of OH with Pt(1 1 1) was also studied by DFT using Pt clusters [128] and found that OH adsorbates prefer high coordination three-fold sites. Clearly, the geometry of OH adsorption on Pt(hkl) surfaces is not yet fully understood.

4.4. Surface structures and energetics of anion adsorption on Pt(h k l) surfaces

The adsorption of anions on metal electrodes has been one of the major topics in surface electrochemistry. Anion adsorption has an important, and generally adverse, effect on the kinetics of fuel cell reactions and thus needs to be understood in some detail. Most of the progress has come recently from conventional electrochemical methods in combination with ex situ and in situ surface sensitive probes. Of particular interest are chemisorbed (also called contact adsorbed or specifically adsorbed) anions, whose adsorption is controlled by both electronic and chemical forces. The major objectives of these studies were to establish: (i) the potential dependence of surface coverage by anions, sometimes erroneously called the "adsorption isotherms"; (ii) the energetics of adsorption; and (iii) the structure of the adsorbed anions. Representative results for the specific adsorption of bisulfate and halide anions on Pt(*h k l*) are summarized in this section.

4.4.1. (Bi)sulfate adsorption

In 1990, Yeager's group [129] provided the first vibrational spectroscopy for sulfuric acid anion adsorption on Pt(1 1 1) in 0.05 M H_2SO_4. In the potential region where there is a corresponding large adsorption pseudo-capacitance, a single absorption band was observed ca. 1200 cm^{-1} whose intensity changes with potential is in correspondence with the integrated charge. This band was assigned to SO_3 asymmetric stretching mode of bisulfate anions which are adsorbed on the Pt(1 1 1) sites via the three unprotonated oxygen atoms, thus *strongly* interacting with the (1 1 1) symmetry of the substrate atoms. The potential dependence of simulated bisulfate adsorption integrated band intensity is shown in Fig. 13. Subsequent studies by other groups have disputed the interpretation of the 1200 cm^{-1} band as being the asymmetric stretching mode of bisulfate. Nichols and Bewick [120], and Ito and co-workers [130] pointed out that the intense 1200 cm^{-1} band can arise from the symmetric SO_3 stretch of bisulfate anions. Nart et al. [131], however, argued that the infrared data can be interpreted as sulfate adsorption via three oxygens presenting a C_{3v} symmetry. To reconcile these two extremes, Faguy et al. [132] suggested that the adsorbate in the anomalous potential region is H_3O^+–SO_4^{2-} species. We consider the issue to be unresolved, and thus refer to the adsorbing species equivocally as (bi)sulfate.

The IRRAS experiments provide only a qualitative measure of the surface coverage of (bi)sulfate (e.g. $\Theta_{(H)SO_4}$) as a function of the electrode potential. Quantitative evaluation of $\Theta_{(H)SO_4}$ versus E relationship came first from radiotracer measurements [133] and later from chronocoulometric measurements of the Gibbs excess of sulfate and bisulfate anions [134]. Both measurements are summarized in Fig. 14. Interestingly, the latter measurements confirm that (bi)sulfate adsorption is indeed two stage process [134], as suggested by us in earlier studies with other methods [135]. At the first plateau in the Γ versus E curves, where Γ is the surface excess, the coverage is ca. 0.33 ML (one anion for three Pt atoms). The second plateau corresponds to a total coverage of 0.4 ML, which was significantly higher from the coverage found by radiotracer method [133] (inset of Fig. 14). The surface concentration at the first plateau correlates very well with the $(\sqrt{3} \times \sqrt{3})$ $R30°$ surface structure in ex situ LEED measurements by Wieckowski and co-workers [136] and Ito and co-workers [137]. Very recently, the potential dependence of the formation of $(\sqrt{3} \times \sqrt{3})$ $R30°$-sulfate overlayer was modeled by analytical Monte Carlo techniques [138]. This model was able to reproduce the so-called complex "butterfly" pattern in the voltammetry, with the sharp peak at ca. 0.6 V corresponding to an order–disorder transition. However, Stimming and his co-workers found by in situ STM that the adsorbed ions (at $\Theta = 0.22$ ML) form an ordered $(\sqrt{3} \times \sqrt{7})$ symmetry in the second stage of adsorption between 0.5 and 0.7 V [139]. This adlattice was interpreted in terms of the co-adsorption of sulfate anions and water. A modified radioactive labeling method used by Kolics and Wieckowski [140] demonstrated that the highest surface concentration of anion is 0.21 ML, which is in agreement with the observed STM structure. The inconsistency between ex situ versus in situ structures was rationalized by Shingaya et al. [137]. They proposed that the loss of water under the UHV conditions causes the $(\sqrt{3} \times \sqrt{7})$ structure to transform to $(\sqrt{3} \times \sqrt{3})$ $R30°$. If this were so the $(\sqrt{3} \times \sqrt{3})$ $R30°$ domains would have to exist in patches to be consistent with the fixed coverage. It is not clear that this consequence is consistent with the LEED pattern observed. Finally, it should be noted that none of the recent results have resolved the issue of bisulfate versus sulfate as the adsorbed species.

The adsorption of (bi)sulfate anions on Pt(1 0 0) and Pt(1 1 0) is much less studied than on Pt(1 1 1). It is worth noting, however, that there are two major differences between bisulfate adsorption on Pt(1 1 1) and

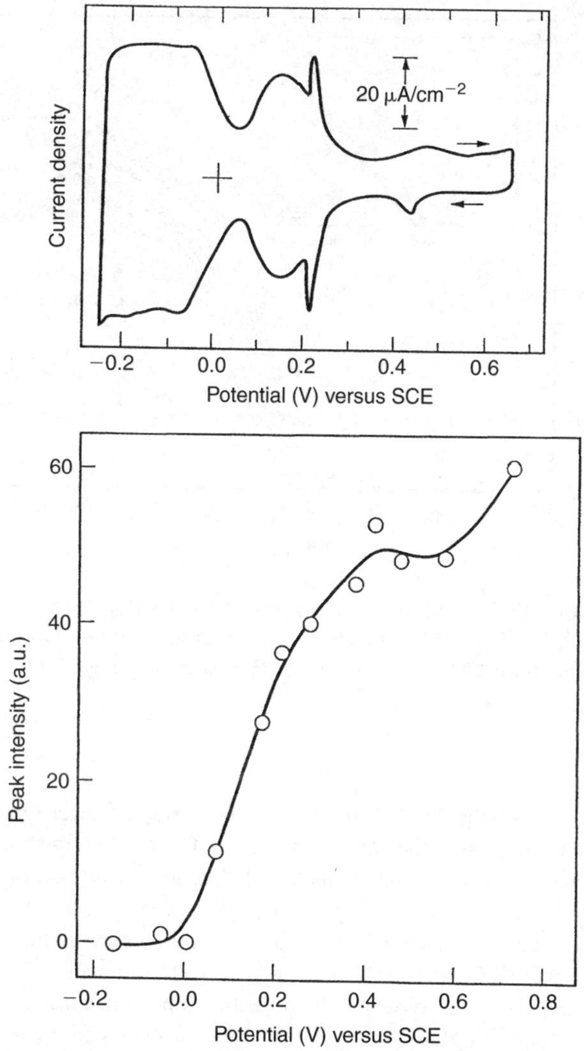

Fig. 13. Upper panel: cyclic voltammogram for flame-annealed Pt(1 1 1) single crystal in 0.05 M H_2SO_4. Lower panel: potential dependence of corrected bisulfate adsorption IR integrated band intensity [129]

the two other low-index single-crystal platinum surfaces. First, due to the lower work function and thus lower potential of zero charge (PZC) of Pt(1 0 0) and Pt(1 1 0), anion adsorption on these surfaces is shifted negatively in potential, and is always concurrent with H_{upd} desorption. Secondly, the orientation for adsorbed bisulfate anions on Pt(1 1 0) and Pt(1 0 0) is a markedly different as compared with the (1 1 1) face [120,141]. Using the symmetry arguments and the surface selection rule, the spectra for the (1 1 0) and (1 0 0) faces have been rationalized in terms of (bi)sulfate bonding to these surfaces through either one or two oxygen atoms. These results in a reduced adsorption strength relative to Pt(1 1 1), and consequently a (bi)sulfate anion interacts more weakly with (1 0 0) and (1 1 0) surfaces than with (1 1 1). No ordered structures of (bi)sulfate anions nor quantitative measurements of coverage have been reported for the (1 0 0) or (1 1 0) surfaces.

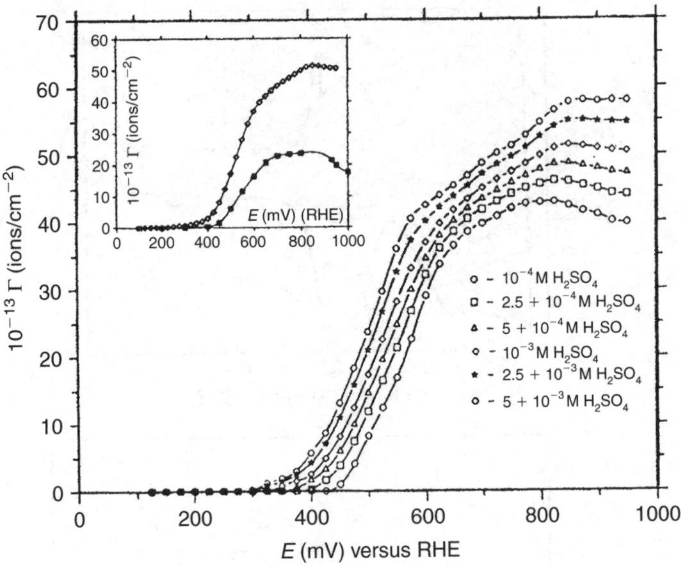

Fig. 14. The Gibbs excess of sulfate/bisulfate adsorbed from 0.1 M $HClO_4$ + *x* M $HClO_4$ solutions with *x* equal to (○) 10^{-4}, (□) 2.5×10^{-4}, (▽) 5×10^{-4} and (★) 2.5×10^{-3}. Inset: comparison of the chronocoulonometric data with radiochemical data for 0.1 M $HClO_4$ + 10^{-3} M H_2SO_4. *E* versus RHE [134]

4.4.2. Halide adsorption

The interaction of halogen ions with Pt(*h k l*) surfaces has been studied extensively at the solid–electrolyte interface, perhaps even more so than at the vacuum interface. The modern in situ structural probes of STM and SXS have played a significant role in understanding halide ion adsorption on Pt and other noble metal surfaces. It is now well-established that the adsorption strength increases in the order $F_{ad} > Cl_{ad} > Br_{ad} > I_{ad}$ for all three low-index surfaces. What we will emphasize in this review is the importance of the surface geometry in the Pt–halide energetics, and thus in the ordering of the adlayers.

(a) *Pt(*h k l*)—chloride system.* In early work, the potential dependence of chloride adsorption (denoted hereafter as Cl_{ad}) on platinum single-crystal surfaces was investigated by the combination of cyclic voltammetry and ex situ UHV surface analytical tools such as AES and LEED [135]. On Pt(1 1 1) and Pt(1 0 0), Cl_{ad} was clearly seen in the AES spectra of the emerged electrodes from electrolytes containing Cl^- anions. Adsorption of Cl^- was found to proceed in two stages, one coupled to H_{upd} desorption, and a second stage of adsorption at more positive potential, which is probably concurrent with OH_{ad} formation. The AES "isotherms" clearly showed that the onset of adsorption starts at a more negative potential on Pt(1 0 0), and that Cl_{ad} coverage increases with increasing potential more dramatically on the (1 0 0) surface than on Pt(1 1 1). These two observations were rationalized by both the more negative PZC of the Pt(1 0 0) surface, and the stronger chemical interaction of Cl_{ad} with (1 0 0) surface geometry then with (1 1 1) sites [135]. Ex situ AES still left one uncertain as to the true coverages at the surface in electrolyte, due to the potentially drastic disturbance of the adlayer by emersion and dehydration in UHV.

Recently, much more reliable measures of Θ_{Cl} versus. *E* for the Pt(1 1 1)–solution interface were obtained by in situ SXS [142]. Although no ordered in-plane structures were found over the entire potential region where completely discharged Cl_{ad} (the electrosorption valency, $\gamma = 1$) is present on the surface, information about the Pt—Cl_{ad} bond lengths and chloride coverages were obtained from analysis of CTR intensities. At 0.7 V, $\Theta_{Cl} = 0.6$ ML (1 ML = 1 Cl_{ad}/Pt) and Pt—Cl ion-core separation ($\Delta_{Pt—Cl}$) of 2.4 Å. These parameters are consistent with the presence of covalently bonded hexagonal closed-packed Cl_{ad}

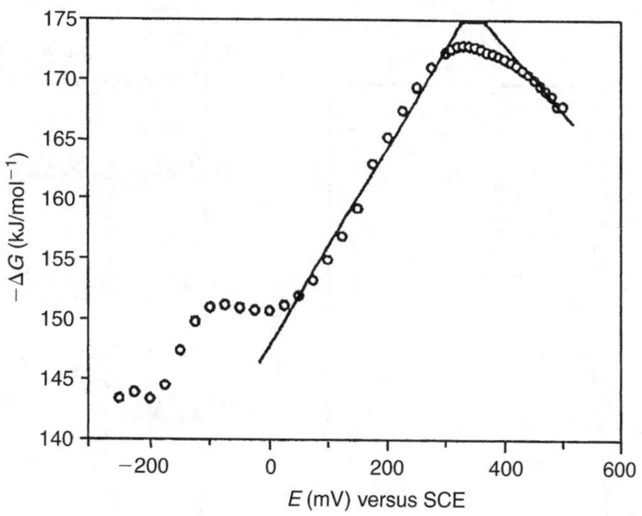

Fig. 15. Plot of the Gibbs energy of chloride adsorption at Pt(1 1 1) versus electrode potential. The straight lines were drawn to determine the average slopes of the Gibbs energy plots for potential region $0 < E < 0.3$ and $E > 0.4$ V, respectively. *E* versus SCE [143]

adlayer with an interatomic Cl_{ad}—Cl_{ad} distance of 3.58 Å. At 0.25 V, the coverage decrease to 0.4 ML, and $\Delta_{Pt—Cl}$ increase slightly to 2.5 Å. The Cl_{ad}—Cl_{ad} distance increases to 4.39 Å.

Using the purely electrochemical Gibbs excess analytical method, Li and Lipkowski [143] reported the potential dependence of surface coverage by Cl_{ad} on Pt(1 1 1) and the Pt—Cl_{ad} bond energy. Fig. 15 shows that the $-\Delta G$ versus E plots have a number of inflection points, indicating that the character of Cl_{ad} adsorption changes with the electrode potential. The maximum surface concentration of Cl_{ad} amounts to about 6.5×10^{14} ions/cm^2, which corresponds to 0.43 ML. The Gibbs free energies of adsorption were also determined in this study, and are plotted against the electrode potential in Fig. 15. The maximum value is ca. 173 kJ/mol. Interestingly, the most weakly bound form of Cl_{ad} has an adsorption free energy, 142 kJ/mol, that is comparable to the OH_{ad} adsorption energy, 136 kJ/mol. This explains why OH_{ad} and Cl_{ad} compete for the Pt surface at potentials near (and positive to) the PZC. The competition between Cl_{ad} and OH_{ad} may frustrate the formation of an ordered Cl_{ad} overlayer, at least with reasonably wide domains that are required for detection by SXS. The potential dependence of the relative coverages of OH_{ad} and Cl_{ad} at potentials away from the PZC depend on the polarizability of the respective bonds, i.e. how the bond responds to the surface electron density. At sufficiently positive potentials (electron deficient), Cl_{ad} is displaced from the electrode surface by OH_{ad}. At more negative potentials (electron excess) OH_{ad} is displaced first by Cl_{ad} and then by H_{upd}.

(b) *Pt*(h k l)*—bromide*. The first studies of bromide adsorption on Pt(1 1 1) by ex situ LEED/AES emersion experiments were reported by the Hubbard group [69,144]. They observed (3 × 3) and (4 × 4) bromide overlayers depending on potential of emersion, the latter having a surface coverage from AES analysis of 0.44 ML. The (3 × 3) coincidence lattice has four Br atoms in the unit cell. The same (3 × 3) Br_{ad} structure was attained in UHV by dosing with either HBr or Br_2. In a more recent ex situ STM study, Bittner et al. [100] prepared Br_{ad} adlayers on low-index platinum single crystals by flame-annealing and subsequent quenching in bromine vapor, an extension of the "iodine method" developed by Wieckowski et al. [74]. They assigned the STM image on Pt(1 1 1) attained in air to a (4 × 4) bromide overlayer, but no in situ STM measurements were reported. The first in situ measurements of bromide adsorption on Pt(1 1 1) were reported by Lucas et al. [83,142] using SXS measurements. In their report the authors showed that an incommensurate hexagonal bromide adlayer ($\Theta_{Br} = 0.44$ ML) is present on the Pt(1 1 1)

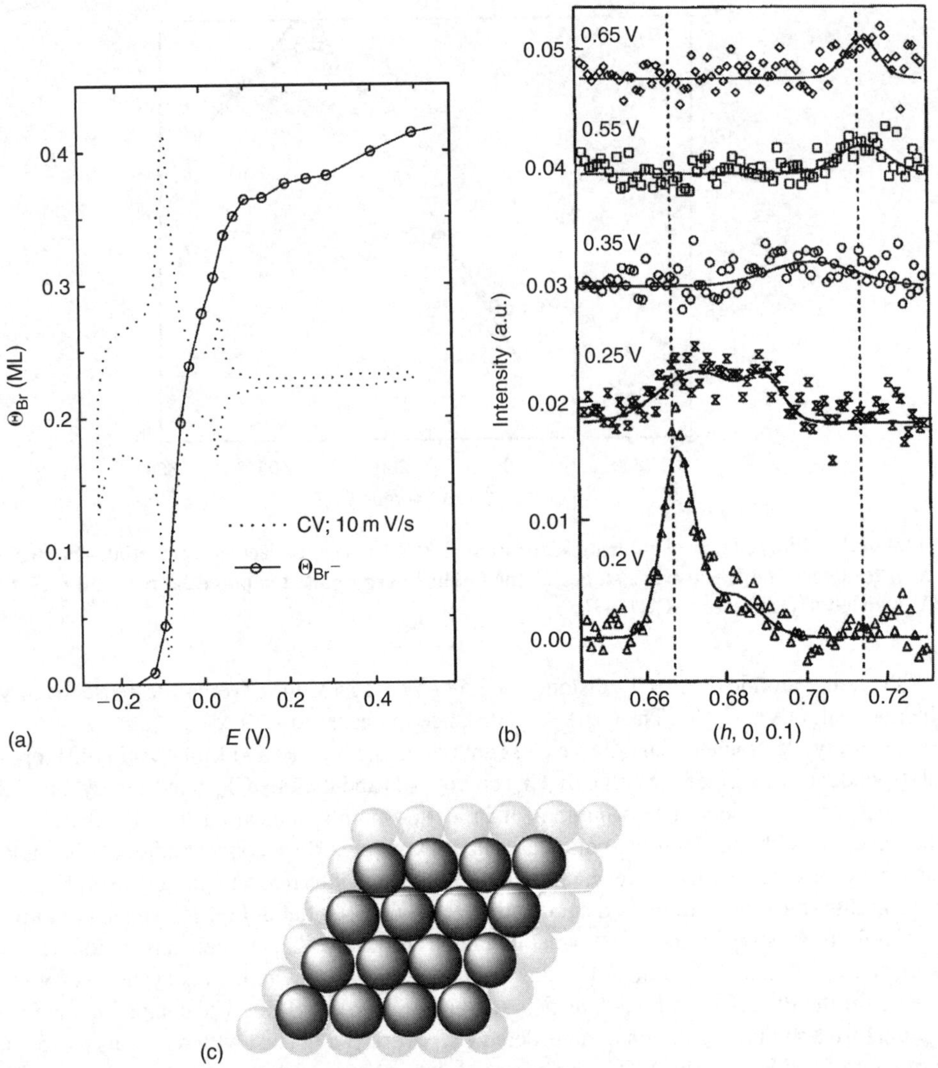

Fig. 16. (a) The dashed line shows the cyclic voltammetry of the Pt(1 1) electrode in 0.01 M KBr/0.1 M $HClO_4$ electrolyte measured in the hanging meniscus geometry. The data points and the solid line correspond to the potential dependent surface coverage by bromide (in ML), obtained from RRDE measurements. (b) Typical scan along (h 0 0.1) for Br^-/Pt(1 1 1) at different electrode potentials. The data have been obtained by subtraction of a background scan performed at −0.2 V. The solid lines are fits in the data. The electrode potential at which each scan was performed are indicated and the dashed line mark the position $h = 0.67$ and 0.71. (c) The real space model in which the Pt substrate atoms are shown as shade circles and the bromide overlayer atoms by dark circles [142]

surface in the potential range 0.05–0.65 V. In contrast, between 0.3 and 0.8 V the commensurate Pt(1 1 1)-(3 × 3) Br_{ad} adlayer was later found by Wang et al. [145] also using SXS. In repeating the SXS measurements, Lucas et al. [142] concluded that Br_{ad} forms a series of high-order commensurate structures on Pt(1 1 1) that are poorly ordered unless the size of the unit cell is small. Typical scans along (h 0 0.1) through the first-order diffraction peak from the hexagonal bromide overlayer at different electrode potentials is shown in Fig. 16b. At 0.2 V, there is a very strong peak due to the (3 × 3) bromide adlayer located exactly at the commensurate (3 × 3) position (domain size, $D \approx 100$ Å) and an additional peak at slightly higher wave vector ($h = 0.68$). Upon stepping to higher potentials, the weak peak gradually emerged at $h = 0.715$ and at 0.65 V the unit cell is a (7 × 7) structure with a 7:5 ratio of the Br and Pt lattice parameters ($\Theta_{Br} = 0.51$ ML).

The surface normal structure of the bromide adlayer on Pt(1 1 1) was ascertained from the CTR measurements. These measurements are sensitive to the surface normal electron density profile of the electrode surface and to adsorbates, even if the adsorbate structure is incommensurate with the underlying lattice [79,80]. The parameters describing the Br_{ad} adlayer gave at 0.2 V $\Theta_{Br} = 0.43$ ML and $\Delta_{Pt—Br} = 2.85$ Å, and at 0.7 V $\Theta_{Br} = 0.48$ ML and $\Delta_{Pt—Br} = 2.61$ Å. The Pt—Br_{ad} layer spacing (≈2.7 Å) is consistent with Br being covalently bonded to Pt [142]. It was proposed that the interaction between Pt and Br_{ad} should be strong enough to minimize a competitive adsorption with OH_{ad} [142], but weak enough to maximize the mobility of the Br_{ad} layer, which is required for the ordering at the Pt(1 1 1) surface.

Quantitative measurements of the coverage by Br_{ad} on Pt($h\,k\,l$) surfaces were obtained by purely electrochemical methods, as described in details in [142]. Briefly, by utilizing the so-called ring shielding properties of the rotating ring disk electrode (RRDE) [147,148], the authors were able to determine the potential-dependent surface coverage by bromide and its electrosorption valence (γ) on both Pt(1 1 1) and Pt(1 0 0) surfaces [146]. The electrosorption valence is a measure of the degree of discharge of the ion upon adsorption, i.e. $\gamma = 1$ for a univalent ion corresponds to complete discharge. The resulting Θ_{Br} versus. E is shown in Fig. 16a. As evident from the shape of curve, the adsorption of Br ($\gamma_{Br} \approx 1$) is also two-step process, the first being a displacement of the H_{upd} state in a narrow interval of potential, the second a *continuous* compression until the close-packed adlayer with maximum surface coverage of 0.44 Br/Pt is reached at ≈0.6 V. In a completely different kind of electrochemical determination, the so-called charge displacements experiments, the coverage reported (only the maximum value is obtained) was very similar, ca. 0.46 Br/Pt [149]. Putting the purely electrochemical measurements of coverage with the structures observed by SXS, a reasonable physical picture of adsorption emerges. With increasing potential, Br_{ad} coverage increases incrementally producing disordered or incommensurate adlattices except at "magic coverages", where a commensurate adlattice is obtained.

The surface structures and potential-dependent coverages for Br_{ad} on Pt(1 0 0) were obtained from the combination of the RRDE (Fig. 17) and SXS measurements [150]. Inset of Fig. 17 shows the Θ_{Br} versus E curve for Pt(1 0 0) deduced from the ring-shielding experiments. It is interesting to note that a small amount of bromide (≈0.01 ML ± 5%) is still on the surface even at the hydrogen evolution potential, which contrasts sharply the Pt(1 1 1) surface, where the desorption of bromide was completed at ≈0.1 V more positive potential. This difference may be attributed to the difference in work function (i.e. the potential of the zero charge) between the (1 0 0) and (1 1 1) surfaces, with the absolute value of the work function of Pt(1 0 0) being lower than that for Pt(1 1 1). In contrast to Br^- adsorption onto Pt(1 1 1), no in-plane diffraction features were found at any potential for the Pt(1 0 0)—Br_{ad} system [150]. The absence of bromide structures with long-range order on the Pt(1 0 0) surface is in agreement with the STM results of Bittner et al. [151]. Information about the surface normal structure and local bonding sites for Br_{ad} was obtained from analyses of the specular and non-specular CTRs [150]. Using the coverages measured in the RRDE experiments, it was proposed that the observed intensity changes at various "anti-Bragg" positions on the CTRs as a function of electrode potential may correspond to presence of a mixture of $c(2 \times 2)$ and $c(\sqrt{2} \times 2\sqrt{2})\,R45°$ adlattices, noting that the $c(\sqrt{2} \times 5\sqrt{2})\,R45°$adlattice was observed in vacuum deposition studies of Br_2/HBr on Pt(1 0 0) [69]. More quantitative information was obtained from modeling the l-dependence of the CTRs, which included relaxation and disorder of the topmost Pt layer. The best-fit to the CTR data gave an average Pt—Br_{ad} vertical separation of ca. 1.9 Å, indicating that the Pt—Br_{ad} bond is covalent in nature, and the bond is considerably stronger than the Pt(1 1 1)—Br_{ad} bond [150]. Very strong bonding on the more corrugated (1 0 0) surface may result in a low mobility of Br_{ad} and may frustrate long-range ordering.

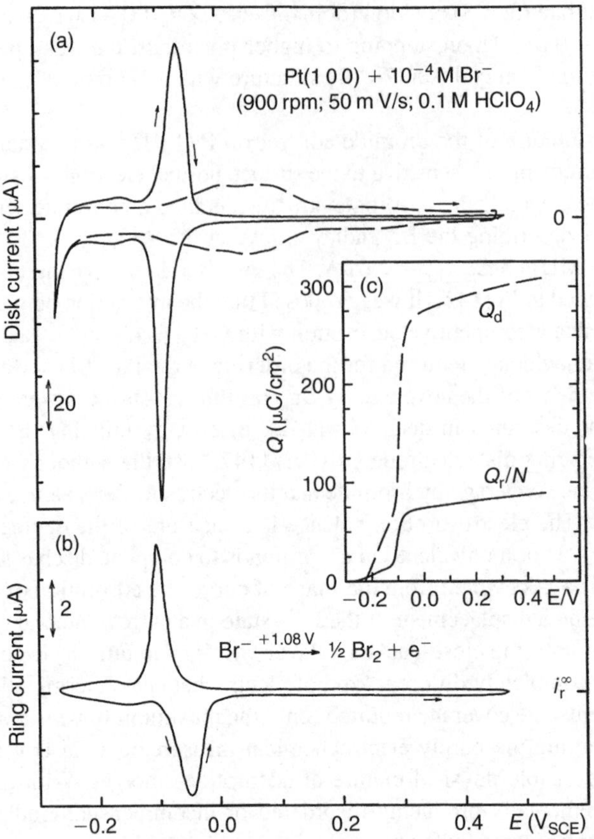

Fig. 17. Potentiodynamic ring-shielding experiment with a Pt(1 0 0) electrode in 0.1 M HClO$_4$ and 8 × 10^{-5} M Br$^-$ at 50 mV/s and 900 rpm. (a) (—) Cyclic voltammogram on the Pt(1 0 0)-disk; (— —) base voltammogram without bromide in solution under otherwise identical conditions. (b) Ring-shielding currents at a ring potential of 1.08 V; the value for the unshielded ring current, i_r^∞, is 19.36 μA. Collection efficiency, $N = 0.21$. (c) Integrated charges on the disk electrode, Q_d, and the ring electrode, Q_r/N. The latter value corresponds to the potential dependent surface coverage by bromide, e.g. $\Theta_{Br} = Q_r/N \times 240\,\mu C\,cm^2$ [150]

(c) *Pt*(h k l)*–I$_{ad}$ system.* The Pt(*h k l*)—I$_{ad}$ system will be discussed only briefly where a comparison is being made to the other halides, or a general conclusion about halide adsorption is being drawn or extended. LEED studies in UHV revealed the presence of three adsorbed structures of iodine on Pt(1 1 1), depending on the method of preparation and the iodine coverage. Dosing of Pt(1 1 1) with iodine in UHV lead to formation of two structures: the $(\sqrt{3} \times \sqrt{3})\,R30°$with the coverage of 1/3, and $(\sqrt{7} \times \sqrt{7})\,R19.1°$ (hereafter termed the $\sqrt{7}$) with a coverage of 0.43 [69]. Adsorption of iodine from KI solution also lead to the ex situ observation of the $\sqrt{7}$ phase and, in addition, a (3 × 3) structure with a slightly higher coverage (0.44 ML). A STM study of Pt(1 1 1)—I$_{ad}$ in air also showed the presence of both the (3 × 3) and $\sqrt{7}$ phases. The proposed $\sqrt{7}$ structure is an hexagonal lattice with an I$_{ad}$—I$_{ad}$ distance of 4.24 Å, with two atoms adsorbed in three-fold hollow steps and one atom at an atop site. The proposed (3 × 3) structure has two slightly different packing arrangements, either with three of the atoms at bridge sites and one atop site, or four atoms in asymmetric position close to atop sites [152–154]. Direct evidence of the true Pt(1 1 1)—I$_{ad}$ structures in situ was obtained from SXS measurements [142]. Two close-packed commensurate structures were observed, a $\sqrt{7}$ phase and a (3 × 3) phase that has slightly higher coverage and is formed

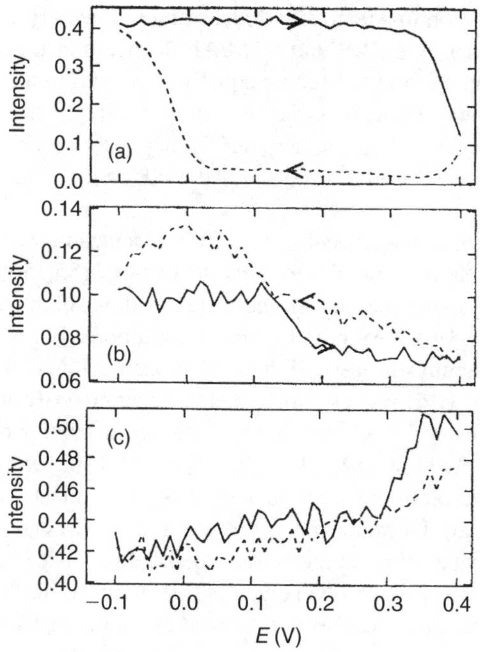

Fig. 18. (a) Dependence of the peak intensity of (1/7, 4/7, 0.15), due to the $\sqrt{7}$ iodine phase, on the Pt electrode potential. (b) Dependence of the peak intensity of (2/3, 2/3, 0.15), due to the (3 × 3) iodine phase, on the Pt electrode potential. (c) Dependence of the intensity at (0, 0, 2.5), a position on the specular CTR, on the Pt electrode potential. In all cases the solid lines are for the negative going sweeps and the background levels, obtained in rocking scans through each peak, have been subtracted from the measured peak intensity [142]

at the positive potentials. The strong Pt—I_{ad} interaction results in a complete suppression of OH_{ad} co-adsorption, and a low adatom mobility that leads to overly slow kinetics of ordering for both structures. The $\sqrt{7}$ phase shows a hysteric effect as a function of the electrode potential (Fig. 18), associated with an order–disorder transition [142]. In contrast, the diffraction peaks from (3 × 3) phase indicate the structure is present on the surface at all potentials, (see Fig. 18). More details about the effects of the electrode history on the potential stability of these two structure can be found in [142].

*4.4.3. Summary of anion adsorption on Pt(*h k l*)*

As evident from the above discussion, the differences between the different anion structures are a consequence of two fundamental properties: the strength of interaction, which on all low-index surfaces increases in the sequence $ClO_4^- < HSO_4^- < Cl^- < Br^- < I^-$; the symmetry of the anions and the geometry of surface, e.g. for tetrahedral (bi)sulfate anions the strength of interaction increases from Pt(1 0 0) ≈ Pt(1 1 0) to Pt(1 1 1), while the spherical halides anions are more strongly adsorbed on (1 0 0) than on (1 1 1) sites. In the following sections we will show how these fundamental properties of the Pt—anion interaction effect the kinetics of the electrocatalytic reactions in different supporting electrolytes.

*4.5. Surface structures of UPD and irreversibly adsorbed metals on Pt(***h k l***) surfaces*

Until recently, UHV prepared surfaces had an enormous advantage over the surfaces prepared by electrodeposition methods, because a UHV system equipped with modern surface sensitive techniques provided a microscopic structural information in a relatively direct fashion. Recently, however, the emergence of surface characterization techniques like SXS and STM/AFM operating under electrochemical conditions have allowed electrodeposition method to become equally important in the synthesis of model bimetallic structures. In this section, we will focus on bimetallic platinum single crystal surfaces which are created by three different solution phase-methods: underpotential deposition (UPD), irreversible adsorption, and electrodeposition. Recall that in Section 3, we discussed model bimetallic systems fabricated by conventional metallurgy.

The UPD of a metal is defined as the reversible deposition of a metal onto a dissimilar metal substrate at a potential that is anodic of the Nernst potential for bulk deposition [155,156]. The deposited adatoms form stable bimetallic surfaces only in the presence of the corresponding soluble cations. In this way, numerous metals have been deposited under a very broad range of conditions. For the purpose of this review, only the UPD of Cu and Pb on platinum surfaces will be summarized, with emphasis on the observed structure of the adatoms. Contrary to the UPD process, some metals spontaneously and irreversibly adsorb onto the substrate surface, with the resulting adlayer being stable in environments lacking the corresponding cation. For example Bi [157], As [158,159], Sb [160,161] and Se [162] exhibit such behavior on Pt single crystals. Of these, Bi-modified platinum electrodes have been the most widely explored, so this system will be described in more detail. Ultrathin metal films, e.g. 1–4 ML, can be prepared electrochemically simply by deposition of the metals close to the Nernst potential [163–168]. As for the irreversible adsorption, for some metals these films are stable in solutions free of corresponding cation. Results for pseudomorphic Pd films on Pt($h\,k\,l$) surfaces will be summarized as a model ultrathin metal film system.

4.5.1. Cu UPD

The UPD of Cu on platinum single crystals in different acidic electrolytes occupies a special position in modern interfacial electrochemistry. However, the electrocatalytic properties of Pt($h\,k\,l$) surfaces modified by Cu_{upd} are generally very poor, particularly with respect to fuel cell reactions, i.e. Cu^{2+} in solution is actually a strongly poisoning species in a fuel cell. The thrust of this relatively short subsection on Pt($h\,k\,l$) surfaces modified by Cu_{upd} will be to understand why this is so.

The interpretation of processes associated with the formation of the Cu monolayer on Pt(1 1 1), and the nature of the Pt(1 1 1)—Cu structure, have been the subject of considerable controversy. Overviews with some different perspectives can be find in [169–178]. The RRDE method [147,179] was successfully applied to investigate the kinetics of Cu^{2+} deposition and to determine the potential dependent surface coverage by Cu_{upd} ($\Theta_{Cu(upd)}$). The fact that the ring current in Fig. 19 mirrors the voltammogram recorded on the disk is a proof that both UPD peaks are related to the deposition of Cu on Pt(1 1 1). The inset of Fig. 19 shows that a near saturation coverage of ≈0.88 ML (420 μC/cm^2) is reached just before the multilayer deposition starts, e.g. complete monolayer does not form and 3D deposition begins (≈0 V in Fig. 18) before the Pt surface is completely covered [173]. Note that 1 ML would correspond to one fully discharged Cu adatom ($\gamma_{Cu} \approx 2$ [173]) per Pt atom, ca. 480 μC/cm^2. Structural details pertaining to Cu UPD on single crystal electrodes have emerged from a combination of classical electrochemical methods with both ex situ [170,172] and in situ [174,180] surface structural probes. These measurements have demonstrated that Cu UPD on platinum single crystals is a multi-stage process which, irrespective of the orientation of surface and/or the nature of specifically adsorbing anions, is governed by a complex interplay of Cu_{upd}—Pt, Cu_{upd}—anion, and the Pt—anion energetics. The structure of the adlayer depends strongly on the anion in the supporting electrolyte. In perchloric acid, the UPD layer on Pt(1 1 1) grows as progressively larger patches of Cu having the Pt lattice constant, i.e. pseudomorphic adlayer [172]. From the electrocatalysis standpoint, the Cu_{upd}—ClO_4^- system is potentially the most interesting, because a true bimetallic surface exists, where both Pt and Cu atoms can do catalysis. In either sulfuric or the halide acids, a multistep deposition occurs with the formation of ordered anion adlattices. A representation of possible

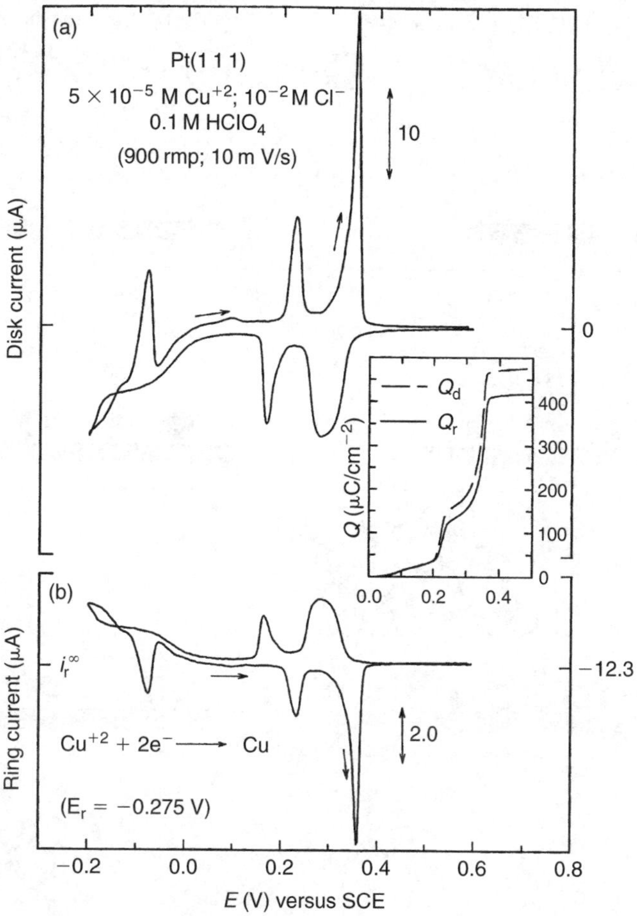

Fig. 19. (a) Cyclic voltammogram for Cu UPD on a Pt(1 1 1) disk electrode. (b) Ring electrode currents recorded with the ring being potentiostated at −0.275 V. Collection efficiency, *N* = 0.22. Inset: integrated charges on the disk, Q_d, and the ring, Q_r electrodes [332]

Pt(1 1 1)—Cu_{upd}—anion and Pt(1 0 0)—Cu_{upd}—anion structural models as a function of Cu_{upd} coverage is shown in Fig. 20. This model is very similar to previously proposed models for Cu UPD on Pt(1 1 1) [174] and Au(1 1 1) [181], and is based on an *enhanced* adsorption of halide and other anions on Cu_{upd}-modified Pt sites. Clearly, Cu_{upd} is either sandwiched between the Pt surface and anions or is in contact with anions adsorbed on the adjacent Pt sites [176]. States 1a and b show a close-packed bisulfate/halide layer which is present on the Pt(1 1 1) and Pt(1 0 0) surfaces at potentials positive of Cu UPD. States 2a and b refer to the initial deposition of Cu_{upd} on Pt(1 1 1) and Pt(1 0 0) surfaces where the Cu^{2+} ion is only partially discharged. At higher surface coverages by Cu_{upd} ($0.5 < \Theta_{Cu} < 1$), anions are either entirely displaced from the surface by Cu_{upd} (State 3a) or Cu_{upd} and anions form two-layer structures in which anions are adsorbed on both Pt as well as Cu_{upd} sites (State 3b in Fig. 20). Note State 3a is representative of an ordered (4 × 4) Cl_{upd}—Cl_{ad} (Br_{ad}) bilayer structure, which is formed in between the Cu UPD peaks in the cyclic voltammetry (Fig. 19), each with a coverage of 0.585 ML [180]. The same structure is observed by LEED [182]. The final step in Cu UPD is the filling-in of the Cu_{upd} monolayer to form a bilayer phase: i.e. a pseudomorphic (1 × 1) Cu_{upd} monolayer with either a disordered anion adlayer on the top of

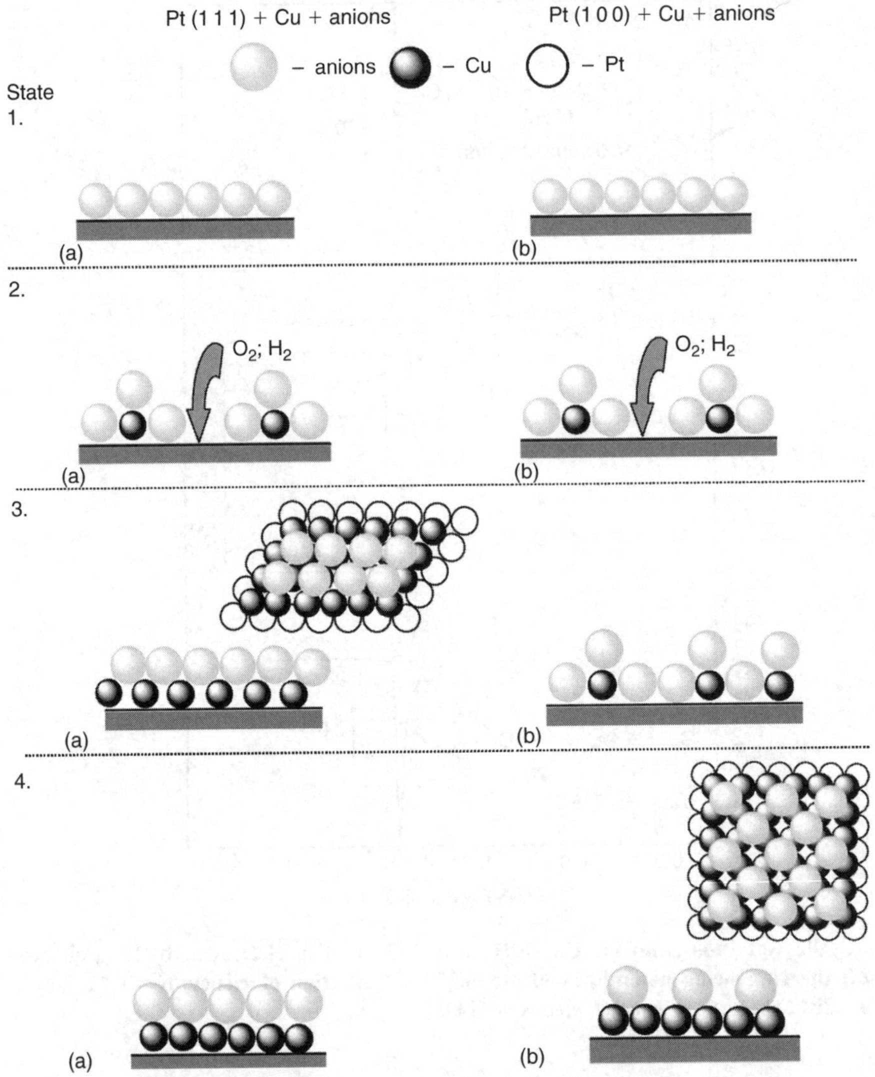

Fig. 20. The representation of the proposed Cu_{upd}-anion structures on (a) Pt(1 1 1) and (b) Pt(1 0 0). ***Note*****: the top view structure in Step 3 is shown for the Pt(1 1)—Cu_{upd}—Cl_{ad} system [180]. The top view structure for the Pt(1 1 1)Cu_{upd}—$SO_{4,ad}$ system can be found in [183,350]**

Pt(1 1 1)—Cu_{upd} [180,183], State 4a, or an ordered $c(2 \times 2)$ bilayer Cu_{upd}—Cl_{ad} (Br_{ad}) structure on the Pt(1 0 0) surface [172], State 4b. The $c(2 \times 2)$ structure consists of an ordered Br lattice which is formed on the top of a pseudomorphic Cu layer. The bilayer structures in Cl^- and Br^- are similar in nature, both being like the (1 1 1) planes in the respective Cu(I)Cl(Br) crystals, which apart from the differences in atomic radii have the identical zinc blende structure. Since Cu UPD in the two halide solutions proceeds via the same intermediate phase, the same physical model applies to both anions, with the differences between the anions accommodated within the same model. In sulfuric acid solution, however, the same state on Pt(1 1 1) is representative of the formation of the honeycomb $(\sqrt{3} \times \sqrt{3})$ $R30°$ Cu_{upd}—HSO_4 structure [183] that consists of both 0.22 ML bisulfate anions and 0.66 ML of Cu_{upd} in the unit cell.

The same structure was proposed for the Au(1 1 1)—sulfate system [184,185]. The formation of State 2 occurs over a very narrow region of potential. As will be shown in Section 5, the existence of bare platinum sites in the State 2 (Fig. 20) is essential for the adsorption of O_2 and H_2 molecules, and this first-stage formation plays the key role in the activity of the Pt—Cu_{upd} interface for the ORR and the HOR.

4.5.2. Pb UPD

As for the Pt($h\,k\,l$)—Cu_{upd} system, most of the early structural details pertaining to Pb UPD on single crystal platinum electrodes have derived from ex situ UHV analysis of emersed surfaces, e.g. by LEED/AES [186–188], XPS [189], and TDS [190]. The recent advances of in situ STM and SXS, in combination with the RRDE method, have provided important new insight into the structural properties of the Pb_{upd} adlayer formed on platinum single crystal surfaces. As demonstrated below, the processes associated with the formation of Pb_{upd} layer at the Pt(1 1 1) and Pt(1 0 0) single crystal surfaces are much better understood, and earlier rather controversial discussions merge into a consistent picture.

For the Pt(1 1 1) surface, the study by El Omar and Durant [191] demonstrated that the Pb UPD on Pt(1 1 1) in 0.1 M $HClO_4$ proceeds as a step-wise process, with four clearly resolved reversible peaks, Fig. 21a. While three of them A_2—C_2, A_4—C_4, A_5—C_5 exhibited a positive shift with an increase of Pb^{2+} concentration, the position of the A_3—C_3 couple was found to be independent of the concentration of Pb^{2+}, but strongly dependent on the pH of solution and/or on the presence of strongly adsorbed anion, such as Cl^- [191]. Closely following the El Omar and Durand work, Clavilier et al. [192] suggested that the A_3–C_3 couple is a surface process associated with a redox reaction between the irreversibly adsorbed lead and OH species. The possible interaction of Pb_{upd} with anionic species, such as O_{ad} or OH_{ad} was first discussed by Borup et al. [189] and subsequently by Adzic et al. [193]. The experimental SXS and STM work by later authors showed that a total coverage of Pb_{upd}, inferred from the rectangular $(3 \times \sqrt{3})$ cell, is $\Theta_{Pb(upd)} = 2/3$. The real space model for the $(3 \times \sqrt{3})$ structure is depicted in Fig. 21b. The $\Theta_{Pb(upd)}$ coverage calculated from this structure is slightly higher than the coverage of 0.63 ML predicted for an hexagonal closed-packed Pb_{upd} monolayer with a nearest-neighbor spacing equal to the Pb—Pb bulk spacing of 3.5 Å. The observed Pb_{upd} layer is uniaxially compressed relative to the *hcp* monolayer. The CTR modeling yielded a Pt—Pb_{upd} layer spacing of 2.34 Å and a ca. $\Theta_{Pb(upd)} = 0.65$ ML. This coverage is in good agreement with the coverage calculated from the ordered structure, but is considerably different from the coverage calculated from the total charge under the voltammetry curve ($\Theta_{Pb(upd)} = 0.48$ ML) [193]. This apparent discrepancy was resolved by using ion-specific current (ISC) measurements with an RRDE [194]. The total amount of Pb deposited by integration of the Pb^{2+} flux was 0.62 ± 5%. The electrosorption valence of Pb_{upd} is $\approx \gamma_{Pb} = 2$, implying that two electrons per Pb_{upd} adatom are exchanged through the interface. A close inspection of Fig. 22 indicates that there is another process ($\approx 0.2 < E < 0.5$ V) in addition to Pb^{2+} deposition that occurs in the UPD potential region. The difference in charge between Pb deposition (ISC) and the total disk charge (TDC) (ca. 100 C/cm^2) was assigned to the induced adsorption of OH_{ad} *onto* Pt atoms adjacent to the Pb_{upd} adatoms. This is an important phenomenon, and will be used later as an argument to explain the surface processes at the Pt($h\,k\,l$)—Bi_{ir} interface. The two small peaks observed at lower potentials are associated with completion of the Pb_{upd} adlayer to form the ordered $(3 \times \sqrt{3})$ close-packed structure [194].

The UPD of Pb on Pt(1 0 0) has been less studied than on the Pt(1 1 1) surface. Most of the earlier studies have focused on ex situ LEED and Auger analysis. Using ex situ LEED to study the UPD of Pb onto Pt(1 0 0) in 0.1 M $HClO_4$, Aberdam et al. [188] reported a $c(2 \times 2)$ superstructure of Pb_{upd} with some diffuse background. In SXS/STM studies by Adzic et al. [193], a $c(2 \times 2)$ superstructure of Pb_{upd} the authors has not been found in in situ experiments. The authors suggested that with loss of potential control in ex situ experiments, evaporation of water and/or cations can effect ordering of Pb adatoms on the Pt(1 0 0) substrate. The most recent SXS studies [195], however, confirmed the existence of an ordered surface

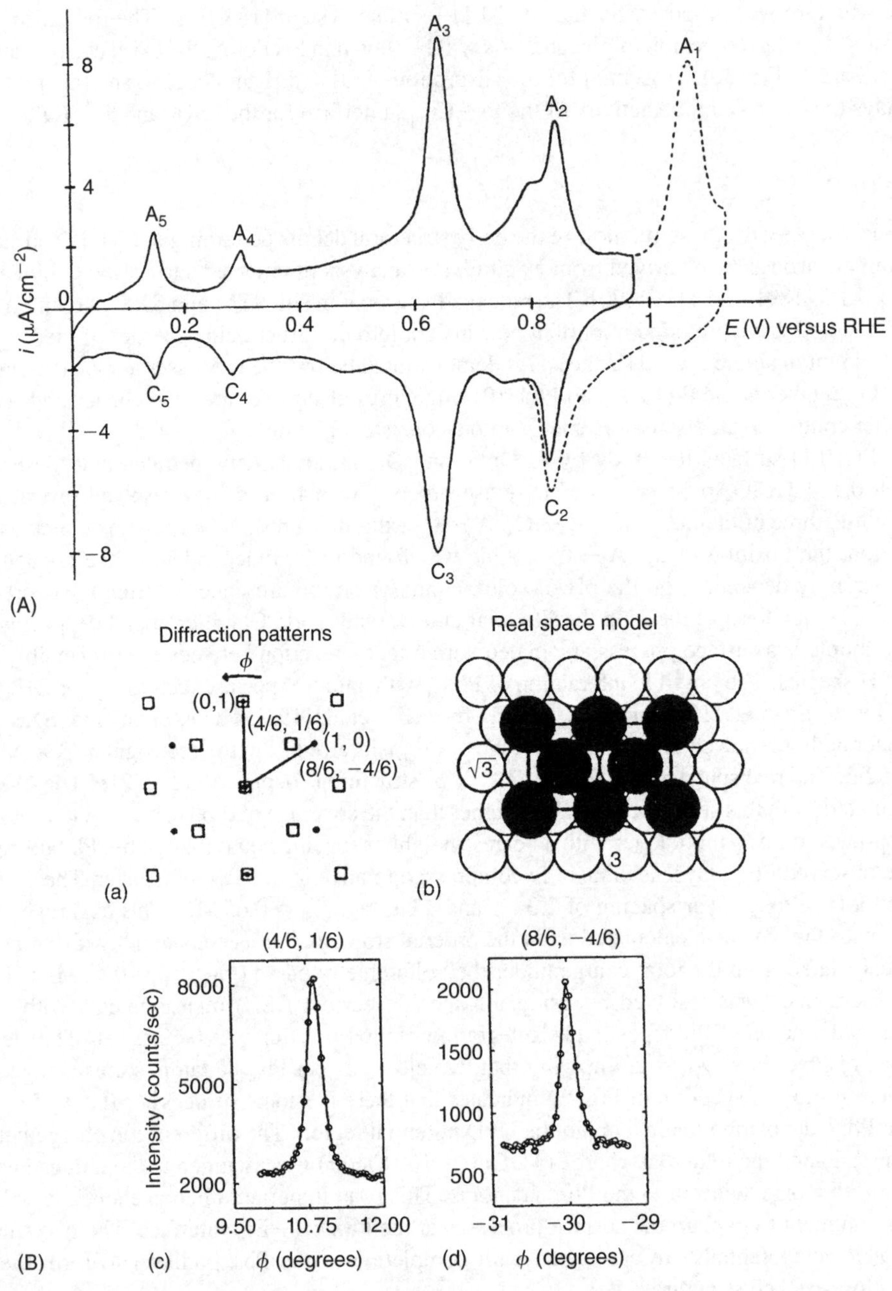

Fig. 21. (A) Cyclic voltammetry of Pt(1 1 1) in 0.1 $HClO_4$ + 6 × 10^{-4}M Pb^{2+} solution. Sweep rate 5 mV/s [191]. (B) (a) In-plain diffraction pattern and (b) real space model obtained from the rectangular (3 × $\sqrt{3}$) Pb monolayer on Pt(1 1 1) at −0.29 V. The solid circles represent the reflections from the Pt(1 1 1) surface and squares show the major diffraction pattern observed from one of the three rotationally equivalent domains of Pb adlayer. Rocking curves at (4/6, 1/6) and (8/6, −4/6) shown in (c) and (d) are plotted when ϕ_0 is along the (1, 0) direction [193]

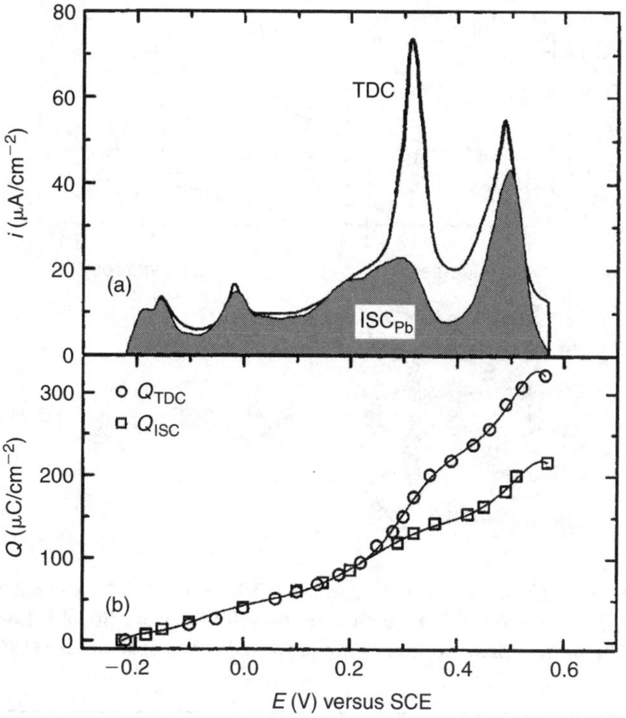

Fig. 22. (a) Comparison of the ion–specific partial currents (ISCs) with the total disk current (TDS) for the Pb UPD in 0.1 M $HClO_4$. (b) Integrated charges during the stripping of Pb_{upd} from the disk electrode, assessed from the disk electrode (Q_{TDC}) and from the ring electrode (Q_{ISC}). Ring electrode currents recorded with the ring being potentiostated at −0.59 V. Collection efficiency, $N = 0.21$ [194].

structure with a $c(2 \times 2)$ unit cell in the potential range $0.15\,V < E < 0.35\,V$, Fig. 23. The total amount of Pb deposited underpotentially in 0.1 M $HClO_4$ is ≈0.62 ± 5% ML (1 ML equals 1 Pb atom per Pt surface atom), nearly equivalent to a close-packed monolayer of fully discharged Pb adatoms, e.g. 0.63 ML.

4.5.3. Irreversibly adsorbed Bi

Clavilier et al. [157] provided the first evidence that Bi can be irreversibly adsorbed (hereafter denoted as Bi_{ir}) on Pt single crystal surfaces. Since then, a number of groups have studied the surface chemistry of the Bi_{ir} adlayer, e.g. [160–162,192,196–207]. Key questions addressed in these studies are: (a) the correlation between the charge under sharp peaks in the cyclic voltammetry of Pt(1 1 1) and Pt(1 0 0), (Fig. 24a and Fig. 25a), and the surface coverage by Bi_{ir}; (b) the valence state of Bi_{ir} before and after the processes that are creating the pseudo-capacitive peaks. In the following, a selective overview of the surface chemistry at the Pt(1 1 1)—Bi_{ir} and Pt(1 0 0)—Bi_{ir} interfaces is presented with the emphasis on results which lead to a determination of the nature of processes under the sharp peaks.

Many groups [197,199,203,208] have followed the original proposition of Clavilier et al. [157] that the pseudo-capacitance at 0.68 V on Pt(1 1 1) and at 0.8 V at Pt(1 0 0) is related to a Bi^{2+}/Bi^{0} surface redox reaction. In recent independent ex situ XPS studies by Hayden et al. [201] (Pt(1 1 0)—Bi_{ir}), Hamm et al. [202] and Schmidt et al. [205] it was found that irrespective of the emersion potential, the irreversibly adsorbed Bi on Pt(1 1 1) does not change its valence state, most notably it remains in a zero valent state over the entire potential range. In Fig. 24, the XPS Bi(4f) core level spectra of Pt(1 1 1)—Bi_{ir} are

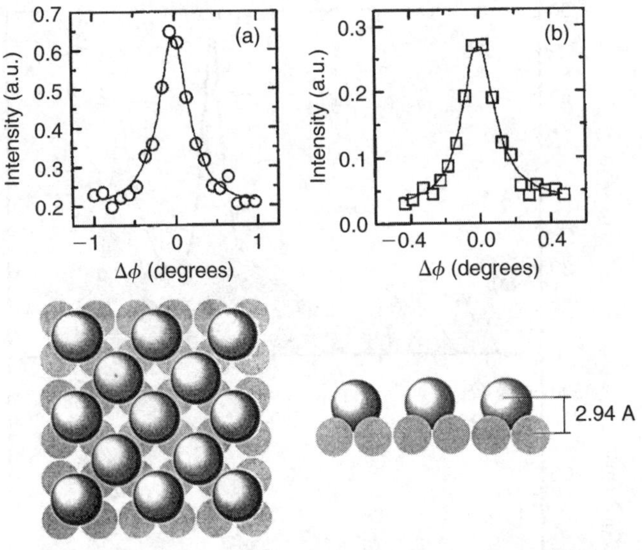

Fig. 23. Rocking curves through (a) (1/2, 1/2, 0.1) and (b) (3/2, −1/2, 0.1) for the $c(2 \times 2)$ structure formed at 0.2 V in 0.1 M $HClO_4$ + 5×10^{-5} M Pb^{2+}. In each case the solid lines are fits of a Lorentzian lineshape to the data. A bottom view: schematic representation of the proposed $c(2 \times 2)$ structure [195]

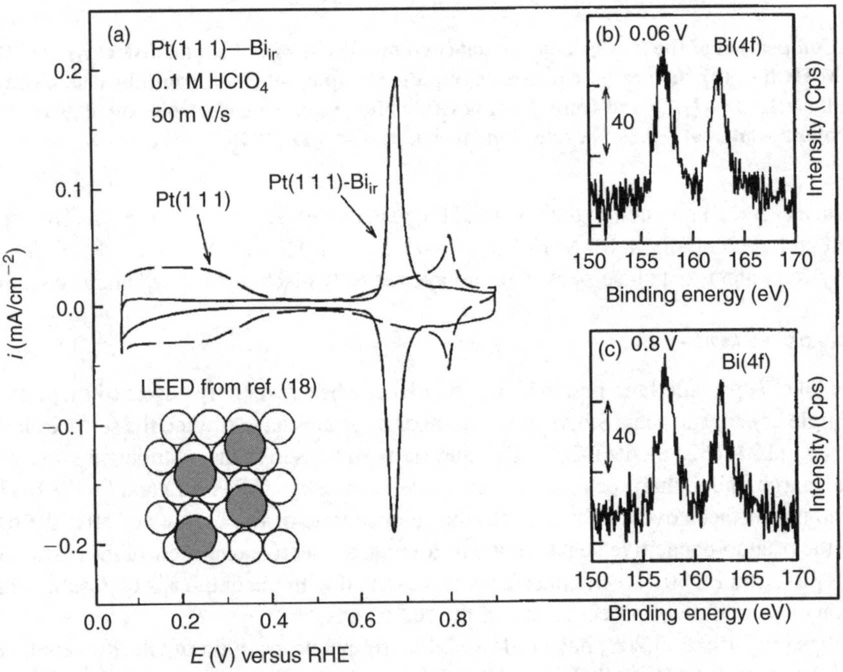

Fig. 24. (a) Base voltammetry of Pt(1 1 1) (dashed line) and Pt(1 1 1)—Bi$_{ir}$ ($\Theta_{Bi_{ir}} \approx 1/3$ ML, solid line) in 0.1 M $HClO_4$ (50 mV/s). XPS-signal for the Bi 4f core level obtained after electrode transfer from the electrochemical cell into the UHV chamber: (b) emersion potential 0.06 V; (c) emersion potential 0.8 V [351] (Ref. [18] in the figure is Ref. [202] in the reference list.)

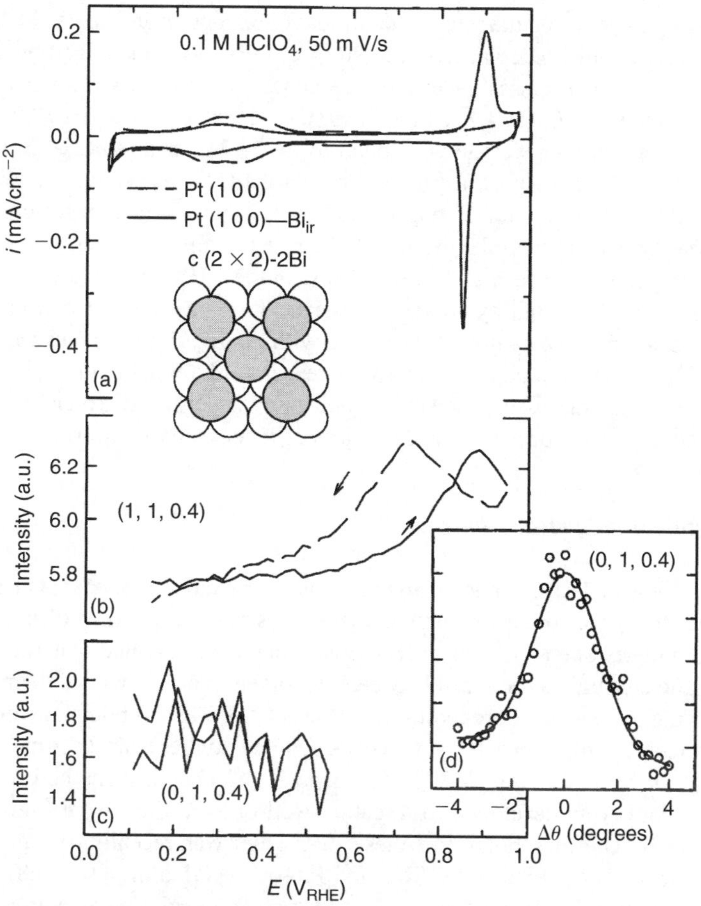

Fig. 25. (a) Base voltammetry of Pt(1 0 0) (dashed line) and Pt(1 0 0)—Bi$_{ir}$ ($\Theta_{Bi_{ir}} \approx 0.5$ ML, solid line) in 0.1 M $HClO_4$ (50 mV/s). (b) Potential dependence of the X-ray intensity at the (1, 1, 0.4) reciprocal space position. (c) Potential dependence of the X-ray intensity at the (0, 1, 0.4) reciprocal space position. (d) Rocking scan through (0.1, 0.4) lattice point, which gives rise to a *c*(2 × 2) Bi$_{ir}$ structure (circles) and fit to Lorentzian lineshape (solid line) [351]

illustrated for emersion potentials of 0.06 and 0.8 V. Consequently, the charge under the sharp peaks in Fig. 24 is *not* due to a Bi^0/Bi^{2+} surface redox reaction, and the Bi$_{ir}$ coverage cannot be determined simply from the charge under the peak, as Clavilier et al. previously proposed. More recently, the Bi surface coverage was determined in situ from the CTR measurements in an SXS study of the Pt(1 1 1)—Bi$_{ir}$ and Pt(1 0 0)—Bi$_{ir}$ interfaces [209]. For Pt(1 1 1)—Bi$_{ir}$, the best-fit of a structural model was obtained for a coverage of $\Theta_{Bi_{ir}} = 0.34 \pm 0.01$ ML, in agreement with the ($\sqrt{3} \times 3$) R30° ex situ LEED structure reported by Hamm et al. [202]. The fact that the surface appears to saturate at a coverage of ~1/3 ML is probably due the large size of the Bi atom, and repulsive forces between the Bi$_{ir}$ atoms, as already observed in UHV studies for vapor deposited Bi on Pt(1 1 1) [210,211]. The Pt—Bi$_{ir}$ spacing was found to be 2.41 Å, indicating that Bi$_{ir}$ is primarily occupying Pt three-fold hollow sites [210,211]. In contrast to ex situ measurements, however, no diffraction features due to an ordered Bi adlayer could be detected in SXS measurements. This discrepancy is possibly explained by the inability of SXS to detect patch-wise structures, i.e. the domain sizes of the ($\sqrt{3} \times \sqrt{3}$) *R*30° may be too small to be detected by SXS.

For Pt(1 0 0)—Bi_{ir}, in the cyclic voltammetry oxide formation/reduction at $E >$ ca. 0.8 V. An asymmetrical peak of 0.8 V is observed, which in comparison to Pt(1 1 1)—Bi_{ir} is characterized by a significant kinetic *irreversibility* ($\Delta E \approx 0.05$ V). The area under the peak centered at ca. 0.9 V amounts to ca. 200 $\mu C/cm^2$. The structure of Bi on the Pt(1 0 0)-(1 × 1) surface was studied with ex situ LEED/XPS by Kizhakevariam and Stuve [212]. The saturation coverage was found to be 0.5 ML with a $c(2 \times 2)$ LEED pattern. In a recent in situ SXS study, the best-fit of a structural model the CTR measurements was obtained for the coverage of $\Theta_{Bi_{ir}} = 0.5 \pm 0.01$ ML. Interestingly, a $c(2 \times 2)$ X-ray diffraction pattern was also observed in-plane, in complete agreement with the ex situ LEED observation, (Fig. 25). The potential stability of the structure was exceptionally wide, e.g. $0 < E < 1.0$ V. In the UHV study [212], the authors concluded that ("strangely") Bi is *not* hydrated by co-adsorbed H_2O. This was the first indication that the surface processes under the sharp peaks may not be associated with the $Bi_{ir}/Bi_{ir}(OH)_2$ surface redox processes as originally proposed by Clavilier et al. [157]. In a very recent work by Schmidt et al. [205], as seen in the Pt($h\,k\,l$)—Pb_{upd} systems just discussed, the surface processes under the sharp and reversible peaks were ascribed to enhanced adsorption of oxygenated species *on* Pt sites adjacent to Bi_{ir}.

4.6. *Surface structure of thin metal films*

The last decade has witnessed a tremendous growth in our understanding of chemical and electronic properties of thin metal films supported on foreign metal substrates. Modern molecular (atomic) surface characterization techniques operating under UHV conditions have revealed that variations in interfacial bonding and energetic constraints produced between monolayer metal films and their substrates provides a means for modifying the chemical properties of surfaces [213–216]. While many different metals have been studied as monolayer films in UHV, in electrochemical studies only thin metal films of Pd have received significant attention. A so-called "forced-deposition" chemical method in addition to purely electrochemical methods were developed for creating well-ordered thin metal films of Pd on Au($h\,k\,l$) and Pt($h\,k\,l$) [163–168]. The morphology of these overlayers was recently examined by in situ STM for Au($h\,k\,l$)—Pd films [217,218] and by SXS for Pt(1 1 1)—Pd films [167,168]. The latter studies confirm the expectation that Pd can be deposited on Pt($h\,k\,l$) from solution as uniform epitaxial metallic layers having the Pt lattice constant, i.e. pseudomorphic growth. Since the latter study is also the most detailed study of the structure of a monolayer Pd film deposited electrochemically, we shall focus on those results here.

Figure 26 (open triangles) shows the (1 0 l) and (0 1 l) CTRs for 1 ML of Pd deposited on Pt(1 1 1) with the electrode potential held at 0.05 V in 0.05 M H_2SO_4 (where nominally a monolayer of H_{upd}, is expected to be adsorbed). The calculated reflectivity from an ideally terminated Pt(1 1 1) surface is shown by the dashed curves. This is typically what is measured for the clean Pt(1 1 1) surface in sulfuric acid electrolyte [87, 118]. The data in Fig. 26, however, differ significantly from the clean Pt model calculations, especially at the "anti-Bragg" positions midway between the bulk Bragg reflections. At this position the scattering from the bulk of the Pt crystal is effectively cancelled and the scattered intensity is due to the topmost Pt atomic layer and any adsorbed species. The decrease in the intensity relative to the clean Pt surface, therefore, results from the destructive interference between X-rays reflected from the Pd layer and the topmost Pt layer. The solid line in Fig. 26 is a calculated best-fit to the data using a structural model in which the Pd coverage (Θ_{Pd}), Pt—Pd distance ($d_{Pt—Pd}$), relaxation of the Pt—Pd surface (Δd_{12}), and roughness parameters (σ_{Pt}, σ_{Pd}) were varied. The best-fit was obtained for Pd adsorbed into the Pt three-fold hollow sites, i.e. effectively continuing the ABC stacking of the substrate, with a coverage of $\Theta_{Pd} = 1.0 \pm 0.1$ ML, $d_{Pt—Pd} = 2.32$ Å, expansion of the Pt lattice $\Delta d_{23} = 0.03 \pm 0.02$ Å, $\sigma_{Pt} = 0.08$ Å, $\sigma_{Pd} = 0.11$ Å. Because the inclusion of a second Pd layer did not improve the fit to the data it was concluded that as-deposited Pd forms a pseudomorphic monolayer on the Pt(1 1 1) electrode.

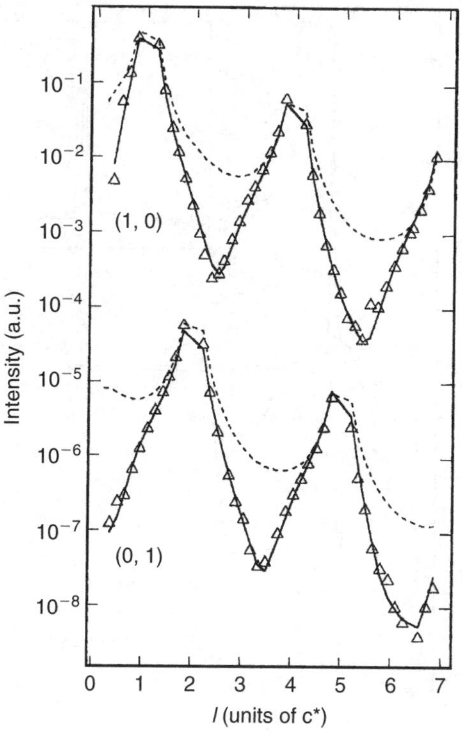

Fig. 26. The (1 0 *l*) and (0 1 *l*) CTRs for the Pt(1 1 1)—Pd surface measured with the electrode potential held at 0.05 V. The dashed lines are calculations for an ideally terminated non-modified Pt(1 1 1) surface. The solid lines are fits to the data [168]

The voltammetric behavior of the Pt(1 1 1)—Pd electrode at 298 K in 0.05 M H_2SO_4 is shown as a solid curve in Fig. 27. For comparison, the base voltammetry of Pt(1 1 1) obtained under the same experimental conditions is shown as a dashed curve. Fig. 27 also shows that at the Pt(1 1 1)—Pd surface the processes of hydrogen adsorption/desorption and bisulfate desorption/adsorption emerge into a single very sharp peak around 0.21 V. The collapse of the H_{upd} peak and the "butterfly" feature into a single sharp peak implies a strong interaction of bisulfate anions with the Pt(1 1 1)—Pd surface. The difference between the total charge in the H_{upd} potential region (≈330 μC/cm^2) and the charge deduced for anion adsorption (ca. 80 μC/cm^2) equals ≈240 μC/cm^2, implying the Pt(1 1 1)—Pd surface is covered by 1 ML of H_{upd} [167, 219]. In the same potential region the maximum charge associated with H_{upd} on Pt(1 1 1) is, however, only 160 μC/cm^2 (0.66 ML). This suggests that the H_{upd} binding energy on the pseudomorphic Pd layer is higher than on Pt(1 1 1) possibly because the lateral repulsion between H_{upd} is reduced relative to the Pt(1 1 1)—H_{upd} system [113]. As will be demonstrated in Section 5.1, this change in the energetics of H_{upd} plays a dominant role in the kinetics of the HER/HOR on Pt(1 1 1)—Pd.

5. ELECTROCATALYSIS AT WELL-DEFINED SURFACES

The term *electrocatalysis* was coined by Kobosev and Monblanova [220] at the beginning of 1930s. However, it has been only in the last 30 years or so that this terminology is commonly employed to describe the study of electrode processes where charge-transfer reactions have a strong dependence on the nature of the electrode material [221,222]. Not surprisingly, virtually every electrochemical reaction,

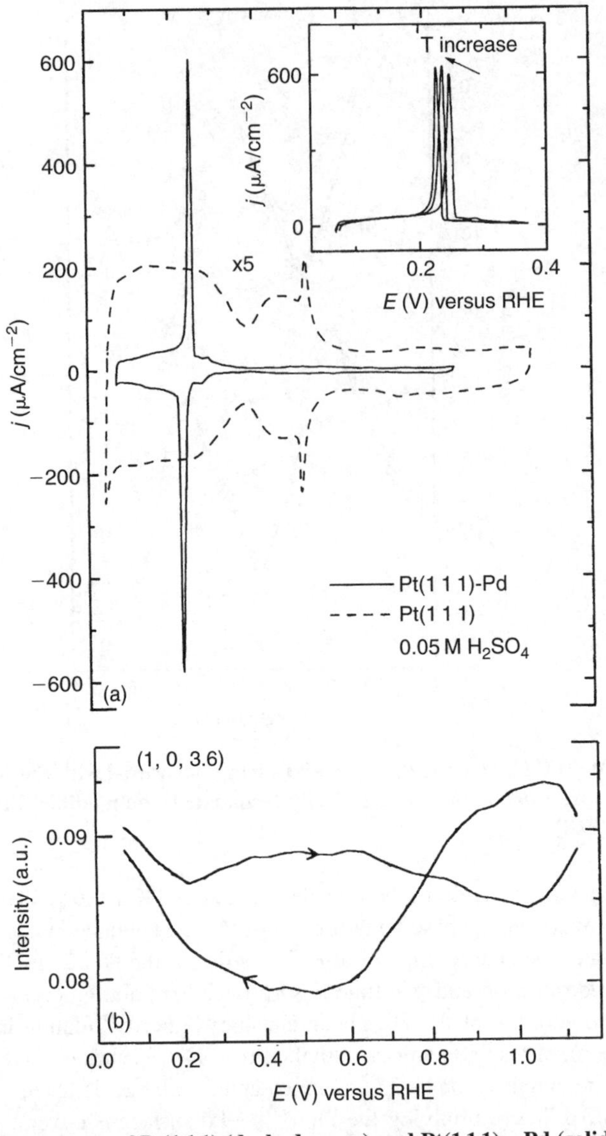

Fig. 27. (a) Cyclic voltammograms of Pt(1 1 1) (dashed curve) and Pt(1 1 1)—Pd (solid curve) in 0.05 M H_2SO_4 at 298 K. Insert: temperature-dependent voltammetry of the Pt(1 1 1)—Pd electrode; sweep rate 50 mV/s. (b) Measurement of the XRV at the (1, 0, 3.6) reciprocal space position for the Pt(1 1 1)—Pd surface; sweep rate 2 mV/s [168]

where chemical bonds are broken or formed is electrocatalytic, and the kinetics vary by many orders of magnitude for different electrode materials. This is true even for the simplest electrochemical reaction where chemical bonds are broken, the hydrogen evolution/oxidation reaction (HER/HOR) and for the more complex reactions such as the oxygen reduction reaction (ORR), and certainly, for bimolecular reactions such as the oxidation of carbon monoxide. While largely different in nature, the kinetics of electrochemical reactions involving either inorganic or organic compounds is governed by the same electrocatalytic law: while the reaction rate passes through a maximum for metals adsorbing reaction *intermediates* moderately, the kinetics is very slow on metals which adsorb intermediates either strongly or weakly, i.e. the Sabatier principle. The establishment of more quantitative relationships between the energetics of intermediates and the rate of electrochemical reaction is somewhat difficult owing in part to the absence of directly measured values for the surface-intermediate bond energy, and in part to the fact that an electrocatalytic reaction occurs on an electrode surface modified by adsorption of "spectator" species rather than on pure metal surfaces. Therefore, for the future development of electrocatalysis as a science, it is essential that these difficulties/complexities be overcome in order to progressively link the atomic/molecular-level properties of the electrochemical interface to the macroscopic kinetic process.

The aim of this section is to summarize recent developments in electrocatalysis on well-defined pure Pt and Pt bimetallic surfaces whose structure and surface chemistry was described in previous sections. The preponderance of electrocatalytic reactions studied on these surfaces are those related to development of fuel cell technology viz the HER/HOR, the ORR, and CO oxidation, since there is no other energy technology whose development into a commercially useful device is so tight to electrocatalysis as is the fuel cell technology. In what follows, the results selected primarily from the authors' laboratory will illustrate the remarkable insight into the effect of electrode structure on the kinetics of electrocatalytic reactions that has been obtained by the study of Pt($h\,k\,l$) surfaces. Furthermore, by focusing on the mechanism of action in bimetallic electrocatalysis, we demonstrate that the ability to make a controlled and well-characterized arrangement of the two elements in the electrode surface and even near-surface region presages a new era of advances in our knowledge of electrocatalysis.

5.1. HER/HOR

The intrinsic kinetic rate of the hydrogen reaction, termed the exchange current density, i_0, is defined as the rate at which the reaction proceeds at the equilibrium (zero net current) potential. Attempts to correlate the exchange current density of the HER with the properties of the electrode substrate date back several decades. A breakthrough with regards to the relationship between log i_0 and physical properties of the electrode substrates was achieved when Conway and Bockris demonstrated a linear relationship between log i_0 and the metal's work function, Φ, and attributed this to the relation between Φ and the metal–hydrogen interaction energy [223]. In the meantime, Parsons [224] and Gerischer [225] independently established from the classical TS theory the relation between log i_0 and the Gibbs energy of adsorbed hydrogen ($\Delta G^0_{\mathrm{H_{ad}}}$). The general form of this relation possesses a "volcano" shape and appears to be valid for both metals and non-metals [226,227]. The exchange current density for the HER varies by 5–6 orders of magnitude depending on the electrode material [227]. Because the highest exchange current density is exhibited by Pt, the major theme in the electrocatalysis of the hydrogen reaction has been to understand the rate dependence on the atomic scale morphology of platinum single crystal surfaces. A number of reviews of this topic that are more detailed than that given here have appeared recently, and should be consulted for further information: these will be cited where appropriate.

*5.1.1. Structure sensitivity on Pt(*h k l*) surfaces*

Early kinetic studies of the HER/HOR were carried out either on polycrystalline platinum electrodes [228] or on platinum single crystals that had poorly defined surface structures [229]. The first reports for the HER on well-ordered Pt($h\,k\,l$) found that the kinetics of the HER in acid solutions was insensitive to the crystallography of the surface [230–232]. However, only recently, it was clearly demonstrated, first by Marković et al. [113,233] and then by Conway and co-workers [234,235], that the HER/HOR on Pt($h\,k\,l$) is indeed a structure-sensitive reaction. The comparison of the polarization curves in Fig. 28 clearly show that in both alkaline as well as acid solution the activity increases in the order (1 1 1) < (1 0 0) < (1 1 0). The order of activity, derived from the Tafel relationships (Fig. 29), reported by Conway and co-workers is slightly different from one shown in Fig. 28a, i.e. Pt(1 0 0) < Pt(1 1 1) < (5 5 1) < (1 1 0) [235]. In any

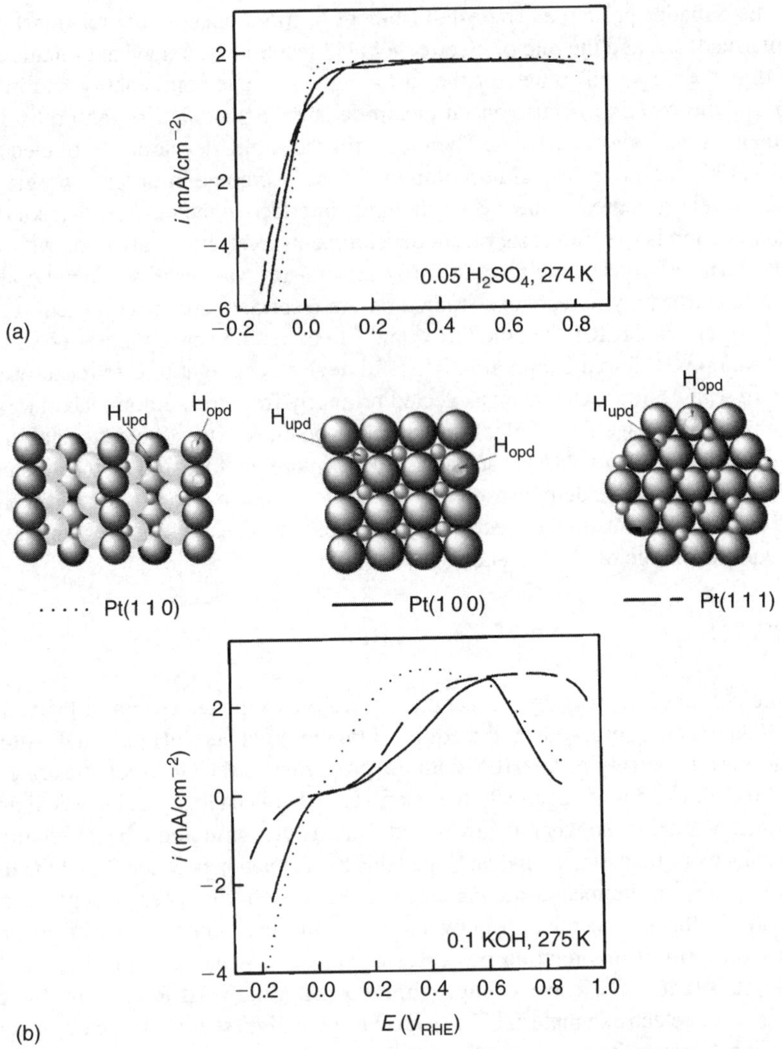

Fig. 28. (a) Polarization curves for the HER and the HOR on Pt(*h k l*) in 0.1 M $HClO_4$ at sweep rate of 20 mV/s [113]. (b) Polarization curves for the HER and the HOR on Pt(*h k l*) in 0.1 M KOH at sweep rate of 20 mV/s [233]. Inserts: ideal models for the Pt(1 1 1)-(1 × 2), Pt(1 0 0)-(1 × 1), and Pt(1 1 1)-(1 × 1) surfaces; small dots represent active sites for H_{upd} and H_{opd}

case, regardless of this rather small disagreement, there is now reasonable consensus that the HER/HOR are structure-sensitive processes. This is confirmed from the values of the apparent activation energy, discerned from the Arrhenius plots [113], which increase in the sequence 9.5 kJ/mol for Pt(1 1 0), 12 kJ/mol for Pt(1 0 0) and 18 kJ/mol for Pt(1 1 1), (Fig. 30a). These differences in the activation energies with crystal face were attributed to structure-sensitive heats of adsorption of the active intermediate, whose physical state is still uncertain. The authors have proposed that the mechanism of the hydrogen reaction is controlled by the interaction between this unknown state and the well-known adsorbed state of hydrogen, H_{upd}, whose adsorption energy is strongly structure-sensitive [126].

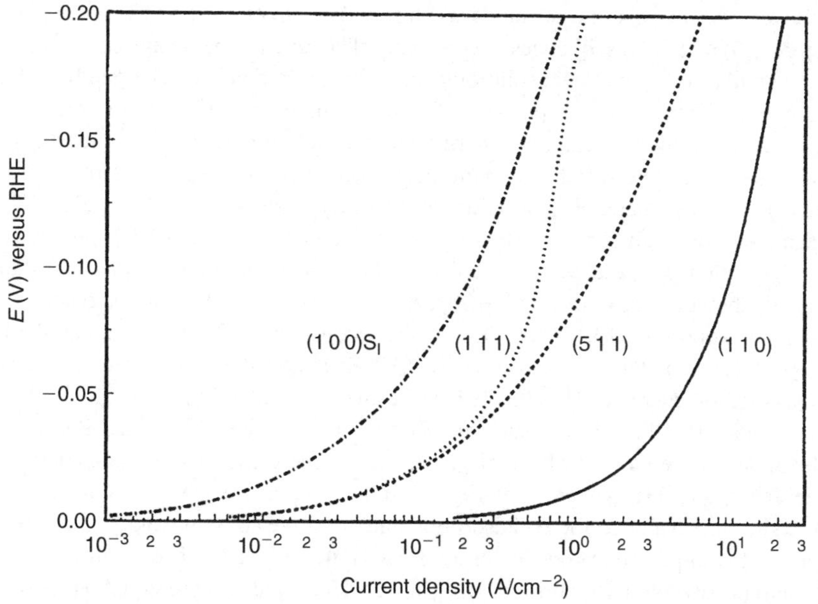

Fig. 29. Simulated Tafel plots, less the diffusion contribution, for the faces (1 0 0)S$_I$, (5 1 1), (1 1 1), and (1 1 0) faces as marked on the plot [235]

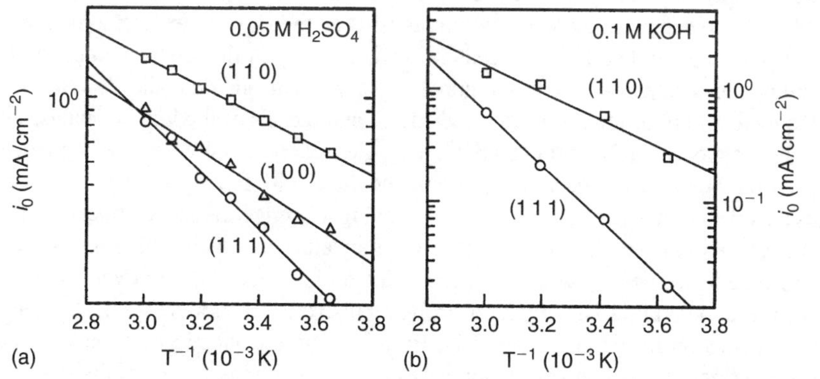

Fig. 30. (a) Arrhenius plots of the exchange current densities (i_0) for the HER/HOR on Pt(hkl) in acid solution [105]. (b) Arrhenius plots of the exchange current densities (i_0) for the HER/HOR on Pt(hkl) in alkaline solution

The mechanism for the HOR on a platinum electrode in acid electrolytes [228,236,237] is usually assumed to proceed by an initial adsorption of molecular hydrogen, which involves either dissociation of H_2 molecules into the atoms, (Eq. (12)), or dissociation into the ion and atom, (Eq. (13)), followed by fast charge-transfer step, (Eq. (14)):

$$H_2 \overset{\text{rds}}{\rightarrow} 2H_{ad} \tag{12}$$

$$H_2 \overset{\text{rds}}{\rightarrow} H^+ + H_{ad} + e^- \tag{13}$$

$$H_{ad} \leftrightarrow H^+ + e^- \tag{14}$$

i.e. historically referred to in some accounts as the Tafel–Volmer (steps (12) and (13)) and Volmer–Heyrovsky (steps (13) and (14)) sequences, respectively. The reaction pathways of the HER/HOR in alkaline solution are similar to those in acid solution except that hydrogen is discharged from H_2O rather than from hydronium ions (H_3O^+). For the purpose of distinguishing the different possible states of adsorbed hydrogen (H_{ad}), it is convenient to employ a thermodynamic notation, referring to the H_{upd} state as the (relatively) strongly adsorbed state which forms on the surface at potentials positive of the Nernst potential, and a more reactive intermediate, H_{opd}, a (relatively) weakly adsorbed state [113,233,238], which forms close to or negative to the Nernst potential. Both Conway and co-workers [239] and ourselves have proposed that the H_{opd} state is *the* reactive intermediate on Pt surfaces, although its physical state is uncertain. The two groups do, however, have different perspectives on the role of H_{upd} in the reaction.

The authors have proposed [113,126,233] that in the vicinity of the Nernst potential, the H_{upd} state may have two modes of action on the kinetics of the HER/HOR: H_{upd} and H_{opd} compete for the same sites; and H_{upd} alters the adsorption energy of H_{opd} on the bare Pt sites adjacent to H_{upd}. This model may seem to conflict with the case of the Pt(1 1 0)-(1 × 2) surface, where the kinetic rate of the HER/HOR is the highest even though the electrode is covered by 1 ML of H_{upd}. The key to resolving this seeming contradiction is in the structure model for H_{upd} and H_{opd} depicted in Fig. 28. If the H_{upd} state is adsorbed only in the troughs of the surface, the top sites are available for H_2 adsorption and bond breaking/making to form H_{opd}. The kinetic model that appears to rationalize the results for the HER/HOR on Pt(1 1 0) is one which follows application of the ideal Langmuir adsorption isotherm for H_{opd}, and the ideal dual-site rate equation for the Tafel–Volmer sequence [113]. The HER/HOR on Pt(1 0 0) also occurs on the surface which is "fully" covered by H_{upd}. On the geometrically homogenous Pt(1 0 0) surface, if there is only one type of site for the adsorbed hydrogen, then a (1 0 0) surface covered by 1 ML of H_{upd} should be completely inactive for the hydrogen reaction. The implication is that unoccupied Pt sites, required for the formation of H_{opd}, can only be created if some amount (unknown) of the H_{upd} adatoms sit in deeper potential wells on the surface, freeing some top-sites to serve as adsorption centers for H_{opd}, and/or the active centers are defect sites. Although the number of these active sites near 0 V is unknown, it appears that this number is very small and that the Pt(1 0 0) is active for the HER/HOR due to very high turnover rate at these sites. On the basis of the kinetic analyses of the polarization curves in the vicinity of the Nernst potential, the single-site Heyrovsky–Volmer sequence was proposed to be operative, with the ion-atom reaction step being the rds [1 1 3]. Finally, on the densely packed Pt(1 1 1) surface H_{upd} is adsorbed *on* the surface (Section 4.3), having a strongly negative effect on the reaction rate of the HER/HOR. As mentioned earlier, H_{upd} on Pt(1 1 1) may alter both the number of Pt-free sites (blocking effect) and/or the interaction of H_{opd} with the surface (electronic effect). The relatively high activation energy for the hydrogen reaction on Pt(1 1 1) implies a strong repulsive interaction between H_{opd} and H_{upd}, as found for H_{upd}—H_{upd} interaction itself, (Section 4.2.1). By analyzing the kinetic parameters for the HER/HOR [113], however, the reaction mechanism cannot be resolved in a definitive way, as discussed in [113].

The HER/HOR was also found to be structure-sensitive in alkaline solution. In fact, the first indication that the rate of the HER/HOR is structure-sensitive arise from the measurements in alkaline solution [233]. At relatively low temperature (273–298 K), the same order of the activity as in acid solution was observed, (Fig. 28b). As for the acid solution, the ordering in the activity is mirrored by the apparent activation energies, which increase from 18 kJ/mol for Pt(1 1 0) to 36 kJ/mol for Pt(1 1 1) [240], Fig. 30b. However, the activation energies in alkaline solution are doubled relative to the values obtained in acid solution, producing exchange current densities on all three surfaces that are dramatically lower in alkaline versus acid solution. A similar pH effect for the HER/HOR kinetics on polycrystalline Pt was reported some time ago by Osetrova and Bagotsky [241]. The reason for this effect has never been adequately explained. Very recently, Schmidt et al. [242] proposed in alkaline solution the kinetics of the HER/HOR even close to the Nernst potential may be affected by OH_{ad}, since OH_{ad} appears to be present deep into the H_{upd} potential region, see 4.2 and 5.3. OH_{ad} may either block active sites required for the adsorption of H_2 molecules and/or change the energetics of the adsorbed states, e.g. change reactive H_{opd} into unreactive H_{upd} [240]. An obvious

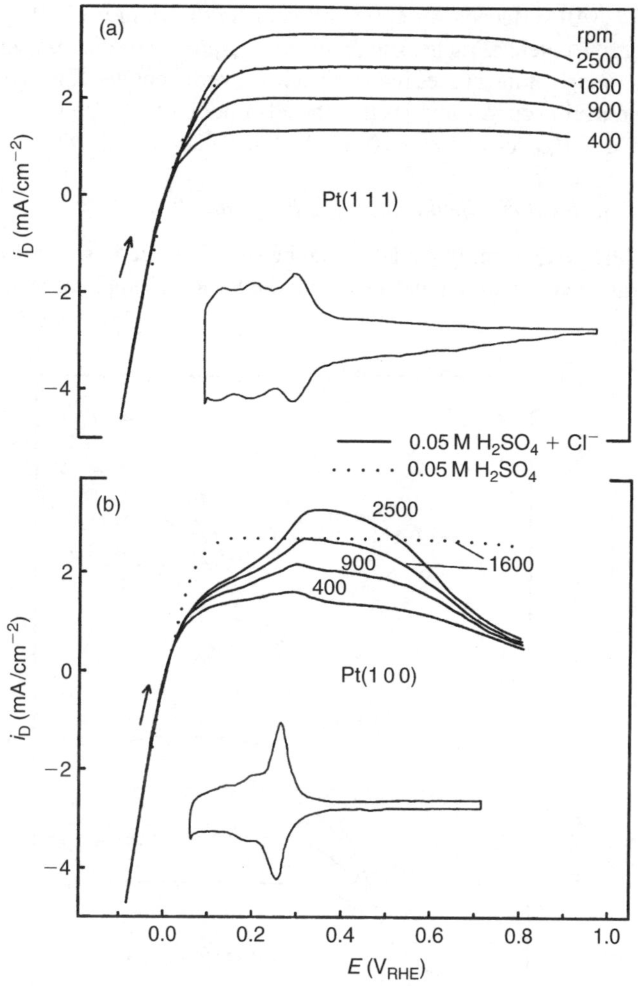

Fig. 31. (a) Polarization curves for the HER and the HOR on Pt(1 1 1) in 0.05 M H_2SO_4 (solid curves), and 0.05 M H_2SO_4 + 10^{-3}M Cl^- (dotted curves) at sweep rate of 20 mV/s. Insert: cyclic voltammetry of Pt(1 1 1) in and 0.05 M H_2SO_4 + 10^{-3}M Cl^-. (b) Polarization curves for the HER and the HOR on Pt(1 0 0) in 0.05 M H_2SO_4 (solid curves), and 0.05 M H_2SO_4 + 10^{-3}M Cl^- (dotted curves) at sweep rate of 20 mV/s. Inset: cyclic voltammetry of Pt(1 0 0) in and 0.05 M H_2SO_4 + 10^{-3}M Cl^- [250]

inhibiting effect of OH_{ad} on the HOR can be seen in the potential region, where OH_{ad} formation is clearly resolved in the cyclic voltammetry, e.g. $E > 0.4$ V. As shown in Fig. 28b, the order of activity in the OH_{ad} potential region, $E > 0.6$ V, increases in order Pt(1 0 0) < Pt(1 1 0) < Pt(1 1 1) and closely follows the nature of the Pt(hkl)—OH_{ad} interaction and the potential dependent surface coverage by OH_{ad} [233].

Adsorption of halide anions in acid solution has also a strong inhibiting effect on the HER/HOR kinetics as OH_{ad} in alkaline. The base voltammogram in Ar-purged solutions and the HER/HOR polarization curves for pure H_2SO_4 and $HClO_4$ with and without added Cl^- are shown in Fig. 31 for Pt(1 1 1) and Pt(1 0 0). The equality of the rate at the Pt(1 1 1) electrode in the Cl^--free and Cl^--containing solutions is consistent with the absence of co-adsorption of Cl^- with H_{upd} on this surface. On the other hand, due to the co-existence of H_{upd} and Cl_{ad} (Cl^- adsorption starts at ca. +30 mV of the Nernst potential), the kinetics of the HOR onto the Pt(1 0 0) surface is strongly inhibited. As shown in Fig. 31b, pure diffusion currents are observed in very narrow potential range, and at higher overpotentials decrease in kinetic activity of the HOR is almost a mirror image of the curves recorded at low overpotentials. It is interesting to note that the "bell-shape" of the polarization curves implies that the adsorption of Cl_{ad} suppresses the catalytic activity in nearly the same way as OH_{ad} does on Pt(1 0 0) in alkaline solution (Fig. 28b).

5.1.2. *HOR on Pt(*hkl*) surfaces modified with* Cu_{upd}, Pb_{upd}, *and* Bi_{ir}

For Cu UPD and Pb UPD systems, only the HOR can be studied since in solutions containing Cu^{2+} and Pb^{2+} bulk metal deposition occurs simultaneously with the HER. A sample of results is shown in Fig. 32.

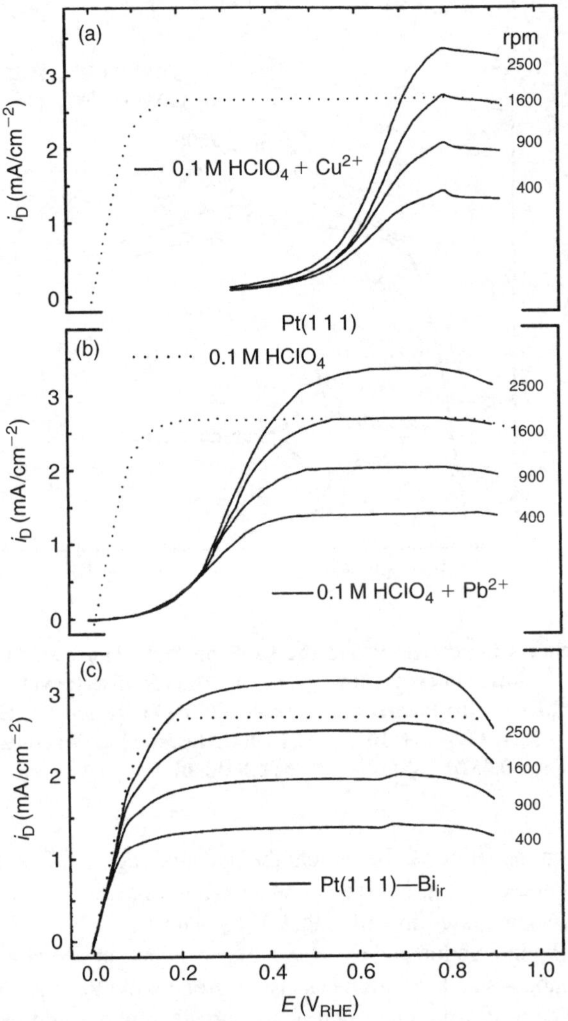

Fig. 32. (a)–(c) Polarization curves for the HOR on Pt(111) in 0.1 M $HClO_4$ (dotted curve), recorded at 1600 rpm. (a) HOR on Pt(111) in 0.1 M $HClO_4$ + 10^{-3} M Cu^{2+} (solid curves) at sweep rate of 20 mV/s. (b) HOR on Pt(111) in 0.1 M $HClO_4$ + 10^{-3} M Pb^{2+} (solid curves) at sweep rate of 20 mV/s. (c) HOR on Pt(111) in 0.1 M $HClO_4$ (dotted curve) and on a Pt(111)—Bi_{ir} electrode in 0.1 M $HClO_4$ at sweep rate of 20 mV/s

One observation evident from this figure is that the HOR is strongly inhibited on all three admetal-modified surfaces. There is, in general, a disinclination to report negative results, except as part of a study of systematic trends in activity with a variation in some characteristic of the catalysts, in which case the results are quite valuable. However, the mechanism of action of these admetals, even though an inhibiting action ("poisoning"), is closely related to one of the central issues of this volume: the role of the energetics of adsorption on the catalytic activity of an electrochemical reaction. Therefore, we proceed with further discussion of these interesting inhibiting effects.

A close inspection of Fig. 32 indicates that at low overpotentials the order of activity of the HOR decreases in sequence: Pt(1 1 1)—Bi_{ir} > Pt(1 1 1)—Pb_{upd} < Pt(1 1 1)—Cu_{upd}. On the other hand, at high overpotentials the most active Pt(1 1 1)—Bi_{ir} surface become the least active for the HOR, due presumably to enhanced formation of OH_{ad} on adjacent Pt sites that are active for the HOR. The most straightforward explanation of this observation is that the activity for the HOR increases in the same order as the availability of Pt sites for the adsorption of molecular H_2, i.e. a purely site blocking effect. At low overpotentials, the Bi_{ir}-modified Pt(1 1 1) surface is the least poisoned because for coverages of $\Theta_{Bi_{ir}} = 0.34$ ML there are still sites available for dissociative adsorption of H_2 to take place. The number of these sites is relatively small, and thus the diffusion limiting currents were not reached on the Pt(1 1 1)—Bi_{ir} electrode. In contrast, the rate of reaction on the UPD-modified Pt(1 1 1) surface mirrors the anodic stripping of the Pb_{upd} and Cu_{upd} on a one-to-one basis. In this case, the admetals do not appear to influence the kinetics via a change in the energetics of the adsorbed hydrogen, but rather with the availability of bare Pt sites. Therefore, due to the strong enhanced adsorption of anions on Cu_{upd}-modified Pt(1 1 1) (Section 4.5.1, Fig. 20, State 2) this surface is the least active for the HOR. Note that in the entire potential range Cu_{upd} is covered by anions, (bi)sulfate in sulfuric acid solution and Cl_{ad} in solution containing Cl^-. For more details the readers are referred to [243] and [350].

5.1.3. HER/HOR on Pt(1 1 1)-modified with a pseudomorphic Pd film

Theoretical studies of the dissociative adsorption of H_2 on transition metals have suggested that electron transfer from the metal into the σ^* anti-bonding orbital of the H_2 molecule plays an important role in the breaking of the H—H bond by lowering the activation energy associated with the process. Transition metals, such as pure Pd, are good electron donors, and dissociation of H_2 on these surfaces occurs readily at room temperature. In contrast, at room temperature no dissociation of H_2 is observed on the Re(0 0 0 1)—Pd, and Nb(1 1 0)—Pd overlayer systems because the charge transfer from the Pd overlayer to the substrate causes the local d-band on Pd to shift down [213]. The development of theoretical tools to describe and understand adsorption on thin metal films has been quite parallel to the experimental developments. A number of concepts have been introduced to describe both "electronic" and "geometrical" effects in gas-phase catalyses of H_2 on ultrathin metal films. For an overview of the key theoretical aspects of adsorption on ultrathin metal surfaces, the readers are referred to the review of Hammer and Norskov [215,216,244]. In general, the results indicate a clear correlation between the d-band center of the overlayer metal atoms and the hydrogen chemisorption energy.

Following the example of these interesting UHV studies, the adsorption and catalytic properties of pseudomorphic Pd films supported on single crystal metal surfaces has also received some attention in surface electrochemistry. Comparison of the HER/HOR at Pt(1 1 1) and Pt(1 1 1)—Pd electrodes in Fig. 33 clearly reveals that the kinetics of the HER/HOR is significantly enhanced on the Pd-modified Pt(1 1 1) surface. The activation energy for the hydrogen reaction obtained from kinetic analyses of polarization curves on Pt(1 1 1)—Pd is ca. 9 kJ/mol, which is half the activation energy obtained for the hydrogen reaction on Pt(1 1 1). In line with discussions of the HER/HOR on Pt(hkl), the kinetics of the hydrogen reaction may be enhanced either by introducing defect sites or by lowering the energy of the transition state on the Pt(1 1 1)—Pd surface compared to the unmodified Pt(1 1 1) surface. Although still

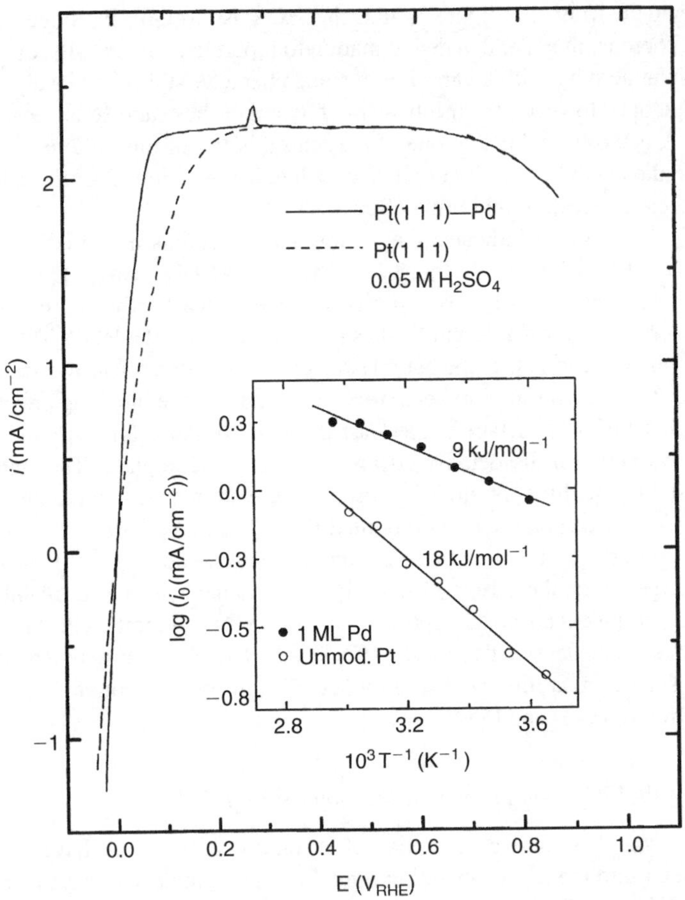

Fig. 33. Comparison of the HER/HOR on Pt(1 1 1)—Pd (solid curve) and Pt(1 1 1) (dashed curve) in 0.05 M H_2SO_4 at 278 K. Rotation rate 1600 rpm. Insert: Arrhenius plots for the HER/HOR on Pt(1 1 1)—Pd and Pt(1 1 1) [168]

uncertain, it appears that the change in the adsorption energetics has the more important role in the enhanced kinetics on the Pd-modified Pt(1 1 1) surface. Close to the Nernst potential, the HER/HOR on Pt(1 1 1)—Pd occurs on a surface, which is completely covered by H_{upd} ($\approx$ 1 H/Pt), in contrast to Pt(1 1 1), which is only partially covered by H_{upd} (0.67 H/Pt). Again, the key to resolving the paradoxical activity of the Pt(1 1 1)—Pd surface for the HER/HOR is to be found in understanding the nature of H_{upd} (Section 4.6) and how this state may effect the formation of H_{opd}. The fact that a full monolayer of H_{upd} can be formed on the Pt(1 1 1)—Pd surface suggests that Pt(1 1 1)—Pd—H_{upd} interaction is much stronger and/or the H_{upd}—H_{upd} repulsive interaction is weaker than for Pt(1 1 1). Consequently, some amount of H_{upd} might sit in deeper potential wells of the three-fold hollow sites, and thus be more *in* the surface than *on* the surface. Along the same lines of argument we used for Pt(1 1 0)-(1 $\times$ 2), the HER/HOR can occur on a geometrically homogenous Pt(1 1 1)—Pd surface fully covered by H_{upd} only if some amount of the H_{upd} indeed is in the subsurface state, allowing adsorption of H_{opd} on free Pd top-sites. An alternative explanation may be that at high coverages by H_{upd}, the hypothetical transition form an unreactive H_{upd} layer into a reactive H_{opd} layer is facilitated on the Pt(1 1 1)—Pd surface due to high coverages by H_{upd}. Nevertheless, the change in the energetics of adsorption would be attributed to the electronic effect of intermetallic bonding between the Pd overlayer and the Pt substrate. Clearly the electronic effect is an extremely important fundamental parameter in electrocatalysis.

5.2. ORR

The ORR is more complex than the HER/HOR, but is an equally important electrocatalytic reaction. In this case, there is no electrode material for which there is even a measurable current from oxygen reduction at the equilibrium potential, 1.23 V versus the normal hydrogen electrode (NHE). Even for the most catalytically active electrode materials, again platinum group metals, measurable currents are obtained only below 1 V. It is customary, therefore, in kinetic studies of the ORR to use the current density at a fixed potential, e.g. at 0.9 V, as a measure of the reaction rate instead of the exchange current density. The experimental relation between the current density, i, and electronic properties of metals known to influence the heat of adsorption of oxygen, e.g. d-orbital vacancies, have also produced "volcano-type" curves. As with the HER, the platinum group metals are at or near the top of the "volcano" plot. The volcano relationship for the ORR appears to be valid for alloys as well as for pure metals, showing variations in reaction rates by 5–6 orders of magnitude [245]. In such plots, however, the scale for the reaction rate is usually a logarithmic scale, where rates differ by orders of magnitude. On a finer scale, additional factors must be accounted for to explain the full range of catalytic response on different surfaces. The optimization of an electrocatalyst for the ORR must, therefore, accommodate many contributing factors, including the rate dependence on the geometry of surface atoms. For the purpose of this review, it will be shown that the elimination of surface heterogeneity and the enhancement of the resolution in the voltammetry that results from a single crystal geometry can bring new insight into our understanding of the reaction pathway of the ORR on platinum electrodes in aqueous electrolytes.

5.2.1. Reaction pathway

The ORR is a multi-electron reaction that may include a number of elementary steps involving different reaction intermediates. Of various reaction schemes proposed for the ORR [246], the simplified version of the scheme given by Wroblowa et al. [247] appears to be the most effective one to describe the complicated reaction pathway by which O_2 is reduced at metal surfaces:

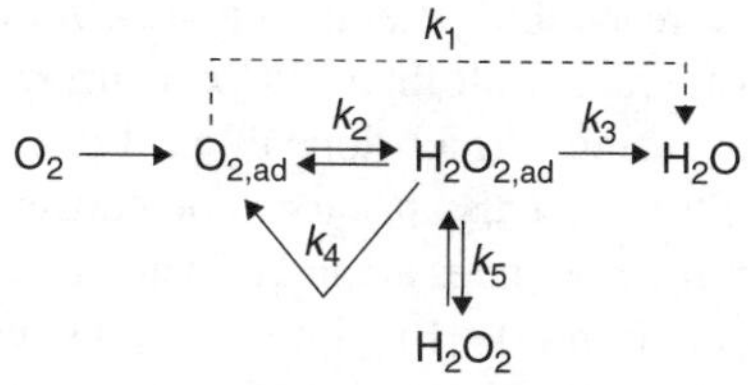

On the basis of this reaction scheme, O_2 can be electrochemically reduced either directly to water with the rate constant k_1 without intermediate formation of $H_2O_{2,ad}$ (the so-called "direct" 4e$^-$ reduction) or to $H_2O_{2,ad}$ with the rate constant k_2 ("series" 2e$^-$ reduction). The adsorbed peroxide can be electrochemically reduced to water with the rate constant k_3 ("series" 4e$^-$ pathway), catalytically (chemically) decomposed on the electrode surface (k_4) or desorbed into the bulk of the solution (k_5). Although a number of important problems pertaining to the interpretation of the reaction pathway for the ORR on Pt($h\,k\,l$) have yet to be resolved, recent studies from our laboratory [248–250] suggest that a series pathway via an $(H_2O_2)_{ad}$ intermediate may be operative on all Pt and Pt bimetallic catalysts. This can be considered as a special case of the general mechanism, where k_1 is essentially zero, i.e. there is no splitting of the O—O bond before a peroxide species is formed. Peroxide, on the other hand may ($k_5 = 0$) or may not ($k_5 \neq 0$) be further reduced to water. In either case, the rate determining step appears to be the addition of the first electron to $O_{2,ad}$. The rate expression is then [249,251,252],

$$i = nFKc_{O_2}(1-\Theta_{ad})^x \exp\left(-\frac{\beta FE}{RT}\right)\exp\left(-\frac{\gamma r\Theta_{OH_{ad}}}{RT}\right) \tag{15}$$

where n is the number of electrons, K the rate constant, c_{O_2} the concentration of O_2 in the solution, Θ_{ad} the total surface coverage by anions (Θ_{anions}) and OH_{ad} ($\Theta_{OH_{ad}}$), x is either 1 or 2 depending on site requirements of the adsorbates, i the observed current, E the applied potential, β and γ are the symmetry factors (assumed to be 1/2), and r a parameter characterizing the rate of change of the apparent standard free energy of adsorption with the surface coverage by adsorbing species, e.g. as in UHV (Section 2.2.2), the adsorption energy decrease with increasing $\Theta_{OH_{ad}}$. In deriving Eq. (15), it is assumed that the reactive intermediates, $(O_2^-)_{ad}$ and $(HO_2^-)_{ad}$, are adsorbed to low coverage, i.e. they are not a significant part of Θ_{ad}. Therefore, the kinetics of O_2 reduction is determined either by the free platinum sites available for the adsorption of O_2 ($1-\Theta_{ad}$ term in Eq. (15)) and/or by the change of Gibbs energy of adsorption of reaction intermediates with $\Theta_{OH_{ad}}$ ($r\Theta_{OH_{ad}}$ term in Eq. (15)). In the following sections, we will use this reaction pathway and rate expression to analyze the effects of various factors on the kinetics of the ORR on Pt($h\,k\,l$) surfaces.

5.2.2. *Structure sensitivity on Pt(*h k l*) surfaces*

Studies of the ORR on Pt($h\,k\,l$) date back to the late 1970s, where essentially the same activity for the ORR was reported on all three low-index single-crystal surfaces [253]. Recently, however, the results obtained by RRDEs showed unambiguously that the kinetics of the ORR on Pt($h\,k\,l$) surfaces vary with the crystal face in a different manner depending on the electrolyte [254,255]. The authors have proposed that not only does structure sensitivity arise because of the sensitivity of the adsorption energy of the reactive intermediates to site geometry, but also from the sensitivity to site geometry of the adsorption of spectator species, such as $HSO_{4(ad)}$ [256], Cl_{ad} [250], and Br_{ad} [249]. Within the limited scope of this volume, it will not be possible to review all of these results. Rather we will show, using selected examples, the kind of information that can be used to improve the interpretation of the role of the local symmetry of platinum surface atoms in the ORR.

There are two general observations concerning the structure sensitivity of the ORR: (i) the structural sensitivity is most pronounced in electrolytes in which there is strong adsorption of anions; (ii) the *same* activation energy in both acid (at the reversible potential ca. 42 kJ/mol [248]) and alkaline solution (at 0.8 V ca. 40 kJ/mol [240]) has been found for all three Pt($h\,k\,l$) surfaces. In solutions containing HSO_4^- [256], Cl^- [250], and Br^- [249] ions, a single Tafel slope of ca. 120 mV/dec is deduced from the kinetic analyses of the results for ORR on Pt($h\,k\,l$). This appears to be consistent support that the standard free energies of adsorption of the reaction intermediates are not affected by the adjacent anions and thus, at the Pt(1 1 1)—$HSO_{4(ad)}$, Pt($h\,k\,l$)—Cl_{ad}, and Pt($h\,k\,l$)—Br_{ad} interfaces, the term $r\Theta_{OH_{ad}}$ becomes negligible. As a consequence, the structure sensitivity in the solution containing bisulfate and halide anions is *completely* determined by the ($1-\Theta_{anions}$) pre-exponential term.

The order of activity of Pt($h\,k\,l$) for the ORR in H_2SO_4 increases in the sequence Pt(1 1 1) < Pt(1 0 0) < Pt(1 1 0), (Fig. 34a). An exceptionally large deactivation is observed at the Pt(1 1 1) surface. This is due to the strong adsorption of the (bi)sulfate anion from the symmetry match between the fcc (1 1 1) face and the C_{3v} geometry of the oxygen atoms of the sulfate anion, (Section 4.4.1). The fact that the activity of all three low-index platinum planes is significantly higher in $HClO_4$ [126] does, however, suggest that adsorption of (bi)sulfate anions effects the kinetics of reduction on all three surfaces, and has a particularly strong effect on Pt(1 1 1) [257]. It should be noted that although bisulfate adsorption onto Pt($h\,k\,l$) surfaces inhibits the reduction of molecular O_2, probably by blocking the initial adsorption of O_2, it does *not* affect the pathway of the reaction, since no H_2O_2 is detected on the ring electrode for any of the surfaces in the kinetically controlled potential region.

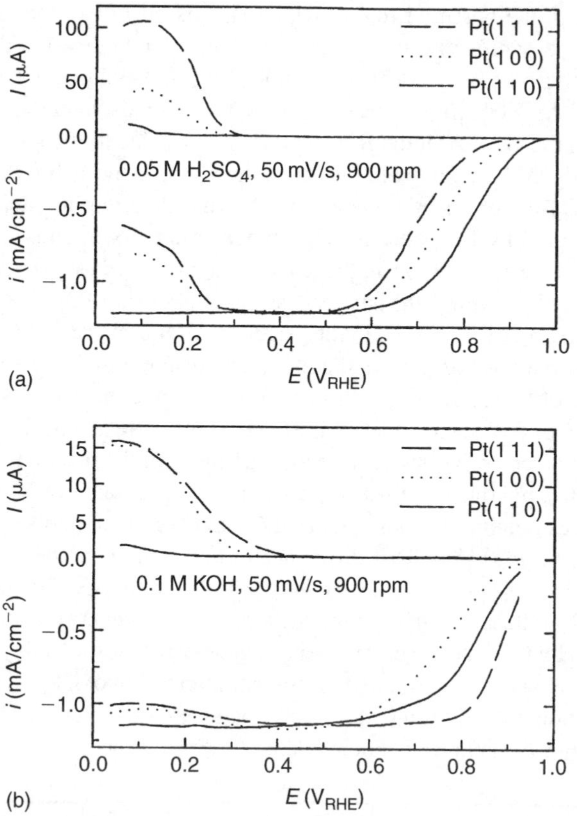

Fig. 34. (a) Disk and ring (I_R) currents during oxygen reduction on Pt(1 1 1) in 0.05 M H_2SO_4 at a sweep rate of 50 mV/s [256]. (b) Disk and ring (I_R) currents during oxygen reduction on Pt(1 1 1) in 0.1 M KOH at a sweep rate of 50 mV/s [258]. Ring potential = 1.15 V. Collection efficiency, $N = 0.22$

The Pt(hkl)/KOH system is the most direct probe of the effects of OH_{ad} on the rate of the electrode reaction, since no other anions are co-adsorbed with hydroxyl species. Fig. 34b shows that in the potential range, where O_2 reduction is under combined kinetic-diffusion control ($E > 0.75$ V), the order of activity of Pt(hkl) in 0.1 M KOH increase in the sequence (1 0 0) < (1 1 0) < (1 1 1). The structure-sensitivity of the ORR in this potential region arises due to the structure-sensitive adsorption of the hydroxyl species, i.e. the most active (1 1 1) surface has the lowest coverage by OH_{ad} and weakest Pt—OH_{ad} interaction. Although at the same overpotential the surface coverage by adsorbed OH_{ad} on Pt(1 1 0)-(1 × 2) is similar, if not slightly higher, than on Pt(1 0 0), the activity of Pt(1 1 0) is higher than the activity of Pt(1 0 0). If on the fcc (1 1 0) surface the adsorption of hydroxyl ion is predominantly in the "trough" positions, as proposed for H_{upd} (Fig. 28), then the top atoms may serve as active centers for the ORR even when the Pt(1 1 0) surface is nominally "fully" covered with OH_{ad} [258]. It is important to note that the Tafel slope on different single crystal surfaces in KOH ($HClO_4$) may deviate significantly from 120 mV/dec [126], similar to what was found in the literature for the O_2 reduction on polycrystalline electrodes [251,252,259]. The change in the slope has been attributed either to the change from Temkin to Langmuirian conditions (the effect of $\Theta_{OH_{ad}}$ term in Eq. (15)) for the adsorption of reaction intermediates, or to a change in the surface coverage by OH_{ad}, e.g. the effect of $(1 - \Theta_{OH_{ad}})$ term in Eq. (15). Very recently, Marković et al. [249] developed a theoretical model which showed that the best-fit of the Tafel slopes in KOH and $HClO_4$ can be obtained simply by introducing both blocking and energetic components in Eq. (15).

As shown in Fig. 34, in the potential region where H_{upd} is adsorbed, the rate of the ORR becomes strongly structure-sensitive, with activity decreasing in the order Pt(1 1 0) > Pt(1 0 0) > Pt(1 1 1). Note that the same order of activity was observed for the HOR, (Fig. 27). Following the discussion for the structure-sensitive kinetics of the HOR in the preceding section, it is reasonable to propose that the structure sensitivity of the ORR in the H_{upd} potential region arises mainly due to the structure-sensitive adsorption of the H_{upd} state [255,256,258]. In acid solution, once the O—O bond is broken, protonation to water is extremely rapid, and H_{upd} is not an intermediate in the rate determining step. However, the ORR on Pt(1 1 1) is strongly inhibited by H_{upd} because H_{upd} blocks the pairs of sites which are required for the O—O bond breaking. In the case of Pt(1 0 0), the active centers for the O—O bond breaking are primarily the defect/step sites created by "lifting" the hex reconstruction. For Pt(1 1 0)-(1 × 2), the active sites are the top-rows of Pt atoms since H_{upd} is adsorbed in the "troughs" of the surface (Fig. 28), thus leaving the top-sites available for O_2 and H_2O_2 adsorption, and consequent cleaving of the O—O bond [256].

The strong adsorption of halide anions, may or may not change the reaction mechanism. Apparently, the Pt(1 1 1)—Cl_{ad} system has the same effect on the ORR as bisulfate anions, i.e. although Cl_{ad} inhibits the initial adsorption of O_2 molecules, it does *not* affect the pathway of the reaction, (Fig. 35a). In contrast, the ORR at the Pt(1 1 1)—Br_{ad} interface is always accompanied quantitatively by H_2O_2 production because the number of Pt pair-sites required for adsorption of O_2 and the breaking of O—O bond are reduced on the Pt(1 1 1)—Br_{ad} surface, (Fig. 35b). This difference in the catalytic activity can be understood simply from the stronger Pt(1 1 1)—Br_{ad} than Pt(1 1 1)—Cl_{ad} interaction, see Section 4.4.2. The adsorption isotherms for Br_{ad} and Cl_{ad} reflect this stronger interaction, with a lower coverage for Cl_{ad} at most potentials, and displacement of Cl_{ad} by OH_{ad} above ca. 0.5 V, i.e. a generally "looser" adlayer with open pair-sites for breaking of O—O bond. However, on Pt(1 0 0), a stronger interaction of Cl_{ad} with (1 0 0) versus the (1 1 1) surface causes the reaction to go through the series pathway via solution phase peroxide (Fig. 35c). Unfortunately, the adsorption isotherm is not available for this surface, but the presumption is that there

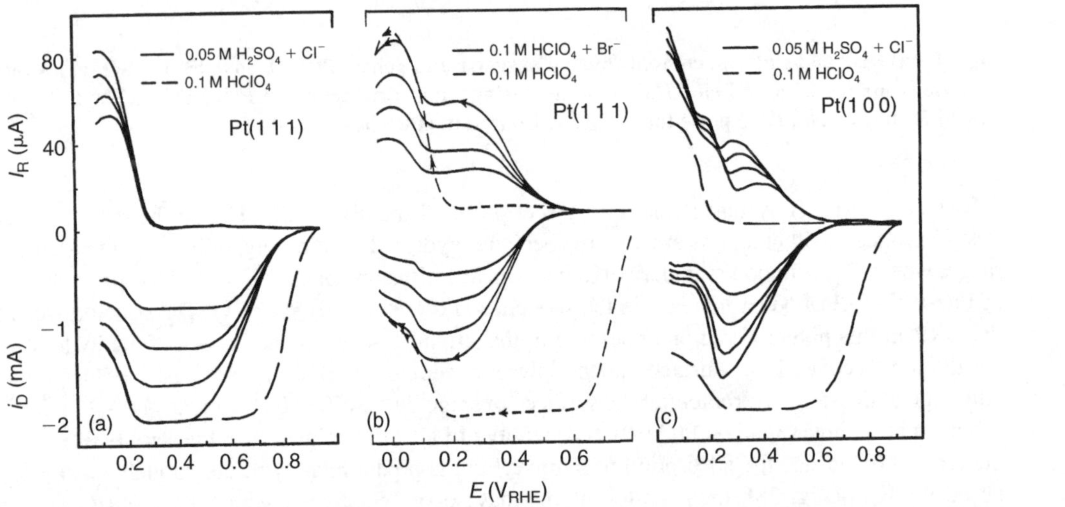

Fig. 35. (a) Polarization curves for the ORR on the Pt(111) disk electrode in 0.1 M $HClO_4$ (dashed curve), 0.05 M H_2SO_4 + 10^{-3}M Cl^- (solid curves), and corresponding peroxide oxidation currents on the ring electrode at a sweep rate of 50 mV/s. (b) Polarization curves for the ORR on the Pt(111) disk electrode in 0.1 M $HClO_4$ (dashed curve), 0.1 M $HClO_4$ + 10^{-4}M Br^- (solid curves) and corresponding peroxide oxidation currents on the ring electrode at a sweep rate of 50 mV/s. (c) Polarization curves for the ORR on the Pt(100) disk electrode in 0.1 M $HClO_4$, 0.05 M H_2SO_4 + 10^{-3}M Cl^-, and corresponding peroxide oxidation currents on the ring electrode at a sweep rate of 50 mV/s [249, 250]

are higher coverages of Cl_{ad} on the (1 0 0) surface and loss of active sites that are required for breaking of the O—O bond [250].

*5.2.3. ORR on Pt(*h k l*) surfaces modified with UPD metals*

Studies of the kinetics of the ORR on platinum single crystal electrodes modified by UPD metal adatoms are of interest primarily from a fundamental science perspective. Since one can start the study from a clean Pt(*h k l*) surface, where the ORR is well-known, the change in both rate and reaction pathway with the addition of another metal to the surface, usually in a highly structured manner, can provide some insight to the ORR on metals that are otherwise difficult to study, e.g. Cu, Pb and Bi. Only the results on Cu_{upd}-modified surfaces are reviewed here. The first result for the ORR on the Pt(1 1 1)—Cu_{upd} electrode was published by Abe et al. [175]. By using STM and hanging meniscus rotating disk electrode (HMRDE) methods, the authors demonstrated a close correlation between the inhibition of the ORR with the microstructure of Cu_{upd} adlayer. In sulfuric acid solution, three-fold hollow platinum sites are still accessible for O_2 adsorption inside the honeycomb $(\sqrt{3} \times \sqrt{3})R30°$ structure that forms at ca. 0.5 V, but the reduced geometry favors $2e^-$ reduction to peroxide versus $4e^-$ reduction to water. More recently, the correlation between the surface structure of Cu_{upd} and the kinetics of the ORR was re-examined by using a combination of RRDE and SXS measurements [243]. In contrast to previous work, these measurements showed that there is *no* correlation between the ORR kinetics and the *ordered* Cu_{upd} surface structures irrespective of the nature of the specifically adsorbing anions [243]. The analyses of the RRDE data (Fig. 36)

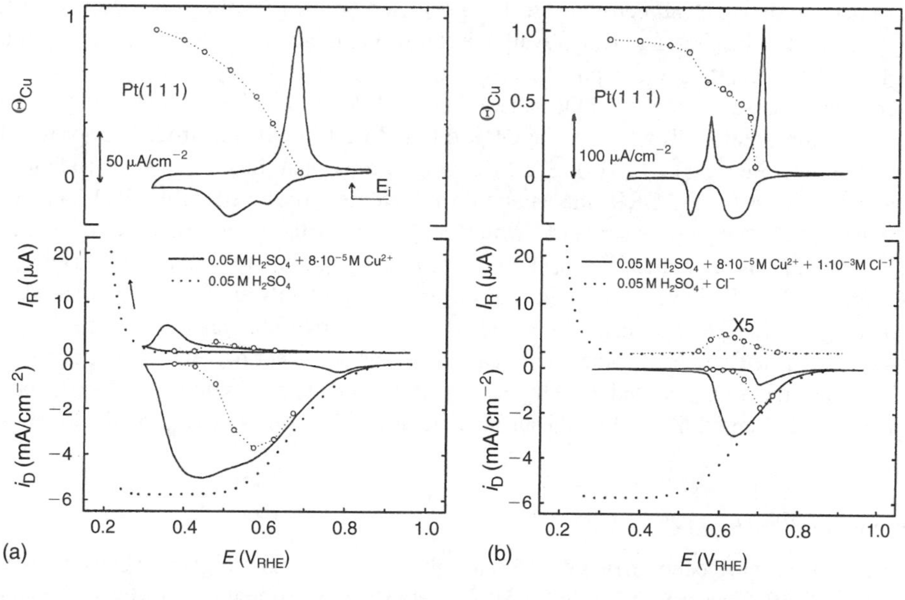

Fig. 36. (a) Upper panel: cyclic voltammetry for Cu UPD on Pt(1 0 0) in 0.05 M H_2SO_4 10^{-3} + 10^{-3}M Cu^{2+} at sweep rate 20 mV/s. Lower panel: polarization curves for the ORR on the Pt(1 0 0) disk electrode and corresponding peroxide oxidation current on the ring electrode in 0.05 M H_2SO_4 + 10^{-3}M Cl^- + 10^{-3}M Cu^{2+} and at sweep rate of 20 mV/s. Close circles: steady-state measurements in 0.05 M H_2SO_4 + 10^{-3}M Cu^{2+}. (b) Upper panel: cyclic voltammetry for Cu UPD on Pt(1 1 1) in 0.05 M H_2SO_4 10^{-3}M Cl^- + 10^{-3}M Cu^{2+} at sweep rate 20 mV/s. Lower panel: polarization curves for the ORR on the Pt(1 1 1) disk electrode and corresponding peroxide oxidation current on the ring electrode in 0.05 M H_2SO_4 + 10^{-3}M Cl^- (dotted curves) and 0.05 M H_2SO_4 + 10^{-3}M Cl^- + 10^{-3}M Cu^{2+} (solid curves) at a sweep rate of 20 mV/s. Close circles: steady-state measurements in 0.05 M H_2SO_4 + 10^{-3}M Cl^- + 10^{-3}M Cu^{2+}. Collection efficiency, N = 0.22 [350]

also revealed that the mechanism for the ORR on Cu_{upd}-modified electrodes is the same as on unmodified Pt(1 1 1) and Pt(1 0 0), i.e. proceeds mostly as a $4e^-$ reaction pathway with negligible solution phase peroxide formation (ca. 3%). It is surprising, however, that a relatively small amount of Cu_{upd} has such a devastating effect on the rate of ORR. This anomalously large inhibition by a very small amount of Cu_{upd} is attributed to an enhanced anion adsorption on platinum atoms adjacent to Cu_{upd} atoms [243], see Section 4.5.1 and Fig. 20. A predominate role of anions in the rate of ORR on Cu_{upd}-modified platinum single crystal surfaces is supported by the fact that the activity of the ORR decreases in the same sequence as the Cu—anion bond strength increases: $Pt—Cu_{upd}—HSO_{4,ad} \gg Pt—Cu_{upd}—Cl^-$. The model which rationalize these results is one where the active sites for the adsorption of molecular O_2 are the small number of platinum "islands" created in State 2 in Fig. 20 [243].

5.2.4. *ORR on the Pt(1 1 1)-modified with a pseudomorphic Pd film*

Figure 37 summarizes a family of polarization curves for the ORR on the Pt(1 1 1)—Pd electrode in O_2–saturated 0.05 M H_2SO_4, 0.1 M $HClO_4$, and 0.1 M KOH solutions along with representative polarization curves recorded in the same solutions but on unmodified Pt(1 1 1). Fig. 37a shows that in 0.05 M H_2SO_4, the ORR is strongly inhibited at the Pt(1 1 1)—Pd surface. The decrease in the rate of the ORR at the disk electrode parallels the increase in the peroxide oxidation currents on the ring electrode, implying that the mechanism on the Pt(1 1 1)—Pd electrode is different from one found for the Pt(1 1 1) electrode. On the other hand, although the activity of the ORR on Pt(1 1 1)—Pd in 0.1 M $HClO_4$ is somewhat lower than on unmodified Pt (Fig. 37b), probably due to stronger interaction of Cl_{ad} with the Pd film than with an unmodified Pt(1 1 1) electrode [167], *no* peroxide has been detected on the ring electrode at low overpotentials, indicating that a pair of Pd sites are always available for the breaking of the O—O bond. This implies that the direct $4e^-$ path is operative in this potential region.

However, somewhat surprisingly the Pt(1 1 1)—Pd electrode has a uniquely high catalytic activity for the ORR in alkaline solution. The rate of the ORR on the Pt(1 1 1)—Pd electrode is improved by factor of 2 relative to unmodified Pt(1 1 1), (Fig. 37c). Keeping in mind that the Pt(1 1 1) electrode in KOH was the most active catalyst for the ORR, the catalytic improvement observed on the Pt(1 1 1)—Pd surface is an extremely important new observation which may help in the quest for developing better ORR catalysts. The fact that the activity of Pt(1 1 1)—Pd electrode is significantly higher in KOH than in $HClO_4$ and H_2SO_4 implies that the lower activity in acid electrolytes probably arises from greater adsorption of Cl_{ad} (an impurity in $HClO_4$) and (bi)sulfate anions on the Pd versus the clean Pt surface. This increased anion adsorption is determined by the modification of surface electrochemical properties (the work function and thus the PZC) caused by charge redistribution upon forming the Pt—Pd bond. More details about the ORR on the Pt(1 1 1)—Pd surface, including the effects of H_{upd} on the kinetics, may be found in [167].

5.2.5. *ORR on Pt alloy surfaces*

Several investigations have been carried out to determine the role of alloying in the electrocatalytic activity of Pt for the ORR (detailed review in [245]). A definitive determination, however, remains elusive. Luczak et al. [142,260,261], e.g. found an increase in mass activity of a factor 1.5–2.5 when using PtCoCr or PtCr instead of pure Pt whereas Beard and Ross [262] and Glass et al. [263] found no change for PtCo and PtCr, respectively. Furthermore, Mukerjee and Srinivasan [264] showed that five binary Pt—M alloy electrocatalysts (M = Cr, Mn, Co, and Ni) supported on carbon produced some enhancement in the kinetics (factor of 3–5) of the ORR relative to "standard" supported Pt catalyst. By utilizing in situ X-ray absorption spectroscopy (XAS) the principle explanation for the enhanced ORR activity could be enumerated as being due one or more of the following effects: (i) modification of the electronic structure of Pt (5d orbital vacancies); (ii) change in the physical structure of Pt (Pt—Pt bond distance and

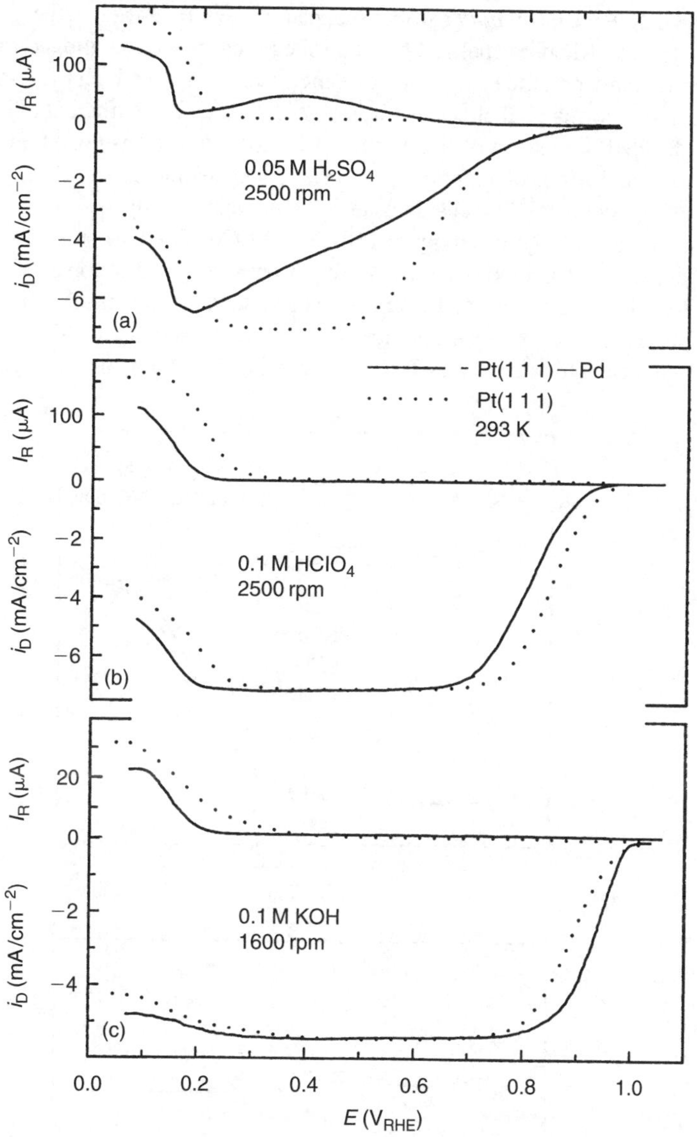

Fig. 37. Polarization curves for the ORR on Pt(1 1 1)—1 ML Pd disk electrode and corresponding peroxide oxidation current on the ring electrode (solid line) in (a) 0.05 M H_2SO_4, (b) 0.1 M $HClO_4$ [167], and (c) 0.1 M KOH. The dotted line represents unmodified Pt(1 1 1) under the same conditions, 50 mV/s, 293 K. Collection efficiency, $N = 0.2$

coordination number); (iii) adsorption of oxygen containing species from the electrolyte onto the Pt or alloying element; and/or (iv) redox type processes involving the first row transition alloying elements.

It is unclear, however, whether there is any alloy of Pt that is more active than Pt itself. Most of the studies reporting improved activity have been done on supported catalysts, where the kinetic measurements themselves are subject to considerable uncertainty. One of the difficulties in determining the effect of alloying components using supported catalysts is that the activity of a pure Pt supported catalysts can have a wide range of values depending on its microstructure and/or method of preparation. The intrinsic activity of Pt for the ORR depends on *both* particle shape and size [105,246], i.e. there is not a single value of the specific activity even when normalized by Pt surface area. Since the alloyed Pt catalysts particles may not have either the same particle size or shape as the Pt catalysts to which they are compared, a simple comparison of activity normalized either by mass or surface area is insufficient to identify a true alloying effect. A more detailed discussion of this point, specifically in the case of Pt—Co catalysts, can be found in [265]. These complexities of supported catalysts reinforced the need for using well-characterized materials to identify the fundamental mechanisms at work in electrocatalysis.

Most recently, the intrinsic catalytic activity of Pt_3Ni and Pt_3Co alloy catalysts was studied in our laboratory with model bulk alloys characterized in UHV (see the summary in Section 3). Figure 38 summarizes results for the ORR on UHV prepared and characterized bulk Pt_3Ni and Pt_3Co alloy catalysts in 0.1 M $HClO_4$. Pure Pt is also shown as a reference. As shown in Fig. 38, the activity at 293 K decreases in the order $Pt_3Co > Pt_3Ni > Pt$, the catalytic activity of Pt_3Co alloy being ca. factor of two higher than on pure Pt.

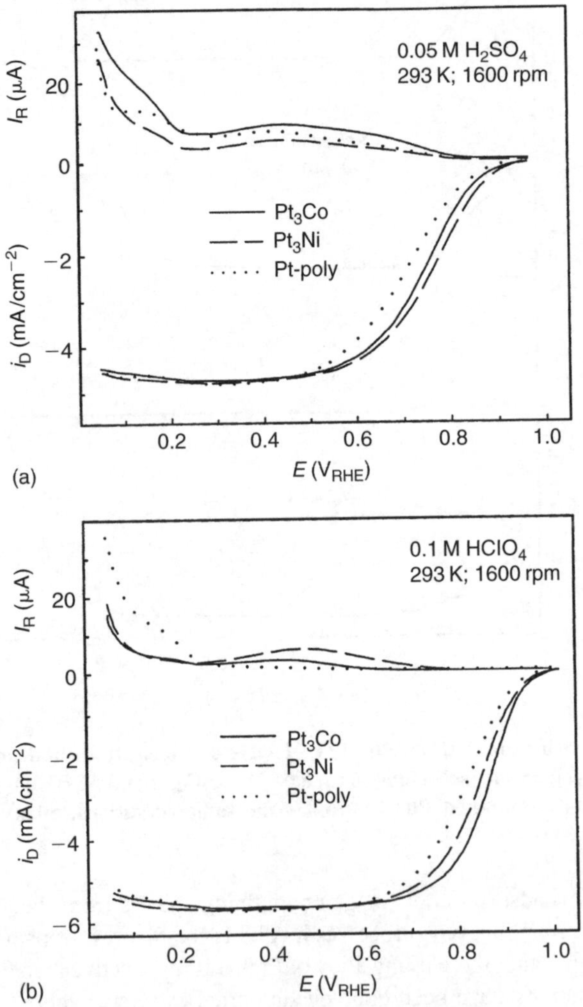

Fig. 38. Polarization curves for the ORR on Pt, Pt_3Co, and Pt_3Ni polycrystalline disk electrodes and corresponding peroxide oxidation current on the ring electrode in (a) H_2SO_4 and (b) $HClO_4$: sweep rate 20 mV/s. Collection efficiency, $N = 0.2$

The order of activity in sulfuric acid solution is slightly different, e.g. $Pt_3Ni > Pt_3Co > Pt$, emphasizing again the importance of anion adsorption on the rate of the ORR. The Tafel slopes and the activation energies (ca. 21–25 kJ/mol at 0.8 V) for Pt_3Ni and Pt_3Co bimetallic surfaces are almost the same as those obtained with pure Pt. This implies that the reaction mechanism on Pt_3Ni and Pt_3Co alloy surfaces is the *same* as one proposed for pure Pt, e.g. a "series" $4e^-$ reduction. As discussed for the ORR on Pt(*h k l*), the constancy of activation energy for the ORR on pure Pt, Pt_3—Ni, and Pt_3Co surfaces may suggest that the small difference in the kinetics of the ORR observed in Fig. 38 is determined with the $(1 - \Theta_{ad})$ term in Eq. (15). Here, Θ_{ad} refers to the coverage by spectator anion and/or OH_{ad} species, and thus a significant improvement of the ORR catalysis on Pt—Ni and Pt—Co alloy surfaces implies an inhibition of the OH_{ad} formation above 0.8 V. The true effects of Ni and Co sites in alloy on the adsorption properties of Pt atoms and the formation OH adlayer are still lacking. In general, Ni and Co may change the distribution of ensembles of Pt which otherwise retain its reactivity (the ensemble effect), but it may also change the local bonding geometry (the structure effects), or directly modify the reactivity of the platinum atoms (the electronic effects). Although it is very difficult to separate these three effects, for the Pt—Ni and Pt—Co alloys it is reasonable to propose that the electronic effects may play the major role in the Pt—OH_{ad} energetics, thus in the reactivity of platinum surface atoms. Theoretical aspects of the relationships between the electronic structure and reactivity are, however, required in order to understand the trends in atomic/molecular chemisorption energies of oxygen containing species on Pt—Ni and Pt—Co bimetallic surfaces.

5.3. *Electrooxidation of CO*

Carbon monoxide is the simplest C_1 molecule that can be electrochemically oxidized in a low temperature fuel cell at a reasonable (although not necessarily practical) potential. It thus serves as an important model "fuel" for fundamental studies of C_1 electrocatalysis. Furthermore, as we shall see in later sections, the oxidation of CO and other C_1 molecules like methanol and formic acid all provide the formation of the same intermediate, adsorbed carbon monoxide, denoted hereafter as CO_{ad}. Thus, the chemistry of CO_{ad} on Pt(*h k l*) electrodes has been a subject of extensive study as a model adsorbate applicable to many important fuel cell reactions. In the first subsection below, we review these studies of CO_{ad}, many very recent, that have exploited the advances in in situ methods. In a second subsection, we discuss the kinetics of the electrochemical oxidation of CO (gas), and present a kinetic model that incorporates the role of CO_{ad} as elucidated from the in situ studies.

5.3.1. *Surface structures of CO_{ad} on Pt(*h k l*) surfaces*

The first ex situ determination of CO_{ad} structure at the Pt(1 1 1) surface was reported by Wieckowski and co-workers [74,266]. Two CO_{ad} structures, $(\sqrt{3} \times \sqrt{3})$rect ($\Theta_{CO} = 0.68$ ML) and $(\sqrt{3} \times \sqrt{5})$rect ($\Theta_{CO} = 0.6$ ML), have been identified by LEED. The important observation from this study is that at the emersed electrode the saturation coverage by CO_{ad} is higher than that observed in the gas phase at room temperature. Very recently, by means of ex situ RHEED measurements, Lin et al. [267] found that CO_{ad} layer exhibits $(\sqrt{3} \times \sqrt{3})R30°$ CO_{ad} layer, hereafter denoted as $\sqrt{3}$ structure, with an average domain diameter of ca. 2.3 nm. Following the early ex situ measurements, an increasing number of in situ studies have continued to appear in this area. STM has gained a certain prominence among the other methods. The first STM structure of compressed adlayer on Pt(1 1 1) in aqueous acidic solution has been documented by Oda et al. [268]. Four different structures were observed depending on the electrode potential and CO_{ad} coverages. At CO_{ad} saturation, $\Theta_{CO} = 0.65$ ML, a complex $\left(\begin{smallmatrix}4 & 1\\ 1 & 5\end{smallmatrix}\right)$ structure and a (3×1) structure $\Theta_{CO} = 0.6$ ML were found at the electrode potential of 0.55 (versus NHE). At lower coverages, the $(\sqrt{3} \times \sqrt{3})R30°$ structure, observed in UHV (Section 2.2.3) and recently in ex situ measurements [267], was found at 0.55 V (versus NHE). A periodic domain structure based on $\sqrt{3}$ symmetry was also observed at 0.3 V (versus NHE). By means of in situ STM Villegas and Weaver [269,270] observed an hexagonal closed packed (2 × 2)-3CO adlayer structure at potentials below 0.25 V (versus SCE), with a CO coverage of $\Theta_{CO} = 0.75$ ML. The *z*-corrugated pattern evident in STM images indicated the presence of two three-fold hollow and one atop CO per unit cell. At potentials above 0 V (up to the onset of CO oxidation) a markedly different adlayer arrangement was formed, having $(\sqrt{19} \times \sqrt{19})R23.4° - 13$CO unit cell, hereafter denoted as $\sqrt{19}$structure, with $\Theta_{CO} = 13/19$. Another CO adlayer structure, having $(\sqrt{7} \times \sqrt{7})R19.1°$ unit cell, hereafter denoted as $\sqrt{7}$ structure, was observed at potentials below 0.2 V after the removal of solution-phase CO.

Besides STM studies, structural information about the Pt(1 1 1)—CO system was obtained by means of SXS [118, 271]. The representative SXS results along with CO stripping voltammetry in 0.05 M H_2SO_4 are summarized in Fig. 39. As for the Pt(*h k l*)—H_{upd} systems, the CTRs yield information about a surface relaxation as the electrode potential is changed and the coverage by CO_{ad} changes, producing what we call an XRV for Pt(*h k l*)—CO_{ad} [118]. The potential dependence of the Pt(1 1 1) surface relaxation induced by CO_{ad} at 0.05 V is represented by the results in Fig. 38b; the XRV measurements for CO_{ad}-free Pt(1 1 1) are also shown as a reference. Recall that the top Pt layer expands ca. ≈2% (0.05 Å) of the lattice spacing away from the second layer when H_{upd} reaches its maximum coverage. Following the adsorption of CO at 0.05 V, in the same potential region the expansion is an even larger ca. 4%. The difference in relaxation of the Pt(1 1 1) surface covered with H_{upd} and CO_{ad} probably arise from the difference in the adsorbate-metal bonding, the Pt(1 1 1)—CO_{ad} interaction being much stronger than the Pt(1 1 1)—H_{upd} interaction [271].

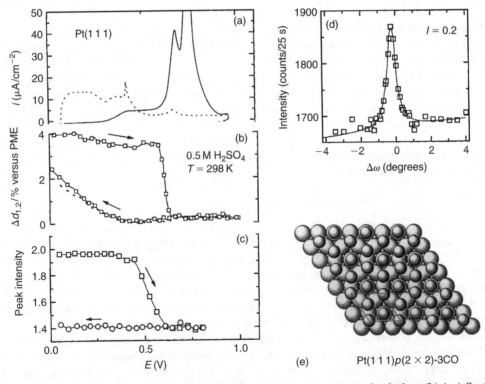

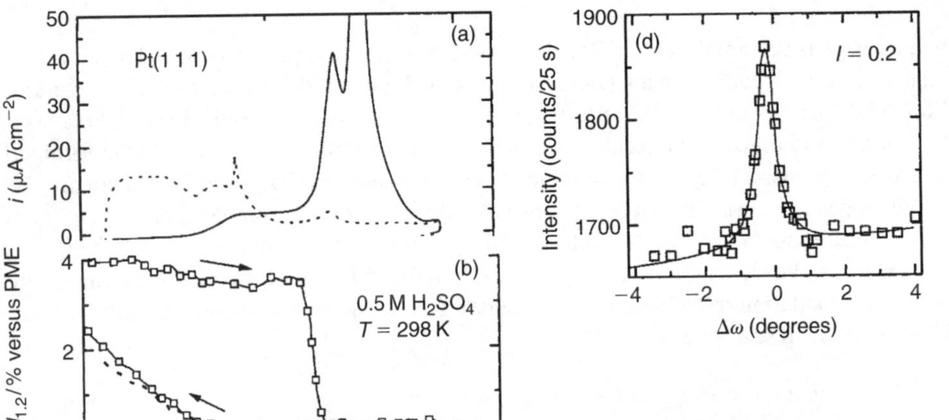

Fig. 39. (a) CO stripping voltammetry on the Pt(1 1 1) surface in argon purged solution. (b) (—) Scattering intensity changes at (1, 0, 3.6) for the surface covered by CO_{ad} and (- - -) free of CO_{ad} (c) Measured X-ray intensity at (1/2, 1/2, 0.2) as a function of electrode potential for the same conditions as in (b). (d) A rocking scan through the (1/2, 1/2, 0.2) position. (e) Ideal model for the $p(2 \times 2)$-3CO structure. Sweep rate 2 mV/s [277]

At 0.05 V, no change in the relaxation of the Pt surface atoms was observed after replacement of CO with nitrogen, indicating that CO_{ad} is indeed irreversibly adsorbed on the Pt(1 1 1) surface. Upon sweeping the potential positively from 0.05 V, the oxidation of CO_{ad} in the so-called pre-oxidation potential region [271, 273] (Fig. 39a) is mirrored with a small contraction of the Pt surface layer. Above ca. 0.6 V, the top layer expansion is reduced significantly, contracting above 0.7 V to the same nearly unrelaxed state the Pt(1 1 1) surface has without CO_{ad}. It is important to note that a significant change in the surface relaxation occurs with the removal of a relatively small fraction of the saturation coverage, ca. 15%. When CO (gas) is present in solution, i.e. when there is CO present to be re-adsorbed, the oxidative removal of CO_{ad} is shifted positively by ≈0.25 V, as shown in [271]. Under this condition the onset of surface contraction is also shifted positively, confirming that relaxation of Pt surface atoms is linked to the *oxidative* removal of CO_{ad}.

Direct information regarding the CO_{ad} structure was obtained by searching in the surface plane of reciprocal space for diffraction peaks characteristic of an ordered adlayer. While holding the potential at 0.05 V and with a continuous supply of CO to the X-ray cell, a diffraction pattern consistent with a $p(2 \times 2)$ symmetry was observed [118]. Once formed the structure was stable even when the CO was replaced by nitrogen. Fig. 39c shows that the potential range of stability of the $p(2 \times 2)$-3CO phase is strongly affected by the oxidation of a *small* fraction (15%) of the saturation coverage. Upon the reversal of the electrode potential at ca. 0.6 V, the $p(2 \times 2)$-3CO structure is not re-formed, confirming that the structure is coverage-dependent and not just potential dependent. However, with a constant overpressure of CO in the X-ray cell, the SXS experiments revealed a *reversible* loss and re-formation of the $p(2 \times 2)$-3CO structure, with the $p(2 \times 2)$-3CO structure re-forming as the potential was slowly (1 mV/s) swept below 0.2 V, (Fig. 40a). After a careful search for diffraction peaks due to $\sqrt{19}$, $\sqrt{7}$, and $\sqrt{3}$ phases of CO_{ad} in 0.1 M $HClO_4$ reported in previous studies [270], such superlattice peaks were not found in SXS measurements at any partial pressure of CO or in any aqueous electrolyte [118 and 270]. Lucas et al. [118] concluded, therefore, that the $p(2 \times 2)$-3CO structure is the only structure present with *long-range order*. Very recently, it was found that the $(2 \times 2) \leftrightarrow \sqrt{19}$ potential transformation in CO-saturated $HClO_4$ solution is sensitive to the defect/step density on Pt(1 1 1) [274]. This is consistent with the conclusion from SXS that on a well-ordered Pt(1 1 1) the $p(2 \times 2)$ structure is the only structure with long-range order. The derived structural model is shown schematically in Fig. 39, which consists of three CO molecules per $p(2 \times 2)$ unit cell. A rocking scan through the (1/2, 1/2, 0.2) position is shown in Fig. 39 together with the fit of a Lorentzian lineshape (solid line) to the data. From the width of this peak and from the result of similar fits to other $p(2 \times 2)$ reflections a coherent domain size in the range of 80–120 Å for the CO adlayer was deduced by Lucas et al. [118,271].

Infrared reflection absorption vibrational spectroscopy (IRRAS) has provided a valuable complement to the structural detail about CO_{ad} on Pt(1 1 1) obtained from STM and SXS. The pioneering studies by Villegas et al. [269,270] are especially noteworthy in this regard. As shown in Fig. 40b and c, the spectra for CO_{ad} on Pt(1 1 1) in CO-saturated perchloric acid solution have three characteristic Pt—CO stretching frequencies, near 2070, 1780, and 1840 cm^{-1} (while the spectra shown are from Rodes and co-workers [275] they do not differ in any significant way from the spectra of Villegas et al. [270]). The frequency at 2070 cm^{-1} is usually assigned to CO_{ad} at the atop sites, that near 1780 cm^{-1} is assigned to multi-coordinated CO_{ad}, and that near 1840 to the bridge site. At saturation coverage, the intensity ratios of atop to multi-coordinate are ca. 2:1. However, examination of the real-space model of the $p(2 \times 2)$-3CO structure in Fig. 39, deduced from the STM and SXS studies, shows that about 1/3 of the CO_{ad} occupy atop sites, and 2/3 are in three-fold hollow sites. The stronger atop signal observed with IR spectroscopy can be explained by "intensity stealing" by which the higher-frequency atop mode gains intensity over the lower frequency hollow-site mode [276]. Thus, changes with IR intensities with potential as seen in the spectra of Fig. 40 must be interpreted with caution. Quantitation of the intensity changes to fractional coverages can only be done with modeling of the dipole coupling, and even then would entail assumptions about the CO_{ad}—CO_{ad} coordination as the coverage changes [276]. Qualitatively, it is clear from Fig. 40b that the

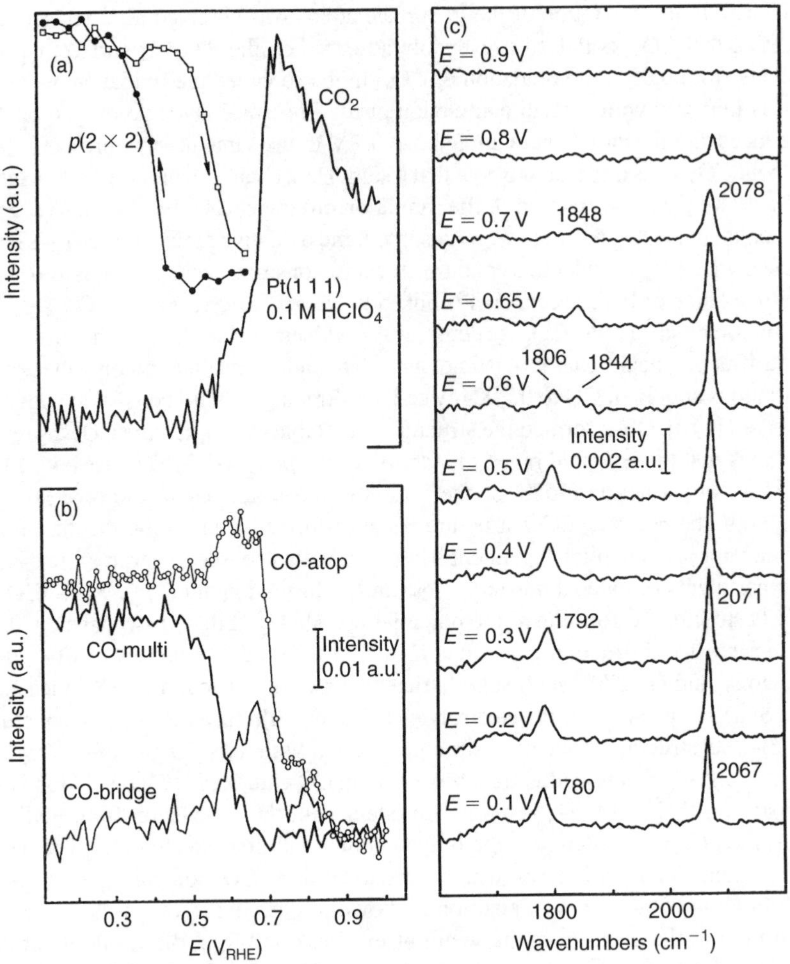

Fig. 40. (a) Potential-dependent stability of the $p(2 \times 2)$-3CO_{ad} structure on Pt(1 1 1) in 0.1 M $HClO_4$ in CO-saturated solution, and CO_2 production as a function of electrode potential (data extracted from FTIR measurement) during the oxidation of CO_{ad}. (b) Integrated intensities for CO_{ad}-atop, CO_{ad}-multi, and CO_{ad}-bridge on Pt(1 1 1) as a function of electrode potential in CO-saturated 0.1 M $HClO_4$ solution. (c) FTIR spectra obtained during progressive oxidation of CO_{ad} on Pt(1 1 1) in 0.1 M $HClO_4$ from the initial potential, ca. E = 0.05 V. Each spectrum, displayed as relative reflectance ($\Delta R/R$), was acquired from 100 interferometer scans at the range potential indicated, rationed to the corresponding spectrum obtained at the final potential, ca. E = 0.9 V [275]

order–disorder transition in the CO adlayer is mirrored by the disappearance of CO_{ad} from the three-fold hollow sites, and the relaxation of the remaining CO_{ad} into a combination of bridge sites and atop sites. Recall that in the SXS studies the disordering transition is governed by electrooxidation of a relatively small amount of $CO_{ad,}$ ca. 15% [271]. Fig. 40a also provides a plot of quantitative measure of CO_2 formation from the asymmetric O—C—O stretch at 2343 cm^{-1}. Compared with Fig. 40b, CO_2 production occurs simultaneously with the binding site occupancy change, indicating that structural transformation in the CO_{ad} layer is triggered by oxidative removal of CO_{ad}, as in Eq. (16). As we shall discuss in Section 5.3.2 the more reactive CO_{ad} adlayer is the one consisting of three CO_{ad} per $p(2 \times 2)$ unit cell located in the atop and three-fold hollow sites. Fig. 40 shows that oxidation of the CO_{ad} in the three-fold hollow

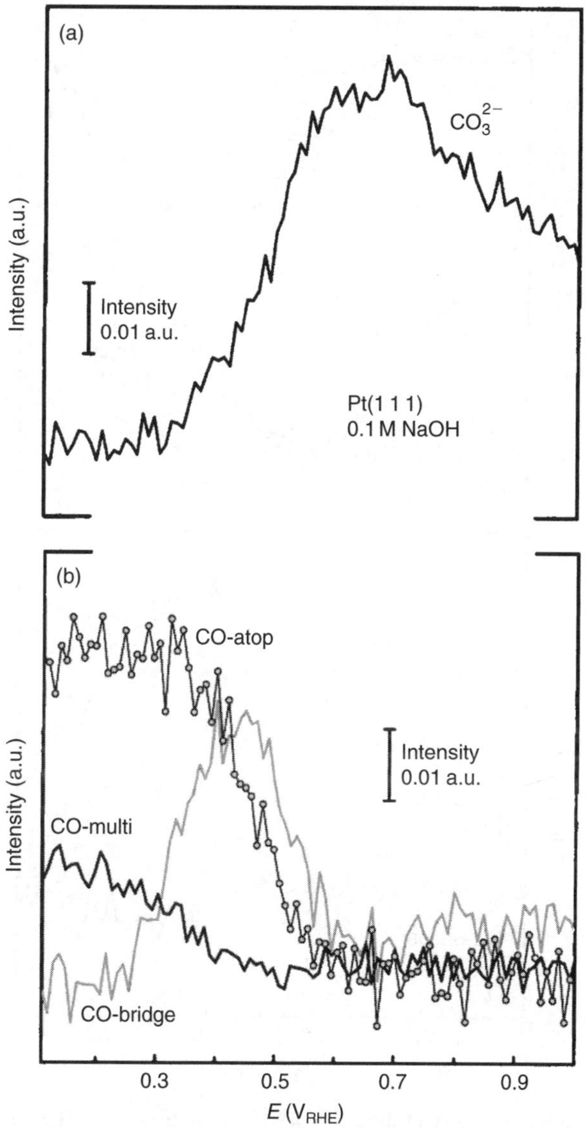

Fig. 41. (a) CO_3^{2-} production as a function of electrode potential (data extracted from FTIR measurement) during the oxidation of CO_{ad} in CO-saturated 0.1 M KOH solution. (b) Integrated intensities for CO_{ad}-atop, CO_{ad}-multi, and CO_{ad}-bridge on Pt(1 1 1) as a function of electrode potential in CO-saturated 0.1 M KOH solution [275]

sites and relaxation of the remaining CO_{ad} into bridge sites and atop sites apparently disrupts the long-range ordering in the remaining adlayer, as the Bragg peak intensity for the $p(2 \times 2)$-3CO structure decreases rapidly in this potential region (Fig. 40a).

Anions in the supporting electrolyte do affect the stability of the $p(2 \times 2)$-3CO structure. Combined SXS/IRRAS data for the Pt(1 1 1)—CO_{ad} system in alkaline solution (Fig. 41) and in perchloric acid solution containing Br^- (Fig. 42) are representative examples. Clearly, in alkaline solution the oxidation of CO (production of CO_3^{2-} in Fig. 41a) begins as low as at 0.3 V, concurrent with the loss of CO_{ad} at three-fold hollow sites and the development of CO_{ad} at bridge sites, (Fig. 41b). The oxidation of CO at

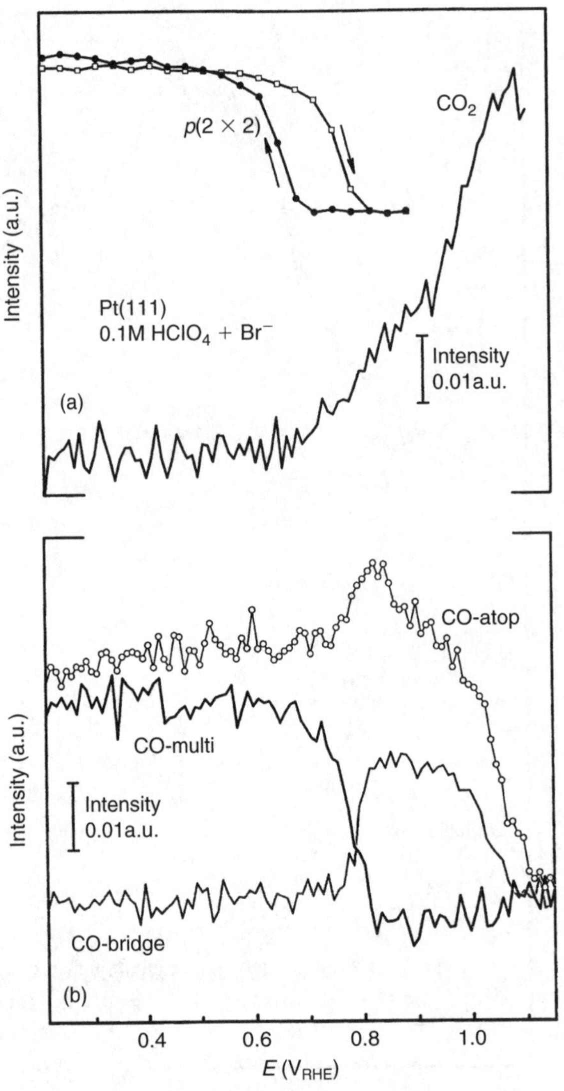

Fig. 42. (a) Potential-dependent stability of the $p(2 \times 2)$-3CO_{ad} structure on Pt(1 1 1) in 0.1 M $HClO_4$ + 10^{-3}M Br^- in CO-saturated solution, and CO_2 production as a function of electrode potential (data extracted from FTIR measurement) during the oxidation of CO_{ad}. (b) Integrated intensities for CO_{ad}-atop, CO_{ad}-multi, and CO_{ad}-bridge on Pt(1 1 1) as a function of electrode potential in CO-saturated 0.1 M $HClO_4$ solution [275]

very low potentials in alkaline solution reduces the potential range of the stability of the $p(2 \times 2)$-3CO structure, 0.05–0.35 V in KOH versus 0.05–0.7 V in $HClO_4$. We shall argue in the section on kinetics below that CO_{ad} is oxidatively removed by OH_{ad} adsorbed at the defect/step sites on the Pt(1 1 1) surface. As emphasized in Section 5.3, the adsorption of OH onto these sites is facile in alkaline solution. In solution containing Br^-, however, these defect/step sites are blocked with Br_{ad}, and consequently the potential stability of the $p(2 \times 2)$-3CO structure is extended to higher potential than even in perchloric acid, up to ca. 0.9 V, (Fig. 42). It is also worth mentioning that besides tuning the fine balance between atop and multi-fold coordinated CO_{ad}, the domain size of the CO_{ad} structure is significantly affected by the nature of

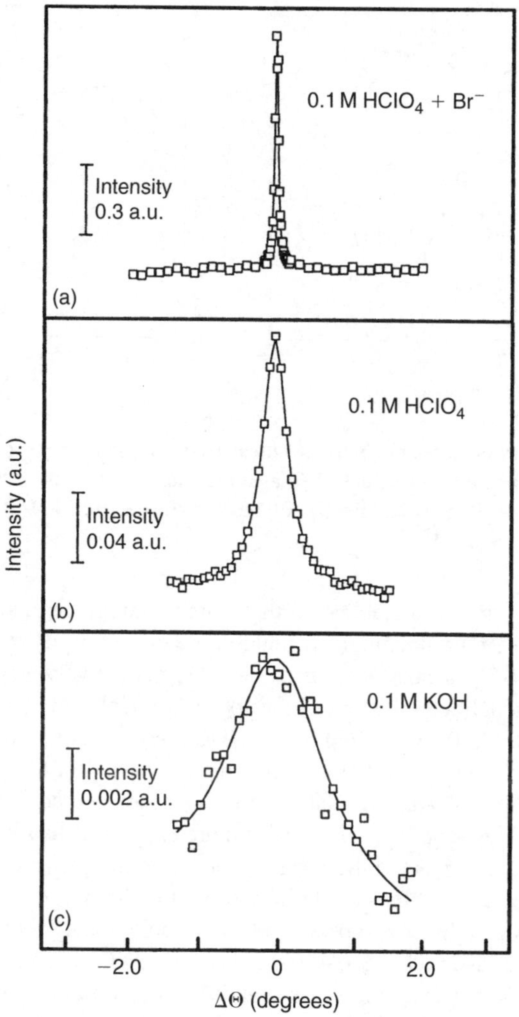

Fig. 43. A rocking scan through the (1/2, 1/2, 0.2) position for the $p(2 \times 2)$-3CO$_{ad}$ structure formed at 0.05 V Pt(1 1 1) in (a) 0.1 M HClO$_4$ + 10^{-3}M Br$^-$, (b) 0.1 M HClO$_4$, and (c) 0.1 M KOH

anions. For example, Fig. 43 shows that the domain size of the $p(2 \times 2)$-3CO structure increases from KOH (ca. 30 Å), to HClO$_4$ (ca. 140 Å) to HClO$_4$ + Br$^-$ (ca. 350 Å), i.e. the less active the surface is toward CO$_{ad}$ oxidation, the larger is ordered domains of the $p(2 \times 2)$-3CO structure. As we shall see in the next section, a self-consistent explanation for this result is that both the stability and domain size of the ordered CO$_{ad}$ adlayer is determined by the competition between OH$_{ad}$ and spectator anions for the defect/step sites on the Pt(1 1 1) surface [114].

In contrast to the Pt(1 1 1) surface, no ordered structures of CO$_{ad}$ were found on either Pt(1 1 0)-(1 × 1), Pt(1 1 0)-(1 × 2), and Pt(1 0 0)-(1 × 1) surfaces. There are some important observations for the interaction of CO with these surfaces, however, which are worth discussing. In situ X-ray scattering (SXS) studies have shown that both the Pt(1 1 0)-(1 × 2) and Pt(1 1 0)-(1 × 1) surfaces are stable in the potential region between 0 and 1.0 V. Adsorption of CO even to full coverage on the (1 × 2) surface in solution does *not* induce the (1 × 2) → (1 × 1) transition that is observed in UHV upon adsorption of CO [47–50]. In fact,

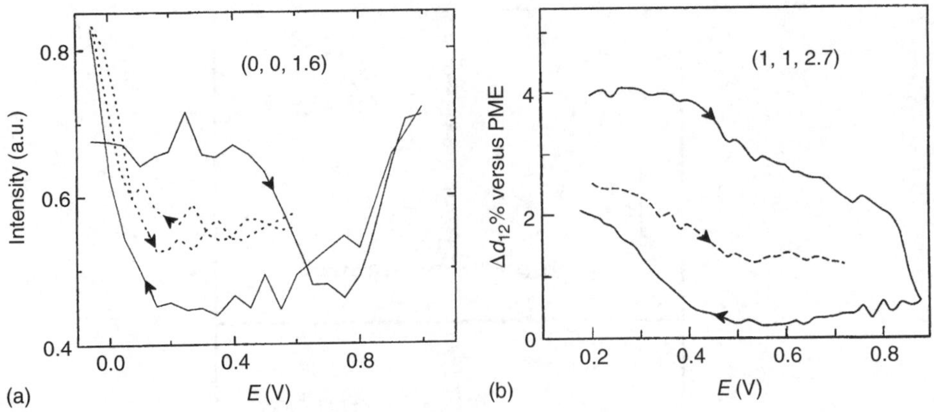

Fig. 44. (a) Scanning intensity changes at (0, 0, 1.6) measured for the (1 × 2) surface covered by CO_{ad} (solid curve) and free of CO_{ad} (dashed curve) at (0, 0, 1.6) as a function of electrode potential in 0.5 M H_2SO_4 [102]. (b) Scattering intensity changes at (1, 1, 2.7) for the Pt(1 0 0) surface in 0.5 M H_2SO_4 covered by CO_{ad} and (- - -) free of CO_{ad}; sweep rate 2 mV/s [277]

the (1 × 2) surface reconstruction was stable over the entire potential range in which CO_{ad} was present on the surface, even at an open circuit potential. Although ordered structures were not observed in-plane on either the Pt(1 1 0) and Pt(1 0 0) surfaces, it appears that CO_{ad} is relatively well-ordered in the surface normal direction because the modeling of the CTR intensity was possible only by inclusion of CO_{ad} layer with a single Pt—CO bond length. In the modeling of the CTR, relaxation of platinum surface atoms was also a very important fitting parameter. In particular, upon the adsorption of CO on (1 1 0)-(1 × 2) surface at 0.05 V, the top layer expansion is *reduced* relative to the surface covered with a monolayer of hydrogen (Fig. 44a), but at about 0.4 V the top layer is actually more expanded than in the absence of CO_{ads} at the same potential [102]. We have explained these relaxations in terms of the structural models for H_{upd} and CO [102], e.g. the H_{upd} sites in the three-fold hollow sites “in” the surface causing significant expansion of the topmost rows of Pt atoms. IRRAS shows that CO_{ad} is only linearly bonded to the topmost rows. At a potential where H_{upd} is not adsorbed, e.g. 0.4 V, the surface with CO_{ad} on it is more expanded than without, (Fig. 44a). Above 0.45 V, however, the expansion decreases continuously up to ≈0.8 V due to oxidative removal of CO_{ad}. Above 0.8 V, the process reverses and the top layer expansion increases with potential, presumably due to accumulation of OH_{ad} (an increase in the expansion also occurs in the same potential region in the solution free of CO). Surprisingly (in light of the UHV observations), the surface maintains (1 × 2) symmetry throughout this 0.05–1.0 V cycle either with or without full coverage of CO_{ad}.

The potential dependence of the Pt(1 0 0) surface relaxation induced by the pre-adsorbed CO_{ad} at 0.05 V and/or H_{upd} is represented by the results shown in Fig. 44b; the effect of H_{upd} on a surface relaxation (no CO present) is also shown as a reference. Following the adsorption of CO at 0.05 V, the expansion is even larger, ca. 4%. Upon sweeping the potential positively from 0.05 V, the oxidation of CO_{ad} layer in nitrogen purged solution is mirrored by a continuous contraction of the Pt surface layer, as shown in Fig. 44b. Above ca. 0.65 V, the top layer expansion is reduced significantly, and the measured X-ray intensity above 0.7 V drops to value below that observed for the unrelaxed Pt(1 0 0) surface without CO_{ad}, which is different from the results observed for the Pt(1 1 1)—CO_{ad} system. This can be explained by some roughening of the Pt(1 0 0) surface caused by the potential excursion to 0.8 V.

*5.3.2. Energetics and kinetics of CO electrooxidation on Pt(*h k l*) surfaces*

Just as in heterogeneous catalysis, the ultimate challenge in electrocatalysis science is to relate the microscopic details of adsorbed states of intermediates to the macroscopic measurement of kinetic rates. There are many strategies that may be employed in this endeavor. The authors have pursued, for the present time at least, a relation via the energetics of adsorption and reaction and classical TS rate theory. Unfortunately, this approach has an intrinsic difficulty. Given that CO adsorption on Pt(*h k l*) at near ambient temperature is an irreversible process, and due to the narrow temperature range available in aqueous solutions, the heat of adsorption of CO cannot be determined at Pt electrodes. However, one can use the values of thermodynamic functions which are obtained from UHV measurements and test them in the electrochemical system for consistency. Two general features in the energetics of the CO_{ad}/Pt system clearly emerge from the UHV studies: that the heat of adsorption of CO on Pt(*h k l*) is strongly coverage-dependent [42]; that the coverage dependence on single crystal surfaces arises primarily from adsorbate–adsorbate repulsive interactions rather than surface heterogeneity [42]; and that the heats of CO adsorption are relatively insensitive to the surface structure of the substrate [25,43]. From these general observations, we suppose that at the Pt(*h k l*)/solution interface the heats of adsorption at saturation coverage of CO_{ad} are close to $\approx$1/3 of the initial value [277], e.g. a heat of adsorption of CO at Pt(*h k l*) aqueous electrolytes may vary from $\approx 150 \pm 15$ kJ/mol at low coverages to $\approx 65 \pm 15$ kJ/mol at saturation. We show below that using only this supposition one can develop a very reasonable relation between the microscopic details of CO_{ad} and the kinetics of CO electrooxidation on Pt(*h k l*) electrodes.

Two forms of CO_{ad} species can be distinguished thermochemically on Pt(1 1 1) in an electrochemical environment [271]: (i) CO_{ad} with a low heat of adsorption is characterized as the "weakly adsorbed" state, and (ii) CO_{ad} with a relatively high enthalpy of adsorption is characterized as the "strongly adsorbed" state. It should be emphasized that at steady-state the weakly and strongly adsorbed states of CO_{ad} as defined do not co-exist on the surface. Therefore, at high coverages ($\Theta_{CO} > 0.65$ ML) *all* CO_{ad} molecules on the surface are in the weakly adsorbed state due to repulsive interactions. When the Θ_{CO} is reduced by oxidative removal of CO_{ad}, the remaining CO_{ad} molecules relax into a CO_{ad} layer, which is characterized as the strongly adsorbed CO adlayer. As we discuss below, these concepts for CO_{ad} energetics, and the transition from the $CO_{ad,w}$ layer into the $CO_{ad,s}$ state, are the keys to understanding the surface electrochemistry and interfacial structures of CO_{ad} on the Pt(*h k l*).

Figure 45 shows representative polarization curves for the oxidation of dissolved CO in solution on platinum single crystal surfaces in 0.05 M H_2SO_4. According to Fig. 45, two potential regions can be distinguished in the current versus potential relationship during CO_b oxidation: a potential region of the electrooxidation via the weakly adsorbed state of CO (the so-called pre-ignition or pre-oxidation potential region [272,273]) and the potential region of the oxidation via the strongly adsorbed state of CO_{ad} (the ignition potential region). The term "ignition potential" is analogous to the term "ignition temperature" in gas-phase oxidation. It is the potential at which the rate becomes entirely mass transfer limited (rate of CO_b diffusion to the surface). In line with the previous section, the macroscopic characterization of a weakly adsorbed state is linked microscopically to a saturated CO adlayer consisting of three CO_{ad} molecules per $p(2 \times 2)$ unit cell located in atop and three-fold hollow sites of the Pt(1 1 1) surface. Clearly, in the pre-ignition potential region the rate of CO oxidation varies with the crystal face, with activity increasing in the order: Pt(1 1 1) < Pt(1 1 0) $\leqslant$ Pt(1 0 0). It is important to note that a reduction of CO partial pressure produces a decrease in the rate of CO_b oxidation in the pre-ignition potential region, insets of Fig. 45, indicative of a *positive* reaction order for CO_b oxidation with respect to the partial pressure of CO. On the other hand, if the partial pressure of CO is reduced, the ignition potential shifts negatively versus that for pure CO, consistent with a *negative* reaction order for the oxidation of CO_b. Although the nature of CO_{ad} changes with electrode potential, the authors have proposed that the mechanism for CO_{ad} oxidation on Pt(*h k l*) is independent of a electrode potential: that CO_{ad} reacts with OH_{ad}, through a Langmuir–Hinshelwood (L—H) type reaction [271]

$$CO_{ad} + OH_{ad} \rightarrow CO_2 + H^+ + e^- \tag{16}$$

CO_{ad} and OH_{ad} may or may not compete for the same sites on the Pt surface. The *competitive* adsorption of CO_{ad} and OH_{ad} on identical sites leads to a negative reaction order in CO, whereas the *non-competitive*

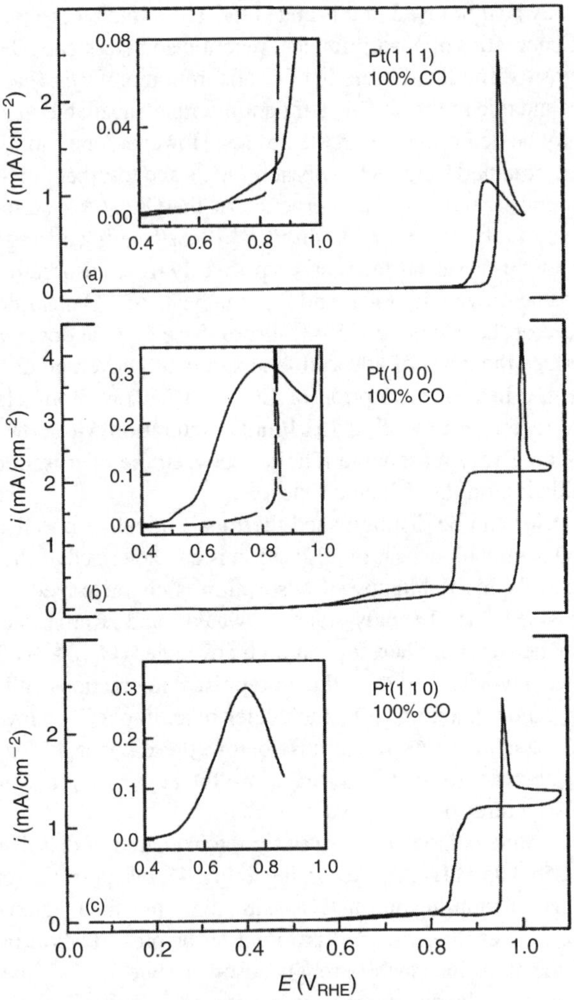

Fig. 45. Potentiodynamic CO_b oxidation current densities on (a) Pt(111)-(1 × 1), (b) Pt(100)-(1 × 1), and (c) Pt(110)-(1 × 2) in 0.5 M H_2SO_4 saturated with CO_b. Insets: magnification of the "pre-ignition region" from (a), (b), and (c), and activities of 2% CO/Ar mixtures. Sweep rate 20 mV/s [277]

adsorption of CO_{ad} and OH_{ad} at uniquely different sites leads to a positive reaction order. In the pre-ignition potential region, the positive reaction order is consistent with *non-competitive* adsorption of $CO_{ad,w}$ and OH_{ad}. At the ignition potential, the negative reaction order is indicative of *competitive* adsorption of $CO_{ad,s}$ and OH_{ad} in this potential region. In the ignition potential region, the coverage by both $CO_{ad,s}$ and OH_{ad} change dramatically with potential, with an inverse potential dependence consistent with *competitive* adsorption.

Other research groups have proposed different models for CO oxidation. By analyzing and modeling the CO current transients, Bergelin et al. [278,279] suggested that in the pre-ignition potential region CO oxidation cannot proceed through the L—H mechanism, but rather through an Eley–Rideal (E—R) mechanism, i.e. reaction between CO_{ad} and "activated" water molecules in the electrical double-layer. Rather than debate the interpretations here, the interested reader is referred to the original references for details.

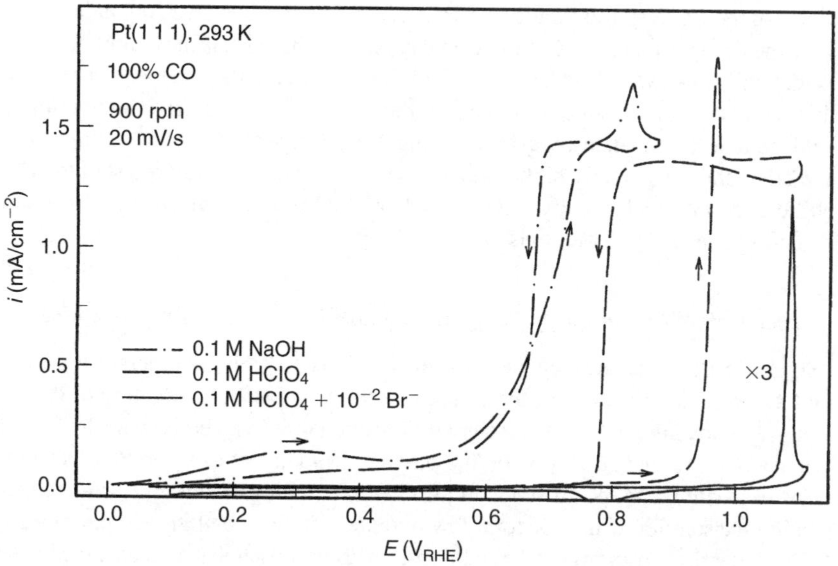

Fig. 46. Potentiodynamic CO_b oxidation current densities on Pt(1 1 1) in 0.1 M $HClO_4$ + 10^{-3} M Br^-, 0.1 M $HClO_4$, and 0.1 M KOH

The role of anions in the supporting electrolyte on the kinetics of CO_b electrooxidation can be seen by comparing the polarization curves for CO_b oxidation on Pt(1 1 1) in different electrolytes, e.g., alkaline solution, perchloric acid solution and perchloric acid solution containing Br^-. Figure 46 shows that the activity of Pt(1 1 1) increases in the sequence: $Br^- \ll HClO_4 \ll NaOH$. In the latter case, the onset of the pre-ignition region is in what is generally considered to be the H_{upd} potential region (!). We had pointed out what a surprising result [114] this is, e.g. how can a mono-metallic surface be both reducing (hydrogenating) and oxidizing at the same potential. If the L—H mechanism is operative, the catalytic activity in the H_{upd} potential region implies that in alkaline solution OH_{ad} is adsorbed even at potentials below ca. 0.2 V (RHE). This would translate into an increase in the Pt—OH_{ad} bond energy by ca. 70–80 kJ/mol, from 136 kJ/mol in acidic solution to ca. 206–216 kJ/mol. We had proposed that adsorption of OH_{ad} with this higher bond energy occurs at the defect/step sites in the Pt(1 1 1) surface [114]. The low defect density on this surface explains the relatively low rate of reaction achieved in this potential region. On the Pt(1 0 0) surface, where the defect density is high due to the lifting of the reconstruction, the activity in the same potential region is much higher than on Pt(1 1 1) [280]. Therefore, our explanation for the remarkable effect of pH on the rate of CO_b oxidation on Pt(*h k l*) is the "pH-dependent" adsorption of OH_{ad} at defect/step sites. The adsorption of OH_{ad} at defect/step sites is in agreement with studies of CO oxidation in UHV which have shown that the oxidative removal of CO_{ad} takes place between oxygen chemisorbed preferentially on the step sites and CO_{ad} on the terrace sites [281]. The reason why OH_{ad} is excluded from active sites in acid solutions was given in the previous section. e.g., the dipole moment at defect/step sites is intrinsically attractive to anions, and in acid solution these anions ($HSO_{4,ad}$, Cl_{ad}, Br_{ad}, etc.) are spectator/blocking species, but in alkaline solution OH_{ad} are in fact the oxidizing species (!) [114]. The supposition that the active centers are the defect/steps sites is fully consistent with the recent results for the electrooxidation of CO on stepped Pt[*n*(1 1 1) × (1 1 1)] electrodes. Lebedeva et al. [282] showed that the onset potential of dissolved CO oxidation increases in the sequence Pt(5 5 3) < Pt(5 5 4) < Pt(1 1 1). For more details regarding the structure sensitivity of CO electrooxidation on Pt(*h k l*) in both pre-ignition and ignition potential region, the readers are referred to details in [277].

From the study of the CO electrooxidation on Pt(hkl) electrodes, we have concluded that there are two key concepts for the design of a new CO oxidation catalyst: (i) there exists a weakly bonded state of CO_{ad} on Pt surfaces that will "turnover," i.e. be oxidized continuously, even though the coverage by CO_{ad} is very high. Since this weakly adsorbed state is created by the CO_{ad}—CO_{ad} repulsive interaction, new Pt-based catalysts should be able to enhance repulsion among the adsorbed CO_{ad}, thus leading to even weaker Pt—CO_{ad} interaction and presumably higher reactivity; (ii) the platinum surface should be modified with a more oxophilic adatom which ideally does not adsorb CO and can insure the formation of reactive hydroxyl adsorbed species at low potentials.

5.3.3. Surface chemistry of CO on Cu_{upd}, Pb_{upd}, Sn_{upd} and Bi_{ir}-modified Pt(hkl) surfaces

Cu_{upd}, Pb_{upd}, Sn_{upd} and Bi_{ir} have been studied in some detail as possible modifiers of Pt(hkl) surfaces that might enhance CO_b oxidation kinetics for the reasons just cited above. The structure of these modified surfaces in solutions not containing CO_b was discussed in Section 4.4. In an early work by Chang and Weaver [197], the influence of Cu_{upd} and Bi_{ir} layers on the binding geometry of CO_{ad} on Pt(1 1 1) and Pt(1 0 0) was studied by means of in situ FTIR spectroscopy in the C—O stretching region. The saturated CO_{ad} layer was formed by sparking the solution with CO_b for a few minutes, followed by nitrogen purging so as to remove the solution CO_b. Under these experimental conditions, it was proposed that an intermixed Bi_{ir}—CO layer is formed throughout the Θ_{CO} range on both Pt(1 1 1) and Pt(1 0 0), where Bi_{ir} occupies primarily bridge sites. By contrast, at saturation CO_{ad} in the presence of a Cu_{upd} adlayer on Pt(1 1 1) were reported to form segregated domains, as evidenced by the an absence of any shift in the C—O stretching frequency due to Cu_{upd}. Corresponding data for CO_{ad} and Cu_{upd} on Pt(1 0 0), however, were consistent with an intermixed structure, as indicated by the marked decreases in the terminal CO_{ad} frequency for larger Cu_{upd} coverages [197]. However, the effect of these adatoms on the kinetics of CO_b oxidation were not reported.

Very recently, RRDE/SXS measurements were performed during the co-adsorption of several UPD metal atoms with CO_b on both the Pt(1 0 0) and Pt(1 0 0) surfaces. The results presented here are selected from a large database in order to highlight three main conclusions drawn from these experiments. The effects of UPD metals on surface (electro)chemistry of CO on Pt(hkl) surfaces were studied for four systems: Cu_{upd}, Pb_{upd}, Bi_{ir}, and Sn_{upd}. This sequence represents an increasing Pt–metal interaction energy, i.e. $Cu_{upd} < Pb_{upd} < Bi_{ir} < Sn_{upd}$, and also incorporates atomic size effects, e.g., Pb, Bi $>$ Cu, Sn. Metal monolayers were formed on the Pt(1 1 1) and Pt(1 0 0) surfaces by holding the electrode potential just positive of the Nernst potential for bulk metal deposition before CO (gas) was introduced to the solution. Both the specific ion flux measurements with the RRDE and the SXS measurements showed that Cu_{up} and Pb_{upd} were nearly completely displaced from the Pt(1 1 1) and Pt(1 0 0) surfaces by CO_{ad}, partial displacement of the Bi_{ir} adlayer was observed, and only for the Sn_{upd} was displacement not observed. Although these results may be somewhat surprising, and would *not* be observed in UHV studies, metal displacement can be understood from a simple thermodynamic analysis by calculating the apparent Gibbs energy change (ΔG) for the component steps of the process. Namely, by calculating ΔG for the UPD process and comparing the value of ΔG for the adsorption of CO_{ad}, it was possible to predict the spontaneous displacement of Cu_{upd} and Pb_{upd} by CO_{ad} from the Pt surface, the driving force being the resulting negative shift in the Gibbs surface free energy [283]. The consequences of the observed displacement phenomenon is that the kinetics of the electrooxidation of CO on single crystal surfaces in Cu^{2+} (Pb^{2+})-containing solution is the *same* as in Cu^{2+} (Pb^{2+})-free solution [283].

Insight into the displacement phenomena is obtained by using SXS to study the structural changes on the microscopic scale. For example, CO adsorption at 0.05 V has a dramatic effect on a $(3 \times \sqrt{3})$ structure of Pb_{upd} (Section 4.5.2). Figure 47 shows a rocking scan through the (4/6, 1/6, 0.2) position, where scattering from the $(3 \times \sqrt{3})$ structure occurs in CO-free solution. The main part of Fig. 47 shows the time dependence of the peak intensity as CO is introduced in solution at $t \approx 200$ s. The presence of CO_b in

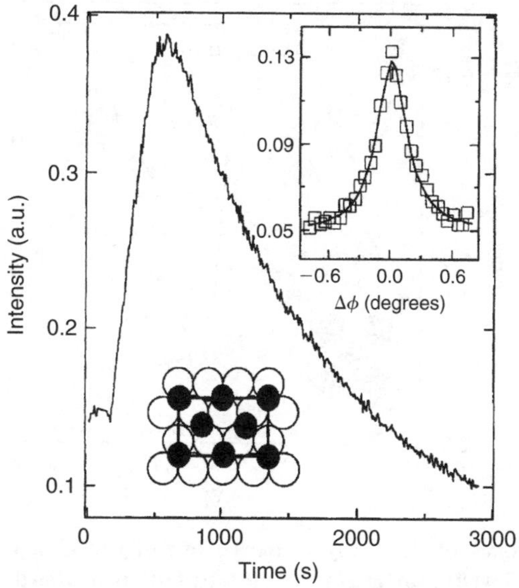

Fig. 47. Time dependence of the X-ray scattering intensity at (4/6, 1/6, 0.1), a Bragg reflection due to the $(3 \times \sqrt{3})$-Pb structure on Pt(1 1 1). CO was introduced to the solution at $t \approx 200$ s. Inset is rocking curve measured at (4/6, 1/6, 0.1) at the beginning of the experiment. The $(3 \times \sqrt{3})$ unit cell is illustrated schematically in the figure where the open circles are surface Pt atoms and the filled circles are Pb adatoms. The Pb adlayer is uniaxially compressed relative to an hexagonal phase [352]

solution initially caused a large increase in intensity due to the $(3 \times \sqrt{3})$ phase. It was proposed that initial increase in the $(3 \times \sqrt{3})$ scattering signal is caused by displacement of Pb_{upd} from Pt defect/step sites by CO_{ad}, and by formation of segregated domains of CO_{ad} and the $(3 \times \sqrt{3})$ structure. Due to a continuous displacement of Pb_{upd} by CO_{ad}, however, the intensity of $(3 \times \sqrt{3})$ peak decreases until, at $t \approx 3000$ s, it has almost disappeared.

Due to a stronger Pt–adatom interaction than in the Cu or Pb cases, in solution containing Bi^{3+} the reversible formation of a Bi $c(2 \times 2)$ adlayer structure (inset of Fig. 48) on the Pt(1 0 0) surface is unaffected by the presence of CO_b. In solutions containing Bi^{3+}, formation of the Bi $c(2 \times 2)$ structure is governed by both Bi_{upd} and Bi_{ir} states. A rocking curve through the (3/2, 1/2, 0.1) reflection is shown in the insert to Fig. 48. The potential range of stability of the $c(2 \times 2)$ structure in was studied by monitoring the scattering signal at (3/2, 1/2, 0.1) as the electrode potential was changed. As demonstrated in Fig. 48, the $c(2 \times 2)$ structure undergoes a potential-dependent order–disorder transition over the range 0.3–0.4 V. Saturating the solution with CO had absolutely no effect on either the reversibly (UPD) or irreversibly adsorbed Bi. By contrast, saturation of the electrolyte with CO and subsequent potential cycling lead to displacement of Bi_{ir} from the Pt(1 1 1) surface in solution free of Bi^{3+} and the appearance of X-ray diffraction due to a $p(2 \times 2)$-3CO adlayer which has previously been observed on the unmodified Pt(1 1 1) electrode, Fig. 49. This observation explains why with time the catalytic activity of CO_b electrooxidation at the Pt(1 1 1)—Bi_{ir} surface approaches the catalytic activity of clean Pt(1 1 1) under the same experimental conditions [205,271].

Unfortunately, the fact that CO_{ad} can displace many metals from the Pt surface strongly suggests that the formation of bimetallic surfaces either by a reversible (UPD) or even an irreversible adsorption from solution is not a practical way to create CO electrooxidation catalysts. From an electrocatalysis perspective, the most positive results from the metal UPD systems studied were with Sn_{upd}. However, much better enhancement is observed with the Pt—Sn alloy system, and these results are presented in the next section.

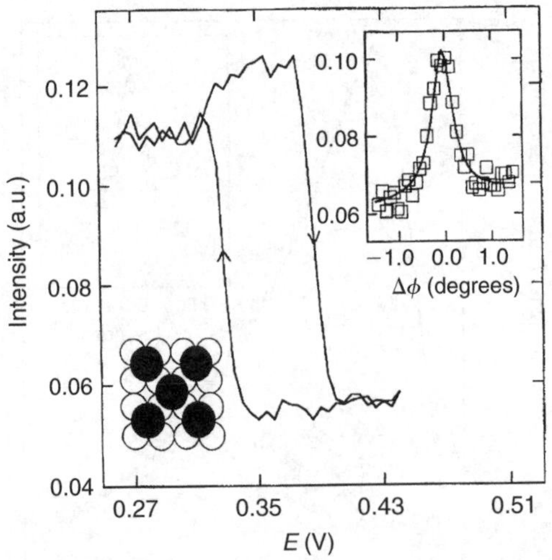

Fig. 48. The potential dependence of the X-ray scattering intensity at (3/2, 1/2, 0.1), a Bragg reflection due to a $c(2 \times 2)$ Bi adlayer on the Pt(1 0 0) surface in CO-saturated solution (sweep rate = 0.15 mV/s). The $c(2 \times 2)$ unit cell is illustrated schematically in the figure, where the open circles are surface Pt atoms and the filled circles are Bi adatoms. The order–disorder transition is fully reversible, even in the presence of CO. Inset: a rocking scan through the (3/2, 1/2, 0.1) position with the Lorentzian lineshape that gives a domain size of ca. 70 Å [119]

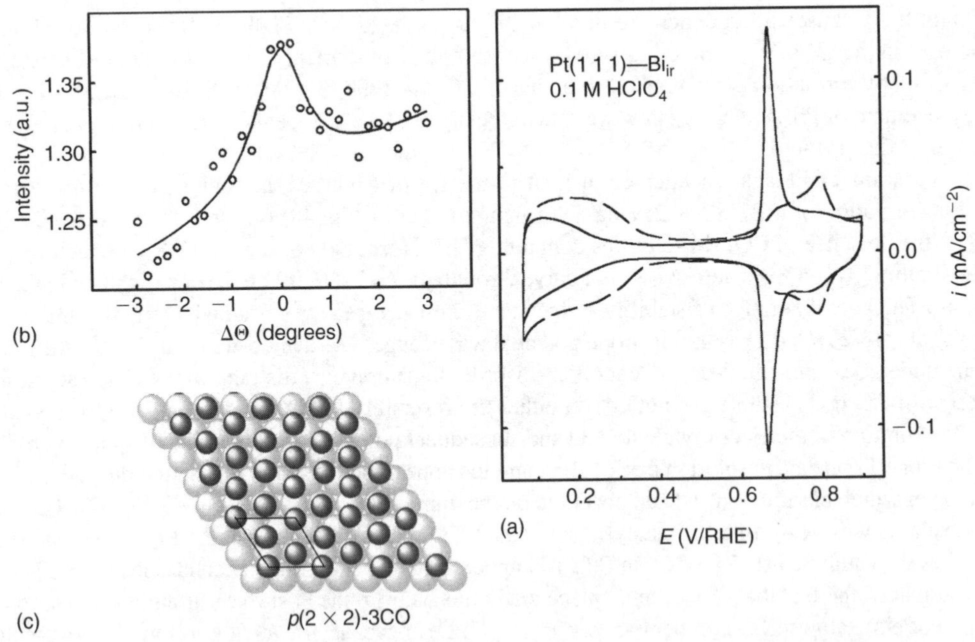

Fig. 49. (a) Base CV of Pt(1 1 1)—Bi_{ir} in Ar-purged 0.1 M $HClO_4$ before (solid line) and after (dashed line) cycling of the electrode in CO-containing electrolyte. (b) Rocking scan through the (1/2, 1/2, 0.2) reciprocal lattice point, which shows the presence of a $c(2 \times 2)$-3CO structure illustrated in (c) [351]

5.3.4. *Surface chemistry of CO on Pt bimetallic alloy surfaces*

In general, the catalytic enhancement by bimetallic surfaces can be ascribed to one or more of the following: (i) *bifunctional* effects, where the second component provides one of the necessary reactive intermediates; (ii) *ligand* (electronic), effects where the promoter alters the electronic properties of the catalytically active metal; (iii) *ensemble* (morphological) effects, where the dilution of the active component with the catalytically inert metal changes the distribution of active sites, thereby opening different reaction pathways. While most studies of electrocatalysis by Pt alloys reported so far invoke the bifunctional effect, information on the corresponding electronic and ensemble effects are more difficult to find. All of these factors may in general operate simultaneously, so that separating these effects and assessing their relative importance in the reaction mechanism is very difficult. For some systems, however, it has been demonstrated that the catalytic effect of the admetal on the kinetics of CO_b oxidation and oxidation of H_2/CO mixtures is consistent with the bifunctional mechanism of action. The concept of bifunctional catalysis was initially established in the field of the gas-phase catalysis [284] and predates the appearance of the concept in the electrochemical community by about two decades. The bifunctional model in electrocatalysis was introduced some time ago by Watanabe and Motoo for the oxidation of bulk (dissolved) CO_b on a Pt surface modified by electrodeposited Ru [285]. This concept is fairly straightforward: modify the surface of Pt with an admetal that is more oxophilic than Pt and shift the potential for OH formation (on the adatom site) negatively by (hopefully) a few tenths of a volt. The practical realization of this concept is, however, not straightforward. For example, the surface composition of alloys is, in general, different from composition of the bulk due to the *surface segregation* that is ubiquitous in bimetallic alloys. A detailed discussion of surface segregation in the Pt alloy systems of interest here was presented in Section 3. In this section, we present selected results for the electrooxidation of CO_b and/or H_2/CO mixtures on Pt bimetallic surface that have been well-characterized in UHV using a variety of surface sensitive tools such as AES, XPS, and ion-scattering spectroscopy (ISS) [286,287]. The surface preparation and characterization in-UHV was followed by clean transfer to a standard electrochemical cell. To simulate the steady state reaction conditions encountered in an envisaged fuel cell application, the flux of reacting gases was controlled in a predictable fashion by use of the rotating disk electrode (RDE) method in different electrolytes and at various temperatures [288–290].

Figure 50 shows the potentiodynamic (1 mV/s) oxidation currents for pure CO_b oxidation on a UHV-prepared $Pt_3Sn(h\,k\,l)$ single crystal alloy surfaces, a Pt—Ru polycrystalline bulk electrode and a pure Pt electrode. Clearly, the ignition potential for CO_b electrooxidation on $Pt_3Sn(1\,1\,1)$ and $Pt_3Sn(1\,1\,0)$ single crystals in acid solution is dramatically reduced in comparison with Pt—Ru and Pt, e.g. by ca. 0.7 V compared to pure Pt(1 1 1). Furthermore, Fig. 50 shows that the ignition potential for CO oxidation on the $Pt_3Sn(1\,1\,1)$ surface is 0.1 V lower than for the sputtered $Pt_3Sn(1\,1\,0)$ surface having (nearly) the same surface composition, demonstrating an unusually large structural effect. The large difference in kinetics between the (1 1 1) and (1 1 0) surfaces was explained by an increase in the mobility of CO_{ad} on the (1 1 1) versus (1 1 0) sites [290], in an analogy with the rate of CO oxidation on pure Pt(1 1 1) and Pt(1 1 0)-(1 × 1) surfaces [102]. Therefore, although there are large differences in activity, the mechanism of action is the same on all these surfaces, viz the L—H mechanism, Eq. (16). On the basis of this reaction mechanism, two classes of catalysts for CO oxidation can be distinguished:

(i) The first class has a pseudo-bifunctional mechanism of action, as in the case of Pt—Ru and Pt—Re [291], where CO is adsorbed on both the Pt and Ru(Re) sites. The second metal does not serve exclusively to adsorb OH_{ad}, e.g. for the Pt—Ru system,

$$\text{Pt—Ru—CO}_{ad} + \text{Ru—OH}_{ad} \rightarrow \text{CO}_2 + \text{H}^+ + \text{e}^- \qquad (17)$$

The kinetics of reaction (17) is strongly dependent on the partial pressure of the dissolved CO. Gasteiger et al. [289] showed that the ignition potential increases with the partial pressure of CO_b in the CO/Ar gas mixtures, reflecting a negative reaction order for Pt—Ru (Fig. 51a). This behavior is consistent

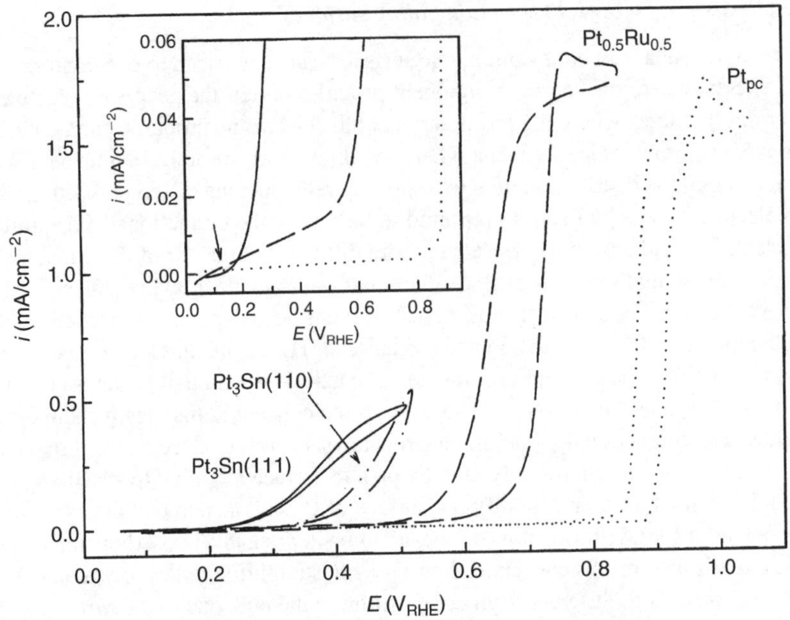

Fig. 50. Anodic oxidation currents on Pt, $Pt_{0.5}$—$Ru_{0.5}$, $Pt_3Sn(110)$ and $Pt_3Sn(111)$ electrodes in an RDE configuration. Inset: magnification of low potential [288, 353]

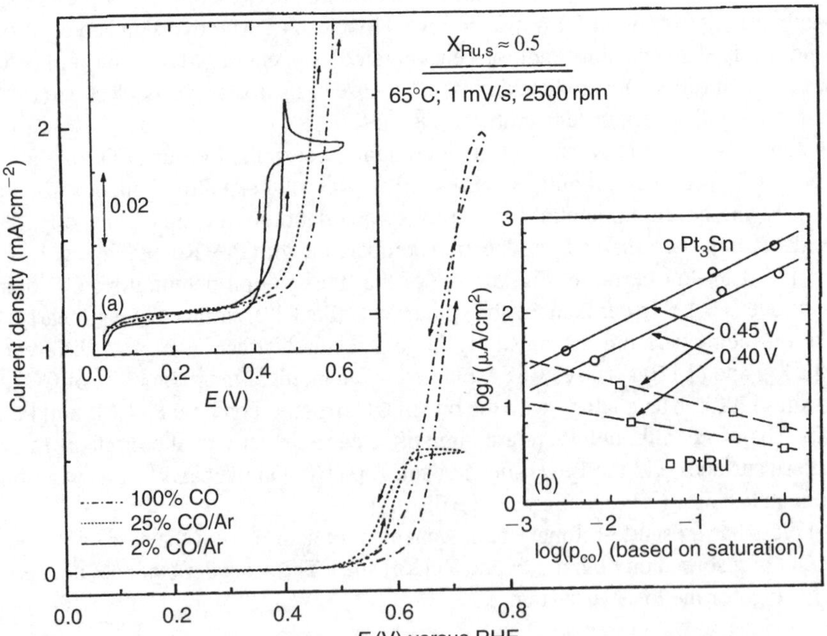

Fig. 51. Comparison of the potentiodynamic oxidation (1 mV/s) of 100% CO, 25% CO/Ar, and 2% CO/Ar on sputter-cleaned $Pt_{0.5}$—$Ru_{0.5}$. Insets: (a) magnification of low potential region; (b) reaction order plot at various electrode potentials for Pt_3Sn and PtRu based on simplified kinetic model with a potential-dependent rate constant, k_E: $i = k_E(p^0_{CO})^n$ [288,353]

with a competitive L—H mechanism, were CO_{ad} and OH_{ad} compete for the same site. Recently, using a simple kinetic model for CO electrooxidation Koper et al. [292] suggested that CO mobility is vital for the Pt—Ru surface to be electrocatalytically more active than the pure elements Pt and Ru. In addition to CO_{ad} mobility, simply from looking at the bifunctional mechanism of action one can see that intermixing of the Pt and admetal atoms in the surface is an extremely important fundamental parameter. Iannielo et al. [293] showed that the IRRAS spectra of CO_{ads} on Pt—Ru alloys have a single vibrational (C—O stretch) band whose frequency (linearly bonded) is in between that on the pure metals. This is rather easily explained as vibrational coupling between identical states of CO_{ad} on individual atoms if they are atomically mixed. On the other hand, if there is clustering of Ru atoms (ca. 3 nm islands), then one expects to see at least two different stretching frequencies for the CO_{ad}. In fact, the latter is exactly what first Friedrich et al. [294] and later Lin et al. [295] reported for Ru electrodeposited on Pt(1 1 1). Both groups found three separate stretching frequencies, corresponding to CO_{ads} on the Ru "islands," on the Pt "ocean", and at the Pt—Ru boundaries (the "beaches"), as shown in Fig. 52 for CO oxidation on Ru, Pt and Ru electrodeposited

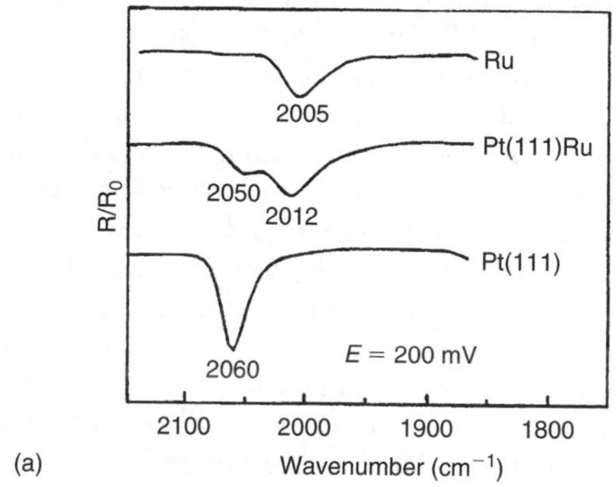

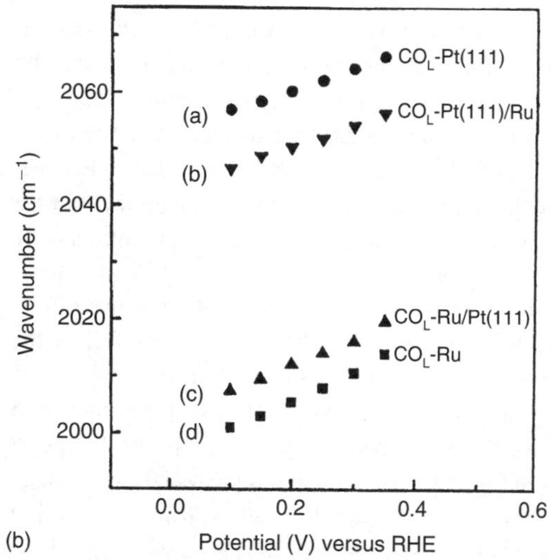

Fig. 52. (A) Comparison of three IR spectra for linear bonded CO (CO_L) on pure Pt(1 1 1), pure Ru, and Ru-modified Pt(1 1 1) with Ru coverage of 0.75 at a given potential of 200 mV vs. RHE. (B) Comparison of C—O stretch wavenumber for the CO_L on Pt(1 1 1) (a), Ru-modified Pt(1 1 1) (b, c; (b) for CO_L on Pt site and (c) for CO_L on Ru site), and Ru (d) electrodes at various potentials. The experimental conditions are described in [295]

on Pt(1 1 1) at nominally 0.75 ML coverage [295]. Thus, one would expect to see fundamentally different catalytic properties of the three differently prepared surfaces of the same Pt—Ru bimetallic system: sputtered bulk alloy, annealed bulk alloy, and submonolayer Ru deposited on Pt. We shall see in the subsequent sections that in fact this is the case for other C_1 reactions.

(ii) The second class has a pure bifunctional mechanism of action, as in the case of the Pt—Sn [146,290] and Pt—Mo [296–299] systems. In the case of Pt—Sn, it is unlikely that CO is adsorbed at Sn sites (it does not do so even in UHV [300]), while it is very likely that here is the formation of OH_{ad} at Sn sites. From ex situ XPS analyses, it is known that there is OH_{ad} on Mo sites at all potentials [298]. Hence, it is reasonable to conclude that CO is exclusively adsorbed on the platinum sites and oxygen containing species are attached to the second element;

$$\text{Pt—CO}_{ad} + \text{Sn(Mo)—OH}_{ad} \rightarrow CO_2 + H^+ + e^- \quad (18)$$

In contrast to Pt—Ru system, the reaction order on Pt_3Sn with respect to the concentration of CO_b in solution on Pt—Sn is positive, as shown in Fig. 51b. The positive reaction order is consistent with a non-competitive L—H reaction mechanism in which there is no competition between OH_{ad} and CO_{ad} for the same adsorption sites [290]. In the case of Pt_3Sn, this system is known to have strong intermetallic bonding that lowers the heat of adsorption of CO at saturation coverage by about 20% [300]. Closely following the assignments of the nature of CO_{ad} at the platinum single crystal surfaces in Section 5.3.2, the uniquely high activity of Pt_3Sn alloy surfaces arise due to the formation of a weakly adsorbed state of CO_{ad} on platinum sites, having a weaker adsorption energy than the weakest bonded $CO_{ad,w}$ formed on platinum. The $CO_{ad,w}$ state formed at the Pt_3Sn interfaces is probably highly mobile on the surface, in particular at the $Pt_3Sn(1\,1\,1)$, and thus is very reactive. Unfortunately, the high reactivity of Pt—Sn alloy surfaces is observed at relatively positive potentials, e.g. at ca. 0.2 V, so the currents for CO_b oxidation are very small below this potential, see inset of Fig. 50. It is interesting that in the same potential region Pt—Ru electrode is more active than Pt—Sn (see arrow in the inset of Fig. 50), suggesting that below 0.1 V small, but continuous oxidation of CO takes place on the Pt—Ru electrode. As we shall see below, it turns out that this small activity is an important property for a CO-tolerant catalysts.

From the practical standpoint, the electrooxidation of pure CO is of less interest than the electrooxidation of an H_2/CO mixture, e.g. produced by steam reforming of a hydrocarbon fuel. Of the various possible compositions, the electrooxidation of H_2 containing 0.1% of CO is used here to provide the benchmark case. As illustrated in Fig. 53, some very dramatic improvements in activity have resulted from the controlled addition of Ru, Re, Sn, and Mo to the platinum surface, with the order of activities (at low overpotentials) increasing in the sequence Pt—Sn < Pt—Re < Pt—Ru ⩽ Pt—Mo. Although there are large differences in activities, the mechanism of action is the same on all these surfaces. Namely, a second metal added to the platinum serves to nucleate OH_{ad} species at lower potentials than on a pure Pt electrode, leading to the oxidative removal of CO_{ad} on Pt sites nearby, freeing Pt sites for the hydrogen oxidation reaction (the non-Pt metals have intrinsic activities for the HOR that are more than an order of magnitude lower than Pt). While the mechanism of action of the admetal may follow this simple concept, the details of the reaction are very complex, with many factors contributing.

Fig. 53 shows that the most promising for electrooxidation of reformate (H_2/CO mixtures) is the $Pt_{77}Mo_{23}$ alloy. Grgur et al. [297] suggested that the superior catalytic properties of the $Pt_{77}Mo_{23}$ surface relative to Pt, or relative to the other catalysts in Fig. 53, is due to the unique reactivity of the MoOOH oxyhydroxides states for the oxidative removal of CO_{ad} in this potential region. More detailed studies on the mechanism of action of Mo atoms in the alloy surface is summarized in [296–298]. There is one outcome from these mechanistic/modeling studies that is important to emphasize here. The only form of CO_{ad} that can be oxidized (turned over) from the Pt sites at low potentials is the weakly adsorbed state, discussed at length in Section 5.3. Unfortunately, this is a small fraction of the total CO_{ad} on the Pt sites and thus only a small number of holes are created for oxidation of the hydrogen even on the best catalyst, the Pt—Mo alloy. From a practical standpoint, this means a large amount of catalyst must be used to achieve the high current densities required

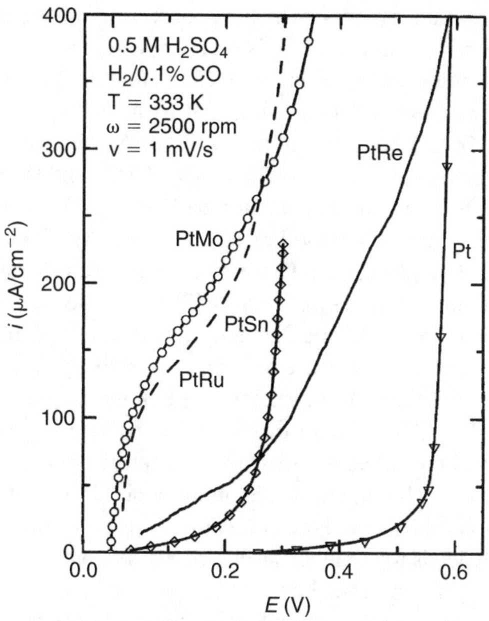

Fig. 53. Polarization curves for the oxidation of H_2 containing 0.1% CO on different catalysts in sulfuric acid solution. *E* is given versus. RHE [304]

in a fuel cell. How much catalyst? That depends on the preparation of high surface area, e.g. nanocrystalline, Pt—Mo catalyst and the catalytic properties of Pt—Mo in this form. In addition to the studies of the well-characterized solid bimetallic surfaces, there have been several studies on the kinetics of electrooxidation of H_2/CO mixtures on Pt nanocluster bimetallic catalysts supported on carbon [301,303]. Additional details, including experimental procedure and the characterization of supported catalysts for the Pt—Sn and Pt—Mo alloys specifically, can be found in two recent review papers by the authors [104,304].

5.4. *Oxidation of formic acid on Pt(***h k l***) and bimetallic surfaces*

The mechanism of formic acid electrooxidation on Pt and selected Pt-group metal surfaces in acid solution is reasonably well-established, via the so-called "dual-pathway" originally suggested by Capon and Parsons [305]. A reaction scheme, expressed as

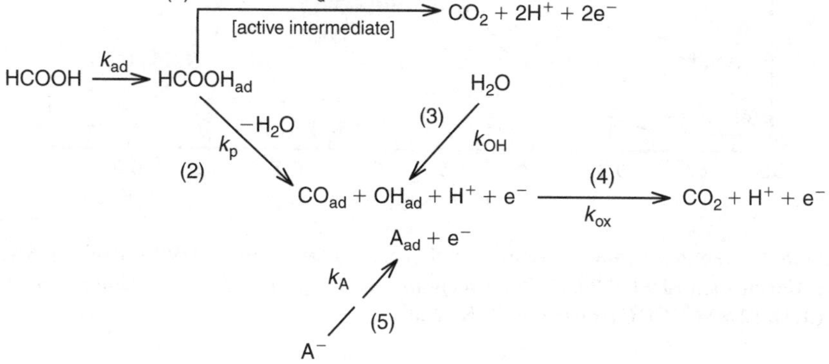

includes most of the known surface process, which are occurring during the electrooxidation of formic acid. While numerous details remain uncertain, this reaction scheme involves the adsorption of HCOOH (k_{ad}), followed by the (non-faradic) dehydration of HCOOH, and the formation of chemisorbed "poison" (reaction (2)) in competition with the direct dehydrogenation path via one or more reactive intermediates (often referred to as $HCOO_{ad}$ or $COOH_{ad}$, reaction (1)) [305]. As we shall see below, the rate of this step is determined by the surface coverage of H_{upd}, anions (A_{ad}), OH_{ad}, and "poisoning" species. It would not be an exaggeration to say that the study of this reaction on Pt by in situ IR spectroscopy is one of the most successful uses of spectroscopic methods in modern electrochemistry. The major "poisoning" species was identified clearly as adsorbed CO [306,307]. These studies are well documented in reviews written by some of the pioneers in this field, e.g. Beden and Lamy [308] and Bewick and Pons [309]. Besides being a "poison," CO_{ad} may also act as an intermediate, where some fraction of the CO_{ad} can be further oxidized to produce CO_2 (reaction (4)). The active surface oxidant is most likely adsorbed OH (or "activated water," H_2O_{ad}) as proposed in Section 5.3 for CO_{ad} oxidation on platinum single crystals. Following the reaction scheme for oxidative removal of CO_{ad}, the adsorption of oxygenated species is in a strong competition with anion adsorption (k_A), and consequently the rate of reaction (4) (k_{ox}) is determined by the delicate balance between the rate constants k_2, k_3, and k_A. Schmidt et al. [240] proposed that the interdependence of the reaction steps in reactions (1)–(4) and the competition for adsorption sites among the reaction partners and intermediates usually leads to complex surface processes, which, under certain experimental conditions may even trigger a transition from linear to non-linear kinetics. The latter systems exhibit an unexpected wealth of dynamic instabilities, often leading to oscillatory behavior. The classical example is electrooxidation of formic acid on the Pt(1 0 0) single crystal, Fig. 54. The discussion of oscillations in electrochemical oxidation is out of scope of this review. An impressive compilation of all the relevant papers up to 1999 as well as the theoretical principles of temporal and spatial pattern formation in electrochemical systems can be found in the review paper by Krischer [310].

The electrochemical oxidation of formic acid on platinum single crystal surfaces is often used as a model system for the oxidation of C_1 oxygenates [311]. The sensitivity of formic acid oxidation to surface

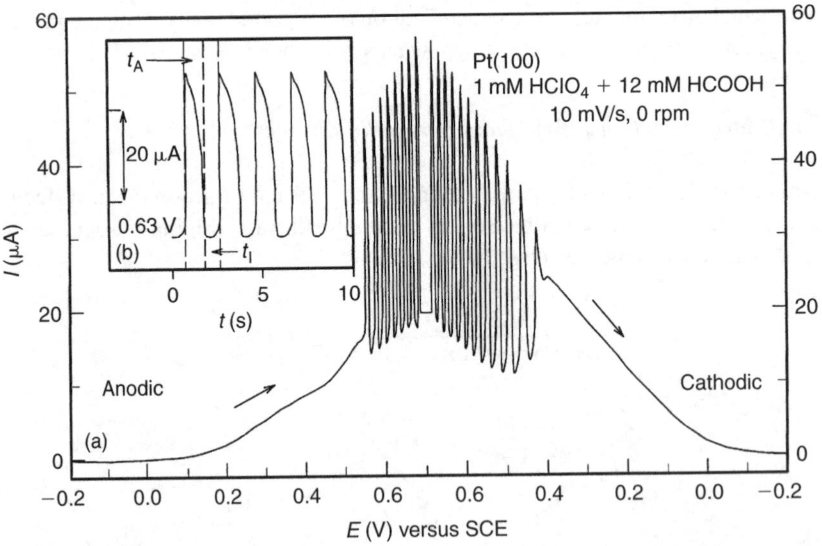

Fig. 54. (a) Unfolded potentiodynamic curve for formic acid oxidation on Pt(1 0 0) (10 mV/s). Sweep reversed at 0.7 V. (b) Potentiostatic transient at 0.63 V in a sweep-and-hold experiment after sweeping from 0.7 V cathodically: 1 mM $HClO_4$; 12 mM HCOOH; 0 rpm; 298 K [240]

structure has been the subject of several papers [312–321], but there is still no consensus on the order of activity. This is due in part to the fact that these activity assignments frequently rely on cyclic voltammetry, which, as we demonstrate below, is only a qualitative measure of reactivity. Nevertheless, regardless of the inconsistencies, it is now well-established that even the most active Pt single crystal faces in Fig. 55 are severely poisoned by the accumulation of CO_{ad} on the surface. Pt(1 1 1) is the least active surface initially, but the activity has a lower rate of decay than the other faces. The self-poisoning behavior of Pt(1 1 1) was nicely documented by in situ FTIR studies [314,316,318]. In particular, the formation of CO_{ad} from HCOOH dehydration (reaction (2)) begins at 0.15 V, followed by monotonic increase of the CO_{ad} coverage up to ca. 0.4 at 0.45 V, where it reaches its maximum coverage of $\Theta_{CO} = 0.35$ CO/Pt [318]. Neither dehydration nor dehydrogenation of HCOOH are possible at potentials below 0.15 V. Schmidt et al. [240] proposed that below ca. 0.15 V the adsorption of HCOOH (k_{ad}) is completely inhibited by H_{upd}, in accord with the proposal by Bagotzky and Vassilyev [322] that the adsorption of small organic molecules on Pt surfaces is inhibited by H_{upd}. Note, the effect of H_{upd} on the adsorption of organic molecules is

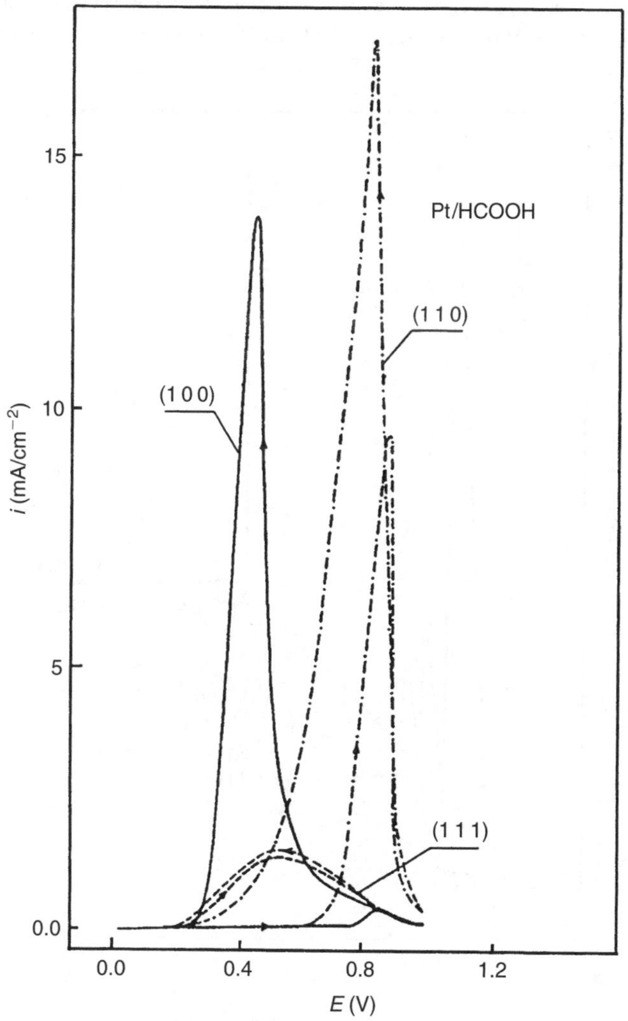

Fig. 55. Oxidation of 0.23 M HCOOH in 1 M $HClO_4$ on low-index Pt surfaces. Sweep rate 50 mV/s [312]

the same as we proposed in 5.1 and 5.2 for the HOR and the ORR, i.e. that H_{upd} is blocking the adsorption of molecular H_2 and O_2. As expected from this premise, the appearance of CO_{ad} parallels the desorption of H_{upd}, so the maximum current is observed on the surface free of H_{upd}. The FTIR observation that CO_{ad} reaches a maximum coverage at ca. 0.45 V is in accord with the supposition that the oxidative removal of CO_{ad} (reaction (4)) starts well below potentials, where the OH adsorption is clearly seen in cyclic voltammetry. At higher potentials, when excess OH_{ad} (not consumed in reaction (3)) accumulates on the surface, OH_{ad} also becomes a "site blocking" species, as was the case for the HOR and the ORR. It appears therefore that the maximum rate of oxidation of HCOOH is obtained at potentials at which the oxidative removal of CO_{ad} is *optimized* by the optimum surface coverage of OH_{ad}. The above discussion one can suggest that: besides being a "poison" at low overpotentials, in the potential region where OH_{ad} is present on the surface, CO_{ad} is also a reaction intermediate in the oxidation of HCOOH; OH_{ad} has two effects in electrocatalysis of HCOOH, a catalytic role at low potential, and an inhibiting at high overpotentials; finally, H_{upd} and anions of supporting electrolytes have only a blocking role, inhibiting the adsorption of both formic acid and OH_{ad}.

The introduction of monoatomic steps in the structure of Pt(1 1 1) decreases its activity (Fig. 56) due to an increase in the "poison" formation at the step sites. Hence, this is not a way to increase the catalytic activity

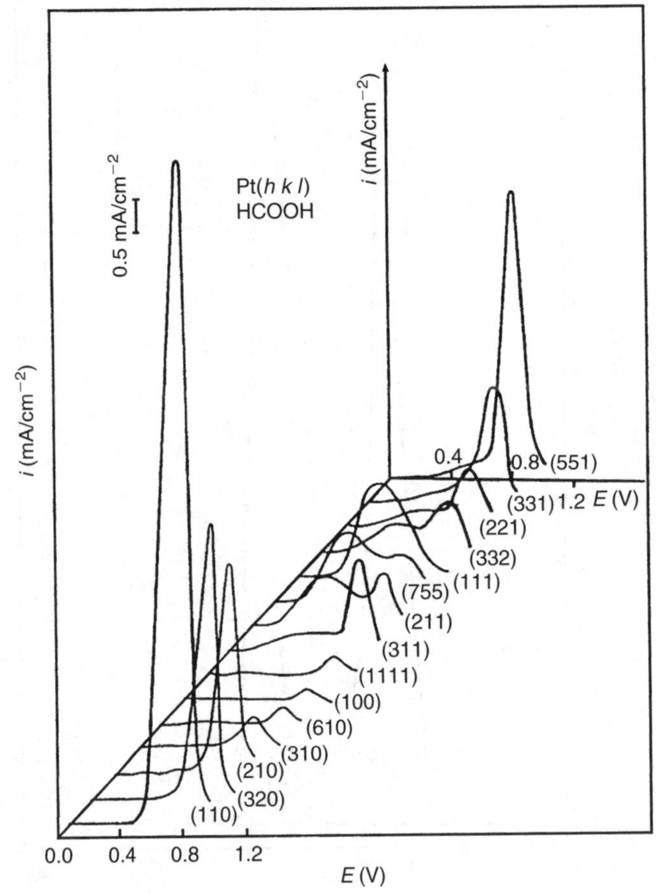

Fig. 56. Anodic sweeps for the oxidation of 0.5 M HCOOH in 0.05 M H_2SO_4 on various single crystal electrodes. Sweep rate 50 mV/s [354]

of Pt(1 1 1). Another generally more successful method is to create bimetallic surfaces that will (ideally) optimize the adsorption of HCOOH while oxidizing CO_{ad} with a minimum surface coverage by OH_{ad}, e.g. the nearly autocatalytic oxidation of CO_{ad}. The most extensively studied bimetallic catalysts for HCOOH oxidation are Pt surfaces modified either by UPD adatoms [312] or by irreversibly adsorbed metal atoms [159,161,206,323,324]. Many examples of these studies (prior to 1988) have already been discussed by Parsons and VanderNoot [325]. In line with previous sections in this volume, we will focus on discussion of representative results from one system, which has been well studied, the kinetics of formic acid on the Bi_{ir}-modifed Pt(1 1 1). The Pt($h\,k\,l$)—Bi_{ir} system occupies a special position in the electrocatalysis of formic acid on Pt-modified surfaces. To our knowledge, Adzic et al. [312] were the first to show that HCOOH oxidation is activated on Pt($h\,k\,l$) surfaces modified with Bi adatoms. Clavilier et al. [159,196, 323,326,327] found that Bi_{ir}-modified platinum single crystal surfaces can also increase the reactivity toward HCOOH oxidation roughly 40 and 20 times on Pt(1 1 1)—Bi_{ir} and Pt(1 0 0)—Bi_{ir} electrodes, respectively. Following this work, many other groups confirmed that HCOOH oxidation is catalyzed on Bi_{ir}-modified platinum single crystal surfaces [204,206,324,328]. A significant catalytic enhancement by Bi_{ir} was interpreted either via a "third-body effect", or via the electronic effect, or sometimes through an interplay of these two mechanisms. The "third-body effect" in (electro)catalysis is neither well-understood nor means the same thing to different researchers. Most explanations of the thirdbody effect in formic acid electrocatalysis are similar to the so-called ensemble effects in hydrocarbon catalysis [329], although the explanations have rarely been put in that form in the electrochemical literature. One of the rare exceptions is the elegant study by Chang et al. [328], who used in situ FTIR spectroscopy to study the formation of CO_{ads} on Pt(1 1 1) and (1 0 0) surfaces modified by Bi adatoms. These studies showed directly that Bi_{ir} reduced the steady-state coverage of CO_{ads}, with the effect being especially dramatic on the (1 0 0) surface, where the coverage by CO_{ads} was essentially nil at the optimum Bi coverage. This result was consistent with an "ensemble effect" by Bi_{ir} on the formic acid adsorption/decomposition reactions, with apparently a larger ensemble of contiguous Pt sites required for dehydration than for dehydrogenation. In contrast to the ensemble effect, Herrero et al. [159] proposed an electronic effect as the reason for the reduced "poison" formation reaction on Pt(1 1 1)—Bi_{ir}. This is consistent with a very simple model for the electrocatalysis of formic acid oxidation on the Pt(1 1 1)—Bi_{ir} electrode recently presented by Leiva et al. [330]. Very recently, Smith and Abruna [204,324] suggested that it is likely that the enhancement of HCOOH oxidation on Bi_{ir}-modified stepped single crystal platinum surfaces is determined by some combination of both ensemble and electronic effects. While the study by Smith and Abruna [204,324] made some progress in clarifying an ensemble effect, interpretation of an electronic effect was still illusive.

Recent work from our laboratory has attempted to clarify and to distinguish the ensemble/electronic effects of Bi_{ir} on Pt(1 1 1). Schmidt et al. [205] proposed two modes of action of Bi_{ir}: (i) an inhibiting effect due to blocking of active Pt sites for adsorption of HCOOH (k_{ad}); (ii) a catalytic effect due to enhanced adsorption of OH_{ad} on Pt sites adjacent to Bi_{ir} and a consequently increased rate of oxidation of the intermediate CO_{ad} (k_{ox}). As discussed in Section 4.5.3, the enhanced adsorption of OH_{ad} arises due to the electronic modification of the Pt surface by adsorbed Bi, i.e. a shift in the local PZC [205]. The opposing action of (i) and (ii) (slowly) leads the system into a steady-state activity that in terms of long-term performance shows little or no difference in the overall activity between Pt(1 1 1) and Pt(1 1 1)—Bi_{ir} (see Fig. 57 for 303 K). Therefore, the Pt(1 1 1)—Bi_{ir} system does not represent a catalytically active system of technological relevance. The same conclusion is true for the other UPD metal-modified Pt surfaces as well.

The studies of oxidation of HCOOH on Pt bimetallic surfaces modified by other than UPD or irreversible adsorption of metal adatoms are relatively scarce. One exception is the study of HCOOH oxidation on UHV-prepared polycrystalline Pt—Ru alloy surfaces. Fig. 58 shows the potentiostatic current densities at 0.4 V on sputter-cleaned Pt—Ru alloys. After a short induction time (t_{ind} in insert of Fig. 58) that was inversely proportional to the Ru surface concentration of the respective alloy electrode, formic acid oxidation current increases by almost 1 order of magnitude. The current density at 0.5 V (RHE) on the

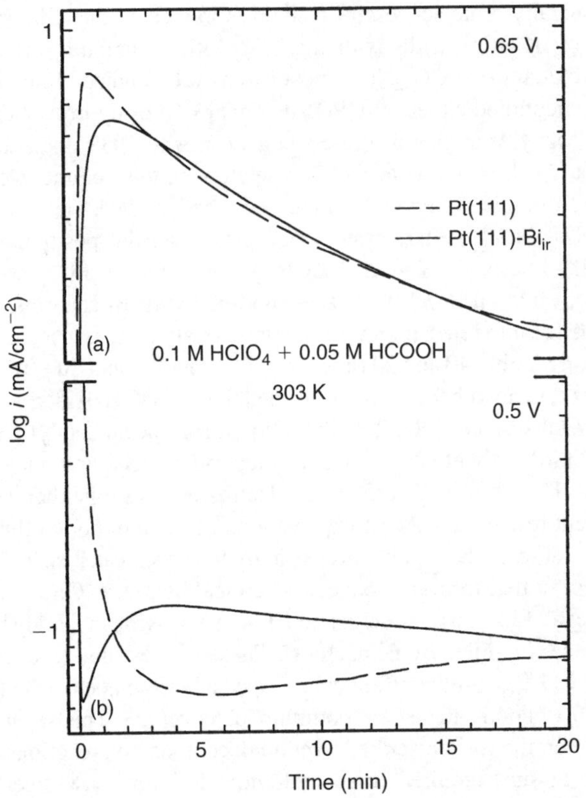

Fig. 57. Comparison of the potentiostatic formic acid oxidation on Pt(1 1 1)- and Pt(1 1 1)-modified with irreversibly adsorbed Bi in 0.1 M $HClO_4$ at 0.5 and 0.65 V. Rotation rate 900 rpm [205]

most active electrode (46 at.% Ru surface composition) was ca. five times larger after 15 min than on a pure Pt electrode. Interestingly, the optimal surface composition for HCOOH oxidation was the same found for CO_{ad} oxidation [287,331,332], implying that the mechanism of action of Ru in these reactions is the same. This mechanism of action of Ru in the Pt surface on HCOOH oxidation can be rationalized by a bifunctional mechanism. As for pure CO_{ad} oxidation, Ru sites nucleate oxygen-containing species at lower potentials than on the pure Pt surface: the adsorbed CO_{ad} is preferentially oxidized at these sites by surface diffusion from sites where adsorption occurs. The enhancement in the rate is due to an enhanced rate (k_{ox}) of CO_{ad} oxidation at Pt—Ru pair sites, which changes CO_{ad} from a mere spectator species (a poison for HCOOH dehydrogenation) to a reaction intermediate [331,332]. The reaction path on the alloy surface occurs consequently via both pathways, a true parallel reaction path with the branching ratio still very high, i.e. Pt-like, but the total rate is accelerated. The principal effect of opening the dehydration channel at steady-state (via the presence of Ru in the surface) is to lower the coverage of CO_{ad} and permit the dehydrogenation path to increase in rate.

*5.5. Oxidation of methanol on Pt(*h k l*) and bimetallic surfaces*

Methanol is probably the most studied of the C_1 compounds because of its potential as a logistical fuel and a feedstock for fuel cells. As for oxidation of formic acid, an analogous dual-pathway was frequently

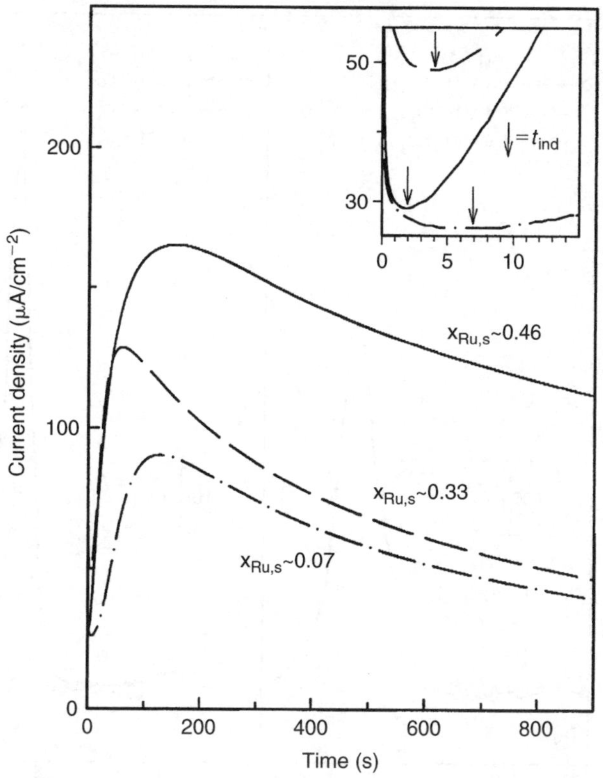

Fig. 58. Potentiostatic oxidation of 0.5 M HCOOH in 0.5 M H_2SO_4 on sputter-cleaned Pt—Ru alloys at 0.4 V after immersion at 0.07 V for 3 min. Ru surface composition are indicated in the figure. The inset shows a magnified scale; induction times, t_{ind}, are marked by an arrow [331]

proposed (see [325,333] for the history) for methanol oxidation, again with CO_{ad} as a spectator species, i.e. a poison, but now from the dehydrogenation of methanol, and an unknown intermediate was responsible for the direct oxidation to CO_2, e.g.

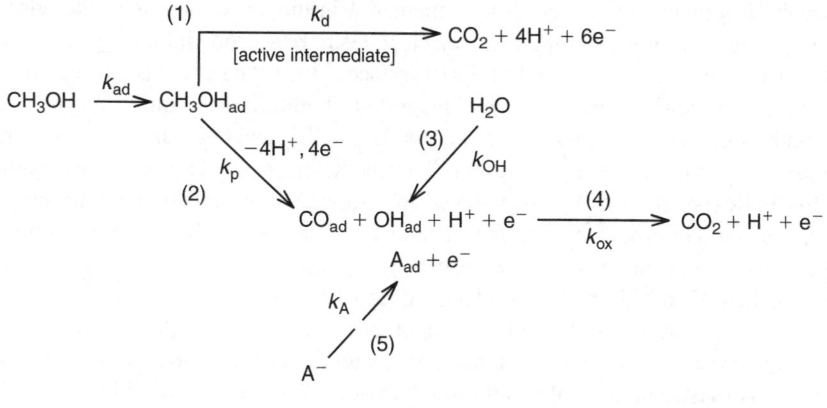

This reaction scheme is almost identical with one proposed for the HCOOH oxidation pathway, again emphasizing the importance of competition between the reactive intermediates and spectator species.

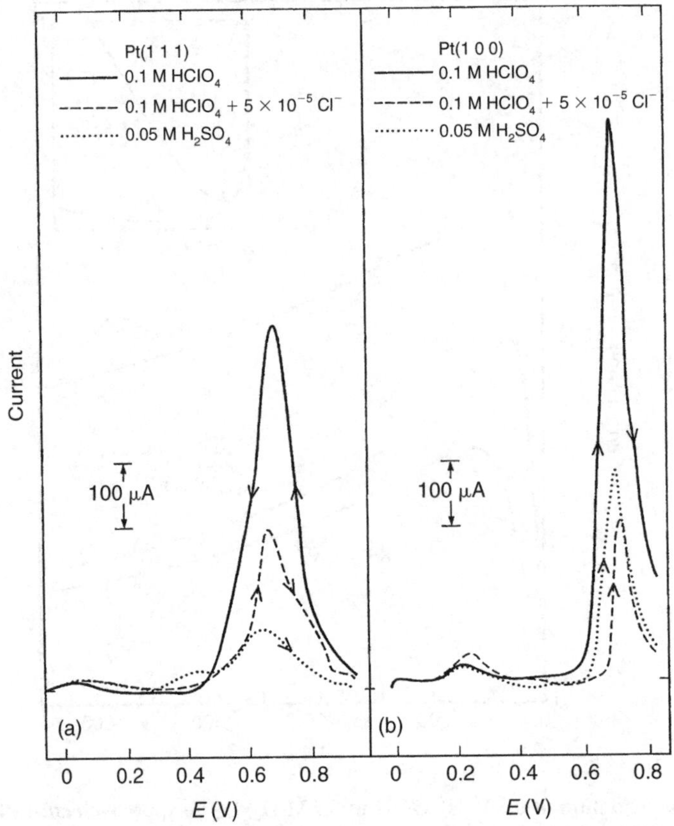

Fig. 59. Cyclic voltammograms for methanol oxidation on flame-annealed platinum single crystals in 0.1 M $HClO_4$, in 0.1 M $HClO_4$ with addition of HCl, and in 0.05 M H_2SO_4 [135]

Much effort, particularly with in situ IR spectroscopy, has been expended in trying to identify the unknown "direct" intermediate, and to date none has been identified. Note that the H_{upd} state is again excluded from the scheme, although H_{upd} blocks the adsorption of methanol in the same way as in the adsorption of HCOOH. As shown in Fig. 59, anions of supporting electrolytes have a strong inhibiting effect on the rate of methanol oxidation on both the Pt(1 1 1) and Pt(1 0 0) surfaces [135]. The effect is particularly strong in Cl^--containing solutions, requiring three orders of magnitude higher concentration of sulfuric acid to achieve the same inhibiting effect. In contrast to anions and H_{upd}, CO_{ad} and OH_{ad} are either spectators (poisoning), blocking the adsorption of methanol (k_{ad}) or "intermediates" which can be further oxidized with rate constant k_{ox}. In situ IR spectroscopy has also played an extremely important role in refining our understanding of this reaction, and in producing a different concept of the role of CO_{ad} in the reaction, namely that of intermediate versus poison. As was the case in gas-phase catalysis, the study of the reaction on alloy catalysts has helped to redefine our understanding of the reaction path and the role of CO_{ad}. There have been many important studies of methanol adsorption on Pt electrodes by in situ IR spectroscopy, which are not discussed in this chapter. These studies have been so extensive, and have such a long history with many twists and turns that a thorough review of the subject would be a full chapter in itself. There have been far fewer studies of methanol adsorption on bimetallic surfaces using in situ IR spectroscopy, and it is those that we will summarize in this chapter.

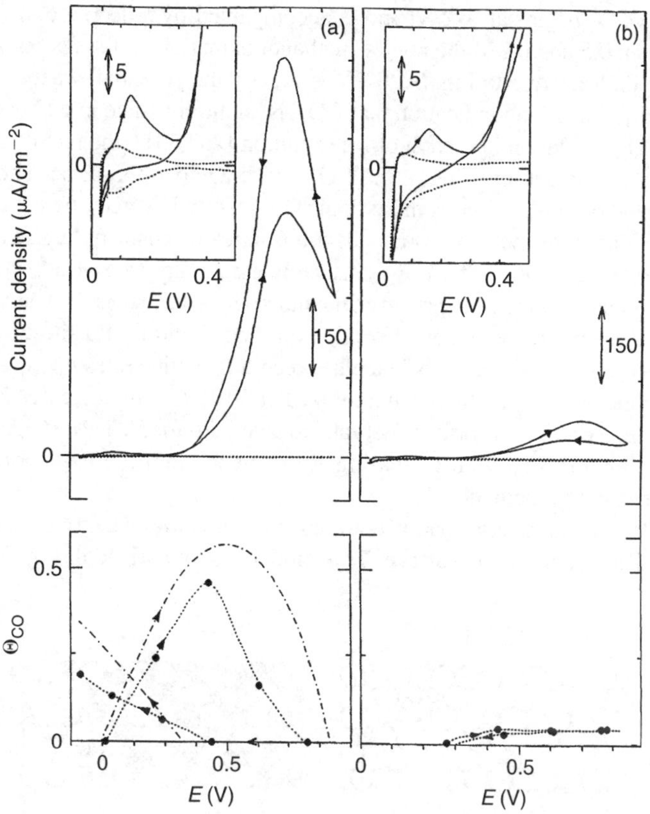

Fig. 60. CO coverages (bottom) assessed from in situ spectroscopy and (top) cyclic voltammetry of UHV-spatter-cleaned Pt—Ru alloys with (a) 10 at.% Ru and (b) 50 at.% Ru surface composition electrodes in 0.1-M $HClO_4$ with 0.05 M CH_3OH. Dotted curves: based voltammogram without CH_3OH. CO coverage on pure Pt (dashed–dotted curve) shown for reference. The electrode were contacted at 0.06 V (versus RHE). The insets show the voltammetry with a magnification of the current in the low-potential region [332]

Fig. 60 shows the CO_{ad} coverages at room temperature on polycrystalline Pt, $Pt_{90}Ru_{10}$, and $Pt_{50}Ru_{50}$ surfaces following immersion at 0.06 V in 0.1 M $HClO_4$ containing 0.05 M CH_3OH. Also shown is the corresponding anodic current as a function of potential. The details are given in [332]. Only linearly bonded CO_{ad}, as evidenced by the C—O stretch feature at 2040–2080 cm^{-1}, was observed on these surfaces and only CO_2, evidenced by the sharp asymmetric stretch at 2343 cm^{-1}, was observed in solution as the oxidation product at all potentials, i.e. there was no detectable amount of adsorbed formyl species nor were any partial oxidation products such as formic acid or methyl formate observed. Only a single C—O stretch was observed even though the frequencies for pure Pt and pure Ru differ by ca. 50 cm^{-1}. This result is similar to that reported by Ianniello et al. [293] and Lin et al. [295] for CO_{ad} from direct CO_b adsorption on the Pt—Ru alloy surfaces, and is attributed to the vibrational coupling between CO_{ad} molecules adsorbed on adjacent atoms. On the Pt and $Pt_{90}Ru_{10}$ surfaces, a bell-shaped functionality of CO coverage versus potential is observed with the onset of the IR signal for CO_{ad} at about 0.1 V, e.g. at the potential, where desorption of H_{upd} is observed in cyclic voltammetry. No CO_2 is observed until about 0.45 V. All of the current on the first sweep from 0.06 V should then correspond to CO_{ad} formation from the dehydrogenation reaction, and integration of this current was consistent with this result (assuming $4e^-$ per CO_{ad}). The onset of methanol oxidation to CO_2 begins at 0.45 V, and is accompanied by a decrease in CO_{ad} coverage. In the potential region between 0.5 and 0.7 V, the rate of methanol oxidation on the $Pt_{90}Ru_{10}$ surface is more than 30 times that for pure Pt. Referring to Fig. 60, 0.5 V is exactly the potential, where CO_{ad} begins to be oxidized on the Pt-rich surfaces, giving indication that CO_{ad} is an intermediate in the reaction. Referring again to the discussion of CO_{ad} oxidation in methanol-free solution [287,331], the rate of this step is maximized at 50% Ru, and thus a significant lowering of the CO_{ads} coverage on this surface should be expected. The CO_{ads} coverage observed on this surface is more than significantly lowered, the coverage is below 0.1 ML (!), suggesting that oxidation of methanol proceeds via a series mechanism, e.g., $k_{ad} \rightarrow k_p \rightarrow k_{ox} \rightarrow CO_2$. Note that in spite of the fact that the 50% Ru surface is essentially free of adsorbed intermediates, i.e. "unpoisoned" in the language of the dual-pathway, the most active surface is 7–10 at.% Ru. If one takes into account that at room temperature methanol adsorption does not occur on Ru sites, then the series pathway helps us understand the way Ru alters the balance between the relative rates k_{ad}, k_p and k_{ox}. Increasing Ru content in the surface increases k_{ox}, which is maximized at 50% Ru, but decreases k_{ad}. The fact that CO_{ad} coverage is relatively high on Pt-rich surfaces but falls to near zero for 50% Ru suggests that there is a transition in the rate determining step with increasing Ru content, from the oxidation of CO_{ad} (k_{ox}) to the adsorption/dehydrogenation of methanol (k_{ad}/k_p).

Gasteiger et al. [334] used this series pathway to develop a quantitative model of the dependence of the oxidation rate on the Ru content in the surface. This model is summarized in Fig. 61. A bifunctional role

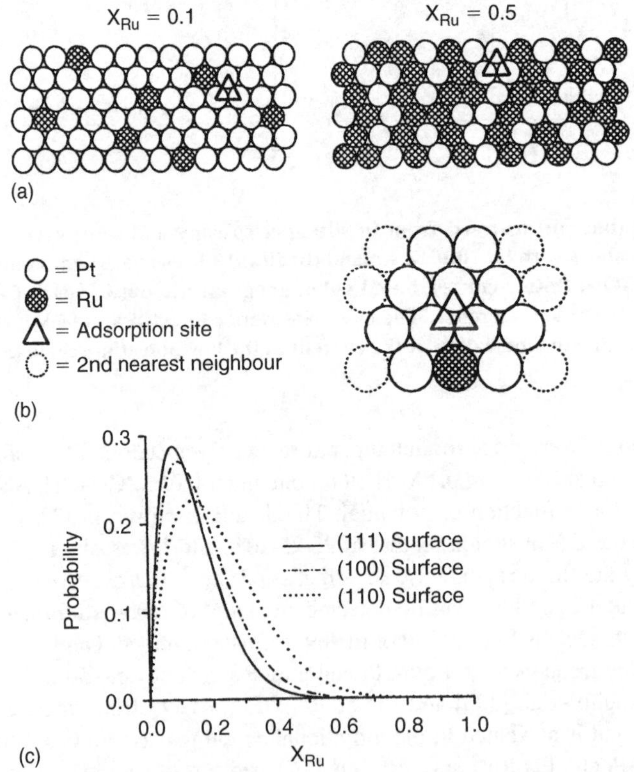

Fig. 61. (a) Schematic representation of sputter Pt—Ru alloy surfaces with 10 and 50 at.% Ru. (b) Geometric arrangement of atoms around a three-fold methanol adsorption site for an hexagonal surface face (fcc (1 1 1) face). (c) Probability distribution for the occurrence of a three-fold Pt site surrounded exactly one Ru atom for different low-index crystal face geometries as a function of the Ru surface composition [334]

of Pt and Ru atoms was assumed, with methanol adsorption/dehydrogenation occurring at an ensemble of Pt atoms and OH_{ads} nucleation occurring at Ru sites. Using statistical analysis, it was shown that the maximum concentration of active ensembles, viz three-fold Pt sites adjacent to exactly one Ru atom, occurs near 10% Ru, exactly where the maximum in rate is observed. Further support for the model was found in experiments conducted at higher temperature [335], where it was found that methanol adsorption/dehydrogenation occurs on Ru sites as well as on Pt sites, and that the maximum in total rate moves to higher Ru content, towards 50%. This shift towards 50% Ru as the most active surface is consistent with the change in ensemble configuration, where now (at higher temperature) any three surface sites can serve to dehydrogenate methanol, and the rate determining step becomes oxidative removal of CO_{ad} (k_4) for all surface compositions.

The effects of thermal activation on the oxidation pathway of methanol on bulk Pt—Ru alloy catalysts has recently been studied by Korzeniewski and co-workers [336] using "high temperature" FTIR spectroscopy. The IR spectra shown in Fig. 62 confirm the previous observation [335] that on Ru-rich surfaces the dissociative chemisorption (dehydrogenation) of methanol is inhibited on Ru sites at temperatures below 330 K; the faradic current for methanol oxidation was low, and only small quantities of adsorbed CO and CO_2 production [336] were detected between 0.2 and 0.8 V. At 353 K, however, strong infrared bands from CO_2 and atop coordinated CO_{ad} appeared at potentials above ca. 0.2 V. These IR results, therefore,

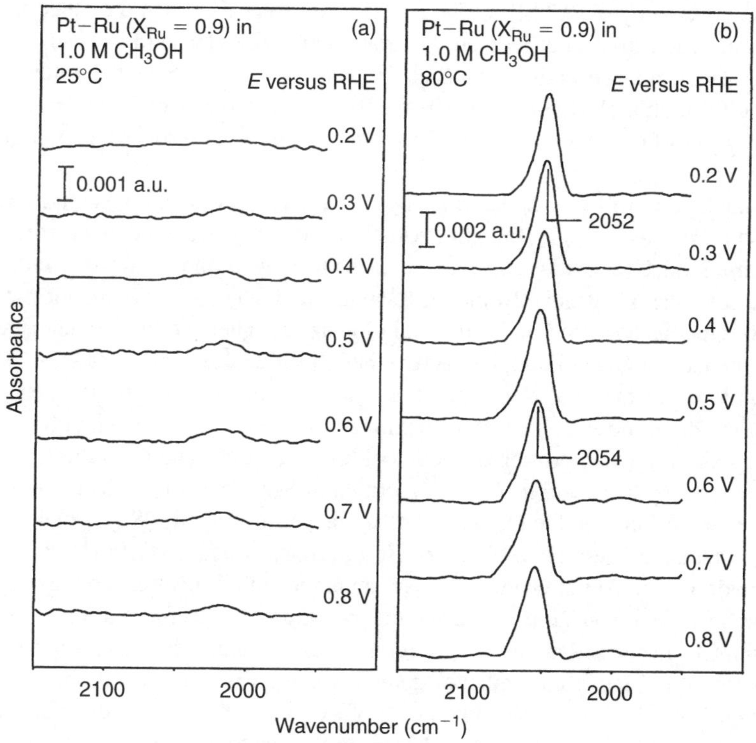

Fig. 62. Potential difference infra red spectra of a Pt—Ru alloy (x_{Ru} = 0.9) electrode in 0.1 M $HClO_4$ containing 1 M CH_3OH at (A) 25 and (B) 70°C. The spectral region for atop coordinated was held at 0.0 V in 0.1 M $HClO_4$ while methanol was added. The cell was maintained at 0.0 V with methanol present in solution for a total of 7 min prior to spectra acquisition. Spectra were recorded in sequence as the potential was stepped positive from 0.0 V. Data acquisition at each potential required (~2.7 min. The background spectra was recorded at 0.0 V [336]

nicely show that methanol adsorption (dehydrogenation) "turns on" at Ru sites at elevated temperatures, shifting the optimum Ru coverage from only 10 to 30 at.% so that methanol adsorption/dehydrogenation is in balance with simultaneous fast oxidative removal of CO_{ad}. While the serial reaction path and the ensemble model of Gasteiger et al. provide a reasonable explanation for the role of Ru in enhancing the activity of Pt for methanol oxidation, there is still much to be learned about this reaction on Pt surfaces, to say nothing of other Pt-group metals and their alloys.

The serial reaction pathway together with current knowledge of the dehydrogenation reaction on Pt provide a sound framework both for further fundamental study and for practical development of methanol oxidation catalysts. With respect to the latter, the Pt—Ru alloy remains the most active catalyst known for methanol electrooxidation in spite of more than 20 years of study of alternative catalysts since the seminal study of Pt bimetallic catalyst by the group at Batelle [337]. These studies have included, at one time or another, the modification of the Pt surface by electrodeposition and/or adsorption from solution of nearly every element in the periodic table that can be deposited in this manner. There are only two of these modifications, which have proven to have any significant stable enhancement of the activity of the Pt surface, and these are Ru and Sn. In the case of Ru-modified surface, it is not clear that one achieves the same result if the Ru atoms are in the surface, as in the bulk alloys, or on the surface. Chrzanowski and Wieckowski [338] and Chrzanowski et al. [339] demonstrated that catalytic activity of either polycrystalline Pt or Pt single crystals toward methanol oxidation was enhanced by irreversibly adsorbed Ru. The most active surface was Pt(1 1 1) covered by ca. 0.2 ML of Ru (or 1 Ru for every 5 Pt). On the other hand, Iwasita et al. [340] compared catalytic activities of methanol oxidation on UHV-cleaned PtRu bulk alloys, UHV-evaporated Ru onto Pt(1 1 1), and irreversibly adsorbed Ru on Pt(1 1 1). At room temperature, it was found that the activity of the PtRu bulk alloys between 10 and 40% is up to factor of 10 higher than activity of the Ru-modified Pt(1 1 1) surfaces. Further study appears to be needed to resolve this important fundamental question.

There has been a kind of folklore in electrocatalysis that a catalyst with high activity for methanol oxidation would also have a high activity for CO_b oxidation and vice versa. One might suspect, from the discussion in Section 3 on CO_b oxidation, that the Pt—Sn alloy is the catalyst of choice for methanol electrooxidation. It is interesting and informative to understand why this turns out not to be the case. Wang et al. [341] showed that Sn adsorbed on Pt from solution has a higher activity for methanol oxidation than the alloy Pt_3Sn, the latter actually being less active than Pt [342,343]. This apparently paradoxical result can be explain in terms of the need to balance the adsorption of methanol on Pt sites and the oxidative removal of CO_{ad}. At 298 K the enhancement is maximized at a very low coverage of Sn, e.g. about 0.1 ML, consistent with an absence of adsorption of methanol on Sn adatoms and blocking effect of Sn on the Pt ensemble needed for dehydrogenation. As we described in Section 3, the surface concentration of Sn in Pt_3Sn is at least 33% Sn (only on the (1 1 1) face, the other faces are 50% Sn), which are too Sn-rich for methanol dehydrogenation. Besides the issue of Sn coverage, Wang et al. [341] also argued that CO_{ad} formed from dehydrogenation of methanol also has an unique inhibiting effect on the rate of methanol oxidation and is not the same as CO_{ad} formed during the oxidation of CO_{ad} formed from adsorption in a solution saturated with CO. In particular, there are differences in the nature (coverages) of CO_{ad} produced from the two sources: the coverage from methanol dehydrogenation is much lower, thus it forms what was assigned in Section 5.3 as the strongly adsorbed state of $CO_{ad,}$ which is a relatively inactive state. With oxidation of dissolved CO in solution, the CO_{ad} adsorbs on Pt sites to high coverage forming the more reactive weakly adsorbed state. In the bulk Pt_3Sn alloy, Sn atoms nucleate OH_{ad} at low potential, and thus the CO_b reaction can proceed even though the Pt sites are completely covered by CO_{ad}.

The fundamental studies of methanol oxidation, while they have not yet produced new catalysts, have revealed some important lessons for catalyst development. The very strong ensemble effect observed with Ru- and Sn-modified surfaces means that one needs to control the surface composition when exploring new systems, which is why the authors put so much emphasis in this chapter on surface analysis and the science of surface enrichment. It is possible that some promising Pt bimetallic systems have been missed because there was too much of the admetal present on the surface.

6. FUTURE DEVELOPMENTS

Recent careful studies of the kinetics of oxygen reduction on pure Pt conclude that the intrinsic activity of Pt for the ORR depends on both the particle shape and size [105,246], i.e. there is not a single value of the specific activity even when normalized by the surface area. While kinetic results from Pt($h\,k\,l$) single crystals identify a structure sensitivity of the reaction, detailed comparison with results from supported catalyst fail to account for the effect of both particle size and shape on the activity. The situation is even more complicated for bimetallic catalysts. In the case of Pt—Mo catalysts for the oxidation of H_2/CO mixtures, detailed comparison of results from well-characterized bulk alloy surfaces with supported catalysts failed to account for the behavior of the supported catalysts [296,297]. A similar conclusion was reached in the cases of supported Pt—Ni and Pt—Co catalysts for the ORR [344]. With supported bimetallic catalysts, in addition to exposing different crystal faces with differing sizes/shapes, each face may have a unique surface composition, as may the edges/points at the intersections of faces. Such possibilities are clearly illustrated in the Monte Carlo simulations of Strohl and King [345] for cubo-octahedral binary clusters of Pt and Group Ib elements. Reactions, where the ensemble effect is very important, such as for methanol and formic acid electrooxidation, would be expected to show very strong dependence on bimetallic particle shape/size. Systematic experimental determinations of the effect of the bimetallic particle shape/size on the kinetics of methanol and formic acid electrooxidation are needed. One important aspect of the Monte Carlo simulations that needs experimental testing is that the {1 0 0} facets of the cluster contain no Pt, i.e. they are composed entirely of the Group Ib element (the surface enriched element). Since both methanol and formic acid electrooxidation kinetics vary significantly with crystal face on Pt($h\,k\,l$) surfaces, the exclusion of the alloying constituent from the {1 0 0} facet would have a profound effect on the activity of the supported catalyst. It is clear that issues of this kind are further evidence that solid surfaces, even when their surfaces are well-characterized, are not adequate models for real catalysts.

What are better model systems for real, i.e. supported, electrocatalysts? There is, we think, no simple answer to this question. Nor is there any single system that will model all the aspects of real catalysts, particularly in the exact configuration they are used in electrolytic cells. Regular arrays of metal particles fabricated by lithographic methods [346,347] will be useful for addressing some issues. Since particles will all have the same size, composition and shape, conventional surface analytical tools can be used to characterize the size/shape/composition and a conventional (macroscopic) kinetic measurement can be made. Unfortunately, the size of the particles created by lithographic methods are limited to relatively large sizes, e.g. >50 nm. How much lower the "feature size" can be reduced in the future is unclear. Chemical methods using "protecting shells" can create monodisperse metallic clusters in solution [348], and in the 1–10 nm size range we need, but putting the clusters on a support and removing the protecting shells produces a distribution of sizes/shapes that reduces their utility as "model" systems. Gas-phase ion cluster beam methods like those pioneered by Haberland et al. [349] may also prove to be a useful synthesis tool with some further improvements. Whatever synthesis technique is chosen, there still remains the challenge to analyze not only the size and shape of the bimetallic particles, but also the surface composition. One wants not only the average or integrated surface composition, but also the variation in composition across the particle, i.e. the composition on each facet. The technique which most obviously comes to mind is individual nanoparticle STM. The challenges to such measurements are obvious. One needs atomic resolution and elemental sensitivity while scanning the facet on a particle that may only be weakly attached to the substrate, i.e. in contact mode will the tip push the particle. Other microscopies and/or spectromicroscopies will probably have to be developed specifically to address these challenges.

The very promising results described here for ultrathin films of Pd on Pt(hkl) substrates suggest possible new directions for electrocatalysts based on a "skin" particle structure, i.e. a monolayer "skin" of one metal covering a core of another. Some binary alloy systems may actually form such 3D structures in nanoparticles as a result of equilibrium surface segregation when there is strong surface enrichment in one metal, e.g. Au—Ni and Au—Pt [345]. More generally, one would have to synthesize non-equilibrium structures to achieve the desired highly structured arrangement of metals. Raft geometries seem attractive for such structures, where one metal selectively wets the substrate (support) and the second metal selectively wets ("paints") the other. Clearly, to begin pursuing any of these highly structured nanoparticle concepts, one needs to develop some model systems, where surface characterization can be applied, and the understanding of the various interactions controlling the structure can be developed.

Finally, it appears certain that quantum chemical and classical Monte Carlo computational methods will play an increasingly important role in the study of model electrocatalyst systems in the future. This will be especially true for the nanocluster types of catalysts. In our view, however, there are important tradeoffs to be considered in the application of these methods in the near future. Because of the shear size of the calculation, it does appear likely one can produce a highly accurate calculation on both sides of the metal/solution interface simultaneously. A calculation, which has a very accurate representation of the solution side, including, e.g. ion–water solvation and the hydrogen bonding in water, cannot also have a highly accurate representation of the metal electronic structure, including simultaneously metal–metal, metal–water and metal–ion interactions. Another way of saying this is that the effect of electrode potential cannot be represented rigorously in current electronic structure calculations of the metal/solution interface. Nonetheless, the computational methods will be important tools in identifying trends in electrocatalytic properties of future novel materials and structures. There are also certain to be surprises in the future, particularly in the area of electrolyte effects in electrocatalysis, which have proven to be difficult to study with the current arsenal of surface analytical tools.

ACKNOWLEDGMENTS

We would like to express our sincere gratitude to a large number of collaborators for contributing to the work described in this review. We would like to thank C. Lucas for his part in performing and discussing the SXS measurements. Our indebtedness also goes to H. Gasteiger, B. Grgur, T. Schmidt, V. Stamenkovic, U. Paulus, and V. Climent, for their contributions in collecting and discussing the electrochemical results. A. Rodes is gratefully acknowledged for his contribution to the FTIR measurements. We thank M. Koper for his careful reading and editing of the manuscript and for suggesting a number of improvements. The authors are pleased to acknowledge long-term support for their research from the US Department of Energy, the Office of Science, Basic Energy Sciences, Materials Sciences Division, and the Office of Advanced Transportation Technologies, Fuel Cell Systems, under Contract DE-AC03-76SF00098.

REFERENCES

1. G.A. Somorjai and M.A. Van Hove. *Prog. Surf. Sci.* **30** (1989) 201.
2. G.A. Somorjai. *Introduction to Surface Chemistry and Catalysis*. Wiley, New York, 1993.
3. P.A. Thiel and P.J. Estrup. In: A.T. Hubbard (Ed.), *The Handbook of Surface Imaging and Visualization*. CRC Press, Boca Raton, FL, 1995.
4. R.I. Masel. *Principles of Adsorption and Reaction on Solid Surfaces*. Wiley, New York, 1996.
5. M.A. Van Hove. *Physics of Covered Solid Surfaces*. Landolt-Boernstein, 1999.
6. K. Griffiths, T.E. Jackman, T.A. Davies and P.R. Norton. *Surf. Sci.* **138** (1984) 113.
7. P.R. Norton, J.A. Davies, D.K. Creber, C.W. Sitter and T.E. Jackman. *Surf. Sci.* **108** (1981) 205.

8. P. Heilmann, K. Heinz and K. Muller. *Surf. Sci.* **83** (1979) 487.
9. J.A. Davies, T.E. Jackman, D.P. Jackson and P.R. Norton. *Surf. Sci.* **109** (1981) 20.
10. U. Korte and G. Meyer-Ehmsen. *Surf. Sci.* **271** (1992) 616.
11. U. Korte and G. Meyer-Ehmsen. *Surf. Sci.* **277** (1992) 109.
12. M.A. Krzyzowski, P. Zeppenfeld, C. Romainczyk, R. David and G. Comsa. *Phys. Rev. B* **50** (1994) 18505.
13. J. Kuntze, M. Huck, S. Bomermann, S. Speller and W. Heiland. *Surf. Sci. Lett.* **355** (1996) L300.
14. E.C. Sowa, M.A. Van Hove and D.L. Adams. *Surf. Sci.* **199** (1988) 174.
15. E. Vlieg, I.K. Robinson and K. Kern. *Surf. Sci.* **233** (1990) 248.
16. K. Christmann. In: J. Lipkowski and P.N. Ross Jr. (Eds.), *Electrocatalysis*. Wiley-VCH, New York, 1998, p. 1.
17. K. Christmann and G. Ertl. *Surf. Sci.* **41** (1976) 365.
18. R.W. McCabe and L.D. Schmidt. *Surf. Sci.* **65** (1977) 169.
19. M. Salmeron, R.J. Gale and G.A. Somorjai. *J. Phys. Chem.* **70** (1979) 2807.
20. A. Baro, H. Ibach and H. Bruchmann. *Surf. Sci.* **88** (1979) 384.
21. R.A. Olsen, G.J. Kroes and E.J. Baerends. *J. Chem. Phys.* **111** (1999) 11155.
22. B. Penneman, K. Oster and K. Wandelt. *Surf. Sci.* **249** (1991) 35.
23. X. Hu and Z. Lin. *Phys. Rev. B* **52** (1995) 11467.
24. A.T. Pasteur, St.J. Dixon-Warren and D.A. King. *J. Chem. Phys.* **103** (1995) 2251.
25. W.T. Lee, L. Ford, P. Blowers, H.L. Nigg and R.I. Masel. *Surf. Sci.* **416** (1998) 141.
26. J.R. Engstrom, W. Tsia and W.H. Weinberg. *J. Chem. Phys.* **87** (1987) 3104.
27. E. Kirsten, G. Parschau, W. Stocker and K.H. Rieder. *Surf. Sci. Lett.* **231** (1990) L183.
28. J.L. Gland, B.A. Sexton and G.B. Fisher. *Surf. Sci.* **95** (1980) 587.
29. C.T. Campbell, G. Ertl, H. Kuipers and J. Segner. *Surf. Sci.* **107** (1980) 220.
30. A. Eichler and J. Hafner. *Phys. Rev. Lett.* **79** (1997) 4481.
31. K. Mortensen, C. Klink, F. Jensen, F. Besenbacher and I. Stensgaard. *Surf. Sci.* **220** (1989) L701.
32. G.N. Derry and P.N. Ross. *Surf. Sci.* **140** (1984) 165.
33. P.R. Norton, P.E. Bindner and K. Griffiths. *J. Vac. Sci. Technol. A* **2** (1984) 1028.
34. X.-C. Guo, J.M. Bradley, A. Hopkinson and D.A. King. *Surf. Sci. Lett.* **292** (1993) L786.
35. R. Ducros and J. Fusy. *Appl. Surf. Sci.* **44** (1990) 59.
36. J. Schmidt, C. Stuhlmann and H. Ibach. *Surf. Sci.* **284** (1993) 121.
37. P.J. Feibelman, S. Esch and T. Michely. *Phys. Rev. Lett.* **77** (1996) 2257.
38. P. Gambardella, Z. Sljivancanin, B. Hammer, M. Blanc, K. Kuhnke and K. Kern. *Phys. Rev. Lett.* **87** (2001) 56103.
39. T. Engel and G. Ertl. *Adv. Catal.* **28** (1979) 1.
40. G. Blyholder. *J. Phys. Chem.* **68** (1964) 2772.
41. B.N.J. Persson, M. Tushaus and A.M. Bradshaw. *J. Chem. Phys.* **92** (1990) 5034.
42. G. Ertl, N. Neumann and K.M. Streit. *Surf. Sci.* **64** (1977) 393.
43. P.A. Thiel, R.J. Behm, P.R. Norton and G. Ertl. *J. Chem. Phys.* **78** (1983) 7448.
44. J.P. Biberian and M.A. Van Hove. *Surf. Sci.* **118** (1982) 443.
45. P. Gardner, C.R. Martin, M. Tushaus and A.M. Bradshaw. *Electron Spectrosc. Relat. Phenom.* **54/55** (1990) 619.
46. T.E. Jackman, J.A. Davies, D.P. Jackson and W.N. Unertl. *Surf. Sci.* **120** (1982) 389.
47. P. Hofmann, S.R. Bare and D.A. King. *Surf. Sci.* **117** (1982) 245.
48. T. Gritsch, D. Coulman, R.J. Behm and G. Ertl. *Phys. Rev. Lett.* **63** (1989) 1086.
49. S. Schwegmann, W. Tappe and U. Korte. *Surf. Sci.* **334** (1995) 55.
50. R.K. Sharma, W.A. Brown and D.A. King. *Surf. Sci.* **414** (1998) 68.
51. C.T. Campbell. *Annu. Rev. Phys. Chem.* **41** (1990) 775.
52. A.P. Dowben and A. Miller. *Surface Segregation Phenomena*. CRC Press, Boca Raton, FL, 1990.
53. W.M.H. Sachtler and R.A. van Santen. *Adv. Catal.* **26** (1977) 69.
54. J.R. Chelikowski. *Surf. Sci.* **139** (1984) L197.
55. P.R. Watson, M.A. Van Hove and K. Herman, Anon. *Atlas of Surface Structure*, Vol. IA, Monograph 5, 1995.
56. W. Heiland and E. Taglauer. In: R.L. Park and M. Lagally (Eds.), *Methods of Experimental Physics*, Vol. 22: *Solid State Physics, Surfaces*. Academic Press, Orlando, FL, 1985, p. 299.
57. H.A. Gasteiger, P.N. Ross Jr. and E.J. Cairns. *Surf. Sci.* **293** (2001) 67.

58. R. Ferrero, R. Capelli, A. Borsese and S. Delfino. *Atti. Accad. Naz. Lincei Red. Cl. Sci. Fis. Mat. Nat.* **54** (1973) 634.
59. A.N. Haner, P.N. Ross and U. Bardi. *Catal. Lett.* **8** (1991) 1.
60. A.N. Haner, P.N. Ross and U. Bardi. *Surf. Sci.* **249** (1991) 15.
61. U. Bardi. *Rep. Prog. Phys.* **57** (1994) 939.
62. L.D. Temmerman, C. Creemers, M.A. Van Hove, A. Neyens, J. Bertolini and J. Messardier. *Surf. Sci.* **178** (1986) 888.
63. Y. Gauthier, R. Baudoing and J. Rundgren. *Phys. Rev. B* **31** (1985) 6216.
64. F.F. Abraham. *Phys. Rev. Lett.* **46** (1981) 546.
65. G. Treglia and B. Legrand. *Phys. Rev. B* **35** (1987) 4338.
66. U. Bardi, B. Beard and P.N. Ross. *J. Catal.* **124** (1990) 22.
67. U. Bardi, A. Atrei, E. Zanazzi, G. Rovida and P.N. Ross Jr. *Vacuum* **41** (1990) 437.
68. Y. Gauthier. *Surf. Rev. Lett.* **3** (2001) 1663.
69. A.T. Hubbard. *Chem. Rev.* **88** (1988) 633.
70. J. Clavilier. *J. Electroanal. Chem.* **107** (1980) 211.
71. J. Clavilier. In: A. Wieckowski (Ed.), *Interfacial Electrochemistry – Theory, Experiment, and Applications*. Marcel Dekker, New York, 1999, p. 231.
72. N. Marković, M. Hanson, G. McDougall and E. Yeager. *J. Electroanal. Chem.* **214** (1986) 555.
73. L.A. Kibler, M. Cuesta, M. Kleinert and D.M. Kolb. *J. Electroanal. Chem.* **484** (2000) 73.
74. M. Wasberg, L. Palaikis, S. Wallen, M. Kamrath and A. Wieckowski. *J. Electroanal. Chem.* **256** (1988) 51.
75. A.T. Hubbard, R.M. Ishikawa and J. Katekaru. *J. Electroanal. Chem.* **86** (1978) 271.
76. A.S. Homa, E. Yeager and B.D. Cahan. *J. Electroanal. Chem.* **150** (1983) 181.
77. F.T. Wagner and P.N. Ross Jr. *J. Electroanal. Chem.* **150** (1983) 141.
78. K. Itaya. *Prog. Surf. Sci.* **58** (1998) 121.
79. M.G. Samant, M.F. Toney, G.L. Borges, K.F. Blurton and O.R. Melroy. *J. Phys. Chem.* **92** (1988) 220.
80. B.M. Ocko, J. Wang, A. Davenport and H. Isaacs. *Phys. Rev. Lett.* **65** (1990) 1466.
81. M.F. Toney and B.M. Ocko. *Synchrotr. Rad. News* **6** (1993) 28.
82. I.M. Tidswell, N.M. Marković and P.N. Ross. *Phys. Rev. Lett.* **71** (1993) 1601.
83. C. Lucas, N.M. Marković and P.N. Ross. *Surf. Sci.* **340** (1996) L949.
84. R.M. Ishikawa and A.T. Hubbard. *J. Electroanal. Chem.* **69** (1976) 317.
85. E. Yeager, W.E. O'Grady and M.Y.C.H.P. Woo. *J. Electrochem. Soc.* **125** (1978) 348.
86. S. Tanaka, S.-L. Yua and K. Itaya. *J. Electroanal. Chem.* **396** (1995) 125.
87. I.M. Tidswell, N.M. Marković and P.N. Ross. *J. Electroanal. Chem.* **376** (1994) 119.
88. B.E. Conway. In: S. Davison (Ed.), *Progress in Surface Science*. Pergamon Press, Fairview Park, NY, 1984, p. 1.
89. F.T. Wagner and P.N. Ross Jr. *J. Electroanal. Chem.* **250** (1988) 301.
90. D.M. Kolb. *Prog. Surf. Sci.* **51** (1996) 109.
91. H. Baltruschat, U. Bringemeier and R. Vogel. *Faraday Discuss.* **94** (1992) 317.
92. M.S. Zei, N. Batina and D.M. Kolb. *Surf. Sci.* **306** (1994) L519.
93. A. Al-Akl, G.A. Attard, R. Price and B. Timothy. *J. Electroanal. Chem.* **467** (1999) 60.
94. K. Sashikata, N. Furuya and K. Itaya. *J. Vac. Sci. Technol. A & B* **9** (1991) 457.
95. I.M. Tidswell, N.M. Marković, C. Lucas and P.N. Ross. *Phys. Rev. B* **47** (1993) 16542.
96. R. Michaelis and D.M. Kolb. *J. Electroanal. Chem.* **328** (1992) 341.
97. D. Armand and J. Clavilier. *J. Electroanal. Chem.* **263** (1989) 109.
98. Y. Kinomoto, S. Watanabe, M. Takahashi and M. Ito. *Surf. Sci.* **242** (2001) 538.
99. E. Morallon, W. Vazquez, R. Duo and A. Aldaz. *Surf. Sci.* **278** (1991) 33.
100. A.M. Bittner, J. Wintterlinn and G. Ertl. *J. Electroanal. Chem.* **388** (1995) 225.
101. C. Lucas, N.M. Marković and P.N. Ross. *Phys. Rev. Lett.* **77** (1996) 4922.
102. N.M. Marković, B.N. Grgur, C.A. Lucas and P.N. Ross. *Surf. Sci.* **384** (1997) L805.
103. G. Jerkiewicz. *Prog. Surf. Sci.* **57** (1998) 137.
104. N.M. Marković, P.N. Ross. *Electrochim. Acta* **45** (2000) 4101.
105. N.M. Marković, H.A. Gasteiger and P.N. Ross. *J. Electrochem. Soc.* **144** (1997) 1591.

106. M.W. Breiter. *Ann. NY Acad. Sci.* **101** (1963) 709.
107. M. Breiter. *Electrochim. Acta* **7** (1962) 25.
108. B.E. Conway, H. Angerstein-Kozlowska and H.P. Dhar. *Electrochim. Acta* **19** (1974) 455.
109. B.E. Conway, H. Angerstein-Kozlowska and W.B.A. Sharp. *J. Chem. Soc., Faraday Trans.* **74** (1978) 1373.
110. B.E. Conway and J.C. Currie. *J. Chem. Soc., Faraday Trans. 1* 74 (1978) 1390.
111. P.N. Ross Jr. In: R. Vanselow and R. Howe (Eds.), *Chemistry, Physics of Solid Surfaces*, Vol. IV, 1st edn. Springer, Berlin, 1982, p. 173.
112. E. Gileadi. *Electrode Kinetics for Chemical Engineers and Material Scientists*. VCH, London, 1993.
113. N.M. Marković, B.N. Grgur and P.N. Ross Jr. *J. Phys. Chem. B* **101** (1997) 5405.
114. N.M. Marković, T.J. Schmidt, B.N. Grgur, H.A. Gasteiger, P.N. Ross Jr. and R.J. Behm. *J. Phys. Chem. B* **103** (1999) 8568.
115. D.D. Wagman, W.H. Evans, V.B. Parker, R.H. Schumm, I. Halow, S.M. Bailey, K.L. Churney and R.L. Nuttall *J. Phys. Chem. Ref. Data* **11** (**Suppl. 2**) (1982).
116. C.H.P. Lupis. *Chemical Thermodynamics of Materials*. North-Holland, New York, 1983, p. 33.
117. Feidenhans'l. *Surf. Sci. Rep.* **10** (1989), p. 105.
118. C.A. Lucas, N.M. Marković and P.N. Ross. *Surf. Sci.* **425** (1999) L381.
119. C.A. Lucas, N.M. Marković, B.N. Grgur and P.N. Ross Jr. *Surf. Sci.* **448** (2000) 65.
120. R.J. Nichols. In: J. Lipkowski and P.N. Ross (Eds.), *Frontiers of Electrochemistry*. Wiley-VCH, New York, 1999, p. 99.
121. H. Ogasawara and M. Ito. *Chem. Phys. Lett.* **221** (1994) 213.
122. A. Bewick and J.W. Russell. *J. Electroanal. Chem.* **132** (1982) 329.
123. A. Peremans and A. Tadjeddine. *J. Chem. Phys.* **103** (1995) 7197.
124. A. Tadjeddine and A. Peremans. *J. Electroanal. Chem.* **409** (1996) 115.
125. A. Le Rille and A. Tadjeddine. *J. Electroanal. Chem.* **467** (1999) 238.
126. N.M. Marković and P.N. Ross Jr. In: A. Wieckowski (Ed.), *Interfacial Electrochemistry – Theory, Experiment, and Applications*. Marcel Dekker, New York, 1999, p. 821.
127. K. Bedurftig, S. Volkening, Y. Wang, J. Winterline, K. Jacoby and G. Ertl. *J. Chem. Phys.* **111** (1999) 11147.
128. A. Michaelides and P. Hu. *J. Chem. Phys.* **114** (2001) 513; M.T.M. Koper and R.A. van Santen. *J. Electroanal. Chem.* **476** (1999) 64.
129. P.W. Faguy, N. Marković, R.R. Adzic, C.A. Fierro and E.B. Yeager. *J. Electroanal. Chem.* **289** (1990) 245.
130. Y. Sawatari, J. Inukai and M. Ito. *J. Electron Spectrosc.* **64–65** (1993) 515.
131. F.C. Nart, T. Iwasita and M. Weber. *Electrochim. Acta* **39** (1994) 961.
132. P.W. Faguy, N.S. Marinkovic and R.R. Adzic. *J. Electroanal. Chem.* **407** (1996) 209.
133. P. Zelenay and A. Wieckowski. *J. Electrochem. Soc.* **139** (1992) 2552.
134. W. Savich, S.G. Sun, J. Lipkowski and A. Wieckowski. *J. Electroanal. Chem.* **388** (1995) 233.
135. N.M. Marković and P.N. Ross. *J. Electroanal. Chem.* **330** (1992) 499.
136. S. Thomas, Y.-E. Sung, H.S. Kim and A. Wieckowski. *J. Phys. Chem.* **100** (1996) 11726.
137. Y. Shingaya, K. Hirota, H. Ogasawara and M. Ito. *J. Electroanal. Chem.* **409** (1996) 103.
138. M.T.M. Koper and J.J. Lukkien. *J. Electroanal. Chem.* **485** (2000) 161.
139. A.M. Funtikov, U. Stimming and R. Vogel. *J. Electroanal. Chem.* **482** (1997) 147.
140. A. Kolics and A. Wieckowski. *J. Phys. Chem. B* **105** (2001) 2588.
141. P.W. Faguy, N. Marković and P.N. Ross Jr. *J. Electrochem. Soc.* **140** (1993) 1638.
142. C.A. Lucas, N.M. Marković and P.N. Ross. *Phys. Rev. B* **55** (1997) 7964.
143. N. Li and J. Lipkowski. *J. Electroanal. Chem.* **491** (2001) 95.
144. G.N. Salaita, D.A. Stern, F. Lu, H. Baltruschat, B.C. Schardt, J.L. Stickney, M.P. Soriaga, D.G. Frank and A.T. Hubbard. *Langmuir* **2** (1986) 828.
145. J.X. Wang, N.S. Marinkovic and R.R. Adzic. *Colloid. Surf. A* **134** (1998) 165.
146. H.A. Gasteiger, N.M. Marković and P.N. Ross. *Langmuir* **12** (1996) 1414.
147. W.J. Albery and M.L. Hitchman. *Ring-disc Electrodes*. Clarendon Press, Oxford, 1971.
148. A.J. Bard and L.R. Faulkner. *Electrochemical Methods*. Wiley, New York, 1980, p. 26.
149. J.M. Orts, R. Gomez, J.M. Feliu, A. Aldaz and J. Clavilier. *J. Phys. Chem.* **100** (1996) 2334.
150. N.M. Marković, C.A. Lucas, H.A. Gasteiger and P.N. Ross. *Surf. Sci.* **365** (1996) 229.

151. A.M. Bittner, B. Wintterline, B. Beran and G. Ertl. *Surf. Sci.* **335** (2001) 291.
152. B.C. Schardt, S.-L. Yua and F. Rinaldi. *Science* **243** (1989) 1050.
153. S.-L. Yua, C.M. Vitus and B.C. Schardt. *J. Am. Chem. Soc.* **112** (1990) 3677.
154. S.C. Chang, S.-L. Yua, B.C. Schardt and M.J. Weaver. *J. Phys. Chem.* **95** (1991) 4787.
155. D.M. Kolb. In: H. Gerischer and C.W. Tobias (Eds.), *Advances in Electrochemistry and Electrochemical Engineering*. Wiley, New York, 1978, p. 125.
156. R.R. Adzic. In: H. Gerischer and C.W. Tobias (Eds.), *Advances in Electrochemistry and Electrochemical Engineering*. Wiley, New York, 1984, p. 159.
157. J. Clavilier, J.M. Feliu and A. Aldaz. *J. Electroanal. Chem.* **243** (1988) 419.
158. A. Fernandez-Vega, J.M. Feliu, A. Aldaz and J. Clavilier. *J. Electroanal. Chem.* **305** (1991) 229.
159. E. Herrero, A. Fernandez-Vega, J.M. Feliu and A. Aldaz. *J. Electroanal. Chem.* **350** (1993) 73.
160. A. Fernandez-Vega, J.M. Feliu, A. Aldaz and J. Clavilier. *J. Electroanal. Chem.* **258** (1989) 101.
161. V. Climent, E. Herrero and J.M. Feliu. *Electrochim. Acta* **44** (1998) 1403.
162. E. Herrero, A. Rodes, J.M. Perez, J.M. Feliu and A. Aldaz. *J. Electroanal. Chem.* **412** (1996) 165.
163. G.A. Attard and A. Bannister. *J. Electroanal. Chem.* **300** (1991) 467.
164. J. Clavilier, M.J. Llorca, J.M. Feliu and A. Aldaz. *J. Electroanal. Chem.* **310** (1991) 429.
165. M. Baldauf and D.M. Kolb. *Electrochim. Acta* **38** (1993) 2145.
166. M. Baldauf and D.M. Kolb. *J. Phys. Chem.* **100** (1996) 11375.
167. V. Climent, N.M. Marković and P.N. Ross. *J. Phys. Chem. B* **104** (2000) 3116.
168. N.M. Marković, C. Lucas, V. Climent, V. Stamenkovic and P.N. Ross. *Surf. Sci.* **465** (2000) 103.
169. J.H. White and H.D. Abruna. *J. Phys. Chem.* **94** (1990) 894.
170. R. Michaelis, M.S. Zei, R.S. Zhai and D.M. Kolb. *J. Electroanal. Chem.* **339** (1992) 299.
171. R. Durand, R. Faure, D. Aberdam, C. Salem, G. Tourillon, D. Guay and M. Ladouceur. *Electrochim. Acta* **37** (1992) 1977.
172. N. Marković and P.N. Ross. *Langmuir* **9** (1993) 580.
173. N. Marković, H.A. Gasteiger and P.N. Ross. *Langmuir* **11** (1995) 4098.
174. R. Gomez, H.S. Yee, G.M. Bommarito, J.M. Feliu and H.D. Abruna. *Surf. Sci.* **335** (1995) 101.
175. T. Abe, G.M. Swain, K. Sashikata and K. Itaya. *J. Electroanal. Chem.* **382** (1995) 73.
176. N.M. Marković, C. Lucas, H.A. Gasteiger and P.N. Ross Jr. *Surf. Sci.* **372** (1997) 239.
177. A.C. Finnefrock, L.J. Buller, K.L. Ringland, J.D. Brock and H.D. Abruna. *J. Am. Chem. Soc.* **119** (1997) 11703.
178. N.M. Marković, B.N. Grgur, C. Lucas and P.N. Ross Jr. *Electrochim. Acta* **44** (1998) 1009.
179. A.J. Bard and L.R. Faulkner. *Electrochemical Methods*. Wiley, New York, 1980, p. 283.
180. I.M. Tidswell, C.A. Lucas, N.M. Marković and P.N. Ross. *Phys. Rev. B* **51** (1995) 10205.
181. S. Wu, Z. Shi, J. Lipkowski, A.P. Hitchcock and T. Tyliszczak *J. Phys. Chem.* **101** (1997), p. 10310.
182. N. Marković and P.N. Ross Jr. *J. Phys. Chem.* **97** (1993) 9771.
183. C.A. Lucas, N.M. Marković and P.N. Ross. *Phys. Rev. B* **56** (1997) 3651.
184. Z. Shi and J. Lipkowski. *J. Electroanal. Chem.* **364** (1994) 289.
185. M.F. Toney, J.N. Howard, J. Richer, G.L. Borges, J.G. Gordon and O.R. Merloy. *Phys. Rev. Lett.* **75** (1995) 4472.
186. B.C. Schardt, J.L. Stickney, D.A. Stern, A. Wieckowski, D.C. Zapien and A.T. Hubbard. *Surf. Sci.* **175** (1986) 520.
187. B.C. Schardt, J.L. Stickney, D.A. Stern, A. Wieckowski, D.C. Zapien and A.T. Hubbard. *Langmuir* **3** (1987) 239.
188. D. Aberdam, S. Traore, R. Durand and R. Faure. *Surf. Sci.* **180** (1987) 319.
189. R.L. Borup, D.E. Sauer and E.M. Stuve. *Surf. Sci.* **293** (1993) 10.
190. R.L. Borup and D.E. Sauer. *Surf. Sci.* **293** (2001) 27.
191. F. El Omar and R. Durand. *J. Electroanal. Chem.* **178** (1984) 343.
192. J. Clavilier, J.M. Orts, J.M. Feliu and A. Aldaz. *J. Electroanal. Chem.* **293** (1990) 197.
193. R.R. Adzic, J. Wang, C.M. Vitus and B.M. Ocko. *Surf. Sci. Lett.* **293** (1993) L876.
194. B.N. Grgur, N.M. Marković and P.N. Ross Jr. *Langmuir* **13** (1997) 6370.
195. N.M. Marković, B.N. Grgur, C. Lucas and P.N. Ross. *J. Chem. Soc., Faraday Trans.* **94** (1998) 3373.
196. J. Clavilier, A. Fernandez-Vega, J.M. Feliu and A. Aldaz. *J. Electroanal. Chem.* **261** (1989) 113.
197. S.C. Chang and M.J. Weaver. *Surf. Sci.* **241** (1991) 11.
198. L. Dollard, R.W. Evans and G.A. Attard. *J. Electroanal. Chem.* **345** (1993) 205.

199. R.W. Evans and G.A. Attard. *J. Electroanal. Chem.* **345** (1993) 337.
200. E. Herrero, J.M. Feliu and A. Aldaz. *J. Catal.* **152** (1995) 264.
201. B.E. Hayden, A.J. Murray, R. Parsons and D.J. Pegg. *J. Electroanal. Chem.* **409** (1996) 51.
202. U.W. Hamm, D. Kramer, R.S. Zhai and D.M. Kolb. *Electrochim. Acta* **43** (1998) 2969.
203. S.P.E. Smith and H.D. Abruna. *J. Phys. Chem. B* **102** (1998) 3506.
204. S.P.E. Smith and H.D. Abruna. *J. Electroanal. Chem.* **467** (1999) 43.
205. T.J. Schmidt, R.J. Behm, B.N. Grgur, N.M. Marković and P.N. Ross Jr. *Langmuir* **16** (2000) 8159.
206. S.P.E. Smith, K.F. Ben-Dor and H.D. Abruna. *Langmuir* **16** (2000) 787.
207. E. Herrero, A. Rodes, J.M. Perez, J.M. Feliu and A. Aldaz. *J. Electroanal. Chem.* **393** (1995) 87.
208. J.M. Feliu, A. Fernandez-Vega, J.M. Orts and A. Aldaz. *J. Chim. Phys.* **88** (1991) 1493.
209. M. Ball, C.A. Lucas, N.M. Marković, B.M. Murphy, P. Steadman, T.J. Schmidt, V. Stamenkovic and P.N. Ross. *Langmuir* **17** (2001) 5943.
210. M.T. Paffett, C.T. Campbell and T.N. Taylor. *J. Vac. Sci. Technol. A* **3** (1985) 812.
211. M.T. Paffett, C.T. Campbell and T.N. Taylor. *J. Chem. Phys.* **85** (1986) 6176.
212. N. Kizhakevariam and E.M. Stuve. *J. Vac. Sci. Technol. A* **8** (1990) 2557.
213. C.T. Campbell, J.A. Rodriguez and D.W. Goodman. *Phys. Rev. B* **46** (1992) 7077.
214. M. Han, P. Mrozek and A. Wieckowski. *Phys. Rev. B* **48** (1993) 8329.
215. B. Hammer, Y. Morikawa and J.K. Norskov. *Phys. Rev. Lett.* **76** (1996) 2141.
216. B. Hammer and J.K. Norskov. In: R.M. Lambert and G. Pacchioni (Eds.), *Chemisorption and Reactivity on Supported Clusters and Thin Films*. Kluwer Academic Publishers, Dordrecht, 1997, p. 285.
217. H. Naohara, S. Ye and K. Uosaki. *J. Phys. Chem. B* **102** (1998) 4366.
218. L.A. Kibler, M. Kleinert, R. Randler and D.M. Kolb. *Surf. Sci.* **443** (1999) 19.
219. B. Alvarez, V. Climent, A. Rodes and J.M. Feliu. *J. Electroanal. Chem.* **497** (2001) 125.
220. N. Kobosev and W. Monblanova. *Acta Physiochem. URSS* **1** (1934) 611.
221. W.T. Grubb. *Nature* **198** (1963) 883.
222. J.O.M. Bockris and A.K.N. Reddy. *Modern Electrochemistry*. Plenum Press, New York, 1970.
223. B.E. Conway and J.O.M. Bockris. *J. Chem. Phys.* **26** (1957) 532.
224. R. Parsons. *Trans. Faraday Soc.* **54** (1958) 1053.
225. H. Gerischer. *Bull. Soc. Chim. Belg.* **67** (1958) 506.
226. S. Trasatti. *J. Electroanal. Chem.* **39** (1972) 163.
227. S. Trasatti. *Surf. Sci.* **335** (1995) 1.
228. K.J. Vetter. *Electrochemical Kinetics*. Academic Press, New York, 1967.
229. S. Schuldiner, M. Rosen and D.R. Flinn. *J. Electrochem. Soc.* **117** (1970) 1251.
230. K. Seto, A. Iannelli, B. Love and J. Lipkowski. *J. Electroanal. Chem.* **226** (1987) 351.
231. H. Kita, S. Ye and Y. Gao. *J. Electroanal. Chem.* **334** (1992) 351.
232. R. Gomez, A. Fernandez-Vega, J.M. Feliu and A. Aldaz. *J. Phys. Chem.* **97** (1993) 4769.
233. N.M. Marković, S.T. Sarraf, H.A. Gasteiger and P.N. Ross. *J. Chem. Soc., Faraday Trans.* **92** (1996) 3719.
234. J.H. Barber, S. Morin and B.E. Conway. *J. Electroanal. Chem.* **446** (1998) 125.
235. J.H. Barber and B.E. Conway. *J. Electroanal. Chem.* **461** (1999) 80.
236. F. Ludwig, R.K. Sen and E. Yeager. *Élektrokhimiya* **13** (1977) 847.
237. B.E. Conway and L. Bai. *J. Electroanal. Chem.* **198** (1986) 149.
238. E. Protopopoff and P. Marcus. *J. Chim. Phys.* **88** (1991) 1423.
239. B.E. Conway. In: A. Wieckowski (Ed.), *Interfacial Electrochemistry – Theory, Experiment, and Applications*. Marcel Dekker, New York, 1999, p. 131.
240. T.J. Schmidt, B.N. Grgur, N.M. Marković and P.N. Ross Jr. *J. Electroanal. Chem.* **500** (2001) 36.
241. N.V. Osetrova and V.S. Bagotzky. *Élektrokhimiya* **9** (1973) 1527.
242. T.J. Schmidt, N.M. Marković and P.N. Ross. *J. Phys. Chem.* **1** (2001) 1.
243. V. Stamenkovic, N.M. Marković and P.N. Ross Jr. (in preparation).
244. A. Ruban, B. Hammer, P. Stoltze, H.L. Skriver and J.K. Norskov. *J. Mol. Catal. A* **115** (1997) 421.
245. A.J. Appleby. *Catal. Rev.* **4** (1970) 221.
246. K. Kinoshita. *Electrochemical Oxygen Technology*. Wiley, New York, 1992.
247. H. Wroblowa, Y.C. Pan and J. Razumney. *J. Electroanal. Chem.* **69** (1976) 195.

248. B.N. Grgur, N.M. Marković and P.N. Ross Jr. *Can. J. Chem.* **75** (1997) 1465.
249. N.M. Marković, H.A. Gasteiger, B.N. Grgur and P.N. Ross. *J. Electroanal. Chem.* **467** (1999) 157.
250. V. Stamenkovic, N.M. Marković and P.N. Ross Jr. *J. Electroanal. Chem.* **500** (2000) 44.
251. M.R. Tarasevich, A. Sadkowski and E. Yeager. In: J.O.M. Bockris, B.E. Conway, E. Yeager, S.U.M. Khan and R.E. White (Eds.), *Comprehensive Treatise in Electrochemistry*. Plenum Press, New York, 1983, p. 301.
252. F. Uribe, M.S. Wilson, T. Springer and S. Gottesfeld. In: D. Scherson, D. Tryk, M. Daroux and X. Xing (Eds.), *Proceedings of the Workshop on Structural Effects in Electrocatalysis and Oxygen Electrochemistry*, PV 92-11, The Electrochemical Society, Pennington, NJ, 1992, p. 494.
253. P.N. Ross Jr. *J. Electrochem. Soc.* **126** (1979) 78.
254. F. El Kadiri, R. Faure and R. Durand. *J. Electroanal. Chem.* **301** (1991) 177.
255. N.M. Marković, R.R. Adzic, B.D. Cahan and E. Yeager. *J. Electroanal. Chem.* **377** (1994) 249.
256. N.M. Marković, H.A. Gasteiger and P.N. Ross. *J. Phys. Chem.* **99** (1995) 3411.
257. N.M. Marković, N.S. Marinkovic and R.R. Adzic. *J. Electroanal. Chem.* **241** (1988) 309.
258. N.M. Marković, H.A. Gasteiger and P.N. Ross. *J. Phys. Chem.* **100** (1996) 6715.
259. A. Damjanovic, M.A. Genshaw and J.O.M. Bockris. *J. Phys. Chem.* **45** (2001) 4057.
260. F.J. Luczak and D.A. Landsman. US Patent 4 447 506 (1984).
261. F.J. Luczak and D.A. Landsman. US Patent 4 677 092 (1987).
262. B. Beard and P.N. Ross Jr. *J. Electrochem. Soc.* **137** (1990) 3368.
263. J.T. Glass, G.L. Cahen and G.E. Stoner. *J. Electrochem. Soc.* **134** (1987) 58.
264. S. Mukerjee and S. Srinivasan. *J. Electroanal. Chem.* **357** (1993) 201.
265. P.N. Ross Jr. In: J. Lipkowski and P.N. Ross Jr. (Eds.), *Electrocatalysis*. Wiley-VCH, New York, 1998, p. 43.
266. D. Zurawski, M. Wasberg and A. Wieckowski. *J. Phys. Chem.* **94** (1990) 2076.
267. W.-F. Lin, M.S. Zei and G. Ertl. *Chem. Phys. Lett.* **312** (1999) 1.
268. I. Oda, J. Inukai and M. Ito. *Chem. Phys. Lett.* **203** (1993) 99.
269. I. Villegas and M.J. Weaver. *J. Chem. Phys.* **101** (1994) 1648.
270. I. Villegas, X. Gao and M.J. Weaver. *Electrochim. Acta* **40** (1995) 1267.
271. N.M. Marković, B.N. Grgur, C.A. Lucas and P.N. Ross. *J. Phys. Chem. B* **103** (1999) 487.
272. A. Wieckowski, M. Rubel and C. Gutiérrez. *J. Electroanal. Chem.* **382** (1995) 97.
273. H. Kita, H. Naohara, T. Nakato, S. Taguchi and A. Aramata. *J. Electroanal. Chem.* **386** (1995) 197.
274. A. Rodes, R. Gomez, J.M. Feliu and M.J. Weaver. *Langmuir* **16** (2001) 811.
275. N.M. Marković, C. Lucas, A. Rodes, V. Stamenkovic and P.N. Ross. *Surf. Sci.* **499** (2002) 149.
276. M.W. Severson, C. Stuhlmann, I. Villegas and M.J. Weaver. *J. Chem. Phys.* **103** (1995) 9832.
277. N.M. Marković, C. Lucas, B.N. Grgur and P.N. Ross. *J. Phys. Chem.* **103** (1999) 9616.
278. M. Bergelin, J.M. Feliu and M. Wasberg. *Electrochim. Acta* **44** (1998) 1069.
279. M. Bergelin, E. Herrero, J.M. Feliu and M. Wasberg. *J. Electroanal. Chem.* **467** (1999) 74.
280. T.J. Schmidt, N.M. Marković, P.N. Ross Jr. *J. Phys. Chem. B* **105** (2001) 12082.
281. J. Xu and J.T. Yates. *J. Chem. Phys.* **99** (1993) 725.
282. N.P. Lebedeva, M.T.M. Koper, E. Herrero, J.M. Feliu and R.A. van Santen. *J. Electroanal. Chem.* **487** (2000) 37.
283. N.M. Marković, B.N. Grgur, C.A. Lucas and P.N. Ross Jr. *Langmuir* **16** (2000) 1998.
284. J.M. Sinfelt. *Bimetallic Catalysts: Discoveries, Concepts and Applications*. Wiley, New York, 1983.
285. M. Watanabe and S. Motoo. *J. Electroanal. Chem.* **60** (1975) 275.
286. H.A. Gasteiger, P.N. Ross and E.J. Cairns. *Surf. Sci.* **293** (1993) 67.
287. H.A. Gasteiger, N. Marković, P.N. Ross and E.J. Cairns. *J. Phys. Chem.* **98** (1994) 617.
288. H.A. Gasteiger, N. Marković and P.N. Ross. *J. Phys. Chem.* **99** (1995) 8290.
289. H.A. Gasteiger, N. Marković and P.N. Ross. *J. Phys. Chem.* **99** (1995) 16757.
290. H.A. Gasteiger, N.M. Marković and P.N. Ross. *J. Phys. Chem.* **99** (1995) 8945.
291. B.N. Grgur, N.M. Marković and P.N. Ross. *Electrochim. Acta* **43** (1998) 3631.
292. M.T.M. Koper, J.J. Lukkien, A.P.J. Jansen and R.A. van Santen. *J. Phys. Chem. B* **103** (1999) 5522.
293. R. Ianniello, V.M. Schmidt, U. Stimming, J. Stumper and A. Wallau. *Electrochim. Acta* **39** (1994) 1863.
294. K.A. Friedrich, K.-P. Geyzers, U. Linke, U. Stimming and J. Stumper. *J. Electroanal. Chem.* **402** (1996) 123.
295. W.-F. Lin, M.S. Zei, M. Eiswirth, G. Ertl, T. Iwasita and W. Vielstich. *J. Phys. Chem. B* **103** (1999) 6968.
296. B.N. Grgur, G. Zhuang, N.M. Marković and P.N. Ross Jr. *J. Phys. Chem. B* **101** (1997) 3910.

297. B.N. Grgur, N.M. Marković and P.N. Ross Jr. *J. Phys. Chem. B* **102** (1998) 2494.
298. B.N. Grgur, N.M. Marković and P.N. Ross Jr. *J. Electrochem. Soc.* **146** (1999) 1613.
299. S. Mukerjee, S.J. Lee, E.A. Ticianelli, J. McBreen, B.N. Grgur, N.M. Marković, P.N. Ross, J.R. Giallombardo and E.S. De Castro. *Electrochem. Sol. Lett.* **2** (1999) 12.
300. A.N. Haner, P.N. Ross, U. Bardi and A. Atrei. *J. Vac. Sci. Technol. A* **10** (1992) 2718.
301. T.J. Schmidt, M. Noeske, H.A. Gasteiger, R.J. Behm, P. Britz, W. Brijoux and H. Bönnemann. *Langmuir* **13** (1997) 2591.
302. T.J. Schmidt, M. Noeske, H.A. Gasteiger, R.J. Behm, P. Britz and H. Bönnemann. *J. Electrochem. Soc.* **145** (1998) 925.
303. T.J. Schmidt, H.A. Gasteiger and R.J. Behm. *J. New Mater. Electrochem. Syst.* **2** (1999) 27.
304. N.M. Marković and P.N. Ross Jr. *Catal. Technol.* **4** (2000) 110.
305. A. Capon and R. Parsons. *J. Electroanal. Chem.* **45** (1973) 205.
306. B. Beden, A. Bewick and C. Lamy. *J. Electroanal. Chem.* **148** (1983) 147.
307. K. Kunimatsu and H. Kita. *J. Electroanal. Chem.* **218** (1987) 155.
308. B. Beden and C. Lamy. In: R.J. Gale (Ed.), *Spectroelectrochemistry, Theory and Practice*. Plenum Press, New York, 1988.
309. A. Bewick, B. Pons, R. Clarke and R. Hester (Eds.). Wiley, Heyden, 1985.
310. K. Krischer. In: B.E. Conway et al. (Eds.), *Modern Aspects of Electrochemistry*. Kluwer Academic Publishers/Plenum Press, New York, 1999, p. 1.
311. R.R. Adzic. In: R.E. White, J.O.M. Bockris and B.E. Conway (Eds.), *Modern Aspects of Electrochemistry*. Plenum Press, New York, 1990, p. 163.
312. R.R. Adzic, A.V. Tripkovic and N.M. Marković. *J. Electroanal. Chem.* **150** (1983) 79.
313. J. Clavilier and S.G. Sun. *J. Electroanal. Chem.* **199** (1986) 471.
314. S.C. Chang, L.W.H. Leung and M.J. Weaver. *J. Phys. Chem.* **94** (1990) 6013.
315. A.V. Tripkovic, K. Popovic and R.R. Adzic. *J. Chim. Phys.* **88** (1991) 1635.
316. C. Lamy and J.M. Leger. *J. Chem. Phys.* **88** (1991) 1649.
317. H. Kita and H.-W. Lei. *J. Electroanal. Chem.* **388** (1995) 167.
318. T. Iwasita, X. Xia, E. Herrero and H.-D. Liess. *Langmuir* **12** (1996) 4260.
319. G.-Q. Ly, A. Crown and A. Wieckowski. *J. Phys. Chem.* **103** (1999) 9700.
320. S.G. Sun and Y.-Y. Yang. *J. Electroanal. Chem.* **467** (1999) 121.
321. M.F. Mrozek, H. Luo and M.J. Weaver. *Langmuir* **16** (2000) 8463.
322. V.S. Bagotzky and Y.B. Vassilyev. *Electrochim. Acta* **12** (1967) 1323.
323. E. Herrero, J.M. Feliu and A. Aldaz. *J. Electroanal. Chem.* **368** (1994) 101.
324. S.P.E. Smith and H.D. Abruna. *Langmuir* **15** (1999) 7325.
325. R. Parsons and T. VanderNoot. *J. Electroanal. Chem.* **257** (1988) 9.
326. J. Clavilier, J.M. Feliu, A. Fernandez-Vega and A. Aldaz. *J. Electroanal. Chem.* **269** (1989) 175.
327. J. Clavilier, A. Fernandez-Vega, J.M. Feliu and A. Aldaz. *J. Electroanal. Chem.* **258** (1989) 89.
328. S.C. Chang, Y. Ho and M.J. Weaver. *Surf. Sci.* **265** (1992) 81.
329. C.T. Campbell, J.M. Campbell, P.J. Dalton, F.C. Henn, J.A. Rodriguez and S.G. Seimanides. *J. Phys. Chem.* **93** (1989) 806.
330. E. Leiva, T. Iwasita, E. Herrero and J.M. Feliu. *Langmuir* **13** (1997) 6287.
331. H.A. Gasteiger, N. Marković, P.N. Ross and E.J. Cairns. *Electrochim. Acta* **39** (1994) 1825.
332. N. Marković, H.A. Gasteiger, P.N. Ross, X. Jiang, I. Villegas and M.J. Weaver. *Electrochim. Acta* **141** (1995) 91.
333. T.D. Jarvi and E.M. Stuve. In: J. Lipkowski and P.N. Ross (Eds.), *Electrocatalysis*. Wiley-VCH, New York, 1998, p. 75.
334. H.A. Gasteiger, N. Marković, P.N. Ross and E.J. Cairns. *J. Phys. Chem.* **97** (1993) 12020.
335. H.A. Gasteiger, N. Marković, P.N. Ross and E.J. Cairns. *J. Electrochem. Soc.* **141** (1994) 1795.
336. D. Kardash, C. Korzeniewski and N. Marković. *J. Electroanal. Chem.* **500** (2001) 518.
337. H. Binder, A. Köhling and G. Sandstede. In: G. Sandstede (Ed.), *From Electrocatalysis to Fuel Cells*. University of Washington Press for Batelle Seattle Research Center, Seattle, 1972, p. 43.
338. W. Chrzanowski and A. Wieckowski. *Langmuir* **14** (1998) 1967.
339. W. Chrzanowski, H. Kim and A. Wieckowski. *Catal. Lett.* **30** (1998) 69.

340. T. Iwasita, H. Hoster, A. Jon-Anacker, W.-F. Lin and W. Vielstich. *Langmuir* **16** (2000) 522.
341. K. Wang, H.A. Gasteiger, N.M. Marković and P.N. Ross. *Electrochim. Acta* **41** (1996) 2587.
342. A.N. Haner and P.N. Ross. *J. Phys. Chem.* **95** (1991) 3740.
343. S.A. Campbell and R. Parsons. *J. Chem. Soc., Faraday Trans.* **88** (1992) 833.
344. U.A. Paulus, G.G. Scherer, A. Wokaun, T.J. Schmidt, V. Stamenkovic, V. Radmilovic, N.M. Marković and P.N. Ross. *J. Phys. Chem. B* **106** (2002) 4181–4191.
345. J.K. Strohl and T.S. King. *J. Catal.* **116** (1989) 540.
346. S. Baldelli, A. Eppler, E. Anderson and Y.R. Shen. *J. Chem. Phys.* **113** (2000) 5432.
347. J. Hulteen and R. Van Duyne. *J. Vac. Sci. Technol. A* **13** (1995) 1553.
348. H. Bönnemann, G. Braun, W. Brijoux, R. Brinkmann, A. Schulze-Tilling, K. Seevogel and K. Siepen. *J. Organomet. Chem.* **520** (1996) 143.
349. H. Huberland, M. Mall, M. Moseler, Y. Quiang, T. Reiners and Y. Thurner. *J. Vac. Sci. Technol. A* **12** (1994) 2925.
350. V. Stamenkovic and N.M. Marković. *Langmuir* **17** (2000) 2388.
351. T.J. Schmidt, N.M. Marković and P.N. Ross. *J. Phys. Chem. B* **67** (2001) 1.
352. C.A. Lucas, N.M. Marković and P.N. Ross Jr. *Surf. Sci.* **448** (2000) 77.
353. H.A. Gasteiger, N.M. Marković and P.N. Ross Jr. *Catal. Lett.* **36** (1996) 1.
354. N.M. Marković, A.V. Tripkovic, N.S. Marinkovic and R.R. Adzic. In: M.P. Soriaga (Ed.), *Electrochemical Surface Science*. American Chemical Society, Washington, DC, 1988, p. 473.

Chapter 17

Proton-conducting polymer electrolyte membranes based on hydrocarbon polymers

M. Rikukawa and K. Sanui

Abstract

This paper presents an overview of the synthesis, chemical and electrochemical properties, and polymer electrolyte fuel cell applications of new proton-conducting polymer electrolyte membranes based on hydrocarbon polymers. Due to their chemical stability, high degree of proton conductivity, and remarkable mechanical properties, perfluorinated polymer electrolytes such as Nafion®, Aciplex®, Flemion®, and Dow membranes are some of the most promising electrolyte membranes for polymer electrolyte fuel cells. A number of reviews on the synthesis, electrochemical properties, and fuel cell applications of perfluorinated polymer electrolytes have also appeared during this period. While perfluorinated polymer electrolytes have satisfactory properties for a successful fuel cell electrolyte membrane, the major drawbacks to large-scale commercial use involve cost and low proton-conductivities at high temperatures and low humidities. Presently, one of the most promising ways to obtain high performance proton-conducting polymer electrolyte membranes is the use of hydrocarbon polymers for the polymer backbone. The present review attempts for the first time to summarize the synthesis, chemical and electrochemical properties, and fuel cell applications of new proton-conducting polymer electrolytes based on hydrocarbon polymers that have been made during the past decade.

Keywords: Proton; Electrolyte; Poly(4-phenoxybenzoyl-1,4-phenylene); Poly(ether-etherketone); Fuel cell; Sulfonation; Conductivity

Article Outline

1. Introduction 376
   1.1. Polymer electrolyte fuel cells 376
   1.2. Solid polymer electrolyte membranes 377
2. Proton-conducting polymer electrolyte membranes based on sulfonated aromatic polymers . . 379
   2.1. Materials 379
   2.2. Thermal stability of sulfonated aromatic polymer electrolyte membranes 382
   2.3. Water uptake in sulfonated aromatic polymer electrolyte membranes 384
   2.4. Proton conductivity of sulfonated aromatic polymer electrolyte membranes 386
3. Proton-conducting polymer electrolyte membranes based on alkylsulfonated aromatic polymers 389
   3.1. Materials 389
   3.2. Thermal stability of alkylsulfonated aromatic polymer electrolyte membranes 391

Fuel Cells Compendium

3.3. Water uptake in alkylsulfonated aromatic polymer electrolyte membranes 392
3.4. Proton conductivity of alkylsulfonated aromatic polymer electrolyte membranes 393
4. Proton-conducting polymer electrolyte membranes based on acid–base polymer complexes . . 396
4.1. Materials 396
4.2. Thermal stability of acid–base polymer complexes 400
4.3. Conductivity of acid–base polymer complex 401
5. Other proton-conducting polymer electrolyte membranes 403
6. Fuel cell applications 408
7. Summary 408
References 409

1. INTRODUCTION

1.1. Polymer electrolyte fuel cells

The idea of using an organic cation exchange membrane as a solid electrolyte in electrochemical cells was first described for a fuel cell by Grubb in 1959. At present the polymer electrolyte fuel cell (PEFC) is the most promising candidate system of all fuel cell systems in terms of the mode of operation and applications. As shown in Fig. 1, a PEFC consists of two electrodes and a solid polymer membrane, which acts as an electrolyte. The polymer electrolyte membrane is sandwiched between two platinum-porous electrodes such as carbon paper and mesh. Some single cell assemblies can be mechanically compressed across electrically conductive separators to fabricate electrochemical stacks. In general, PEFCs require humidified gases, hydrogen and oxygen (or air) as a fuel for their operation. The electrochemical reactions that occur at both electrodes are as follows:

$$\text{Anode:}\quad H_2 \rightarrow 2H^+ + 2e^-$$

$$\text{Cathode:}\quad 1/2O_2 + 2H^+ + 2e^- \rightarrow H_2O$$

$$\text{Overall:}\quad H_2 + 1/2O_2 \rightarrow H_2O + \text{Electrical Energy} + \text{Heat Energy}$$

In recent years, PEFCs have been identified as promising power sources for vehicular transportation and for other applications requiring clean, quiet, and portable power. Hydrogen-powered fuel cells in general

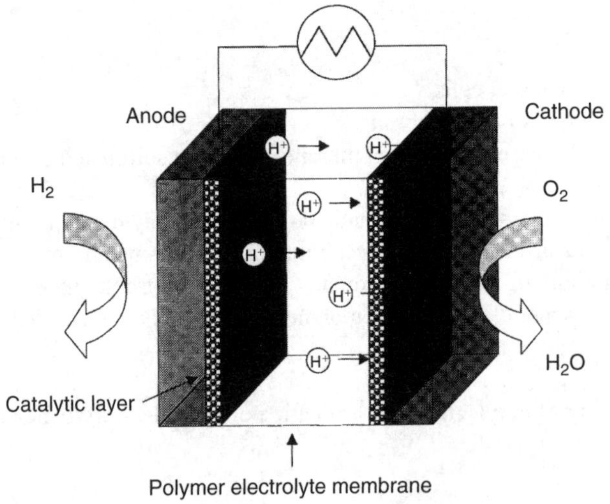

Fig. 1. PEFC schematic

have a high–power density and are relatively efficient in their conversion of chemical energy to electrical energy. Exhaust from hydrogen-powered fuel cells is free of environmentally undesirable gases such as nitrogen oxides, carbon monoxide, and residual hydrocarbons that are commonly produced by internal combustion engines. Carbon dioxide, a greenhouse gas, is also absent from the exhaust of hydrogen-powered fuel cells. Thus, transportation uses, especially fuel cell electric vehicles (FCEV), are on attractive and effective application because of not only clean exhaust emissions and high-energy efficiencies but also effective solution to the coming petroleum shortage. While FCEV might provide the greatest societal benefits, its total impact would be small if only a few FCEVs are sold due to lack of fueling infrastructure or due to high vehicle cost. The major obstacles for the commercial use of FCEV are expensive materials and low performances at high temperatures (over 100°C) and low humidities.

1.2. Solid polymer electrolyte membranes

In general, proton-conducting polymers are usually based on polymer electrolytes, which have negatively charged groups attached to the polymer backbone. These polymer electrolytes tend to be rather rigid and are poor proton conductors unless water is absorbed. The proton conductivity of hydrated polymer electrolytes dramatically increases with water content and reaches values of 10^{-2}–$10^{-1}\,S\,cm^{-1}$.

The first PEFC used in an operational system was the GE-built 1 kW Gemini power plant [1]. This system was used as the primary power source for the Gemini spacecraft during the mid-1960s. The performances and lifetimes of the Gemini PEFCs were limited due to the degradation of poly(styrene sulfonic acid) membrane employed at that time. The degradation mechanism determined by GE was generally accepted until the present time. It was postulated that HO_2 radicals attack the polymer electrolyte membrane. The second GE PEFC unit was a 350 W module that powered the Biosatellite spacecraft in 1969. An improved Nafion® membrane manufactured by DuPont was used as the electrolyte. Figure 2 shows the chemical structures of Nafion® and other perfluorinated electrolyte membranes. The performance and lifetime of PEFCs have significantly improved since Nafion® was developed in 1968. Lifetimes of over 50,000 h have been achieved with commercial Nafion®120.

Nafion® 117 and 115 have equivalent repeat unit molecular weights of 1100 and thicknesses in the dry state of 175 and 125 μm, respectively. Nafion® 120 has an equivalent weight of 1200 and a dry state thickness of 260 μm. Ballard Technologies Corporation showed the possibility of the application of PEFC for electric vehicles by using experimental perfluorinated membranes developed by Dow Chemical. Development of PEFC has been accelerated year by year after the report of Ballard Technologies Corporation. The Dow membrane has an equivalent weight of approximately 800 and a thickness in the wet state of 125 μm. In addition, Flemion® R, S, T, which have equivalent repeat unit molecular weights of 1000 and dry state thicknesses of 50, 80, 120 μm, respectively, were also developed by Asahi Glass

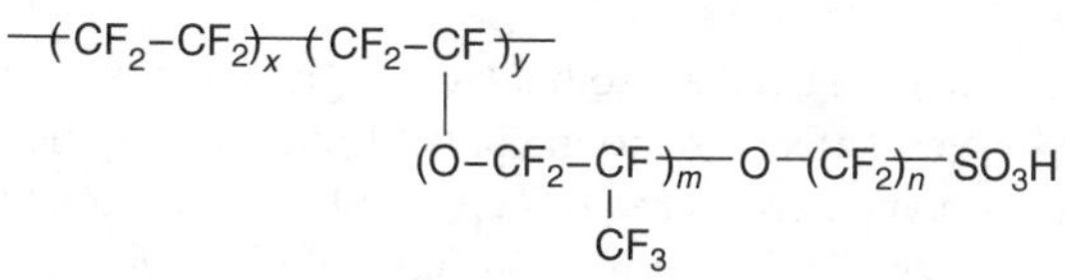

Nafion® 117	$m \geqslant 1$; $n = 2$; $x = 5$–13.5; $y = 1000$
Flemion®	$m = 0$; 1; $n = 1$–5
Aciplex®	$m = 0$, 3; $n = 2$–5; $x = 1.5$–14
Dow membrane	$m = 0$; $n = 2$; $x = 3.6$–10

Fig. 2. Chemical structures of perfluorinated polymer electrolyte membranes

Company [2]. Asahi Chemical Industry manufactured a series of Aciplex®-S membranes, which have equivalent repeat unit molecular weights of 1000–1200 and dry state thicknesses of 25–100 μm.

These perfluorinated ion exchange membranes including Neosepta-F® (Tokuyama) and Gore-Select® (W. L. Gore and Associates, Inc.) have been developed for chlor-alkali electrolysis. The water uptake and proton transport properties of this type of membrane have significant effects on the performance of PEFCs. These membranes have water uptakes of above 15 H_2O/–SO_3H, and maximizing membrane water uptake also maximizes the proton conductivity. In general, conductivities can reach values of 10^{-2}–10^{-1} S cm^{-1}. All of these membranes possess good thermal, chemical, and mechanical properties due to their perfluorinated polymer backbones.

A limiting factor in PEFCs is the membrane that serves as a structural framework to support the electrodes and transport protons from the anode to the cathode. The limitations to large-scale commercial use include poor ionic conductivities at low humidities and/or elevated temperatures, a susceptibility to chemical degradation at elevated temperatures and finally, membrane cost. These factors can adversely affect fuel cell performance and tend to limit the conditions under which a fuel cell may be operated. For example, the conductivity of Nafion® reaches up to 10^{-2} S cm^{-1} in its fully hydrated state but dramatically decreases with temperature above the boiling temperature of water because of the loss of absorbed water in the membranes. Consequently, the development of new solid polymer electrolytes, which are cheap materials and possess sufficient electrochemical properties, have become one of the most important areas for research in PEFC and FCEV.

Proton-conducting polymer electrolyte membranes for high performance PEFCs have to meet the following requirements, especially for electric vehicle applications [3]:

1. low cost materials;
2. high proton conductivities over 100°C and under 0°C;
3. good water uptakes above 100°C;
4. durability for 10 years.

The challenge is to produce a cheaper material that can satisfy the requirements noted above. Some sacrifice in material lifetime and mechanical properties may be acceptable, providing cost factors are commercially realistic. Good electrochemical properties over a wide temperature range may help the early marketing of PEFCs. Presently, one of the most promising routes to high-performance proton-conducting polymer electrolyte membranes is the use of hydrocarbon polymers for polymer backbones. The use of hydrocarbon polymers as polymer electrolytes was abandoned in the initial stage of fuel cell development due to the low thermal and chemical stability of these materials. However, relatively cheap hydrocarbon polymers can be used for polymer electrolytes, since the lifetime of electrolytes required in fuel cell vehicles are shorter when compared to use in space vehicles. Also, catalyst and fuel cell assembly technologies have improved and brought advantages to the lifetimes of PEFCs and related materials.

There are many advantages of hydrocarbon polymers that have made them particularly attractive:

1. Hydrocarbon polymers are cheaper than perfluorinated ionomers, and many kinds of materials are commercially available.
2. Hydrocarbon polymers containing polar groups have high water uptakes over a wide temperature range, and the absorbed water is restricted to the polar groups of polymer chains.
3. Decomposition of hydrocarbon polymers can be depressed to some extent by proper molecular design.
4. Hydrocarbon polymers are easily recycled by conventional methods.

Based on numerous works by other authors and our own research group, we review the syntheses of new proton-conducting polymer electrolyte membranes based on hydrocarbon polymers. The characteristics of these new materials which determine their potential applications, are discussed in detail. A review of electrochemical properties, water uptake, and thermal stability makes possible a comprehensive understanding of the proton conduction mechanism and physical state of absorbed water in these systems.

2. PROTON-CONDUCTING POLYMER ELECTROLYTE MEMBRANES BASED ON SULFONATED AROMATIC POLYMERS

2.1. Materials

Over the last decade new proton-conducting polymer electrolyte membranes have been developed. These new membrane concepts include partially fluorinated membranes, composite membranes, and also aromatic polymer membranes. This section will give a brief overview on sulfonated, aromatic polymer membranes.

Poly(styrene sulfonic acid) is a basic material in this field. In practice, poly(styrene sulfonic acid) and the analogous polymers such as phenol sulfonic acid resin and poly(trifluorostyrene sulfonic acid), were frequently used as polymer electrolytes for PEFCs in the 1960s. Chemically and thermally stable aromatic polymers such as poly(styrene) [4,5], poly(oxy-1,4-phenyleneoxy-1,4-phenylenecarbonyl-1,4-pheneylene) (PEEK) [6–9], poly(1,4-phenylene) [10–13], poly(oxy-1,4-phenylene) [14], poly(phenylene sulfide) [15], and other aromatic polymers [16–19], can be employed as the polymer backbone for proton-conducting polymer electrolytes. These chemical structures are illustrated in Fig. 3. These aromatic polymers are easily sulfonated

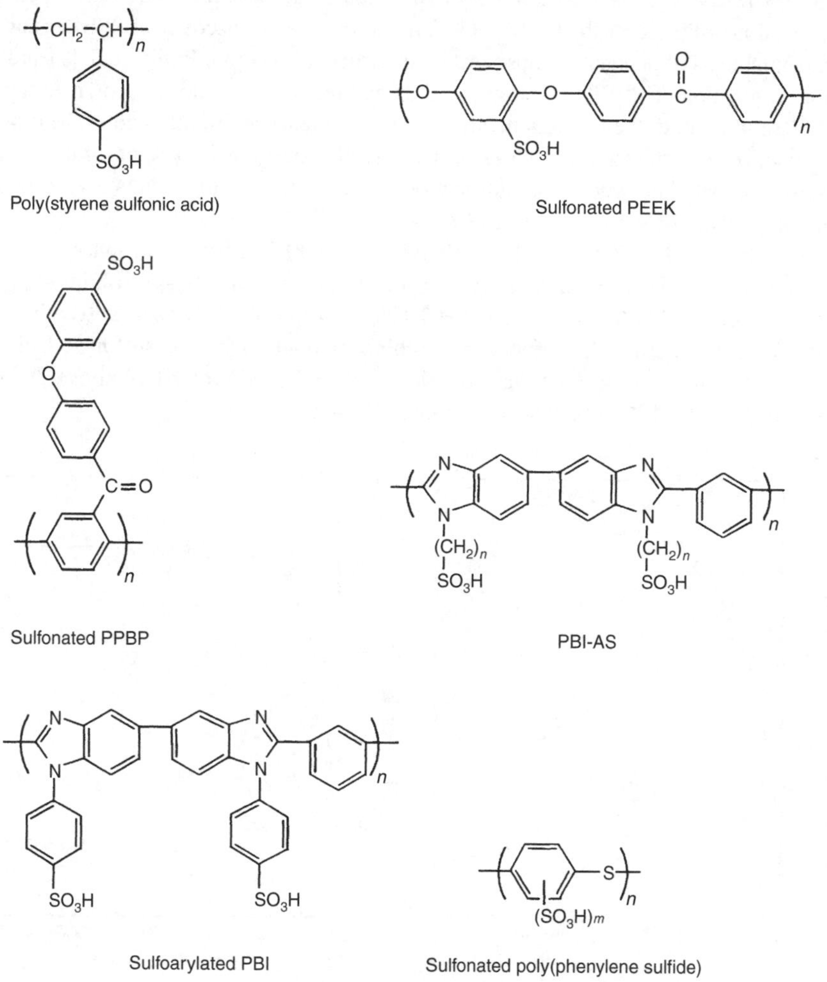

Fig. 3. Chemical structure of polymer electrolyte membranes based on hydrocarbon polymers

by concentrated sulfuric acid [20], by chlorosulfonic acid [21], by pure or complexed sulfur trioxide [16,22, 23], and by acetylsulfate [24]. Sulfonation with chlorosulfonic acid or fuming sulfuric acid sometimes causes chemical degradation in these polymers. According to Bishop et al. [25], the sulfonation rate of PEEK in sulfuric acid can be controlled by changing the reaction time and the acid concentration, and can thereby provide a sulfonation range of 30–100% without chemical degradation or cross-linking reactions [26]. However, this direct sulfuric acid procedure cannot be used to produce truly random copolymers at sulfonation levels of less than 30%, because dissolution and sulfonation in sulfuric acid occur in a heterogeneous environment due to the increase of viscosity of reactant solutions. For this reason, the dissolution process was kept short, for less than 1 h, in order to produce a more random copolymer. Sulfonation levels in PEEK and poly(4-phenoxybenzoyl-1,4-phenylene) (PPBP) [13] as a function of reaction time at room temperature are presented in Fig. 4. Since sulfonation is an electrophilic reaction, its application depends on the substituents present on the ring. Electron-donating substituents will favor the reaction, whereas electron-accepting substituents will not. Indeed, with PPBP the terminal phenyl ring of the substituent can be sulfonated under mild conditions as good as PEEK, although poly(4-benzoyl-1,4-phenylene) containing an electron accepting substituent in the ring cannot be sulfonated under the same conditions. The sulfonation of PPBP and PEEK occurred to almost 80 mol% within 100 h. The level of sulfonation of PPBP became saturated at 85 mol% per repeat unit, whereas the level of sulfonation of PEEK reached 100 mol%. This is because the viscosity of PPBP in sulfuric acid was so high that entanglements of pendant groups prevented further sulfonation. Bailly et al. [9] and Jin et al. [20] showed that the sulfonation of PEEK in sulfuric acid cannot produce a random copolymer especially below sulfonation of 30 mol% due to the heterogeneous nature of the reaction. In our studies, heterogeneous reactions were also confirmed in the case of both PPBP and PEEK. Since it is important for fuel cell applications to produce highly sulfonated polymers, this sulfonation method with sulfuric acid is one of the most suitable routes for preparing polymer electrolytes of PEFCs.

In the case of sulfonated PEEK (S-PEEK), S-PEEK having 30% or more sulfonation was soluble in *N*, *N*-dimethylformamide (DMF), dimethyl sulfoxide (DMSO), or *N*-methyl-2-pyrrolidinone (NMP); above 70%, the polymer were soluble in methanol and at 100%, in water. PPBP was dissolved in common chlorine containing solvents such as chloroform and dichloromethane, whereas sulfonated PPBP (S-PPBP) was insoluble in the same solvents, but was soluble in DMF, DMSO, or NMP above 30% sulfonation. Above 65% sulfonation, S-PPBP swelled in methanol and water.

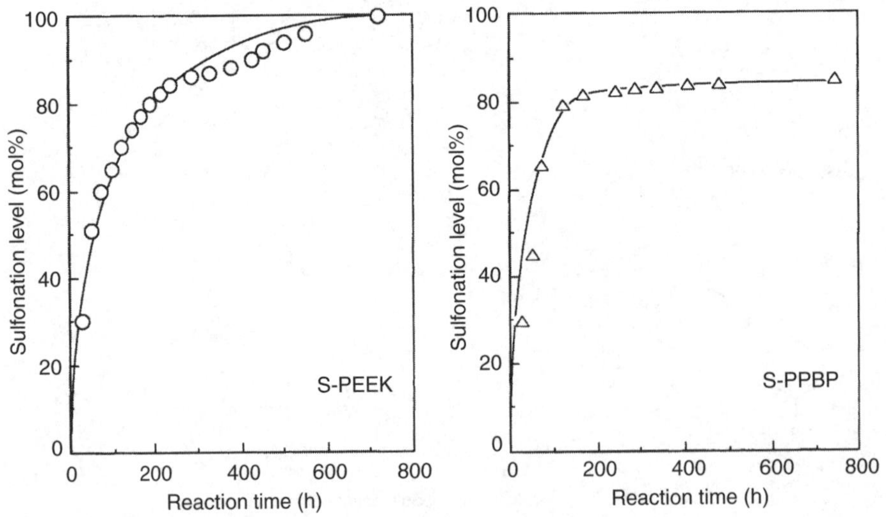

Fig. 4. Levels of sulfonation of PEEK and PPBP as a function of reaction time at room temperature

Fourier Transform Infrared (FT-IR) spectra for PEEK, PPBP, and their sulfonated polymers are shown in Fig. 5. The sulfonation of PEEK at room temperature in concentrated sulfuric acid places a limit of one sulfonic group per repeat unit [21,23,27]. It is also clear from the FT-IR data that PEEK is sulfonated at one position on the phenylene ring between the ether groups. For S-PPBP, the intensity of the absorption band at 1070 cm^{-1} assigned to the mono-substituted benzene ring decreased with an increasing sulfonation level, while the intensities of absorption bands at 1020 and 730 cm^{-1} attributed to a *para*-substituted benzene ring, and the S–O stretching vibration, respectively, also increased with increasing sulfonation level. These FT-IR spectral data indicate that the sulfonation of PPBP in sulfuric acid takes place at the *para*-position of the terminal phenoxy group.

Tsuchida et al. [18,19] have synthesized poly(thiophenylenesulfonic acid) possessing up to 2.0 sulfonic acid groups per phenylene ring, as shown in Fig. 3. They polymerized 4-(methylsulfinyl) diphenyl sulfide

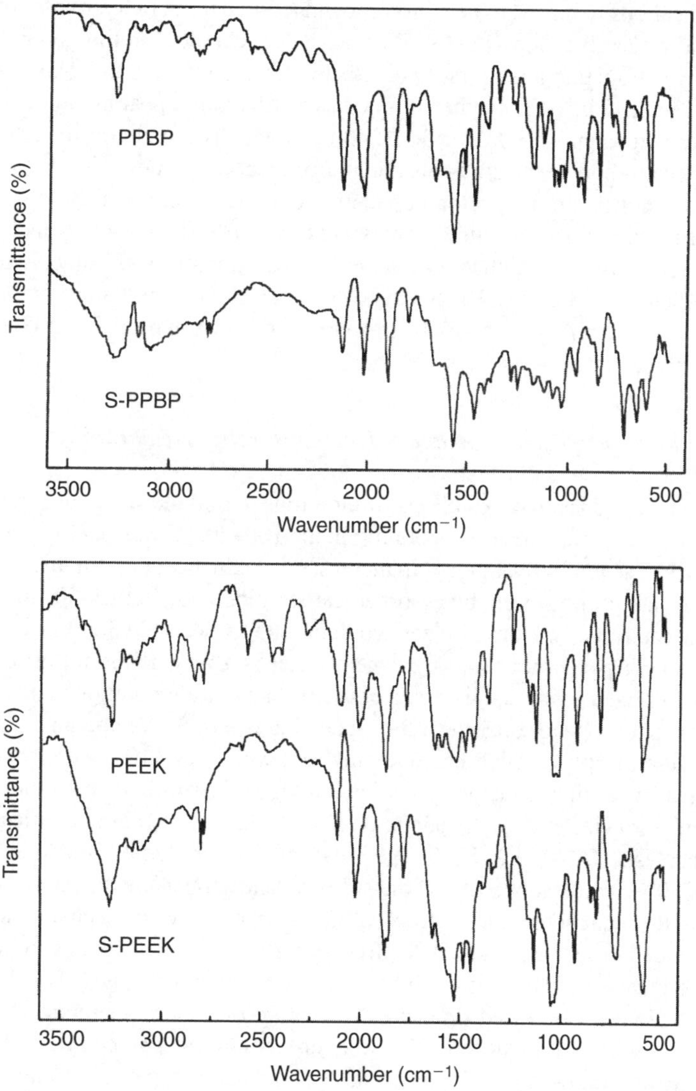

Fig. 5. FT-IR spectra of PEEK, PPBP, and their sulfonated polymers (S-PEEK and S-PPBP)

in sulfuric acid upon heating or in the presence of SO_3 to yield a sulfonated poly(sulfonium cation), which can be converted in to the corresponding sulfonated poly(thiophenylene). The sulfonation can be controlled by the reaction time, the temperature, and/or the addition of SO_3. The resulting polymer electrolytes are soluble in water and methanol and can form transparent films.

Composite polymer electrolytes of Teflon-FEP films and poly(styrenesulfonic acid) were also prepared by simultaneous radiation grafting of styrene onto Teflon-FEP films and subsequent sulfonation to provide low cost proton-conducting polymers [4]. They immersed a Teflon-FEP film (DuPont, $50 \pm 2\,\mu m$) into styrene monomer at 60°C and irradiated at a dose rate of $0.5\,Gy/min^{-1}$ for variable times up to 15 h with γ-rays from a ^{60}Co source. The radicals created by radiation in the system induce the polymerization, and styrene monomers were grafted to the perfluorinated matrix. The radiation-grafted films were sulfonated with chlorosulfonic acid. A linear dependence of degree of grafting on time was observed for grafting times up to 15 h. This sulfonated grafted-film was thermally stable up to 310°C as determined by thermogravimetric analysis.

Some patent applications in regard to new proton-conducting polymer electrolytes have been submitted by several industrial research groups [28,29]. Iwasaki et al. claimed proton-conducting polymer electrolytes for PEFC by sulfonating a precursor polyethersulfone having a $0.6–1.5\,dl\ g^{-1}$ reduced viscosity (in 1% w/v DMF). The sulfonated polyethersulfone has a structural repeating unit containing benzene or naphthalene and has equivalent weights of $500–2500\,g\ mol^{-1}$. The fuel cell measurements using these materials were carried out in order to study the electrochemical properties.

Konishi et al. also claimed similar proton-conducting polymer electrolytes based on polyethersulfone. The protons of the sulfonic acid groups in their polymers are partly exchanged by metal ions such as magnesium, titanium, aluminum, and lanthanum ions to improve durability. Their polymers have high conductivities of more than $10^{-2}\,S\ cm^{-1}$. The results of fuel cell tests using their membranes showed a cell voltage of 550 mV and a current density of $700\,mA\ cm^{-2}$ under 1 atm humidified H_2/O_2 at 70°C. No significant loss of performance was seen after 1000 h.

2.2. *Thermal stability of sulfonated aromatic polymer electrolyte membranes*

Polymer electrolyte membranes that exhibit fast proton transport at elevated temperatures are needed for PEFCs and other electrochemical devices operating in the 100–200°C range. Operation of a PEFC at elevated temperatures has several advantages. It increases the kinetic rates for the fuel cell reactions, it reduces problems of catalyst poisoning by absorbed carbon monoxide in the 150–200°C range, it reduces the use of expensive catalysts, and it minimizes problems due to electrode flooding. Thus, the thermal stability of proton-conducting polymer electrolyte membranes is a very important factor for fuel cell applications. Several studies have addressed the issue of thermal stability for Nafion® and analogous polymers. Chu et al. used IR spectroscopy to study the effect of heating Nafion® at temperatures of 22, 110, 160, 210, and 300°C [30]. These samples, which had an equivalent weight of 1100, were coated onto a platinum film, and heated in air. The study concluded that Nafion® loses its sulfonic acid groups after being heated at 300°C for 15 min. Surowiec et al. [31] conducted a study using thermogravimetric analysis (TGA), differential thermal analysis (DTA), and FT-IR spectroscopy . Their studies concluded that Nafion® loses only water below 280°C, while above 280°C the sulfonic acid groups are also lost. Wilkie et al. reported using TGA and FT-IR spectroscopy that Nafion® produced water as well as small amounts of sulfur dioxide and carbon dioxide between 35 and 280°C. Between 280 and 355°C, the evolution of sulfur dioxide and carbon dioxide increased while the evolution of water decreased. At 365°C, the amounts of sulfur dioxide and carbon dioxide decreased dramatically, and above this temperature absorbencies in FT-IR spectra of evolution gas were thought to be due to hydrogen fluoride, silicon fluoride, and carbonylfluorides. Another study was conducted by Samms et al. in simulated fuel cell environments [32]. To simulate the conditions in a fuel cell, Nafion® samples were loaded with fuel cell grade platinum black and heated under atmospheres of nitrogen, 5% hydrogen, or air in a thermogravimetric analyzer. The products of

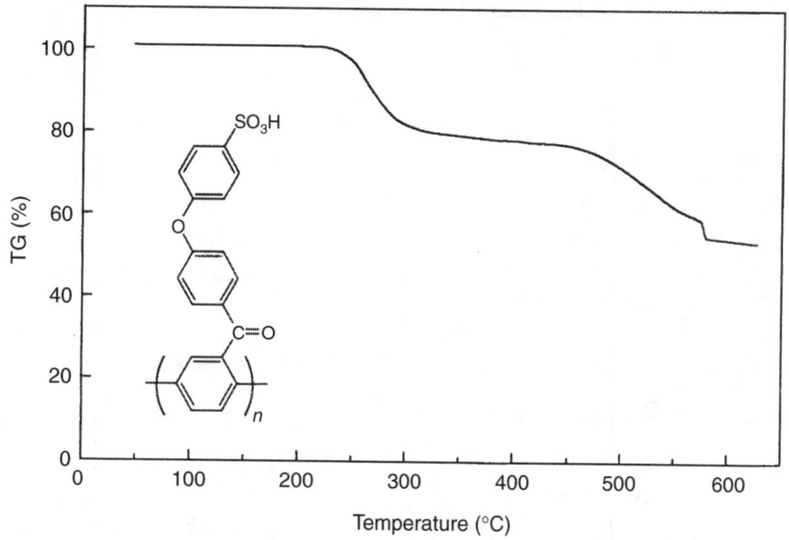

Fig. 6. TG-DTA curve of S-PPBP with 80 mol% sulfonation level

decomposition were taken directly into a mass spectrometer for identification. In all cases, Nafion® was found to be stable up to 280°C, at which temperature the sulfonic acid groups began to decompose. They concluded that the addition of platinum to the polymer electrolyte had no detectable effect on the temperature at which the sulfonic acid groups decomposed. The long-term stability in PEFCs cannot be accurately predicted by these results.

In our group, the thermal stabilities of our polymer electrolyte membranes were studied by heating samples in a TGA and analyzing the resulting residues by elemental analysis. For example, data for S-PPBP investigated by TGA at a heating rate of 10°C min^{-1} under nitrogen are displayed in Fig. 6. In the TGA data, S-PPBP showed an initial weight loss of about 20% of the original weight between 250 and 400°C, which corresponds to a loss of sulfonic acid groups. The thermal stabilities of dry S-PEEK and S-PPBP are shown in Fig. 7, which shows thermal decomposition temperatures (T_d) as a function of the level of sulfonation. In contrast to what was observed for PEEK and PPBP, degradation steps were observed for the sulfonated polymers between 350 and 250°C. The T_d of S-PEEK and S-PPBP decreased from 500 to 300°C for S-PEEK and 250°C for S-PPBP with increasing levels of sulfonation. Elemental analysis of the residues indicated that the sulfur contents of S-PEEK and S-PPBP decreased from 7.62 to 0.86% after heating at temperatures above 400°C. These results suggest that thermal decomposition occurs by desulfonation. However, such thermal decomposition was not detected below 200°C. Sufficient thermal stabilities for fuel cell applications is retained in these polymer electrolytes even if the T_d values drop with sulfonation levels.

Other proton-conducting polymer electrolytes based on sulfonated aromatic polymers also show thermal degradation between 200 and 400°C. As arylsulfonation is known to be a reversible reaction, desulfonation of arylsulfonic acid easily occurs on heating acidic aqueous solution of arylsulfonic acid at 100–175°C. Thus, the desulfonation of arylsulfonic acid provides fatal limitations of the thermal stability of sulfonated aromatic polymer electrolytes. The substituents present on the phenyl ring can be somewhat effective at depressing the thermal degradation process.

According to Tsuchida et al. [18], highly sulfonated poly(phenylene sulfide) possesses higher thermal stability than other sulfonated aromatic polymer electrolytes. They studied the thermal stabilities of poly(phenylene sulfide sulfonic acid) with different sulfonation levels using TGA. The decomposition

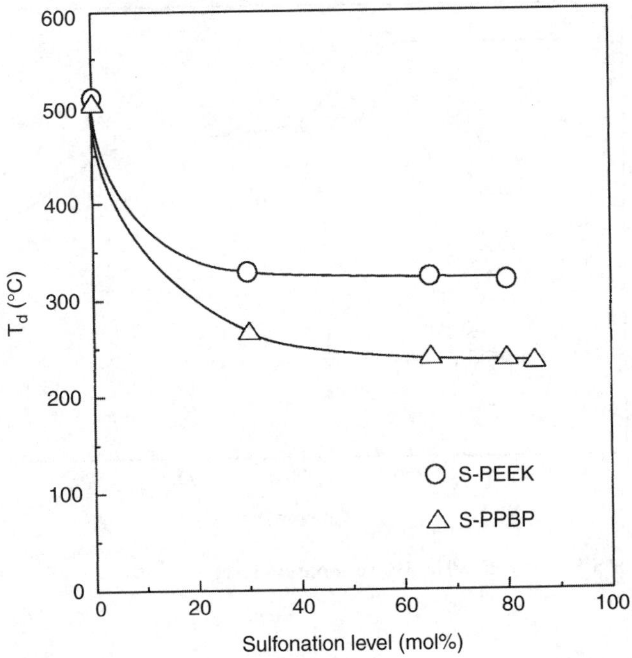

Fig. 7. Degradation temperature of S-PPBP and S-PEEK as a function of sulfonation level

temperature of highly sulfonated polymer (degree of sulfonation per repeat unit; $m = 2.0$) is 265 and 125°C higher than that for low-sulfonated polymer ($m = 0.6$). They suggest that the C–S bond of the highly sulfonated polymer is stronger due to the two electron withdrawing sulfonic acid substituents on one phenyl ring. In addition, the initial weight loss for the higher sulfonated polymer between 265 and 380°C is only 13%, which corresponds to the loss of two H_2O molecules per repeat unit. Thus, the desulfonation reaction in this polymer is depressed by substitution of electron withdrawing groups.

2.3. Water uptake in sulfonated aromatic polymer electrolyte membranes

Water is carried into the fuel cell via the humidified H_2/O_2 gas streams entering the gas diffusion electrodes. Some combination of water vapor and liquid water passes through each electrode to the electrode/electrolyte interface. Water crossing this interface assists in the hydration of the electrolyte membrane. An additional source of water involves oxygen reduction occurring at the cathode. Water in the membrane is transported in two main ways: electro-osmotic drag of water by proton transport from anode to cathode and diffusion down concentration gradients that build up. Thus, adequate hydration of electrolyte membranes is critical to fuel cell operation. If the electrolyte membrane is too dry, its conductivity falls, resulting in reduced cell performance. An excess of water in the fuel cell can lead to cathode flooding problems, also resulting in less than optimal performances.

In the case of perfluorinated polymer electrolytes, the structure of the electrolyte membrane in various states of hydration, the water uptake characteristics when exposed to liquid- or vapor- phase water, and the properties of water in the electrolyte membrane have all been investigated in detail. The extreme hydrophobicity of the perfluorinated polymer backbone and the extreme hydrophilicity of the terminal sulfonic acid groups lead to a spontaneous hydrophilic/hydrophobic nano-separation [33,34]. In the presence of water only the hydrophilic domain of the nano structure is hydrated. The absorbed water also acts as a

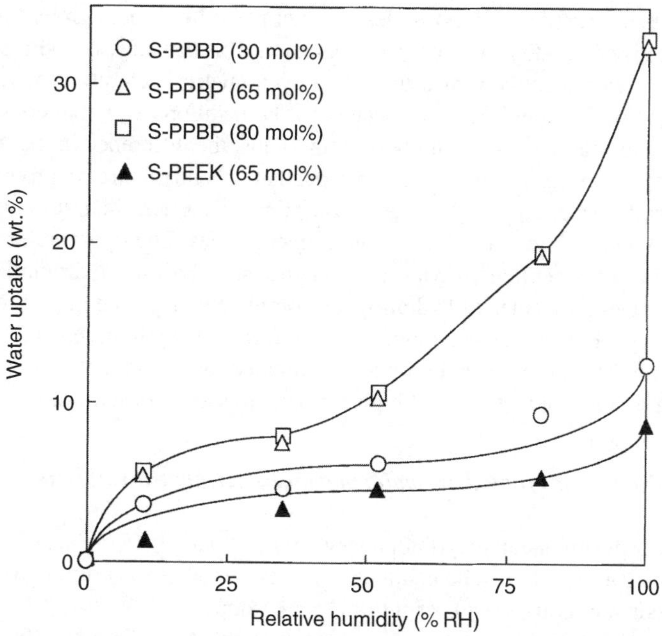

Fig. 8. Water uptake of S-PEEK and S-PPBP at room temperature as a function of relative humidity

plasticizer, which leads to further phase separation. Maximizing the membrane water uptake maximizes the proton conductivity, while the hydrophobic domains provide relatively good mechanical properties even in the presence of water. The activation enthalpy of water diffusion in such membranes is almost identical to that in pure water for water contents higher than $H_2O/SO_3H = 3$. The perfluorinated polymer electrolyte acts as a nano-porous inert sponge for the absorbed water, which in fact shows very little interaction with the polymer except for the first three water molecules per sulfonic acid group.

Water uptake from the vapor phase is likely to be the principal mode of external hydration for the membrane in PEFCs. We have investigated the absorption of water vapor in S-PEEK and S-PPBP. S-PEEK and S-PPBP films were placed under various humidities to measure their equilibrium water contents. It is quite interesting to compare the water uptake and proton conductivities of S-PEEK and S-PPBP in terms of the structural effect of hydration on the sulfonic acid moieties because PEEK is a structural isomer of PPBP. The water uptake of S-PEEK and S-PPBP at room temperature is shown in Fig. 8. The results are similar to those reported with Nafion® membranes [1]. Over the entire range of relative humidity, the activity coefficient of water in the polymer is greater than unity if one assumes as the ideal behavior a Rault's law relationship between water activity and membrane water content. The equilibrium water contents of S-PEEK and S-PPBP increased sigmoidally with increasing levels of sulfonation. Two regions are discriminated in the following way. Relatively little increase of water uptake is observed for the region of relative HUMIDITY = 0–50%, and a significantly greater increase of water uptake is noted with increasing relative humidities in the range of 50–100%. The former region corresponds to water uptake due to solvation of the proton and sulfonate ions. The latter region corresponds to water involved in polymer swelling. In the former region, the water in the polymer is engaged in strong interactions with ionic components of the polymer, and these interactions overcome the strong tendency of polymer to exclude water due to its hydrophobic nature and resistance to swelling. In the latter region, these hindrances to water uptake are surmounted once the water in the bathing gas has sufficient chemical potential. The water content of S-PPBP with 65 mol% sulfonation was larger than that of S-PEEK with the same sulfonation over the entire relative humidity range, and these values increased to 8.7 molecules of water per sulfonic acid group for S-PPBP and 2.5 for S-PEEK at 100% relative humidity and room temperature. The densities of S-PEEK and S-PPBP with 65 mol% sulfonation level, which were measured by a pycnometric method, were 1.373 for S-PPBP and 1.338 for S-PEEK. The surfaces and fracture surfaces for both sulfonated polymers were also found to be nearly the same by using scanning electron microscopy. Thus, the difference in the water uptake between S-PEEK and S-PPBP might be due to the flexible pendant phenoxybenzoyl groups in S-PPBP, which enhance plasticity, water penetration, and water absorption in the terminal sulfonic groups. The water uptakes of S-PPBP are nearly equal to that of Nafion® membranes. The hydrophilic domains only absorb water in perfluorinated polymer electrolytes that have a spontaneous hydrophilic/hydrophobic nano-separation, while the water molecules of hydration are completely dispersed in polymer electrolytes based on hydrocarbon polymers. In fact, the water molecules in sulfonated hydrocarbon polymers show relatively strong interactions with their sulfonic acid groups as estimated by differential thermal analysis and consequently result in high proton conductivities at high temperatures and low humidities.

2.4. *Proton conductivity of sulfonated aromatic polymer electrolyte membranes*

The proton conductivity of sulfonated poly(thiopheneylene) is about 10^{-5} S cm^{-1} at 30% RH and it increases exponentially with relative humidity. The conductivity at 94% RH reaches up to 10^{-2} S cm^{-1}, where the water content of the polymer is 10.3 H_2O per sulfonic acid group.

Films of S-PEEK and S-PPBP were placed in various relative humidities (0–100%) in order to absorb water into the films. The proton conductivities of the films containing absorbed equilibrium water were measured as a function of the relative humidities. As shown in Fig. 9, the conductivity increased with a sigmoidal curve with increasing relative humidities and water uptake, and it reached almost 10^{-2} S cm^{-1}

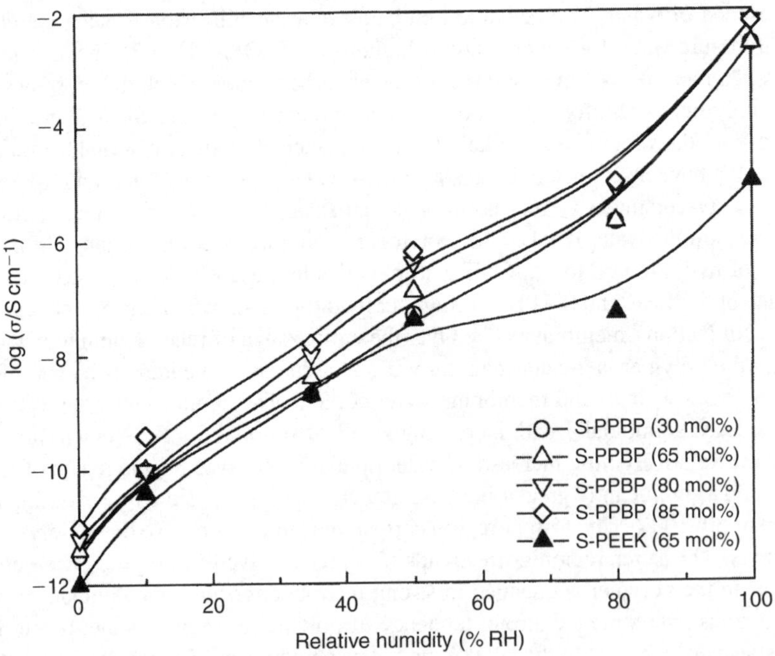

Fig. 9. Proton conductivity of S-PPBP and S-PEEK with various sulfonation levels as a function of relative humidity at room temperature

for S-PPBP with 65 mol% sulfonation, while that of S-PEEK for the same sulfonation level was $10^{-5}\,S\,cm^{-1}$. A comparison of the proton conductivities of S-PEEK and S-PPBP for the same sulfonation level (65 mol%) is shown in Fig. 10, where the films were saturated at 100% RH. It is clear that the proton conductivities and the water absorption of S-PPBP are much higher than those of S-PEEK. Moreover, the proton conductivities of S-PEEK dropped sharply at temperatures above 100°C, while those of S-PPBP exhibited less temperature dependence.

These stable proton conductivities of sulfonated poly(thiophenylene) and S-PPBP at elevated temperatures are an interesting feature worthy of further applications such as high temperature and non-humidifying PEFCs, compared with Nafion® membranes which show a sharp decrease in proton conductivities at temperatures above 100°C due to dehydration. The high water uptake and the strong interaction between absorbed water and polymer chains might be maintained at elevated temperatures so that these polymers do not show a measurable reduction in proton conductivity.

The conductivity of S-PPBP was also measured using a galvanostatic four-point-probe electrochemical impedance spectroscopy technique [35]. A four-point-probe cell with two platinum foil outer current-carrying electrodes and two platinum wire inner potential-sensing electrodes was mounted on a Teflon plate. The schematic view of the cell is illustrated in Fig. 11. Membrane samples were cut into strips that were approximately 1.0 cm wide, 5 cm long, and 0.01 cm thick prior to mounting in the cell. The cell was placed in a thermo-controlled humidic chamber to measure the temperature and humidity dependence of proton conductivity. With this method, a fixed AC current is passed between two outer electrodes, and the conductance of the material is calculated from the ac potential difference observed between the two inner electrodes. The method is relatively insensitive to the contact impedance at the current-carrying electrode, and is therefore well suited for measuring proton conductances. This open cell is also suited for studying the humidity dependence of conductivity for proton-conducting polymer electrolytes because of the low interfacial resistance.

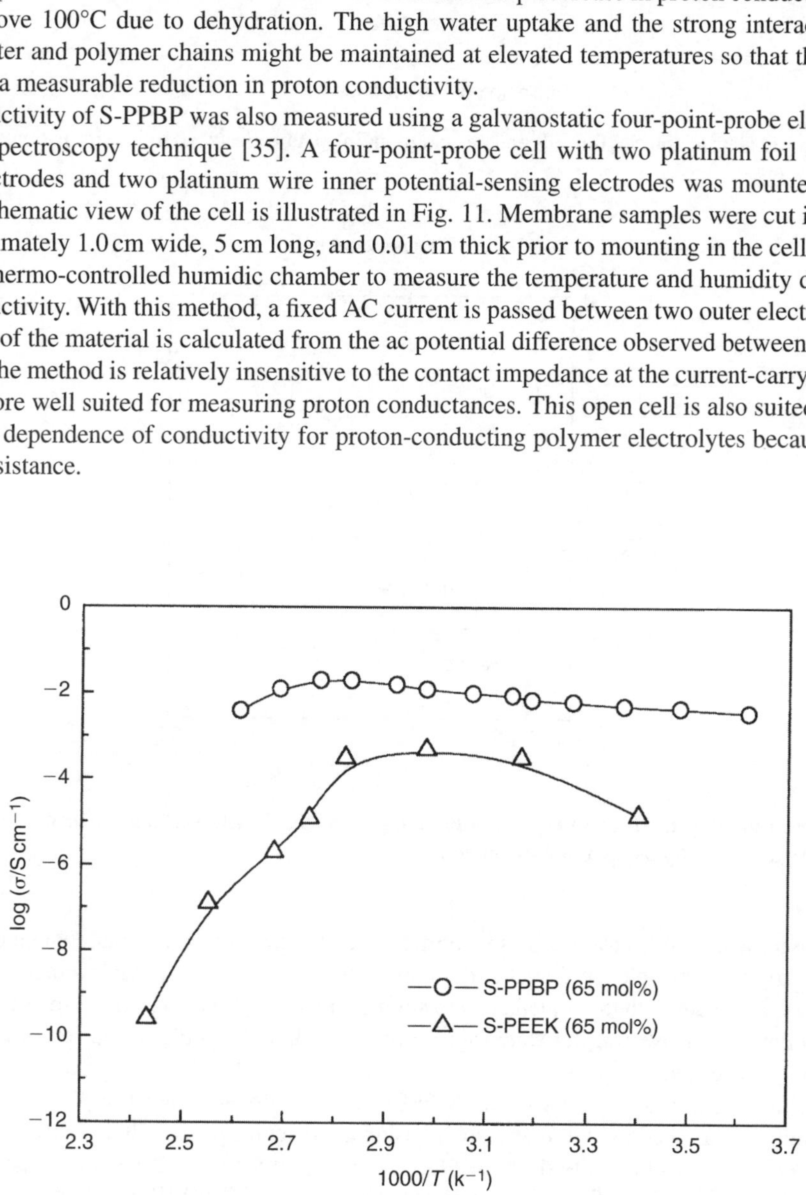

Fig. 10. Comparison of proton conductivities of S-PEEK and S-PPBP for the same sulfonation level (65 mol%)

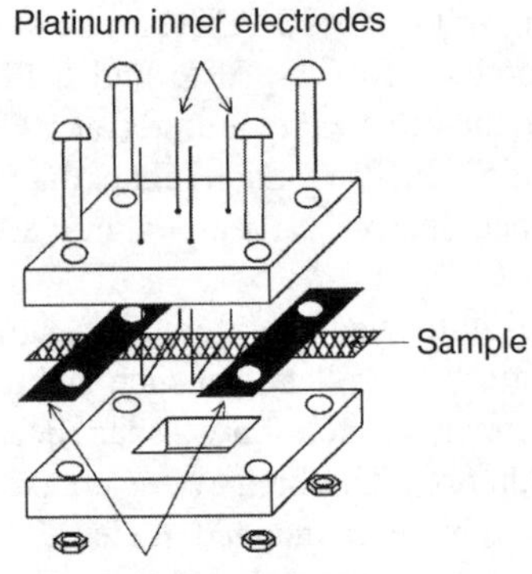

Fig. 11. Schematic view of the cell for galvanostatic four-point-probe electrochemical impedance spectroscopy technique

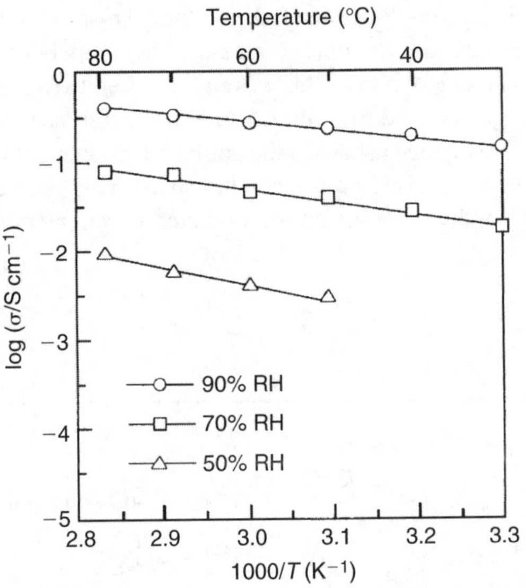

Fig. 12. Temperature dependence of proton conductivity for S-PEEK with sulfonation level of 85 mol% under various relative humidity by using a four-point cell

In many cases, sealed cells have been used to investigate temperature and water content dependence of proton conductivity for samples hydrated under certain humidities prior to measurements. Conductivity measurements carried out with the sealed cell are simple and useful for conductivity measurements above 100°C and below 0°C. However, the water uptake of the hydrated sample in the sealed cell sometimes changes during measurements.

Figure 12 shows the proton conductivity of S-PPBP with a sulfonation level of 85 mol% under various relative humidities as a function of temperature by using a four-point-probe cell. S-PPBP exhibits a strong decrease in conductivity as humidity decreases like Nafion® membranes. This strong dependence of proton conductivity on humidity reflects in part a strong tendency of S-PPBP to absorb water vapor. This behavior in conductivity under various relative humidity is attributed to a liquid-like proton conductivity mechanism, where protons are transported as free hydronium ions through water-filled ionic pores in the membrane [36].

Usually perfluorinated polymer electrolytes exhibit a remarkable decrease in conductivity with increasing temperature, such that conductivity at 80°C is diminished by more than 10 times relative to that at 60°C. Perfluorinated polymer electrolyte membranes become less conductive under high temperature conditions because, they dry out, causing the channels to collapse and proton transport to become more difficult. However S-PPBP exhibits little increase in conductivity with increasing temperature under each relative humidity (90–50%). This phenomenon is attributable to the strong interaction between the sulfonic acid group and the absorbed water molecules in S-PPBP. Proton-conducting polymer electrolyte membranes based on hydrocarbon polymers such as S-PPBP and sulfonated poly(thiopheneylene) have a tendency to include relatively higher amounts of bound water, resulting in higher conductivities at high temperatures and/or low humidities, as determined by differential scanning calorimetry.

3. PROTON-CONDUCTING POLYMER ELECTROLYTE MEMBRANES BASED ON ALKYLSULFONATED AROMATIC POLYMERS

3.1. Materials

Sulfonation with sulfuric acid, chlorosulfonic acid, sulfur trioxide, and acetylsulfate easily provides proton-conducting polymer electrolytes, but these sulfonated polymer electrolytes decompose on heating around 200–400°C due to desulfonation. The introduction of alkylsulfonic substituents onto the backbone of aromatic polymers leads to thermally stable proton-conducting polymer electrolytes whose electrochemical properties can be controlled by the content of the substituent and the length of alkyl chain. These alkylsulfonic substituents can induce water uptake and proton conductivities similar to sulfonic acid groups without sacrificing desirable properties such as thermal stabilities, mechanical strengths, and chemical resistance. Gieselman et al. have synthesized poly(*p*-phenyleneterephthalamido-*N*-propanesulfonate), poly(*p*-phenyleneterephthalamido-*N*-methylbenzenesulfonate), poly[2,2′-*m*-phenylene-5,5′-bibenzimidazolyl-*N*-propanesulfonate], based on all-aromatic parent polymers, via a general method for attachment of alkylsulfonate or arenesulfonate side chains onto polymers containing reactive N–H sites [37–39]. The alkylsulfonation and arenesulfonation of aromatic polymers are promising routes to the preparation of proton-conducting polymer electrolytes. Recently, Glipa et al. and our group reported the synthesis and electrochemical characterization of arylsulfonated [40] and alkylsulfonated polymer electrolytes [41–43]. For example, a solution of PBI in dimethylacetoamide (DMAc) is treated LiH or NaH, which deprotonates the nitrogen of the benzimidazole rings in the polymer backbone, and the addition of propanesulfonate side chains onto PBI was accomplished by the reaction of 1,3-propanesultone and the PBI anion, as shown in Fig. 13. Subsequent ion-exchange with H_3PO_4 or HCl provides proton-conducting polymer electrolytes based on PBI. The alkylsulfonation of PBI improved its solubility in organic solvents, where the solubility depended on the alkylsulfonation level. Alkysulfonated PBI was more soluble in polar aprotic solvents like methanol, DMAc, and DMSO when compared to the parent polymer. The relationship between the feed ratio of 1,3-propanesultone and alkylsulfonation level is shown in Fig. 14. The alkylsulfonation level for the NH groups in PBI was estimated by elemental analysis and ^{1}H NMR. The alkylsulfonation level was easily controlled by varying the feed ratio of 1,3-propanesultone and led to almost 60 mol% with the feed ratio of 5.0 (1,3-propanesultone/PBI repeat unit) [44].

We also tried to synthesize ethylphosphrylated PBI via the same derivatization procedure for the reactive N–H sites in PBI, as shown in Fig. 15. The substituted polymer was successfully synthesized, but the resulting

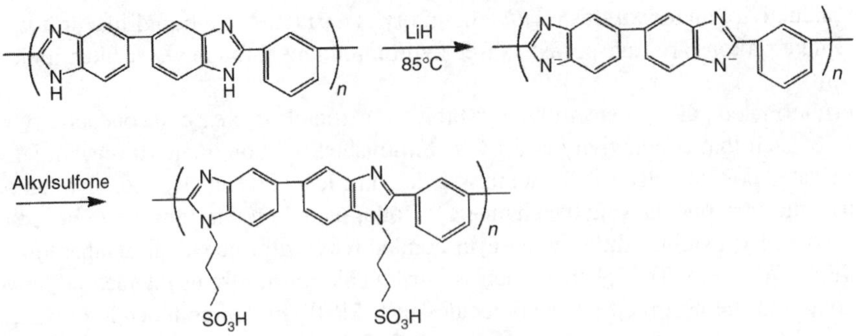

Fig. 13. Synthesis of alkylsulfonated PBI

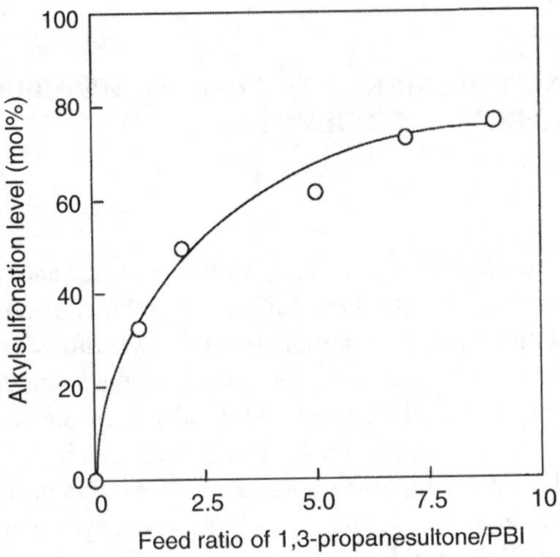

Fig. 14. Relationship between the feed ratio of 1,3-propanesultone and alkylsulfonation level

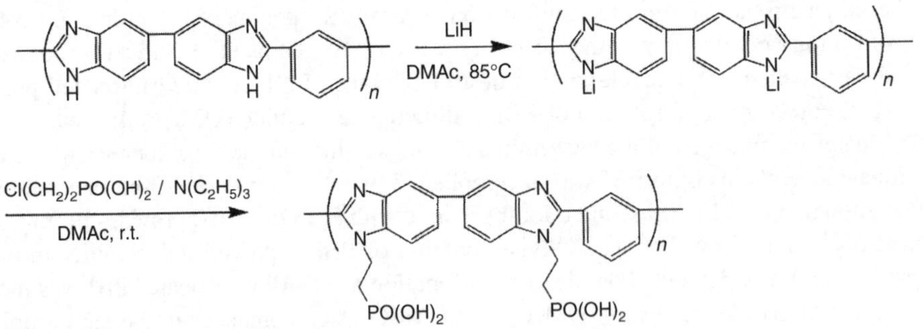

Fig. 15. Synthesis of phosphoethylated PBI

polymer was insoluble in organic solvents. It can be presumed that the phosphoric acid groups aggregate with each other during the substitution process. The ethylphosphrylated PBI showed a high proton conductivity of 10^{-3} S cm^{-1} even as a compressed pellet. It is found that phosphoric acid groups are an effective polar group for proton-conducting polymer electrolytes.

3.2. Thermal stability of alkylsulfonated aromatic polymer electrolyte membranes

The arenesulfonated and alkylsulfonated polymers based on all-aromatic backbones maintain high thermal stabilities when compared to the parent materials. The aim of derivatization of the parent polymers is to improve water absorption and to promote proton conductivity but retain a significant fraction of their high thermal stabilities. The TGA of these polymers were examined in both inert and oxidative atmospheres by Gieselman et al. [38]. PBI is extremely thermally stable, as expected for an all-aromatic polymer. In an inert atmosphere, it exhibits an onset of degradation around 650°C and 5 wt.% mass loss at 700°C and retains more than 80% of its mass at 800°C. The onset of degradation temperatures for the substituted polymers in an inert atmosphere is significantly lower than PBI, as expected for the addition of pendant groups that are not conjugated with the main chains. The degradation of poly[2,2′-*m*-phenylene-5,5′-bibenzimidazolyl-*N*-methylbenzenesulfonate] at 22% substitution begins at about 480°C, and poly[2,2′-*m*-phenylene-5,5′-bibezimidazolyl-*N*-propanesulfonate] at 54% substitution shows mass loss beginning at 450°C. After the removal of the side chains, the degradation is gradual with high residual masses (50–60%) at 800°C. The onset of degradation of PBI in an atmosphere of dry air begins at a significantly lower temperature relative to that in an inert atmosphere. PBI begins to degrade at approximately 520°C, a full 100°C below the degradation temperature in an inert atmosphere. The substituted PBI shows little difference in its onset of degradation temperature in changing from an inert to an oxygen-containing atmosphere. The major difference in the behavior in air compared to nitrogen is that the magnitudes of the mass losses are much higher and the percentages of residual char at high temperatures are much lower in the oxidative atmosphere. This is apparently the result of the lower thermal stability of PBI in dry air, rather than a consequence of the addition of the pendant side chain.

Poly(*p*-phenylene terephthalamide) (PPTA) is also thermally stable. The onset of degradation dose not occur until above 550°C, and a sharp decrease in mass begins at 600°C with a residue of 50% of the original mass. After derivatization with propanesulfonate side chains (66%), the polymer shows retention of thermal integrity until approximately 400°C, followed by a steady degradation to 40 wt.% at 800°C. The methylbenzenesulfonated derivative (66%) is more thermally stable; degradation initiates at about 470°C. After an initial mass loss, a decrease in mass to a residue of 50% occurs at 800°C. Thermal analyses of PPTA and its derivatives in an atmosphere of dry air were also carried out by the same research group. The PPTA is much less stable in air with degradation beginning about 70°C lower than in nitrogen. Once degradation begins, the entire mass of the polymer is volatilized. The TGA traces of the methylbenzenesulfonated PPTA (66%) run in air and in N_2 indicate a relatively small decrease in the onset of degradation temperature. The major difference in behavior is that the mass decrease is steeper after the initial mass loss, and the amount of the residue at high temperatures is much lower due to oxidative degradation of the polymer backbone. The addition of pendant groups to aromatic polymers results in a lowering of thermal stability due to the lower temperature at which the side chain cleaves, regardless of the nature of the atmosphere. This is the expected behavior because the side chain, especially sulfonic acid groups is not stabilized by conjugation to the polymer backbone. It has been concluded that the methylbenzenesulfonate side chain is more stable than the propanesulfonate side chain, regardless of the type of polymer. This result suggests that the point of cleavage of the side chain is not exclusively at the N–C bond. Glipa et al. also investigated the thermal stability of methylbenzenesulfonated PBI (substitution 75%) using TGA in air with a heating rate 1°C min^{-1} [40]. They reported that the addition of the metylbenzenesulfonate group reduced the thermal stability of the polymer electrolyte. The onset of degradation begins at 370°C with a residue of 60% of the original mass. They concluded that the mass loss between 370 and 420°C was attributable to the degradation of the sulfonic acid groups. The degradation process and mechanism is likely to be more complex. Residual water, impurities, substitution levels, and measuring conditions significantly affect in the TGA data for these polymer electrolytes.

As mentioned above, Glipa et al. also reported that arylsulfonated PBI is stable up to 350°C, while Gieselman et al. concluded that the methylbenzenesulfonated PBI was stable up to 500°C in air. It is difficult to compare these data due to the difference in substitution levels. It can be presumed that highly

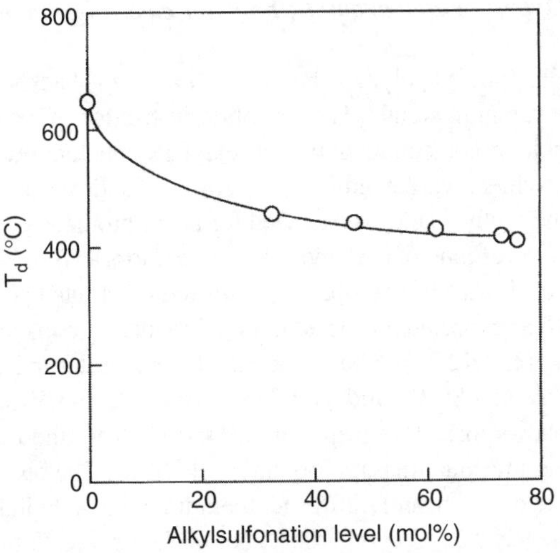

Fig. 16. Decomposition temperature (T_d, the temperature at the onset of mass loss) of PBI-PS as a function of alkylsulfonation level

methylbenzenesulfonated PBI has a lower stability than propanesulfonated PBI due to the weak aryl–S bond. Actually, the degradation temperature of methylbenzenesulfonated PBI is almost the same as polymer electrolytes that were sulfonated with sulfuric acid, etc.

The thermal stability of anhydrous PBI-PS was also investigated by using TGA at a heating rate of 5°C min^{-1} under a nitrogen atmosphere by our research group. All samples were vacuum oven-dried at 60°C for at least 2 days prior to analysis. The PBI-PS is hygroscopic and quickly reabsorbs water after vacuum drying. Due to the difficulty encountered in drying and keeping the samples dry, an in situ drying method followed by immediate thermal analysis was used. Samples were loaded into a TGA instrument, heated to 160°C, and held for 1 h to dry the sample. After cooling under a N_2 atmosphere, the sample was analyzed without removing it from the instrument. Figure 16 shows the thermal decomposition temperature (T_d; the temperature at the onset of mass loss) as a function of the alkylsulfonation level. In contrast to what was observed for PBI, degradation steps were observed for PBI-PS between 450 and 400°C. Although the T_d of PBI-PS decreased to 400°C with increasing alkylsulfonation level, the degradation temperature of PBI-PS is higher than those for perfluorinated polymer electrolytes that exhibit T_d points around 280°C. Elemental analysis and FT-IR measurements of PBI-PS were performed in order to investigate the decomposition process. In practice, the sulfur contents and the intensities of S–O stretching vibrations for PBI-PS heated above 400°C for 1 h decreased. This result is similar to the result of Gieselman et al. and indicates that the decomposition of PBI-PS can be attributed to desulfonation. Consequently, alkylsulfonated PBI is thermally more stable when compared with sulfonated aromatic polymer electrolytes whose degradation temperatures are in range of 200–350°C. This thermal stability in alkylsulfonated polymer electrolytes is attributable to the strong chemical bond between the alkyl chain and the sulfonic acid groups. The addition of alkysulfonic acid groups to thermostable polymers with alkylsultone is one of the most important routes to synthesize thermally stable proton-conducting polymer electrolytes. To design thermostable polymer electrolytes with hydrocarbon polymers, the addition of polar groups, as well as the choice of thermostable polymer backbones, are important factors.

3.3. Water uptake in alkylsulfonated aromatic polymer electrolyte membranes

The addition of arylsulfonate and alkylsulfonate groups onto aromatic polymers induces water absorption and makes the polymer more hygroscopic. The water uptake of PBI-PS was determined by measuring the

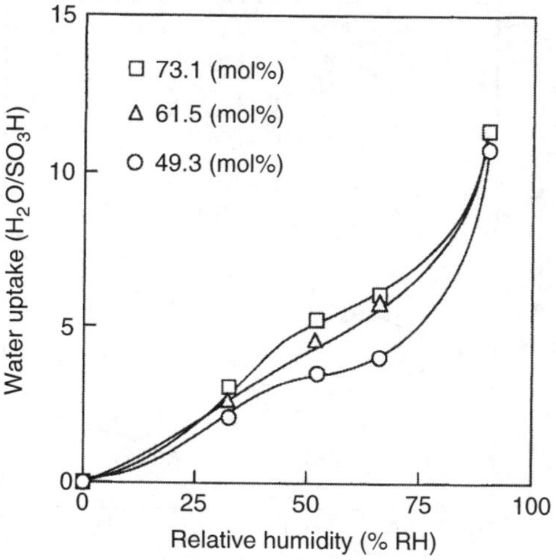

Fig. 17. Water uptake of PBI-PS as a function of relative humidity

change in the mass before and after hydration. The results are shown in Fig. 17 in terms of the moles of water absorbed per mole of sulfonic acid in the polymer. These results show that water uptake varies with relative humidities. The equilibrium water uptake of PBI-PS films increased with increasing the relative humidity and alkylsulfonation level. This behavior and the water absorption level are similar to perfluorinated polymer electrolyte membranes. Two regions with high and low increases in water uptake of PBI-PS were observed over the entire range of relative humidity. The water uptake of PBI-PS with 73.1 mol% alkylsulfonation achieved 11.3 H_2O/SO_3H at 90% RH and room temperature, while Nafion® 117 showed a water uptake of 11 H_2O/SO_3H under the same conditions. We also synthesized butanesulfonated and methylpropanesulfonated PBI (PBI-BS and PBI-MPS) by the same procedure with butanesultone and methylpropanesultone. These polymers showed different features in their water uptakes compared with PBI-PS. The water uptakes of PBI-BS and PBI-MPS were 19.5 and 27.5 H_2O/SO_3H at 90% RH, respectively. These values for alkylsulfonated PBI were affected by the length of alkyl chains and chain branching, and hence the long alkyl chain and chain branching induced its water uptake. It is thought that this phenomenon is attributed to the greater flexibility of long alkyl chains and larger water absorbed cavity formed by the branching chains.

Several studies have examined the specific role of absorbed water in polymer electrolytes, and the physical states of absorbed water have been characterized by several techniques such as IR spectroscopy [45] and low temperature ^{1}H–NMR relaxation time measurements [46]. We tried to elucidate the physical states of the absorbed water in hydrous PBI-PS films by means of DSC. Figure 18 shows a DSC curve of a hydrous PBI-PS (73.1 mol%) film containing 11.3 H_2O per sulfonic acid group. While anhydrous PBI-PS showed no peak, hydrous PBI-PS showed two sharp phase transitions of absorbed water at −36.6 and −21.6°C, which are attributable to the freezing and melting temperature of the absorbed water, respectively. On the other hand, hydrous Nafion® mostly showed a phase transition temperature around 0°C under the same condition. These results suggest that the absorbed water in PBI-PS is more restricted by sulfonic acid groups on the polymer chains, as compared with Nafion® membranes. Therefore, the absorbed water tends to be retained in PBI-PS even at elevated temperatures.

3.4. Proton conductivity of alkylsulfonated aromatic polymer electrolyte membranes

The electrical properties of wet PBI-PS films were investigated by both DC measurements and the complex impedance method. When direct current was applied to PBI-PS films, the DC conductivity suddenly

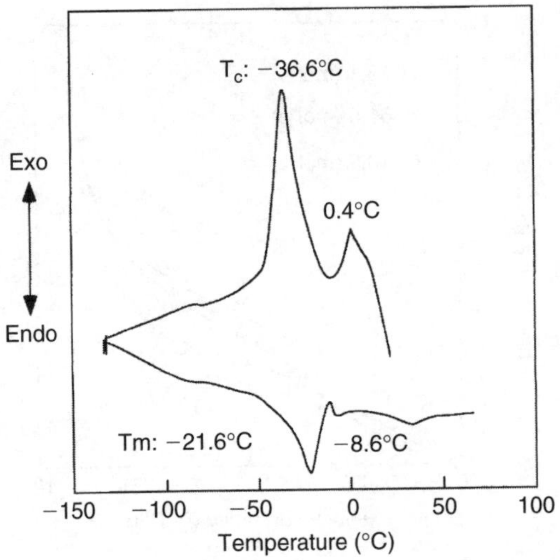

Fig. 18. DSC curve of hydrous PBI-PS (73.1 mol%) film containing 11.3 H_2O per sulfonic acid group

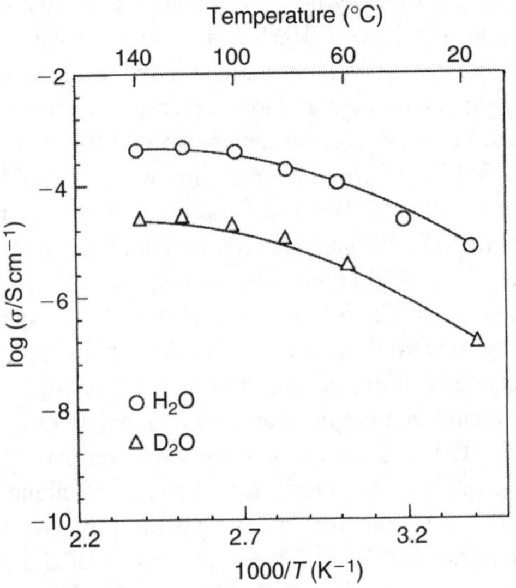

Fig. 19. Conductivity of PBI-PS (61.5 mol%) films containing H_2O or D_2O

decreased. This result demonstrated that there is no electron conduction in PBI-PS, though the main chains are conjugated. In order to clarify the carrier ion in hydrous PBI-PS films, the conductivity of PBI-PS films containing H_2O or D_2O was measured, and the result is shown in Fig. 19. The conductivity of PBI-PS containing H_2O increased with increasing water uptake and was higher than that of PBI-PS containing D_2O in the entire temperature range. This result indicates that the carrier ion is a proton (hydronium ion) for hydrous PBI-PS. Although the conventional theory for H/D isotope effects supports that the mobility of a proton is 1.4 times higher than that of deuterium ion, PBI-PS containing H_2O showed much higher conductivity than that of PBI-PS containing D_2O. Further studies regarding the differences in conductivity from theoretical values are now in progress.

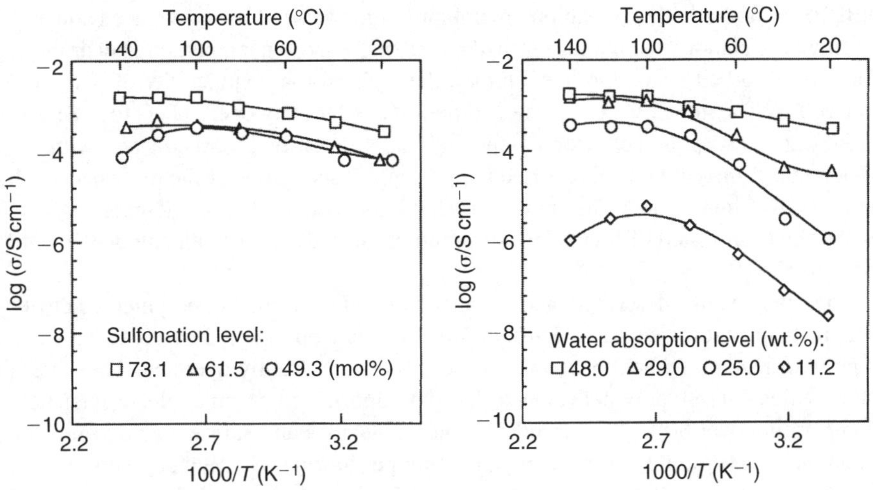

Fig. 20. Temperature dependence of conductivity for PBI-PS. Right: water uptakes of PBI-PS films are almost 48 wt.%. Left: sulfonation level of PBI-PS films is 73.1 mol%

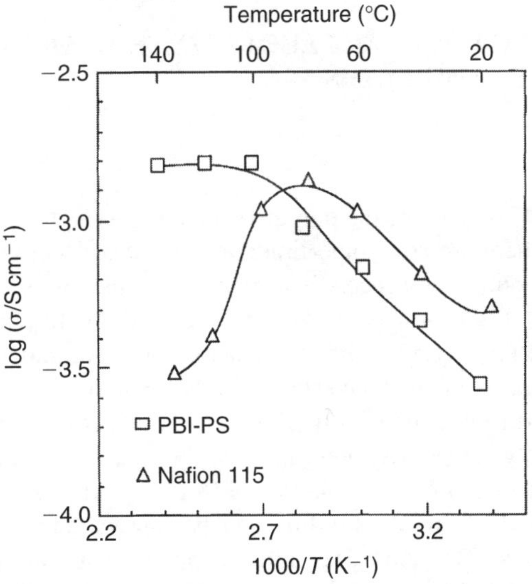

Fig. 21. A comparison of Arrhenius plots of conductivity between PBI-PS and Nafion® membrane

The proton conductivity of PBI-PS containing equilibrium water was measured as a function of temperature by the complex impedance method. As shown in Fig. 20, hydrous PBI-PS exhibits a high proton conductivity at room temperature. The conductivity of a PBI-PS film containing 3.1 H_2O/SO_3H reached $10^{-5}\,S\,cm^{-1}$ at 80°C and decreased slightly at higher temperatures due to a small loss of water (about 10 wt.%). The conductivity of PBI-PS containing more than 5.2 H_2O/SO_3H increased with an increase in temperature, and a conductivity of the order of $10^{-3}\,S\,cm^{-1}$ was maintained over 100°C. The PBI-PS film containing 11.3 H_2O/SO_3H showed high proton conductivities on the order of $10^{-3}\,S\,cm^{-1}$ and an Arrhenius type dependence of conductivity with the activation energy $E_a = 9.4\,kJmol^{-1}$. A comparison of Arrhenius plots for conductivities between PBI-PS and Nafion® membranes is shown in Fig. 21. The water uptake of PBI-PS is similar to that of Nafion® membrane when these films were placed under 90% RH. The proton conductivity of Nafion® membrane reached 10^{-3} S cm^{-1} at room temperature but dropped over 100°C due to the loss of absorbed water. On the contrary, the high proton conductivity of hydrous PBI-PS was retained over 100°C. The higher water uptake and proton conductivity over 100°C for PBI-PS is attributed to the different water absorption behavior and the physical state of the absorbed water within PBI-PS.

On the other hand, Glipa et al. [40] reported the proton conductivities of methylbenzenesulfonated PBI under various relative humidities. The proton conductivity increased with increases in the substitution level. Methylbenzenesulfonated PBI with 75% substitution showed a high conductivity of about 10^{-2} S cm^{-1} at 40°C under 100% RH.

Judging from the results described above, alkylsulfonated aromatic polymer electrolytes possess enough thermal stability for use as an electrolyte in a fuel cells operating near 80°C. This is the temperature where perfluorinated polymer electrolytes based PEFC are usually operated. The water uptakes and proton conductivities of these polymers are similar to perfluorinated polymer electrolytes below 80°C but are better temperature over 80°C. The absorbed water molecules and sulfonic acid groups in these polymers have relatively strong interactions, compared with perfluorinated polymer electrolytes. It is thought that this difference is closely related to the water absorption mechanism and physical state of the absorbed water between these polymers and perfluorinated polymer electrolytes. Further studies on the proton conduction mechanism in these polymers are required.

4. PROTON-CONDUCTING POLYMER ELECTROLYTE MEMBRANES BASED ON ACID–BASE POLYMER COMPLEXES

4.1. Materials

The development of novel proton-conducting polymers for various electrochemical applications is attracting considerable interest. There are still some limitations on the electrochemical properties and water uptake in proton-conducting polymers, especially at elevated temperatures. Complexes of basic polymers, such as poly(ethylene oxide) (PEO) [47], poly(ethylene imine) (PEI) [48], poly(acrylamide) (PAAM) [49, 50], and poly(vinylalchol) (PVA) [51,52], with strong acids have been shown to possess high proton conductivities both in the dehydrated and hydrated states [53–58]. For these systems the mechanism of proton conduction has been inferred from conductivity, FT-IR, and nuclear magnetic resonance measurements. These complexes are relatively inexpensive and can be easily processed as thin films for applications such as hydrogen sensors [59], electrochromic displays, and PEFC systems. For example, complexes of PVA with orthophosphoric acid in the form of thin films have been prepared by a solution cast technique with different stoichiometric ratios. The ionic transference number of mobile ions was estimated from Wagner's polarization method and the value was reported to be $t_{ion} = 0.97$. At 30°C the conductivity of PVA/H_3PO_4 complexes (molar ratio H^+/PVA = 0.77) is about 10^{-3} S cm^{-1} [51]. According to Lassegues et al. [58], acid–base polymer complexes of PAAM and strong acids such as H_3PO_4 or H_2SO_3 exhibit a high proton conductivity in the range of 10^{-4}–10^{-3} scm^{-1} at ambient temperatures. The proton conductivity increases with temperature to about 10^{-2} S cm^{-1} at 100°C. However, the mechanical and chemical stability of these complexes is relatively poor, and chemical degradation is often observed after humidification. Anhydrous acid–base polymer complexes (PEI/xH_2SO_4 and PEI/xH_3PO_4) can also be obtained by protonation of branched commercial poly(ethylene imine) with H_2SO_4 or H_3PO_4. For partially protonated PEI ($x \leqslant 0.35$), only SO_4^{2-} or HPO_4^{2-} anions are present and the conduction seems to occur by proton exchange between protonated and unprotonated amine groups. Above $x \approx 0.35$, HSO_4^- or $H_2PO_4^-$ anions appear and the conductivity, likely to occur along the SO_4^{2-}/HSO_4^- or $HPO_4^{2-}/H_2PO_4^-$ anionic hydrogen-bonded chains, increases up to values of about 10^{-4} S cm^{-1} for PEI/0.5H_2SO_4 and about 10^{-5} S cm^{-1} for

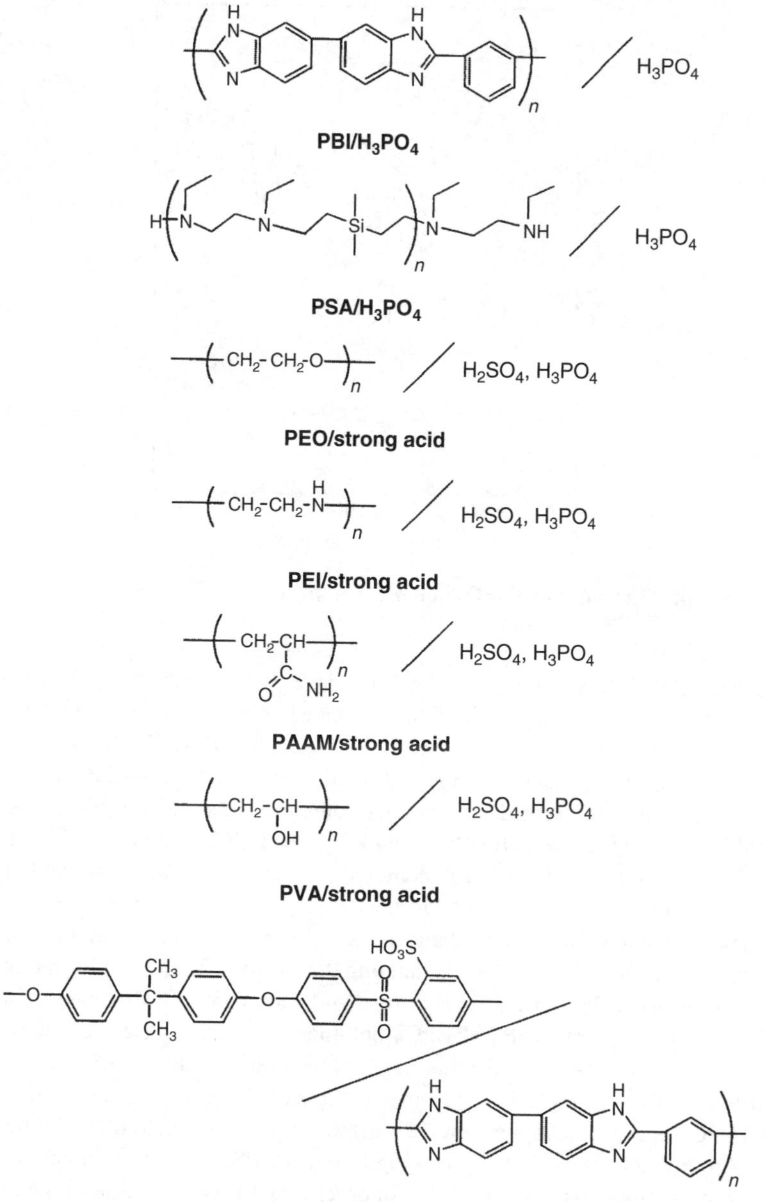

Fig. 22. Chemical structures of polymer electrolytes based on acid–base polymer complexes

PEI/0.5H_3PO_4. Thin transparent films can be produced for $x \leqslant 0.35$, and they lose their mechanical properties and become very hygroscopic above this value because of the presence of excess acid. Figure 22 shows some of the chemical structures for these materials.

New proton-conducting polymers based on poly(silamine) (PSA) were synthesized by mixing PSA and H_3PO_4 by our research group [60]. The reaction of *N,N′*-diethylethylenediamine (DEDA) and

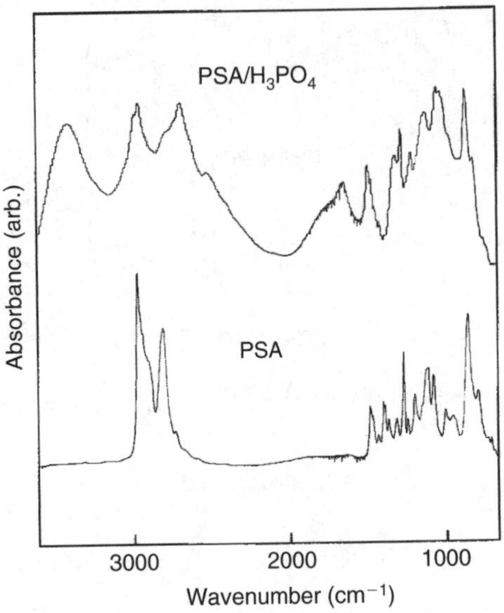

Fig. 23. FT-IR spectra of PSA and PSA/H_3PO_4 polymer complex

dimethyldivinylsilane (DVS) in the presence of *n*-BuLi gave PSA, which was a clear yellow and viscous liquid. The results of ^{1}H NMR analysis for PSA indicated that both ends of the PSA chains carried *N*-ethylamino groups and that the molecular weight was about 2000. PSA/H_3PO_4 complexes were obtained by mixing PSA and H_3PO_4 in methanol. To a H_3PO_4 methanol solution, liquid PSA was added dropwise, and then a white precipitate of the resulting polymer complex was obtained. FT-IR measurements were carried out in order to prove the interaction between PSA and H_3PO_4 molecules (Fig. 23). The spectrum of the polymer complex exhibits a broad peak centered at 3000 cm^{-1} and a CN band at 1469 cm^{-1}. The N–H band appears at 3393 cm^{-1} due to the hydrogen bond interaction between the N atoms of PSA and the phosphoric acid molecules. Because of the presence of the ethylenediamine units in the main chain, PSA shows a two-step protonation process on changing the pH [61]. These results and previous data indicate that PSA is protonated with phosphoric acid to afford a solid complex containing the double chelating structure. The H_3PO_4 contents in PSA/H_3PO_4 polymer complexes were controlled by changing the feed ratio. The PSA absorbed H_3PO_4 of 0.8 mol/unit at a feed ratio (H_3PO_4/PSA unit) of 2. Over a H_3PO_4 content of 0.8 mol/unit, these PSA/H_3PO_4 polymer complexes became viscoelastomers. TGA measurements for PSA/H_3PO_4 polymer complexes were carried out to investigate their thermal stabilities. The results are shown in Fig. 24. The onset of the weight loss for PSA/H_3PO_4 polymer complexes began at 200°C, while PSA was stable up to about 300°C. In order to identify the origin of weight loss, elemental analysis and FT-IR measurements were carried out before and after the heat treatment of the polymer complexes at 200°C for 1 h. These data indicated that the weight loss at 200°C results from the change from orthophosphoric acid to pyrophosphoric acid. Thus, these PSA/H_3PO_4 complexes were found to be quite stable up to at least 200°C.

Recently, new proton-conducting polymer electrolyte membranes based on poly(benzimidazole) (PBI) have been proposed for use in fuel cells [62–65]. The main advantage compared with perfluorinated polymer electrolytes and other acid–base polymer complexes, is that the polymer is conductive even when the activity of water is low and has a high thermal stability. These materials are expected to operate from ambient to high temperatures in humid or dry gas. These PBI complex films are fabricated by immersing PBI cast films into phosphoric acid solutions.

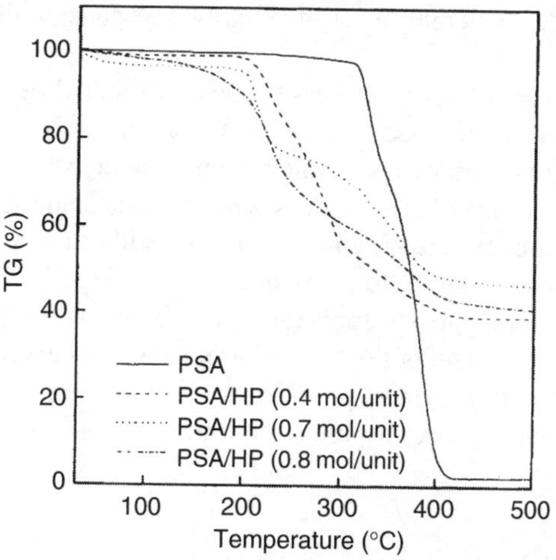

Fig. 24. Thermogravimetric analysis of PSA and PSA/H_3PO_4 polymer complexes

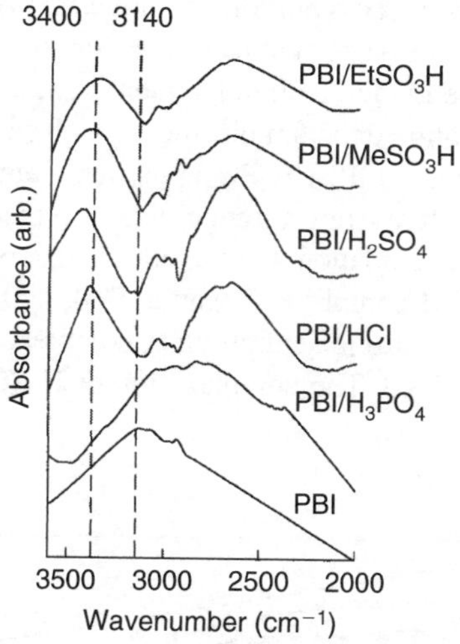

Fig. 25. FT-IR spectra of PBI and PBI/strong acid polymer complexes

In our case, the PBI/strong acid polymer complexes were easily prepared by immersing PBI films into a strong acid/methanol solution [66,67]. The absorption level of strong acid molecule increased with increasing the concentration of the strong acid, and the highest absorption level of 2.9 mol/unit was observed for PBI/H_3PO_4 polymer complexes. In order to clarify the nature of the interaction between PBI and strong acid molecules, FT-IR measurements were carried out. As shown in Fig. 25, the interactions between strong acids and imidazole groups were observed in the range from 2000 to 3600 cm^{-1}. PBI showed a self-associated NH stretching band at 3139 cm^{-1}. Although NH^+ stretching bands were observed around 2900 cm^{-1} for PBI/H_2SO_4, CH_3SO_4, and $C_2H_5SO_3H$ complexes, PBI/H_3PO_4 did not show the characteristic absorptions of NH^+ groups. These data indicate that the acid molecules, except for H_3PO_4, protonate the N atom in the imidazole ring. H_3PO_4 does not protonate imidazole groups in PBI but interacts with it via strong hydrogen–bonding interaction of OH and NH groups. Savinel et al. [62–65] prepared PBI complex films doped with phosphoric acid by immersion in a phosphoric acid aqueous solution for a least 16 h. Equilibration in an 11 M H_3PO_4 solution yielded a doping level of roughly five phosphoric acid molecules per polymer repeat unit.

In general, sulfonated aromatic polymers are very brittle in the anhydrous state, which can happen in the fuel cell application under intermittent conditions. Kerres et al. have discovered new materials, which are not as brittle as anhydrous sulfonated polymers using polymer blend techniques [68]. These materials were synthesized by combining polymeric N-bases, such as *ortho*-sulfone diamine polysulfone, poly(4-vinylpyridine), PBI, and poly(ethyleneimine), and polymeric sulfonic acids such as sulfonated poly(etheretherketone) and *ortho*-sulfone-sulfonated polysulfone, as illustrated in Fig. 22. These blend polymers showed high proton conductivities and moderate swelling values combined with high thermal stabilities. The specific interaction of the SO_3H groups and of the basic N groups was observed by FT-IR measurements. The acid–base interaction between both polymers provided suitable mechanical strengths and thermal stabilities. Their thermal and mechanical properties are similar to those of other acid–base polymer complexes described later.

4.2. Thermal stability of acid–base polymer complexes

The thermal stability of PBI/strong acid polymer complexes was investigated by TG-DTA analysis. Figure 26 shows TG-curves of PBI and PBI/strong acid polymer complexes. PBI is extremely thermally stable over the entire measuring temperature range. The small mass loss for all samples up to 200°C is attributable to losses of water and solvent, which tend to remain in the membranes. A lowering of degradation temperature was expected with the complexation of acid molecules, which easily corrode and oxidize the polymer backbone. However, the degradation of PBI/H_3PO_4 polymer complexes in N_2 atmosphere was not observed over the entire measuring temperature range, where typical proton-conducting polymer electrolytes show considerable thermal degradation. On the other hand, onsets of thermal decompositions for PBI/H_2SO_4, CH_3SO_4, and $C_2H_5SO_3H$ complexes began at 330, 240, and 220°C, respectively. After the thermal degradation of these polymer/strong acid polymer complexes between 220 and 400°C, a residue with 50% of the original mass remained. The complexation of H_2SO_4, CH_3SO_4, and $C_2H_5SO_3H$ to PBI

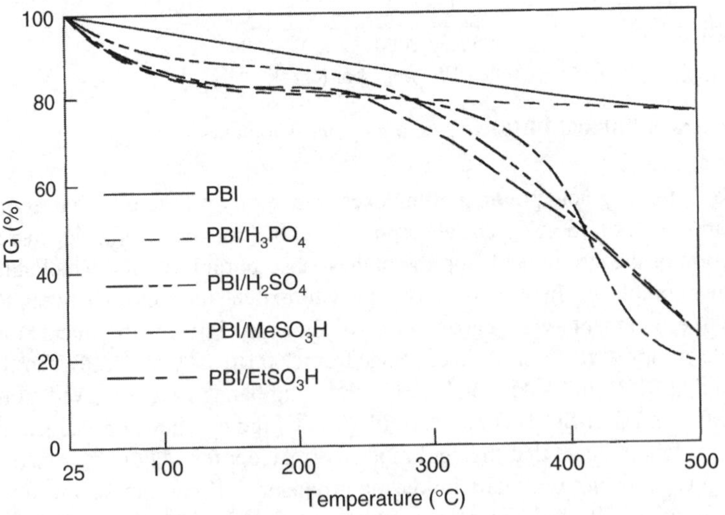

Fig. 26. TG-curves of PBI and PBI/strong acid polymer complexes

results in a loss of thermal stability. This decomposition is mainly due to the elimination of acid molecules from the polymer/strong acid polymer complexes as determined by elemental analysis. Over 400°C, the PBI polymer backbone gradually decomposes due to the presence of strong acids and high temperatures.

PBI/H_3PO_4 complexes, however, were thermally stable up to 500°C and so also PBI. The sulfonation of PBI via acid treatment has been shown to stabilize PBI, and Powers et al. reported that doping PBI with 27 wt.% phosphoric acid also increased thermal stability [69]. They attributed the increased stability to the formation of benzimidazonium cations.

Savinell's group also investigated the thermal stability of PBI/H_3PO_4 polymer complexes in simulated fuel cell environments [63]. They found that PBI/H_3PO_4 polymer complexes have promising properties for use as polymer electrolytes in hydrogen/air and direct methanol fuel cells. To simulate the conditions present in a high–temperature fuel cell, PBI/H_3PO_4 polymer complexes were loaded with fuel cell grade platinum black, doped with ca. 480 mol% phosphoric acid (4.8 H_3PO_4 molecules per PBI repeat unit) and heated under atmospheres of either nitrogen, 5% hydrogen, or air in a thermogravimetric analyzer. The products of decomposition were taken directly into a mass spectrometer for identification. In all cases weight loss below 400°C was found to be due to the loss of water. They concluded that the PBI/H_3PO_4 polymer complexes loaded with fuel cell grade platinum were quite stable up to 600°C.

4.3. Conductivity of acid–base polymer complex

Savinell's group studied the conductivity of PBI/H_3PO_4 polymer complexes as a function of water vapor activity, temperature, and acid doping levels [65]. The conductivity of PBI/H_3PO_4 polymer complexes at the higher doping level (500 mol%) was roughly twice that for the film doped to 338 mol% under similar conditions of temperature and humidity. For example, the conductivity of PBI doped with ca. 500 mol% (5 H_3PO_4 molecules per PBI repeat unit) at 190°C and at water vapor activity of 0.1 was 3.5×10^{-2} Scm^{-1}. The conductivity increased with temperature and water vapor activity regardless of the doping level of H_3PO_4. In addition, they found that the methanol cross-over with PBI/H_3PO_4 polymer complexes was one order of magnitude smaller than that with perfluorinated polymer electrolytes. The mechanical strength is three orders of magnitude greater than that of Nafion® membranes.

We also investigated the proton conductivity of PBI/strong acid polymer complexes prepared by our method with methanol solutions of strong acids. The temperature dependences of conductivity for anhydrous PBI/strong acid polymer complexes at an acid concentration of about 1.9 mol/unit is shown in Fig. 27. All of the anhydrous PBI/strong acid polymer complexes exhibited proton conductions of 10^{-6}–10^{-9} S cm^{-1} at 100°C. The conductivity of PBI/H_3PO_4 polymer complex reached up to 10^{-5} S cm^{-1} at 160°C, while other PBI/strong acid polymer complexes showed a decrease in conductivity over 80°C. This result also reflects the good thermal stability of PBI/H_3PO_4 polymer complexes. The films of these PBI/strong acid polymer complexes were placed in a 90% RH vessel for 3 days to prepare hydrous films. The water uptakes of these complexes were estimated to be near 13–26 wt.%. As shown in Fig. 28, the proton conductivities of the hydrous PBI/strong acid polymer complexes were one order of magnitude higher than those of anhydrous polymer complexes due to the improvement of carrier generation by the absorbed water. Especially, remarkable increases in proton conductivities were observed at ambient temperatures. Figure 29 shows the temperature dependence of conductivity for anhydrous PBI/H_3PO_4 complexes with different acid contents. The conductivity of the PBI/H_3PO_4 polymer complexes increased with increasing contents of H_3PO_4. The conductivity of PBI/H_3PO_4 complex ($\chi = 1.4$) showed a different behavior, having relatively low conductivities over the entire temperature range. This phenomenon suggests that two H_3PO_4 molecules interact quantitatively with a PBI unit containing two imidazole groups and consequently an excess of H_3PO_4 to imidazole groups is necessary to give sufficient proton conductivity. In order to clarify the interaction between H_3PO_4 and PBI, the characteristic absorption of H_3PO_4 molecules

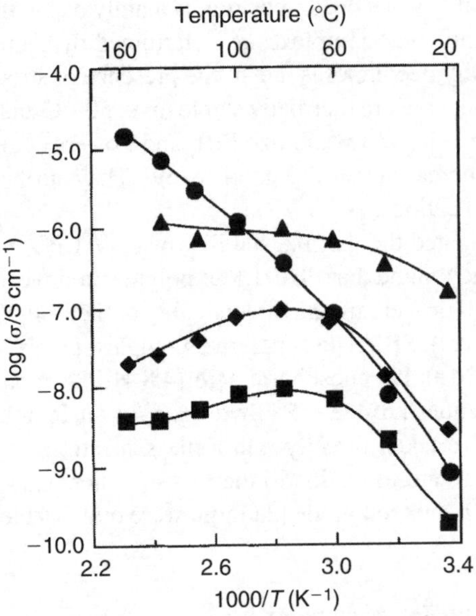

Fig. 27. Temperature dependence of conductivity for anhydrous PBI/strong acid complexes. (●) PBI/H_3PO_4 (2.0 mol/unit); (▲) PBI/H_2SO_4 (1.8 mol/unit); (■) $MeSO_3H$ (2.0 mol/unit); (◆) $EtSO_3H$ (1.9 mol/unit)

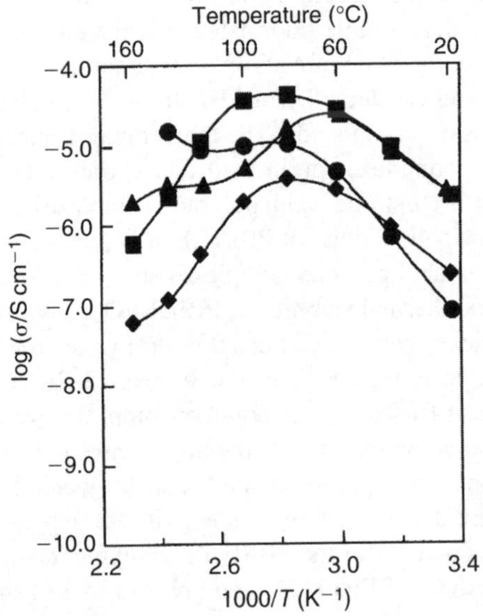

Fig. 28. Temperature dependence of conductivity for hydrous PBI/strong acid complexes. (●) PBI/H_3PO_4 (2.0 mol/unit, H_2O 13 wt.%); (▲) PBI/H_2SO_4 (1.8 mol/unit, H_2O 19 wt.%); (■) $MeSO_3H$ (2.0 mol/unit, H_2O 26 wt.%); (◆) $EtSO_3H$ (1.9 mol/unit, H_2O 20 wt.%)

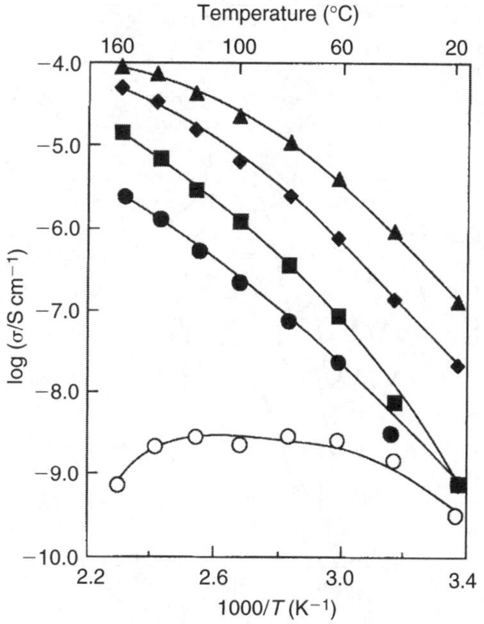

Fig. 29. Temperature dependence of conductivity for anhydrous PBI/H_3PO_4 complexes with different acid contents. H_3PO_4 contents (mol/unit) (▲) 2.9; (◆) 2.7; (■) 2.3; (●)2.0; (○) 1.4

in the PBI/H_3PO_4 polymer complex was investigated by FT-IR. The results are shown in Fig. 30. Three characteristic absorptions of the HPO_4^{2-} P–OH, and $H_2PO_4^-$ groups for PBI/H_3PO_4 polymer complexes appear at 1090, 1008, and 970 cm^{-1}, respectively [49,50,53,70]. The intensity of absorption band of HPO_4^{2-} increases with increase in the concentration of H_3PO_4. The presence of HPO_4^{2-} and $H_2PO_4^-$ anions implies that proton conduction may occur according to the Grotthus mechanism, which involves an exchange of protons between H_3PO_4 and HPO_4^{2-} or $H_2PO_4^-$.

The PSA/H_3PO_4 polymer complexes also exhibit high proton conductions in the hydrated state as well as the dehydrated state. Figure 31 shows the temperature dependence of conductivity in the dehydrated state for PSA/H_3PO_4 polymer complexes with different H_3PO_4 contents. The conductivities of the PSA/H_3PO_4 polymer complexes increased with increasing H_3PO_4 contents and did not diminish over 100°C. The PSA/H_3PO_4 polymer complex possesses a conductivity of 10^{-5} S cm^{-1} at 160°C at a concentration of 0.8 H_3PO_4 per repeat unit. In order to identify the carrier ion, the conductivities of PSA/H_3PO_4 and PSA/D_3PO_4 polymer complexes were measured under the same conditions, and the results are shown in Fig. 32. The temperature dependence of conductivity for the PSA/D_3PO_4 polymer complex was similar to that of the PSA/H_3PO_4 polymer complex, but the conductivity of PSA/H_3PO_4 complex was higher than the conductivity of the PSA/D_3PO_4 polymer complex at every temperature. The isotope effect for the conductivity of these polymer complexes proves that the carrier ions are protons generated from strong acid molecules [71].

5. OTHER PROTON-CONDUCTING POLYMER ELECTROLYTE MEMBRANES

As described above, many kinds of proton-conducting polymer electrolyte membranes based on hydrocarbon polymers have been developed not only for PEFC and DMFC applications but also for electrochromic display

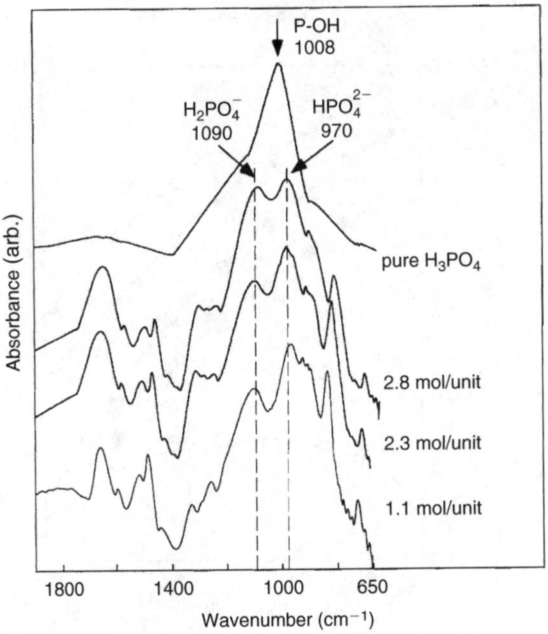

Fig. 30. FT-IR spectra of PBI/H_3PO_4 and H_3PO_4

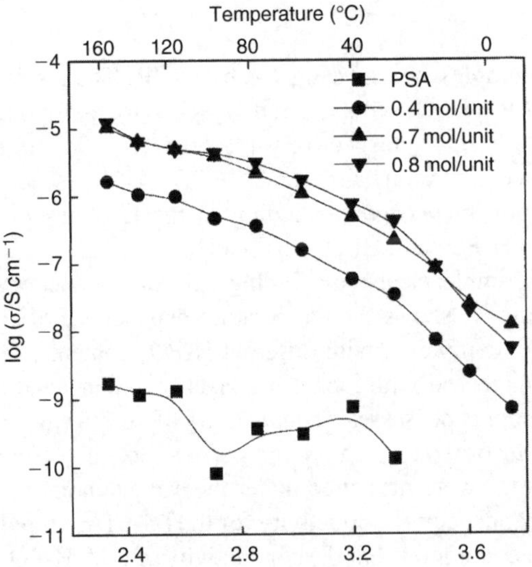

Fig. 31. Temperature dependence of conductivity for anhydrous PSA/H_3PO_4 complexes with different H_3PO_4 contents

applications and sensor devices. There are several reports and patents concerning new proton-conducting polymer electrolytes having different chemical structures and concepts from conventional perfluorinated polymer electrolytes and hydrocarbon materials described above. These materials involve phosphoric acid–based polymer electrolytes, silicone based-electrolytes, and other inorganic polymer electrolytes.

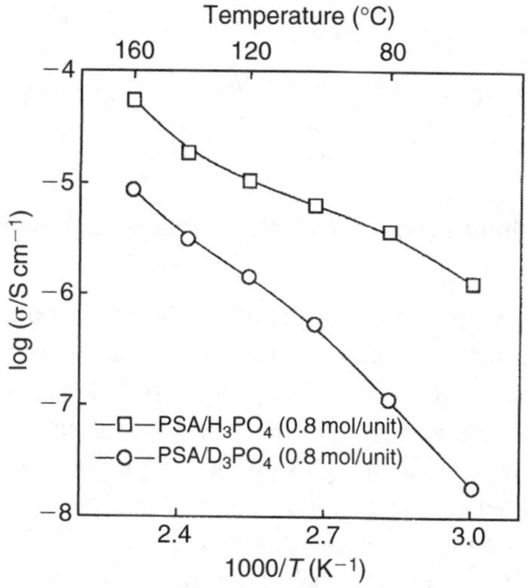

Fig. 32. Comparison of temperature dependence of conductivity for PSA/H_3PO_4 and PSA/D_3PO_4 complexes

Proton-conducting polymer electrolyte membranes for possible use in H_2/O_2 and direct methanol fuel cells have been fabricated from poly[bis(3-methylphenoxy)phosphazene] by Quo et al. [72]. These polymer electrolytes were sulfonated with SO_3 and cast from solution. The ion exchange capacity of the membrane was 1.4 mmol/g^{-1}. The polymer cross-linking was carried out by dissolving a benzophenone photoinitiator in the casting solution and exposing the resulting cast films to UV light. The sulfonated and cross-linked polyphosphazene membrane swelled less than Nafion 117® membranes in both water and methanol. The proton conductivity of water equilibrated polymers was similar to Nafion® membranes between 25 and 65°C. A Japanese company also claimed gas diffusion electrodes for fuel cells comprised of polyphosphazene [73].

Gautier-Luneau et al. prepared an organic–inorganic proton-conducting polymer electrolyte to be used as a membrane for DMFC [74]. They synthesized the electrolytes by hydrolysis–co-condensation of three different alkoxy silanes; benzyltriethoxysolane, *n*-hexyltrimethoxysilane, and triethoxysilane. After co-condensation and solvent evaporation, sulfonation was achieved in dichloromethane by adding chlorosulfonic acid in a stoichiometric amount, with respect to the benzyl groups. The cross-linking was performed in THF by hydrosilyation of the silane groups with divinylbenzene using a divinyltetramethyldisiloxane platinium complex as catalyst. The TGA data showed one broad endothermic peak with a maximum at 103°C due to the loss of water. They reported that these polymer electrolytes were thermally stable until 250°C and that the carbonization of the organic chains started at 375°C. The proton conductivity of these materials was about 1.6×10^{-2} Scm^{-1}.

Another alkoxy silanes based polymer electrolyte was reported by Honma et al. [75]. Their organic–inorganic hybrid electrolyte membranes have been synthesized by the sol–gel process. Poly(ethyleneglycols) ($M_w = 200, 300, 400, 600, 2000$) were reacted with 3-isocyanatopropyltriethoxy silane in an N_2 atmosphere at 70°C to obtain alkoxy end-capped poly(ethyleneoxide) precursors. The hybrid electrolyte membranes were fabricated by hydrolyzing the end-capped precursor with monophenyltriethoxysilane. A composition of the monophenyltriethoxysilane of 20% was found to be the most flexible and chemically stable composition for the hybrid electrolyte membranes. The TGA showed that decomposition of the hybrid membrane started near 300°C. They reported that the thermal stability of the electrolyte membranes were enhanced enormously compared with poly(ethyleneoxide), which was the base material of the hybrid membranes due to structural confinement of the poly(ethyleneoxide) chain between the nano-sized silicate domains. Proton conductivity within the hybrid membranes was provided by incorporating acid molecules

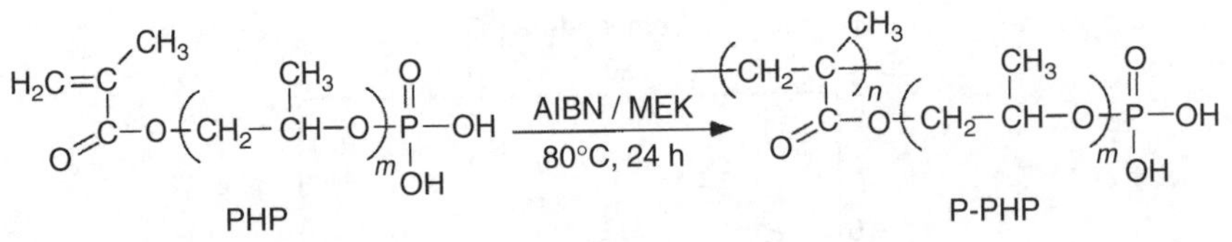

Fig. 33. Chemical structure of soluble polymer electrolyte containing phosphoric acid group

such as monododecylphosphate and phosphotungstic acid. The proton conductivity as a function of humidifier temperature against the constant cell temperature at 80°C was measured for the electrolyte membrane doped with 20 wt.% of monododecylphosphate. The proton conductivity was very small at low humidifier temperatures and increased with temperature and reached a maximum value of 10^{-3} S cm^{-1} above 80°C. The hybrid electrolyte membranes needed almost saturated humidities to afford maximum proton conductivities. They also attempted to measure the conductivity of hybrid electrolyte membranes containing phosphotungstic acid. This system showed a specific conductivity of 0.17 S cm^{-1} at room temperatures. The maximum conductivity of 5×10^{-3} Scm^{-1} was observed for the hybrid electrolyte membrane incorporating phosphotungstic acid of 10 wt.% at a humidifier temperature of 80°C.

Polymer electrolytes prepared from hydrocarbon ionomers or perfluorinated ionomers generally use sulfonic acid groups for the hydrophilic part, which absorbs water and generates water clusters and proton carriers. Except for sulfonic acid groups, phosphoric acid, and boric acid groups can be used as a polar groups for proton-conducting polymer electrolytes. Our research group has synthesized new proton-conducting polymer electrolytes containing phosphoric acid groups. In many cases, polymer electrolytes containing phosphoric acid groups are insoluble in organic solvents and water due to the strong aggregation and condensation of the phosphoric acid groups. It is, hence, difficult to synthesize polymer electrolyte membranes from monomers containing phosphoric acid groups. We tried to synthesize soluble polymer electrolytes with various vinyl monomers containing phosphoric acid groups. However, most of the resulting polymer electrolytes were insoluble in organic solvents and water. A soluble polymer electrolyte containing phosphoric acid groups has been synthesized by modifying the polymerization method and substituting methyl groups onto the polymer side chains in the neighborhood of the phosphoric acid groups. The chemical structure of this soluble polymer electrolyte is shown in Fig. 33. Vinyl acrylate phosphoric acid monomer (PHP) and azobisisobutyronitrile (AIBN) were first dissolved in methylethylketone. To methylethylketone heated at 80°C, the monomer solution was added dropwise, and then the mixing solution was maintained at 80°C for at least 24 h. The polymer was obtained and purified by reprecipitation. Copolymers with phenyl mareimide, styrene, or acrylonitrile were also synthesized by the same polymerization method with AIBN. The mechanical properties, proton conductivities, and solubilities were modified by changing the types of co-monomers and their contents. The thermal stability of these polymer electrolytes was studied using TGA in nitrogen atmosphere. These polymers and copolymers were thermally stable up to 200°C. A first mass loss began between 200 and 300°C in N_2 at a heating rate of 10°C min^{-1}. After the first degradation, 60 wt.% of the original mass remained. These results suggest that the dehydration and degradation of P-PHP main chains occur at that temperature. It can be therefore presumed that these polymer electrolytes have sufficient thermal stabilities for most PEFC applications, which are usually operated around 100°C. The equivalent weight of P-PHP is about 500. Figure 34 shows the water uptake of P-PHP containing phosphoric acid. The water absorption behavior of the polymer electrolyte was similar to conventional perfluorinated polymer electrolytes, which showed the Brunauer–Emmett–Teller (BET) absorption of water. Although P-PHP absorbed $12H_2O$ molecules per phosphoric acid at 100% RH, less than $4H_2O$ molecules per phosphoric acid were absorbed below 60% RH. Its water uptake is very sensitive to humidity. The proton conductivity of P-PHP showed great dependence on water uptake and temperature. Figure 35 displays the temperature dependence of conductivity for P-PHP with various water contents. The conductivity of hydrous P-PHP increased with increasing water content. The P-PHP material containing $12H_2O$ molecules

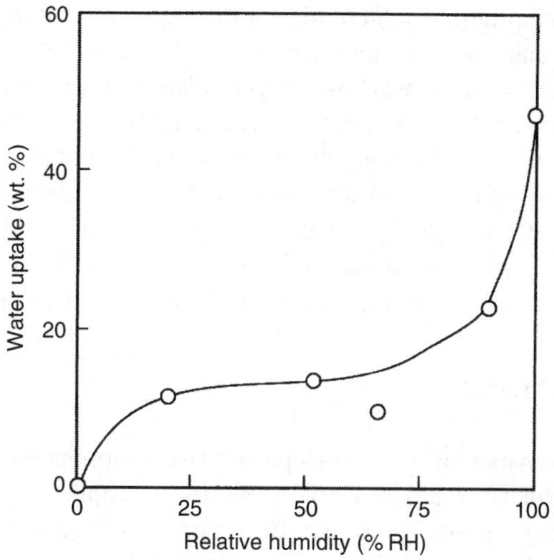

Fig. 34. Water uptake of soluble polymer electrolyte containing phosphoric acid group

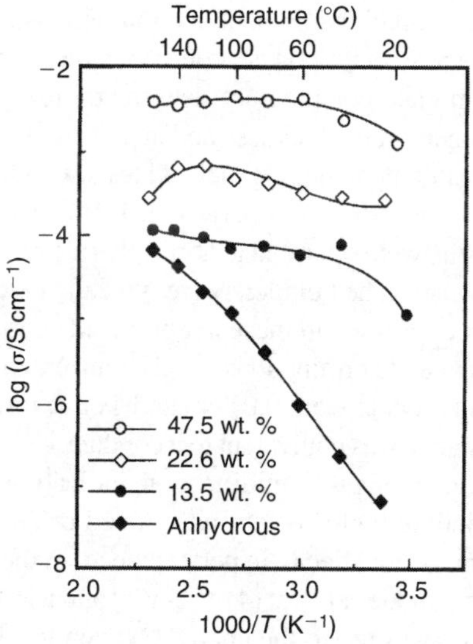

Fig. 35. Temperature dependence of conductivity for soluble polymer electrolyte containing phosphoric acid group with various water contents

per phosphoric acid group showed 20 times higher conductivity than that of the P-PHP containing only 3.4 H_2O molecules per phosphoric acid group. All of these samples showed little decreases in conductivity up to 100°C. This result indicates that the absorbed water molecules in P-PHP have a strong interaction with the polymer back bones and hence remain in the membranes even at elevated temperatures. Judging from these results, phosphoric acid groups can serve as polar groups with sufficient water absorption and proton conductivities for polymer electrolytes in PEFC applications. Since phosphoric acid groups may have different water absorption mechanisms from that of sulfonic acid groups, further studies on the water uptake and physical state of water in these materials are desired.

Doyle et al. [76] demonstrated that perfluorinated polymer electrolyte membranes such as Nafion® membrane can be swollen with room temperature molten salts giving composite free-standing membranes having excellent stabilities and proton conductivities in the high temperature range while retaining a low volatility of the ionic liquid. Ionic conductivities in excess of 0.1 S cm^{-1} at 180°C have been demonstrated using a room temperature molten salt, 1-butyl 3-methyl imidazolium trifluoromethane sulfonate.

On the other hand, a new material, which is permeable to gases and water with efficient proton and electron conductivity, has been developed. Such a material could replace both carbon and Nafion® in the catalyst layer and should provide enhanced performance on PEFCs. Lefebvre et al. reported the use of polypyrrole/poly(styrene-4-sulfonate), poly(3,4-ethylenedioxythiophene)/poly(vinylsulfate), and poly(3,4-ethylenedioxythiophene)/poly(styrene-4-sulfonate) in such a role [10,15,77]. They showed high electron- and proton conductivities, which facilitated rapid electron chemical reaction rates in thick layers of catalyst.

6. FUEL CELL APPLICATIONS

Two of the acid–base blend polymer electrolytes (type 1: 90 wt.% sulfonated PEEK and 10 wt.% PBI, type 2: 95 wt.% sulfonated poly(ethersulfone) and 5 wt.% PBI) were applied in a H_2/O_2 PEFC. Kerres et al. reported that the current/voltage curves of the acid–base blend polymers in the fuel cell were comparable with that of Nafion 112® membranes [68]. These fuel cell tests were performed up to 300 h.

Our research group carried out H_2/O_2 fuel cell tests with sulfonated PPBP membranes. Several cells have been operated at 80°C using E-TEK Pt/C electrodes with a Pt loading of 0.8 mg cm^{-2}. The fuel cells were operated at 4 atm, with the reactant gases humidified by bubbling through distilled water. Both O_2 and H_2 bubblers were typically retained at 80°C. The maximum power observed in these unoptimized cells was 0.3 W cm^{-2} at 800 mA cm^{-2}. The electrolyte membrane conductivity determined using the current interrupt method was found to be 3×10^{-3} S cm^{-1}. The membrane thickness and area were 0.01 cm and 3.15 cm^2, respectively.

Savinell's group has done H_2/O_2 and methanol/O_2 fuel cell tests with PBI/H_3PO_4 polymer complexes. For the case of H_2/O_2 fuel cell tests, several cells were operated at 150°C using E-TEK Pt/C electrodes with a Pt loading of 0.5 mg cm^{-2}. The fuel cells were operated at atmospheric pressure, with the reactant gases humidified by bubbling through distilled water. The bubblers were typically held at 48°C (hydrogen side) and 28°C (oxygen side). The maximum power observed in these unoptimized cells was 0.25 W cm^{-2} at 700 mA cm^{-2}. The electrolyte membrane resistance determined using the current interrupt method was found to be 0.4 ohm. The membrane thickness and area were 0.01 cm and 1 cm^2, respectively, and the doping level was 500 mol%. The measured cell resistance was equivalent to a conductivity of 0.025 S cm^{-1}. The cell resistance was found to be essentially independent of gas humidification, indicating that the water produced at the cathode is sufficient to maintain conductivity in the electrolyte. A cell of this type was operated continuously at 200 mA cm^{-2} for over 200 h with no overall decay in performance. In the case of methanol/O_2 fuel cell tests, the fuel cell electrodes were a Pr–Ru anode catalyst (4 mg cm^{-2}) and a Pt black cathode catalyst (4 mg cm^{-2}). A 4:1 water:methanol vapor mixture was fed to the anode. Oxygen for the cathode was humidified at room temperature. Operating at 200°C and atmospheric pressure, the cell produced over 0.1 W cm^{-2} for current densities between 250 and 500 mA cm^{-2}. The conductivities of the membranes in the test cells did not change between 30 and 140°C under the same conditions.

7. SUMMARY

This review concerning new proton-conducting polymer electrolytes was devoted to the description of numerous synthesis of materials, thermal stabilities, water uptakes, proton conductivities, and fuel cell applications. The numerous advantages of these materials have been critically reviewed, together with synthesis methods and characterizations.

Perfluorinated polymer electrolyte membranes such as Nafion® and Flemion® have been extensively used as polymer electrolytes for fuel cells. These polymer electrolytes have sufficient electrochemical properties, mechanical properties, and chemical and thermal stabilities. However, PEFCs constructed with these perfluorinated polymer electrolyte membranes tend to be expensive and have several problems especially for use in motor vehicles. To overcome these problems and their high costs, the development of new proton-conducting polymer electrolytes is necessary for extensive applications. Proton-conducting polymer electrolytes based on hydrocarbon polymers or inorganic polymers are one of the most promising materials for the development of new PEFCs. These polymer materials have great variety with regard to chemical structure and can be modified chemically at very low cost. As mentioned in this review, these polymer electrolytes possess high water absorptions and proton conductivities at high temperatures and low humidities with sufficient thermal and chemical stabilities. These materials can be identified as a remarkable family of proton-conducting polymer electrolytes, which provide for new high performance PEFCs that can be operated at high temperatures without humidification. Further work will be required to develop materials with sufficient long-term stabilities and mechanical strengths, and to further optimize the fuel cell performances.

REFERENCES

1. T.A. Zawodzinski, C. Derouin, S. Radzinski, R.J. Sherman, V.T. Smith, T.E. Springer and S. Gottesfeld. *J. Electrochem. Soc.* **140** (1993) 1041–1047.
2. D.S. Watkins. In: L.J.M.J. Blomen and M.N. Mugerwa (Eds.), *Fuel Cell Systems*. Plenum Press, New York, 1993, pp. 493–530.
3. M. Higuchi, N. Minoura and T. Kinoshita. *Chem. Lett.* (1994) 227–230.
4. B. Gupta, F.N. Buchi and G.G. Scherer. *Solid State Ionics* **61** (1993) 213–218.
5. S.D. Flint and R.C.T. Slade. *Solid State Ionics* **97** (1997) 299–307.
6. J.L. Bredas, R.R. Chance and R. Silbey. *Phys. Rev. B* **26** (1982) 5843.
7. H. Kobayashi, H. Tomita and H. Moriyama. *J. Am. Chem. Soc.* **116** (1994) 3153–3154.
8. F. Wang and J. Roovers. *Macromolecules* **26** (1993) 5295–5302.
9. C. Bailly, D.J. Williams, F.E. Karasz and W.J. Macknight. *Polymer* **28** (1987) 1009–1016.
10. Z. Qi and P.G. Pickup. *J. Chem. Soc. Chem. Commun.* (1998) 15.
11. T.I. Wallow and B.M. Novak. *J. Am. Chem. Soc.* **113** (1991) 7411.
12. A.D. Child and J.R. Reynolds. *Macromolecules* **27** (1994) 1975–1977.
13. T. Kobayashi, M. Rikukawa, K. Sanui and N. Ogata. *Solid State Ionics* **106** (1998) 219–225.
14. A.J. Chalk and A.S. Hay. *J. Polym. Sci. A* **7** (1968) 691.
15. Z. Qi, M.C. Lefebvre and P.G. Pickup. *J. Electroanal. Chem.* **459** (1998) 9.
16. B.C. Johnson, I. Ylgor, M. Iqbal, J.P. Wrightman, D.R. Lliyd and J.E. McGrath. *J. Polym. Sci Polym. Chem. Ed.* **22** (1984) 72.
17. X.L. Wei, Y.Z. Wang, S.M. Long, C. Bobeczko and A.J. Epstein. *J. Am. Chem. Soc.* **118** (1996) 2545–2555.
18. K. Miyatake, E. Shouji, K. Yamamoto and E. Tsuchida. *Macromolecules* **30** (1997) 2941–2946.
19. K. Miyatake, H. Iyotani, K. Yamamoto and E. Tsuchida. *Macromolecules* **29** (1996) 6969–6971.
20. X. Jin, M.T. Bishop, T.S. Ellis and F.E. Karasz. *Br. Polym. J.* **17** (1985) 4.
21. J. Lee and C.S. Marvel. *J. Polym. Sci. Polym. Chem. Ed.* **22** (1984) 295.
22. A. Noshay and L.M. Robeson. *J. Appl. Polym. Sci.* **20** (1976) 1885.
23. M.I. Litter and C.S. Marvel. *J. Polym. Sci.: Polym. Chem. Ed.* **23** (1985) 2205.
24. H. Makowsky, R.D. Lundberg and G.H. Singahl. US Patent 3 870 841 (1975).
25. M.T. Bishop, F.E. Karasz, P.S. Russo and K.H. Langley. *Macromolecules* **18** (1985) 86.
26. J. Devaux, D. Delimoy, D. Daoust, R. Legras, J.P. Mercier, C. Strazielle and E. Neild. *Polymer* **26** (1985) 1994.
27. T. Ogawa and C.S. Marvel. *J. Polym. Sci. Polym. Chem. Ed.* **23** (1985) 1231.
28. K. Iwasaki, A. Terahara and H. Harada. Japanese Patent H11-116679 (1999).
29. M. Konishi, H. Murata and F. Yamamoto. Japanese Patent H11-67224 (1999).
30. D. Chu, D. Gervasio, M. Razaq and E.B. Yeager. *J. Appl. Electrochem.* **20** (1990) 157.
31. J. Surowiec and R. Bogoczek. *J. Therm. Anal.* **33** (1988) 1097.
32. S.R. Samms, S. Wasmus and R.F. Savinell. *J. Electrochem. Soc.* **143** (1996) 1498–1504.

33. T.D. Gierke. *J. Electrochem. Soc.* **134** (1977) 319c.
34. K.D. Kreuer. *Solid State Ionics* **97** (1997) 1–15.
35. J.J. Sumner, S.E. Creger, J.J. Ma and D.D. DesMarteau. *J. Electrochem. Soc.* **145** (1998) 107–110.
36. K.D. Kreuer. *Chem. Mater.* **8** (1996) 610.
37. M.B. Gieselman and J.R. Reynolds. *Macromolecules* **25** (1992) 4832–4834.
38. M.B. Gieselman and J.R. Reynolds. *Macromolecules* **26** (1993) 5633–5642.
39. M.B. Gieselman and J.R. Reynolds. *Macromolecules* **23** (1990) 3188.
40. X. Glipa, M.E. Haddad, D.J. Jones and J. Rozière. *Solid State Ionics* **97** (1997) 323–331.
41. K. Tsuruhara, K. Hara, M. Kawahara, M. Rikukawa, K. Sanui and N. Ogata. *Electrochim. Acta* **45** (2000) 1223–1226.
42. M. Kawahara, M. Rikukawa and K. Sanui. *Polym. Adv. Technol.* **11** (2000) 1–5.
43. M. Kawahara, M. Rikukawa, K. Sanui and N. Ogata. *Proceedings of the Sixth International Symposium on Polymer Electrolytes*, Extended Abstracts, 1998, p. 98.
44. M. Kawahara, M. Rikukawa, K. Sanui and N. Ogata. *Solid State Ionics* **136–137** (2000) 1193–1196.
45. H. Yoshida, T. Hatakeyama and H. Hatakeyama. *Polymer* **31** (1990) 693–698.
46. M. Folk. *Can. J. Chem.* **58** (1980) 1495.
47. P. Donoso, W. Gorecki, C. Berthier, F. Dfendini, C. Poinsignon and M.B. Armand. *Solid State Ionics* **28–30** (1988) 969–974.
48. G.K.R. Senadeera, M.A. Gareem, S. Skaarup and K. West. *Solid State Ionics* **85** (1996) 37–42.
49. J.R. Stevens, W. Wieczorek, D. Raducha and K.R. Jeffrey. *Solid State Ionics* **97** (1997) 347–358.
50. W. Wieczorek and J.R. Stevens. *Polymer* **38** (1997) 2057–2065.
51. P.N. Gupta and K.P. Singh. *Solid State Ionics* **86–88** (1996) 319–323.
52. M.A. Vargas, R.A. Vargas and B.-E. Mellander. *Electrochim. Acta* **44** (1999) 4227–4232.
53. R. Tanaka, H. Yamamoto, S. Kawamura and T. Iwase. *Electrochim. Acta* **40** (1995) 2421–2424.
54. D. Rodrigz, C. Jegat, O. Trinquet, J. Grondin and J.C. Lassègues. *Solid State Ionics* **61** (1993) 195–202.
55. D. Weng, J.S. Wainright, U. Landau and R.F. Savinell. *J. Electrochem. Soc.* **143** (1996) 1260–1263.
56. J.C. Lassegues, B. Desbat, O. Trinquet, F. Cruege and C. Poinsignon. *Solid State Ionics* **28–30** (1988) 969.
57. S. Petty-Weeks, J.J. Zupancic and J.R. Swedo. *Solid State Ionics* **31** (1988) 117.
58. J.C. Lassegues. In: P. Colomban (Ed.), *Proton Conductors: Solids, Membranes and Gels*. Cambridge University Press, Cambridge, UK, 1992, pp. 311–328.
59. R. Bouchet, E. Siebert and G. Vitter. *J. Electrochem. Soc.* **144** (1997) L95–L97.
60. K. Tsuruhara, M. Rikukawa, K. Sanui, N. Ogata, Y. Nagasaki and M. Kato. *Electrochim. Acta* **45** (2000) 1391–1394.
61. Y. Nagasaki and K. Kataoka. *Trend. Polym. Sci.* **4** (1996) 59–64.
62. J.-T. Wang, S. Wasmus and R.F. Savinell. *J. Electrochem. Soc.* **143** (1996) 1233–1239.
63. S.R. Samms, S. Wasmus and R.F. Savinell. *J. Electrochem. Soc.* **143** (1996) 1225–1232.
64. R.F. Savinell, E. Yeager, D. Tryk, U. Landau, J.S. Wainright, D. Weng, K. Lux, M. Litt and C. Rogers. *J. Electrochem. Soc.* **141** (1994) L46–L48.
65. J.S. Wainright, J.-T. Wang, D. Weng, R.F. Savinell and M. Lit. *J. Electrochem. Soc.* **142** (1995) L121–L123.
66. M. Kawahara, J. Morita, M. Rikukawa, K. Sanui and N. Ogata. *Electrochim. Acta* **45** (2000) 1395–1398.
67. M. Rikukawa, J. Morita, K. Sanui and N. Ogata. *Proceedings of the Fifth International Symposium on Polymer Electrolytes*, Abstracts, 1996, p. 32.
68. J. Kerres, A. Ullrich, F. Meier and T. Haring. *Solid State Ionics* **125** (1999) 243–249.
69. E.D. Powers and G.A. Serad. *High Performance Polymers: Origin and Development*. Elsevier, New York, 1986, p. 355.
70. M.F. Daniel, B. Destbat, F. Cruege, O. Trinquet and J.C. Lassegues. *Solid State Ionics* **28–30** (1988) 637.
71. K.-D. Kreuer, A. Fuchs and J. Maier. *Solid State Ionics* **77** (1995) 157.
72. Q. Quo, P.N. Pintauro, H. Tang and S. O'Connor. *J. Membr. Sci.* **154** (1999) 175–181.
73. T. Saito. Japanese Patent H11-3715 (1999).
74. I. Gautier-Luneau, A. Denoyelle, J.Y. Sanchez and C. Poinsignon. *Electrochim. Acta* **37** (1992) 1615–1618.
75. I. Honma, Y. Takeda and J.M. Bae. *Solid State Ionics* **120** (1999) 255–264.
76. J.-C. Liu, H.R. Kunz, M.B. Cutlip and J.M.Fenton. *Proceeding of the 31st Mid-Atlantic Industrial and Hazardous Waste Conference*, 1999, p. 656–662.
77. M.C. Lefebvre, Z. Qi and P.G. Pickup. *J. Electrochem. Soc.* **146** (1999) 2054–2058.

Chapter 18

Advanced materials for improved PEMFC performance and life

Dennis E. Curtin, Robert D. Lousenberg, Timothy J. Henry, Paul C. Tangeman and Monica E. Tisack

Abstract

Physical and functional attributes are reviewed for recently developed Nafion® products that satisfy emerging fuel cell requirements – including stronger, more durable membranes, and polymer dispersions of higher quality and consistency for catalyst inks and film formation. Size exclusion chromatography (SEC) analysis has confirmed that dispersion viscosity is related to an "apparent" molar mass, resulting from a molecular aggregate structure. Membranes produced with solution-casting and advanced extrusion technologies exhibit improved water management and mechanical durability features, respectively. Additionally, DuPont has shown that experimentally modified Nafion® polymer exhibits 56% reduction in fluoride ion generation, which is considered a measure of membrane lifetime.

Keywords: Nafion®; Proton exchange membrane; Polymer dispersion; Fuel cell

Article Outline

1. Introduction 412
   1.1. Nafion® PFSA polymer 412
   1.2. Nafion® PFSA membranes 412
   1.3. Nafion® PFSA polymer dispersions 413
   1.4. Polymer chemical stability 413
2. Experimental 414
   2.1. Polymer chemical stability measurements 414
   2.2. Viscosity measurements 414
   2.3. Size-exclusion chromatography (SEC) 415
   2.4. Dynamic mechanical analysis (DMA) 415
   2.5. Surface tension 415
   2.6. Contact angle 415
   2.7. Electrical shorts tolerance 416
   2.8. Accelerated lifetime 416
3. Results and discussion 416
   3.1. Polymer chemical stability 416
   3.2. Nafion® PFSA polymer dispersions 418

Fuel Cells Compendium

3.3. Nafion® PFSA solution-cast membranes 419
3.4. Nafion® ST membranes 421
4. Conclusions 423
Acknowledgments 423
References 423

1. INTRODUCTION

1.1. Nafion® PFSA polymer

DuPont introduced Nafion® perfluorinated polymer [1] in the mid-1960s. Nafion® is a copolymer of tetrafluoroethylene or "TFE", and perfluoro(4-methyl-3,6-dioxa-7-octene-1-sulfonyl fluoride) or "vinyl ether", as shown in Fig. 1. Nafion® polymer is a thermoplastic resin that can be melt-formed into typical shapes such as beads, film, and tubing. The perfluorinated composition of the copolymer imparts chemical and thermal stability rarely available with non-fluorinated polymers. The ionic functionality is introduced when the pendant sulfonyl fluoride groups (SO_2F) are chemically converted into sulfonic acid (SO_3H). The copolymer's acid capacity is related to the relative amounts of co-monomers specified during polymerization, and can range from 0.67 to 1.25 $meq \cdot g^{-1}$ (1500–800 EW, respectively).

The unique functional properties of Nafion® PFSA polymer have enabled a broad range of applications. Initially, Nafion® membranes were used for spacecraft fuel cells; however, by the early-1980s, membrane electrolysis production of chlorine and sodium hydroxide from sodium chloride emerged as the largest application for Nafion® membranes. Other important industrial applications include production of high purity oxygen and hydrogen, recovery of precious metals, and dehydration/hydration of gas streams. In addition, Nafion® super-acid catalysts are used to produce fine chemicals. Starting in 1995, DuPont began a series of process and product development programs specific to PEM fuel cell applications.

1.2. Nafion® PFSA membranes

The traditional extrusion-cast membrane manufacturing process was developed for "thick" films, typically greater than 125 μm. The extruded polymer film must be converted from the SO_2F into the SO_3K form using an aqueous solution of potassium hydroxide and dimethyl sulfoxide, followed by an acid exchange with nitric acid to the final SO_3H form [2].

Technical advances in fuel cell design and performance have increased demand for thin membranes produced at production rates that will meet the lower conversion cost goals required for fuel cell applications. Furthermore, there is a growing demand for larger production lot sizes, increased roll lengths, and improved physical appearance. To meet this need, DuPont developed a solution-casting process for supplying high-volume, low-cost membrane to the fuel cell industry that was planning automated processes for membrane electrode assemblies [3,4].

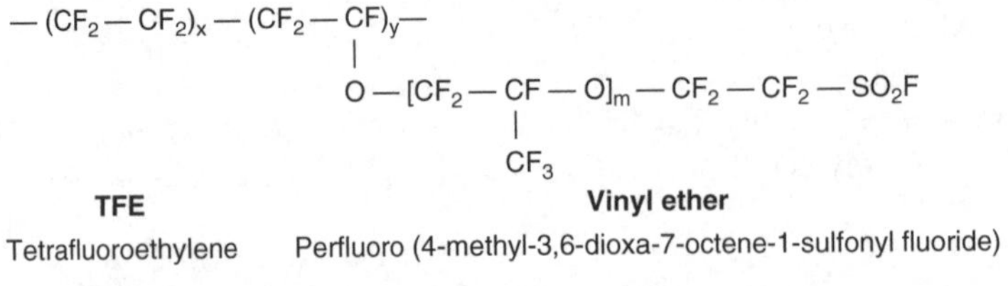

Fig. 1. Nafion® polymer structure before conversion into the sulfonic acid form

DuPont's new membrane process uses typical solution-casting technology and equipment, as shown in Fig. 2. A base film (1) is unwound and measured for thickness (2). Polymer dispersion is applied (3) to the base film, and both materials enter a dryer section (4). The composite membrane/backing film is measured for total thickness (5), with the membrane thickness the difference from the initial backing film measurement. The membrane is inspected for defects (6), protected with a coversheet (7), and wound on a master roll (8). The membrane is produced in a clean room environment [9]. Master rolls are slit into product rolls, which are individually sealed and packaged for shipment.

This process has several key advantages: (1) pre-qualification of large dispersion batches for quality (e.g., free of contamination) and expected performance (e.g., acid capacity); (2) increased overall production rates for H^+ membrane from solution-casting as compared to polymer extrusion followed by chemical treatment; and (3) improved thickness control and uniformity, including the production capability of very thin membranes (e.g., 12.7 μm).

1.3. Nafion® PFSA polymer dispersions

Two patented high-pressure processes, solvent-based [5] and water-based [6], are used to convert Nafion® polymer (sulfonic acid form) into polymer dispersions having solids contents ranging from 5% to 20% by weight. These dispersions are formulated into carbon inks and catalyst coatings, and used either "as supplied" or with modifiers [7], and/or reinforcement materials to fabricate electrode coatings and membranes [8–10].

The manufacture of polymer dispersions has undergone considerable change since first introduced by DuPont, with the recent "second generation" dispersions exhibiting more stable and consistent viscosity, improved acid capacity, and reduced metal ion content. These features enable more predictable coating formulations, consistent processing, and improved fuel cell performance. A "third generation" dispersion is in the final R&D stages, and will provide broader formulation capabilities for both solvent and polymer content. It will also allow further process simplification for preparing coatings, casting membranes and fabrication of membrane electrode assemblies.

1.4. Polymer chemical stability

The useful lifetime of a membrane is related to the chemical stability of the ionomer. While Nafion® PFSA polymer has demonstrated highly efficient and stable performance in fuel cell applications, evidence of

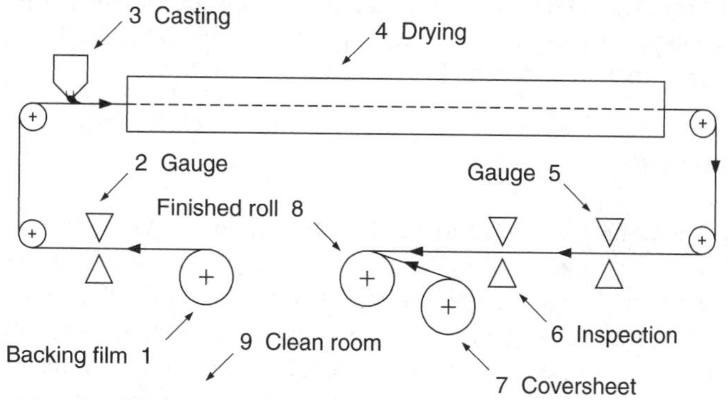

Fig. 2. Solution-casting process for Nafion® membranes

membrane thinning and fluoride ion detection in the product water indicates that the polymer is undergoing chemical attack. The fluoride loss rate is considered an excellent measure of the health and life expectancy of the membrane [11]. Peroxide radical attack on polymer endgroups [12] with residual H-containing terminal bonds is generally believed to be the principal degradation mechanism.

In this degradation mechanism, cross-over oxygen from the cathode side, or air bleed on the anode side, provides the oxygen needed to react with hydrogen from the anode side and produce H_2O_2, which can decompose to give $^{\bullet}OH$ or $^{\bullet}OOH$ radicals. These radicals can then attack any H-containing terminal bonds present in the polymer. Peroxide radical attack on H-containing endgroups is generally believed to be the principal degradation mechanism. This form of chemical attack is most aggressive in the presence of peroxide radicals at low relative humidity conditions and temperatures exceeding 90°C.

Hydroxy or peroxy radicals resulting from the decomposition of hydrogen peroxide in the fuel cell attack the polymer at the endgroup sites and initiate decomposition. The reactive endgroups can be formed during the polymer manufacturing process and may be present in the polymer in small quantities. An example of attack on an endgroup such as CF_2X, where X = COOH, is shown below.

Several proposed mechanisms include the following sequential reactions: abstraction of hydrogen from an acid endgroup to give a perfluorocarbon radical, carbon dioxide and water (step 1). The perfluorocarbon radical can react with hydroxy radical to form an intermediate that rearranges to an acid fluoride and one equivalent of hydrogen fluoride (step 2). Hydrolysis of the acid fluoride generates a second equivalent of HF and another acid endgroup (step 3).

$$R_f\text{—}CF_2COOH + {}^{\bullet}OH \rightarrow R_f\text{—}CF_2^{\bullet} + CO_2 + H_2O \tag{1}$$

$$R_f\text{—}CF_2^{\bullet} + {}^{\bullet}OH \rightarrow R_f\text{—}CF_2OH \rightarrow R_f\text{—}COF + HF \tag{2}$$

$$R_f\text{—}COF + H_2O \rightarrow R_f\text{—}COOH + HF \tag{3}$$

2. EXPERIMENTAL

2.1. *Polymer chemical stability measurements*

A sample of **Nafion®** membrane is treated in a solution of 30% hydrogen peroxide containing 20 ppm iron (Fe^{2+}) salts at 85°C for 16–20 h. The resulting solution is checked for fluoride ion content using a fluoride-specific ion electrode. The same membrane sample is treated two additional times, each treatment using fresh peroxide and iron. The results are recorded as the "total milligrams of fluoride per gram of sample" generated during the three treatment cycles. For membrane operating in fuel cells, the fluoride loss rate is reported as "micromoles fluoride ion per gram of sample per hour".

2.2. *Viscosity measurements*

Dispersion viscosity is measured using a Brookfield (Middleboro, MA) digital viscometer model LVD-VIII+ employing a wide gap concentric cylinder geometry. Using the SC4 series spindles and small sample adapter connected to a temperature controlled water bath (VWR), samples were equilibrated at 25.0 ± 0.1°C before measurements were taken. For those samples that showed shear-thinning behavior, observed viscosities and shear rates were corrected at the spindle wall using established power law relationships for viscosity, shear rate, and geometry [13]. Dispersion viscosity is reported at $40\,s^{-1}$ shear rate.

2.3. *Size-exclusion chromatography (SEC)*

Size-exclusion chromatography (SEC) molecular characterization was performed using a size-exclusion chromatograph equipped with a Waters (Bedford, MA) 2410 refractive index (RI) detector, Waters 515 HPLC pump, Waters column heater, and Rheodyne injector with 200 μl sample loop. A Precision Detectors (Franklin, MA) PD2020 light scattering (LS) detector with static light scattering at 15° and 90° ($\lambda = 800$ nm) and dynamic light scattering (DLS) at 90° was installed within the 2410 for constant temperature control (40°C). Samples were eluted through two Polymer Laboratories (Amherst, MA) SEC columns (Plgel 10 μm MIXED-B LS) maintained at 50°C. The LS detector millivolt output relative to ΔR_θ and inter-detector volume were calibrated from an average of six injections (100 μl of ~2 mg ml^{-1}) of a narrow polystyrene standard (Aldrich, product/lot # 330345, MW = 44 000 g mol^{-1}) using dimethyl formamide (DMF) as the mobile phase (0.6 ml min^{-1}).

2.4. *Dynamic mechanical analysis (DMA)*

The Dynamic mechanical analysis (DMA) responses were measured using a TA Instruments Model 2980. The test measurements used thick specimens prepared by pressure laminating eightlayers of the 2 mil Nafion® membrane at 1000 psi, 80°C between DuPont Kapton® polyimide film. Film specimens approximately 15 × 7 × 0.4 mm^3 were cut from multilayer membrane laminate for tensile DMA measurements. The test specimen was equilibrated at ambient temperature and humidity, then clamped in the tensile fixture, which is flushed with dry nitrogen and cooled to −100°C. The test chamber's relative humidity was not controlled during the DMA tests. Small-amplitude oscillatory stresses were applied at a frequency of 10 Hz, while measurements were made of the storage modulus E' and loss modulus E'' (plus the derived loss tangent, tan delta = E''/E') as a function of temperature. Each DMA test included a number of sequential heating cycles, where the maximum temperature for each cycle progressively extended to a higher maximum temperature.

2.5. *Surface tension*

The air-liquid (surface) tension for Nafion® polymer dispersions was measured at 23°C using the Wilhelmy platinum plate method. The platinum plate is pre-cleaned by flame treatment. The sample liquid is placed in a clean glass vessel, free of contaminates that may effect the surface tension of the liquid. The platinum plate (40 mm wide × 0.2 mm thick × 10 mm high) is attached to a force measuring device (Kruss K100 Tensiometer) and bought down into contact with the surface of the liquid being measured, along the 40 mm × 0.2 mm bottom edge. The plate first is submerged below the surface of the liquid to a depth of 2.0 mm to wet the plate, and then pulled back to within 10 μm of the liquid's surface. The force of the liquid pulling down on the plate (the liquid's Wilhelmy force) is measured 60 s after the plate has stopped moving. The surface tension is the Wilhelmy force divided by the wetted length of the plate (its perimeter of 80.4 mm). The cited surface tensions are averages of three measurements, and reported in milliNewton per meter (mN m^{-1}).

2.6. *Contact angle*

The contact angle data for Nafion® PFSA Membranes was obtained with a Kruss Automated Goniometer DSA10, using an environmental chamber equipped with a dew point sensor to monitor and control conditions.

Dew point was set at 12°C for measurements at 23°C, which yielded relative humidity of 50% at 23°C. For each drop of water placed on a sample membrane, contact angles were measured every 5 s for 30 min. The reported "average" contact angle was based on the average of contact angle measurements for five drops of water placed on a particular sample, as a function of time after droplet placement.

2.7. *Electrical shorts tolerance*

A resistivity test measures the membrane's electrical shorts tolerance caused by penetration of surface fibers from the gas diffusion layer. The test is performed in a constant humidity, constant room temperature with the samples conditioned at least 24 h before testing. The test apparatus consists of the inner elements of a single fuel cell, namely, the top and bottom electrode plates with lead wires and the top and bottom Pocco graphite flow fields. These elements sit on a rigid base in a constant rate of extension (CRE) test machine and are compressed with a 25.4 mm-diameter ball mounted in the center of a 6.35 mm thick rigid steel plate. The ball is pushed by a rod attached to the load cell, so the plate remains parallel to the assembly. The DC resistance of the stack is measured across the electrode plates using an ohmmeter. A square, 50.8 mm × 50.8 mm, of a gas diffusion backing (GDB) material is placed on the bottom flow field with the microporous layer facing up. The membrane is placed over the GDB and a second GDB piece is placed with the microporous layer facing down, over the membrane. The top flow field is then placed on top. The stack is centered over the bottom electrode plate and covered with the top electrode plate with the insulated side up. The CRE machine is closed so that the load cell just begins to measure load. The ohmmeter is attached to the electrode plates and the stack assembly is left to reach equilibrium as the "capacitor" charges. The test begins once the resistance reading is stable. The CRE machine is closed at a rate of 0.635 mm min^{-1}. The resistance is recorded as a function of the applied load and the pressure causing an electrical short is reported. A "failure" occurs when the electrical resistance drops below 1000 Ω.

2.8. *Accelerated lifetime*

The "time to failure" in hours is measured using a single fuel cell apparatus and proprietary testing protocols. This test is used to compare membranes and MEA designs against each other in a simulated fuel cell environment.

3. RESULTS AND DISCUSSION

3.1. *Polymer chemical stability*

Previously, DuPont had determined that fluoropolymer endgroup reactions could be minimized during extrusion processes by pre-treating the polymer with elemental fluorine [14 and 15] to remove reactive endgroups and impart greater thermal stability. When Nafion® polymer was treated in a similar manner, the number of measurable endgroups was reduced by 61%, thus providing a good candidate for chemical stability testing. Using the peroxide stability test, this treated polymer was compared with a sample of the same polymer before treatment. After >50 h of exposure, there was a 56% decrease in total fluoride ion generated per gram of treated polymer, versus the non-treated polymer, as shown in Fig. 3.

Recently, DuPont has developed proprietary protection strategies that substantially reduce both the number of polymer endgroup sites and their vulnerability to attack. Using ex situ accelerated degradation protocols, Fig. 4 shows that membranes made the modified polymer (type A and type B) exhibited 10 to 25× reduction in fluoride ion emissions when compared to the standard Nafion® N-112 membrane. The

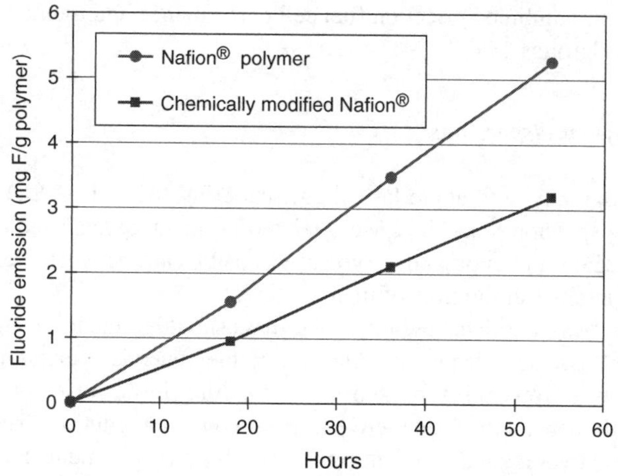

Fig. 3. Fluoride emissions for membranes made using standard and chemically modified Nafion® polymer

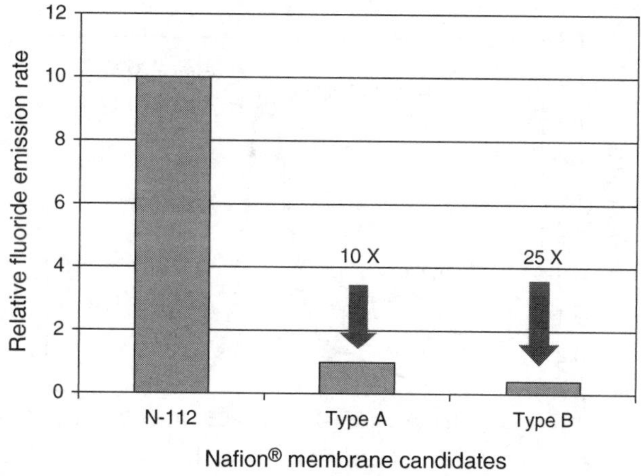

Fig. 4. Reduction in fluoride emissions for developmental Nafion® membranes made using DuPont's proprietary protection strategies

reduction in fluoride ion release was consistent with the reduction in the number of reactive polymer endgroups. It confirms that reactive endgroups are the vulnerable sites, and that the polymer can be effectively protected using DuPont's proprietary methods.

It should be noted that the ex situ accelerated degradation tests do not necessarily correlate to fuel cell accelerated degradation results. In one case, when membranes prepared from treated and non-treated polymer were tested in a fuel cell configuration using accelerated operating conditions, both membranes generated similar amounts of fluoride ion. Furthermore, clear relationships between accelerated test protocols and real-time fuel cell operating conditions have yet to be resolved satisfactorily.

DuPont continues to investigate those conditions present during PEM fuel cell operation which initiate F^- formation, including both initial and long-term release rates. The analysis includes identifying PFSA polymer and membrane features susceptible to attack, and subsequent modifications to minimize and/or eliminate polymer stress and degradation. In addition, the scope and range of fuel cell operating conditions are being assessed for their combined effect on fuel cell performance and membrane durability, including polymer structure and endgroups.

3.2. *Nafion® PFSA polymer dispersions*

Size-exclusion chromatography–low angle laser light scattering (SEC–LALLS) has been used to show that observed aqueous dispersion viscosities were related to an "apparent" molar mass, as a result of a process dependent aggregation phenomenon. Typical viscosities are between 4 and 5 cP for the nominal 10% solids aqueous dispersions at the time of manufacture.

As seen in Fig. 5, there was a noticeable high molar mass shoulder due to the aggregate structure in the mass distribution of the "as made" dispersion. On heating the aqueous dispersions to high temperatures, the aggregate structure was irreversibly broken down resulting in narrower mono-modal distributions, which were similar for all dispersions. The high temperature heating reduced viscosities to approximately 2 cP. Interestingly, limited two-angle dependent light scattering measurements had indicated a linear relationship between the radius of gyration and molar mass for the aggregate portion of the "as made" distribution.

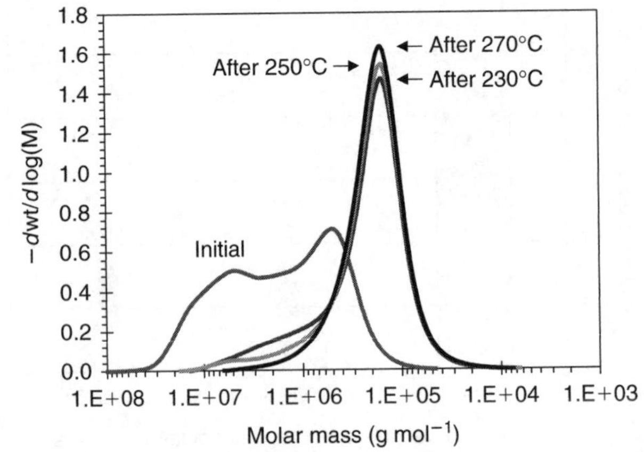

Fig. 5. Change in molar mass distributions after heating dispersions at progressively higher temperatures

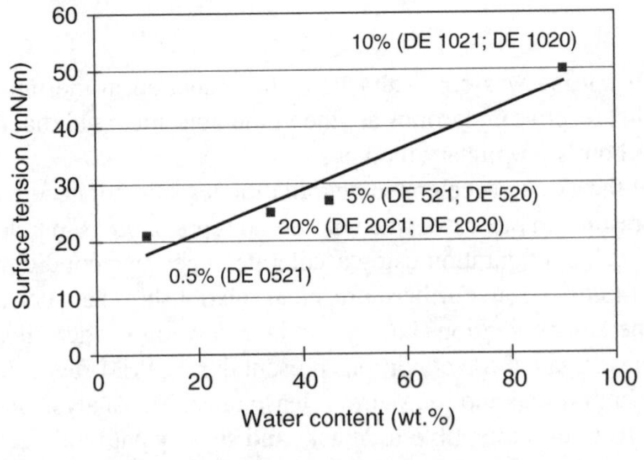

Fig. 6. Surface tension measurements Nafion® polymer dispersions

This was consistent with previous research, which concluded that the dispersion particle shape was anisotropic, possibly rod and/or ribbon form.

This evidence suggested a model for Nafion® dispersions where elongated, charge stabilized particles exist on a three-dimensional lattice with the particle centers of mass at the lattice points. Furthermore, for dispersions that had even greater aggregate distributions, viscosity-shear thinning might result as a consequence of particle overlap and lattice deformation in a shear field.

Based on this work and other process developments, DuPont has introduced a new generation of polymer dispersions offering increased acid capacities, reduced metal ion content, and improved color and viscosity stability. The new dispersions are available in 5%, 10% and 20% polymer content, two acid capacity levels, and mixed 1-propanol/water and water-only dispersions.

The surface tension of a polymer dispersion is an important consideration for optimization of catalyst coatings and efficiencies, coating adhesion, and membrane formation from dispersion. The surface tensions for the Nafion® polymer dispersions are reported in Fig. 6. The "percent (%)" value next to each data point indicates the dispersion's polymer content, which when added to the wt.% water (indicated on the graph) and wt.% alcohol (inferred) equal 100%. As expected, the aqueous dispersions exhibit the highest surface tensions ($50\,mN\,m^{-1}$); while increasing alcohol content depresses the surface tension. For example, DE 2021 (containing 20% polymer, 34% water, and 46% 1-propanol) has a surface tension of $25\,mN\,m^{-1}$. The range of possible surface tensions offers broad formulation capabilities for MEA fabrication.

3.3. Nafion® PFSA solution-cast membranes

The tensile storage modulus exhibits three distinct relaxation modes (α, β, γ) as a function of temperature, and all are sensitive to water content. Kyu and Eisenberg [16] attribute the α-relaxation to the glass relaxation of the hydrophilic phase domains, and the β-relaxation to the glass relaxation for the fluorocarbon phase domains. In Fig. 7, the DMA response for solution-cast membrane shows the dominant α-relaxation (T_g) shifting to lower values with successive heating cycles (and decreasing water content). The T_g

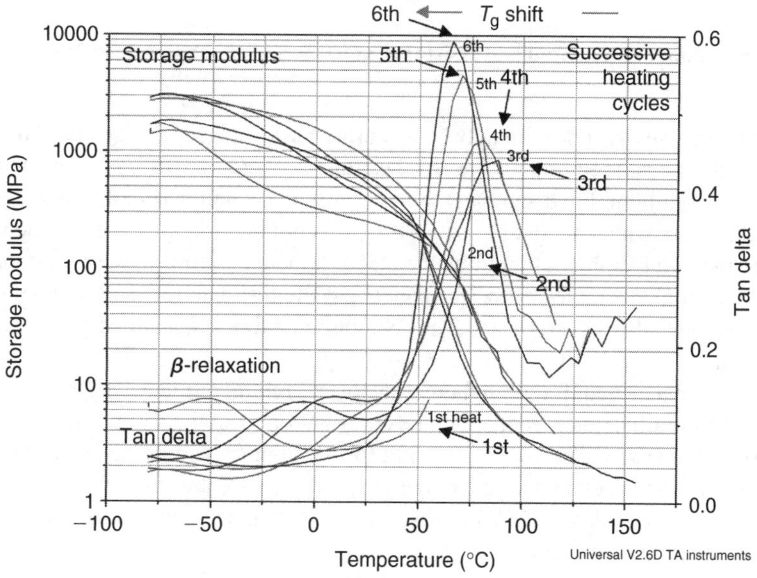

Fig. 7. DMA response data for NR-112 solution-cast membrane

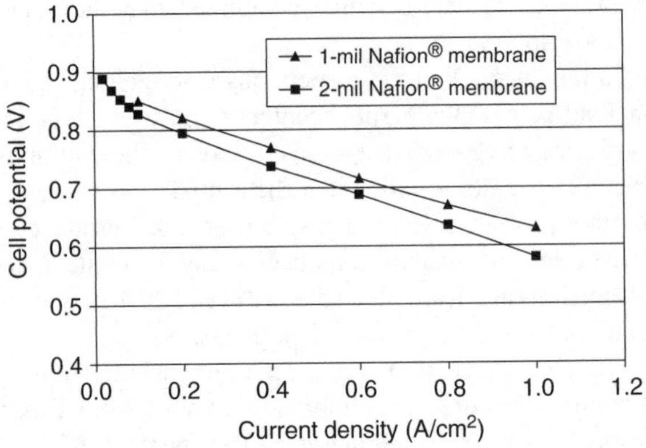

Fig. 8. Single-cell MEA performance at 65°C, 100% RH, 0 psig, H_2/air (80/60%), 0.7 mg_{Pt} cm^{-2} total loading

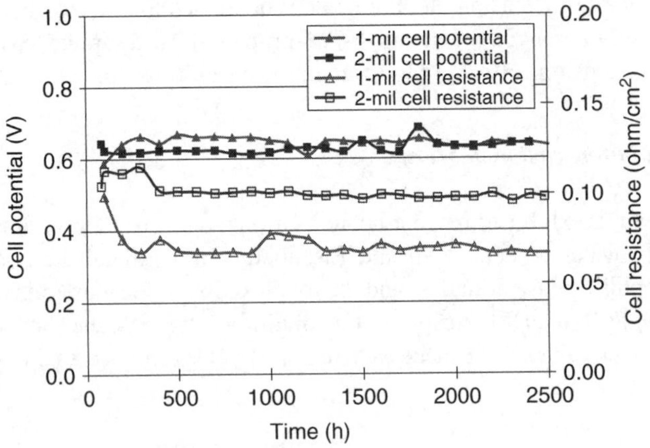

Fig. 9. MEA life test data: cell potential and resistance at 0.8 A cm^{-2} unchanged for 2500 h

starts at 85°C and decreases to approximately 65°C during the six successive cycles. This response is similar to that of extrusion-cast membrane. The β-relaxation (-50°C to 10°C) is stronger for solution-cast membrane (starting from humidified samples), but the difference attenuates after heating above 100°C in dry N_2. The γ-relaxation, which occurs below -80°C, was not measured. When the membrane sample is re-equilibrated to the initial ambient humidity, the DMA responses return to their original starting values.

Figures 8 and 9 document single-cell performance and life test data for DuPont Fuel Cells three-layer MEAs fabricated with catalyst-coated 1 and 2 mil Nafion® membranes, and using commercially available gas diffusion media.

The fuel cell performance gain for 1 mil membrane over 2 mil membrane at high current density operation under reduced humidification is more than the contribution from the cell resistance differential alone. This gain represents enhanced water back-diffusion for the thinner membrane [17]. The 1 mil membrane experiences a lower voltage decline over the range of reduced anode and cathode humidification levels, as shown in Fig. 10. This provides a 10% increase in fuel cell power output for MEAs using the 1 mil membrane versus the 2 mil membrane.

Fabrication requirements, such as lamination, reinforcement, surface coatings, and other membrane-related processes rely heavily on interfacial strength, which can be optimized by matching cohesive energies

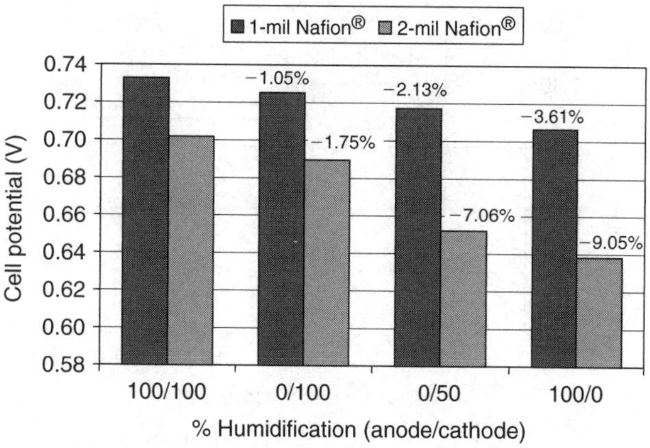

Fig. 10. Humidification effect on MEA performance at 80°C, 25 psig, H_2/air (50/50%), 0.8 A cm^{-2}, 0.7 $mg_{Pt}cm^{-2}$ total loading

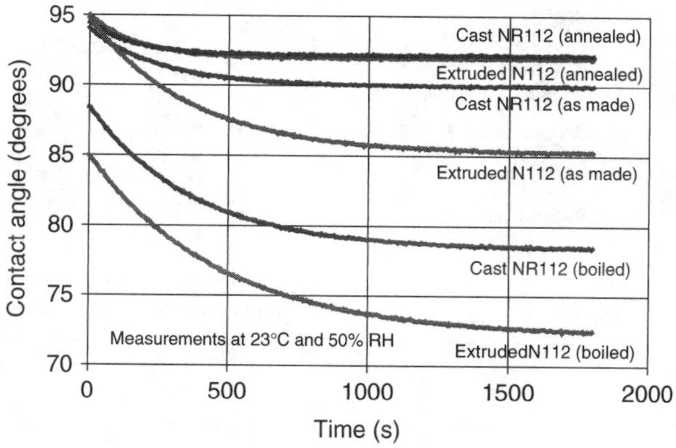

Fig. 11. Water contact angle for "as made" and treated 2-mil Nafion® membranes

of adjacent layers. This attribute for Nafion® membranes can be estimated by measuring water contact angles. For this evaluation, membrane samples were tested "as made", "annealed" at 130°C for 30 min in a dry nitrogen purged environment, and "boiled" in D.I. water for 30 min and blotted dry prior to testing.

Figure 11 illustrates the degree of change in surface characteristics achievable by various membrane treatments; as well as how the surface characteristics change with time for the various treatments. The water contact angle responses for the extruded (N-112) and solution-cast (NR-112) membrane samples were identical for the "annealed" state, but very different for the "as made" and "boiled" states. This behavior may be influenced by the membrane's prior thermal history (melt extrusion versus solution-cast) and water content, which is higher for the solution-cast (NR) membrane in the "as made" and "boiled" states; but very similar for both membranes in the "annealed" state.

3.4. *Nafion® ST membranes*

DuPont has developmental programs focused on improving membrane mechanical durability as measured by several physical property and performance indicators. These include mechanical durability, as measured

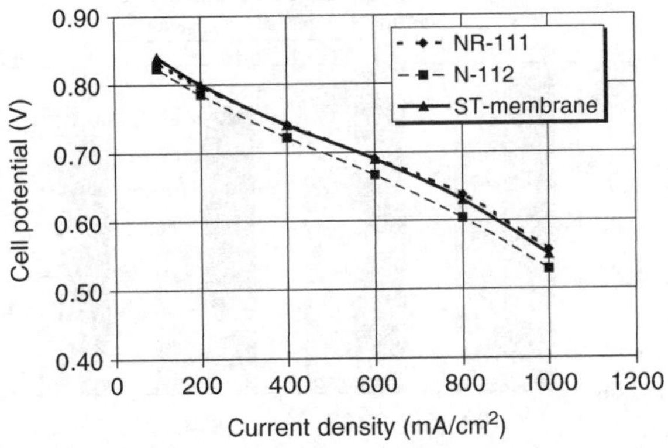

Fig. 12. Single cell MEA performance at 65°C, 100% RH, 0 psig, H_2/air (80/60%), 0.7 mg_{Pt} cm^{-2} total loading

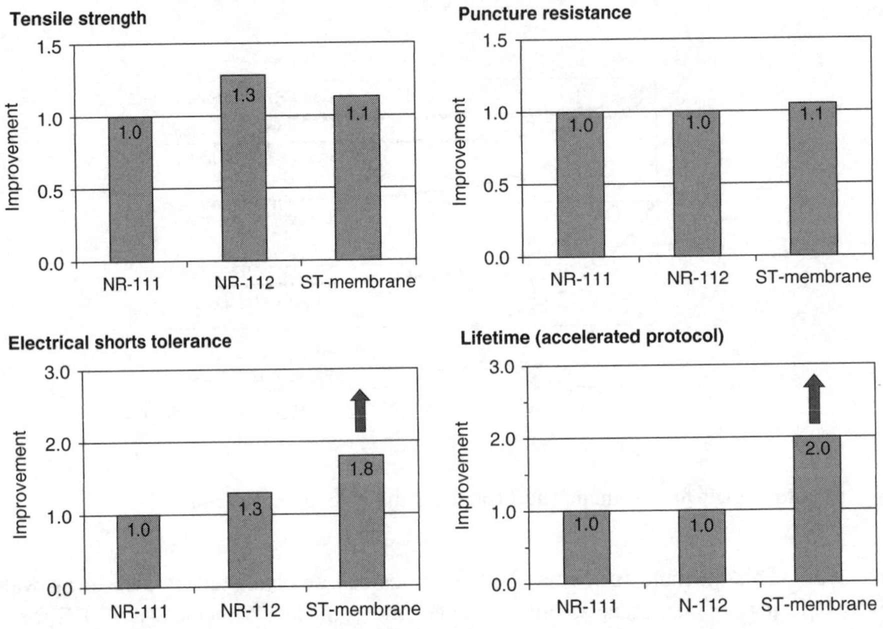

Fig. 13. Improvement comparisons between solution-cast and strengthened membranes

by tensile strength, reduced dimensional change, puncture resistance, electrical short tolerance, and lifetime (voltage over time using accelerated test protocols).

MEAs made with Nafion® ST membrane by DuPont Fuel Cells have demonstrated improved mechanical durability and lifetime, with fuel cell performance similar to Nafion® NR-111 as shown in Fig. 12.

Tensile strength and puncture resistance performance were similar for both solution-cast and ST Membrane. However, compared to NR-111, the ST membrane has 50–80% improvement in electrical shorts tolerance and 2× extended lifetime as measured using DuPont's accelerated test protocol. Figure 13 summarizes these comparisons.

The reported polymer and membrane improvements are undergoing validation and qualification in several commercial applications. Using the "voice of the customer", DuPont is determining what additional membrane improvements are needed for MEA fabrication and fuel cell applications based on customer evaluations and durability feedback. The improvements will incorporate chemical stability features as they are developed for both the Nafion® polymer and membrane.

4. CONCLUSIONS

DuPont Fuel Cell's polymer, dispersion, solution-cast and extrusion-cast membrane technologies provide the fuel cell industry with more efficient and flexible production capabilities, specialized membrane features, and reduced overall MEA fabrication costs. Today, DuPont is operating large-scale, thin membrane production facilities to provide the long-term projected membrane volumes at automotive quality standards and customer performance targets.

Ongoing polymer and membrane improvement programs are focused on providing the required performance, mechanical durability, and chemical stability necessary for successful PEM fuel cell applications. Fundamental research has enabled processing advancements for Nafion® polymer dispersions as well as formulation choices for improved fuel cell membrane fabrication and performance. Developmental ST membrane has shown favorable response based on our current mechanical durability indicators. The pipeline of new products and features is evidence of DuPont Fuel Cell's commitment and leadership in supplying membranes and components to the fuel cell industry.

ACKNOWLEDGMENTS

DuPont CR&D: David Londono and Steve Mazur; and DuPont Fuel Cells: Jayson Bauman, Gonzalo Escobedo, Kim Raiford, Eric Teather, Elizabeth Thompson and Mark Watkins. Nafion® is a DuPont registered trademark for its brand of perfluorinated polymer products made and sold only by E.I. du Pont de Nemours and Company.

REFERENCES

1. D.J. Connolly and W.F. Gresham. *Fluorocarbon Vinyl Ether Polymers*. US Patent 3 282 875 (1 November 1966).
2. R.A. Smith. *Coextruded Multilayer Cation Exchange Membranes*. US Patent 4 437 952 (20 March 1984).
3. C. Preischl, P. Hedrich and A. Hahn. *Continuous Method for Manufacturing a Laminated Electrolyte and Electrode Assembly*. US Patent 6 291 091 B1 (18 September 2001).
4. J. Kohler, K.-A. Starz, S. Wittphal and M. Diehl. *Process for Producing a Membrane Electrode Assembly for Fuel Cells*. US Patent 2002/0064593 A1 (30 May 2002).
5. W.G. Grot. *Process for Making Liquid Composition of Perfluorinated Ion Exchange Polymer, and Product Thereof*. US Patent 4 433 082 (21 February 1984).
6. D.E. Curtin and E.G. Howard Jr. *Compositions Containing Particles of Highly Fluorinated Ion Exchange Polymer*. US Patent 6 150 426 (21 November 2000).
7. W.G. Grot and G. Rajendran. *Membranes Containing Inorganic Fillers and Membrane and Electrode Assemblies and Electrochemical Cells Employing same*. US Patent 5 919 583 (6 July 1999).
8. W.G. Grot. *Process for Making Articles Coated with a Liquid Composition of Perfluorinated Ion Exchange Resin*. US Patent 4 453 991 (12 June 1984).
9. S. Banerjee. *Fuel Cell Incorporating a Reinforced Membrane*. US Patent 5 795 668 (18 August 1998).
10. J.E. Spethmann and J.T. Keating. *Composite Membrane with Highly Crystalline Porous Support*. US Patent 6 110 333 (29 August 2000).
11. R. Baldwin, M. Pham, A. Leonida, J. McElroy and T. Nalette. Hydrogen-oxygen proton-exchange membrane fuel cells and electrolyzers. *J. Power Sources* **29** (1990) 399–412.

12. M. Pianca, E. Barchiesi, G. Esposto and S. Radice. End groups in fluoropolymers. *J. Fluorine Chem.* **95** (1999) 71–84.
13. H.A. Barnes, J.F. Hutton and K. Walters. *An Introduction to Rheology*. Elsevier, New York, 1989.
14. R.A. Morgan and W.H. Sloan. *Extrusion Finishing of Perfluorinated Copolymers*. US Patent 4 626 587 (2 December 1986).
15. J.F. Imbalzano and D.L. Kerbow. *Stable Tetrafluoroethylene Copolymers*. US Patent 4 743 658 (10 May 1988).
16. T. Kyu and A. Eisenberg. In: A. Eisenberg and H.L. Yeager (Eds.), *Mechanical Relaxations in Perfluorosulfonate Ionomer Membranes, Perfluorinated Ionomer Membranes*. ACS Symposium Series 180, American Chemical Society, Washington, DC, 1982 (Chapter 6).
17. T.J.P. Freire and E.R. Gonzalez. Effect of membrane characteristics and humidification conditions on the impedance response of polymer electrolyte fuel cells. *J. Electroanal. Chem.* **503** (2001) 57–68.

Chapter 19
Polymer–ceramic composite protonic conductors

B. Kumar and J.P. Fellner

Abstract

This paper reviews emerging polymer–ceramic composite protonic conductors in the context of their usefulness as membrane material for fuel cells. These composite protonic conductors appear to exhibit a superior propensity to retain water, enhanced conductivity, superior thermal and mechanical robustness, and reduced permeability of molecular species.

Keywords: Fuel cells; Composite membrane; PEFC

Article Outline

1. Introduction 425
2. Prior work: chemistry, processing, and properties 426
3. Discussion 427
   3.1. Water retention 427
   3.2. Polymer–ceramic particle interaction and microstructure 427
   3.3. Transport of charged species in a composite material 428
4. Thermal and mechanical robustness 430
5. Permeability of molecular species 430
6. Summary and conclusions 431
Acknowledgments 432
References 432

1. INTRODUCTION

Composites are an important class of materials. They are comprised of two or more phases mixed in predetermined proportions to obtain superior performance as compared to any of the pure, single-phase, solid components. They have found widespread application as materials of construction in structural components requiring superior mechanical and thermal properties. However, the application of composites as electrical conductors, both electronic and ionic, is in its infancy. It has been shown that the dispersion of fine, electronically conducting particles into an insulator matrix and insulating particles into an ionically conducting matrix leads to enhancements in electrical conductivity by several orders of magnitude. In both cases, a significant concentration (10–30 vol%) of filler particles is required to achieve the optimum conductivity. To explain the composite effect on electrical conductivity, a unified model has recently been proposed for the two types of composite conductors [1].

Fuel Cells Compendium

The use of solid polymer–ceramic composite materials as protonic conductors has recently attracted significant interest [2–5]. The motivation for the interest is a commercial application as high conductivity and thermally stable membrane material for polymer electrolyte fuel cells (PEFCs). Preliminary investigations have shown that these polymer–ceramic composites are associated with attributes, such as enhanced protonic conductivity, improved water retention, increased cell operating temperature, higher carbon monoxide tolerance threshold, and reduced permeability of molecular species. The improved thermomechanical stability of these membranes in a practical fuel cell is expected to provide them with long-term structural integrity. The polymer–ceramic composite protonic conductors are now a subset of the general class of solid protonic conductors and possess the potential to provide membrane materials for commercial applications.

The history of the composite ionic conductors may be traced to the work of Liang [6]. In a pioneering work, Liang [6] investigated polycrystalline lithium iodide doped with aluminum oxide and reported that lithium iodide doped with 35–45 mol% aluminum oxide exhibited conductivity on the order of 10^{-5} S cm^{-1} at 25°C, about three orders of magnitude higher than that of the LiI conductivity. However, no significant amount of aluminum oxide was determined to be soluble in LiI; thus, the conductivity enhancement could not be explained by the classical doping mechanism and creation of Schottky defects such as in the LiI–CaF_2 system. Subsequently, a number of investigations have reported enhanced conductivity of silver in the AgI–Al_2O_3 system [7], copper in the CuCl–Al_2O_3 system [8], fluorine in the PbF_2–SiO_2 and PbF_2–Al_2O_3 systems [9], and lithium in polymer–ceramic composite electrolytes [10]. Two review papers [1,11] also document the developmental history and general characteristics of these fast ionic conductors. Analyses of these reviews point out that a new conduction mechanism evolves, which augments the bulk conductivity of single-phase ionic conductors. The new conduction mechanism makes use of interfacial and/or space charge regions between the two primary components.

The purpose of this paper is to review the literature on composite protonic conductors in the context of their usefulness as a membrane material, specifically their ability to retain water, enhance conductivity, augment thermal and mechanical robustness, and suppress the permeability of molecular species in a practical, commercially-viable fuel cell.

2. PRIOR WORK: CHEMISTRY, PROCESSING, AND PROPERTIES

Since Nafion® has been the mainstay of protonic conductors, a number of studies on composite protonic conductors have been conducted using it as a matrix for reinforcing ceramic particles. Mauritz et al. [12] reported a processing method for producing nanocomposites of Nafion® and silica using the sol–gel reactions. They observed that the polar/nonpolar nanophase-separated morphological template exists despite incorporation of silicon oxide phase in the composite. The highest silicon oxide concentration was observed near the surface and decreased to a minimum in the center of the specimen. Watanabe et al. [2] extended the work of Mauritz et al. [12] toward the applicability of the nanocomposite for a fuel cell membrane. Watanabe et al. [2] characterized these nanocomposites as "self-humidifying," as colloidal silica (silica gel) possesses an inherent capacity to absorb and retain water. They also reported that in addition to water retention capability, silica particles also suppressed H_2 and O_2 cross-over through the membrane. The use of the nanocomposite membrane also facilitated cold starts of the cell. Antonucci et al. [3] reported the use of Nafion® and silica nanocomposite in direct methanol fuel cells (DMFCs) at 145°C. The favorable humidification conditions in the nanocomposite allowed high operating temperatures and enhancement of methanol oxidation kinetics. A peak power density of 240 mWcm^2 for the oxygen-fed fuel cell was obtained. Park and Nagai [13] fabricated fast protonic organic–inorganic hybrid nanocomposites from the hydrolysis and condensation reaction of 3-glycidoxypropyltrimethoxy silane and tetraethyleorthosilicate. The protonic conductivity of the composite increased up to 1.0×10^{-1} S cm^{-1} by the addition of silicotungstic acid. The sol–gel derived Nafion®/silica composite membrane was also investigated by Miyake et al. [4] as a membrane material for fuel cells. These membranes exhibited higher water contents at 25 and 120°C but not at 150 and 170°C. Despite the higher water content, the protonic conductivity of the membranes were lower or equal to unmodified Nafion® membranes. Miyake et al. [14] also investigated uptake and permeability of methanol in liquid and vapor phases as a function of temperature. They concluded that the Nafion®/silica hybrid membranes with high silica content (≅20 wt.%) are potentially useful as membranes for direct methanol fuel cells using either liquid- or vapor-feed fuels. Adjemian et al. [5] also investigated silicon oxide/Nafion® composite membranes in hydrogen/oxygen proton-exchange membrane fuel cells from 80 to 140°C. All composite membranes had a silicon oxide content of less that or equal to 10 wt.%. The silicon oxide enhanced the water retention of the composite membranes and contrary to the report of Miyake et al. [4] these membranes exhibited increased protonic conductivity at elevated temperatures. They also exhibited impressive current densities – four times greater than unmodified Nafion® at 130°C and a pressure of 3 atm. Furthermore, these membranes were structurally and mechanically robust in comparison to unmodified Nafion®, which degraded after a higher operating temperature and thermal cycling. Uchida et al. [15] reported attributes of titanium dioxide/Nafion® composites prepared by the sol–gel reactions. They reported increased water absorbability and self-humidifying characteristics by dispersing only 2 wt.% TiO_2.

Analyses of the aforementioned investigations appear to suggest that the incorporation of a ceramic phase in the polymer matrix provides some benefits such as superior water retention, higher operating temperature, and enhanced thermal stability. However, a scientific and quantitative basis for the formulation and optimization of the composites is lacking. The qualitative approach of composite formulation and nonstandard processing techniques may also account for the deviations in properties from the aforementioned general observations.

3. DISCUSSION

3.1. *Water retention*

Silicates and aluminosilicates are known to be associated with various concentrations of water. Silica gel, for example, is a prominent desiccant. It can adsorb water and also react to form silicic acid. The substitution of silicon by aluminum in the silica network leads to the formation of a variety of aluminosilicate minerals (felspar, clay, zeolites, and mica) in nature [16]. These minerals can be associated with various concentrations of water. For example, clay ($Al_2O_3 \cdot 2SiO_2 \cdot 2H_2O$) is a layered silicate and the chemically held water is retained up to 560°C. Similarly, a cage-type structure of zeolites can physically hold significant concentrations of water.

A composite membrane also possesses a very high concentration of polymer–ceramic interfaces. These interfaces possess a defect structure and free volume, which can accommodate significant concentration of water. The interfaces can serve as a water reservoir, and it is likely that non-silicates and non-aluminosilicates can also improve water retention.

In and around a fuel cell membrane, water formation, retention, and movement at elevated temperatures must be regulated in a precise manner. The incorporation of a ceramic component is expected to facilitate the water management issue. However, water that is physically held in voids and interfaces may be lost around its boiling point. The structural or chemically held water may be useful for protonic conductivity up to its decomposition temperature.

3.2. *Polymer–ceramic particle interaction and microstructure*

An interaction between a polymer chain and a ceramic particle is schematically illustrated in Fig. 1. The length of the polymer chain will depend upon its molecular weight. The size of the ceramic particle may

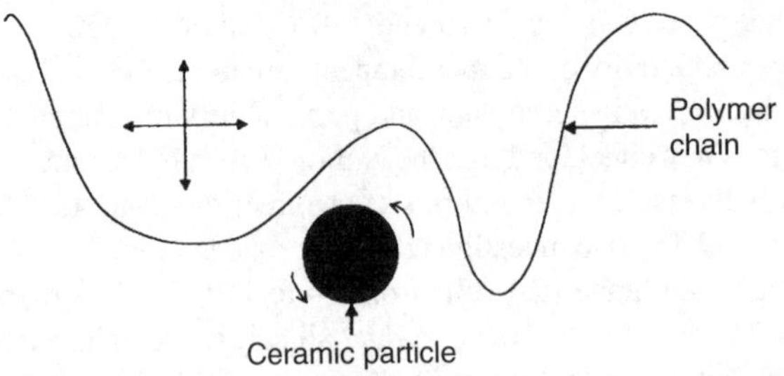

Fig. 1. Schematic representation of a polymer chain segment and ceramic particle interaction

vary from a few nanometers to several micrometers. The shape of the ceramic particle may also vary. The extent of the polymer chain–ceramic particle interaction will depend upon the polymer chain length and the ceramic particle size. For example, if the polymer chain length (generally of the order of microns) and weight percent of ceramic particles are maintained constant and the particle size of the ceramic phase is reduced from 10 μm to 10 nm, the number of polymer–ceramic interaction sites is increased by a factor of 10^9 and the distance between the ceramic particles is reduced from 4.12 μm to 4.12 nm. These approximations are based on the assumptions that the agglomeration of ceramic particles is absent and particles are uniformly distributed.

A difference in the dielectric constants of the polymer and ceramic phases may lead to a chemical interaction between them resulting in the formation of chemical bonds. In fact, the formation of chemical bonds in PEO:$LiBF_4$–MgO, a lithium ion composite material, has been demonstrated [17].

It has been shown that in the PEO:$LiBF_4$–MgO system, lithium ion conductivity is enhanced initially by reducing the degree of polymer crystallinity and subsequently by facilitating the polymer chain–ceramic particle interaction. The interaction is further enhanced by reducing the particle size and mass of the ceramic particle [10]. A similar interaction and microstructure are expected in composite protonic conductors, which may facilitate the protonic conduction.

There is a similarity in the microstructure of polymer–ceramic composites and Nafion® – the mainstay of solid polymer protonic conductors. Nafion® is produced by attaching a side chain to polytetrafluoroethylene (PTFE). The end of the side chain is sulfonated, which is highly hydrophyllic. The side chain molecules of Nafion® tend to cluster, thus creating hydrophyllic regions within a generally hydrophobic material, PTFE. The hydrophyllic regions allow for the absorption of large quantities of water – up to 50 wt.% of dry Nafion®. The inhomogeneous nature of the microstructure and regions for storage of water resemble characteristics of the two classes of materials.

3.3. Transport of charged species in a composite material

A composite material in which metallic particles are dispersed in an insulating matrix displays the electronic conductivity enhancement as a function of weight fraction of the metallic filler particle as depicted in Fig. 2a. Initially, there is little influence on conductivity as the metallic particles are introduced, but around 20 wt.% of the metallic component, there is a sharp jump – about 10 orders of magnitude – in conductivity. At this concentration level of the metal phase, the microstructure allows steady-state percolation of electrons and thus it is defined as the percolation threshold. The percolation threshold varies (shown by arrows in Fig. 2a) depending upon the nature of the metal and insulating phases. The steady-state conductivity of the composite beyond the percolation threshold is comparable to the conductivity of the metal phase.

Figure 2b schematically shows the effect of insulating particle reinforcement on the conductivity of the ionically conducting matrix. Unlike Fig. 2a, the ionic conductivity of the composite in this case gradually increases and reaches a peak at around 20 wt.% of the insulating doping phase. Further increases of the

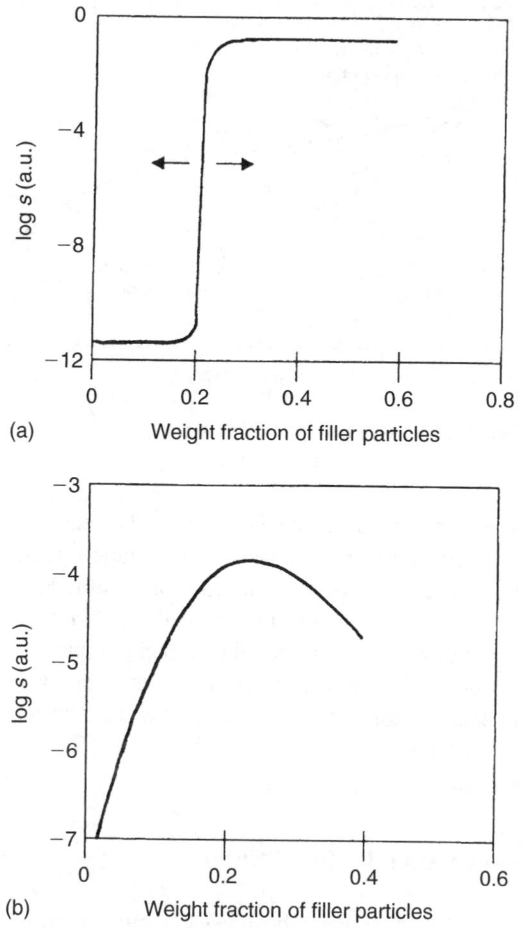

Fig. 2. Schematic representation of the effect of reinforcement on conductivity. (a) Electronic conductivity of insulating matrix reinforced with metallic particles; (b) ionic conductivity of ionically conducting matrix reinforced with insulating particles

dopant decreases the conductivity as it impedes the transport of charged species. Again in this case, steady-state percolation occurs around 20 wt.% of the insulating dopant phase. The percolation threshold may vary depending upon the matrix and dopant chemistries, particle sizes, and processing parameters. This composite effect has been demonstrated in a number of diverse ionic conductors [6–11].

The schematic data shown in Figs 2a and b are based on a number of theoretical and experimental investigations. A review of these investigations can be found in [1,11]. The intent of presenting the electronic and ionic conductivity of composite materials in Figs 2a and b is to impress upon readers the role of microstructures on the transport of charged species. In a fuel cell membrane, the lowest electronic conductivity (preferably zero) is desired. However, for electrode materials, mixed (electronic and ionic) conductivity is required. It is anticipated that the transport of protons should also be facilitated in the composite microstructure. However, authentic experimental evidence correlating the protonic conductivity and composite microstructure is lacking.

In the fuel cell literature, one of the widely accepted protocols for evaluating a membrane material is to obtain polarization curves (*V–j* plots) as schematically shown in Fig. 3. The figure illustrates regions in

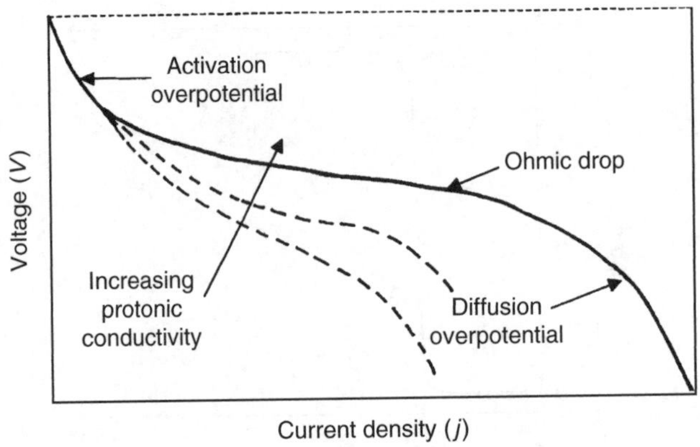

Fig. 3. Schematic of cell voltage (*V*) vs. current density, (*j*) of a typical cell

which various types of voltage losses occur. From Fig. 3 it can be seen that at low current densities, the major contribution to the losses originate from activation polarization, which is characterized by a sharp drop in voltage with increasing current. As the current increases, ohmic loss emerges, as exhibited by linearity in the central region of Fig. 3. At high currents, the cell resistance is controlled by mass–transport limitation (diffusion overpotential), resulting in a rapid decline in voltage.

An increase in the protonic conductivity of the membrane is reflected by a change in the slope of the linear ohmic region. The composite protonic conductors are expected to have lower slope in the ohmic region. In fact, there is evidence of this kind of behavior in Nafion 115 and 6% SiO_2 composite membranes investigated by Adjemian et al. [5].

4. THERMAL AND MECHANICAL ROBUSTNESS

Improvements in the thermal and mechanical robustness of commercial polymers have been a topic of considerable interest. A number of different types of fillers such as clay, carbon, and mica have been incorporated in polymer matrices to improve wear resistance, retard flammability, and enhance the heat distortion temperature of industrial polymeric products. In recent times, the heat resistance of a commercial Nylon 6 has been improved by incorporating clay [18–20]. The Nylon 6 begins to soften at 60°C; however, the addition of 3–5 wt.% of clay raises the softening temperature to 140°C. Furthermore, the clay additive improved dimensional stability, enhanced barrier properties, and retarded flammability. In general, the addition of a ceramic phase to a polymer matrix raises the glass transition temperature, T_g [21]. The modulus of elasticity also increases with the incorporation of a ceramic phase in polymers [22].

It is evident that ceramic additives impart large positive influences on the mechanical and thermal properties of polymers. These benefits are of enormous interest in developing robust, durable, and high performance membrane materials for fuel cells.

5. PERMEABILITY OF MOLECULAR SPECIES

The permeation of molecular species, either from anode or cathode side through the electrolyte, is detrimental to the electrochemical performance of a fuel cell. The permeability of the species through the membrane must be reduced to a minimum, preferably zero. The ceramic additives in a polymer matrix have been found to be effective in reducing permeability of molecular species. Miyake et al. [14] reported

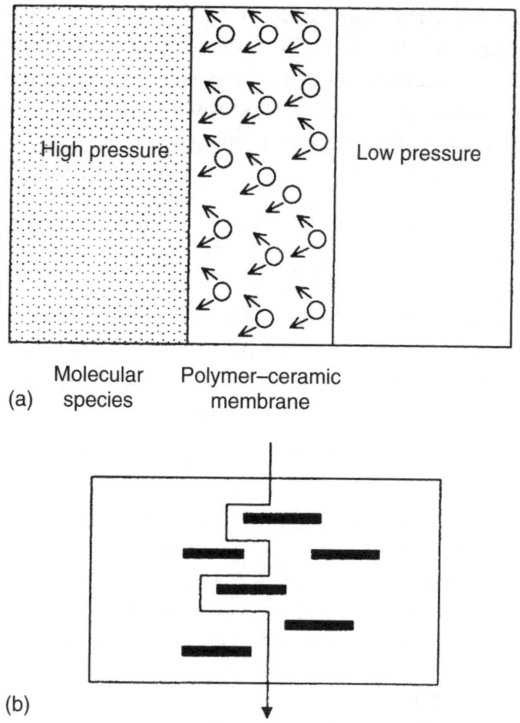

Fig. 4. A schematic of a polymer–ceramic membrane showing (a) backscattering, and (b) tortuous pathway of molecular species

that a hybrid membrane containing 20 wt.% silica in a Nafion® matrix exhibited significantly lower methanol permeation rates.

The mechanism for reducing the permeability of gases through the membrane is believed to originate from rigid scattering sites and tortuous pathways that a permeant must encounter to transverse the composite material. Fig. 4a schematically illustrates backscattering of the molecular species by rigid spherical ceramic particles in a polymer matrix, whereas Fig. 4b depicts a tortuous pathway, as proposed earlier by Yano et al. [23] in a composite membrane containing a platelet type ceramic phase, for example, clay in a polymer.

6. SUMMARY AND CONCLUSIONS

A review of the state-of-the-art polymer–ceramic protonic conductors for application as membranes in fuel cells have been presented and discussed. The discussion focused on the key attributes of the membrane such as water retention, polymer–ceramic particle interaction, transport of charged species through a composite structure, thermal and mechanical properties, and permeability of molecular species. An analysis of a broader range of ceramic fillers reveals that while silicates and aluminosilicates have an inherent capacity to retain water, non-silicates and non-aluminosilicates can enhance water retention by providing polymer–ceramic interfacial regions as water storage channels. The polymer and ceramic phases can also chemically interact, leading to nanostructures and microstructures beneficial for protonic conductivity and mechanical and thermal properties. The microstructure of solid composites facilitates transport of charged species provided that there is a large difference in the electronic properties and structure of the components of the composite. The observation is valid for an electron and also for a number of ionic species. Thus, it is expected that the structure of polymer–ceramic composites should enhance protonic conductivity; however, definitive experimental evidence is yet to emerge. A number of reports in the literature support the claim that a ceramic phase in the composite enhances mechanical and thermal properties. It has also been suggested that the polymer–ceramic composites should exhibit suppressed permeability of molecular species.

ACKNOWLEDGMENTS

One of the authors (B. Kumar) gratefully acknowledges the financial support provided by the Propulsion Directorate, Air Force Research Laboratory, under Contract no. F33615-98-D-2891, DO 18. The authors also express their gratitude to Dr. S.J. Rodrigues for reading the manuscript.

REFERENCES

1. A. Mikrajuddin, G. Shi and K. Okuyama. *J. Electrochem. Soc.* **147(8)** (2000) 3157–3165.
2. M. Watanabe, H. Uchida, Y. Seki and M. Emori. *J. Electrochem. Soc.* **143(12)** (1996) 3847–3852.
3. P.L. Antonucci, A.S. Arico, P. Creti, E. Ramunni and V. Antonucci. *Solid State Ionics* **125** (1999) 431–437.
4. N. Miyake, J.S. Wainright and R.F. Savinell. *J. Electrochem. Soc.* **148(8)** (2001) A898–A904.
5. K.T. Adjemian, S.J. Lee, S. Srinivasan, J. Benziger and A.B. Bocarsly. *J. Electrochem. Soc.* **149(3)** (2002) A256–A261.
6. C.C. Liang. *J. Electrochem. Soc.* **120** (1973) 1289.
7. K. Shahi and J.B. Wagner. *Solid State Ionics* **3–4** (1981) 295.
8. T. Jow and J.B. Wagner. *J. Electrochem. Soc.* **126** (1979) 1963.
9. K. Hariharan and J. Maier. *J. Electrochem. Soc.* **142(10)** (1995) 3469.
10. B. Kumar, S.J. Rodrigues and L.G. Scanlon. *J. Electrochem. Soc.* **148(10)** (2001) A1191.
11. P. Knauth. *J. Electroceram.* **5(2)** (2000) 111–125.
12. K.A. Mauritz, I.D. Stefanithis, S.V. Davis, R.W. Scheetz, R.K. Rope, G.L. Wilkes and H. Huang. *J. Appl. Polym. Sci.* **55** (1995) 181–190.
13. Y. Park and M. Nagai. *J. Electrochem. Soc.* **148(6)** (2001) A616–A623.
14. N. Miyake, J.S. Wainright and R.F. Savinell. *J. Electrochem. Soc.* **148(8)** (2001) A905–A909.
15. H. Uchida, Y. Ueno, H. Hagihara and M. Wantanabe. *J. Electrochem. Soc.* **150(1)** (2003) A57–A62.
16. W.D. Kingery, H.K. Bowen and D.R. Uhlmann. *Introduction to Ceramics*, Wiley, New York, 1976, pp. 77–80.
17. B. Kumar, S.J. Rodrigues and R.J. Spry. *Electrochim. Acta* **47** (2002) 1275–1281.
18. Y. Kojima, A. Usuki, M. Kawasumi, A. Okada, Y. Fukushima, T. Kuranchi and O. Kamigaito. *J. Mater. Res.* **8** (1993) 1185–1189.
19. Y. Kojima, A. Usuki, M. Kawasumi, A. Okada, T. Kuranchi and O. Kamigaito. *J. Appl. Polym. Sci.* **49** (1993) 1259–1264.
20. Y. Kojima, K. Fukumori, A. Usaki, A. Okada and T. Kuranchi. *J. Mater. Sci. Lett.* **12** (1993) 889–890.
21. B. Kumar and L.G. Scanlon. *J. Power Sources* **52** (1994) 261–268.
22. F. Rodriguez. *Principles of Polymer Systems*, Hemisphere Publishing Corporation, New York, 1989, p. 254.
23. K. Yano, A. Usuki, A. Okada, T. Kuranchi and O. Kamigaito. *J. Polym. Sci., Part A: Polym. Chem.* **31** (1993) 2493–2498.

Chapter 20

Recent developments in high-temperature proton conducting polymer electrolyte membranes

Patric Jannasch

Abstract

Progress in the area of proton conducting polymer electrolyte membranes is intimately linked with the development of polymer electrolyte membrane fuel cells, and is today largely driven by the insufficient properties of humidified Nafion® membranes at temperatures above 100°C. Recent developments in the field include new ionomers and hybrid membranes containing inorganic nanoparticles to control morphology and enhance water retention, as well as improved systems based on the complexation of basic polymers with oxo-acids. In addition, the molecular design and synthesis of completely new all-polymeric electrolytes that rely entirely on structure diffusion of the protons holds great promise in the long perspective.

Keywords: Ionomers; Acid–base complexes; Organic–inorganic hybrids; Proton conductivity; Proton exchange membrane fuel cells

Article Outline

1. Introduction . . . 433
2. Ionomers and ionomer membranes . . . 434
3. Organic–inorganic hybrid membranes . . . 435
4. Membranes based on polymers and oxo-acids . . . 437
5. All-polymeric electrolytes . . . 438
6. Theoretical studies . . . 438
7. Conclusions . . . 439
Acknowledgements . . . 439
References and recommended reading . . . 439

1. INTRODUCTION

Conventional proton conducting polymer electrolyte membranes (PEMs) are based on hydrated ionomers in their protonated form. These materials are typically phase separated into a percolating network of hydrophilic nanopores embedded in a hydrophobic polymer-rich phase domain. The hydrophilic nanopores contain water and the acidic moieties, and conductivity occurs via transport of dissociated protons by the dynamics of the water. The hydrophobic phase domain provides mechanical strength by stabilizing the morphology of the membrane.

Fuel Cells Compendium

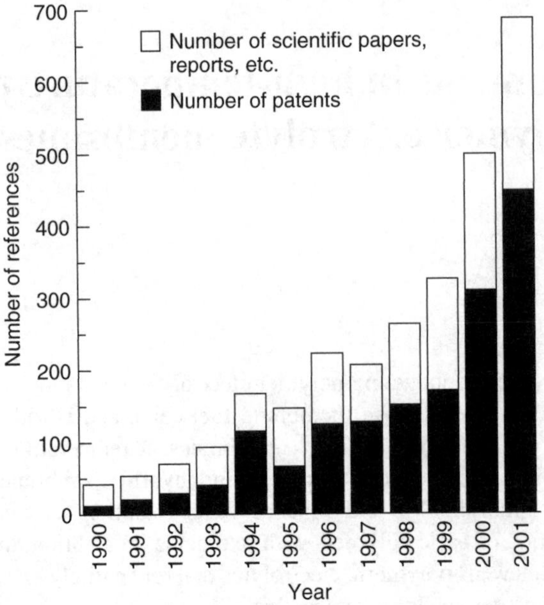

Fig. 1. The number of references found in the scientific database "CAplus" by using the ACS search facility "SciFinder® Scholar 2000" and a combination of the words: "polymer", "fuel", and "cell"

Proton conducting PEMs attract considerable attention because they are key-components in PEM fuel cells (PEMFCs), which are promising environmentally friendly and efficient power sources for a wide range of different applications [1*]. During normal H_2/O_2 PEMFC operation, anodic dissociation of H_2 produces protons that are transported through the hydrated PEM to the cathode, where reduction of O_2 produces water. Potentially, this type of fuel cell gives no emissions, and the use of ozone-depleting petroleum based fuels can thus be avoided. The research activities on PEMFCs have increased progressively during the last decade, and are today rather extensive as seen by the accelerating number of yearly publications (Fig. 1).

The primary demands on the hydrated PEM are high proton conductivity (at least above 0.01 S cm^{-1}), low fuel and O_2 permeability, and high chemical, thermal and mechanical stability. Conventional PEMFCs typically operate with Nafion® membranes, which offers quite good performance below 90°C. However, to decrease the complexity and increase the efficiency and CO-tolerance of the PEMFC system, there is today a strong need for PEMs capable of sustained operation above 100°C. Unfortunately, the proton conductivity of Nafion® suffers greatly at temperatures above 90°C due to loss of water. Also, the barrier properties of this membrane are usually insufficient when methanol is used as fuel. These factors, in addition to the high cost of Nafion®, has triggered an extensive research for alternative PEM materials, some relying on other species than water for proton conduction [2–4*]. The present review focuses on the development of different types of high-temperature proton conducting PEMs during the last 2 years.

2. IONOMERS AND IONOMER MEMBRANES

Ionomer-based membranes intended for high-temperature PEMFCs should preferably retain a high conductivity at low levels of humidification. There is thus a need to improve water retention at high temperatures and to improve performance at low water contents, while simultaneously giving special attention to chemical as well as morphological stability to resist excessive water swelling. The membrane morphology is important for the performance, and is linked to the nature of the ionomer and the membrane formation process in a quite complex manner. It typically depends strongly on the water content, and on the concentration and distribution of the acidic moieties [4*,5,6]. For example, Kreuer has shown that hydrated membranes based on sulfonated polyetherketone have a less pronounced separation into hydrophilic/hydrophobic domains, as well as a larger distance between the acidic moieties, as compared to the Nafion® membrane [4*].

The majority of the new ionomers developed currently are based on different arylene main-chain polymers, which are characterized by excellent thermal, chemical, and mechanical properties. Some of these ionomers are shown in Fig. 2. Several research groups are working with different sulfonated polymers containing diaryl-sulfone units [7*,8–11]. For example, Wang et al. have prepared high molecular weight polysulfones containing randomly distributed disulfonated diarylsulfone units [7*]. Analysis of the membrane morphology by atomic force microscopy revealed hydrophilic phase domains that increased in size, from 10 to 25 nm, with increasing degree of sulfonation. The membranes were stable up to 220°C in air, and highly sulfonated ones showed conductivities of 0.17 S cm^{-1} at 30°C in water [7*]. Poppe et al. have produced flexible PEMs based on carboxylated and sulfonated poly(arylene-co-arylene sulfone)s [8]. As expected, the carboxylated materials showed lower water uptake and lower conductivity in comparison with the sulfonated ones. Sulfonated polysulfones have also been blended with basic polymers such as polybenzimidazole (PBI) and poly(4-vinyl pyridine) in order to improve the performance in direct methanol fuel cells [11].

Different sulfonated aromatic polyimides are also under investigation [12,13*,14,15]. These ionomers typically reach high levels of conductivity, but the hydrolytic stability is reported to be very sensitive to the chemical structure of the polyimide main-chain [13*,14]. A membrane-electrode assembly based on a sulfonated polyimide was recently evaluated in a fuel cell at 70°C, and was found to have a performance similar to Nafion® [15].

Sulfonated PBI has been investigated by Kawahara et al. [16 and 17] and Asensio et al. [18]. At low water contents, PEMs of PBI grafted with sulfopropyl units showed a proton conductivity in the order of 10^{-3} S cm^{-1} in the temperature range from 20 to 140°C, which is superior to Nafion® under the same conditions [17]. The performance of these PEMs have also recently been investigated in fuel cells at temperatures up to 150°C under fully humidified conditions [19].

As the operation temperature of the PEMs is increased to temperatures above 100°C, desulfonation, i.e. loss of the sulfonic acid unit though hydrolysis, becomes an increasingly important problem. Acidic moieties having higher stability include phosphonic acid and sulfonimides. The latter is a significantly stronger acid compared to sulfonic acid, which may be especially advantageous at low water contents. Allcock et al. have prepared different poly(aryloxyphosphazene)s functionalized with phenyl phosphonic acid units with the intended use in direct methanol fuel cells [20]. Just recently, the same authors also report on the preparation of poly(aryloxyphosphazene)s having sulfonimide units [21]. Blending and radiation cross-linking have been investigated as means to reduce water swelling and methanol permeation of poly(aryloxyphosphazene) ionomers [22].

3. ORGANIC–INORGANIC HYBRID MEMBRANES

The incorporation of various hygroscopic, and often proton conducting, inorganic nanoparticles has shown to significantly improve the high-temperature performance of several types of PEMs [23*,24–29*,30*]. A number of research groups are developing hybrid membranes based on Nafion® to improve its high-temperature performance [23*,24–27,31]. Nafion®-silica hybrid membranes were produced by Miyake et al. via a sol–gel process [31]. Although containing more water, the conductivity of the hybrid membranes decreased with increased silica content, and was lower than that in the unmodified membrane under all conditions investigated. Bocarsly and co-workers have prepared Nafion® PEMs containing silicon oxide [24],

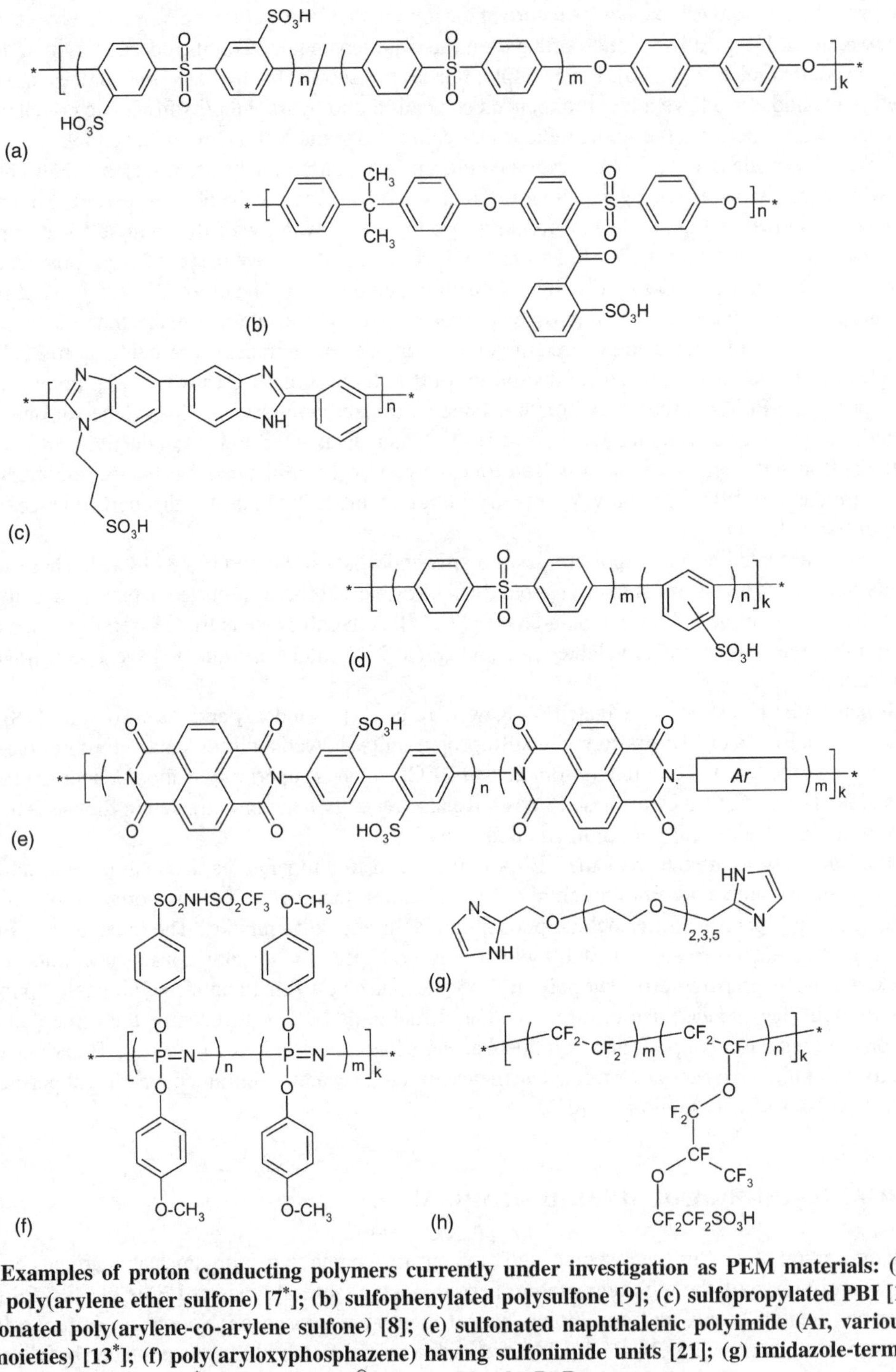

Fig. 2. Examples of proton conducting polymers currently under investigation as PEM materials: (a) sulfonated poly(arylene ether sulfone) [7*]; (b) sulfophenylated polysulfone [9]; (c) sulfopropylated PBI [16,17]; (d) sulfonated poly(arylene-co-arylene sulfone) [8]; (e) sulfonated naphthalenic polyimide (Ar, various aromatic moieties) [13*]; (f) poly(aryloxyphosphazene) having sulfonimide units [21]; (g) imidazole-terminated ethylene oxide oligomers [46*]; and (h) Nafion® marketed by the DuPont company

as well as zirconium phosphate [23*,25] particles. They found, for example, that the silicon oxide modified PEMs showed improved robustness and water retention, which resulted in high conductivities at 130°C during at least 50 h. Staiti et al. have investigated Nafion®-silica membranes doped with phosphotungstic acid and silicotungstic acid [26]. The authors claim that the heteropolyacid-modified Nafion®-silica recast membranes showed suitable properties for operation at 145°C in a direct methanol fuel cell. Tazi and Savadogo [27] have prepared membranes based on Nafion®, silicotungstic acid and thiophene. The modified membranes are reported to have a higher water uptake and conductivity than the unmodified membrane, resulting in improved fuel cell characteristics.

Hybrid membranes based on different arylene main-chain polymers have also been investigated [28,29*,30*,33*]. For example, nanocomposite PEMs based on phosphotungstic acid in sulfonated polysulfones have been prepared by Hickner et al. [28]. Interestingly, the presence of the nanoparticles was found to increase the proton conductivity, while at the same time decreasing the water absorption. In addition, the mechanical modulus of the material was improved after the addition of the particles. Genova-Dimitrova et al. [29*] incorporated phosphatoantimonic acid particles into sulfonated polysulfone and obtained PEMs with improved mechanical properties and conductivities close to Nafion®, while avoiding excessive water swelling at 80°C. Bonnet et al. [30*] have studied the properties of hybrid membranes based on sulfonated polyetheretherketone and particles of amorphous silica, zirconium phosphate sulfophenylphosphate, and zirconium phosphate as a function of temperature and humidity. In all cases the presence of the particles led to increased conductivities at 100°C. Staiti [32] has attached silicotungstic acid on SiO_2-support particles, and then used PBI as a binder to prepare membrane films. The materials are reported to be thermally stable with a conductivity of 10^{-3} S cm^{-1} at 160°C and 100% relative humidity. The use of a phosphonated PBI gave membranes with twice the conductivity at the same operating conditions [32].

A somewhat different approach has been pursued by Honma et al. [33*,34] who prepared different organic–inorganic hybrid materials by forming networks containing nanoparticles covalently linked by oligoether segments. After doping the networks with various heteropolyacids, they reported proton conductivities of 10^{-3}–10^{-2} S cm^{-1} in the temperature range 20–140°C under fully humidified conditions [34]. Also, the thermal stability of the oligoethers was greatly improved after formation of the hybrids. Stangar et al. have shown that a similar material based on silica functionalized by poly(propylene glycol) and doped with a heteropolyacid showed better results than Nafion® in a methanol fuel cell, mostly due to a lower methanol cross-over [35].

In conclusion, the incorporation of various nanoparticles seems to be very promising, although a great deal remains to be understood when it comes to, for example, interactions, synergy and long-term stability in these rather complex hybrid materials.

4. MEMBRANES BASED ON POLYMERS AND OXO-ACIDS

Several research groups are currently developing high-temperature PEMs based on complexes of strong acids, such as H_3PO_4 and H_2SO_4, with different basic polymers, especially PBI [36–40*,41*,42,43*,44]. Typically, the mechanical properties of these PEMs are favored by low levels of acid-doping, while the conductivity is favored by high doping levels and increasing water concentrations. Various polymers beside PBI have been evaluated for use in these types of membranes [36–39]. Bozkurt and Meyer have for example investigated poly(4-vinylimidazole)–H_3PO_4 complexes and found by thermogravimetry that they were stable up to 150°C [36]. When increasing the concentration of H_3PO_4, they found that the materials became softer and that the conductivity at ambient temperature increased to reach approximately 10^{-4} S cm^{-1} at 2 mol H_3PO_4 per mol imidazole unit. Lassègues et al. [37] found that complexes of an amorphous polyamide with H_3PO_4 showed high conductivities, but had poor mechanical properties and poor chemical stability above 90°C.

Pu et al. [38] have used dielectric spectroscopy to investigate the proton conduction mechanism in poly(4-vinylimidazole) and PBI [40*] doped with H_3PO_4 and H_2SO_4. The results showed that the proton transport in the PBI-based materials mainly is controlled by proton hopping and diffusion, rather than by segmental motions of the polymer chains. For the poly(4-vinylimidazole)-based materials in the glassy state, the proton transport was controlled by a hopping mechanism, while the segmental motions and diffusion also contributed above the glass transition temperature.

The concept of the PBI—H_3PO_4 system has been further exploited. For example, Qingfeng et al. [41*] found that the conductivity of PBI—H_3PO_4 complexes was insensitive to humidity, but strongly dependent on the acid content, reaching values of 0.13 S cm^{-1} at 160°C and high acid-doping levels. It was also shown that the water drag due to proton transport was almost zero in these PEMs. Asensio et al. [18] have synthesized and compared the performance of different sulfonated and unsulfonated PBIs and their complexes with H_3PO_4. They found that the complexes were stable up to 400°C, and that sulfonation of a PBI promoted the conductivity of the complex in comparison to when the unsulfonated polymer was used. Furthermore, Hasiotis et al. have prepared blends of sulfonated polysulfones and PBI which were doped with H_3PO_4 [42,43*]. These PEMs showed improved mechanical properties and conductivities above 10^{-2} S cm^{-1} at 160°C and 80% relative humidity, which was higher than for acid-doped PBI membranes under the same conditions. Initial work has also indicated the suitability of the blend membranes in fuel cells [44]. A membrane-electrode assembly based on PBI–H_3PO_4 complexes has recently been evaluated in a H_2/O_2 fuel cell [44]. Finally, the Celanese company has announced the introduction of a commercial membrane-electrode assembly based on the same system.

5. ALL-POLYMERIC ELECTROLYTES

Kreuer et al. [45*] have outlined a very interesting approach to obtain proton conducting polymeric systems based on nitrogen-containing heterocycles, such as imidazole, benzimidazole and pyrazole. These heterocycles form hydrogen bonded networks similar to that found in water, and also their transport properties are similar to that of water with proton transfer occurring via structure diffusion. An important advantage of the heterocycles over water is that they can be covalently incorporated into polymer structures to obtain all-polymeric proton conductors, thus avoiding any volatile low molecular weight species. It is, however, important that the incorporation is accomplished in such a way that the heterocyclic groups retain a high mobility. Recently, Schuster et al. [46*] showed that imidazole-terminated ethylene oxide oligomers can reach conductivities of up to 10^{-5} S cm^{-1} at 120°C. The conductivity was further enhanced after acid-doping. In another study, Yoon et al. [47] prepared a polyurethane having imidazole units in the main-chain which reached conductivities of 10^{-4} S cm^{-1} at 140°C. Notably, these levels of conductivity were obtained in the complete absence of water.

6. THEORETICAL STUDIES

Several theoretical studies have recently been carried out to increase the knowledge concerning the mechanism of proton transport in ionomer membranes [48–54]. Studies that take into account molecular structure and membrane morphology may play a vital role in the development of new ionomers and membranes.

Eikerling et al. [48] were able to predict experimental values of PEM conductivities by using a model based on a heterogeneous membrane structure, and addressing relevant experimental parameters such as the concentration of acidic moieties and the level of hydration. In another study, the same authors carried out computations to evaluate the proton dissociation of various acidic moieties at different levels of hydration [49]. They found that the sulfonimide moieties have higher degrees of proton dissociation at low water contents as compared to triflic acid, CF_3SO_3H, which has a higher tendency towards contact-ion pair formation. Paddison and coworkers have taken morphological parameters obtained from SAXS data into account when calculating proton diffusion coefficients [51,52]. Coefficients were obtained at different hydration levels and at different distances from the pore walls. Also in this study, the computed values were close to experimentally measured values. Li et al. [54] have studied interactions of the hydronium ion with water and model Nafion® structures using ab initio, density functional theory, and molecular dynamics simulations. The results indicated, for example, that the flexible sulfonated perfluorinated side chain is stretched in the aqueous phase, and that the sulfonate–hydronium contact ion pair is very stable.

Ab initio molecular dynamics simulations have also been performed by Münch et al. [55] to investigate the diffusion process of an excess proton in hydrogen bonded imidazole chains. The diffusion mechanism was described by a Grotthus mechanism involving a proton transfer step and a rate-determining molecular reorientation step.

7. CONCLUSIONS

It seems reasonable that operational temperatures up to approximately 130°C may be reached in PEMFCs by using, for example, well-designed hydrated organic–inorganic hybrid PEMs. However, in order to attain operational temperatures above 150°C, focus has most probably to be on different modes of proton conduction using durable non-volatile components. In this context, PBI–H_3PO_4 complexes have already been evaluated with good results, even if the critical long-term stability remains to be proven. The perspective of all-polymeric electrolytes based on imidazoles is appealing and initial results are encouraging. A future challenge is to develop useful PEMs based on this concept, combining high conductivity, and high thermal, chemical, and mechanical stability. Also in these systems, the use of suitable nanoparticles may prove fruitful. For nearly all PEM systems there is a further need for carefully designed polymers to control structure and morphology in order to manipulate interactions and processes taking place on the molecular scale.

ACKNOWLEDGEMENTS

I thank L.E. Karlsson for helpful discussions and for reading the proofs. Support from the Swedish Research Council and the Swedish Foundation for Strategic Environmental Research is gratefully acknowledged.

REFERENCES AND RECOMMENDED READING

(Asterisk * denotes articles of special interest)

*1. P. Costamanga and S. Srinivasan. Quantum jumps in the PEMFC science and technology from the 1960s to the year 2000. Part I. Fundamental scientific aspects. *J. Power Sources* **102** (2001) 242–252.
2. B.C.H. Steele and A. Heinzel. Materials for fuel-cell technologies. *Nature* **414** (2001) 345–352.
3. J.A. Kerres. Development of ionomer membranes for fuel cells. *J. Membr. Sci.* **185** (2001) 3–27.
*4. K.D. Kreuer. On the development of proton conducting membranes for hydrogen and methanol fuel cells. *J. Membr. Sci.* **185** (2001) 29–39.
5. J. Ding, C. Chuy and S. Holdcroft. Solid polymer electrolytes based on ionic graft polymers: effect of graft chain length on nano-structured, ionic networks. *Adv. Funct. Mater.* **12** (2002) 389–394.
6. H. Tang and P.N. Pintauro. Polyphosphazene membranes IV. Polymer morphology and proton conductivity in sulfonated polybis(3-methylphenoxy)phosphazene. films. *J. Appl. Polym. Sci.* **79** (2001) 49–59.
*7. F. Wang, M. Hickner, Y.S. Kim, T.A. Zawodzinski and J.E. McGrath. Direct polymerization of sulfonated poly(arylene ether sulfone) random (statistical) copolymers: candidates for new proton exchange membranes. *J. Membr. Sci.* **197** (2002) 231–242.

8. D. Poppe, H. Frey, K.D. Kreuer, A. Heinzel and R. Mülhaupt. Carboxylated and sulfonated poly(arylene-co-arylene sulfone)s: thermostable polyelectrolytes for fuel cell applications. *Macromolecules* **35** (2002) 7936–7941.
9. B. Lafitte, L.E. Karlsson and P. Jannasch. Sulfophenylation of polysulfones for proton conducting fuel cell membranes. *Rapid Macromol. Commun.* **23** (2002) 896–900.
10. Y.Z. Meng, S.C. Tjong, A.S. Hay and S.J. Wang. Synthesis and proton conductivities of phosphoric acid containing poly-(arylene ether)s. *J. Polym. Sci. Polym. Chem.* **39** (2001) 3218–3226.
11. L. Jönissen, V. Gogel, J. Kerres and J. Garche. New membranes for direct methanol fuel cells. *J. Power Sources* **105** (2002) 267–273.
12. X. Guo, J. Fang, T. Watari, K. Tanaka, H. Kita and K. Okamoto. Novel sulfonated polyimides as polyelectrolytes for fuel cell application. 2. Synthesis and proton conductivity of polyimides from 9,9-bis(4-aminophenyl) fluorene-2,7-disulfonic acid. *Macromolecules* **35** (2002) 6707–6713.
*13. C. Genies, R. Mercier, B. Sillion, N. Cornet, G. Gebel and M. Pineri. Soluble sulfonated naphthalenic polyimides as materials for proton exchange membranes. *Polymer* **42** (2001) 359–373.
14. C. Genies, R. Mercier, B. Sillion et al. Stability study of sulfonated phthalic and naphthalenic polyimide structures in aqueous medium. *Polymer* **42** (2001) 5097–5105.
15. S. Besse, P. Capron, O. Diat et al. Sulfonated polyimides for fuel cell electrode membrane assemblies (EMA). *J. New Mater. Electrochem. Syst.* **5** (2002) 109–112.
16. M. Kawahara, M. Rikukawa, K. Sanui and N. Ogata. Synthesis and proton conductivity of sulfopropylated poly(benzimidazole) film. *Solid State Ionics* **136–137** (2000) 1193–1196.
17. M. Kawahara, M. Rikukawa and K. Sanui. Relationship between absorbed water and proton conductivity in sulfopropylated poly(benzimidazole). *Polym. Adv. Technol.* **11** (2000) 544–547.
18. J.A. Asensio, S. Borrós and R. Gómez. Proton conducting polymers based on benzimidazole and sulfonated benzimidazoles'. *J. Polym. Sci. Polym. Chem.* **40** (2002) 3703–3710.
19. J.M. Bae, I. Honma, M. Murata, T. Yamamoto, M. Rikukawa and N. Ogata. Properties of selected sulfonated polymers as proton-conducting electrolytes for polymer electrolyte fuel cells. *Solid State Ionics* **147** (2002) 189–194.
20. H.R. Allcock, M.A. Hofmann, C.M. Ambler et al. Phenyl phosphonic acid functionalized poly(aryloxyphosphazenes) as proton-conducting membranes for direct methanol fuel cells. *J. Membr. Sci.* **201** (2002) 47–54.
21. M.A. Hofmann, C.M. Ambler, A.E. Maher et al. Synthesis of polyphosphazenes with sulfonimide side groups. *Macromolecules* **35** (2002) 6490–6493.
22. R. Carter, R. Wycisk, H. Yoo and P.N. Pintauro. Blended polyphosphazene/polyacrylonitrile membranes for direct methanol fuel cells. *Electrochem. Solid State Lett.* **5** (2002) A195–A197.
*23. C. Yang, P. Costamagna, S. Srinivasan, J. Benziger and A.B. Bocarsly. Approaches and technical challenges to high temperature operation of proton exchange membrane fuel cells. *J. Power Sources* **103** (2001) 1–9.
24. K.T. Adjemian, S.J. Lee, S. Srinivasan, J. Benziger and A.B. Bocarsly. Silicon oxide Nafion composite membranes for proton-exchange membrane fuel cell operation at 80–140°C. *J. Electrochem. Soc.* **149** (2002) A256–A261.
25. P. Costamagna, C. Yang, A.B. Bocarsly and S. Srinivasan. Nafion® 115/Zirconium phosphate composite membranes for operation of PEMFCs above 100°C. *Electrochim. Acta* **47** (2002) 1023–1033.
26. P. Staiti, A.S. Aricò, V. Baglio, F. Lufrano, E. Passalacqua and V. Antonucci. Hybrid Nafion-silica membranes doped with heteropolyacids for application in direct methanol fuel cells. *Solid State Ionics* **145** (2001) 101–107.
27. B. Tazi and O. Savadogo. Parameters of PEM fuel-cells based on new membranes fabricated from Nafion®, silicotungstic acid and thiophene. *Electrochim. Acta* **45** (2000) 4329–4339.
28. M. Hickner, Y.S. Kim, F. Wang, T.A. Zawodzinski and J.E. McGrath. Proton exchange membrane nanocomposites for fuel cells. *Int. SAMPE Tech. Conf.* **33** (2001) 1519–1532.
*29. P. Genova-Dimitrova, B. Baradie, D. Foscallo, C. Poinsignon and J.Y. Sanchez. Ionomeric membranes for proton exchange membrane fuel cell (PEMFC): sulfonated polysulfone associated with phosphatoantimonic acid. *J. Membr. Sci.* **185** (2001) 59–71.
*30. B. Bonnet, D.J. Jones, J. Roziere et al. Hybrid organic–inorganic membranes for a medium temperature fuel cell. *J. New Mater. Electrochem. Syst.* **3** (2000) 87–92.
31. N. Miyake, J.S. Wainright and R.F. Savinell. Evaluation of a sol–gel derived Nafion/silica hybrid membrane for proton electrolyte membrane fuel cell applications. I. Proton conductivity and water content. *J. Electrochem. Soc.* **148** (2001) A898–A904.

32. P. Staiti. Proton conductive membranes constituted of silicotungstic acid anchored to silica-polybenzimidazole matrices. *J. New Mater. Electrochem. Syst.* **4** (2001) 181–186.
*33. I. Honma, S. Nomura and H. Nakajima. Protonic conducting organic/inorganic nanocomposites for polymer electrolyte membrane. *J. Membr. Sci.* **185** (2001) 83–94.
34. H. Nakajima, S. Nomura, T. Sugimoto, S. Nishikawa and I. Honma. High temperature proton conductive organic–inorganic nanohybrids for polymer electrolyte membrane. Part II. *J. Electrochem. Soc.* **149** (2002) A953–A959.
35. U.L. Stangar, N. Groselj, B. Orel, A. Schmitz and P. Colomban. Proton-conducting sol–gel hybrids containing heteropoly acids. *Solid State Ionics* **145** (2001) 109–118.
36. A. Bozkurt and W.H. Meyer. Proton conducting blends of poly(4-vinylimidazole) with phosphoric acid. *Solid State Ionics* **138** (2001) 259–265.
37. J.C. Lassègues, J. Grondin, M. Hernandez and B. Marée. Proton conducting polymer blends and hybrid organic inorganic materials. *Solid State Ionics* **145** (2001) 37–45.
38. H. Pu, W.H. Meyer and G. Wegner. Proton conductivity in acid-blended poly(4-vinylimidazole). *Macromol. Chem. Phys.* **202** (2001) 1478–1482.
39. A. Bozkurt and W.H. Meyer. Proton-conducting poly(vinylpyrrolidon)-phosphoric acid blends. *J. Polym. Sci. Polym. Phys.* **39** (2001) 1987–1994.
*40. H. Pu, W.H. Meyer and G. Wegner. Proton transport in polybenzimidazole blended with H_3PO_4 or H_2SO_4. *J. Polym. Sci. Polym. Phys.* **40** (2002) 663–669.
*41. L. Qingfeng, H.A. Hjuler and N.J. Bjerrum. Phosphoric acid doped polybenzimidazole membranes: physico-chemical characterization and fuel cell applications. *J. Appl. Electrochem.* **31** (2001) 773–779.
42. C. Hasiotis, V. Deimede and C. Kontoyannis. New polymer electrolytes based on blends of sulfonated polysulfones with polybenzimidazole. *Electrochim. Acta* **46** (2001) 2401–2406.
*43. C. Hasiotis, L. Qingfeng, V. Deimede, J.K. Kallitsis, C.G. Kontoyannis and N.J. Bjerrum. Development and characterization of acid-doped polybenzimidazole/sulfonated polysulfone blend polymer electrolytes for fuel cells. *J. Electrochem. Soc.* **148** (2001) A513–A519.
44. O. Savadogo and B. Xing. Hydrogen/oxygen polymer electrolyte membrane fuel cell (PEMFC) based on acid-doped polybenzimidazole (PBI). *J. New Mater. Electrochem. Syst.* **3** (2000) 345–349.
*45. K.D. Kreuer, A. Fuchs, M. Ise, M. Spaeth and J. Maier. Imidazole and pyrazole-based proton conducting polymers and liquids. *Electrochim. Acta* **43** (1998). 1281–1288.
*46. M. Schuster, W.H. Meyer, G. Wegner et al. Proton mobility in oligomer-bound proton solvents: imidazole immobilization via flexible spacers. *Solid State Ionics* **145** (2001) 85–92.
47. C.B. Yoon, W.H. Meyer and G. Wegner. New functionalized polyurethane with proton conductivity. *Synthetic Met.* **119** (2001) 465–466.
48. M. Eikerling, A.A. Kornyshev, A.M. Kuznetsov, J. Ulstrup and S. Walbran. Mechanisms of proton conductance in polymer electrolyte membranes. *J. Phys. Chem. B* **105** (2001) 3646–3662.
49. M. Eikerling, S.J. Paddison and T.A. Zawodzinski. Molecular orbital calculations of proton dissociation and hydration of various acidic moieties for fuel cell polymers. *J. New Mater. Electrochem. Syst.* **5** (2002) 15–23.
50. M. Eikerling and A.A. Kornyshev. Proton transfer in a single pore of a polymer electrolyte membrane. *J. Electroanal. Chem.* **502** (2001) 1–14.
51. S.J. Paddison and R. Paul. The nature of proton transport in fully hydrated Nafion®. *Phys. Chem. Chem. Phys.* **4** (2002) 1158–1163.
52. S.J. Paddison, R. Paul and K.D. Kreuer. Theoretical computed proton diffusion coefficients in hydrated PEEKK membranes. *Phys. Chem. Chem. Phys.* **4** (2002) 1151–1157.
53. S.J. Paddison, R. Paul and T.A. Zawodzinski. Proton friction and diffusion coefficients in hydrated polymer electrolyte membranes: computations with a non-equilibrium statistical mechanical method. *J. Chem. Phys.* **115** (2001) 7753–7761.
54. T. Li, A. Wlaschin and P.B. Balbuena. Theoretical studies of proton transfer in water and model polymer electrolyte systems. *Ind. Eng. Chem. Res.* **40** (2001). 4789–4800.
55. W. Münch, K.D. Kreuer, W. Silvestri, J. Maier and G. Seifert. The diffusion mechanism of an excess proton in imidazole molecule chains: first results of an ab initio molecular dynamics study. *Solid State Ionics* **145** (2001) 437–443.

Chapter 21
PEM fuel cell electrodes

S. Litster and G. McLean

Abstract

The design of electrodes for polymer electrolyte membrane fuel cells (PEMFC) is a delicate balancing of transport media. Conductance of gas, electrons, and protons must be optimized to provide efficient transport to and from the electrochemical reactions. This is accomplished through careful consideration of the volume of conducting media required by each *phase* and the distribution of the respective conducting network. In addition, the issue of electrode flooding cannot be neglected in the electrode design process. This review is a survey of recent literature with the objective to identify common components, designs and assembly methods for PEMFC electrodes. We provide an overview of fabrication methods that have been shown to produce effective electrodes and those that we have deemed to have high future potential. The relative performances of the electrodes are characterized to facilitate comparison between design methodologies.

Keywords: Polymer electrolyte membrane fuel cell (PEMFC); Catalyst layer; Gas diffusion layer; Thin-film electrodes; PTFE-bound electrodes; Sputtered electrodes

Article Outline

1. Introduction 444
   1.1. Catalyst layer 446
   1.2. Gas diffusion layer 446
   1.3. Electrode designs 446
2. PTFE-bound methods 447
   2.1. Nafion impregnation 447
3. Thin-film methods 448
   3.1. Nafion loading 451
   3.2. Organic solvents 451
   3.3. Pore formers in the catalyst layer 452
   3.4. Thermoplastic ionomers 453
   3.5. Colloidal method 454
   3.6. Controlled self assembly 455
4. Vacuum deposition methods 455
   4.1. Graded catalyst deposition 457
   4.2. Multiple layer sputtering 458
5. Electrodeposition methods 459
   5.1. Effect of current control 460
   5.2. Membrane layer 460

Fuel Cells Compendium

6. Impregnated catalyst layer . . . 461
7. Catalyst supports . . . 461
7.1. Pt/C weight ratio . . . 462
7.2. Binary carbon catalyst supports . . . 462
7.3. Conducting polymer catalyst supports . . . 463
7.4. Carbon nanohorn catalyst supports . . . 463
8. Gas diffusion layer development . . . 463
8.1. Polytetrafluoroethylene (PTFE) content . . . 464
8.2. Influence of carbon powder . . . 464
8.3. Thickness . . . 464
8.4. Composite gas diffusion layer . . . 465
8.5. Pore formers in the gas diffusion layer . . . 465
9. Conclusion . . . 466
References . . . 466

1. INTRODUCTION

The first application of a proton exchange membrane (PEM), also referred to as a polymer electrolyte membrane, in a fuel cell was in the 1960s as an auxiliary power source in the Gemini space flights. Subsequently, advances in this technology were stagnant until the late 1980s when the fundamental design underwent significant reconfiguration. New fabrication methods, which have now become conventional, were adopted and optimized to a high degree. Possibly, the most significant barrier that PEM fuel cells had to overcome was the costly amount of platinum required as a catalyst. The large amount of platinum in original PEM fuel cells is one of the reasons why fuel cells were excluded from commercialization. Thus, the reconfiguration of the PEM fuel cell was targeted rather directly on the electrodes employed and, more specifically, on reducing the amount of platinum in the electrodes. This continues to be a driving force for further research on PEM fuel cell electrodes.

A PEM fuel cell is an electrochemical cell that is fed hydrogen, which is oxidized at the anode, and oxygen that is reduced at the cathode. The protons released during the oxidation of hydrogen are conducted through the PEM to the cathode. Since the membrane is not electrically conductive, the electrons released from the hydrogen travel along the electrical detour provided and an electrical current is generated. These reactions and pathways are shown schematically in Fig. 1.

At the heart of the PEM fuel cell is the membrane electrode assembly (MEA). The MEA is pictured in the schematic of a single PEM fuel cell shown in Fig. 1. The MEA is typically sandwiched by two flow field plates that are often mirrored to make a bipolar plate when cells are stacked in series for greater voltages. The MEA consists PEM, catalyst layers, and gas diffusion layers. Typically, these components are fabricated individually and then pressed together at high temperatures and pressures.

As shown in Fig. 1, the electrode is considered herein as the components that span from the surface of the membrane to gas channel and current collector. A schematic of an electrode is illustrated in Fig. 2. Though the membrane is an integral part of the MEA, a review of the design and fabrication of polymer electrolyte membranes is beyond the scope of this paper. However, the interface between the membrane and the electrode is critical and will be given its due attention. Current collectors and gas channels, typically in the form of bipolar plates, will not be reviewed herein.

An effective electrode is one that correctly balances the transport processes required for an operational fuel cell, as shown in Fig. 2. The three transport processes required are the transport of

1. protons from the membrane to the catalyst;
2. electrons from the current collector to the catalyst through the gas diffusion layer; and
3. the reactant and product gases to and from the catalyst layer and the gas channels.

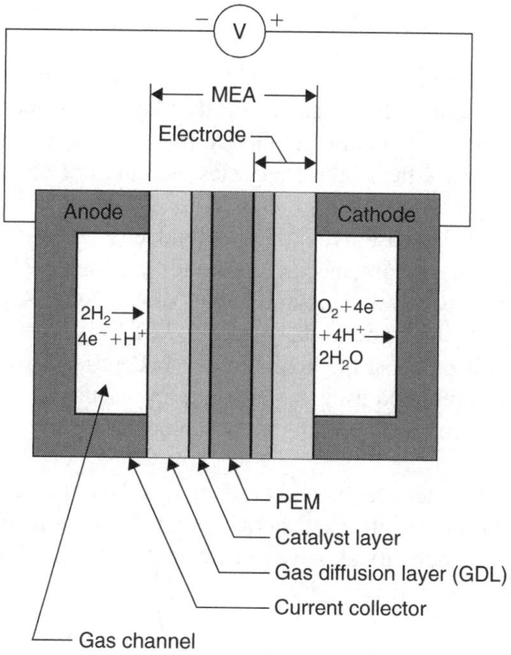

Fig. 1. Schematic of a single typical PEM fuel cell

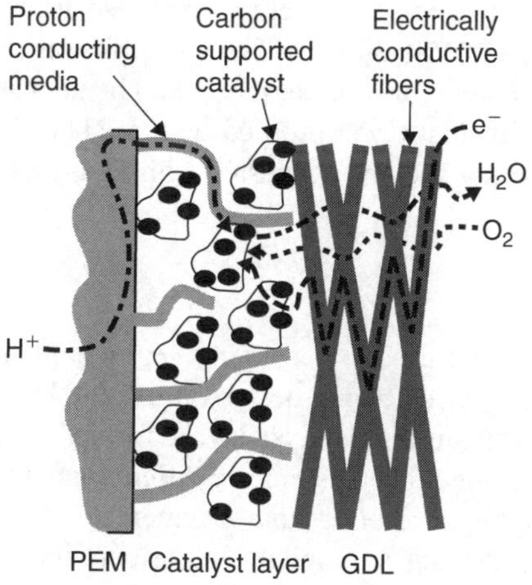

Fig. 2. Transport of gases, protons, and electrons in a PEM fuel cell electrode

Protons, electrons, and gases are often referred to as the three phases found in a catalyst layer. Part of the optimization of an electrode design is the attempt to correctly distribute the amount of volume in the catalyst layer between the transport media for each of the three phases to reduce transport losses. In addition, an intimate intersection of these transport processes at the catalyst particles is vital for effective operation of a PEM fuel cell. Each portion of the electrode will now be introduced.

1.1. Catalyst layer

The catalyst layer is in direct contact with the membrane and the gas diffusion layer. It is also referred to as the active layer. In both the anode and cathode, the catalyst layer is the location of the half-cell reaction in a PEM fuel cell. The catalyst layer is either applied to the membrane or to the gas diffusion layer. In either case, the objective is to place the catalyst particles, platinum or platinum alloys (shown as black ellipses in Fig. 2), within close proximity of the membrane.

The first generation of PEMFC used Polytelrafluoroethylene (PTFE)-bound Pt black electrocatalysts that exhibited excellent long-term performance at a prohibitively high cost [1]. These conventional catalyst layers generally featured expensive platinum loadings of 4 mg/cm^2. A generous amount of research has been directed at reducing Pt loading below 0.4 mg/cm^2 [2,3]. This is commonly achieved by developing methods to increase the utilization of the platinum that is deposited. Recently, platinum loadings as low as 0.014 mg/cm^2 have been reported using novel sputtering methods [4,5]. As a consequence of this focused effort, the cost of the catalyst is no longer the major barrier to the commercialization of PEM fuel cells.

In addition to catalyst loading, there are a number of catalyst layer properties that have to be carefully optimized to achieve high utilization of the catalyst material: reactant diffusivity, ionic and electrical conductivity, and the level of hydrophobicity all have to be carefully balanced. In addition, the resiliency of the catalyst is an important design constraint [1].

1.2. Gas diffusion layer

The porous gas diffusion layer in PEM fuel cells ensures that reactants effectively diffuse to the catalyst layer. In addition, the gas diffusion layer is the electrical conductor that transports electrons to and from the catalyst layer. Typically, gas diffusion layers are constructed from porous carbon paper, or carbon cloth, with a thickness in the range of 100–300 μm. The gas diffusion layer also assists in water management by allowing an appropriate amount of water to reach, and be held at, the membrane for hydration. In addition, gas diffusion layers are typically wet-proofed with a PTFE (Teflon) coating to ensure that the pores of the gas diffusion layer do not become congested with liquid water.

1.3. Electrode designs

Proven and emerging methods that are used to construct integrated membrane electrodes are illuminated in this review. Two widely employed electrode designs are the PTFE-bound and thin-film electrodes. Emerging methods include those featuring catalyst layers formed with electrodeposition and vacuum deposition (sputtering). In general, electrode designs are differentiated by the structure and fabrication of the catalyst layer. As well, we highlight recent accomplishments in the development of gas diffusion layers. However, most commercial PEM fuel cells and the majority of those reported herein still employ conventional carbon cloth or paper. There has been a significant amount of research conducted on producing composite gas diffusion layers with graded porosity and wet-proofing, as well as the optimization of carbon and PTFE loading in the gas diffusion layer. This report also includes a section describing some recent advances in increasing the surface area of the catalyst by optimization of catalyst supports.

It is evident throughout the report that the most common electrode design currently employed is the thin-film design. The thin-film design is characterized by the thin Nafion film that binds carbon-supported catalyst particles. The thin Nafion layer provides the necessary proton transport in the catalyst layer. This is a significant improvement over its predecessor, the PTFE-bound catalyst layer, which requires the less effective impregnation of Nafion. However, one fault of the Nafion thin-film method is its reduced resiliency. Methods of increasing this resiliency, such as using a thermoplastic form of the ionomer, have been found and are reported herein. Sputter deposited catalyst layers have been shown to provide some of the lowest catalyst loadings, as well as the thinnest layers. The short conduction distance of the thin sputtered layer dissipates the requirement of a proton-conducting medium, which can simplify production. The performance of the state of the art sputtered layer is only slightly lower than that of the present thin-film convention.

The performances of many of the electrodes reviewed are reported to accommodate comparison between designs. The performances are provided in the form of power densities at 200 mA/cm^2 and 0.6 V. These power densities are benchmarked because they typically represent two characteristics of the electrode. At 200 mA/cm^2, the losses can be associated to activation overpotential (the losses associated with the irreversibilities of the chemical reaction). The 0.6 V benchmark depicts the resistive components of the cell and its ability to provide adequate transport of gases, electrons, and protons to the catalyst sites. Together, these two benchmarks provide an overall picture of a PEM fuel cell's electrode performance. However, when comparing electrode designs it is important to weigh the operating characteristics such as temperatures, pressures, and the purity of the gases as they can have an overriding effect on the fuel cell performance.

2. PTFE-BOUND METHODS

Before the development of the thin-film catalyst layer [3], PTFE-bound catalyst layers were the convention [6–9]. In these catalyst layers, the catalyst particles were bound by a hydrophobic PTFE structure commonly cast to the diffusion layer. This method was able to reduce the platinum loading of prior PEM fuel cells by a factor of 10; from 4 to 0.4 mg/cm^2 [9]. In order to provide ionic transport to the catalyst site, the PTFE-bound catalyst layers are typically impregnated with Nafion by brushing or spraying. However, platinum utilization in PTFE-bound catalyst layers remains approximately 20% [8,10]. Nevertheless, researchers have continued to work on developing new strategies for Nafion impregnation [7].

Some of the original low-platinum loading PEM fuel cells featuring PTFE-bound catalyst layers were fabricated by Ticianelli et al. [9] at the Los Alamos National Laboratory. Cheng et al. [10] fabricated conventional PTFE-bound catalyst layer electrodes for direct comparison with the current thin-film method. The process employed for forming the PTFE-bound catalyst layer MEA in their study is detailed below.

1. 20 wt.% Pt/C catalyst particles were mechanically mixed for 30 min in a solvent.
2. PTFE emulsion was added until it occupied 30% of the mixture.
3. A bridge-builder and a peptization agent were added, followed by 30 min of stirring.
4. The slurry was coated onto the wet-proofed carbon paper using a coating apparatus.
5. The electrodes were subsequently dried for 24 h in ambient air, and then baked at 225°C for 30 min.
6. The electrodes were rolled and then sintered at 350°C for 30 min.
7. A 5 wt.% Nafion solution was brushed onto the electrocatalyst layer (2 mg/cm^2).
8. The Nafion-impregnated electrodes were placed in an oven at 80°C and allowed to dry for an hour in ambient air.
9. Once dry, the electrodes were bonded to the H^+ form of the polymer electrolyte membrane through hot pressing at 145°C for 3 min at a pressure of 193 atm to complete the MEA.

2.1. Nafion impregnation

Lee et al. [7] investigated the effect of Nafion impregnation on commercial low-platinum loading PEMFC electrodes. The researchers employed a conventional MEA with PTFE-bound catalyst layers featuring platinum loadings of 0.4 mg/cm^2. Nafion was impregnated in electrode structures, with the Nafion loadings varying from 0 to 2.7 mg/cm^2, by a brushing method. The results presented by Lee et al. depict a non-linear relationship between performance and Nafion loading. In addition, the polarization curves showed the effect that the oxidant composition has on the optimum amount of Nafion loading. When the oxidant was air, there was a sharp increase in performance as the Nafion loading was increased to 0.6 mg/cm^2. However, performance dropped as additional Nafion was added. The researchers found that 0.6 mg/cm^2 was the ideal Nafion loading when operating on air. When pure oxygen was employed as the oxidant, the performance increased with Nafion loading up to 1.9 mg/cm^2. This difference is due to mass transport being the limiting rate when air is the oxidant, as the partial pressure of oxygen is much lower. Without the addition of some Nafion, the majority of the catalyst sites were inactive. However, as more Nafion is added the porosity of the composite decreases and limits mass transfer. The same phenomenon was originally presented by Ticianelli et al. [9].

3. THIN-FILM METHODS

The present convention in fabricating catalyst layers for PEM fuel cells is to employ thin-film methods. In his 1993 patent, Wilson [3] described the thin-film technique for fabricating catalyst layers for PEM fuel cells with catalyst loadings less than 0.35 mg/cm^2. In this method the hydrophobic PTFE traditionally employed to bind the catalyst layer is replaced with hydrophilic perfluorosulfonate ionomer (Nafion). Thus, the binding material in the catalyst layer is composed of the same material as the membrane. Even though PTFE features effective binding qualities and imparts beneficial hydrophobicity in the gas diffusion layers, there is no particular benefit to its presence in the catalyst layer [11]. The resilient binding of PTFE catalyst layer is traded for the enhanced protonic conductivity of a Nafion-bound thin-film catalyst layer. Thin-film catalyst layers have been found to operate at almost twice the power density of PTFE-bound catalyst layers. This correlates with an active area increase from 22% to 45.4% when a Nafion-impregnated and PTFE-bound catalyst layer is replaced with a thin-film catalyst layer [10]. Moreover, thin-film MEA manufacturing techniques are more established and applicable to stack fabrication [6]. However, an active area of 45% suggests there is still significant potential for improvement.

The procedure for forming a thin-film catalyst layer on the membrane, according to Wilson's 1993 patent [3], is as follows:

1. Combine a 5% solution of solubilized perfluorosulfonate ionomer (such as Nafion) and 20 wt.% Pt/C support catalyst in a ratio of 1:3 Nafion/catalyst.
2. Add water and glycerol to weight ratios of 1:5:20 carbon–water–glycerol.
3. Mix the solution with ultrasound until the catalyst is uniformly distributed and the mixture is adequately viscous for coating.
4. Ion-exchange the Nafion membrane to the Na^+ form by soaking it in NaOH, then rinse and let dry.
5. Apply the carbon–water–glycerol ink to one side of the membrane. Two coats are typically required for adequate catalyst loading.
6. Dry the membrane in a vacuum with the temperature of approximately 160°C.
7. Repeat Steps 5 and 6 for the other side of the membrane.
8. Ion-exchange the assembly to the protonated form by lightly boiling the MEA in 0.1 M H_2SO_4 and rinsing in de-ionized water.
9. Place carbon paper/cloth against the film to produce a gas diffusion layer.

Alternatively, the catalyst layer can be applied using a transfer printing method in which the catalyst layer is cast to a PTFE blank. The catalyst layer is then decaled on to the membrane. This process is mainly used to ease fabrication in a research laboratory [1]. Moreover, direct coating methods (catalyst layer is cast directly onto the membrane) have been shown to provide higher performance because they offer better

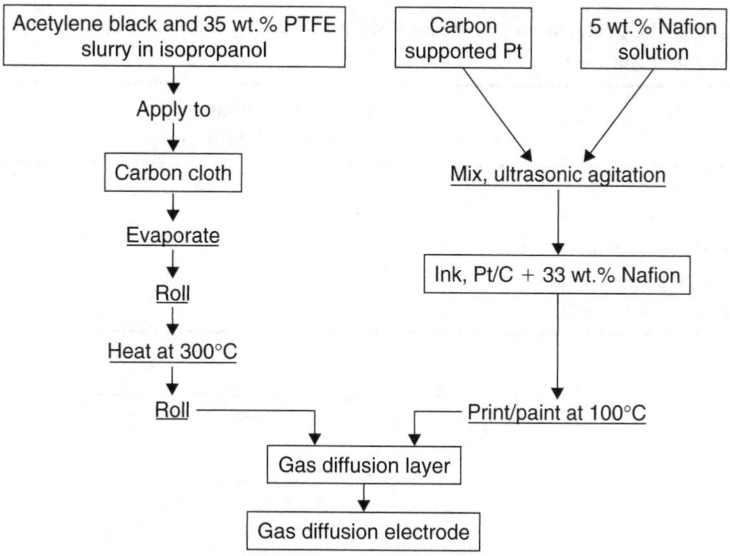

Fig. 3. Preparation of a gas diffusion electrode employing the thin-film methodology. Reproduced from [13]

ionic connection between the membrane and the ionomer in the catalyst layer [6]. To improve platinum utilization, Qi and Kaufman (QK) [12] boiled or steamed the electrodes as the last step in the fabrication of their thin-film electrodes. Their paper presented a significant increase in the performance over the entire spectrum of current densities when the electrodes were steamed or boiled for 10 min. Gamburzev and Appleby (GA) [13] also applied a thin-film catalyst layer to the carbon cloth gas diffusion layer. Their methodology is schematically described in Fig. 3.

Paganin et al. [14] documented the results of a fuel cell with a thin-film catalyst layer. Alternatively to Wilson's 1993 patent [3], the research group brushed the catalyst slurry (containing isopropanol instead of glycerol to achieve the desired viscosity) onto the gas diffusion layer rather than the membrane. The group was able to achieve good performance with platinum and Nafion loadings of 0.4 and 1.1 mg/cm^2, respectively, using 20 wt.% Pt/C catalyst particles.

The DLR research group in Germany [15,16] has developed a dry layer preparation method for fabricating catalyst layers bound by either PTFE or Nafion. Their method consists of dry mixing either supported or unsupported catalyst with PTFE or Nafion powder and spraying the atomized dry mixture in a nitrogen stream onto either the porous carbon diffusion layer or the membrane. Subsequently, the assembly would be hot-pressed or rolled. Some of the benefits of the dry layer technique are its simplicity because of the lack of evaporation steps, and its ability to create graded layers with multiple mixture streams. In addition, the platinum loading in the electrode fabricated is reported to be as low as 0.08 mg/cm^2. The cell performance results presented by the DLR group depict a preparation method with good future potential for use in MEA mass production.

Qi and Kaufman (QK) [17] of the H Power Corporation reported in 2003 on low Pt loading high performance cathodes for PEM fuel cells. The method used was of the thin-film variety, in which supported catalyst was mixed into a Nafion and water solution without the addition of organic solvents. The viscous solution was then applied to an ELAT gas diffusion layer and dried at moderate temperatures. The carbon-supported catalysts were purchased from E-TEK. QK achieved some of the highest power densities reported in this review (0.72 W/cm^2 at 75°C) [17]. All of QK's test cells featured Nafion 112 membranes.

The power densities listed in Table 1 demonstrate the contrast between the performance of thin-film catalyst layers and PTFE-bound catalyst layers in the study performed by Chun et al. [6]. There is a dramatic

Table 1. Comparison of the performance of various MEAs (Nafion 115 membrane, H_2/O_2 pressure = 1/1 atm, H_2/O_2 feed rate = 8.5/3.8 l/min) [6]

Type of electrode	Power density at 200 mA/cm²(mW/cm²)	Power density at 0.6 V(mW/cm²)
Commercial	140	233
PIFE-bound	114	93
Thin-film, direct membrane coating	145	200
Thin-film, transfer printing (20% Pt/C)	129	147
Thin-film, transfer printing (40% Pt/C)	123	132

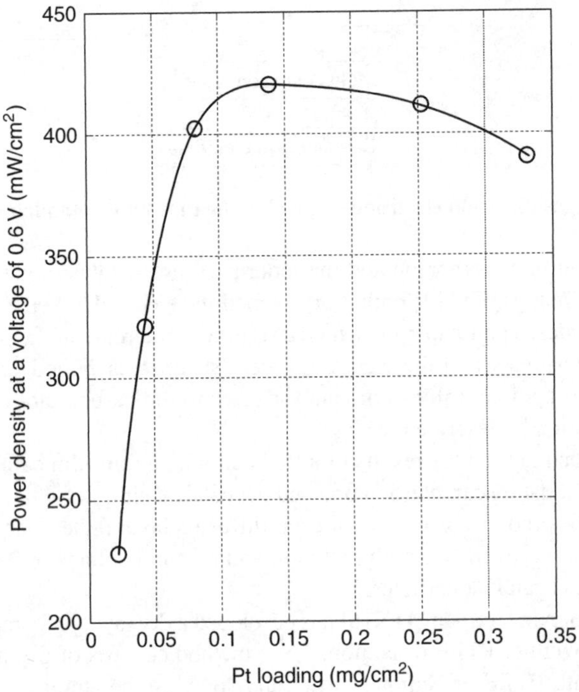

Fig. 4. Effect of Pt loading on performance for electrodes made using E-TEK 20% Pt/C, 35/45/45°C [17]

increase in the performance when the catalyst layer is formed via the thin-film method. In addition, the polarization curves show the increase in performance obtained when the catalyst layer is directly coated onto the membrane instead of transfer printing with a PTFE blank. This enhancement is a result of the intimate interface formed between the Nafion in the catalyst ink and the membrane when the direct coating of the membrane is employed.

Paganin et al. [14] investigated the effect of platinum loading on their thin-film cell with a Pt/C weight ratio 20 wt.%. They found that the performance in the cathode improved significantly when the loading was increased from 0.1 to 0.3 mg/cm^2. This is an effect of the increased active area. Conversely, there was a slight reduction in performance when the loading was increased to 0.4 mg/cm^2. No explanation was given for this response, but it could be caused by reduced reactant transport to the areas closest to the membrane. They compared loadings of 0.1 and 0.4 mg/cm^2 in the anode and found the lower loading to provide better performance. QK [17] found the highest performance with a Pt/C weight ratio 20 wt.% and a platinum loading of 0.20 $\pm$ 0.05 mg/cm^2. The influence of platinum loading on QK's cell is presented in Fig. 4.

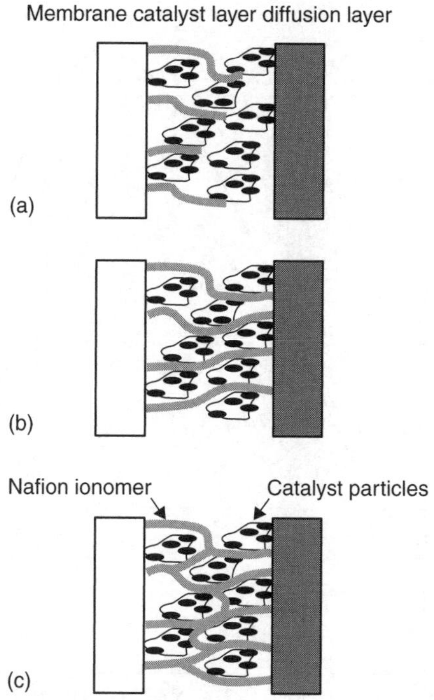

Fig. 5. Schematic planar representation of the catalyst layer. (a) Content of Nafion too low: not enough catalyst particles with ionic connection to membrane. (b) Optimal Nafion content: electronic and ionic connections well balanced. (c) Content of Nafion too high: catalyst particles electronically isolated from diffusion layer. Reproduced from [18]

3.1. Nafion loading

Paganin et al. [14] ascertained that, in their thin-film catalyst layer, when the Nafion loading was increased from 0.87 to 1.75 mg/cm^2 the performance improved significantly. Moreover, the performance deteriorated at higher current densities when the Nafion loading was increased beyond 2.2 mg/cm^2, which is equivalent to an optimum Nafion percentage of 33% of the catalyst layer weight. These values have been supported by several other recent studies [13,17,18]. Fig. 5 is a schematic of the catalyst layer that depicts the effect of Nafion loading. The effect of Nafion loading, as found by QK [17], is presented in Fig. 6.

The effect of Nafion loading on performance is aptly depicted by the results of Song et al. [19], who varied the Nafion content from 0.2 to 2.0 mg/cm^2 in a thin-film catalyst layer featuring a platinum loading 0.4 mg/cm^2. Table 2 shows that the increase in Nafion loading from 0.2 to 0.8 mg/cm^2 dramatically increased the power density in both spectrums. This indicates an increased utilization of the platinum. However, when the Nafion concentration was further increased to 2.0 mg/cm^2, the only change was a sharp drop in the power density at higher current densities. This is because additional Nafion is blocking reactant gases and the hydrophilic Nafion is likely trapping water in the catalyst layer.

3.2. Organic solvents

Organic solvents such as Glycerol are typically added to the ink mixture to improve paintability. Chun et al. [6] investigated the effect of glycerol in the catalyst ink on the performance of their thin-film electrodes.

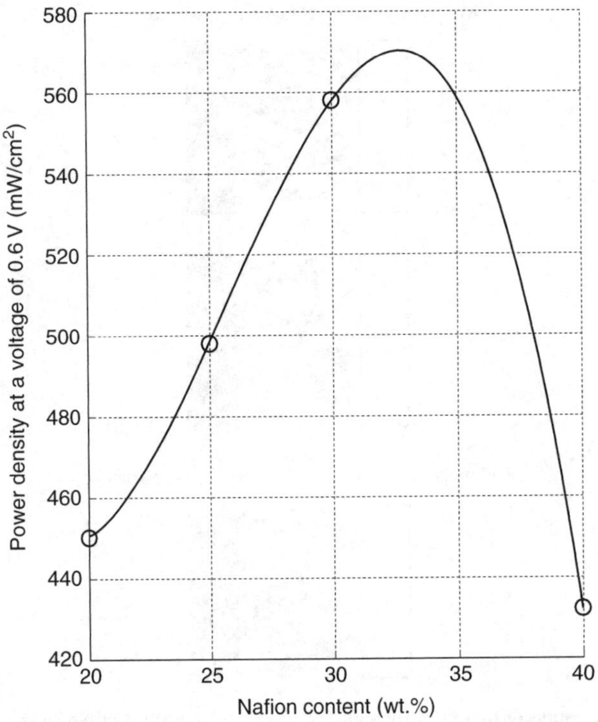

Fig. 6. Effect of Nafion content in the catalyst layer on performance. E-TEK 20% Pt/C, 35/45/45°C [17]

Table 2. Power densities of thin-film electrodes, with a Pt loading of 0.4 mg/cm^2, containing various Nafion ionomer concentrations in the catalyst layer [19]

Nafion ionomer concentration (mg/cm^2)	Power density at 200 mA/cm^2(mW/cm^2)	Power density at 0.6 V(mW/cm^2)
0.2	110	72
0.8	144	240
2.0	140	204

They found that high glycerol content (3:1 glycerol-5% Nafion solution) in the catalyst ink causes a significant drop in performance at current densities above 350 mA/cm^2. Chun et al. suggest that high glycerol loading reduces the contact area between the catalyst and the Nafion, and limits the charge transfer.

3.3. Pore formers in the catalyst layer

Fischer et al. [20] investigated the effect of additional porosity in the catalyst layers of thin-film catalyst layer fuel cells. They constructed their electrodes using a hot spray method, in which catalyst slurry containing catalyst and Nafion was sprayed onto the gas diffusion layer. To create additional porosity, several types of pore formers were added to the slurry, including:

- low temperature, decomposable (ammonium carbonate);
- high temperature, decomposable (ammonium oxalate); and
- soluble additives (lithium carbonate).

Without pore formers, the porosity of the catalyst layer was 35%. Ammonium carbonate and ammonium oxalate increased the porosity to 42% and 48%, respectively. With the addition of lithium carbonate, the porosity increased to 65%. However, it was shown that the electrical conductivity decreased (from 1.64 to 0.44 S/cm^2) with these increases in porosity. This change in the conductivity was shown to have little influence. It was found that the addition of pore formers made a negligible difference on performance when the cell is supplied with oxygen. However, there was a significant performance improvement when the oxidant was air (reactant transport became the limiting factor).

GA [13] documented an enhanced gas transport and fuel cell efficiency by the addition of proprietary pore formers in the catalyst layer. The pore former is mixed with the ink that forms the catalyst layer. Following the painting of the catalyst layer, the pore former material is dissolved and pores are formed. Calcium carbonate is a typical pore former [21]. Increased efficiency was found with the addition of pore former until the ink held 44 wt.% pore former. At this point, the efficiency was lower than that of the no pore former case. With more pore former present, the mass transfer rate increases. However, the electron and proton transport rates readily decrease when pore former is added. GA found the optimum pore former content to be 33 wt.%.

Contrary to the previously discussed increase in performance with the addition of pore former in the catalyst layer, Yoon et al. [22] found a decrease in performance with the addition of either 27 or 60 wt.% ethylene glycol as a pore forming agent. However, no explanation was given for this and it can only be assumed that ethylene glycol is a poor choice for a pore forming agent.

3.4. Thermoplastic ionomers

Wilson et al. [1] introduced a method of using thermoplastic ionomers into the catalyst layer to counter the steadily decaying performance of fuel cells with poorly bound catalyst layers. This work was prompted by the discovery that Nafion can be converted into a thermoplastic form by ion-exchanging the Nafion with large hydrophobic counter ions such as tetrabutylammonium (TBA^+). In the thermoplastic form, the ionomer can be processed in a melted phase, which leads to the possibility of fabricating ionomer structures by molding and extruding. The fabrication of the thermoplastic catalyst layer is similar to the thin-film method described by Wilson [3] (with the inclusion of TBA^+ in the mixing of the ink). However, ion-exchanging the MEA to the protonated (H^+) was hindered by the hydrophobic TBA^+. Consequently, the thermoplastic ionomer requires a more rigorous ion-exchanging process than conventional thin-film catalyst layers. The results presented depict a fuel cell with an adequate power density and low Pt loadings (0.12 mg/cm^2). In addition, the power density decreases by only 10% after 4000 h of operation. Chun et al. [6] also ion-exchanged the catalyst layer ionomer to the TBA^+ form during the catalyst layer preparation. However, no conclusions on molding or extruding techniques were made. The effect of the ion-exchange on the resiliency of the catalyst layer was not explored in this paper.

Yoon et al. [22] recently experimented with the effect of ion-exchanging the catalyst layer Nafion ionomer to the TBA^+ form. This was achieved through the addition of tetrabutylammonium hydroxide (TBAOH) to the original catalyst slurry containing electrocatalyst, Nafion ionomer, and water. Both the conventional and TBA^+ versions of thin-film catalyst layers were sprayed on the gas diffusion layers at various droplet sizes. It was visible in scanning electron micrographs that the grain size in the catalyst

Table 3. Power densities of thin-film electrodes prepared with conventional catalyst ink and with the addition of TBAOH to produce the thermoplastic form of the Nafion ionomer [22]

Ink preparation	Power density at 200 mA/cm^2(mW/cm^2)	Power density at 0.6 V(mW/cm^2)
Conventional ink	142	252
Addition of TBAOH	146	288

increases two-fold with the addition of TBAOH. The change in performance due to the addition of TBAOH is depicted in Table 3. In addition, it can be seen that the optimal case of the thermoplastic form of the ionomer performs moderately better than the non-thermoplastic case between current densities of 200–500 mA/cm^2. At high current densities the thermoplastic form of ionomer performs markedly better. Yoon et al.'s results also show that too fine or too coarse of droplet size can have a significant effect on the performance.

3.5. Colloidal method

An alternative method to conventional thin-film techniques is the colloidal method. Typically, the catalyst layers are applied as a solution. It is well understood that Nafion forms a solution in solvents with dielectric constants greater than 10. A typical solvent is isopropyl alcohol, which has a dielectric constant of 18.3. When normal-butyl acetate, which has a dielectric constant of 5.01, is employed as the solvent, a colloid forms in lieu of a solution. Shin et al. [23] suggested that in the conventional solution method the catalyst particles could be excessively covered with ionomer, which leads to under-utilization of platinum. In addition, it was proposed that in the colloidal method the ionomer colloid absorbs the catalyst particles and larger Pt/C agglomerates are formed. The colloidal method is known to cast a continuous network of ionomer that enhances proton transport.

Shin et al. [23] prepared colloidal catalyst ink with a method similar to the conventional thin-film approach. A mixture of Pt/C powder and Nafion ionomer was dripped drop by drop into the normal-butyl acetate solvent to form the ionomer colloids. The ink was then treated ultrasonically to allow the colloids to absorb the Pt/C powder. The ink was then sprayed via air brushing onto the carbon paper, which was to be used as the gas diffusion layer. It was stated that the colloidal method is the preferred ink for spraying methods, as it forms larger agglomerates. Small agglomerates formed by the solution method have a tendency to penetrate too far into the gas diffusion layer, blocking pores needed for gas transport.

The thickness of a catalyst layer that Shin et al. [23] formed by the colloidal ink was twice that of the 0.020 mm thick layer formed with solution ink. In addition, the size of Pt/C agglomerates increased from 550 to 736 nm with the introduction of the colloidal method. The catalyst layers formed with colloidal ink were hot-pressed to a Nafion 115 membrane and tested in a single cell test apparatus, along with a similar cell featuring catalyst layers formed by the solution method. The colloidal method dramatically outperformed the solution method at high current densities (see Table 4 for power densities). This is attributed to a significant increase in the proton conductivity, as well as a moderate enhancement of the mass transport in the catalyst layer formed with the colloidal ink. Shin et al. quantified these improvements by inserting resistance layers, formed from inactive catalyst layers, between either the membrane and active catalyst layer or the active catalyst layer and the gas diffusion layer. The increase in proton conductivity is due to the continuous network of ionomer in the colloidal catalyst layer. The increased mass transport is a product of the larger agglomerates of Pt/C in the colloidal catalyst layer, which translates to a higher porosity, allowing a greater flux of the reactant and product gases.

Table 4. Effect of the catalyst ink preparation method on the cell performance (Pt loading = 0.4 mg/cm^2, H_2/O_2 = 1/1 atm, and T = 80°C) [23]

Ink preparation	Power density at 200 mA/cm^2(mW/cm^2)	Power density at 0.6 V(mW/cm^2)
Ionomer solution method	153	417
Ionomer colloidal method	157	516

3.6. Controlled self assembly

Middelman [24] of Nedstack fuel cell technology reported on the development of a catalyst layer that features a controlled morphology to enhance performance. A fabrication method to create a highly oriented catalyst morphology was revealed as an alternative to conventional methods that typically create a random morphology. To create highly oriented structures, Middelman increased the mobility of the catalyst layer with high temperatures and chemical additives. Then an electric field was employed as the driving force to orient the strands. Middelman suggests that this method could increase Pt utilization to almost 100%, and states that increases in voltages of 20% are obtained with this process.

4. VACUUM DEPOSITION METHODS

Common vacuum deposition methods include chemical vapor deposition, physical or thermal vapor deposition, and sputtering. Sputtering is commonly employed to form catalyst layers and is known for providing denser layers than the alternative evaporation methods [25]. The sputtering of catalyst layers consists of a vacuum evaporation process that removes portions of a coating material (the target) and deposits a thin and resilient film of the target material onto an adjacent substrate. A schematic of the sputtering apparatus is shown in Fig. 7. In the case of sputtered catalyst layers, the target material is the catalyst material and the substrate can be either the gas diffusion layer or the membrane. Sputtering provides a method of depositing a thin catalyst layer (onto either the membrane or the gas diffusion layer) that delivers high performance combined with a low Pt loading. The entire catalyst layer is in such intimate contact with the membrane that the need for ionic conductors in the catalyst layer is resolved [4]. Moreover, platinum and its alloys are easily deposited by sputtering [26]. The success of the sputtering method on reducing platinum loading depends heavily on the reduction in the size of catalyst particles below 10 nm. State of the art thin-film electrodes feature Pt loading of 0.1 mg/cm^2 [4]. A 5 nm sputtered platinum film amounts to a platinum loading of 0.014 mg/cm^2. However, the performance of a fuel cell with a sputtered catalyst layer can vary by several orders of magnitude depending on the thickness of the sputtered catalyst layer [5].

According to Weber et al. [26], fuel cells with sputter deposited catalyst layers were first investigated by Cahan and Bockris in the late 1960s. Half a decade later, the method was further refined by Asher and Batzold, but without adequate power density. In 1987, Weber et al. continued to explore the belief that electrode fabrication could be significantly streamlined if platinum was applied directly to a wet-proofed substrate by vacuum deposition (i.e. sputtering). In their study, they sputtered platinum onto wet-proofed, porous substrates that were then used as hydrogen and oxygen electrodes. Their early results showed that the hydrogen electrodes were limited by the rate of mass transfer at low current densities (5–20 mA/cm^2). However, their oxygen electrodes performed considerably better; to current densities as high as 500 mA/cm^2. They found that the performance of their sputtered electrodes depended more on the substrate preparation than on the sputtering process. Substrate preparation includes the impregnation of PTFE and carbon powder into the porous substrate. Typical Pt loadings in this study were 0.15 mg/cm^2 and reached as high as 0.6 mg/cm^2.

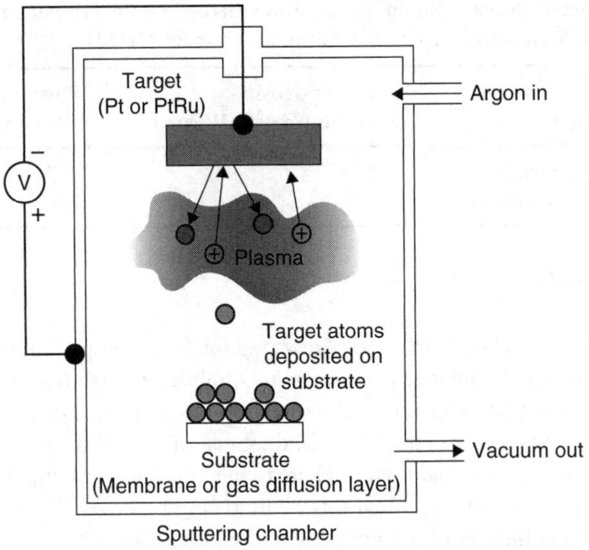

Fig. 7. Schematic of sputtering in argon plasma for fabricating PEM fuel cell electrodes

In 1997, Hirano et al. [27] documented their study of sputter-deposited catalyst layers in high performance PEM fuel cells. The cathode featured an uncatalyzed E-TEK electrode with a sputtered catalyst layer featuring loadings between 0.04 and 0.1 mg Pt/cm^2. The anode they employed was always a commercial E-TEK electrode. The anode and cathode catalyst layers both featured a Nafion loading of 0.6 mg/cm^2 to improve protonic conductivity. The performance of the cells with a sputtered platinum loading of 0.1 mg/cm^2 was nearly equivalent to that of the commercial variant. However, there was a visible drop in performance at very high current densities. One case, which featured a platinum loading of 0.04 mg/cm^2, suffered a dramatic increase in resistances because of the low active area. However, this low loading could be effective in low-power portable applications because at 200 mA/cm^2 the cell featured a respectable power density of 160 mW/cm^2.

O'Hayre et al. [5] of the Rapid Prototyping Lab at Stanford University reported in 2002 on their development of a catalyst layer with ultra-low platinum loading. Their paper suggests that they are developing these electrodes for use in micro-fuel cells since it was stated that the sputtering process is compatible with many other integrated circuit fabrication techniques. O'Hayre et al. also suggest the future ability to apply the gas diffusion layer with a sputtering process. In their study, they deposited a single sputtered platinum layer with a nominal thickness of 2–1000 nm to Nafion 117. After the catalyst layer was applied to both sides of the membrane, the catalyst layers were covered by carbon cloth. They were not hot-pressed or fixed by any other method. The operating conditions used in their performance tests were dry oxygen and dry hydrogen at the ambient temperature and pressure. They found that films only 5–10 nm thick produced the best performance when the catalyst was applied to smooth Nafion. This corresponds to platinum loadings of 0.01–0.02 mg/cm^2.

The performance of the O'Hayre et al. [5] fuel cell dropped dramatically when the thickness of the catalyst layer was less than 5 nm or greater than 10 nm. Using scanning electron microscopy (SEM), they depicted the reasoning for this balance. When the layer is very thin, there are only islands of the catalyst material. Once a nominal thickness of 4 nm is reached, the islands coalesce into a single film. At this thickness, a fine crack structure can be found in the film, which provides gas access to the reaction sites closest to the membrane and increases catalyst utilization. When the thickness is increased further, the crack structure becomes coarser as the film increases in mechanical strength. Thus, gas transport to the reaction

Table 5. Comparison of the power density of a commercial MEA with 0.4 mg/cm^2 platinum loading to that of a 15 nm sputtered platinum MEA with 0.04 mg/cm^2 platinum loading

Catalyst layer	Power density at 200 mA/cm^2(mW/cm^2)	Power density at 0.6 V(mW/cm^2)
Commercial MEA, 0.4 mg Pt/cm^2	NA	34
15 nm thin-film Pt, 0.04 mg Pt/cm^2	NA	17

Cell measurements were taken at room temperature using dry (non-humidified) H_2/O_2 at 1 atm [5].

sites most intimate with the membrane is limited. The effect of roughening the membrane surface prior to sputtering was also investigated. They found that the roughened surface delayed the coalescence of the platinum. The delayed coalescence on the roughened membrane reduces the maximum performance attainable, but reduces the performance sensitivity to the catalyst layer thickness. This can be beneficial since it is difficult to control the thickness of the sputtered layer [5]. A sputtered MEA (thickness of 15 nm, Pt loading of 0.04 mg/cm^2 on Nafion 115) was compared to a conventional MEA (Pt loading of 0.4 mg/cm^2 on Nafion 115) purchased from Electrochem. Inc. The power densities obtained in this comparison are presented in Table 5. The conventional MEA produced a maximum power of 50 mW/cm^2 and the sputtered MEA produced a maximum power of 33 mW/cm^2. Therefore, 3/5 of the power was produced with 1/10 the catalyst loading.

4.1. Graded catalyst deposition

A graded or composite catalyst layer refers to a variety of catalyst layers that are produced with multiple deposition methods. A typical form is a supported catalyst layer, PTFE-bound or thin-film electrode, with an additional sputtering of platinum on the surface of the membrane or electrode. The objective of this method is to reduce the thickness of the supported catalyst layer and increase the catalyst concentration at the interface between the electrode and polymer electrolyte membrane. Reducing the catalyst layer's thickness is vital for PTFE-bound catalyst layers as the depth that Nafion can be impregnated is limited to 10 μm [28]. Catalyst beyond 10 μm is unreachable by the protons and is therefore inactive. Fig. 8 illustrates the distribution of catalyst concentration in a graded catalyst layer.

Ticianelli et al. [28] of the Los Alamos National Laboratory published a paper in 1988 on the sputtering of catalyst onto electrodes with or without a preformed PTFE-bound layer present. They conducted their study by observing the performance of a conventional PTFE-bound electrode and comparing that with fuel cells employing a composite catalyst layer that combined a 50 nm thick sputtered layer and a PTFE-bound layer. They found the addition of a sputtered layer had the capability of increasing the power density 100–150%. At a current density of 1.0 A/cm^2, the single cell voltage increased from 0.42 to 0.54 V when a 50 nm layer of platinum was sputtered between a 20 wt.% Pt/C PTFE-bound catalyst layer with a Pt loading of 0.45 mg/cm^2. Mukerjee et al. [29] undertook a similar comparison, published in 1993, with observations concentrated on the oxygen reduction reaction kinetics. They found that the electrochemically active area was greater by a factor of two for a PTFE-bound electrode with the addition of a sputtered layer. As well, the oxygen reduction reaction overpotential was shown to be lower in the cell featuring the sputtered component.

In a recently granted (2001) patent, held by Cavalca et al. [25] (assigned to Gore Enterprise Holdings Incorporated), a method for fabricating electrodes with catalyst grading in the catalyst layer was divulged. The inventors combined thin-film methods and vacuum deposition techniques, such as electron beam-physical

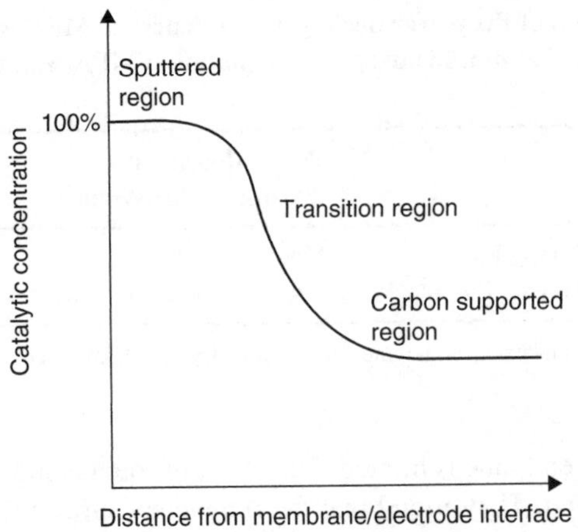

Fig. 8. Change in catalyst concentration with respect to distance from membrane. Reproduced from [25]

Table 6. Power densities of a single cell with a graded catalyst layer and a reference electrode that does not feature catalyst grading [25]

Catalyst layer	Power density at 200 mA/cm^2(mW/cm^2)	Power density at 0.6 V(mW/cm^2)
Reference MEA	156	510
MEA with graded catalyst layer	160	696

vapor deposition (EB-PVD) and dc magnetron sputtering, to fabricate a catalyst layer with progressive loading. This is not an entirely unique concept. Commercial PTFE-bound catalyst layers have often featured a 50 nm sputtering of platinum to enhance performance [11]. The preparation of the catalyst layer began by mixing a common thin-film ink that contained carbon supported platinum, Nafion solution, and solvents, which was then brushed onto a PTFE blank for transfer-printing. Subsequently, a layer of catalyst, single metal or bimetallic, was deposited via EB-PVD or sputtering onto either the thin-film catalyst, layer or the polymer electrolyte membrane. The inventors preferred method of vacuum deposition was EB-PVD because it exhibited greater surface texture, which aids the reaction kinetics. Thus, this method produces a dense pure catalyst layer directly adjacent to the membrane and places dispersed platinum further from the membrane with ionic transport provided by the impregnated Nafion. Table 6 presents the power densities for a single cell that features catalyst grading on the cathode side. This single cell is compared with an identical MEA that does not feature catalyst grading, and thus does not feature the EB-PVD region. It is visible in the polarization curves presented by Cavalca et al. that there is no significant benefit in the range of 0–400 mA/cm^2. However, at high current densities the performance of the graded cell is far superior. A 30% increase in power density at 1000 mA/cm^2 is presented. The membrane represented in the polarization curves is a 20 μm thick Gore Select membrane.

4.2. Multiple layer sputtering

Cha and Lee (CL) [4] presented a novel strategy for depositing the catalyst layer onto the membrane (Nafion 115) of a PEM fuel cell. The process consisted of multiple short sputterings separated by an application of carbon-Nafion ink. The process was carried out on both sides of the membrane. After each sputtering, the newly formed film was brushed with a Nafion solution and then again with a Nafion-XC-72 carbon powder-isopropyl mixture. The addition of the carbon powder increases the electrical conductivity in the intermediate Nafion layer. CL found that after enough catalyst had accumulated on the surface, additional sputtering of platinum does not contribute to the amount of active area. A single sputtering thickness of 5 nm was found to be ideal. However, when the Nafion-carbon powder-alcohol mixture was applied between additional 5 nm thick sputterings the performance increased considerably. But, the marginal increase in performance was negligible after five sputterings. CL found that if the same amount of catalyst was used, the case of multiple thin layers performed markedly better than a single thick layer. The best performance was found when the ratio of Nafion to carbon in the applied mixture was 1:1. The result of these electrodes is comparable to conventional carbon supported catalyst electrodes when the sputtering is repeated four times at a thickness of 5 nm and a Nafion-carbon mixture ratio of 1:1 is used. The conventional electrode had a Pt loading level of 0.4 mg/cm^2, whereas the sputtered electrode featured a loading of only 0.043 mg/cm^2, resulting in a ten times increase in platinum utilization.

5. ELECTRODEPOSITION METHODS

The first disclosure of electrodeposition of the catalytic layer in PEM fuel cells was in the form of Vilambi Reddy et al.'s 1992 US patent [30]. This patent detailed the fabrication of electrodes featuring low platinum loading in which the platinum was electrodeposited into their uncatalyzed carbon substrate in a commercial plating bath. The uncatalyzed carbon substrate consisted of a hydrophobic porous carbon paper that was impregnated with dispersed carbon particles and PTFE. Nafion was also impregnated onto the side of the carbon substrate that was to be catalyzed. The typical Nafion loading was 1.5 mg/cm^2.

The Nafion coated carbon paper was placed in a commercial platinum acid-plating bath, along with a platinum counter electrode. The face of the substrate that was not coated with Nafion was most likely masked with some form of a non-conducting film. This step would have been taken to ensure that platinum would only be deposited in regions impregnated with Nafion. Thus, when an interrupted dc current was applied to the electrodes in the plating bath, catalyst ions would pass through the Nafion to the carbon particles and successfully be deposited only where protonic and electronic conduction coexists. This method was able to produce electrodes featuring platinum loadings of 0.05 mg/cm^2. This is a significant reduction in loading from the state of the art thin-film electrode.

The performance of the aforementioned half-cell was revealed in a subsequent publication [31] by the same research group. Their results showed a definite increase in utilization due to the platinum catalyst only being deposited where the three-phase zone is located. Their electrodeposited catalyst layer electrode, with a Pt loading of 0.05 mg/cm^2, provided equivalent performance to the then state of the art PTFE-bound electrode with a Pt loading of 0.5 mg/cm^2.

In the following years, additional research on electrodeposition of platinum onto porous substrates was continued by Verbrugge [32]. According to Verbrugge, a distinguishing difference between his study and the aforementioned patent is the larger amount of sulfuric acid employed by Vilambi Reddy et al. [30]. Another distinguishing feature of Verbrugge's study is the employment of a membrane instead of a Nafion impregnated layer. Using the area provided by the deposition channel, platinum was selectively electrodeposited through the membrane and into the membrane-electrode interfacial region. Verbrugge suggested that this method has the potential to increase platinum utilization because of the concentrated platinum found at the membrane-electrode interface. However, he did not provide the results of these electrodes implemented in a functional fuel cell.

The objective of studies by Hogarth et al. [33], and later by Gloaguen et al. [34], was to improve the reaction kinetics for the oxidation of methanol using electrodes fabricated with electrodeposition. Hogarth

Table 7. Effect of frequency (on/off time) for a duty cycle of 25% and deposition current density of 50 mA/cm^2, H_2/O_2, $T = 70°C$ [35]

Frequency (Hz)	Power density at 0.6 V(mW/cm^2)
Dc	24
0.25	135
2.5	276
25	252

et al. placed electrodes in a plating bath that contained 0.02 M chloroplatinic acid and exposed only 1 cm^2 of the PTFE impregnated carbon cloth electrode face by using a water seal. In this study, neither a Nafion layer nor a membrane film was applied to the carbon substrate prior to electrodeposition. The Gloaguen et al. study focused on the oxygen reduction reaction kinetics of electrodes formed with the electrodeposition of platinum on carbon supports that were bound by Nafion onto a glassy carbon stick. One of the most significant conclusions of the study was that Pt activity is less related to particle size and more to the fine structure of the platinum surface.

5.1. *Effect of current control*

Recently (1998), a study was undertaken by Choi et al. [35] that investigated electrodeposition of platinum for producing electrodes in PEM fuel cells. Their specific interest was the effect of the current density, duty cycle, and frequency used during the electrodeposition process on the performance of their electrodes. Their uncatalyzed electrode was fabricated by applying a mixture of glycerin, PTFE, carbon black, and isopropyl alcohol to a carbon cloth backing. No Nafion or membrane film was applied to the carbon paper prior to the electrodeposition of the catalyst. A pulse generator was used to control the magnitude of the current passing through the electrodes that were placed in a bath of $H_2PtCl_6 \cdot 6H_2O$.

During the electrodeposition they varied the current density from 10 to 50 mA/cm^2, the duty cycle from 15 to 50%, and the frequency from 0 to 20 Hz. The group found that the optimum current density was 25 mA/cm^2. At higher values dendritic crystals form and performance falls. However, electrodes were fabricated most efficiently at a current density of 50 mA/cm^2 by applying an optimized duty cycle and frequency. The power density found for a fuel cell operating at 70°C with pure oxygen as the oxidant, where the catalyst layers were fabricated with a current density of 50 mA/cm^2, a duty cycle of 25%, and a frequency 2.5 Hz, was shown to be approximately 276 mW/cm^2 at a fuel cell voltage of 0.6 V (see Table 7).

5.2. *Membrane layer*

The highest performance publicized for a fuel cell featuring an electrodeposited catalyst layer was depicted in a US patent granted to Stäb and Urban [36] in 2001. In contrast to the Vilambi Reddy et al. patent [30], this group's uncatalyzed electrode did not feature a Nafion impregnated surface, but rather a thin membrane film. The membrane was approximately 10 μm thick and was applied to the carbon substrate, which was 75 μm thick. The cathodes were fabricated with the addition of a carbon powder sublayer. In the electrodeposition process, the back of the electrode was covered to avoid electrodeposition in locations far from the future membrane-electrode interface. Thus, during fabrication, the electrocatalyst passes from the electrolyte solution through membrane and is deposited where it encounters the electrically conductive carbon. The process deposits catalyst only where both protonic and electronic conduction

Table 8. Power densities of a cell fabricated by Stäb and Urban, which features a cathode produced by electrodeposition [36]

Catalyst layer	Power density at 200 mA/cm^2(mW/cm^2)	Power density at 0.6 V(mW/cm^2)
Produced by electrodeposition	158	408

Table 9. Power densities a MEA made by pressing a perfluorosulfonyl fluoride sheet with Pt/C catalyst at two different temperatures (0.2 mg Pt/cm^2, H_2/O_2 = 1/1 atm) [37]

Temperature	Power density at 200 mA/cm^2(mW/cm^2)	Power density at 0.6 V(mW/cm^2)
70	126	144
80	134	180

is possible. Using this methodology they were able to obtain some of the lowest platinum loading ever reported for an operational PEM fuel cell (0.0073 mg/cm^2 in the anode). The resulting power densities of this cell are listed in Table 8. Little detail is given to the conditions and components used to obtain these results. Nevertheless, the results do indicate a promising potential for this method. In this particular case, the cathode is formed with the electrodeposition method and the anode is a standard electrode with a Pt loading of 4 mg/cm^2. The power density reported at a current density of 200 mA/cm^2 is 158 mW/cm^2.

6. IMPREGNATED CATALYST LAYER

The ability to use fabrication techniques that require meltable materials, such as molding and extruding, would be extremely valuable in the production of MEAs. The conventional perfluorosulfonate acid membranes are not melt-processable because of side chain entanglement and the ionic interactions between the functional groups [37]. Kim et al. [37] have been working on a melt-processable membrane and the encapsulating MEA, which is formed out of perfluorosulfonyl fluoride copolymer (PFSF). PFSF is melt-processable and can be fabricated into many forms. Kim et al. formed membrane sheets by hot-pressing PFSF powder at 200–250°C. Since the membrane can be melted, the catalyst ink no longer requires Nafion ionomer. Kim et al. applied a catalyst ink that contained only Pt/C, glycerol, and water to both sides of the PFSF preformed sheet. Subsequently, they hot-pressed the catalyzed sheet at 200°C, which imbedded the catalyst layer into the softened surface of the membrane. This forms a composite membrane-electrode. The MEA is reported to have effective three-phase boundaries and good adhesion between the membrane and the catalyst layer. Though the performance depicted in the polarization curves of a test cell lags the present state of the art electrodes, the method does show promise (Table 9).

7. CATALYST SUPPORTS

The most common supported catalyst is platinum supported by high surface area carbon and is used in both the cathode and anode. When CO is present in the fuel stream because of reforming, the platinum is alloyed with other materials such as Ruthenium to reduce poisoning of the fuel cell and retain the performance. Electrocatalysts are commonly prepared by solution precipitation, which is followed by reduction of platinum salt in either the gas or liquid phase [13]. Though platinum and platinum alloys are employed in virtually all PEM fuel cell electrodes, other noble metals have been evaluated and there has been development of methods to synthesize non-noble metal catalysts, such as the pyrolysis of iron [38].

Catalyst metals employed in PEM fuel cells will not be reviewed herein as it is beyond the scope of the review. The metal chosen can have a profound effect on fuel cell performance, but does not alter the general design or fabrication of the electrode. However, in the following section it will be shown that the supporting material can influence the general design and fabrication of PEM fuel cell electrodes. The catalyst supports discussed in this section are not relevant to the vacuum deposition and electrodeposition methods of constructing the catalyst layer. In these methods, the catalyst metal is deposited directly onto the membrane or gas diffusion layer and no intermediate support is necessary.

7.1. *Pt/C weight ratio*

The most common support for the catalyst metal is carbon powder. The platinum to carbon weight ratio (Pt/C) is the ratio of the weight of platinum deposited onto the carbon support to the weight of the carbon support itself. Paganin et al. [14] found that their thin-film cell's performance was approximately unchanged when the Pt/C weight ratio was varied from 10 to 40 wt.% with a platinum loading of 0.4 mg/cm^2. However, the performance deteriorated as the weight ratio was increased beyond 40 wt.%. Paganin et al. suggested that this indicates a negligible change in the active catalyst area for weight ratios between 10 and 40 wt.%, and that the active area markedly decreases beyond these values. The Qi and Kaufman [17] electrodes performed marginally better when the platinum to carbon ratio (Pt/C) was increased from 20% to 40% with a platinum loading of 0.2 mg/cm^2.

7.2. *Binary carbon catalyst supports*

Wang et al. [39] reported on the use of binary carbon support catalysts. Their thin-film layer was catalyzed with platinum supported on Vulcan XC-72 and Black pearl 2000 carbon powders, featuring surface areas of 254 and 1475 m^2/g and particle sizes of 30 and 15 nm, respectively. The catalysts were mixed in a ratio of 9:1. The group constructed their thin-film catalyst layers according to Wilson's descriptions [3] with a platinum loading of 0.20 mg/cm^2. The results found using cyclic voltammetry showed the binary supported catalyst layer had a greater active area and increased the utilization efficiency by 14%. In addition, the higher power density found at 200 mA/cm^2 (Table 10) reveals an increase in the oxygen reduction reaction kinetics.

Table 10. Power densities for single- and binary-support electrodes

Support type	Power density at 200 mA/cm^2(mW/cm^2)	Power density at 0.6 V(mW/cm^2)
Binary	150	312
Single	136	246

Pt loading in cathode and anode are 0.2 and 0.35 mg/cm^2, respectively. Operating conditions: cell temperature, 60°C; atmospheric pressure; O_2 and H_2 flow rate at stoichiometry of 5 [39].

7.3. Conducting polymer catalyst supports

Qi et al. [40] prepared supported catalyst by depositing platinum onto protonically and electronically conductive polymers. The conducting polymers were fabricated from a composite of polypyrrole and polystyrenesulfonate. These supports were developed to provide a catalyst support that could fulfill the roles of both the electron conducting carbon powder and the Nafion in conventional catalyst layers. The development of these supports dissolves the need for Nafion layers or impregnation.

To fabricate electrodes, the conducting polymer supported catalyst was mixed with a 15% PTFE solution and applied to the carbon paper gas diffusion layer. The researchers attained maximum current densities of 100 mA/cm^2. Even though this does not compare well with state of the art electrodes, it does suggest future possibilities for this concept if the procedure for the synthesis of the conducting support composite is optimized.

7.4. Carbon nanohorn catalyst supports

Yoshitake et al. [41] deposited platinum catalyst onto high surface area single-wall carbon nanohorns for use in the catalyst layers of PEM fuel cells. The catalyst supports were prepared by CO_2 laser ablation. The platinum was deposited via a colloidal method. The platinum to support ratio in typical PEM fuel cells is 20% and ratios of 20–40% were obtained for the nanohorn-supported catalyst particles. The size of particles was 2 nm. The research group compared a fuel cell they prepared with catalyst supported by the usual carbon black to that of a fuel cell with catalyst supported by single-wall carbon nanohorn supports. The latter was shown to have better performance.

8. GAS DIFFUSION LAYER DEVELOPMENT

The gas diffusion layer has many roles to fulfill. Firstly, it is the electronic conductor between the current collecting bipolar plates and the catalyst layers. Thus, thin gas diffusion layers with a high conductivity is desired for electrical efficiency. Secondly, the gas diffusion layer is fabricated in the form of porous media to allow the passage of the reactant and product flows. To improve mass transport, gas diffusion layers can be made more porous at the cost of increased electrical resistance. In addition, the porous gas diffusion layer is often employed as the base substrate for the deposition of the catalyst layer. Another important function of the gas diffusion layer in an MEA is to reject liquid water from the internals of the MEA. If water collects near, or in, the catalyst layer, a large fraction of the catalyst will not be utilized. Typically, PTFE (Teflon) is applied through various methods to the gas diffusion layer in order to eject the water. However, Teflon is not an electric conductor and reduces the porosity, which hinders the transport of reactant gases. Thus, Teflon should be applied with careful measure. The common Nafion content in the solution applied to the gas diffusion layer is 33%.

Paganin et al. [14] fabricated a typical Teflonated gas diffusion layer by filtering a PTFE and carbon powder suspension onto both sides of a carbon cloth. The layer had to be dried for 30 min at 280°C to remove the dispersion agent from the PTFE suspension. The layer was then sintered at 330°C. Prior to use, the composite gas diffusion layers were cleansed by heat and chemical treatments.

A recent development in gas diffusion layers is the use of carbon aerogels to form the porous substrate. Glora et al. [42] documented their integration of carbon aerogels in PEM fuel cells. The 300 μm-thick layers featured micron-thin fine-structured layers on both sides of the gas diffusion layer. These fine layers are incorporated to decrease contact resistances between the electrode and the membrane, as well as the current collecting bipolar plate. The highest level of electrical conductivity attained in their study was 28 S/cm in an 80% porous sample. The largest pore sizes were in the range of several microns. The group

Table 11. Power for PEM fuel cells operating at 85°C

Gas diffusion layer	Power density at 200 mA/cm^2(mW/cm^2)	Power density at 0.6 V(mW/cm^2)
Carbon paper/Vulcan	152	402
Carbon cloth/Vulcan	156	552
Carbon cloth/Shawiningan	166	744

H_2/O_2 pressures = 2/2 atm for the electrodes with Vulcan on carbon paper or carbon cloth and 2/5 atm for the electrode with Shawinigan on carbon cloth in the gas diffusion layer [43].

implemented the gas diffusion layers in a test cell and found a power density only 1/6 that of a typical PEM fuel, but this was attributed to poor catalyst layer preparation and not the gas diffusion layer.

8.1. Polytetrafluoroethylene (PTFE) content

Paganin et al. [14] found that the PTFE content of the gas diffusion layer in their thin-film cell was optimal at a value of 15%. However, no drop in voltage, at any current density, was found when the content was varied from 10% to 40%.

8.2. Influence of carbon powder

Antolini et al. [43] studied the influence of the powder placed in both the gas diffusion and catalyst layers of PEM fuel cells. The two carbon powder types investigated were oil-furnace carbon black and acetylene-black, namely Vulcan XC-72R and Shawinigan carbon powder, respectively. The group found that an electrode featuring Shawinigan carbon powder performed substantially better than an electrode containing Vulcan carbon powder. The resulting power densities for each formulation of the gas diffusion layer are listed in Table 11. In addition, even higher levels of performance were attainable when combinations of the two carbon powders were used and the fuel and oxidant streams were sufficiently pressurized. The ideal combination found was to use Vulcan carbon powder in the catalyst layer, because of Vulcan's high surface area, and to impregnate the gas diffusion side of the electrode with the Shawinigan carbon powder.

8.3. Thickness

Paganin et al. [14] showed that the performance of their thin-film electrodes increases considerably when the gas diffusion layer thickness is increased from 15 to 35 μm. According to the researchers, this increase in performance is because the very thin layers provide a poor electrical contact between the catalyst layer and the current collecting plate. It is also possible that the thinner gas diffusion layers did not possess the mechanical strength to resist the compression of the bipolar plate. This compression would force the collapse of the pores in the gas diffusion layer underneath the ribs of the bipolar plate.

They also found that the marginal increase in performance was negligible when the thickness was further increased from 35 to 50 μm. In addition, they found a drop in performance at higher current densities when the diffusion layer thickness was increased to 60 μm. This polarization can be attributed to an increase in the diffusion distance and electrical resistance.

8.4. *Composite gas diffusion layer*

A composite gas diffusion layer commonly consists of Teflonated carbon cloth/paper backing with a microporous hydrophobic sublayer sandwiched between the carbon backing and the catalyst layer. The role of the hydrophobic sublayer is to improve the transport mechanisms across the porous backing and catalyst layer interface, and to enhance water management. QK [12] speculated that the improved water management is caused by the microporosity of the sublayer. The macropore size is largely determined by the particle size of the carbon aggregate. A common carbon particle size is 30 nm. QK suggest that stable water particles may not be able to form in such small and hydrophobic pores and is, therefore, less likely to flood.

Lufrano et al. [44] documented the results of thin-film electrodes with the introduction of an intermediate hydrophobic carbon layer. This sublayer was composed of carbon and PTFE that was sprayed onto the carbon paper backing. They found that the optimal PTFE content was 20 wt.%, though no significant change in performance was found when varying the content from 10 to 60 wt.%. However, when the oxidant stream consisted of air, the cell performance became more influenced by the PTFE content at elevated current densities.

Song et al. [19] formed a microporous sublayer, consisting of PTFE and carbon powder, that was sandwiched between a wet-proofed carbon paper and a thin-film catalyst layer. They employed ac impedance measurement techniques to optimize their thin-film electrodes. The researchers found that the optimal loading of the microporous sublayer was 3.5 mg/cm^2, with a PTFE concentration of 30 wt.%.

GA [13] from the Center for Electrochemical Systems and Hydrogen Research (CESHR) at Texas A&M University presented a paper in 2002 describing recent progress in performance improvement of PEM fuel cells at their center. The CESHR-developed gas diffusion layer consists of a 0.65:0.35 acetylene black-PTFE mixture spread and rolled at 3 mg/cm^2 onto carbon cloth. The porosity of CESHR gas diffusion layer was much greater than that of the standard ELAT gas diffusion layer that was also used in their study: 60% in comparison to 45%. Moreover, the thickness of the CESHR gas diffusion layer was 0.08 mm thinner than the standard 0.38 mm ELAT gas diffusion layer. Thus, a shorter diffusion path is a likely factor in the augmented performance of the CESHR gas diffusion layer over the standard gas diffusion layer.

Qi and Kaufman (QK) [45] documented a study where a microporous sublayer was inserted between the carbon cloth gas diffusion layer and the catalyst layer to improve water management. The microporous layer reduced the difference in performance when different batches of carbon paper were used. QK presented results that prove carbon paper supplied by the same manufacturer with identical specifications can provide significantly different performances. The best performing sublayer consisted of 35% PTFE and 65% Vulcan XC-72 carbon powder with a carbon loading of 2.0 mg/cm^2. The carbon paper, to which the sublayer was applied, contained 20% PTFE.

8.5. *Pore formers in the gas diffusion layer*

Kong et al. [46] documented an examination of the effect of pore size distribution in the gas diffusion layer. In order to study this effect pore former was added in varying amounts during the fabrication of the gas diffusion layer. The gas diffusion layer was formed from a viscous mixture containing carbon powder, isopropyl alcohol, and lithium carbonate (the pore former), that was applied to a carbon cloth. The catalyst layer used was thin-film and a Nafion 115 membrane was employed. Kong et al. also compared the pores formed by heat-treating versus the pore former additive. When the PTFE in the gas diffusion layer is heated to 350 K, the PTFE melts and transforms into a fibrous phase, increasing the porosity. They found that a combination of heat-treating and pore former addition produced the highest porosity and power density. The effect of each was visualized with scanning electron microscopy images presented by Kong et al.

[46]. Using mercury-intrusion porosimetry measurements they determined that heat-treating increased the pore volume of pores with diameters between 0.03 and 0.07 μm, whereas the addition of pore former increased the pore volume of pores with much larger diameters ranging between 2 and 13 μm. The optimum amount of pore former in the diffusion layer was determined to be 7 mg/cm^2 with 5 mg/cm^2 carbon loading in the diffusion layer and 0.4 mg/cm^2 platinum loading in the catalyst layer. At a voltage of 0.6 V their optimized cell produced a power density of 174 mW/cm^2, and with a current density of 200 mA/cm^2 the cell produced a power density of 136 mW/cm^2.

9. CONCLUSION

This report outlined major advances made in the fabrication of electrodes for PEM fuel cells from the PTFE-bound catalyst layers of almost 20 years ago to the present investigation of membrane-impregnated catalyst layers. It was found that the most common form of electrode is that featuring a thin-film catalyst layer. This is a common selection because of the ample proton conductivity provided by the binding Nafion film. This method is shown to significantly increase performance and reduce the level of platinum loading required. It is evident in the report that one of the main resistances in the catalyst layer is the membrane-catalyst layer interface contact resistance. The ability of the interface to conduct protons from the membrane into catalyst layer and the deposited catalyst is crucial. Thin sputtered layers have shown promise for low catalyst loading with adequate performance. This is achieved by the reduced thickness of the sputter deposited catalyst layer. Such short conduction distances alleviate the need for a proton-conducting medium. Electrodeposition methods were introduced and the suggestions of researchers that this method deposits catalyst only where electronic and protonic conduction exists were investigated. This and the ability to mass-produce electrodes in a commercial plating bath are considered to be the key advantages of electrodeposition. Integrated membrane-electrodes were highlighted to show the edge of the research envelope for electrode design. In addition, the continued development of gas diffusion layers and catalyst supports was investigated.

REFERENCES

1. M.S. Wilson, J.A. Valerio and S. Gottesfeld. Low platinum loading electrodes for polymer electrolyte fuel cells fabricated using thermoplastic ionomers. *Electrochim. Acta* **40** (1995) 355–363.
2. G.S. Kumar, M. Raja and S. Parthasarathy. High performance electrodes with very low platinum loading for polymer electrolyte fuel cells. *Electrochim. Acta* **40** (1995) 285–290.
3. M.S. Wilson. *Membrane Catalyst Layer for Fuel Cells*. US Patent 5 234 777 (1993).
4. S.Y. Cha and W.M. Lee. Performance of proton exchange membrane fuel cell electrodes prepared by direct decomposition of ultrathin platinum on the membrane surface. *J. Electrochem. Soc.* **146** (1999) 4055–4060.
5. R. O'Hayre, S.J. Lee, S.W. Cha and F.B. Prinz. A sharp peak in the performance of sputtered platinum fuel cells at ultra-low platinum loading. *J. Power Sources* **109** (2002) 483–493.
6. Y.-G. Chun, C.-S. Kim, D.-H. Peck and D.-R. Shin. Performance of a polymer electrolyte membrane fuel cell with thin film catalyst electrodes. *J. Power Sources* **71** (1998) 174–178.
7. S.J. Lee, S. Mukerjee, J. McBreen, Y.W. Rho, Y.T. Kho and T.H. Lee. Effects of Nafion impregnation on performances of PEMFC electrodes. *Electrochim. Acta* **43** (1998) 3693–3701.
8. O.J. Murphy, G.D. Hitchens and D.J. Manko. High power density proton-exchange membrane fuel cells. *J. Power Sources* **47** (1994) 353–368.
9. E.A. Ticianelli, C.R. Derouin, A. Redondo and S. Srinivasan. Methods to advance technology of proton exchange membrane fuel cells. *J. Electrochem. Soc.* **135** (1988) 2209–2214.
10. X. Cheng, B. Yi, M. Han, J. Zhang, Y. Qiao and J. Yu. Investigation of platinum utilization and morphology in catalyst layer of polymer electrolyte fuel cells. *J. Power Sources* **79** (1999) 75–81.

11. M.S. Wilson and S. Gottesfeld. Thin-film catalyst layers for polymer electrolyte fuel cell electrodes. *J. Appl. Electrochem.* **22** (1992) 1–7.
12. Z. Qi and A. Kaufman. Enhancement of PEM fuel cell performance by steaming or boiling the electrode. *J. Power Sources* **109** (2002) 227–229.
13. S. Gamburzev and A.J. Appleby. Recent progress in performance improvement of the proton exchange membrane fuel cell (PEMFC). *J. Power Sources* **107** (2002) 5–12.
14. V.A. Paganin, E.A. Ticianelli and E.R. Gonzalez. Development and electrochemical studies of gas diffusion electrodes for polymer electrolyte fuel cells. *J. Appl. Electrochem.* **26** (1996) 297–304.
15. E. Guzlow and T. Kaz. New results of PEFC electrodes produced by the DLR dry preparation technique. *J. Power Sources* **106** (2002) 122–125.
16. E. Gulzow, M. Schulze, N. Wagner, T. Kaz, R. Reisser, G. Steinhilber and A. Schneider. Dry layer preparation and characterization of polymer electrolyte fuel cell components. *J. Power Sources* **86** (2000) 352–362.
17. Z. Qi and A. Kaufman. Low Pt loading high performance cathodes for PEM fuel cells. *J. Power Sources* **113** (2003) 37–43.
18. E. Passalacqua, F. Lufrano, G. Squadrito, A. Patti and L. Giorgi. Nafion content in the catalyst layer of polymer electrolyte fuel cells: effect on structure and performance. *Electrochim. Acta* **46** (2001) 799–805.
19. J.M. Song, S.Y. Cha and W.M. Lee. Optimal composition of polymer electrolyte fuel cell electrodes determined by the ac impedance method. *J. Power Sources* **94** (2001) 78–84.
20. A. Fischer, J. Jindra and H. Wendt. Porosity and catalyst utilization of thin layer cathodes in air operated PEM-fuel cells. *J. Appl. Electrochem.* **28** (1998) 277–282.
21. V. Mehta and J.S. Cooper. Review and analysis of PEM fuel cell design and manufacturing. *J. Power Sources* **114** (2003) 32–53.
22. Y.-G. Yoon, G.-G. Park, T.-H. Yang, J.-N. Han, W.-Y. Lee and C.-S. Kim. Effect of pore structure of catalyst layer in a PEMFC on its performance. *Int. J. Hydrogen Energ* **28** (2003) 657–662.
23. S.-J. Shin, J.-K. Lee, H.-Y. Ha, S.-A. Hong, H.-S. Chun and I.-H. Oh. Effect of the catalytic ink preparation method on the performance of polymer electrolyte membrane fuel cells. *J. Power Sources* **106** (2002) 146–152.
24. E. Middelman. Improved PEM fuel cell electrodes by controlled self-assembly. *Fuel Cells Bulletin* November 2002 9–12.
25. C.A. Cavalca, J.H. Arps and M. Murthy. *Fuel Cell Membrane Electrode Assemblies with Improved Power Outputs and Poison Resistance*. US Patent 6 300 000 (2001).
26. M.F. Weber, S. Mamiche-Afare, M.J. Dignam, L. Pataki and R.D. Venter. Sputtered fuel cell electrodes. *J. Electrochem. Soc.* **134** (1987) 1416–1419.
27. S. Hirano, J. Kim and S. Srinivasan. High performance proton exchange membrane fuel cells with sputter-deposited Pt layer electrodes. *Electrochim. Acta* **42** (1997) 1587–1593.
28. E.A. Ticianelli, C.R. Derouin and S. Srinivasan. Localization of platinum in low catalyst loading electrodes to attain high power densities in SPE fuel cells. *J. Electroanal. Chem.* **251** (1988) 275–295.
29. S. Mukerjee, S. Srinivasan and J. Appleby. Effect of sputter film of platinum on low platinum loading electrodes on electrode kinetics of oxygen reduction in proton exchange membrane fuel cells. *Electrochim. Acta* **38** (1993) 1661–1669.
30. N.R.K. Vilambi Reddy, E.B. Anderson and E.J. Taylor. *High Utilization Supported Catalytic Metal-Containing Gas-Diffusion Electrode, Process for Making it, and Cells Utilizing it*. US Patent 5 084 144 (1992).
31. E.J. Taylor, E.B. Anderson and N.R.K. Vilambi. Preparation of high-platinum-utilization gas diffusion electrodes for proton-exchange-membrane fuel cells. *J. Electrochem. Soc.* **139** (1992) L45–L46.
32. M. Verbrugge. Selective electrodeposition of catalyst within membrane-electrode structures. *J. Electrochem. Soc.* **141** (1994) 46–53.
33. M.P. Hogarth, J. Munk, A.K. Shukla and A. Hamnett. Performance of carbon-cloth bound porous-carbon electrodes containing an electrodeposited platinum catalyst towards the electrooxidation of methanol in sulphuric acid electrolyte. *J. Appl. Electrochem.* **24** (1994) 85–88.
34. F. Gloaguen, J.M. Leger and C. Lamy. Electrocatalytic oxidation of methanol on platinum nanoparticles electrodeposited onto porous carbon substrates. *J. Electrochem. Soc.* **27** (1997) 1052–1060.
35. K.H. Choi, H.S. Kim and T.H. Lee. Electrode fabrication for proton exchange membrane fuel cells by pulse electrodeposition. *J. Power Sources* **75** (1998) 230–235.

36. G.D. Stäb and P. Urban. *Process for the Manufacture of an Electrode for a Solid Polymer Fuel Cell*. US Patent 6 258 239 (2001).
37. C.S. Kim, Y.G. Chun, D.H. Peck and D.R. Shin. A novel process to fabricate membrane electrode assemblies for proton exchange membrane fuel cells. *Int. J. Hydrogen Energy* **23** (1998) 1045–1048.
38. G. Faubert, R. Cote, J.P. Dodelet, M. Lefevre and P. Bertrand. Oxygen reduction catalyst for polymer electrolyte fuel cells from pyrolysis of Fe(II) acetate adsorbed on 3,4,9,10-perylenetetracarboxylic dianhydride. *Electrochim. Acta* **44** (1999) 2589–2603.
39. X. Wang, I.M. Hsing and P.L. Yue. Electrochemical characterization of binary carbon supported electrode in polymer electrolyte fuel cells. *J. Power Sources* **96** (2001) 282–287.
40. Z. Qi, M.C. Lefebvre and P.G. Pickup. Electron and proton transport in gas diffusion electrodes containing electronically conductive proton-exchange polymers. *J. Electroanal. Chem.* **459** (1998) 9–14.
41. T. Yoshitake, Y. Shimakawa, S. Kuroshima, H. Kimura, T. Ichihashi, Y. Kubo, D. Kasuya, K. Takahashi, F. Kokai, M. Yudasaka and S. Iijima. Preparation of fine platinum catalyst supported on single-wall carbon nanohorns for fuel cell application. *Physica B* **323** (2002) 124–126.
42. M. Glora, M. Wiener, R. Petricevic, H. Probstle and J. Fricke. Integration of carbon aerogels in PEM fuel cells. *J. Non-Cryst. Solids* **285** (2001) 283–287.
43. E. Antolini, R.R. Passos and E.A. Ticianelli. Effects of the carbon powder characteristics in the cathode gas diffusion layer on the performance of polymer electrolyte fuel cells. *J. Power Sources* **109** (2002) 477–482.
44. F. Lufrano, E. Passalacqua, G. Squadrito, A. Patti and L. Giorgi. Improvement in the diffusion characteristics of low Pt-loaded electrodes for PEFCs. *J. Appl. Electrochem.* **29** (1999) 445–448.
45. Z. Qi and A. Kaufman. Improvement of water management by a microporous sublayer for PEM fuel cells. *J. Power Sources* **109** (2002) 38–46.
46. C.S. Kong, D.Y. Kim, H.K. Lee, Y.G. Shul and T.H. Lee. Influence of pore-size distribution of diffusion layer on mass-transport problems of proton exchange membrane fuel cells. *J. Power Sources* **108** (2002) 185–191.

Chapter 22

Review and analysis of PEM fuel cell design and manufacturing

Viral Mehta and Joyce Smith Cooper

Abstract

Design and manufacturing alternatives for Proton Exchange Membrane (PEM) fuel cells are described and analysed within the context of vehicle applications. Specifically, following a review of many alternatives, 16 polymer electrolyte membranes, two types of gas diffusion layers (GDL), eight types of anode catalysts, four types of cathode catalysts and over 100 bipolar plate designs are recommended for further study. This work not only reviews membrane electrode assembly manufacturing options and synthesis processes for many of the membranes and for the gas diffusion layers, but also adds to the bipolar plate fabrication options described in literature. This work is intended to facilitate material and process selection through the consideration of the variety of design and manufacturing alternatives prior to capital investment for wide-scale production.

Keywords: PEM fuel cells; Membrane electrode assembly; Bipolar plate

Article Outline

1. Introduction 470
2. Review and analysis of membrane electrode assembly design and manufacturing 471
   2.1. MEA design 471
      2.1.1. Membrane design 471
      2.1.2. Catalyst layer design 474
      2.1.3. Gas diffusion layer design 481
   2.2. MEA manufacturing 481
      2.2.1. Membrane and GDL fabrication 481
      2.2.2. MEA assembly 485
3. Review and analysis of bipolar plate design and manufacturing 490
   3.1. Bipolar plate design 490
      3.1.1. Non-porous graphite plates 491
      3.1.2. Coated metallic plates 491
      3.1.3. Composite plates 493

Fuel Cells Compendium

3.2. Bipolar plate manufacturing .494
3.2.1. Non-porous graphite plate fabrication .494
3.2.2. Coated metallic plate fabrication .494
3.2.3. Composite plate fabrication .494
4. Discussion .496
Acknowledgments .499
References .499

1. INTRODUCTION

On 9 January 2002, the US Secretary of Energy Spencer Abraham and executives of Ford Motor Company, General Motor Corporation, and Daimler Chrysler announced a new cooperative automotive research partnership between the US Department of Energy and the US Council for Automotive Research (USCAR) called *FreedomCAR*. The partnership, which replaces the partnership for a New Generation of Vehicles program, focuses on the development of fuel cell vehicle technologies. Fuel cell vehicle technologies are those that enable mass production of affordable hydrogen-powered fuel cell vehicles and the hydrogen-supply infrastructure to support them. Among the vehicle technology options, proton exchange membrane (PEM) fuel cells, also referred to as solid polymer fuel cells, are favored for use in automobiles ([1,2], and many others). This preference is due to the high power density, relatively quick start-up, rapid response to varying loads, and low operating temperatures provided by PEM fuel cells.

Figure 1 depicts the key components of PEM fuel cells in which the oxidative and reductive half reactions are kept separate (i.e. in which the bipolar plates to be impervious to the reactants). As shown, a single PEM cell is comprised of three types of components: a membrane–electrode assembly (MEA), two bipolar (a.k.a. flow field or separator) plates, and two seals. In its simplest form, the MEA consists of a membrane, two dispersed catalyst layers, and two gas diffusion layers (GDL). The membrane separates the half reactions allowing protons to pass through to complete the overall reaction. The electron created on the anode side is forced to flow through an external circuit thereby creating current. The GDL allows direct and uniform access of the fuel and oxidant to the catalyst layer, which stimulates each half reaction. In a fuel cell stack, each bipolar plate supports two adjacent cells. The bipolar plates typically have four

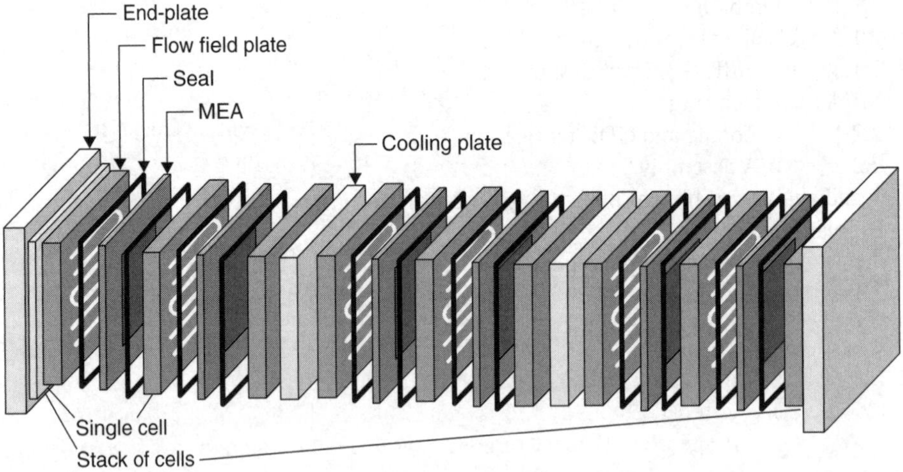

Fig. 1. PEM fuel cell stack hardware

functions: (1) to distribute the fuel and oxidant within the cell, (2) to facilitate water management within the cell, (3) to separate the individual cells in the stack, and (4) to carry current away from the cell. In the absence of dedicated cooling plates, the bipolar plates also facilitate heat management. Individual cells are combined into a fuel cell stack of the desired power. End plates and other hardware (bolts, springs, intake/exhaust pipes and fittings, etc. not shown in Fig. 1) are needed to complete the stack.

Previous works summarizing PEM fuel cell design alternatives are provided by Larminie and Dicks [3], EG&G Services [2], and Gottesfeld and Zawodzinski [1]. Specifically, Larminie and Dicks and EG&G Services provide textbooks on emerging fuel cell technologies. Their discussions of PEM fuel cell design include very general descriptions of materials use and configurations, the advantages and disadvantages of each design, stack performance relationships related to thermodynamics, water management, operating temperatures and pressures, and fuel and oxidant composition, and potential applications issues. Gottesfeld and Zawodzinski [1] provide a more research-oriented, electrochemistry-based discussion of fuel cell design when compared to these textbooks.

More specific discussions of materials and topologies for design alternatives can be found for specific components, typically accompanying related research or an analysis of that component. In particular, summaries of membrane materials have been published by Glipa and Hogarth from Johnson Matthey Technology Center, UK [4] and Rikukawa and Sanui from Sophia University, Japan [5]. Also, analysis of some bipolar plate materials is presented by Borup and Vanderborgh [6].

Similarly, PEM fuel cell manufacturing information can be found for specific components, especially for novel designs. Unlike PEM fuel cell design, current literature does not include summaries of manufacturing alternatives. Also, little analysis of fabrication options for more typical designs is available.

This paper, based on [7], reviews and extends existing PEM fuel cell design and manufacturing literature within the context of vehicle propulsion. We provide a comprehensive review of design and manufacturing alternatives described in literature for MEAs and bipolar plates. We also critique and broaden this set of alternatives based on a functional analysis of design, the application of process selection techniques with respect to component design features, and analyses of process inputs and outputs.

2. REVIEW AND ANALYSIS OF MEMBRANE ELECTRODE ASSEMBLY DESIGN AND MANUFACTURING

Figures 2 and 3 provide classifications of MEA material and manufacturing alternatives, described as follows.

2.1. MEA design

Again, an MEA consists of a membrane, a dispersed catalyst layer, and a GDL. The membrane separates the reduction and oxidation half reactions. It allows the protons to pass through to complete the overall reaction while forcing the electrons to pass through an external circuit. The catalyst layer stimulates each half reaction. The GDL further improves the efficiency of the system by allowing direct and uniform access of the fuel and oxidant to the catalyst layer. Design and manufacturing alternatives for each of these three components are reviewed and analyzed as follows.

2.1.1. Membrane design

Gottesfeld and Zawodzinski [1] suggest that perfluorosulfonic acid (PFSA) is the most commonly used membrane material for PEM fuel cells. PFSA consists of three regions: (1) a polytetrafluoroethylene (PTFE, a.k.a. DuPont's Teflon™)-like backbone, (2) side chains of –O–CF_2–CF–O–CF_2–CF_2– which connect the molecular backbone to the third region, and (3) ion clusters consisting of sulfonic acid ions [8].

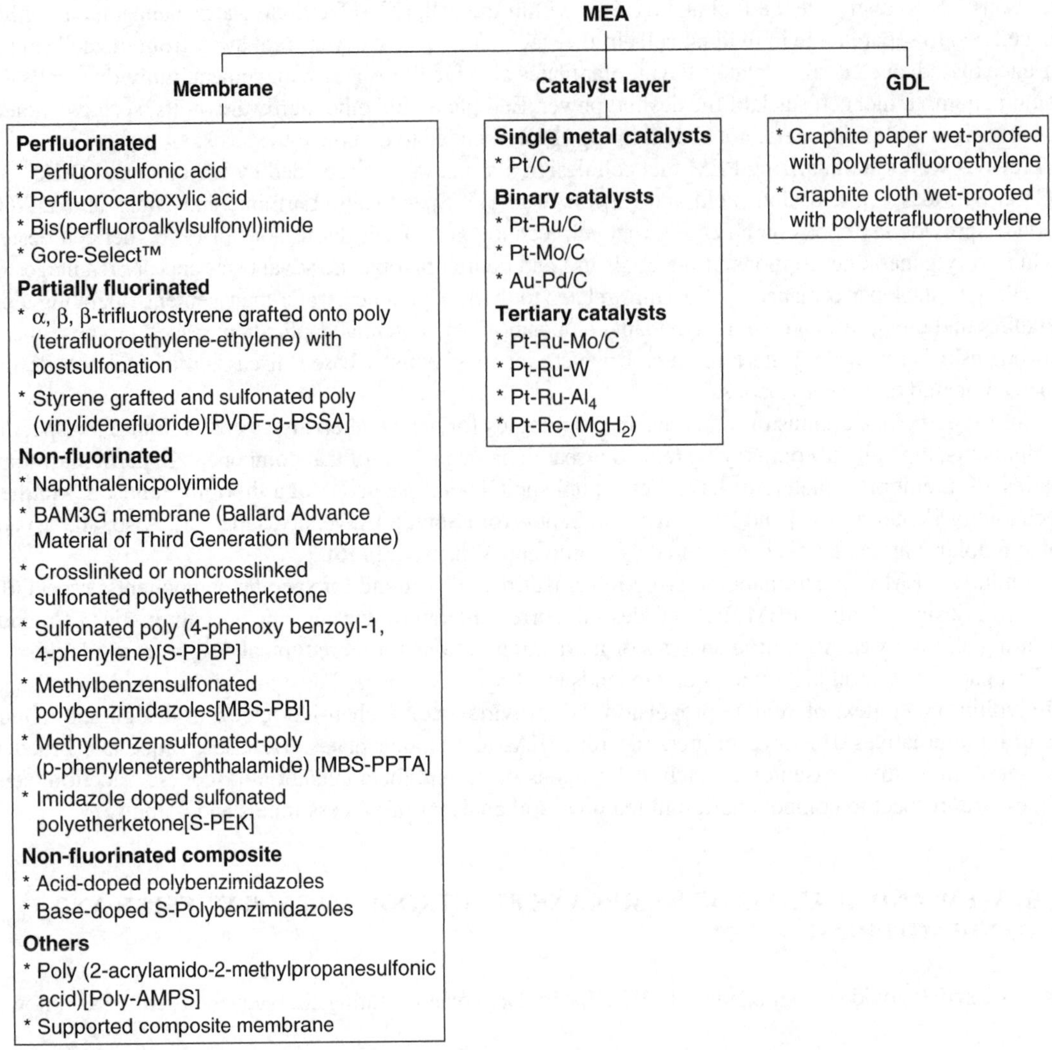

Fig. 2. Classification of MEA materials

When the membrane becomes hydrated, the hydrogen ions in the third region become mobile by bonding to the water molecules and moving between sulfonic acid sites.

There are two advantages to the use of PFSA membranes in PEM fuel cells. First, because the structure is based on PTFE backbone, PFSA membranes are relatively strong and stable in both oxidative and reductive environments. In fact, durability of 60 000 h has been reported [4]. Second, the protonic conductivities achieved in a well-humidified PFSA membrane can be as high as 0.2 S/cm at PEM fuel cell operating temperatures. This translates to a cell resistance as low as $0.05\,\Omega\,cm^2$ for a $100\,\mu$ thick membrane with voltage loss of only 50 mV at 1 A/cm^2 [1].

Given these advantages, there are several disadvantages to the use of PFSA membranes in PEM fuel cells. In addition to the membrane material being expensive (currently averaging US$25/kW [4]), disadvantages can be categorized as those related to safety, supporting equipment requirements, and temperature-related limitations. First, safety concerns arise from toxic and corrosive gases liberated at temperatures above 150°C [4,9].

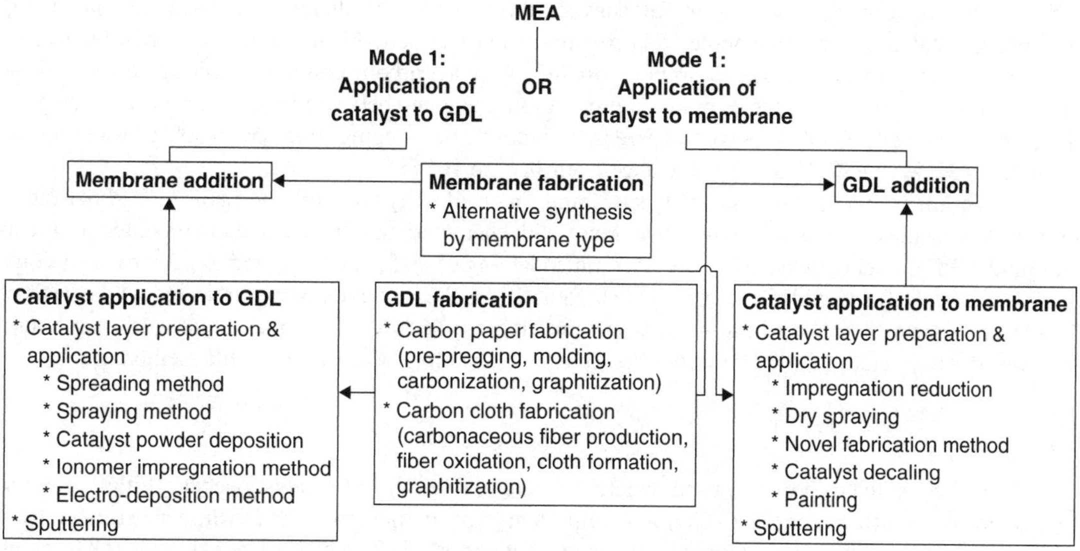

Fig. 3. Classification of MEA manufacturing alternatives

Decomposition products could be a concern during manufacturing emergencies or vehicle accidents and could limit fuel cell recycling options. Second, extensive supporting equipment requirements for use with PFSA membranes are described by Glipa and Hogarth [4] and Crawford [10]. Among the equipment needed, the hydration system adds considerable cost and complexity to the vehicle powertrain. Third, at elevated temperatures PFSA membrane properties degrade. For example, the conductivity at 80°C is diminished by more than 10 times relative to that at 60°C [5]. Also, phenomena like membrane dehydration, reduction of ionic conductivity, decreased affinity for water, loss of mechanical strength via softening of the polymer backbone and increased parasitic losses through high fuel permeation are observed at temperature above 80°C [4]. Making the temperature problems seem worse, Rikukawa and Sanui [5] note that operation of PEM fuel cells improves at elevated temperatures. Specifically, operation at elevated temperatures increases the rates of reaction, reduces problems related to catalyst poisoning by absorbed carbon monoxide in the 150–200°C range, reduces the use of expensive catalysts, and minimizes problems due to electrode flooding. Because PFSA membranes must be kept hydrated to retain proton conductivity, the operating temperature must be kept below the boiling point of water. Some increase in operating temperature, up to 120°C, may be possible at the expense of operation under pressurized steam. This alternative will however shorten the life of the cell.

Because of the disadvantages of PFSA membranes, an extensive literature review was done to identify alternatives. Much of the literature is summarized by Glipa and Hogarth from Johnson Matthey Technology Center, UK [4] and Rikukawa and Sanui from Sophia University, Japan [5]. Particularly, Rikukawa and Sanui suggest the foremost challenge is to produce materials that are cheaper than PFSA. They note that some sacrifice in material lifetime and mechanical properties may be acceptable, providing the cost factors are commercially realistic.

Among the different alternatives, Rikukawa and Sanui suggest the use of hydrocarbon polymers even though they had been previously abandoned due to low thermal and chemical stability. Hydrocarbon membranes provide some definite advantages over PFSA membranes. First, they are less expensive. Second, many types are commercially available. Third, polar groups can be formed to have high water uptakes over a wide temperature range with the absorbed water restricted to the polar groups of polymer chains. Fourth, decomposition of hydrocarbon polymers can be depressed to some extent by proper molecular design. Finally, it is possible membranes made from hydrocarbon polymers will be recyclable by conventional methods.

Glipa and Hogarth [4] extend upon Rikukawa and Sanui's list of alternatives. Their final taxonomy includes five categories of membranes: (1) perfluorinated, (2) partially fluorinated, (3) non-fluorinated (including hydrocarbon), (4) non-fluorinated (including hydrocarbon) composite, and (5) others. These authors also note the wide range of material properties among and between membranes in each category. Specifically, they cite membranes with degradation temperatures ranging from 250 to 500°C, water uptake from 2.5 to 27.5 H_2O/SO_3H, and conductance from 10^{-5} to 10^{-2} S/cm.

Together, Glipa and Hogarth and Rikukawa and Sanui identify over 60 alternatives to PFSA membranes. Among these, we identified 46 membranes with characteristics that make them ill-suited for use as automotive PEM fuel cells based on the recommendations of and personal communications with Glipa [11] Rikukawa [12] and with DesMarteau [13]. Table 1 lists these 46 membranes, rejected on the basis of 13 reasons shown as column headings. After removing the 46 "ill-suited" membranes, 16 membranes remain for further study. Table 2 provides design information for these 16 acceptable membranes.

2.1.2. Catalyst layer design

In PEM fuel cells, the type of fuel used dictates the appropriate type of catalyst needed. Within this context, tolerance to carbon monoxide (CO) is an important issue, particularly when hydrogen is formed from methanol by steam reforming. Methanol reformate contains as much as 25% carbon dioxide (CO_2) along with a small amount (1%) of carbon monoxide (CO). It has been proven that PEM fuel cell performance drops with a CO concentration of only several parts per million. This is due to the strong chemisorption force of CO onto the catalyst [25].

There are two techniques to counter the problem of CO poisoning: fuel reforming or catalyst alloying. First, the fuel can be reformed to reduce the CO level in fuel. If using on-board fuel reforming, it has been determined that the PEM fuel cell must be capable of tolerating a CO concentration of at least 100 ppm in order to reduce the size of the reformer unit. Reforming techniques include [2,26]:

- *Selective oxidation*: Selective oxidation is usually the preferred method for CO removal because of the parasitic system loads and energy required by the other methods. In selective oxidation, the reformed fuel is mixed with air or oxygen either before the fuel is fed into the cell or within the stack itself. Another approach involves the use of a selective oxidation catalyst that is placed between the fuel stream inlet and the anode catalyst. Current selective oxidation technologies can reduce CO levels to $<$10 ppm, but this is difficult to maintain under actual operating conditions.
- *Catalysis*: Ballard Power Systems has demonstrated that the CO level in fuel cell can be significantly reduced (to 100 ppm) by passing reformed methanol and small amount of oxygen over a Pt on aluminum catalyst.
- *Hydrogen peroxide bleeding*: The use of hydrogen peroxide (H_2O_2) in an anode humidifier successfully mitigated 100 ppm CO in an H_2 rich feed [27]. It was reported that mitigation appears to be provided by an unintended O_2 bleed produced by the decomposition of H_2O_2 in the humidifier rather than by H_2O_2 vapors transported from the humidifier to the anode.

When alloying the catalyst to counter the problem of CO, one (a binary catalyst) or sometimes two elements (a ternary catalyst) are added to the base catalyst. Table 3 lists 26 anode catalyst alloys. As shown, binary and ternary anode catalysts are typically, but not always, Pt-based and supported on carbon (or "/C"). It can be summarized that for hydrogen contaminated with CO there are at least seven Pt-based catalysts that give performance equal or similar to that given by Pt/C with pure hydrogen cell: Pt-Ru/C, Pt-Mo/C, Pt-W/C, Pt-Ru-Mo/C, Pt-Ru-W/C, Pt-Ru-Al_4, and Pt-Re-(MgH_2).

Table 3 lists 12 binary catalysts. Specifically, Iwase and Kawatsu [25] investigated 10 of these catalysts: Pt-Ru/C, Pt-Ir/C, Pt-V/C, Pt-Rh/C, Pt-Cr/C, Pt-Co/C, Pt-Ni/C, Pt-Fe/C, Pt-Mn/C, and Pt-Pd/C. Each catalyst was made of a 20 wt.% alloy on carbon with a Pt loading rate of 0.4 mg/cm^2 in a 5 wt.% PFSA solution.

Table 1. Membranes ill-suited for automotive applications

Membrane no.		1 No FC data (conceptual)	2 Expensive	3 Fire risk	4 Low durability	5 Degradation	6 Poor performance property	7 Water soluble	8 Shrinks (low flexibility)	9 Low conductivity	10 Low thermal stability	11 Low stability	12 Depoly-merization	13 No longer in production
1	Dow membrane													X
2	Watanabe self-humidifying membrane		X	X										
3	Poly(tetrafluoroethylene)-G-polystyrene sulfonic acid					X								
4	Poly(tetrafluoro-co-hexafluoropropylene)-G-polystyrene sulfonic acid					X								
5	Poly(ethylene-alt-tetrafluoroethylene)-G-polystyrene sulfonic acid					X								
6	Sulfonated phthalic polyimide				X		X							
7	Sulfonated poly(phenylqunox-aline)-(SPPQ or Ballard Advance Materials of first Generation-BAM1G)				X									
8	Sulfonated poly(2,6-diphenyl 1,4-phynylen oxide) or (Ballard Advance Materials of Second Generation-BAM2G)				X									
9	Sulfonated poly(aryl ether sulfone) or (Ballard Advance Materials of Second Generation BAM2G)				X									
10	Sulfonated styrene/ethylene-butadiene/styrene triblok copolymer					X						X	X	
11	Sulfonated polysulphone with SO_3H on bisphenol-A part							X						
12	Sulfonated polysulphone with SO_3H on diarylsulphone part							X						
13	Cross-linked sulfonated polysulphone with								X					
14	Sulfonated polyethersulphone cross-linked with diamines	X												
15	Benzylsulfonate-G-polybenzimidazoles								X					

(Continued)

Table 1. (*Continued*)

Membrane no.		1 No FC data (conceptual)	2 Expensive	3 Fire risk	4 Low durability	5 Degradation	6 Poor performance property	7 Water soluble	8 Shrinks (low flexibility)	9 Low conductivity	10 Low thermal stability	11 Low stability	12 Depoly-merization	13 No longer in production
16	Cross- or non-cross-linked sulfonated poly[bis(3-methylphenoxy)phosphazene] (*S*-PP membranes)						X							
17	Macrocomposites-aromatic polymers	X												
18	SiO_2/poly(ethylene oxide) doped with monododecylphosphate or phosphotungstic						X							
19	Sulfonated polysulphone doped with phosphoantimonic acid						X							
20	Phosphoric acid doped poly(ethylene oxide)-(PEO/H_3PO_4)							X			X			
21	Phosphoric acid doped poly(vinyl alcohol)-(PVA/H_3PO_4)							X			X			
22	Phospheric acid or sulfuric acid doped poly(acrylamide) PAAM/H_3PO_4/H_2SO_4							X			X			
23	Phospheric acid or sulfuric acid doped poly(vinylpyrrolidone)-(PVP/H_3PO_4/H_2SO_4)							X			X			
24	Phospheric acid or sulfuric acid doped poly(2-vinylpyridine)-(P_2VP/H_3PO_4/H_2SO_4)							X			X			
25	Phospheric acid or sulfuric acid doped poly(4-vinylpyridine)-(P_4VP/H_3PO_4)							X			X			
26	Phospheric acid or sulfuric acid doped linear poly(ethyleneimine)-(LPEI/H_3PO_4/H_2SO_4)							X			X			
27	Phospheric acid or sulfuric acid doped branched poly(ethyleneimine)-(BPEI/H_3PO_4/H_2SO_4/HCl)							X			X			
28	Phospheric acid doped poly(diallyldimethylammonium-dihydrogenophosphate) (PAMA+/H_3PO_4)							X			X			

Table 1. (*Continued*)

Membrane no.		1 No FC data (conceptual)	2 Expensive	3 Fire risk	4 Low durability	5 Degradation	6 Poor performance property	7 Water soluble	8 Shrinks (low flexibility)	9 Low conductivity	10 Low thermal stability	11 Low stability	12 Depoly-merization	13 No longer in production
29	KOH or NaOH or LiOH doped polybenzimidazoles (PBI/(KOH, NaOH, LiOH))	X												
30	Sulfonated polyethersulfone doped polybenzimidazoles (*S*-PSU/PBI)				X									
31	Sulfonated polyetherether ketone doped polybenzimidazoles (*S*-PEEK/PBI)				X									
32	Sulfonated polyphenylene oxide blended with poly(vinylidene fluoride) (*S*-PPO/PVDF)				X									
33	Polystyrene sulfonic acid (PSSA)					X								
34	Phenol sulfonic acid	X												
35	Poly(trifluorostyrene sulfonic acid)	X												
36	Poly(styrene)	X												
37	Poly(1,4-phenylene)	X												
38	Poly(oxy-1,4-phenylene)	X												
39	Poly(phenylene sulfide)	X												
40	Propane sulfonated polybenzimidazoles (PBI-PS)						X							
41	Butane sulfonated polybenzimidazoles (PBI-PS)						X							
42	Methylpropane sulfonated polybenzimidazoles (PBI-MPS)						X							
43	Propane sulfonated poly (*p*-phenylene terephthalamide) (PPTA-PS)						X							
44	Phosphoethylated polybenzimidazoles	X												

Table 2. Possible alternatives to PFSA membranes

Membrane no.	Membrane type (category)	Design information
1	α,β,β-Trifluorostyrene grafted membrane (partially fluorinated)	This membrane is based on grafting of α,β,β-trifluorostyrene and PTFE/ethylene copolymers [1]
2	Acid-doped polybenzimidazoles-(PBI) membrane (non-fluorinated composite)	This membrane is based on PBI and acids like phosphoric acid. PBI is a basic polymer ($pK_a = 5.5$) which can readily be complexed with strong acids. The immersion of a PBI film in aqueous phosphoric acid leads to a membrane which has high conductivity and thermal stability [14]
3	BAM3G membrane (Ballard Advance Material of Third Generation Membrane) (non-fluorinated)	This membrane is based on polymerization of α,β,β-trifluorostyrene and includes monomer(s) selected from a group of substituted α,β,β-trifluorostyrene. The polymers possess favorable properties, such as high heat stability, chemical resistance and favorable mechanical properties, such as tensile strength, compared to the homopolymeric material formed from α,β,β-trifluorostyrene (TFS) alone [15]
4	Base-doped *S*-polybenzimidazoles membrane (non-fluorinated composite)	This membrane is based on the introduction of organic or inorganic Bronsted bases to sulfonated PBI [4]
5	Bis (perfluoroalkylsulfonyl)imide membrane (perfluorinated)	Bis (perfluoroalkylsulfonyl) imide is based on the copolymerization of sodium 3,6-dioxa-Δ′-4-trifluoromethyl perfluorooctyl trifluoromethyl with tetrafluoroethylene (TFE). This membrane is thermally stable to nearly 400°C in the acid form. It has excellent conductivity and its water uptake is typically 40% by weight [13]
6	Cross- or non-cross-linked sulfonated polyetheretherketone membrane (non-fluorinated)	This membrane is based on polyetheretherketone. Direct sulfonation of polyetheretherketone results in materials with wide range of equivalent weights. The initial results obtained with the cross- and non-cross-linked *S*-PEEK membranes show very good thermal stability, proton conductance and water uptake compared to PFSA at even elevated temperature [16]
7	Gore-Select™ membrane (perfluorinated)	This is an ultra-thin integral composite membrane, which includes a base material and an ion exchange material or ion exchange resin with 0.025 mm thickness. The preferred base material is an expanded-polytetrafluoroethylene (e-PTFE) membrane with thickness of less than 0.025 mm and a porous microstructure. The ion exchange resin substantially impregnates the membrane. Suitable ion exchange materials include perfluorinated sulfonic acid resin, perfluorinated carboxylic acid resin, polyvinyl alcohol, divinyl benzene, styrene-based polymers, and metal salts with or without a polymer. A surfactant is preferably employed with the ion exchange material to ensure impregnation of the interior volume of the base material. Alternatively, the composite membrane may be reinforced with a woven or non-woven material bonded to one side of the base material. Suitable woven materials may include, scrims made of woven fibers of expanded porous polytetrafluoroethylene; webs made of extruded or oriented polypropylene or polypropylene netting [17]

Table 2. (*Continued*)

Membrane no.	Membrane type (category)	Design information
8	Imidazole doped sulfonated polyetherketone (*S*-PEK) membrane (non-fluorinated)	Sulfonated poly(arylether ketone) membranes and in particular sulfonated polyetherketone (*S*-PEK) exhibit high proton conductivities when in their hydrated forms. *S*-PEK can be complexed with imidazole to give membranes with high proton conductivities around 2×10^{-2} S/cm at a high temperature of 200°C [4,18]
9	Methylbenzensulfonated polybenzimidazoles membrane (non-fluorinated)	These alkylsulfonated aromatic polymer electrolyte posses very good thermal stability even above 80°C. Water uptake and proton conductivity are also reported to be higher than PFSA membranes above 80°C [5]
10	Methylbenzensulfonate poly (*p*-phenylene terephthalamide) membrane (non-fluorinated)	These alkylsulfonated aromatic polymer electrolyte posses very good thermal stability even above 80°C. Water uptake and proton conductivity are also reported to be higher than PFSA membranes above 80°C [5]
11	Perfluorocarboxylic acid membrane (perfluorinated)	Perfluorocarboxylic acid is based on a copolymer of tetrafluoroethylene and perfluorovinyl ether having a carboxylated group instead of a sulfonated group. The molar ratio of functional perfluorovinyl ether to tetrafluoroethylene in the copolymer is directly related to ion exchange capacity of resulting polymeric acid. Copolymerization of tetrafluoroethylene and functional perfluorovinyl ether is carried out by using a radical initiator [19]
12	Poly(2-acrylamido-2-methylpropanesulfonic acid) (poly-AMPS) membrane (other)	This membrane is made from polymerization of AMPS® monomer. AMPS® monomer is made from acrylonitrile, isobutylene and sulfuric acid [20]
13	Styrene grafted and sulfonated poly(vinylidene fluoride) membranes (PVDF-G-PSSA) (partially fluorinated)	This membrane is based on the pre-irradiation grafting of styrene onto a matrix of poly(vinylidene fluoride) (PVDF) after electron beam irradiation. It can be cross-linked with divinylbenzene (DVB) or bis (vinylphenyl) ethane (BVPE). The proton conductivity of membrane is influenced by degree of cross-linking [21]
14	Sulfonated naphthalenic polyimide (non-fluorinated)	This membrane is based on sulfonated aromatic diamines and dihydrides. It gives a performance very similar to PFSA membranes [4]
15	Sulfonated poly(4-phenoxybenzoyl-1, 4-phenylene) (*S*-PPBP) (non-fluorinated)	This membrane is based on poly(4-phenoxybenzoyl-1, 4-phenylene). This material is a poly(*p*-phenylene) derivative and is structurally similar to PEEK. The direct sulfonation of PPBP is reported to give a membrane that gives water absorption and proton conductance better than *S*-PEEK membranes [23]
16	Supported composite membrane (other)	Composite membrane is made of ion conducting polymer (ICP) and poly-*p*-phenylene benzobisoxazole (PBO) substrates [24]

They found that only the Pt-Ru catalyst showed cell performance equivalent to that of pure hydrogen cell with a single metal Pt/C catalyst when exposed to reformate gas with 100 ppm of CO. Also, they found that Ru in the binary catalyst absorbs water and facilitates the oxidation of CO. Although adequate CO tolerance can be obtained over a Ru-range of 15–85%, the optimum ratio of Pt/Ru was determined by Iwase and Kawatsu to be 50:50.

Table 3. Anode catalyst materials [22,25,28–34]

	Single metal catalyst	Binary catalyst	Tertiary catalyst
Pt/C	X		
Pt-Co/C		X	
Pt-Cr/C		X	
Pt-Fe/C		X	
Pt-Ir/C		X	
Pt-Mn/C		X	
Pt-Mo/C		X	
Pt-Ni/C		X	
Pt-Pd/C		X	
Pt-Rh/C		X	
Pt-Ru/C		X	
Pt-V/C		X	
Au-Pd/C		X	
Pt-Ru-Al_4			X
Pt-Ru-Mo/C			X
Pt-Ru-Cr/C			X
Pt-Ru-Ir/C			X
Pt-Ru-Mn/C			X
Pt-Ru-Co			X
Pt-Ru-Nb/C			X
Pt-Ru-Ni/C			X
Pt-Ru-Pd/C			X
Pt-Ru-Rh/C			X
Pt-Ru-W/C			X
Pt-Ru-Zr/C			X
Pt-Re-(MgH_2)			X

Other researchers add Pt-Mo/C and a non-Pt-based alloy Au-Pd/C to the list of possible binary catalysts. Specifically, Bauman et al. [28] found Pt-Mo/C to achieve high tolerance to low levels (10–20 ppm) of CO in reformate without the need of an air bleed. However, at CO levels above 20 ppm, the benefit of this catalyst is lessened. Although Pinheiro et al. [35] also found Pt-Ru/C to outperform Pt-Mo/C, Bauman et al. [28] found better performance with Pt-Mo/C as compared to Pt-Ru/C catalyst. Finally, Lawrence Berkeley researchers [33] have developed a non-platinum-based binary catalyst. They reported a three-fold improvement in electro-oxidation of CO/H_2 with their Au-Pd catalyst as compared to a Pt-Ru catalyst.

Tertiary catalysts are typically based on a Pt-Ru alloy. The largest number of tertiary catalysts along with some binary catalyst has been investigated by scientists at ECI Laboratories [29] and performances were compared to pure Pt/C catalyst performance. They investigated Pt-Ru alloys with Ni, Pd, Co, Rh, Ir, Mn, Cr, W, Zr, and Nb. They found that out of all the catalyst investigated, in the presence of CO, the binary catalysts $Pt_{0.53}$–$Ru_{0.47}$ and $Pt_{0.82}$–$W_{0.18}$ were far superior to pure platinum. Of the two, Pt-Ru was better in the low potential region while Pt-W proved superior in the plateau region except at very high current densities. But the performance of ternary $Pt_{0.53}$–$Ru_{0.32}$–$W_{0.15}$ alloy exceeded both binaries in the low potential and potential plateau regions. Similarly, Pinheiro et al. [35] analyzed the performance of Pt-Ru, Pt-Mo, and Pt-Ru-Mo/C and found the tertiary catalyst to have the best performance.

In another ternary catalyst development, Denis et al. [30] investigated the ternary electrocatalyst of Pt–Ru–Al_4 with no carbon support. Their results show that an unsupported Pt-Ru-Al_4 catalyst produced by high-energy ball milling gives equal performance to Pt-Ru/C when exposed to reformate gas with 100 ppm of CO. Using similar kind of ball milling technique, Dodelet et al. [31] produced a ternary catalyst Pt-Re-(MgH_2) without carbon support that performed better than Pt-Ru/C when exposed to reformate gas with 100 ppm of CO.

Little information was found on cathode catalysts for PEM fuel cells, which do not have to be CO tolerant. Notably, in addition to the use of Pt/C, Ross et al. [33] at Lawrence Berkeley National Laboratory report the use of Pt-Ni/C and Pt-Co/C as cathode catalyst. Also, Faubert et al. [34] produced a special, non-platinum-based cathode catalyst. The catalyst is produced by pyrolysis of iron acetate adsorbed on perylenetetracarboxylic dianhydride in Ar:H_2:NH_3 under ambient conditions. Also, at the National Renewable Energy Laboratory [32], a "rapid throughput" system has been developed to identify catalysts for oxygen reduction. This study investigates 1200 bimetallic complexes. Approximately 20 complexes were found suitable for fuel cells although detailed information about what these complexes was not included in the report.

2.1.3. Gas diffusion layer design

The GDLs, one next to the anode and the other next to the cathode, are usually made of a porous carbon paper or carbon cloth, typically 100–300 μm thick. The porous nature of the GDL material ensures effective diffusion of each reactant gas to the catalyst on the membrane/electrode assembly. The structure allows the gas to spread out as it diffuses so that the gas will be in contact with the entire surface area of the catalyzed membrane [8,36].

The GDL also assists in water management during the operation of the fuel cell. A GDL that allows the appropriate amount of water vapor to reach the membrane/electrode assembly keeps the membrane humidified and improves the efficiency of the cell. The GDL allows the liquid water produced at the cathode to leave the cell so it does not flood. The GDL is typically wet-proofed to ensure that at least some, and hopefully most, of the pores in the carbon cloth or paper do not become clogged with water, which would prevent the rapid gas diffusion necessary for a good rate of reaction to occur at the electrodes [8,36]. PTFE is the wet-proofing agent used for carbon-based PEM GDLs by several research groups [1,37,38].

A literature review did not reveal any research group who has studied both carbon paper and carbon cloth with the specific objective of identifying the most favorable among these two in a PEM fuel cell. In a study of water management, Ralph et al. [39] found that carbon cloth offered a distinct advantage at high current densities in Ballard Mark V cells. In fact, the slope of the pseudolinear region of the cell potential versus current density plot was lowered from 0.27 to 0.21 $\Omega\,cm^2$ and the limiting current was substantially raised by the use of the carbon cloth. Also, the cloth was found to enhance mass transport properties at the cathode derived from improved water management and enhanced oxygen diffusion rates. Finally, the surface porosity and hydrophobicity of the cloth substrate are more favorable for the movement of the liquid water.

2.2. MEA manufacturing

2.2.1. Membrane and GDL fabrication

Whereas the catalyst layer is typically prepared and applied during MEA assembly, the membrane and GDL are fabricated prior to assembly. Considering membranes first, a variety of polymerization processes are used in the fabrication of PFSA membranes and the alternatives listed in Table 2. Table 4 presents the processing steps and the primary inputs and outputs for many of these membranes. Notably, the processing steps include many chemical processes and a number of energy-intensive heating and drying steps. Process flow diagrams and additional synthesis information is available in [7].

Like the membrane, the GDL is fabricated prior to assembly. Carbon paper is fabricated in four steps: prepregging (continuous strands are aligned with spools and a surface treatment is followed by a resin bath and

Table 4. Analysis of membrane synthesis methods

Processing steps	Primary process inputs	Primary process outputs
Synthesis of PFSA membranes [40]		
1. Partial fluorination	Hydrogen fluoride (HF), antimony fluoride, chloroform	Chlorodifluoromethane
2. Pyrolysis	Chlorodifluoromethane	Tetrafluoroethylene (TFE)
3. Pyrolysis [290–370°C]	Diacyl fluoride	Perfluorinated vinyl ether
4. Copolymerization	TFE, perfluorinated vinyl ether	PFSA solution
5. Casting	PFSA solution	PFSA film
6. Sulfonation	PFSA film, sodium/potassium hydroxide	Sulfonated PFSA membrane
Synthesis of perfluorocarboxylic acid membranes [19]		
1. Reaction	Oleum, 1,4-diiodo-perfulorobutane	Cyclic lactone
2. Reaction	Methanol, cyclic lactone	3-Methoxycarbonyl perfluoro-propionyl fluoride
3. Pyrolysis	3-Methoxycarbonyl perfluoro-propionyl fluoride, HFPO	Carboxylated perfluorovinyl ethers
4. Partial fluorination	HF antimony fluoride, chloroform	Chlorodifluoromethane
5. Pyrolysis	Chlorodifluoromethane	TFE
6. Copolymerization	Carboxylated perfluorovinyl ethers, TFE	Perfluorocarboxylic acid (PFCA) solution
7. Casting	PFCA solution	PFCA film
8. Sulfonation	PFCA film, sodium/potassium hydroxide	Sulfonated PFCA membrane
Synthesis of poly-AMPS membranes [1]		
1. Reaction	Acrylonitrile, Isobutylene, Sulfuric acid	AMPS monomer
2. Addition (40°C)	Distilled water, AMPS monomer	Dissolved AMPS monomer
3. Addition	Dissolved AMPS monomer, ammonium persulfate in distilled water	AMPS with ammonium persulfate
4. Polymerization (60°C)	AMPS with ammonium persulfate, sodium metabisulfite in distilled water	Poly-AMPS
5. Casting	Poly-AMPS	Poly-AMPS membrane
α,β,β-Trifluorostyrene grafted onto poly(tetrafluoroethylene-ethylene) with post sulfonation membranes [1]		
1. Fluorination at α, β position	HF, vinylbenzene	Fluorinated vinylbenzene
2. Grafting	Fluorinated vinylbenzene, PTFE/ethylene	PTFE grafted fluorinated vinylbenzene
3. Sulfonation	PTFE grafted fluorinated vinylbenzene, sulfuric acid	α,β,β-Trifluorostyrene grafted poly(tetrafluoroethylene-ethylene) membrane
Styrene grafted and sulfonated PVDF membranes [21]		
1. Electron beam irradiation	PVDF films	Irradiated films
2. Mixing	Styrene, DVB or BVPE, toluene	Monomer
3. Grafting (immersion into monomer)	Monomer, irradiated films	Styrene grafted membranes
4. Sulfonation	Styrene grafted membranes, sulfuric acid	PVDF-G-PSSA membrane

Table 4. (***Continued***)

Processing steps	Primary process inputs	Primary process outputs
Synthesis of bis(perfluoroalkylsulfonyl) imide membranes [13]		
1. Polymerization	Monomer, TFE, $C_8F_{17}CO_2NH_4$, $(NH_4)_2S_2O_8/NHSO_3$	Emulsion
2. Filtration	Emulsion, acid	Bulk polymer
3. Drying	Bulk polymer	Dried polymer
4. Sonication	Dried polymer, DMF	Sonicated polymer
5. Removal of DMF in vacuum oven	Sonicated polymer	DMF-free polymer
6. Annealing at 220–250°C	DMF-free polymer	Annealed polymer
7. Boiling	HNO_3, annealed polymer	Bis(perfluoroalkylsulfonyl) imide membrane
Synthesis of Gore-Select™ membranes [17]		
1. Roll lamination/ultrasonic lamination/adhesive lamination	Polypropylene woven fabric, e-PTFE	Laminated e-PTFE sheets
2. Mixing	HC/FC based surfactant, PFSA/PFCA	Solution
3. Roll coating/Reverse roll coating/Gravure coating	Laminated e-PTFE sheets	Coated sheets
4. Removal of excess solution	Coated sheets, solution	Excess solution-free sheets
5. Oven drying	Excess solution-free sheets	Dried sheets
6. Repeat steps 3–5 several times		
7. Soaking	Water/H_2O_2/CH_3OH, dried sheets	Surfactant-free membrane
8. Boiling in swelling agent	Surfactant free membrane	Gore-Select™ membrane
Synthesis of BAM3G membranes [15]		
1. Mixing (35–96°C, 24–74 h, inert atmosphere)	α,β,γ-Trifluorostyrene monomer, substituted-α,β,β-trifluorostyrene comonomers, free radical initiator, emulsifying agent	Base copolymer
2. Dissolve in solvent	Base copolymer, dichloroethane, tetrachloroethylene and chloroform	Dissolved copolymer
3. Sulfonation	Dissolved copolymer, chlorosulfonic acid	BAM3G membrane
Synthesis of cross- or non-cross-linked sulfonated PEEK membranes [16]		
1. Polymerization	EEK monomer	PEEK
2. Sulfonation 60%	PEEK, 95–96.5% sulfuric acid	Sulfonated PEEK
3.1. Conversion of sulfonic acid group into sulfonyl chloride	Sulfonated PEEK, PC15/thionyl chloride	Sulfonyl PEEK
3.2. Crosslinking by high-energy radiation or heat	Sulfonyl PEEK, aliphatic/aeromatic amine, chloroform or dichloroethane	Cross-linked *S*-PEEK
OR		
3.1. Intra/inter chain polymerization (120°C, under vacuum)	Sulfonated PEEK	Cross-linked *S*-PEEK
4. Casting	Cross-linked *S*-PEEK	*S*-PEEK membrane

(*Continued*)

Table 4. (*Continued*)

Processing steps	Primary process inputs	Primary process outputs
Synthesis of sulfonated PPBP membranes [23]		
1. Mixing	Triphenylphosphine, nitrogen bis (triphenylphosphine) nickel(II) chloride, sodium iodide, zinc dust, anhydrous *N*-methylpyrrolidinone (NMP)	Solution
2. Stirring overnight	Solution	Solution
3. Coagulation in acetone bath	Solution	Coagulated solution
4. Blending	Coagulated solution	Crude polymer
5. Removal of excess zinc	Crude polymer, hydrochloric acid in ethanol	Excess zinc-free crude polymer
6. Filtration	Excess zinc-free crude polymer	Filtrate
7. Washing	Filtrate, water/acetone	Washed filtrate
8. Dissolve	Methylene chloride, washed filtrate	Dissolved solution
9. Filtration with polypropylene membrane	Dissolved solution	Filtrate
10. Coagulation	Acetone, filtrate	Coagulated polymer
11. Filtration	Coagulated polymer	Filtrate
12. Drying	Filtrate	Poly(4-phenoxybenzoyl-1,4-phenylene) (PPBP)
13. Dissolve	Chloroform, PPBP	Dissolved PPBP
14. Reprecipitation	Dissolved PPBP, methanol	Solid polymer
15. Drying	Solid polymer	Dried polymer
16. Pulverization	Dried polymer	Dried PPBP
17. Dissolve in H_2SO_4	Dried PPBP, H_2SO_4	Dissolved PPBP
18. Add	Dissolved PPBP, water	Water mixed PPBP
19. Filtration	Water mixed PPBP	Precipitate
20. Washing	Precipitate, water	Washed precipitate
21. Pulverization	Washed precipitate	Pulverized precipitate
22. Washing	Pulverized precipitate	Pulverized precipitate
23. Dialyzation	Pulverized precipitate, distilled water	Sulfonated polymer
24. Dissolve	NMP, sulfonated polymer	Dissolved polymer
25. Reprecipitation in tetrahydrofuran	Dissolved polymer, tetrahydrofuran	Re-precipitated polymer
26. Drying	Re-precipitated polymer	Dried polymer
27. Dissolve	NMP, dried polymer	2% wt. Solution
28. Casting	2% wt. Solution	Cast films
29. Drying	Cast films	Dried films
30. Washing	Dried films, methanol	Washed films
31. Vacuum drying	Washed films	Sulfonated PPBP membrane
Synthesis of acid-doped PBI membranes [41]		
1. Casting	Dimethylacetamide	Cast films
2. Boiling	Water, cast films	Films
3. Doping	Film, phosphoric acid	Acid doped PBI membrane
OR		
1. Casting	PBI	Cast films
2. Boiling	Water, cast films	Films
3. Doping	Film, phosphoric acid	Acid doped PBI membrane

Table 4. ***(Continued)***

Processing steps	Primary process inputs	Primary process outputs
Supported composite membranes [24]		
1. Water quench of substrate	Bi-axially oriented PBO extruded polymers, water	Hydrated substrate
2. Sulfonation of ion conducting polymer	PFSA, H_2SO_4	Sulfonated PFSA
3. Solvent exchange	Hydrated substrate, sulfonated PFSA	PFSA rich substrate
4. Tension drying	PFSA loaded substrate	Dried substrate
5. Degassing for complete solvent removal	PFSA loaded dried substrate	Composite membrane
6. Hot pressing	Composite membrane	Supported composite membrane

formation of a layered structure), molding, carbonization, and graphitization [24]. Carbon cloth is also fabricated in four steps: carbonaceous fiber production (made from mesophase pitch spun by melt spinning, centrifugal spinning, blow spinning, etc.), fiber oxidation, cloth formation by weaving or knitting, and graphitization [42]. Finally, the carbon cloth or paper is wet-proofed, typically using PTFE. Specifically, Bevers et al. [37] describe their wet-proofing process in which a carbon/PTFE suspension is applied to both sides of the carbon cloth or paper substrate. Application of the carbon/PTFE mixture flattens out any roughness of the cloth or paper and improves the gas and water transport properties.

2.2.2. MEA assembly

As shown in Fig. 4, there are two modes of MEA assembly: (1) application of the catalyst layer to the GDL followed by membrane addition or (2) application of the catalyst layer to the membrane followed by GDL addition. No matter the mode of assembly, the catalyst layer can be prepared and applied in two separate steps (catalyst preparation and application) or using a single sputtering process. As described later, several manufacturing options exist within these two modes of MEA manufacturing.

For either mode, early catalyst preparation methods were based on the use of platinum black. Later, Raistrick [43] used 10% carbon-supported platinum (Pt/C, 2 nm size particles) and a 100 μm thick catalyst layer instead of platinum black. The obvious advantage was a higher degree of platinum dispersion. Raistrick impregnated the Pt/C/PTFE catalyst layer on carbon cloth with a solution of PFSA, in order to fill it, or at least a significant part of it, with recast ionomer prior to hot pressing the impregnated electrode onto the membrane. This process overcame cell performance problems related to the lack of protonic access to the majority of catalyst sites not in intimate contact with membrane. Ticianelli et al. [44] further improved cell performance by optimizing the percentage of PFSA impregnant. They replaced a 10% Pt/C-100 m catalyst layer with a 20% Pt/C-50 m catalyst layer. Although this work was considered a major breakthrough by Gottesfeld and Zawodzinski [1], not all methods use ionomer impregnation, as follows. As described later, spreading method, spraying method, and catalyst powder deposition method do not use ionomer impregnation.

For mode 1, we identified five methods for catalyst preparation and application to prepare a GDL/catalyst assembly.

- *Spreading*: The spreading method described by Srinivasan et al. [45] consists of preparing a catalyzed carbon and PTFE dough by mechanical mixing and spreading it on a wet-proofed carbon cloth using a heavy stainless steel cylinder on a flat surface. This operation leads to a thin and uniform active layer on the GDL/catalyst assembly for which the Pt loading is directly related to the thickness.

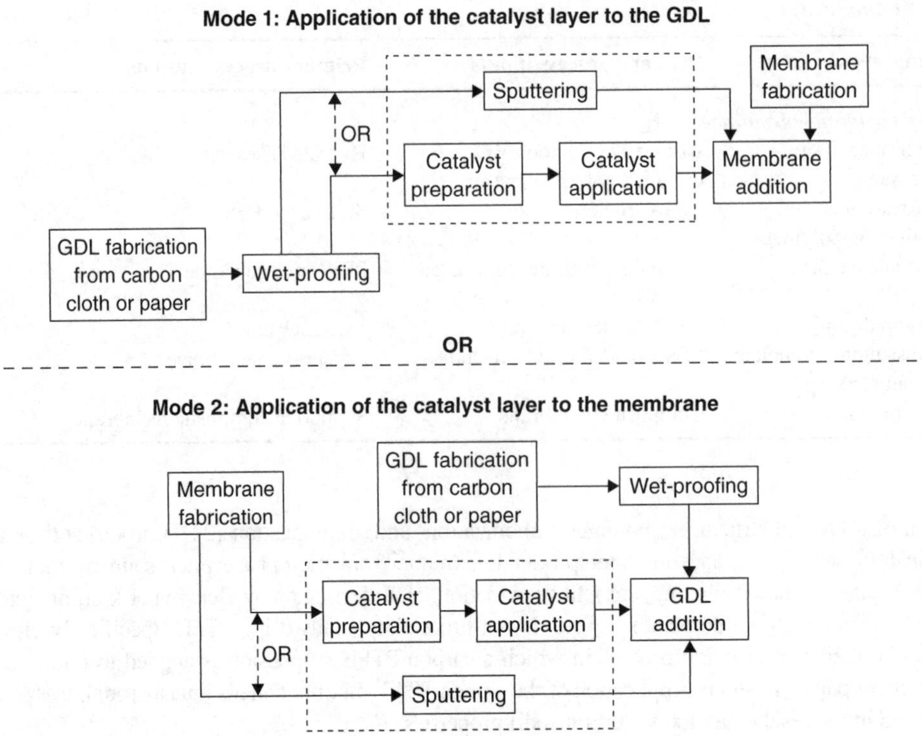

Fig. 4. Modes of MEA assembly

- *Spraying*: In the spraying method described by Srinivasan et al. [45], the electrolyte is suspended in a mixture of water, alcohol, and colloidal PTFE. This mixture is then repeatedly sprayed onto wet-proofed carbon cloth. Between each spraying, the electrode is sintered in order to prevent the components from re-dissolving in the next layer. The last step is rolling of the electrode. This operation has been found to produce a thin layer of uniform thickness and of low porosity on the GDL/catalyst assembly.
- *Catalyst powder deposition*: In catalyst powder deposition described by Bevers et al. [37], the components of the catalytic layer (Vulcan XC-72, PTFE powder, and a variety of Pt/C loadings) are mixed in a fast running knife mill under forced cooling. This mixture is then applied onto a wet-proofed carbon cloth. Also applying a layer of carbon/PTFE mixture flattens out the roughness of the paper and improves the gas and water transport properties of the MEA.
- *Ionomer impregnation*: In the ionomer impregnation method described by Gottesfeld and Zawodzinski [1], the catalytically active side of GDL is painted with solubilized PFSA in a mixture of lower aliphatic alcohols and water. To improve reproducibility of the GDL/catalyst assembly, the catalyst and ionomer are premixed before the catalyst layer is deposited, rather then ionomer impregnation of Pt/C/PTFE layer.
- *Electro-deposition*: Electro-deposition as described by Gottesfeld and Zawodzinski [1] and Taylor et al. [46] involves impregnation of the porous carbon structure with ionomer, exchange of the cations in the ionomer by a cationic complex of platinum and electrodeposition of platinum from this complex onto the carbon support. This results in deposition of platinum only at sites that are accessed effectively by both carbon and ionomer.

In mode 2, we identified six methods for catalyst application to prepare a membrane/catalyst assembly.

- *Impregnation reduction*: In impregnation reduction (a.k.a. electroless deposition) as described by Foster et al. [47] and Fedkiw and Her [48], the membrane, ion exchanged to the Na^+ form is equilibrated with an aqueous solution of $(NH_3)_4PtCl_2$ and a co-solvent of H_2O/CH_3OH. Following impregnation, vacuum-dried PFSA in the H^+ form is exposed on one face to air and the other to an aqueous reductant $NaBH_4$. The method has been found to produce metal loadings of the order of 2–6 mg Pt/cm^2 on the membrane/catalyst assembly.
- *Evaporative deposition*: In evaporative deposition as described by Foster et al. [47] and Fedkiw and Her [48], $(NH_3)_4PtCl_2$ is evaporatively deposited onto a membrane from an aqueous solution. After deposition of the salt, metallic platinum is produced by immersion of the entire membrane in a solution of $NaBH_4$. The method has been found to produce metal loadings of the order of $\leqslant 0.1$ mg Pt/cm^2 on the membrane/catalyst assembly.
- *Dry spraying*: In the dry spraying method described by Gulzow et al. [49], reactive materials (Pt/C, PTFE, PFSA powder and/or filler materials) are mixed in a knife mill. The mixture is then atomized and sprayed in a nitrogen stream through a slit nozzle directly onto the membrane. Although adhesion of the catalytic material on the surface is strong, in order to improve the electric and ionic contact, the layer is fixed by hot rolling or pressing. Depending upon the degree of atomization, a completely, uniformly covered reactive layer with thickness down to 5 μm can be prepared with this technique.
- *Novel fabrication method*: In a novel method suggested by Matsubayashi et al. [50], PFSA solution is mixed with the catalyst and dried in a vacuum. Then, the PFSA coated catalyst is mixed with a PTFE dispersion, calcium carbonate used to form pores, and water. The mixture is passed through a filter and the filtrate is formed into a sheet. The sheet is then dipped in nitric acid to remove any calcium carbonate. The sheet is then dried and PFSA solution is applied to one side of the electrode catalyst layer. Finally catalyst layer is applied to the membrane.
- *Catalyst decaling*: In the catalyst decaling method described by Gottesfeld and Wilson [51,52] and Gottesfeld and Zawodzinski [1], Pt ink is prepared by thoroughly mixing the catalyst and solubilized PFSA. The protonated form of PFSA in the ink is next converted into the TBA+ (tetrabutylammonium) form by the addition of TBAOH in methanol to the catalyst and PFSA solution. The paintability of the ink and the stability of the suspension can be improved by the addition of glycerol. Membranes are catalyzed using a "decal" process in which the ink is cast onto PTFE blanks for transfer to the membrane by hot pressing. When the PTFE blank is peeled away, a thin casting layer of catalyst is left on the membrane. In the last step, the catalyzed membranes are rehydrated and ion-exchanged to the H^+ form by immersing them in lightly boiling sulfuric acid followed by rinsing in deionized water.
- *Painting*: In the painting method described by Gottesfeld and Wilson [51,52] and Gottesfeld Zawodzinski [1], Pt ink is prepared as described for the decaling method. A layer of ink is painted directly onto a dry membrane in the Na^+ form and baked to dry the ink. When using thinner membranes or heavy ink applications, there will be considerable amount of distortion of the painted area. The distortion is managed through drying on a specially heated and fixtured vacuum table. Also, the bulk of the solvent is removed at a lower temperature to alleviate cracking and the final traces of solvent are rapidly removed at higher temperatures. In the last step, the catalyzed membranes are rehydrated and ion-exchanged to the H^+ form by immersing them in lightly boiling sulfuric acid followed by rinsing in deionized water.

In modes 1 and 2, sputtering can also be used as a single step option to catalyst preparation and application. In mode 1, Srinivasan et al. [45] describe a method in which a ~5 μm layer is sputter deposited on the wet-proofed GDL. Also, a remarkable improvement in performance was achieved when the catalyzed GDL was over-coated with a very thin layer of sputter deposited catalyst at the anode. Alternatively, they did not find over-coating improved performance at the cathode. In mode 2, Cha and Lee [38] describe a method in which the catalyst is sputtered onto both sides of the membrane. To enhance the performance,

Table 5. Analysis of catalyst preparation and application for MEA assembly mode 1

Processing steps	Primary process inputs	Primary process outputs
Catalyst preparation and application: spreading method [45]		
1. Dough making	Carbon supported metal-catalyst, PTFE	Dough
2. Spreading	Dough, wet-proofed GDL	Coated GDL
3. Rolling	Coated GDL	GDL/catalyst assembly
Catalyst preparation and application: spraying method [45]		
1. Composite mixture making	Carbon supported metal-catalyst, water, PTFE, alcohol	Mixture for spraying
3. Spraying	Mixture, wet-proofed GDL	Coated GDL
4. Sintering	Coated GDL	Coated and sintered GDL
5. Rolling	Coated and sintered GDL	GDL/catalyst assembly
Catalyst application: catalyst powder deposition method [37]		
1. Mixing	Carbon supported metal-catalyst, PTFE	Reactive powder
2. Powder application		
2.1. Using a line funnel under gravity	Powder, wet-proofed GDL	Coated GDL
OR		
2.1. Horizontal powder application	Powder, wet-proofed GDL	Coated GDL
3. Rolling	Coated GDL	GDL/catalyst assembly
Catalyst preparation and catalyst application: ionomer impregnation method [1]		
1. Mixing	Carbon supported metal-catalyst, PTFE, PFSA in aliphatic alcohols and water	Catalyst solution
2. Painting	Catalyst solution, wet-proofed GDL	GDL/catalyst assembly
Sputtering [45]		
1. Sputtering	Wet-proofed GDL, metal-catalyst	Sputter deposited paper
2. Rolling	Sputter deposited paper	GDL/catalyst assembly

a mixture of PFSA solution, carbon powder, and isopropyl alcohol is brushed on the catalyzed surfaces of membrane/catalyst assembly. The assembly is then dried in a vacuum chamber to remove any residential solvent. Sputtering and application of the ink is repeated to form a second layer of catalyst.

Tables 5 and 6 present processing steps and the primary inputs and outputs for catalyst preparation and application in two steps or by sputtering. Again, process flow diagrams and additional manufacturing information is available in [7].

The final step in modes 1 and 2 is the addition of the membrane and GDL respectively. Hot pressing is used in both modes. During the hot pressing procedure, the membrane will dry out but becomes re-hydrated adequately after insertion in the stack with humidified gases. Also, in mode 1, Ticianelli et al. [44] suggest that prior to hot pressing, the membrane be treated with a H_2O_2/H_2O solution heated to the boiling point, rinsed in deionized water, immersed in hot dilute sulfuric acid, and treated several times in boiling water. The process oxidizes organic impurities, removes any metallic impurities, and removes traces of acid from the finished MEA.

No matter the mode of assembly, the primary challenge in the assembly of MEAs is to achieve good contact between the membrane, the GDL, and the catalyst layers. Good contact maximizes catalyst utilization during cell operation. Gottesfeld and Zawodzinski [1] suggest good reproducibility is gained by hot pressing Pt/C/ionomer layers (that is Pt/C impregnated with premixed catalyst and ionomer) to the membrane as opposed to ionomer-impregnation of the GDL. This is because ionomer-impregnation is hard to achieve due to hydrophobic nature of GDL. Alternatively, Stuve [53] suggests that pressing a catalyzed GDL against the

Table 6. Analysis of catalyst preparation and application for MEA assembly mode 2

Processing steps	Primary process inputs	Primary process outputs
Catalyst preparation and application: impregnation reduction method (electroless deposition method) [48]		
1. Platinum salt impregnation	Membrane, $(NH_3)_4PtCl_2(aq)$	Membrane loaded with metal salt
2. Reduction of $(NH_3)_4Pt^{2+}$	Membrane loaded with metal salt, $NaBH_4(aq)$	Membrane/catalyst assembly
Catalyst preparation and application: evaporative deposition [47]		
1. Metal salt evaporation	Membrane, heat, metal salt such as $(NH_3)_4PtCl_2(aq)$	Membrane impregnated with metal salt
2. Reduction of metal ion	Membrane impregnated with metal salt, $NaBH_4(aq)$	Membrane/catalyst assembly
Catalyst preparation and application: dry spraying method [49]		
1. Composite powder making	Metal-catalyst supported C, PTFE, membrane material powder	Composite powder
2. Atomization	Composite powder	Atomized powder
3. Dry spraying	Atomized powder, membrane	Coated membrane
4. Hot rolling	Coated membrane	Membrane/catalyst assembly
Catalyst preparation and application: novel fabrication method [50]		
1. Catalyst preparation	PFSA, metal-catalyst	PFSA coated catalyst
2. Vacuum drying	PFSA coated catalyst	Dried catalyst
3. Mixing catalyst with other elements	Catalyst, PTFE, $CaCO_3$, water	Composite mixture
4. Filtration	Composite mixture	Filtrate
5. Rolling	Filtrate	Rolled sheet
6. $CaCO_3$ removal	HNO_3, sheet	$CaCO_3$ free sheet
7. Drying	Sheet	Electrode catalyst assembly
8. Hot pressing at 150°C	Electrode catalyst assembly, membrane	Membrane/catalyst assembly
Catalyst preparation and application: catalyst decaling method [1]		
1. Mixing	Metal-catalyst supported C, solubilize ionomer	Metal ink
2. Conversion of ink to TBA+ form	Metal ink, TBAOH	TBA+ ink
3. Apply TBA+ ink to PTFE blank	TBA+ ink, PTFE blank	Coated blank
4. Drying	Coated blank	Dried blank
5. Hot press to Na^+ membrane	Membrane, dried blank, heat	Membrane/blank assembly
6. Peel of blank	Membrane/blank assembly	Coated membrane
7. Protonation	Coated membrane, boiling H_2SO_4	Protonated membrane
8. Rinsing	Deionized water, membrane	Clean, protonated membrane
9. Air drying	Clean, protonated membrane	Membrane/catalyst assembly
Catalyst preparation and application: painting method [1]		
1. Painting of ink on Na^+ polymer electrolytemembrane	TBA+ ink, Na^+ membrane	Coated membrane
2. Oven dry in a vacuum table	Heat, coated membrane	Semi-dried coated membrane
3. Rapid heating	Semi-dried membrane	Dried coated membrane
4. Protonation	Coated membrane, boiling H_2SO_4	Protonated membrane
5. Rinsing	Deionized water, membrane	Clean, protonated membrane
6. Air drying	Clean, protonated membrane	Membrane/catalyst assembly
Sputtering [38]		
1. Mixing	PFSA solution, carbon powder, Isopropyl alcohol	Ion conducting polymer ink
2. Sputtering	Membrane	Sputtered membrane
3. Brushing	Ion conducting polymer ink, sputtered membrane	Coated membrane
4. Repeat 2 and 3		Membrane/catalyst assembly

membrane might cause some ionomer particles to embed in the electrode structure and thereby improve protonic access.

3. REVIEW AND ANALYSIS OF BIPOLAR PLATE DESIGN AND MANUFACTURING

Figures 5 and 6 provide classifications of bipolar plate material and manufacturing alternatives, described as follows.

3.1. Bipolar plate design

Bipolar plates perform a number of functions within the PEM fuel cell. They have been used to distribute the fuel and oxidant within the cell, separate the individual cells in the stack, carry current away from each cell, carry water away from each cell, humidify gases, and keep the cells cool. Plate topologies and materials facilitate these functions. Topologies can include straight, serpentine, or inter-digitated flow fields, internal manifolding, internal humidification, and integrated cooling. Materials have been proposed on the basis of chemical compatibility, resistance to corrosion, cost, density, electronic conductivity, gas diffusivity/impermeability, manufacturability, stack volume/kW, material strength, and thermal conductivity

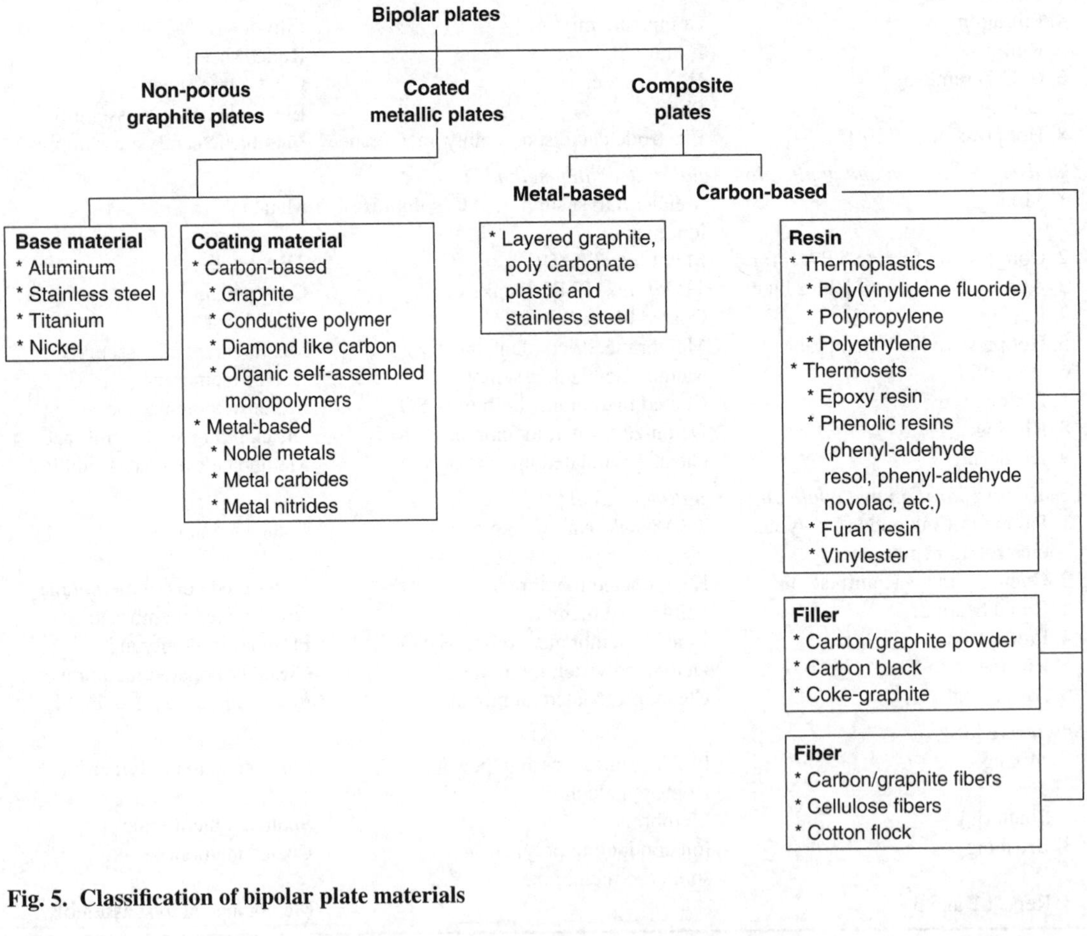

Fig. 5. Classification of bipolar plate materials

[6,8,36]. Given the criteria found in literature, non-porous graphite, a variety of coated metals, and a number of composite materials have been suggested for use in bipolar plates, as described later.

3.1.1. Non-porous graphite plates

Long experience has shown that non-porous graphite has the chemical stability to survive the fuel cell environment. Natural as well as synthetic graphite has been used to make non-porous bipolar plate material for PEM fuel cells.

3.1.2. Coated metallic plates

Aluminum, stainless steel, titanium, and nickel are considered possible alternative materials for the bipolar plate in PEM fuel cells. Although typically metallic plates are based on a single piece of metal, Allen [54] developed a modular metallic bipolar plate. The design provides for parallel flow of coolants within each sub-section arranged perpendicular to the direction of manufacture. The plate design combines nested subplates. The difference in depth of the nested subplates produces a center-cooling chamber with the interior surfaces of the subplates not in contact. Flat wire current collectors are bonded to the diffusion electrode or to the flow channels of the bipolar plate.

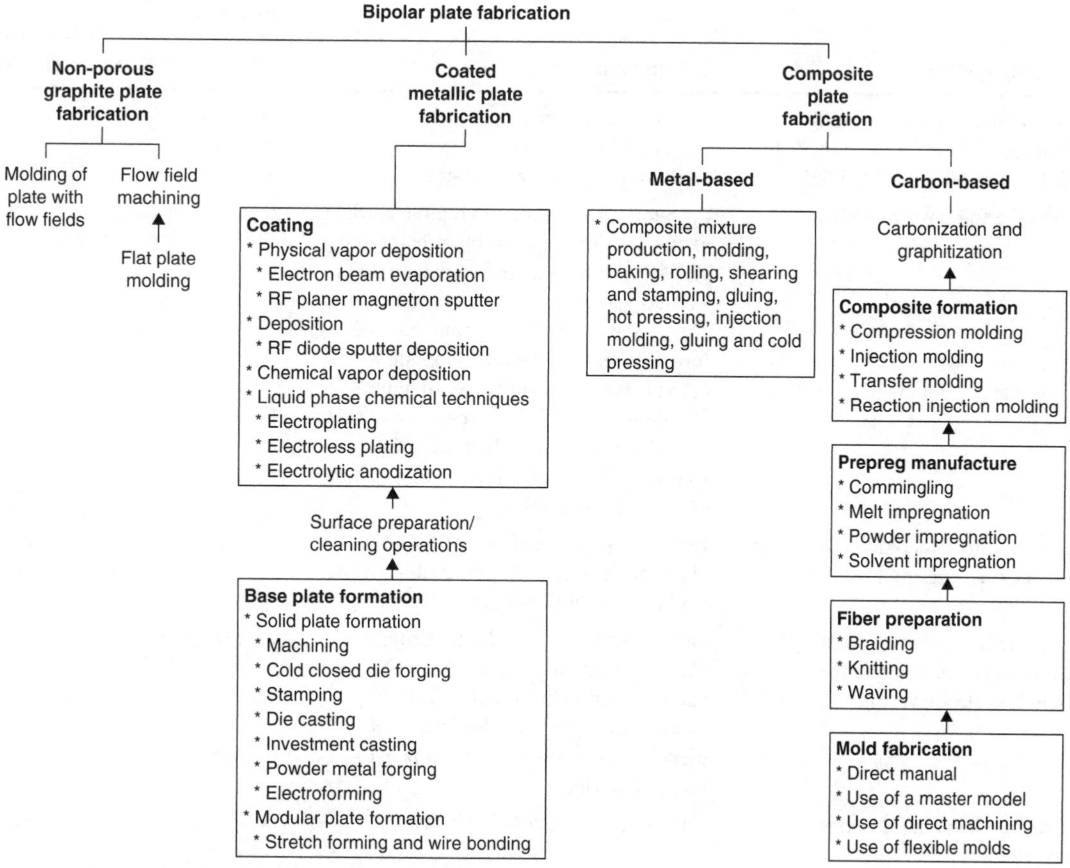

Fig. 6. Classification of bipolar plate manufacturing alternatives

No matter the plate configuration, because bipolar plates are exposed to an operating environment with a pH of 2–3 at high temperatures, if not designed properly, dissolution or corrosion of the metal will occur. Specifically, when the metal plate is dissolved, the dissolved metal ions diffuse into the membrane and are trapped at ion exchange cites, resulting in a lowering of ionic conductivity. In addition, a corrosion layer on the surface of a bipolar plate increases the electrical resistance in the corroded portion and decreases the output of the cell.

Because of these issues, metallic bipolar plates are designed with protective coating layers. Borup and Vanderborgh [6] suggest that coatings for bipolar plates should be conductive and adhere to the base material properly to protect the substrate from the operating environment. They present an overview of carbon-based and metallic bipolar plate coating materials. Carbon-based coatings include: (1) graphite, (2) conductive polymer, (3) diamond like carbon, (4) organic self-assembled monopolymers. Metal-based coatings include: (5) noble metals, (6) metal nitrides, (7) metal carbides. Table 7 lists bipolar plate coatings suggested by Borup and Vanderborgh and others. The coating techniques for these methods can be found in Table 10.

Woodman et al. [55] studied the corrosion failure mechanism for coated bipolar plates. They concluded that the coefficient of thermal expansion (CTE), corrosion resistance of coating, and micro-pores and

Table 7. Coating materials for metallic bipolar plates

		Applicable base plate materials			
Coating method	**Coating materials**	**Al**	**SS**	**Ti**	**Ni**
Conductive polymers coating [6]	Conductive polymers	Not specified			
Diamond-like carbon coating [6]	Diamond-like carbon	Not specified			
Gold topcoat layering [55,56]	Gold over nickel over copper	x			
Graphite foil layering [57]	(1) Sublayer – sonicated graphite particles in an emulsion, suspension or paint (e.g. graphite particles in an epoxy resin thinned by an organic solvent, such as toluene); (2) topcoat – exfoliated graphite in the form of sheets of flexible, graphite foil	x		x	x
Graphite topcoat layering [57]	(1) Sublayer – titanium over titanium–aluminum-nitride; (2a) overcoat – transient metal sublayer of Cr (Ti, Ni, Fe, Co) followed by sulfuric/chromic acid OR; (2b) topcoat – graphite	x	x	x	x
Indium-doped tin oxide layering [58]	Indium doped tin oxide ($Sn(In)O_2$)		x		
Lead oxide layering [58]	(1) Sublayer – lead; (2) topcoat – lead oxide (PbO/PbO_2)		x		
Organic monopolymer coating [6]	Organic self-assembled monopolymers	Not specified			
Silicon carbide layering [58]	(1) n-type silicon carbide (SiC); (2) gold		x		
Stainless-steel layering [59]	(1) Sublayer – chromium/nickel/ molybdenum-rich stainless steel OR nickel-phosphorus alloy; (2) topcoat – titanium nitride	x	x	x	
Titanium–aluminum nitride layering [58]	Titanium–aluminum-nitride layer	x			
Titanium nitride layering [58]	Titanium nitride (TiN) layer			x	

micro-cracks play a vital role in protecting the bipolar plate from the hostile PEM fuel cell environment. Woodman et al. also state that even though PEM fuel cells typically operate at temperatures less than 100°C, vehicle service would impose frequent startup and shut down conditions, and temperature differentials of 75–125°C would be expected. This is an important consideration for metal-coated bipolar plates because the two metals may expand and contract at different rates. The resulting micro-pores and micro-cracks may lead to failure if the base metal becomes exposed to the acidic fuel cell environment. A large difference in the CTE of the substrate and coating materials may lead to coating layer failure. One technique to minimize the CTE differential is to add intermediate coating layers with CTEs between that of adjacent layers.

3.1.3. Composite plates

Composite plates can be categorized as metal- or carbon-based. A metal-based composite bipolar plate has been developed by Los Alamos National Laboratory [24]. This design combines porous graphite, polycarbonate plastic and stainless steel in an effort to leverage the benefits of the different materials. Since producing porous graphite plates is not as time consuming or expensive as producing non-porous graphite plates, it can be used while impermeability is provided by the stainless steel and polycarbonate parts. Stainless steel also provides rigidity to the structure while the graphite resists corrosion. The polycarbonate provides chemical resistance and can be molded to any shape to provide for gaskets and manifolding. The layered plate appears to be a very good alternative from stability and cost standpoints.

Extensive literature was found citing the use of carbon-based composite bipolar plates in fuel cells. Table 8 summarizes notable citations. As shown, carbon composite bipolar plates have been made using thermoplastic (polypropylene, polyethylene, poly(vinylidene fluoride)) or thermosetting resins (phenolics, epoxies, and vinyl esters) with fillers and with or without fiber reinforcement. In the early years of development, thermoplastics were preferred as the resin material. However, in the late 1980s the preference changed to thermosets. Wilson [60] suggests this preference was due to short processing cycles. However, more recently select bipolar plate manufacturers, such as Micro Molding Technology [61], are producing bipolar plates using thermoplastic resin because they are more readily recyclable.

Table 8. Summary of carbon composite materials used in bipolar plates [62–71]

Resin	Type of resin	Filler	Fiber
Poly(vinylidene fluoride)	Thermoplastic	Carbon/graphite particles	
Poly(vinylidene fluoride)	Thermoplastic	Carbon/graphite particles	Carbon/graphite fibers
Polypropylene	Thermoplastic	Carbon black, graphite powder	
Mixture of an epoxy resin and an aromatic amine hardener	Thermoset	Graphite powder	
Phenyl-aldehyde resol OR phenyl-aldehyde novolac	Thermoset	Graphite powder	Graphite fibers or whiskers
Phenyl-aldehyde resol OR phenyl-aldehyde novolac	Thermoset	Coke-graphite particles	
Reichhold 24-655 phenolic resin	Thermoset	Graphite powder	Cellulose fibers (but not rayon and cellulose acetate)
Phenol resin or furan resin	Thermoset	Graphite powder	Cellulose fibers (but not rayon and cellulose acetate)
Phenolic resin	Thermoset		Carbon fibers (PAN-based)
Vinyl ester	Thermoset	Graphite powder	Cotton flock (graphite/carbon, glass, cotton, and polymer)

3.2. Bipolar plate manufacturing

3.2.1. Non-porous graphite plate fabrication

Woodman et al. [55] states that state-of-the-art PEM fuel cells contain graphite bipolar plates that are machined or molded with flow fields. Although no other citation was found that discussed machined graphite bipolar plates, compression molding is discussed by Meissner [72]. In this method, graphite mixtures that contain crystalline graphite with additives and/or binders are compression molded and preferably subjected to a heat treatment in the absence of oxygen. Suitable additives include aluminum oxide, zircon dioxide, silicon dioxide, titanium dioxide, silicon carbide, and powdered coke. Suitable binders are cokable from 300 to 800°C and include carbohydrates such as fructose, glucose, galactose, and mannose and oligosaccharides such as sucrose, maltose, and lactose.

3.2.2. Coated metallic plate fabrication

The process for fabricating coated metallic bipolar plates includes formation of the base plate, surface preparation and cleaning operations, and coating processes. For the more typical solid metallic bipolar plate designs, formation of the base plate by machining or stamping is suggested by Woodman et al. [55]. Mehta [7], in an effort to extend Woodman's suggestions, applied the process selection system, *Cambridge Engineering Selector* by Granta Designs to identify additional solid plate formation processes. Using design information such as materials, part shape and size, tolerance, surface finish, production quantity and a "quality factor" (used to select among processes prone to defects), Mehta added five fabrication alternatives to Woodman et al.'s original two: cold closed die forging, die casting, investment casting, powder metal forging, and electroforming. The result of Mehta's analysis, which assumed production quantities above 10 000, is presented in Table 9 for two basic plate designs. The analysis captures changes in quality (factors 5–10), tolerance (from 0.05 to 0.1 mm) and surface finish (10–100 μm) for the two basic designs. In general, the number of manufacturing options increases as tolerances increase and as surface finish and quality requirements decrease. No differences were noted above these ranges, and literally no processes fall below them. A significant observation was that more processes were identified for the larger faced plate. Specifically, investment casting and powder metal forging and often die-casting were not compatible to produce the smaller faced design.

For Allen's less typical modular metallic bipolar plate [54] discussed in Section 3.1.2, plate formation is more specialized. In plate formation, flow channels and manifolds are "stretched formed" into finite subsections by progressive tooling. The tooling is designed such that the tool may be closed to any engagement to produce any depth for the flow ribs within the elongation constraints of the material being formed. As such, a subplate stretch formed to the maximum extent will nest with a subplate stretch formed to less than the maximum extent. Finally, parallel flat wire current collectors are continuously slit from coiled material and bonded to the diffusion electrode or to the flow channels of the bipolar plate prior to assembly.

Coating processes for solid or modular metallic bipolar plates include a variety of deposition processes as listed in Table 10. Processes include physical vapor deposition techniques like electron beam evaporation, sputtering and glow discharge decomposition, chemical vapor deposition technique, and liquid phase chemical techniques like electro- and electroless deposition, chemical anodization/oxidation overcoating, and painting.

3.2.3. Composite plate fabrication

For the layered metal composite bipolar plates developed at Los Alamos [24] described in Section 3.1.2, a multi-step process is used in fabrication. First, a stainless steel layer is produced using shearing and stamping methods. Next, graphite powder and resin are mechanically mixed and molded to the required

Table 9. Analysis of process options for solid metallic bipolar plates

	Basic design												
	Smaller faced, thicker plate[a]						Larger faced, thinner plate[b]						
Quality factor[c]	5–6	7–8	9–10	8	8	8	5–6	7–8	9	10	8	8	8
Design tolerance (mm)	0.1	0.1	0.1	0.05	0.08	0.1	0.1	0.1	0.1	0.1	0.05	0.08	0.1
Surface finish (μm)	10	10	10	10	10	50–100	10	10	10	10	10	10	50–100
Aluminum													
Machining	x	x	x	x	x	x	x	x	x	x	X	x	x
Cold closed die forging	x	x	x	–	–	x	x	x	x	x	–	–	x
Stamping	x	x	–	–	x	x	x	x	–	–	–	x	x
Die casting	x	–	–	–	–	–	x	–	–	–	–	–	–
Investment casting	–	–	–	–	–	–	x	x	–	–	–	–	x
Powder metal forging	–	–	–	–	–	–	x	x	x	x	X	x	x
Stainless steel													
Machining	x	x	x	x	x	x	x	x	x	x	X	x	x
Cold closed die forging	x	x	x	–	–	x	x	x	x	x	–	–	x
Stamping	x	x	–	–	x	x	x	x	–	–	–	x	x
Die casting	–	–	–	–	–	–	–	–	–	–	–	–	–
Investment casting	–	–	–	–	–	–	x	x	–	–	–	–	x
Powder metal forging	–	–	–	–	–	–	x	x	x	–	X	x	x
Titanium													
Machining	x	x	x	x	x	x	x	x	x	x	X	x	x
Cold closed die forging	x	x	x	–	–	x	x	x	x	x	–	–	x
Stamping	x	x	–	–	x	x	x	x	–	–	–	x	x
Die casting	–	–	–	–	–	–	–	–	–	–	–	–	–
Investment casting	–	–	–	–	–	–	x	x	–	–	–	–	x
Powder metal forging	–	–	–	–	–	–	x	x	x	–	X	x	x
Nickel													
Machining	x	x	x	x	x	x	x	x	x	x	x	x	x
Cold closed die forging	x	x	x	–	–	x	x	x	x	x	–	–	x
Stamping	x	x	–	–	x	x	–	–	–	–	–	–	–
Die casting	–	–	–	–	–	–	–	–	–	–	–	–	–
Investment casting	–	–	–	–	–	–	x	x	–	–	–	–	x
Powder metal forging	–	–	–	–	–	–	x	x	x	–	x	x	x
Electroforming	x	x	x	x	x	x	–	–	–	–	–	–	–

[a] Maximum surface area 100 × 150 mm; thickness 5.4 mm; flow channel width 1.6 mm; maximum hole diameter 6.4 mm; minimum section 1.6 mm; aspect ratio 30.

[b] Maximum surface area 250 × 600 mm; thickness 4.7 mm; flow channel width 1.6 mm; maximum hole diameter 6.4 mm; minimum section 1.6 mm; aspect ratio 49.

[c] The quality factor is assessed on a numerical scale of 1–10 such that processes that are prone to produce defects receive a 1 and processes that minimize defects receive a 10.

shape by conventional compression or injection molding. The resulting graphite plate is baked in an oven. Next, a conductive adhesive is applied to the graphite plate using a screen-printing method. Next, a hot-press is used to join the stainless steel and graphite plates. Finally, to form the polycarbonate plates, polycarbonate resin is injection molded to the required shape and adhesive is applied to bond the plate with the stainless steel/graphite plate assembly using cold pressing.

For carbon-based composite bipolar plates, fabrication includes mold fabrication (direct manual, fabrication using a master model, fabrication through direct machining, fabrication through flexible molds),

Table 10. Coating processes for metallic plates

Coating method[a]	Coating processes
Gold topcoat layering	Pulse current electrodeposition
Stainless steel layering	Physical vapor deposition (PVD) (e.g. magnetron sputtering), or chemical vapor deposition (CVD), and electroless deposition for Ni–Ph alloy
Graphite topcoat layering	PVD (closed-field, unbalanced, magnetron sputter ion plating) and chemical anodization/oxidation overcoating
Graphite foil layering	Painting OR pressing
Titanium nitride layering	RF-diode sputtering
Indium doped tin oxide layering	Electron beam evaporation
Lead oxide layering	Vapor deposition and sputtering
Silicon carbide layering	Glow discharge decomposition and vapor deposition
Titanium aluminum nitride layering	RF-planar magnetron (sputtering)

[a] See also Table 7.

Table 11. Analysis of process options for carbon composite bipolar plate designs

	Basic design				
	Smaller faced, thicker plate[a]				
Quality factor	5–6	7–10	8	8	8
Design tolerance (mm)	0.1	0.1	0.05	0.08	0.1
Surface finish (μm)	10	10	10	10	50–100
Compression molding	x	–	–	–	X
Injection molding	x	–	–	x	X
Transfer molding	x	–	–	x	X
Reaction injection molding	x	–	–	x	X

[a] Maximum surface area 100 × 150 mm; thickness 5.4 mm; flow channel width 1.6 mm; maximum hole diameter 6.4 mm; minimum section 1.6 mm; aspect ratio 30.

fiber preparation (braiding, knitting, weaving), prepreg manufacture (commingling; melt, powder, or solvent impregnation), and composite formation. Post-processing steps include carbonization and graphitization. In carbonization, the resin is converted into carbon by controlled decomposition of the carbon–hydrogen bonds. In graphitization, the structure gets denser.

Mehta [7] again applied *Cambridge Engineering Selector* to identify composite plate formation options. The result of Mehta's analysis, which again assumed production quantities above 10 000, is presented in Table 11 for the smaller faced plate design. Again, the number of manufacturing options increase as tolerances increase and as surface finish and quality requirements decrease. Again no differences were noted above the quality, tolerance, and surface finish ranges analyzed, and no processes fall below them or were recommended for production of the larger-faced plate.

4. DISCUSSION

For use in PEM fuel cells, this work identified 16 polymer electrolyte membranes, 2 types of GDLs, and over 100 bipolar plate designs that are promising and require further study for use in PEM fuel cells. This

Table 12. Areas for future research

Area for future research	Recommended basis for comparison of alternatives
Membrane materials	Fuel cell performance, cost, safety, supporting equipment requirements, synthesis issues (such as hazardous materials used in processing), and temperature-related limitations
Catalyst materials	Fuel cell performance, cost, abundance of materials and related social and political concerns, recyclability
GDL materials	Density, thickness, pore-size distribution, electrical conductivity
Catalyst preparation and application methods (in one or two steps)	Fuel cell performance, metal loadings of the order of $\leqslant 0.1\ mg/cm^2$; degree of metal dispersion; energy use, hazardous material use and waste
Bipolar plate materials	Fuel cell performance, chemical compatibility, resistance to corrosion, cost, density, resistance to dissolution (for metallic plates), electronic conductivity, gas diffusivity/impermeability, manufacturability, recyclable/recycled, stack volume/kW, strength, surface finish requirements, thermal conductivity, and tolerance requirements
Bipolar plate fabrication	Fuel cell performance, tolerance and surface finish capability, capacity relative to mass production needs, and flexibility relative to incorporating unique features (seals, manifolding, cooling systems, etc.)

work also reviews synthesis processes for many of the membranes and GDLs and added several processes to the bipolar plate fabrication options described in literature.

The review presented here combines the work of many researchers. As such, the designs and manufacturing methods can for the most part not be directly compared based on the information presented because there is no reason to believe fuel cell operating conditions or manufacturing process parameters are comparable. For example, for the process steps listed in the input–output tables (Tables 4–6), the level of abstraction presented is dependent upon that provided by the citations listed. This, and the lack of consistent data on input and output quantities, makes quantitative comparison impossible and qualitative comparison of processes difficult. But, we can recommend a basis for future comparisons, as provided in Table 12.

MEA design is dominated by a variety of materials issues. For membrane materials, issues relate to the dependence on complicated polymers (fluorinated, sulfonated or otherwise) that solve one problem while having repercussions (safety, temperature limitations, etc.) at the MEA, stack, and systems (supporting equipment) levels. For catalyst materials, tertiary alloys seem to offer the best performance when CO poisoning is of concern. However, in spite of environmental, social, and political concerns surrounding the use of Pt and other more rare and valuable metals in fuel cells, the use of non-Pt-based alloys has been limited to investigation of only one binary catalysts. Finally, for the GDL, although the carbon cloth offered superior performance in the Ballard Mark V cell study, no corroborating study was found. Also, properties like density, thickness, pore-size distribution, electrical conductivity should also be investigated.

For MEA manufacturing, catalyst preparation and application dominates research opportunities. With several processes able to prepare a catalyst layer with thickness down to 5 μm, the ability to maximize fuel cell performance and still meet the speed requirements of mass production will be the basis for final selection. With respect to performance, there is still disagreement on the preferred mode of application. Also, although sputter deposition is considered as one of the better approaches to apply catalyst to either of GDL or membrane, it is quite a slow process.

Table 13. Design criteria for bipolar plate materials [6–8,36]

S. no.	Material selection criteria	Limit
1	Chemical compatibility	Anode face must not produce disruptive hydride layer; cathode face must not passivate and become non-conductive
2	Corrosion	Corrosion rate $< 0.016\,mA/cm^2$
3	Cost	Material + fabrication $<$ US\$0.0045/cm^2
4	Density	Density $< 5\,g/cm^3$
5	Dissolution	Minimization of dissolution (for metallic plates)
6	Electronic conductivity	Plate resistance $< 0.01\,\Omega\,cm^2$
7	Gas diffusivity/impermeability	Maximum average gas permeability $< 1.0 \times 10^{-4} cm^3/s\,cm^2$
8	Manufacturability	Cost of fabrication (see 3) should be low with high yield
9	Recyclable	Material can be recycled during vehicle service, following a vehicle accident, or when the vehicle is retired
10	Recycled	Made from recycled material
11	Stack volume/kW	Volume $< 1\,g/kW$
12	Strength	Compressive strength $> 22\,lb/in.^2$
13	Surface finish	$>50\,\mu m$
14	Thermal conductivity	Material should be able to remove heat effectively
15	Tolerance	$>0.05\,mm$

Bipolar plate design is dominated by the management of cost and stack mass and durability in the low pH, high temperature fuel cell operating environment. Bipolar plate design criteria, compiled from this and other research, are presented in Table 13. Among the bipolar plate materials presented in Section 3.1, none meet all the design criteria. Specifically, when compared to metal bipolar plates, non-porous graphite provides high corrosion resistance without the need for coating ($<15\,\mu m$/year) and thermal conductivity ($\sim 4\,W/cm\,K$) at a low density. Although tending to be lower in electrical conductivity, compressive strength, and recyclability, graphite plates have been preferred for space applications over other materials. Also, graphite plates are expensive, from both material and processing standpoints [24]. Composite plates, although processing can include many steps, have the advantage of being flexible for incorporating seals, manifolding, cooling systems, and other features.

For bipolar plate manufacturing, two non-porous graphite plate formation options, eight metal plate formation processes and nine coating processes, and five composite formation processes are presented. Although forming non-porous graphite plates can be a time-consuming and therefore expensive process, hazardous materials costs related to purchasing, handling/training, and the management of wastes in all of the composite fabrication processes and most of the coating processes investigated could balance costs.

The research described here is part of a larger project analyzing the environmental life cycle of fuel cells for transportation applications. The project applies Life Cycle Assessment (LCA), a protocol for assessing the environmental aspects (for example, oil and platinum consumption, greenhouse gas emissions, etc.) of a product from various points in their life cycle: from raw materials acquisition through production, use, and disposal [73–75]. Understanding product design is important to the development of a bill-of-materials for the identification of materials used in the product for the application of LCA. Because the bills-of-materials used in LCA need to represent equivalent products (for example, a variety of fuel cells capable of moving the same automobile), linking the design alternative above to fuel cell performance is an important research need. Also, understanding manufacturing alternatives allows the production phase and ultimately materials acquisition to be modeled. For both design and manufacturing, understanding the variety of alternatives that exist prior to wide-scale production facilitates product design and process selection based on environmental criteria prior to large capital investment.

ACKNOWLEDGMENTS

The Ford Motor Company of Dearborn, Michigan provided financial support for this work. The authors are grateful to Drs John Sullivan and Dennis Schutzle of the Ford Motor Company; Per Reinhall and Mamidala Ramulu, Professors of Mechanical Engineering at the University of Washington; and Eric Stuve, Professor and Chair of Chemical Engineering at the University of Washington.

REFERENCES

1. S. Gottesfeld and T. Zawodzinski. Polymer electrolyte fuel cells. *Adv. Electrochem. Sci. Eng.* **5** (1997) 195–301.
2. EG&G Services. *Fuel Cell Handbook*, 5th edn. Parsons Inc., October 2000. Available from World Wide Web: http://www.fuelcells.org/fchandbook.pdf, last retrieved on 2 January 2002.
3. J. Larminie and A. Dicks. *Fuel Cell Systems Explained*. Wiley, New York, 2000.
4. X. Glipa and M. Hogarth. Department of Trade and Industry (UK) Homepage, 2001. Available from World Wide Web: http://www.dti.gov.uk/renewable/pdf/f0200189.pdf, last retrieved on 2 November 2001.
5. M. Rikukawa and K. Sanui. Proton conducting polymer electrolyte membranes based on hydrocarbon polymers. *Prog. Polym. Sci.* **25** (2000) 1463–1502.
6. R. Borup and N. Vanderborgh. Design and testing criteria for bipolar plate materials for PEM fuel cell applications. In: *Proceedings of the Materials Research Society Symposium on the Materials for Electrochemical Energy Storage and Conversion I – Batteries, Capacitors and Fuel Cells*, Vol. 393, 1995, pp. 151–155.
7. V. Mehta. Analysis of design and manufacturing of proton exchange membrane fuel cells. M.S. Thesis. University of Washington, Washington, DC, 2002.
8. Los Alamos National Laboratory Home page, 31 July 2000. Available from World Wide Web: http://education.lanl.gov/resources/fuelcells/, last retrieved on 2 November 2001.
9. Ion Power Homepage, Nafion® Material Safety Data Sheet (MSDS). Available from World Wide Web: http://ion-power.com/nafion/naf001.html, last retrieved on 2 November 2001.
10. J. Crawford. Towards a PEM Fuel Cell Bill of Materials Model. M.S. Thesis. University of Washington, Washington, DC, 2001.
11. X. Glipa. E-mail to author, 31 October 2001.
12. M. Rikukawa. E-mail to author, 31 October 2001.
13. D. DesMarteau. E-mail to author, 31 October 2001.
14. J.S. Wainright, J.J. Fontanella, M.C. Wintersgill, R.F. Savinell and M. Litt. High pressure electrical conductivity studies of acid doped polybenzimidazole. *Electrochim. Acta* **43** (1998) 1289–1294.
15. J. Wei, C. Stone and A. Steck. US patent 5 422 411 (6 June 1995).
16. F. Helmer-Metzman, F. Osan, A. Schneller, H. Ritter, K. Ledjeff, R. Nolte and R. Thorwirth. US patent 5 438 082 (August 1995).
17. B. Bahar, A. Hobson, J. Kolde and D. Zuckerbrod. US patent 5 547 551 (20 August 1996).
18. K.D. Kreuer. On the development of proton conducting materials for technological applications. *Solid State Ionics* **97** (1997) 1–15.
19. H. Miyake. The design and development of Flemion® membranes. *Modern Chlor-Alkali Technology* Vol. 5, pp. 59–67.
20. A. Clark. Lubrizol Corporation, Wickliffe, OH. E-mail to author, 28 August 2001.
21. S. Hietala, M. Paronen, S. Holmberg, J. Nasman, J. Juhanoja, M. Karjalainen, R. Serimaa, M. Toivola, T. Lehtinen, K. Parovuori, G. Sundholm, H. Ericson, B. Mattsson, L. Torell and F. Sundholm. Phase separation and crystallinity in proton conducting membranes of styrene grafted and sulfonated poly(vinyldenefluoride). *J. Polym. Sci.* **37** (1999) 1741–1753.
22. G. Holleck. ECI Laboratories. E-mail to author, 4 September 2002.
23. N. Ogata and M. Rikukawa. US patent 5 403 675 (4 April 1995).
24. Los Alamos National Laboratory Home page (1998). Available from World Wide Web: http://www.ott.doe.gov/pdfs/contractor.pdf, last retrieved on 5 November 2001.

25. M. Iwase and S. Kawatsu. Optimized CO tolerant electrocatalysts for polymer electrolyte fuel cells. In: *Proceedings of the First International Symposium on Proton Conducting Membrane Fuel Cells*, Vol. 1, 1995, pp. 12–18.
26. L.J.M.J. Blomen and M.N. Mugerwa. *Fuel Cell Systems*. Plenum Press, NewYork, 1993.
27. R.J. Bellows, E. Marucchi-Soos and R.P. Reynolds. Proposed Mechanism of CO mitigation in PEMFC's, using dilute H_2O_2 in the anode humidifier. In: *Proceedings of the Second International Symposium on the Proton Conducting Membrane Fuel Cells*, Vol. 2, 1998, pp. 121–126.
28. J. Bauman, T. Zawodzinski, T. Rockward, P. Haridoss, F. Uribe and S. Gottesfeld. Enhanced CO tolerance in polymer electrolyte fuel cells with Pt—Mo anodes. In: *Proceedings of the Second International Symposium on the Proton Conducting Membrane Fuel Cells*, Vol. 2, 1998, pp. 200–204.
29. G.L. Holleck, D.M. Pasquariello and S.L. Clauson. Carbon monoxide tolerant anodes for proton exchange membrane (PEM) fuel cells. II. Alloy catalyst development. In: *Proceedings of the Second International Symposium on Proton Conducting Membrane Fuel Cells*, Vol. 2, 1998, pp. 150–155.
30. M.C. Denis, G. Lalande, D. Guay, J.P. Dodelet and R. Schulz. Performing CO tolerant anode catalyst for polymer electrolyte fuel cells. In: *Proceedings of the Second International Symposium on the Proton Conducting Membrane Fuel Cells*, Vol. 2, 1998, pp. 209–271.
31. J.P. Dodelet, M.C. Denis, P. Gouerec, D. Guay and R. Schulz. In: *Proceedings of the Fuel Cell Program and Abstracts on the CO Tolerant Anode Catalysts for Fuel Cells made by High Energy Ball-Milling*, 2000, pp. 51–54.
32. D. Dubois and J. Ohi. Cathode Catalyst, Fuel Cell for Transportation, 1999 Annual Progress Report Energy Conversion Team, pp. 107–110. Also available from World Wide Web: http://www.ott.doe.gov/pdfs/fuelcellpdf.pdf, last retrieved on 2 January 2002.
33. P.N. Ross, N.M. Markovic, T.J. Schmidt and V. Stamenkovic. New Electrocatalysts for Fuel Cells, Transportation Fuel Cell Power System, 2000 *Annual Progress Report*, 2000, pp. 115–118.
34. G. Faubert, R. Cote and J.P. Dodelet. A new method to prepare non-noble metal based catalysts for the reduction of oxygen in polymer electrolyte fuel cells. In: *Proceedings of the Second International Symposium on the Proton Conducting Membrane Fuel Cells*, Vol. 2, 1998, pp. 31–38.
35. A.L.N. Pinheiro, A.O. Neto, E.C. De Souza, J. Perez, V.A. Paganin, E.A. Ticianelli and E.R. Gonzalez. In: *Proceedings of the Fuel Cell – Program and Abstracts on the Electro-Catalysis of PEM Fuel Cell Reaction on Noble Metal and Noble Metal Alloys*, 2000, pp. 63–66.
36. K. Stroh. Los Alamos National Laboratory, Los Alamos, NM, E-mail to author, 2 January 2001.
37. D. Bevers, N. Wagner and M. Bradke. Innovative production procedure for low cost PEFC electrodes and electrode membrane structures. *Int. J. Hydrogen Energy* **23** (1998) 57–63.
38. S. Cha and W. Lee. Performance of proton exchange membrane fuel cell electrodes prepared by direct deposition of ultra thin platinum on the membrane surface. *J. Electrochem. Soc.* **146** (1999) 4055–4060.
39. T. Ralph, G. Hards and J. Keating. Low cost electrodes for proton exchange membrane fuel cells. *J. Electrochem. Soc.* **144** (1997) 3845–3857.
40. J.G. Drobny. *Technology of Fluropolymers*. CRC Press, Boca Raton, 2001.
41. J.S. Wainright, J.J. Fontanella, M.C. Wintersgill, R.F. Savinell and M. Litt. High pressure electrical conductivity studies of acid doped polybenzimidazole. *Electrochim. Acta* **43** (1998) 1289–1294.
42. D. Schulz. US patent 4 014 725 (29 March 1977).
43. I.D. Raistrick. Modified gas diffusion electrode for proton exchange membrane fuel cells. In: *Proceedings of the Symposium on the Diaphragms, Separators and Ion exchange Membrane*, 1986, pp. 172–177.
44. E. Ticianelli, C. Derouin, S. Srinivasan and A. Redondo. Methods to advance technology of proton exchange membrane fuel cells. *J. Electrochem. Soc.* **135** (1998) 2209–2214.
45. S. Srinivasan, A. Ferreira, R. Mosdale, S. Mukerjee, J. Kim, S. Hirano, S. Lee, F. Buchi and A. Appleby. In: *Proceedings of the Fuel Cell – Program and Abstracts on the Proton Exchange Membrane Fuel Cells for Space and Electric Vehicle Application*, 1994, pp. 424–427.
46. E. Taylor, E. Anderson and N. Vilambi. Preparation of high platinum untilization gas diffusion electrode for proton exchange membrane fuel cells. *J. Electrochem. Soc.* **139** (1992) L45–L46.
47. S. Foster, P. Mitchell and R. Mortimer. In: *Proceedings of the Fuel Cell – Program and Abstracts on the Development of a Novel Electrode Fabrication Technique for Use in Solid Polymer Fuel Cells*, 1994, pp. 442–443.

48. P. Fedkiw and W. Her. An impregnation reduction method to prepare electrodes on Nafion SPE. *J. Electrochem. Soc.* **136** (1989) 899–900.
49. E. Gulzow, M. Schulze, N. Wagner, T. Kaz, R. Reissner, G. Steinhilber and A. Schneider. Dry layer preparation and characterization of polymer electrolyte fuel cell components. *J. Power Sources* **86** (2000) 352–362.
50. T. Matsubayashi, A. Hamada, S. Taniguchi, Y. Miyake and T. Saito. In: *Proceedings of the Fuel Cell – Program and Abstracts on the Development of the High Performance Electrode For PEFC*, 1994, pp. 581–584.
51. S. Gottesfeld and M. Wilson. High performance catalyzed membrane of ultra-low Pt loading for polymer electrolyte fuel cells. *J. Electrochem. Soc.* **139** (1992) L28–L30.
52. S. Gottesfeld and M. Wilson. Thin film catalyst layers for polymer electrolyte fuel cell electrodes. *J. Appl. Electrochem.* **22** (1992) 1–7.
53. E. Stuve. Chair and Professor of Chemical Engineering, University of Washington, Personal Communication, 2001.
54. J. Allen. In: *Proceedings of the Fuel Cell – Program and Abstracts on the Bipolar Separator and Current Collector Design and Manufacture*, 1994, pp. 424–427.
55. A. Woodman, E. Anderson, K. Jayne and M. Kimble. Physical Science Inc. Home Page, 1999, Development of corrosion-resistant coatings for fuel cell bipolar plates. Available at World Wide Web: http://www.psicorp.com/html/pubs/Subjects/FuelCell.htm, last retrieved on 5 November 2001.
56. K.D. Jayne. Physical Science Laboratory, Andover, MA. E-mail to author, 7 August 2001.
57. I. Zafar, J. Guiheen, N. Dave and R. Timothy. World Patent WO00 128 019 (19 April 2001).
58. T. Matsumoto, J. Niikura, H. Ohara, M. Uchida, H. Gyoten, K. Hatoh, E. Yasumoto, T. Kanbara, K. Nishida and Y. Sugawara. European Patent EP1094535 (25 April 2001).
59. Y. Li, W. Meng, S. Swathirajan, S. Harris and G. Doll. US patent 5 624 769 (20 April 1997).
60. M.S. Wilson. Los Alamos National Laboratory, Los Alamos, NM. E-mail to author, 4 February 2002.
61. P. Blaine. Micro Molding Technology LLC, Boynton Beach, FL. E-mail to author, 3 October 2001.
62. R. Lawrence. US patent 4 214 969 (29 July 1980).
63. E. Balko and R. Lawrence. US patent 4 339 322 (13 July 1982).
64. K. Ledjeff-Hey, F. Mahlendorf, O. Niemzig, A. Trautmann and G. Mercator. In: *Proceedings of the Fuel Cell – Program and Abstracts on the Electronically conducting composite materials as bipolar plates for PEM fuel cells*, 1998, pp. 570–573.
65. A. Pellegri and P. Spaziante. US patent 4 197 178 (8 April 1980).
66. R. Emanuelson, W. Luoma and W. Taylor. US patent 4 301 222 (17 November 1981).
67. W. Taylor. US patent 4 592 968 (3 June 1986).
68. Stewart Jr. and C. Robert. US patent 4 670 300 (2 June 1987).
69. T. Uemura and S. Murakami. US patent 4 737 421 (12 April 1988).
70. T. Besmann, J. Klett and T. Burchell. Carbon composite for a PEM fuel cell bipolar plate. In: *Proceedings of the Materials Research Society Symposium on the Materials for Electrochemical Energy Storage and Conversion-II – Batteries, Capacitors and Fuel Cells*, Vol. 496, 1998, pp. 243–248.
71. D. Busick and M. Wilson. US patent 6 248 467 (19 June 2001).
72. R. Meissner, M. Irgang, K. Eger, P. Weidlich and H. Dreyer. US patent 5 736 076 (7 April 1998).
73. M.A. Curran. *Life Cycle Assessment.* Amazon Hill, 1996.
74. International Organization for Standardization. ISO14040: *Life Cycle Assessment – Principles and Framework*, 1997.
75. W. Klöpffer and O. Hutzinger. *LCA Documents: Life Cycle Assessment – State-of-the-Art and Research Priorities*. Eco-Informa Press, 1997.

Chapter 23

Aging mechanisms and lifetime of PEFC and DMFC

Shanna D. Knights, Kevin M. Colbow, Jean St-Pierre and David P. Wilkinson

Abstract

This paper provides an overview of several operating conditions which can have a significant effect on the durability of polymer electrolyte fuel cells (PEFCs) and direct methanol fuel cells (DMFCs), including: low reactant flows, high and low humidification levels, and high and low temperatures. The possible effects of these conditions, along with possible mitigating strategies, are discussed. Data from various tests are presented demonstrating lifetimes from 1000 h to greater than 13 000 h for various conditions and applications.

Keywords: Polymer electrolyte fuel cells; Direct methanol fuel cells; Lifetime; Aging mechanisms; Failure; Mitigation strategies

Article Outline

1. Introduction 503
2. Experimental 504
3. Discussion 505
   3.1. Low reactant flow 505
   3.2. Low humidification 506
   3.3. High humidification 509
   3.4. Low temperature 512
   3.5. High temperature 513
4. Conclusions 515
Acknowledgements 515
References 515

1. INTRODUCTION

Fuel cell systems, when properly designed, can be a reliable and durable method to produce efficient and environmentally friendly energy for various applications. As polymer electrolyte fuel cells (PEFCs) approach commercialization, significant progress is being made towards producing systems that achieve the optimum balance of cost, efficiency, reliability, and durability.

Fuel cell lifetime requirements vary significantly, ranging from 3000 to 5000 operating hours for car applications, up to 20 000 operating hours for bus applications and up to 40 000 operating hours for stationary

Fuel Cells Compendium

applications. There can be various lifetime requirements for other types of applications, such as portable power, uninterrupted power supply (UPS), etc. Degradation rate requirements are normally set based on beginning-of-life (BOL) performance, end-of-life (EOL) performance requirements, and lifetime durability requirements in terms of operational hours and/or stand-by hours. The degradation range of 2–10 μV/h is common for most applications.

The most common fuel for the PEFC is hydrogen, which can be either in an essentially pure gas stream or in the form of reformate produced from various fuels, such as methane, methanol, and gasoline. A special case of PEFCs is that of direct methanol fuel cells (DMFC), which convert methanol directly into electrical energy, without the use of a reformer. Wasmus [1] provides an overview of DMFC technology. Such systems offer many advantages in terms of system simplicity and will likely be preferred for a number of niche markets, but development lags hydrogen-based PEFCs. Degradation rates of DMFCs are generally higher than that of hydrogen PEFCs, and depend on the application, but are typically in the range of 10–25 μV/h.

The ability of the fuel cell to operate under a wide range of operating conditions with different system characteristics is described by the term, "fuel cell operational flexibility." Optimum fuel cell operational flexibility must take into account both specified and an estimated amount of unexpected, or "out-of-specification" conditions over the fuel cell target lifetime. Some of the conditions to consider include: reactant flow rates and composition, operating and environmental temperature, operating and environmental pressure, humidification levels, peak load requirements and turn-down ratios, duty-cycle characteristics (including percentage of time at different load points), and required rate of transient responses.

Due to the absence of moving parts, the fuel cell is an inherently reliable system, but can be prone to material degradation from the presence of reactants; various materials, including catalyst; significant electrical potential and current density; and various operating conditions, including temperature and pressure ranges. The management of the fuel cell stack for lifetime is dependent on how these components and interacting conditions are designed and managed. As a very general statement, the lifetime achieved in a fuel cell can often be traded-off against another characteristic, such as operating regime, cost, and power density. Targeted lifetime and failure testing can be conducted during the development stage, to provide the basis to understand potential failure mechanisms and develop the necessary technology to mitigate such mechanisms. Therefore, a fuel cell system may be custom designed to meet the needs of a particular application, including a target lifetime requirement.

This paper will provide an overview of some operational conditions that can affect fuel cell lifetime. The conditions to be discussed include: low reactant flows, high and low humidification levels, and high and low temperatures. Each of these conditions and associated failures can be discussed in much more depth, but is beyond the scope of this paper. For a discussion on the effects of water management, please refer to the paper by St-Pierre et al. [2]. For a literature review of PEFC durability, please refer to the discussion by Wilkinson and St-Pierre [3].

2. EXPERIMENTAL

Various standard Ballard fuel cells and fuel cell stacks were used to generate the data presented in this paper, with active areas ranging from 49 to 1280 cm^2. Tests were conducted on single-cells or short-stacks (typically in the range of 4–20 cells). The conditions used varied with the application and are noted with the data when relevant. Standard Ballard test equipment, including humidifier, test station, gas mixer, and electronic load, was used for control of operating conditions.

Fuel starvation was induced by replacing the fuel with humidified nitrogen. An external power supply was used to simulate the stack power to drive the current through the cell or short-stack. External humidifiers were used for humidification experiments, with control of humidification level through adjustment of dew point temperature. Lifetime tests were generally conducted by applying a constant load (unless otherwise stated) over a fixed period of time. During the lifetime, voltage performance points were recorded continuously.

Diagnostic tests were generally conducted throughout the test period, but are not shown on the charts. Only operational hours are included in the lifetime calculations. Freeze–thaw data was collected by running a cell through a repetitive cycle, consisting of operation at two different load points, removal of the load and gases, cool down and purging of the cell, freezing, then warm-up and re-introduction of gases and load.

3. DISCUSSION

3.1. Low reactant flow

High reactant utilization is generally required for most applications in order to maximize fuel efficiency and reduce system parasitic load, size, and weight that may be associated with the oxidant and fuel delivery and/or storage systems. Transient operation, particularly under the demanding conditions of automotive applications, introduces greater challenges due to rapid load changes and the resulting wide range of conditions.

During high overall stack utilization, uneven flow sharing between cells can result in partial fuel and/or air starvation conditions in individual cells. This situation can be exacerbated by the presence of liquid water in channels or other blockages, resulting in further flow sharing difficulties that in extreme cases can lead to complete starvation conditions. One example of such a condition is sub-zero start-up or operation. As long as the stack temperature remains below zero, the cells are prone to ice formation and subsequent flow channel blockage. Although the stack and system operation can be designed to reduce these occurrences, it is generally accepted that rapid heating of the stack to minimize ice formation is desirable.

In the case of oxidant starvation, the protons passing through the membrane will combine, in the absence of oxygen, to form hydrogen, and the cell essentially acts as a hydrogen pump. The cathode potential drops due to the lack of oxygen and the presence of hydrogen, and the cell voltage generally drops to very low levels or may even become negative.

In the case of fuel starvation, if hydrogen is no longer available to be oxidized, the anode potential will rise to that required to oxidize water, assuming water is available, resulting in the evolution of oxygen and protons at the anode, according to

$$H_2O \rightarrow \frac{1}{2}O_2 + 2H^+ + 2e^- \tag{1}$$

The protons will pass through the membrane and combine with oxygen at the cathode in the normal reduction reaction to produce water (reverse of reaction (1)).

A polarization of a complete fuel starvation (no hydrogen) with humidified nitrogen flowing on the anode is presented in Fig. 1 for an anode with a 4 mg Pt/cm^2 loading of Pt black catalyst on the anode. The cell voltage, as measured from the cathode to the anode, drops due to a rise in anode potential, which reaches levels sufficient for water oxidation, i.e. >1.23 V. The values of the cathode and anode potential are not individually known, but the presence of oxygen on the anode due to reaction (1) was confirmed by gas chromatographic (GC) analysis, indicating that high anode potentials are being achieved.

Most current technology development is focussed on the use of platinum (or platinum and ruthenium) supported on carbon particles, in order to reduce the amount of platinum required on the anode down to 0.05–0.45 mg/cm^2, and as described in [4]. These types of anodes are prone to degradation during fuel starvation due to reaction (2), the oxidation of carbon, which is catalyzed by the presence of platinum [5]. This reaction proceeds at an appreciable rate at the electrode potentials required to electrolyze water in the presence of platinum (greater than approximately 1.4 V [6]):

$$C + 2H_2O \rightarrow CO_2 + 4H^+ + 4e^- \tag{2}$$

This is shown schematically in Fig. 2. The catalyst support is converted in to CO_2, and Pt and/or Ru particles may be lost from the electrode, resulting in the loss of performance.

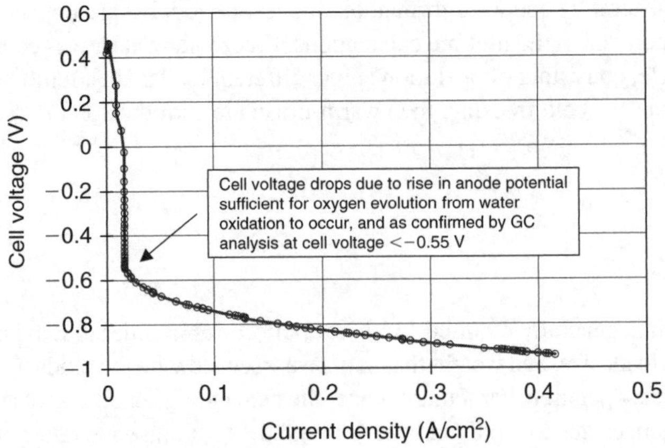

Fig. 1. Fuel starvation polarization. Humidified anode/cathode feed streams: nitrogen/air. 3 bar, 75°C, 4 mg/cm^2 Pt on each of cathode and anode

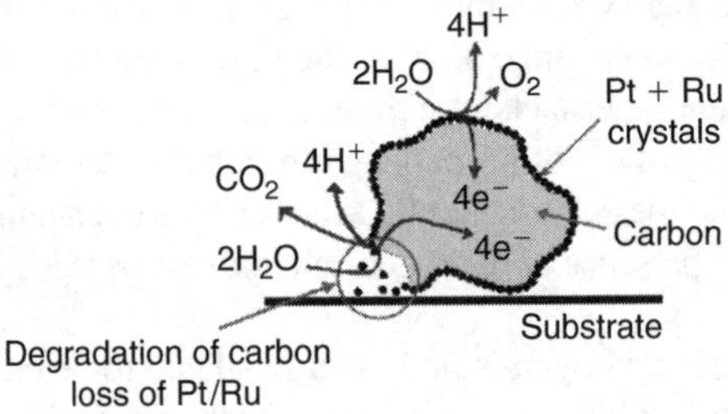

Fig. 2. Schematic representation of degradation of carbon catalyst support during operation in the absence of fuel

Reduced degradation can be achieved through modification of the anode structure to favour oxidation of water over carbon. Some strategies to accomplish this include: enhanced water retention on the anode (e.g. through modifications to PTFE and/or ionomer, and addition of water blocking components such as graphite); use of improved catalysts to reduce the required anode potential for water electrolysis and thus the associated carbon oxidation (e.g. additional Ru on the anode); use of a more robust catalyst support (e.g. more graphitic carbon or alternative support materials); and increased catalyst coverage on the support to reduce contact of carbon with reactants (e.g. higher weight percentage Pt on the carbon) [7–10].

Figure 3 shows the cell voltage response over time of four different four-cell stacks with different anode designs. Each stack was subjected to fuel starvation conditions through an equivalent number of cycles. Each stack was then finally starved of hydrogen and allowed to go into voltage reversal until an average cell voltage of −2 V was reached. The length of time the cells operated prior to reaching −2 V is a measure of robustness to fuel starvation.

3.2. Low humidification

Water management is key to optimum reliability and durability of fuel cells. Inadequate water content, either globally within the stack or locally at certain locations within the unit cell, results in reduced conductivity in the membrane and in any ionomer present in the catalyst layer. This results in increased ohmic

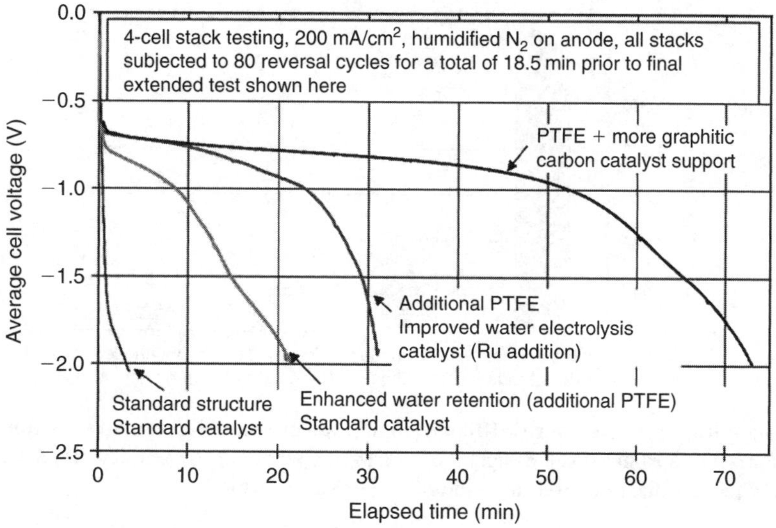

Fig. 3. Comparison of different anode structures in severe failure testing. Each cell has equivalent cathode (~0.7 mg/cm^{2}Pt, supported on carbon). Testing conducted at 200 mA/cm^2, fully humidified nitrogen on anode. Anode loading at ~0.3 mg/cm^2 Pt supported on carbon (varied materials and compositions). Each curve represents the results from a four-cell stack with each cell in the stack of identical composition. Four separate stack tests were run to generate the curves

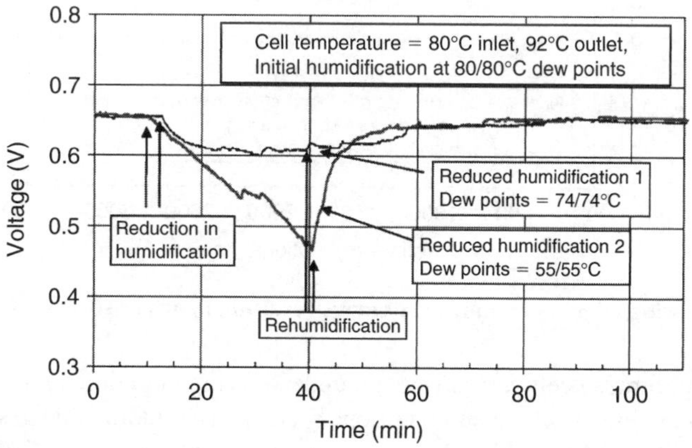

Fig. 4. Effect of reduced humidification levels on non-optimized cell design. Air/hydrogen operation with specified humidification levels. Current density: 1 A/cm^2

losses and a drop in cell voltage. This effect can be demonstrated by conducting controlled tests on a cell design that has not been optimized for low humidification conditions. The results of such a test are shown in Fig. 4. The cell is initially running fully humidified with a constant performance. Once a step-change is made through a reduction in humidification level, the performance quickly drops. The drier the conditions, the greater the loss in performance. When the humidification level is subsequently increased, the performance quickly recovers.

Inadequate water content can also accelerate membrane physical degradation, and can ultimately result in membrane holes and reactant gas cross-over. This effect was observed through a series of tests in which

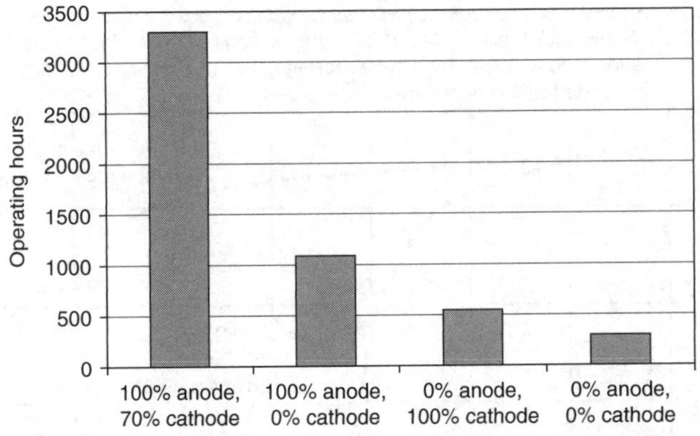

Fig. 5. Time to significant gas crossover (>10 cm^3/min at 2 bar pressure differential) as a function of inlet gas humidification for a non-optimized cell design. Air/hydrogen operation, current density: 540 mA/cm^2, coolant temperature: 75°C, humidification levels of reactant streams as indicated

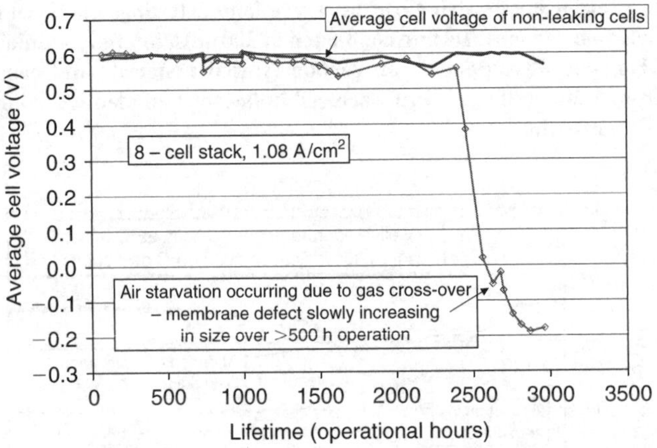

Fig. 6. Lifetime test showing effect of membrane damage resulting in gas cross-over. Air/hydrogen operation

the same type of non-optimized cell design was run under varying humidification conditions. The results of these tests are shown in Fig. 5, where, as conditions become drier, failure due to significant reactant gas cross-over occurs after shorter lifetimes.

It is possible to continue running a cell with membrane holes present, but the performance becomes very sensitive to pressure differentials across the membrane. In this case, the reactants are able to transfer across the membrane, and dilute or consume the opposite reactant. When this occurs, either fuel or air starvation can result, depending on the direction of the pressure differential. Figure 6 shows a lifetime plot for an eight-cell stack during which a large cross-over leak developed in one cell. The stack was operated with the fuel pressure slightly higher than the air pressure, resulting in air starvation in that cell. It can be observed that the cell voltage dropped to negative values, but continued operation for approximately 500 h prior to the end of the lifetime test. The remaining cells in the stack continued to operate with stable voltages for the entire planned 3000 h test duration.

Improved water management can be achieved through various strategies. One challenge faced by cell and stack designers is how to effectively distribute the product water within the cell. Unless the inlet gas

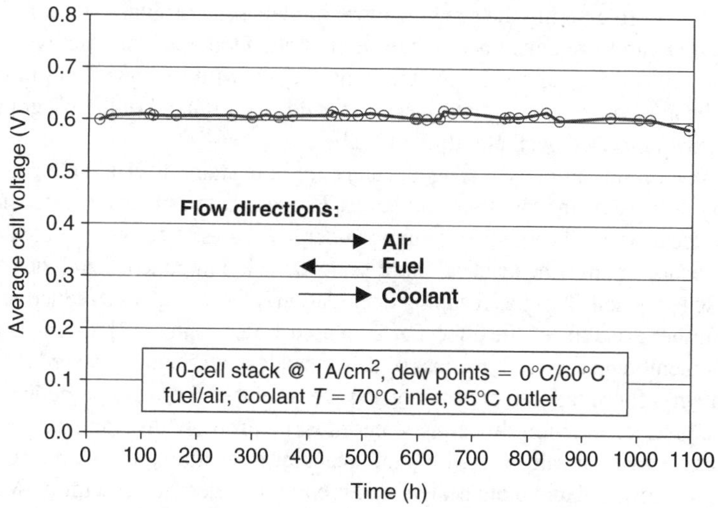

Fig. 7. Stable lifetime performance achieved using counter-flow operation. Ten-cell stack, hydrogen/air operation

streams are fully humidified, the inlet region will have low humidity. High oxygen concentration at the inlet can also result in high catalytic activity in this region, which, when combined with the reduced water for heat transfer removal, can result in increased local temperatures. The combined effects of low humidification and locally increased temperatures can result in a region prone to membrane failure. Conversely, as the product water accumulates, the back portion of the cell may be fully saturated, leading to the possibility of two-phase flow and potential mass transport losses, contamination, and flow sharing issues.

One of the easiest strategies to implement to redistribute product water within the cell is through operation with countercurrent reactant flows, essentially using the MEA as an in-cell humidifier [2,11,12]. A study was conducted to determine the optimum flow configuration for drier conditions (no air humidification, fuel humidification at 10°C lower than the inlet coolant temperature). The strategies tested included co- and counter-flow strategies for the three fluid streams: air, fuel, and coolant. In this case, the cell was operated with a 15°C temperature differential from inlet to outlet in the coolant stream. Operation with fuel counter-flow to air and coolant achieved the most stable performance and the lowest cell resistance. Figure 7 shows a lifetime test run under these conditions, with no degradation observed. The other configurations tested resulted in a steady decrease in cell voltage and the accompanying increase in cell resistance, indicating the effect was due to membrane drying, resulting in proton conductivity loss.

3.3. High humidification

While insufficient humidification can introduce problems for durability and reliability, excess humidification can also cause many potential problems. Excess humidification may result from several scenarios. During transient operation, rapid electrical load changes may be demanded of the fuel cell. The change in operating conditions may lag behind that of the load and the conditions can become temporarily non-optimized. This situation may cause flooding or drying conditions to exist in the stack. Stack and system design must be carefully optimized to ensure uniform temperatures throughout the stack. Any local regions of lower temperature have the potential to cause associated water condensation and/or pooling. The use of reformed fuel is generally associated with a large fuel water content which in some cases may result in greater than 100% saturation upon being cooled to stack temperature if the moisture is not adequately removed prior to entering the stack. A more fundamental problem with the fuel cell is the production of water as the gases are consumed, resulting in the tendency for product water to accumulate in the back half of the cell. In most applications, the fuels are run with low gas flows to minimize parasitic losses, which may be inadequate to completely vaporize and/or entrain all of the water produced, although various strategies exist to reduce oversaturation [12,13].

Excess water content, if not properly managed, can result in reactant diffusion blockages, particularly on the cathode, causing an increase in mass transport losses. Excess water can also increase flow-sharing issues, which becomes particularly significant when running under low flows. A further effect of excess water is the increase in contaminants that may be leached out of system/stack components, and the increased opportunities to transfer these to the cell. This has resulting degradation effects, such as a reduction in hydrophobicity in the cell, which further exacerbates the effects of increased water content [2]. Other contamination effects may include loss of membrane conductivity, loss of active catalyst sites, and increased mass transport losses.

Many of the strategies for mitigation of excess water are similar to those used to reduce the impact of drying conditions. The product and humidification water must be properly distributed throughout the cell and between cells. Strategies which can be used include reactant flow strategies (such as counter-flow design discussed previously), flowfield and plate design, and membrane–electride assembly (MEA) design [14].

One example of MEA design to mitigate the product water accumulation is the use of a non-uniform in-plane electrode structure to improve water retention in the drier inlet regions, and enhance water removal in the wetter outlet regions [2,14,15]. This strategy minimizes both dry and flooded regions. This may be achieved through a localized increase in electrode gas diffusion layer (GDL) substrate porosity in areas where water accumulates, such as the oxidant outlet. Figure 8 shows one example of electrode design and performance improvements achieved at high current densities [14,15]. In this case, the GDL contains added perforations, with increased density of perforations in the wetter outlet region. The performance is equivalent in the low current density region, but shows a significant improvement to conventional GDL design in the region of greater than 1 A/cm^2, which is typically limited by mass transport losses.

A particular case where water management is of high importance is that of DMFCs. Although the cathode reactions in a DMFC are the same as that of hydrogen based fuel cells, the fuel consists of a methanol and water mixture, usually introduced into the anode in the liquid phase. DMFCs do not suffer from issues with dry regions due to the liquid fuel, but are particularly prone to cathode flooding. A significant portion of the liquid water in the fuel stream will ultimately cross over to the cathode. This results in increased cathode mass transport losses due to the difficulty of the reactants to diffuse through water vapour and/or liquid present which must be removed by the cathode gas stream.

As a result of liquid fuel feed, DMFCs require different water management strategies than hydrogen-based PEFCs. One strategy that can be used to recover performance loss is load cycling. When operating under constant load, the cells tend to suffer high performance degradation due to an increase in gas diffusion limitations as water builds up on the cathode. This can be seen in the polarization curves of Fig. 9. A cell operating for only 16 h under constant current load has lost significant performance at high current densities, indicative of mass transport losses due to gas diffusion limitations. Quantification of this increased loss can be accomplished through testing of the cell in the high current density region on each of two different oxidants, air and a mixture, termed "helox," consisting of 21% oxygen and 79% helium. The voltage response as a function of current density and oxidant used provides information on gas diffusion limited performance, with an increasing gap at high current densities indicating increased losses [2,16]. The change in gas diffusion limited performance of a DMFC run continuously for only 16 h is compared in Fig. 10 against one run under load cycling for almost 2000 h. These cells suffered a change in gas diffusion limited performance of ~370% and ~35%, respectively, an order of magnitude improvement in degradation. The load cycling strategy used in this case consists of removing the load from the cell for 30 s during every 30 min operational period. As can be observed given the results in Fig. 10, this has proven to be very effective in maintaining low degradation rates. In Fig. 11, a DMFC lifetime plot at 0.2 A/cm^2, during which the

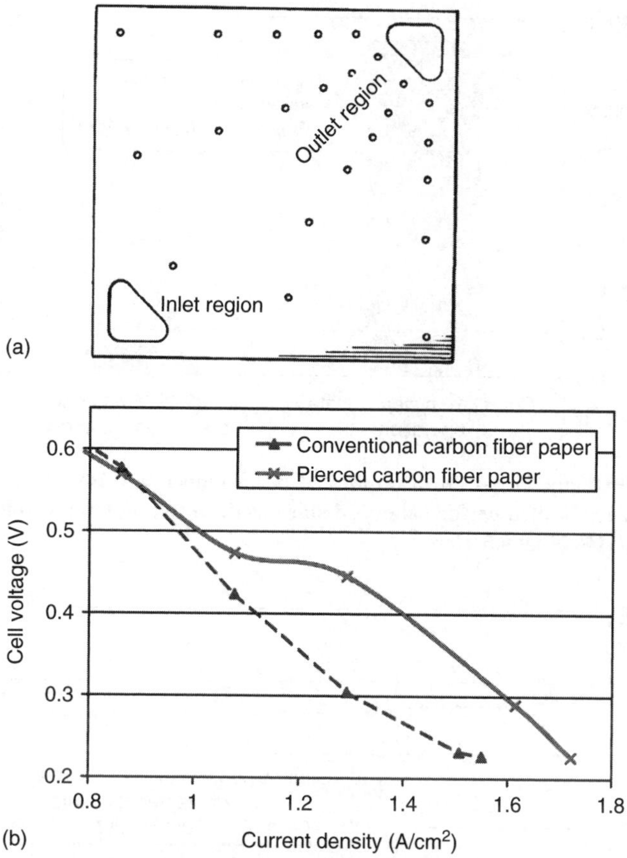

Fig. 8. Water management through MEA design. (a) Plan view of MEA showing perforations in GDL with increased perforation density in wetter outlet region. (b) Performance of MEA at increased current density compared to a conventional, non-pierced MEA design

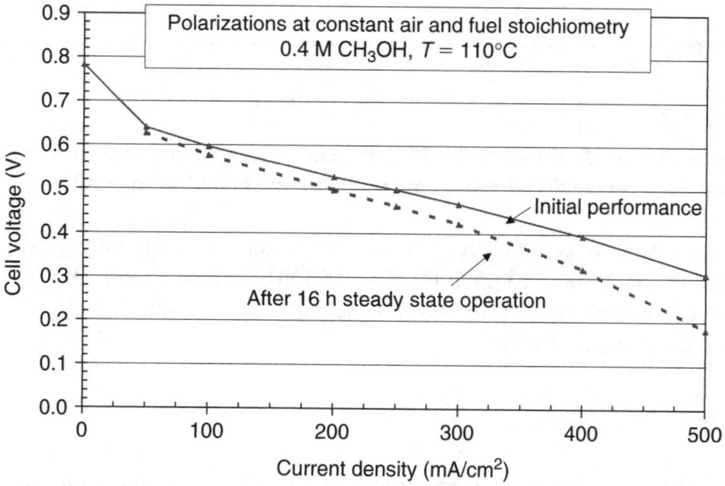

Fig. 9. Performance of a DMFC after 16 h of steady-state operation compared to initial performance

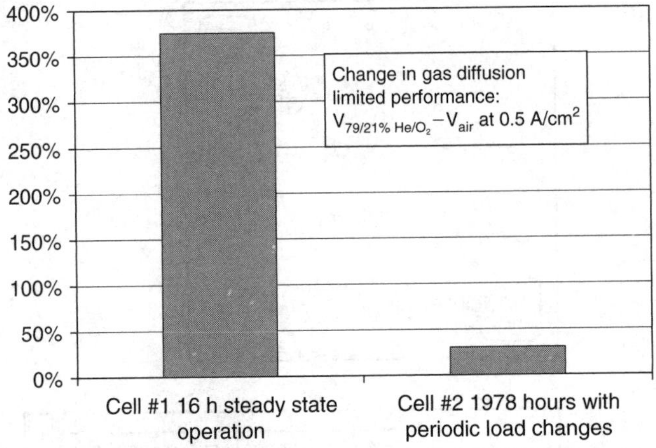

Fig. 10. Comparison of change in gas diffusion limited performance in a DMFC depending on operational strategy used. Gas diffusion limited performance is defined as the difference in cell voltage on air as compared to a mixture of 79% O_2/21% N_2 at 0.5 A/cm^2

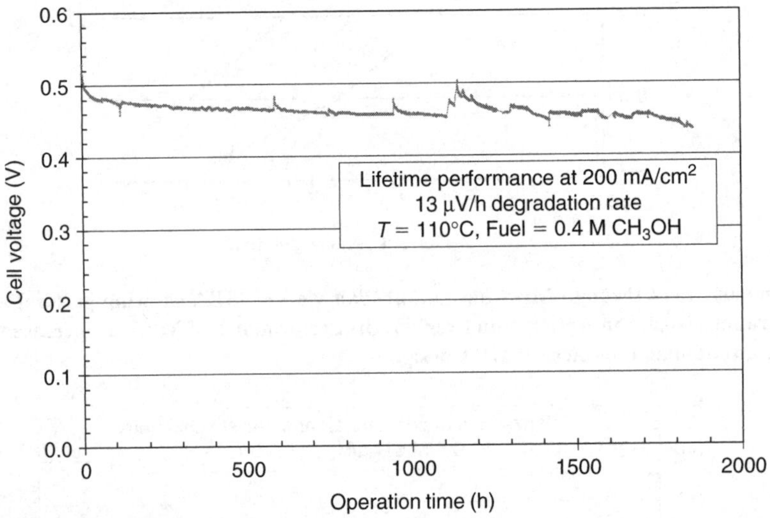

Fig. 11. Lifetime performance of a direct methanol fuel cell using load cycling strategy to reduce mass transport-related degradation: 30 s of open-circuit voltage operation/30 min of on-load operation

load cycling strategy was used, is presented. The lifetime performance was very good for this type of fuel cell, with a degradation rate of only 13 μV/h for almost 2000 h with no failures. It is important to note that many DMFC applications can also achieve these low degradation rates simply due to the dynamic nature of load conditions, including on–off cycles.

3.4. Low temperature

A wide range of operating temperatures is required for fuel cells, depending on the application. For automotive applications, this may include sub-zero operation upon start-up in freezing temperatures [17].

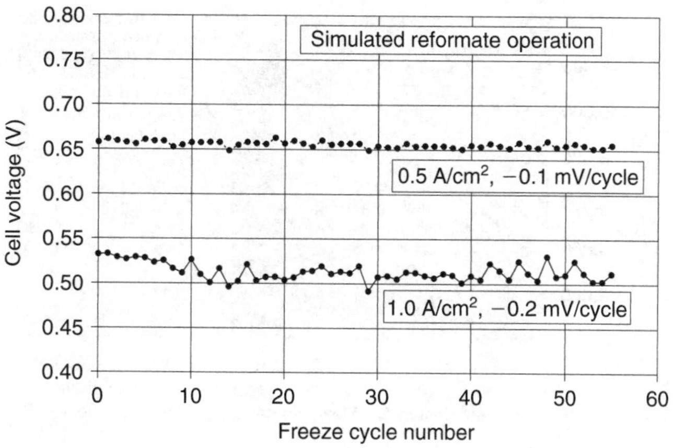

Fig. 12. Effect of freeze–thaw cycles on fuel cell performance. Simulated reformate operation. Operation and purging conducted between each purge/thaw cycle

Regardless of the start-up requirements, the stack and system will likely be required to withstand many freeze–thaw cycles. Possible low temperature effects can include: reduced material degradation rates; increased contamination effects, e.g. carbon monoxide, which is generally present in reformed fuel, is adsorbed much more strongly at lower temperatures [18]; lower performance due to increased kinetic, ohmic, and reactant transport losses; and increased importance of water management. Water management is particularly important for sub-zero operation, as water blocking channels may freeze, resulting in partial or complete reactant starvation in some cells.

One mitigation strategy that can be used for operation when sub-zero temperatures are expected, is purging of flow channels. This is an effective technique to clear the channels of water prior to freezing [19]. Figure 12 shows the performance of a cell after each of 55 freeze–thaw cycles, with operation and purging during each cycle. No significant performance loss is observed at 0.5 A/cm^2, and at 1.0 A/cm^2 a loss of only 0.2 mV/cycle is observed. This is consistent with data reported in [20], in which no degradation was observed when the cell was cooled to −78°C, and with that reported in [21], in which three freeze–thaw cycles to −10°C were completed with no observable degradation.

3.5. High temperature

Operation of the fuel cell at increased temperature has several advantages. Higher temperatures result in reduced cooling requirements, which is particularly important for automotive applications in which engine size can be limited by radiator capacity [22]. Higher temperatures are preferred for co-generation of heat and electricity, which is particularly advantageous for residential applications. Contaminants tend to adhere less strongly to the catalyst and other fuel cell components as the temperature is increased. For example, carbon monoxide present in reformed fuel has less poisoning effect at higher temperatures. However, the disadvantages can include: increased material degradation rates and associated contaminant levels; and, a reduction in membrane and ionomer moisture content for a given set of operational conditions [23]. The last effect can result in loss of performance due to decreased proton conductivity and increased membrane degradation leading to holes and/or thin spots.

The use of reformed fuel provides an example of a localized high temperature effect that has the potential to reduce lifetime of the fuel cell. The use of a small stream of air bleed into the fuel is often used to recover the CO induced anode catalyst poisoning by oxidizing the CO to CO_2 [24]. However, the presence

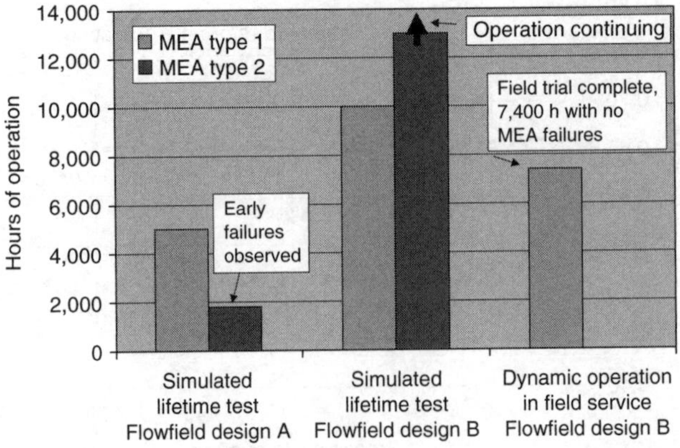

Fig. 13. Comparison of lifetimes achieved in 250 kW natural gas power plant hardware. Air/simulated reformate operation with 2% air bleed

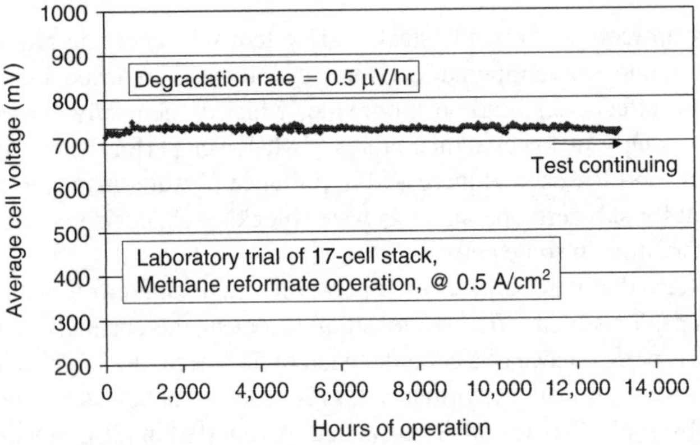

Fig. 14. Lifetime test of 17-cell stack comprised of 250 kW natural gas power plant hardware. Coolant operating temperature: 82°C at stack outlet. Design of cell optimized to minimize localized high temperature effects at membrane in fuel inlet. Actual temperature at membrane is not measured

of this air bleed can result in localized hot spots generally close to the fuel flow channel inlet area, depending on the cell design. Failure analysis of the problem region is important to detect any link between the failure and specific features of the flow channel design [25]. The use of modeling techniques can be further used to determine the most likely contributing factors, such as coolant channel location, plate conductivity, air and coolant channel locations, interactions with MEA design, etc. In this way, the cell can be designed to prevent the occurrence of hot spots and extend the lifetime of the fuel cell under these conditions.

An example of the improvement achievable using this approach is represented in Fig. 13. In this case, the application was the Ballard 250 kW Natural Gas Power plant. In an early design, the cells began to fail very early for this type of application; at 5000 h for MEA type 1 and in less than 2000 h for MEA type 2. In this case, the MEA type 2 was the preferred design in all other respects. Thus, the strategy mentioned above of failure analysis and modeling was accomplished to highlight key contributors to the early failures. Subsequent redesign and testing resulted in very significant improvements to durability, particularly with the target MEA type 2. In laboratory simulation trials, a short stack (17-cells) operated for over 13 000 h without failure, with an average cell degradation rate of only 0.5 μV/h, as presented in Fig. 14. In a field trial under dynamic operation, the full power plant operated for 7400 h, to the end of the field trials, with no MEA failures.

4. CONCLUSIONS

An overview of the effect of various operational conditions on the durability of the fuel cell was presented. Lifetime can be extended both through control of these conditions, within application and system abilities, and through optimized design to reduce the impact of these conditions. Fundamental understanding of failure mechanisms and development of mitigating technology is facilitated by directed lifetime and failure testing throughout the technology development cycle. Various lifetime plots were presented showing very stable performance. Although DMFC is at an earlier stage of development, significant progress has been made in achieving good durability, and it is a prime candidate for niche markets with moderate lifetime requirements.

ACKNOWLEDGEMENTS

The authors would like to acknowledge the following groups within Ballard Power Systems for generating and providing much of the data shown and discussed: R&D Department, Product Development Reliability Group, and Field Testing Personnel.

REFERENCES

1. S.K.A. Wasmus. *J. Electroanal. Chem.* **461** (1999) 14–31.
2. J. St-Pierre, D.P. Wilkinson, S. Knights and M.L. Bos. *J. New Mater. Electrochem. Sys.* **3** (2000) 99–106.
3. D.P. Wilkinson and J. St-Pierre. Durability. In: W. Vielstich, H. Gasteiger and A. Lamm (Eds.), *Handbook of Fuel Cells: Fundamentals, Technology and Applications*, Vol. 3. Wiley, New York, 2003, pp. 611–626 (Chapter 47).
4. P. Costamagna and S. Srinivasan. *J. Power Sources* **102** (2001) 242–252.
5. E. Passalacqua, M. Vivaldi, N. Giordano, P.L. Anotonucci and K. Kinoshita. In: *Proceedings of the 27th Intersociety Energy Conversion Engineering Conference*, Vol. 929294, 1992, pp. 3425–3431.
6. R.E. Billings. *The Hydrogen World View*. American Academy of Science, 1991.
7. S.D. Knights, D.P. Wilkinson, S.A. Campbell, J.L. Taylor, J.M. Gascoyne and T.R. Ralph. PCT WO 01/15247 A2, March 1, 2001.
8. J.L. Taylor, D.P. Wilkinson, D.S. Wainwright, T.R. Ralph and S.D. Knights. PCT WO 01/15249 A2, March 1, 2001.
9. S.D. Knights, J.L. Taylor, D.P. Wilkinson and S.A. Campbell. PCT WO 01/15254 A2, March 1, 2001.
10. S.D. Knights, J.L. Taylor, D.P. Wilkinson and D.S. Wainwright. PCT WO 01/15255 A2, March 1, 2001.
11. D.P. Wilkinson, H.H. Voss, N.J. Fletcher, M.C. Johnson and E.G. Pow. US Patent 5 773 160 (30 June 1998).
12. J. St-Pierre, D.P. Wilkinson, H. Voss and R. Pow. In: O. Savadogo and P.R. Roberge (Eds.), *New Materials for Fuel Cell and Modern Battery Systems II, Ecole Polytechnique*, Montreal, Canada, 1997, pp. 318–329.
13. N.J. Fletcher, C.Y. Chow, E.G. Pow, B.M. Wozniczka, H.H. Voss, G. Hornburg and D.P. Wilkinson. US Patent 5 547 776 (20 August 1996).
14. D.P. Wilkinson and J. St-Pierre. *J. Power Sources* **113** (2003) 101–108.
15. M.C. Johnson, D.P. Wilkinson, C.P. Asman, M.L. Bos and R.J. Potter. US Patent 5 840 438 (24 November 1998).
16. Y.W. Rho, S. Srinivasan and Y.T. Kho. *J. Electrochem. Soc.* **141** (1994) 2084–2096.
17. R. Sims and I.F. Kuhn Jr. *Cold Weather Operational Considerations for Direct Hydrogen*, Automotive Fuel Cell Systems.
18. T.A. Zawodzinski Jr., C. Karuppaiah, F. Uribe and S. Gottesfeld. In: J. Mc Breen, S. Mukherjee and S. Srinivasan (Eds.), *Proceedings of the First International Symposium of the Electrochemical Society, Electrode Materials and Processes for Energy Conversion and Storage IV*, PV 97-13, Montreal, Canada, May, 1997. The Electrochemical Society, Pennington, NJ, pp. 139–149.

19. J. Roberts, J. St-Pierre, M. van Der Geest, A. Atbi and N. Fletcher. WO 01/24296, 5 April, 2001.
20. S.F. Simpson, C.E. Salinas, A.J. Cisar and O.J. Murphy. Factors affecting the performance of proton exchange membrane fuel cells. In: A. Landgrebe, S. Gottesfeld and G. Halpert (Eds.), *Proceedings of the First International Symposium of the Electrochemical Society, First International Symposium on Proton Conducting Membrane Fuel Cells*, PV 95-23, Hardbound, Chicago, Illinois. The Electrochemical Society, Pennington, NJ, pp. 182–192 (English).
21. M.S. Wilson, J.A. Valerio and S. Gottesfeld. *Electrochim. Acta* **40** (1995) 355–363.
22. M. Sadler, A.J. Stapleton, R.P.G. Heath and N.S. Jackson. *Fuel Cell Power for Transportation*, SP-1589, SAE, Vol. 81–90, March 2001.
23. A. Sen, K.E. Leach and R.D. Varjian. Determination of water content and resistivity of perfluorosulfonic acid fuel cell membranes. In: D.H. Doughty, B. Vyas, T. Takamura and J.R. Huff (Eds.), *Proceedings of the Materials Research Society Symposium on the Materials for Electrochemical Energy Storage and Conversion: Batteries Capacitors and Fuel Cells*, Vol. 393, April 17, 1995, pp. 157–162 (English).
24. S. Gottesfeld and J. Pafford. *J. Electrochem. Soc.* **135** (1988) 2651–2652.
25. G. Lamont and D. Wilkinson. US Patent 5 763 765 (9 June 1998).

Chapter 24
Materials for hydrogen storage

Andreas Züttel

Abstract

Hydrogen storage is a materials science challenge because, for all six storage methods currently being investigated, materials with either a strong interaction with hydrogen or without any reaction are needed. Besides conventional storage methods, i.e. high-pressure gas cylinders and liquid hydrogen, the physisorption of hydrogen on materials with a high specific surface area, hydrogen intercalation in metals and complex hydrides, and storage of hydrogen based on metals and water are reviewed.

Article Outline

1. Introduction . . . 517
2. Storing hydrogen as a gas . . . 518
   2.1. High-pressure gas cylinders . . . 519
3. Liquid-hydrogen storage . . . 520
4. Physisorption of hydrogen . . . 521
5. Metal hydrides . . . 524
6. Complex hydrides . . . 527
7. Storage via chemical reactions . . . 528
8. Conclusion . . . 528
Acknowledgments . . . 529
References . . . 529

1. INTRODUCTION

The goal is to pack hydrogen as close as possible, i.e. to reach the highest volumetric density by using as little additional material as possible. Hydrogen storage implies the reduction of an enormous volume of hydrogen gas. At ambient temperature and atmospheric pressure, 1 kg of the gas has a volume of 11 m^3. To increase hydrogen density, work must either be applied to compress the gas, the temperature decreased below the critical temperature, or the repulsion reduced by the interaction of hydrogen with another material.

The second important criterion for a hydrogen storage system is the reversibility of uptake and release. Materials that interact with hydrogen, therefore, as well as inert materials, are important. The reversibility criterion excludes all covalent hydrogen–carbon compounds because hydrogen is only released if they are heated to temperatures above 800°C, or if the carbon is oxidized. Basically, six methods of reversible hydrogen storage with a high volumetric and gravimetric density are known today, listed in Table 1. This article reviews the various hydrogen storage methods and illustrates their advantages and the material challenges.

Fuel Cells Compendium

Table 1. The six basic hydrogen storage methods and phenomena. The gravimetric density ρ_m, the volumetric density ρ_v, the working temperature T, and pressure p are listed. RT stands for room temperature (25°C)

Storage method	ρ_m (Mass%)	ρ_v (kg H_2 m^{-3})	T (°C)	p (Bar)	Phenomena and remarks
High-pressure gas cylinders	13	<40	RT	800	Compressed gas (molecular H_2) in light weight composite cylinders (tensile strength of the material is 2000 MPa)
Liquid hydrogen in cryogenic tanks	Size dependent	70.8	−252	1	Liquid-hydrogen (molecular H_2) continuousloss of a few % per day of hydrogen at RT
Adsorbed hydrogen	≈2	20	−80	100	Physisorption (molecular H_2) on materials, e.g. carbon with a very large specific surface area, fully reversible
Absorbed on interstitial sites in a host metal	≈2	150	RT	1	Hydrogen (atomic H) intercalation in host metals, metallic hydrides working at RT are fully reversible
Complex compounds	<18	150	>100	1	Complex compounds ($[AlH_4]^-$ or $[BH_4]^-$), desorption at elevated temperature, adsorption at high pressures
Metals and complexes together with water	<40	>150	RT	1	Chemical oxidation of metals with water and liberation of hydrogen, not directly reversible?

2. STORING HYDROGEN AS A GAS

Three isotopes of hydrogen are known, hydrogen or protium (H), deuterium (D), and the unstable tritium (T). All the isotopes of hydrogen form covalent molecules like H_2, D_2, and T_2, respectively, because of the single electron in the atom. Hydrogen has an ambivalent behavior toward other elements, occurring as an anion (H^-) or cation (H^+) in ionic compounds, forming covalent bonds, e.g. with carbon, or even behaving like a metal to form alloys or intermetallic compounds at ambient temperature.

The phase diagram of the hydrogen molecule H_2 is shown in Fig. 1. At low temperatures, hydrogen is a solid with a density of 70.6 kg m^{-3} at −262°C, and a gas at higher temperatures with a density of 0.089886 kg m^{-3} at 0°C and a pressure of 1 bar. Hydrogen is a liquid in a small zone between the triple and critical points with a density of 70.8 kg m^{-3} at −253°C. At ambient temperature (298.15 K), hydrogen gas is described by the Van der Waals equation

$$p(V) = \frac{nRT}{V - nb} - a\frac{n^2}{V^2} \tag{1}$$

where p is the gas pressure, V the volume, T the absolute temperature, n the number of moles, R the gas constant, a the dipole interaction or repulsion constant, and b the volume occupied by the hydrogen molecules [1]. The strong repulsive interaction between hydrogen is responsible for the low critical temperature (T_c = 33 K) of the gas.

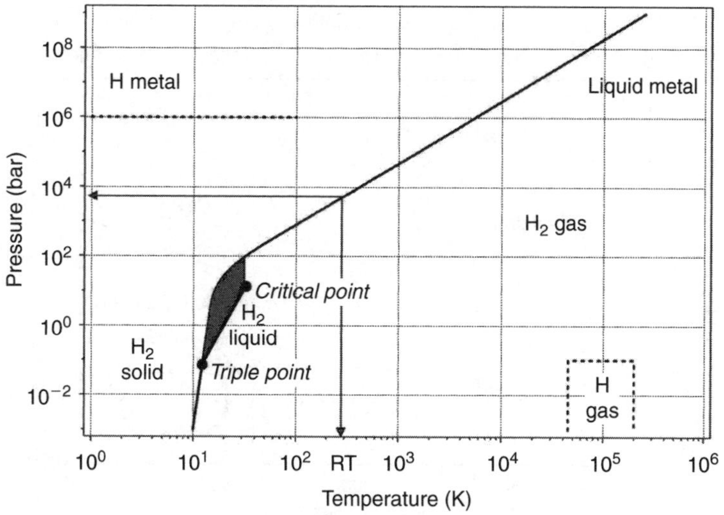

Fig. 1. Primitive phase diagram for hydrogen [46]. Liquid hydrogen only exists between the solidus line and the line from the triple point at 21.2 K and the critical point at 32 K

2.1. High-pressure gas cylinders

The most common storage system is high-pressure gas cylinders, which are operated at a maximum pressure of 20 MPa. The wall thickness of a cylinder capped with two hemispheres is given by the following equation:

$$\frac{d_W}{d_O} = \frac{\Delta p}{2\sigma_V + \Delta p} \tag{2}$$

where d_w is the wall thickness, d_o the outer diameter of the cylinder, Δp the overpressure, and σ_V the tensile strength of the material. The tensile strength of materials varies from 50 MPa for Al to more than 1100 MPa for high quality steel. Other materials like B have a tensile strength of up to 2410 MPa and a density of only 2370 kg m^{-3}. New lightweight composite cylinders have been developed that are able to withstand pressures up to 80 MPa, so that hydrogen can reach a volumetric density of 36 kg m^{-3}, approximately half as much as in its liquid form at normal boiling point. The ideal material for a high-pressure cylinder has a very high-tensile strength (not necessarily isotropic), a low density, and does not react with hydrogen or allow hydrogen to diffuse into it. Most pressure cylinders to date have used austenitic stainless steel (e.g. AISI 316 and 304 and AISI 316L and 304L above 300°C to avoid carbon grain-boundary segregation [2]), Cu, or Al alloys, which are largely immune to hydrogen effects at ambient temperatures. Figure 2 shows the volumetric density of hydrogen inside the cylinder and the ratio of the wall thickness to the outer diameter of the pressure cylinder for stainless steel with a tensile strength of 460 MPa. The volumetric density increases with pressure and reaches a maximum above 1000 bar, depending on the tensile strength of the material. However, the gravimetric density decreases with increasing pressure, and the maximum gravimetric density is found for zero overpressure! Therefore, the increase in volumetric storage density is sacrificed with the reduction in the gravimetric density in pressurized gas systems.

The safety of pressurized cylinders is a concern, especially in highly populated regions. It is envisaged that future pressure vessels will consist of three layers: an inner polymer liner over-wrapped with a carbon-fiber

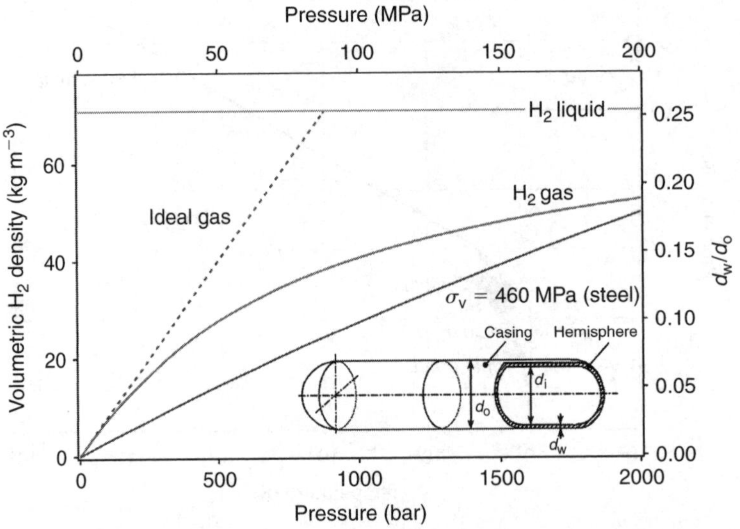

Fig. 2. Volumetric density of compressed hydrogen gas as a function of gas pressure, including the ideal gas and liquid hydrogen. The ratio of the wall thickness to the outer diameter of the pressure cylinder is shown on the right-hand side for steel with a tensile strength of 460 MPa. Inset: A schematic drawing of the pressure cylinder

composite (which is the stress-bearing component) and an outer layer of an aramid-material capable of withstanding mechanical and corrosion damage. Industry has set itself a target of a 110 kg, 70 MPa cylinder with a gravimetric storage density of 6 mass% and a volumetric storage density of 30 kg m^{-3}.

Hydrogen can be compressed using standard, piston-type mechanical compressors. The theoretical work for the isothermal compression of hydrogen is given by the equation:

$$\Delta G = RT \ln\left(\frac{p}{p_0}\right) \tag{3}$$

where R is the gas constant, T the absolute temperature, p and p_0 the end pressure and the starting pressure, respectively. The error of the work calculated with Eq. (3) in the pressure range of 0.1–100 MPa is less than 6%. The isothermal compression of hydrogen from 0.1 to 80 MPa therefore consumes 2.21 kWh kg^{-1}. In a real process, the work consumption is significantly higher because compression is not isothermal. Compression ratios of greater than 20:1 are possible [3] with final pressures >100 MPa.

The relatively low hydrogen density together with the very high gas pressures in the system are important drawbacks of this technically simple and, on the laboratory scale, well-established high-pressure storage method.

3. LIQUID-HYDROGEN STORAGE

Liquid hydrogen is stored in cryogenic tanks at 21.2 K at ambient pressure. Because of the low critical temperature of hydrogen (33 K), the liquid form can only be stored in open systems, as there is no liquid phase existent above the critical temperature. The pressure in a closed storage system at room temperature (RT) could increase to ~10^4 bar.

The simplest liquefaction cycle is the Joule–Thompson cycle (Linde cycle). The gas is first compressed and then cooled in a heat exchanger, before it passes through a throttle valve where it undergoes an isenthalpic

Joule–Thomson expansion, producing some liquid. The cooled gas is separated from the liquid and returned to the compressor via the heat exchanger [4]. The Joule–Thompson cycle works for gases, such as nitrogen, with an inversion temperature above *RT*. Hydrogen, however, warms upon expansion at *RT*. For hydrogen to cool upon expansion, its temperature must be below its inversion temperature of 202 K. Hydrogen is usually precooled using liquid nitrogen (78 K), therefore, before the first expansion step occurs. The free enthalpy change [5] between gaseous hydrogen at 300 K and liquid hydrogen at 20 K is 11640 kJ kg^{-1}. The necessary theoretical energy (work) to liquefy hydrogen from *RT* is $W_{th} = 3.23$ kWh kg^{-1}, the technical work [6] is about 15.2 kWh kg^{-1}, almost half of the lower heating value of hydrogen combustion.

The boil-off rate of hydrogen from a liquid storage vessel because of heat leaks is a function of its size, shape, and thermal insulation. Since boil-off losses as a result of heat leaks are proportional to the surface-to-volume ratio, the evaporation rate diminishes as the storage tank size increases. For double-walled, vacuum-insulated spherical dewars, boil-off losses are typically 0.4% per day for those with a storage volume of 50 m^3, 0.2% for 100 m^3 tanks, and 0.06% for 20 000 m^3 tanks.

The large amount of energy necessary for liquefaction and the continuous boil-off of hydrogen limit the possible use of liquid-hydrogen storage systems to applications where the cost of hydrogen is not an issue and the gas is consumed in a short time, e.g. air and space applications.

4. PHYSISORPTION OF HYDROGEN

Resonant fluctuations in charge distributions, which are called dispersive or van der Waals interactions, are the origin of the physisorption of gas molecules onto the surface of a solid. In this process, a gas molecule interacts with several atoms at the surface of a solid. The interaction is composed of two terms: an attractive term, which diminishes with the distance between the molecule and the surface to the power of –6, and a repulsive term, which diminishes with distance to the power of −12. The potential energy of the molecule, therefore, shows a minimum at a distance of approximately one molecular radius of the adsorbate. The energy minimum [7] is of the order of 0.01–0.1 eV (1–10 kJ mol^{-1}). Because of the weak interaction, significant physisorption is only observed at low temperatures ($<$273 K).

Once a monolayer of adsorbate molecules is formed, gaseous molecules interact with the surface of the liquid or solid adsorbate. The binding energy of the second layer of adsorbate molecules is, therefore, similar to the latent heat of sublimation or vaporization of the adsorbate. Consequently, a single monolayer is adsorbed at a temperature equal to or greater than the boiling point of the adsorbate at a given pressure [8]. To estimate the quantity of adsorbate in the monolayer, the density of the liquid adsorbate and the volume of the molecule is required. If the liquid is assumed to consist of a close-packed, face-centered cubic structure, the minimum surface area, S_{ml}, for 1 mole of adsorbate in a monolayer on a substrate can be calculated from the density of the liquid, ρ_{liq}, and the molecular mass of the adsorbate, M_{ads}:

$$S_{ml} = \frac{\sqrt{3}}{2}\left(\sqrt{2N_A}\,\frac{M_{ads}}{\rho_{liq}}\right)^{\frac{2}{3}} \tag{4}$$

where N_A is the Avogadro constant. The monolayer surface area for hydrogen is $S_{ml}(H_2) = 85\,917$ m^{2}mol^{-1}. The amount of adsorbate, m_{ads}, on a substrate material with specific surface area, S_{spec}, is given by $m_{ads} = M_{ads}S_{spec}/S_{ml}$. In the case of carbon as the substrate and hydrogen as the adsorbate, the maximum specific surface area of carbon is $S_{spec} = 1315$ m^{2}g^{-1} (single-sided graphene sheet) and the maximum amount of adsorbed hydrogen is $m_{ads} = 3.0$ mass%. From this approximation, we may conclude that the amount of adsorbed hydrogen is proportional to the specific surface area of the adsorbent with $m_{ads}/S_{spec} = 2.27 \times 10^{-3}$ mass% m^{-2}g, and can only be observed at very low temperatures.

Materials with a large specific surface area like activated or nanostructured carbon and carbon nanotubes (CNTs) are possible substrates for physisorption. The main difference between CNTs and high surface area graphite (HSAG) is the curvature of the graphene sheets and the cavity inside the tube. In microporous solids with capillaries, which have a width of less than a few molecular diameters, the potential fields from opposite walls overlap so that the attractive force acting upon adsorbate molecules is increased compared with that on a flat carbon surface [9]. This phenomenon is the main motivation for the investigation of the hydrogen–CNT interaction.

Most work on the theoretical absorption of hydrogen in carbon nanostructures uses the Feynman (semi-classical) effective potential approximation to calculate the adsorption potential [10] or the grand canonical Monte Carlo simulation [11,12]. The adsorption potential is 9 kJ mol^{-1} (0.093 eV) for hydrogen molecules inside (13,0) zigzag CNTs with a diameter of 1.018 nm at 50 K – about 25% higher than the flat surface of graphite. The amount of absorbed hydrogen depends on the surface area of the sample, the maximum is 0.6 mass% (at $p = 6$ MPa, $T = 300$ K). The investigation of hydrogen absorption inside CNTs has shown that it is energetically more favorable for hydrogen atoms to recombine and form molecules, [13]. Molecular dynamics simulations of hydrogen atom implantation [14] with an energy of 20 eV through the sidewalls of a (5,5) single-walled carbon nanotube (SWNT) consisting of 150 atoms and a diameter of 0.683 nm has been performed. The hydrogen atoms recombine to form molecules inside the tube and arrange themselves in a concentric pattern. The hydrogen pressure inside the SWNT increases with the number of injected atoms and reaches 35 GPa for 90 atoms (5 mass%). This simulation does not show condensation of hydrogen inside the nanotube. The critical temperature [15] of hydrogen is 33.25 K. If we assume that hydrogen behaves in a similar way to nitrogen [16], it should only form one monolayer of liquid at the carbon surface at temperatures above its boiling point. Geometrical considerations of CNTs lead to the specific surface area and to the maximum amount of condensed hydrogen in a surface monolayer. Figure 3 shows the maximum amount of hydrogen in mass% for the physisorption of hydrogen on CNTs [17]. The theoretical maximum amount of adsorbed hydrogen is 3.0 mass% for SWNTs with a specific surface area of 1315 m^2 g^{-1} at a temperature of 77 K.

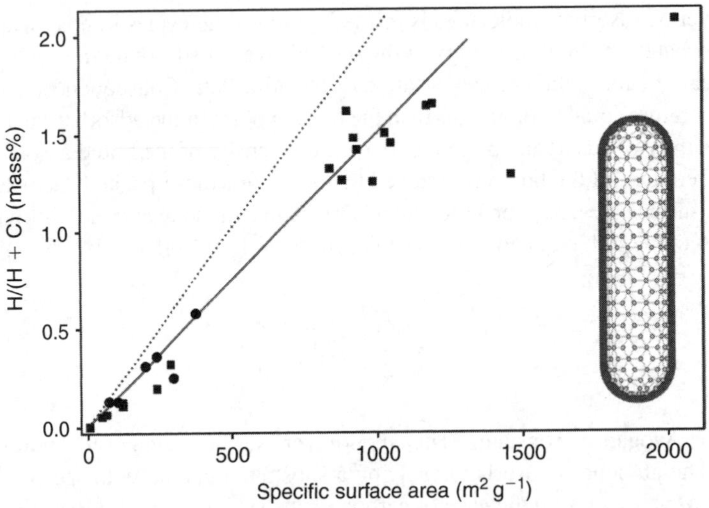

Fig. 3. Reversible amount of hydrogen adsorbed (electrochemical measurement at 298 K) versus the surface area (black circles) of a few CNT samples including two measurements on HSAG samples together with the fitted line. Hydrogen gas adsorption measurements at 77 K from Nijkamp et al. [19] (black squares) are included. The dotted line represents the calculated amount of hydrogen in a monolayer at the surface of the substrate

A large variety of different nanostructured carbon samples have been investigated using a high-pressure microbalance [18,19] at 77 K, electrochemical galvanostatic measurements at *RT* [20–23], and volumetric (mass flow) gas-phase measurements at 77 K. From adsorption–desorption experiments, it is evident that reversible physisorption takes place with all samples. The amount of adsorbed hydrogen correlates with the specific surface area of the sample (Fig. 3). The electrochemical hydrogen absorption is reversible. The maximum discharge capacity measured at 298 K is 2 mass% with a very small discharge current. A few electrochemical measurements are shown in Fig. 3 compared with calculated values. It is remarkable that measurements of hydrogen uptake in the gas phase at 77 K have the same value as electrochemical measurements at 298 K.

To summarize, the reversible hydrogen sorption process is based on physisorption. The amount of adsorbed hydrogen is proportional to the surface area of the nanostructured carbon sample. The amount of adsorbed hydrogen from the gas phase at 77 K and electrochemically at *RT* is 1.5×10^{-3} mass% m^{-2} g. Together with the maximum specific surface area of carbon (1315 $m^2 g^{-1}$), the maximum measured absorption capacity of the nanostructured material is 2 mass%. The experimental results are in good agreement with theoretical estimations if we take into account that the measurements were carried out at a temperature of 77 K. This is still far above the critical temperature of hydrogen, which means that the monolayer of hydrogen is not complete. There is no evidence that the geometric structure of nanostructured carbon influences the amount of adsorbed hydrogen. It is obvious that the curvature of nanotubes influences the adsorption energy, but not the amount of absorbed hydrogen. All attempts to open nanotubes and absorb hydrogen inside do not show any increased absorption. Theoretical studies beyond the well-known physisorption routes lead to various maximum hydrogen absorption capacities. Most of the experiments were performed under special conditions, e.g. at zero K or high-energy hydrogen atom implantation. No evidence has been found for a higher density of hydrogen in and on carbon nanostructures compared with liquid hydrogen at ambient conditions.

Besides carbon nanostructures, other nanoporous materials have been investigated for hydrogen absorption. Zeolites of different pore architecture and composition, e.g. A, X, and Y, have been analyzed [24] in the temperature range 293–573 K and at pressures of 2.5–10 MPa. In this work, hydrogen was absorbed at the desired temperature and pressure. Samples were cooled to RT and then evacuated. Hydrogen release upon heating of the samples to the absorption temperature was measured. The amount of hydrogen absorbed increased with temperature and absorption pressure. The maximum amount of desorbed hydrogen was found to be 0.08 mass% for a sample loaded at a temperature of 573 K and a pressure of 10 MPa. This behavior indicates that absorption is caused by a chemical reaction rather than physisorption. At liquid nitrogen temperatures (77 K), zeolites physisorb hydrogen in proportion to the specific surface area of the material. A maximum of 1.8 mass% of adsorbed hydrogen was found [25] for a zeolite (NaY) with a specific surface area of 725 $m^2 g^{-1}$. The low-temperature physisorption (type I isotherm) of hydrogen in zeolites is in good agreement with the model for nanostructured carbon. The desorption isotherm follows the same path as adsorption [25], indicating that no pore condensation is occurring.

Recently, a microporous metal-organic framework of the composition Zn_4O (1,4-benzenedicarboxylate)$_3$ was proposed as a hydrogen storage material [26]. The material was shown to absorb hydrogen at a temperature of 298 K in proportion to the applied pressure. The slope of the linear relationship between the gravimetric hydrogen density and the hydrogen pressure was found to be 0.05 mass% bar^{-1}. No saturation of the hydrogen absorption was found, which is very unlikely for any kind of hydrogen absorption process. At 77 K, the amount of adsorbed hydrogen was 3.7 mass% at very low hydrogen pressures and showed an almost linear increase with pressure. This behavior is not a type I isotherm, as the authors claim, and the results should be taken with care.

The big advantages of physisorption for hydrogen storage are the low operating pressure, the relatively low cost of the materials involved, and the simple design of the storage system. The rather small gravimetric and volumetric hydrogen density on carbon, together with the low temperatures necessary, are significant drawbacks.

5. METAL HYDRIDES

Hydrogen reacts at elevated temperatures with many transition metals and their alloys to form hydrides. The electropositive elements are the most reactive, i.e. Sc, Yt, lanthanides, actinides, and members of the Ti and Va groups. The binary hydrides of the transition metals are predominantly metallic in character and are usually referred to as metallic hydrides. They are good conductors, have a metallic or graphite-like appearance, and can often be wetted by Hg.

Many of these compounds (MH_n) show large deviations from ideal stoichiometry ($n = 1, 2, 3$) and can exist as multiphase systems. The lattice structure is that of a typical metal with hydrogen atoms on the interstitial sites, and for this reason they are also called interstitial hydrides. This type of structure is limited to the compositions MH, MH_2, and MH_3, with the hydrogen atoms fitting into octahedral or tetrahedral holes in the metal lattice, or a combination of the two. The hydrogen carries a partial negative charge, depending on the metal, but an exception [27] is $PdH_{0.7}$. Pt and Ru are able to adsorb considerable quantities of hydrogen, which becomes activated. These two elements, together with Pd and Ni, are extremely good hydrogenation catalysts, although they do not form hydrides [28].

Especially interesting are the metallic hydrides of intermetallic compounds, in the simplest case the ternary system AB_xH_n, because the variation of the elements allows the properties of these hydrides to be tailored. Element A is usually a rare earth or an alkaline earth metal and tends to form a stable hydride. Element B is often a transition metal and forms only unstable hydrides. Some well-defined ratios of B:A, where $x = 0.5, 1, 2, 5$, have been found to form hydrides with a hydrogen/metal ratio of up to 2.

The reaction of hydrogen gas with a metal is called the absorption process and can be described in terms of a simplified one-dimensional potential energy curve [29] (Fig. 4). The hydrogen atoms contribute their

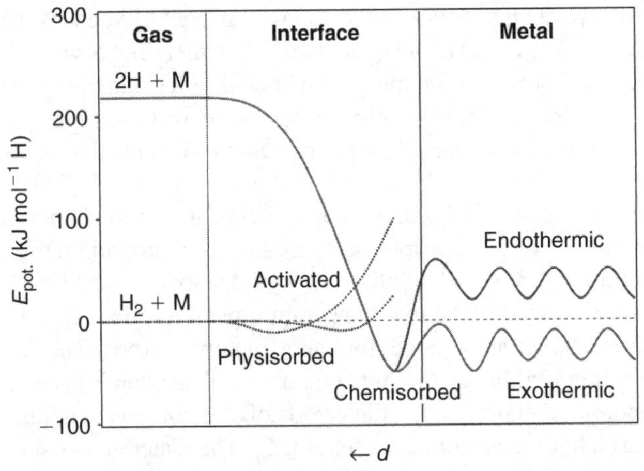

Fig. 4. Far from the metal surface, the potential of a hydrogen molecule and of two hydrogen atoms are separated by the dissociation energy ($H_2 \rightarrow 2H$, $E_D = 435.99\,kJ\,mol^{-1}$). The first attractive interaction of the hydrogen molecule approaching the metal surface is the Van der Waals force leading to the physisorbed state ($E_{Phys} \approx 10\,kJ\,mol^{-1}$) approximately one hydrogen molecule radius ($\approx$0.2 nm) from the metal surface. Closer to the surface, the hydrogen has to overcome an activation barrier for dissociation and formation of the hydrogen metal bond. The height of the activation barrier depends on the surface elements involved. Hydrogen atoms sharing their electron with the metal atoms at the surface are then in the chemisorbed state ($E_{Chem} \approx 50\,kJ\,mol^{-1}\,H_2$). The chemisorbed hydrogen atoms may have a high surface mobility, interact with each other, and form surface phases at sufficiently high coverage. In the next step, the chemisorbed hydrogen atom can jump in the subsurface layer and finally diffuse on the interstitial sites through the host metal lattice

electron to the band structure of the metal. At a small hydrogen/metal ratio (H/M < 0.1), the hydrogen is exothermically dissolved in the metal (solid-solution, α-phase). The metal lattice expands proportional to the hydrogen concentration by approximately 2–3 Å^3 per hydrogen atom [30].

At greater hydrogen concentrations in the host metal (H/M > 0.1), a strong hydrogen–hydrogen interaction becomes important because of the lattice expansion, and the hydride phase (β-phase) nucleates and grows. The hydrogen concentration in the hydride phase is often found to be H/M = 1. The volume expansion between the coexisting α and β-phases corresponds, in many cases, to 10–20% of the metal lattice. At the phase boundary, therefore, a large stress builds up and often leads to a decrepitation of brittle host metals such as intermetallic compounds. The final hydride is a powder with a typical particle size of 10–100 μm.

The thermodynamic aspects of hydride formation from gaseous hydrogen are described by pressure–composition isotherms (Fig. 5). When solid solution and hydride phases coexist, there is a plateau in the isotherms, the length of which determines the amount of hydrogen stored. In the pure β-phase, the hydrogen pressure rises steeply with the concentration. The two-phase region ends in a critical point, T_c, above which the transition from the α- to β-phase is continuous. The equilibrium pressure, p_{eq}, is related to the changes ΔH and ΔS in enthalpy and entropy, respectively, as a function of temperature by the Van't Hoff equation

$$\ln\left(\frac{p_{eq}}{p_{eq}^0}\right) = \frac{\Delta H}{R}\frac{1}{T} - \frac{\Delta S}{R} \tag{5}$$

As the entropy change corresponds mostly to the change from molecular hydrogen gas to dissolved solid hydrogen, it is approximately the standard entropy of hydrogen ($S_0 = 130\,\mathrm{J\,K^{-1}mol^{-1}}$) and is, therefore, $\Delta S_f \approx -130\,\mathrm{J\,K^{-1}mol^{-1}}\,H_2$ for all metal–hydrogen systems. The enthalpy term characterizes the stability of the metal–hydrogen bond. To reach an equilibrium pressure of 1 bar at 300 K, ΔH should amount to $39.2\,\mathrm{kJ\,mol^{-1}}\,H_2$. The entropy of formation term for metal hydrides leads to a significant heat evolution $\Delta Q = T\Delta S$ (exothermal reaction) during hydrogen absorption. The same heat has to be provided to the

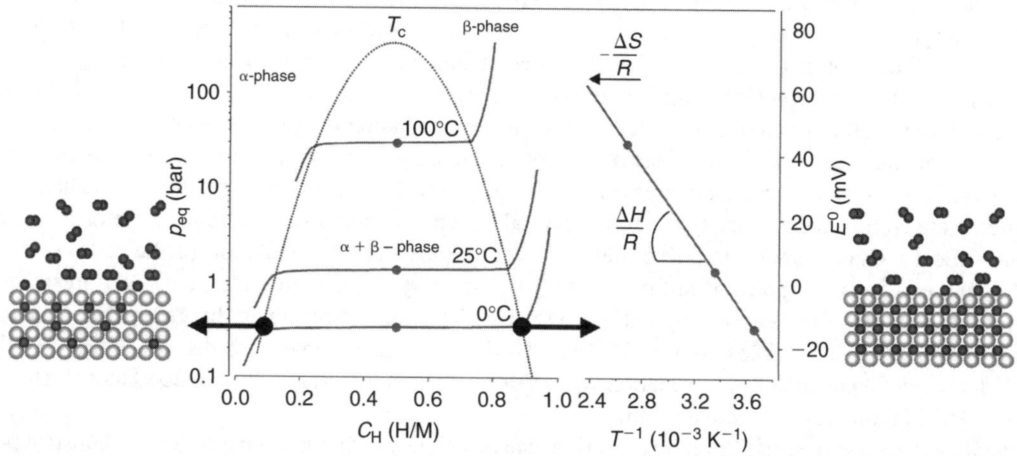

Fig. 5. Pressure composition isotherms for hydrogen absorption in a typical intermetallic compound on the left-hand side. The solid solution (α-phase), the hydride phase (β-phase) and the region of the coexistence of the two phases are shown. The coexistence region is characterized by the flat plateau and ends at the critical temperature T_c. The construction of the Van't Hoff plot is shown on the right-hand side. The slope of the line is equal to the enthalpy of formation divided by the gas constant and the intercept is equal to the entropy of formation divided by the gas constant

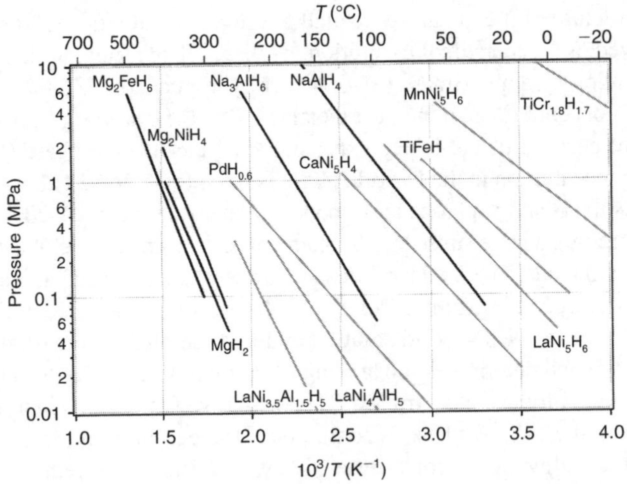

Fig. 6. Van't Hoff plots of some selected hydrides. The stabilization of the hydride of $LaNi_5$ by the partial substitution of Ni with Al in $LaNi_5$ is shown, as well as the substitution of La with mischmetal (e.g. 51% La, 33% Ce, 12% Nd, 4% Pr)

metal hydride to desorb the hydrogen (endothermal reaction). If the hydrogen desorbs below *RT*, this heat can be delivered by the environment. However, if the desorption is carried out above *RT*, the necessary heat has to be delivered from an external source, such as the combustion of hydrogen. For a stable hydride like MgH_2, the heat necessary for the desorption of hydrogen at 300°C and 1 bar is ≈25% of the higher heating value of hydrogen.

Several empirical models allow the estimation of the stability and concentration of hydrogen in an intermetallic hydride. The maximum amount of hydrogen in the hydride phase is given by the number of interstitial sites in the intermetallic compound [31,32]. As a general rule, it can be stated that all elements with an electronegativity in the range of 1.35–1.82 do not form stable hydrides (hydride gap) [33]. Here, the stability of a hydrogen atom on an interstitial site is the weighted average of the stability of the corresponding binary hydrides of the neighboring metallic atoms [34]. More general is the Miedema model: the more stable an intermetallic compound is the less stable the corresponding hydride and vice versa [35].

These semiempirical models allow estimation of the stability of binary hydrides as long as rigid band theory can be applied. However, the interaction of hydrogen with the electronic structure of the host metal is often more complicated. In many cases, the crystal structure of the host metal and, therefore, the electronic structure change upon the phase transition and the theoretical calculation of the stability of the hydride becomes very complex, if not impossible. The stability of metal hydrides is usually presented in the form of Van't Hoff plots according to Eq. (5) (Fig. 6). The most stable binary hydrides have enthalpies of formation of $\Delta H_f = -226\,\text{kJ mol}^{-1}\ H_2$, e.g. HoH_2. The least stable hydrides are $FeH_{0.5}$, $NiH_{0.5}$, and $MoH_{0.5}$ with enthalpies of formation of $\Delta H_f = +20\,\text{kJ mol}^{-1}\ H_2$, $\Delta H_f = +20\,\text{kJ mol}^{-1}\ H_2$, and $\Delta H_f = +92\,\text{kJ mol}^{-1}\ H_2$, respectively [36].

Metal hydrides, because of this phase transition, can absorb large amounts of hydrogen at a constant pressure, i.e. the pressure does not increase with the amount of hydrogen absorbed. The characteristics of hydrogen absorption and desorption can be tailored by partial substitution of the constituent elements in the host lattice. Some metal hydrides absorb and desorb hydrogen at ambient temperature and close to atmospheric pressure. Several families of intermetallic compounds listed in Table 2 are interesting for hydrogen storage. They all consist of an element with a high affinity to hydrogen, element A, and a low affinity one, element B. The latter is often at least partially Ni, since it is an excellent catalyst for hydrogen dissociation.

Table 2. The most important families of hydride-forming intermetallic compounds, including the prototype and the structure. Element A has a high affinity to hydrogen and element B has a low affinity to hydrogen

Intermetallic compound	Prototype	Structure
AB_5	$LaNi_5$	Haucke phases, hexagonal
AB_2	ZrV_2, $ZrMn_2$, $TiMn_2$	Laves phase, hexagonal or cubic
AB_3	$CeNi_3$, YFe_3	Hexagonal, $PuNi_3$-typ
A_2B_7	Y_2Ni_7, Th_2Fe_7	Hexagonal, Ce_2Ni_7-typ
A_6B_{23}	Y_6Fe_{23}	Cubic, Th_6Mn_{23}-typ
AB	TiFe, ZrNi	Cubic, CsCl-or CrB-typ
A_2B	Mg_2Ni, Ti_2Ni	Cubic, $MoSi_2$-or Ti_2Ni-typ

One of the most interesting features of metallic hydrides is the extremely high volumetric density of hydrogen atoms present in the host lattice. The highest volumetric hydrogen density reported to date is $150\,kg\,m^{-3}$ in Mg_2FeH_6 and $Al(BH_4)_3$. Both hydrides belong to the complex hydrides family (see below). Metallic hydrides can reach a volumetric hydrogen density of $115\,kg\,m^{-3}$, e.g. $LaNi_5$. Most metallic hydrides absorb hydrogen up to a hydrogen/metal ratio of $H/M = 2$. Greater ratios up to $H/M = 4.5$, e.g. for $BaReH_9$, have been found [37]. However, all hydrides with a hydrogen to metal ratio of more than 2 are ionic or covalent compounds and belong to the complex hydrides group.

Metal hydrides are very effective at storing large amounts of hydrogen in a safe and compact way. All the reversible hydrides working around ambient temperature and atmospheric pressure consist of transition metals; therefore, the gravimetric hydrogen density is limited to less than 3 mass%. It remains a challenge to explore the properties of the lightweight metal hydrides.

6. COMPLEX HYDRIDES

Group 1, 2, and 3 light metals, e.g. Li, Mg, B, and Al, give rise to a large variety of metal-hydrogen complexes. They are especially interesting because of their lightweight and the number of hydrogen atoms per metal atom, which is 2 in many cases. The main difference between the complex and metallic hydrides is the transition to an ionic or covalent compound upon hydrogen absorption. The hydrogen in the complex hydrides is often located in the corners of a tetrahedron with B or Al in the center. The negative charge of the anion, $[BH_4]^-$ and $[AlH_4]^-$, is compensated by a cation, e.g. Li or Na. The hydride complexes of borane, the tetrahydroborates $M(BH_4)$, and the tetrahydroaluminates $M(AlH_4)$ are interesting storage materials. They are known to be stable and decompose only at elevated temperatures, often above the melting point of the complex.

In 1996, Bogdanovic and Schwickardi [38] showed, for the first time, adsorption and desorption pressure–concentration isotherms for catalyzed $NaAlH_4$ at temperatures of 180°C and 210°C. The isotherms, which have a nearly horizontal pressure plateau, do not show hysteresis. Furthermore, the catalyzed system reversibly absorbs and desorbs hydrogen of up to 4.2 mass%. The mechanism of the two-step reaction was also described. A more detailed study of $NaAlH_4$ with an improved catalyst has been conducted more recently [39]. A desorption hydrogen pressure of 2 bar at 60°C was found and the enthalpy for the dissociation reaction was determined to be 37 and $47\,kJ\,mol^{-1}$ for the first and second dissociation steps of Ti-doped $NaAlH_4$, respectively, according to the reactions

$$3NaAlH_4 \rightarrow Na_3AlH_6 + 2Al + 3H_2 \quad (3.7wt\%H)$$

$$Na_3AlH_6 \rightarrow 3NaH + Al + 3/2H_2 \quad (3.0wt\%H)$$

The equilibrium hydrogen pressure at *RT*, therefore, is approximately 1 bar. Furthermore, the reaction is reversible, a complete conversion to product was achieved at 270°C under 175 bar hydrogen pressure in 2–3 h [40].

The compound with the highest gravimetric hydrogen density at *RT* known today is $LiBH_4$ (18 mass%). This complex hydride could, therefore, be the ideal hydrogen storage material for mobile applications. $LiBH_4$ desorbs three of the four hydrogens in the compound upon melting at 280°C and decomposes into LiH and B. The desorption process can be catalyzed by adding SiO_2 and significant thermal desorption has been observed [41], starting at 100°C. The stability of metal tetrahydroborides has been discussed in relation to their percentage ionic character, and those compounds with less ionic character than diborane are expected to be highly unstable [42]. Steric effects have also been suggested to be important in some compounds [43,44]. The special feature exhibited by covalent metal hydroborides is that the hydroboride group is bonded to the metal atom by bridging hydrogen atoms, similar to the bonding in diborane, which may be regarded as the simplest of the so-called 'electron-deficient' molecules. Such molecules possess fewer electrons than apparently required to fill all the bonding orbitals, based on the criterion that a normal bonding orbital involving two atoms contains two electrons. The molecular orbital bonding scheme for diborane has been discussed [44].

Complex hydrides open a new field of hydrogen storage materials. While the alanates have been investigated extensively during the last 6 years, there is a whole field of new compounds ready to be explored. The borides are especially interesting because of their very high gravimetric and volumetric hydrogen density.

7. STORAGE VIA CHEMICAL REACTIONS

Hydrogen can be generated by reacting metals and chemical compounds with water. The common experiment, seen in many chemistry classes, where a piece of Na floating on water produces hydrogen, demonstrates the process.

The Na transforms to NaOH in this reaction. The reaction is not directly reversible, but NaOH can be removed and reduced in a solar furnace back to metallic Na. Two Na atoms react with two H_2O molecules and produce one hydrogen molecule. The hydrogen molecule produces a H_2O molecule in combustion, which can be recycled to generate more hydrogen gas. However, the second H_2O molecule necessary for the oxidation of the two Na atoms has to be added. Therefore, Na has a gravimetric hydrogen density of 3 mass%. The same process carried out with Li leads to a gravimetric hydrogen density of 6.3 mass%. The major challenge of this storage method is reversibility and control of the thermal reduction process in order to produce the metal in a solar furnace. The process has been successfully demonstrated with Zn [45].

8. CONCLUSION

The materials science challenge of hydrogen storage is to understand the interaction of hydrogen with other elements better, especially metals. Complex compounds like $Al(BH_4)_3$ have to be investigated and new compounds of lightweight metals and hydrogen will be discovered. Hydrogen production, storage, and conversion has reached a technological level, although plenty of improvements and new discoveries are still possible.

Six different hydrogen storage methods have been described here. Alongside well-established, high-pressure cylinders for laboratory applications and liquid-hydrogen storage methods for air and space applications, metal hydrides and complex hydrides offer a safe and efficient way to store hydrogen. Further research and technical development will lead to higher volumetric and gravimetric hydrogen density. The best materials known today show a volumetric storage density of 150 kg m^{-3}, which can still be

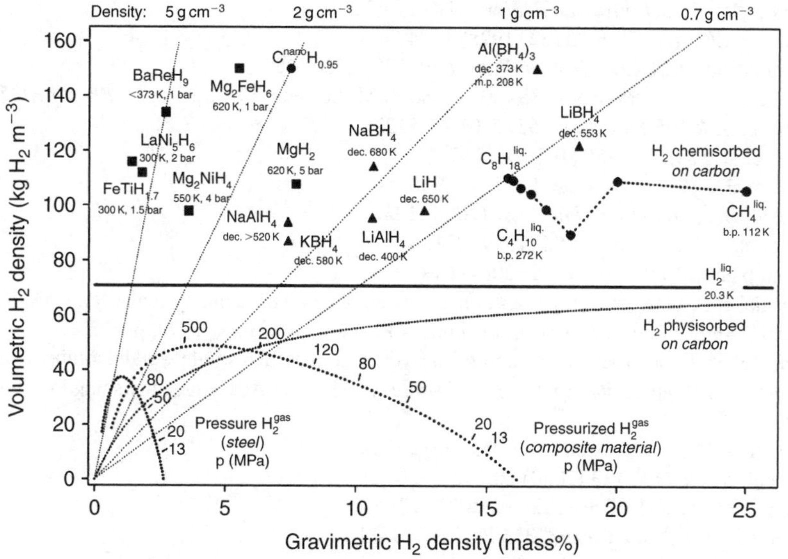

Fig. 7. Volumetric and gravimetric hydrogen density of some selected hydrides. Mg_2FeH_6 shows the highest known volumetric hydrogen density of 150 kg m^{-3}, which is more than double that of liquid hydrogen. $BaReH_9$ has the largest H/M ratio of 4.5, i.e. 4.5 hydrogen atoms per metal atom. $LiBH_4$ exhibits the highest gravimetric hydrogen density of 18 mass%. Pressurized gas storage is shown for steel (tensile strength $\sigma_v = 460$ MPa, density 6500 kg m^{-3}) and a hypothetical composite material ($\sigma_v = 1500$ MPa, density 3000 kg m^{-3})

improved by approximately 50% according to theoretical estimations. Figure 7 shows the volumetric versus gravimetric hydrogen density for the various materials reviewed in this article.

ACKNOWLEDGMENTS

This work was supported by the Swiss Federal Office of Energy (Bundesamt für Energie, BfE) in contract with the International Energy Agency (IEA), the Swiss Federal Office of Education and Science (BBW), the European Commission (Project FUCHSIA), and the Science Faculty of the University of Fribourg in Switzerland.

REFERENCES

1. R.C. Weast. *Handbook of Chemistry and Physics*, 57th edn. CRC Press, 1976.
2. L. Schlapbach. In: L. Schlapbach (Ed.), *Hydrogen in Intermetallic Compounds II*. Springer, Heidelberg, 1988, p. 219.
3. E.L. Huston. A liquid and solid storage of Hydrogen. In: T.N. Veziroqlu and J.B. Taylor (Eds.), *Proceedings of the 5th World Hydrogen Energy Conference*,Vol. 3. Pergamon Press, Oxford, 1984.
4. T.M. Flynn. A liquification of gases. In: S.P. Parker (Ed.), *McGraw-Hill Encyclopedia of Science & Technology*, Vol. 10, 7th edn. McGraw-Hill, New York, 1992, p. 106.
5. Gary Chen and Samim Anghaie based on NASA/NIST databases, http://www.inspi.ufl.edu/data/h_prop_package.html
6. M. von Ardenne et al. *Effekte der Physik*, Verlag Harri Deutsch, Frankfurt am Main, 1990, p. 712.
7. F. London. *Z. Physik.* **63** (1930) 245; *Z. Physik. Chem.* **11** (1930) 222.
8. S. Brunauer et al. *J. Am. Chem. Soc.* **60** (1938) 309.
9. S.J. Gregg and K.S.W. Sing. *Adsorption, Surface Area and Porosity*. Academic Press, London and New York, 1967.

10. G. Stan and M.W. Cole. *J. Low Temp. Phys.* **110** (1998) 539.
11. M. Rzepka et al. *J. Phys. Chem.* **B102** (1998) 10849.
12. K.A. Williams and P.C. Eklund. *Chem. Phys. Lett.* **320** (2000) 352.
13. S.M. Lee et al. *J. Korean Phys. Soc.* **38** (2001) 686; S.M. Lee and Y.H. Lee. *Appl. Phys. Lett.* **76** (2000) 2879.
14. Y. Ma et al. *Phys. Rev. B Solid State* **63** (2001) 115422.
15. W.B. Leung et al. *Phys. Lett.* **56** (1976) 425.
16. R.A. Beebe et al. *J. Am. Chem. Soc.* **69** (1947) 95.
17. A. Züttel et al. *Int. J. Hydrogen Energ.* **27** (2002) 203.
18. R. Ströbel et al. *J. Power Sources* **84** (1999) 221.
19. M.G. Nijkamp et al. *Appl. Phys.* A **72** (2001) 619.
20. Ch. Nützenadel et al. Electronic properties of novel materials. In: H. Kuzmany et al. (Eds.), *Science and Technology of Molecular Nanostructures*. American Institute of Physics, New York, 1999, p. 462.
21. Ch. Nützenadel et al. Electrochemical storage of hydrogen in carbon single wall nanotubes. In: D. Tománek and R.J. Enbody (Eds.), *Sciences and Application of Nanotubes*. Kluwer Academic Publishing/Plenum Press, Dordrecht, 2000, p. 205.
22. A. Züttel et al. *J. Metastable Nanocryst. Mater.* **11** (2001) 95.
23. S.M. Lee et al. *Synth. Met.* **113** (2000) 209.
24. J. Weitkamp et al. *Int. J. Hydrogen Energ.* **20** (1995) 967.
25. H.W. Langmi et al. *J. Alloy. Compd.* **356** (2003) 710–715.
26. N.L. Rosi et al. *Science* **300** (2003) 1127.
27. G.R. Pearson. *Chem. Rev.* **85** (1985) 41.
28. W.M. Mueller et al. (Eds.). *Metal Hydrides*. Academic Press, New York, 1968.
29. J.E. Lennard-Jones. *Trans. Faraday Soc.* **28** (1932) 333.
30. Y. Fukai. *Z. Phys. Chem.* **164** (1989) 165.
31. A.C. Switendick. *Z. Phys. Chem. Neue Fol.* **117** (1979) 89.
32. D.J. Westlake. *J. Less-Common Met.* **91** (1983) 275.
33. P. Rittmeyer and U. Wietelmann. Hydrides. In: M. Bohnet et al. (Eds.), *Ullmann's Encyclopedia of Industrial Chemistry*, Vol. A13, 5th edn. Wiley-VCH, Weinheim, 1996, p. 199.
34. A.R. Miedema. *J. Less-Common Met.* **32** (1973) 117.
35. H.H. Van Mal et al. *J. Less-Common Met.* **35** (1974) 65.
36. R. Griessen and T. Riesterer. Heat of formation models. In: L. Schlapbach (Ed.), *Hydrogen in Intermetallic Compounds I*. Springer, Berlin 1988, p. 219.
37. K. Yvon. *Chimia* **52** (1998) 613.
38. B. Bogdanovic and M. Schwickardi. *J. Alloy. Compd.* **253–254** (1997) 1.
39. B. Bogdanovic et al. *J. Alloy. Compd.* **302** (2000) 36.
40. T.N. Dymova et al. *Dokl. Akad. Nauk. USSR* **215** (1974) 1369.
41. A. Züttel et al. *J. Power Sources* **118** (2003) 1.
42. G.N. Schrauzer. *Naturwissenschaften* **42** (1955) 438.
43. S.J. Lippard and D.A. Ucko. *Inorg. Chem.* **7** (1968) 1051.
44. W.N. Lipscomb. *Boron Hydrides*. W.A. Benjamin, New York, 1963.
45. A. Steinfeld. *Int. J. Hydrogen Energ.* **27** (2002) 611.
46. W.B. Leung et al. *Phys. Lett. A* **56** (1976) 425.

Chapter 25

Fuel economy of hydrogen fuel cell vehicles

Rajesh K. Ahluwalia, X. Wang, A. Rousseau and R. Kumar

Abstract

On the basis of on-road energy consumption, fuel economy (FE) of hydrogen fuel cell (FC) light-duty vehicles is projected to be 2.5–2.7 times the FE of the conventional gasoline internal combustion engine vehicles (ICEV) on the same platforms. Even with a less efficient but higher power density 0.6 V per cell than the base case 0.7 V per cell at the rated power point, the hydrogen FC vehicles are projected to offer essentially the same FE multiplier. The key to obtaining high FE as measured on standardized urban and highway drive schedules lies in maintaining high efficiency of the FC system at low loads. To achieve this, besides a high-performance FC stack, low parasitic losses in the air management system (i.e., turndown and part load efficiencies of the compressor–expander module) are critical.

Keywords: Fuel economy; Hydrogen fuel cell; Light-duty vehicles

Article Outline

1. Introduction . . . 531
2. Fuel cell system . . . 532
3. Performance of fuel cell system . . . 533
   3.1. Air management system . . . 534
   3.2. PEFC stack . . . 535
   3.3. PEFC system performance . . . 535
4. Performance of fuel cell vehicles . . . 537
   4.1. Fuel economy . . . 539
   4.2. Effects of fuel cell system parameters . . . 540
   4.3. Effects of improved H_2-FCV parameters . . . 541
   4.4. Onboard hydrogen storage requirements to yield 320-mile range . . . 542
5. Conclusions . . . 542
Acknowledgements . . . 543
References . . . 543

1. INTRODUCTION

Almost all of the major automobile manufacturers around the world are actively engaged in developing prototype fuel cell (FC) vehicles to meet the future transportation needs of people in developed and developing

Fuel Cells Compendium

countries. Hydrogen-fueled fuel cell vehicles (H_2-FCVs) are also an essential component of the hydrogen economy – a vision of clean, sustainable energy for future generations. For developed countries, such FCVs hold the promise of greatly reduced urban pollution and decreased dependence on imported petroleum. For developing countries, H_2-FCVs also offer an attractive alternative to vastly increasing their petroleum imports, refining, and distribution infrastructure. Underlying these projected benefits is the higher efficiency of FC systems compared to conventional gasoline-fueled internal combustion engines (ICEs). In this paper, we present and discuss the results of an analytical study to examine the potential for fuel economy gains by FCVs over the conventional gasoline-fueled passenger cars. Such analyses also help to determine the amounts of hydrogen such vehicles will need to carry onboard to achieve the desired driving range between refuelings.

We quantify the potential gain in fuel economy (FE) in terms of a multiplier which is defined as the ratio of the miles per gallon gasoline equivalent (mpgge) achieved by the hydrogen FCV on standardized drive schedules to the miles per gallon gasoline achieved by the reference gasoline internal combustion engine vehicle (ICEV) on the same drive cycles. In calculating the multiplier, hydrogen is converted into an equivalent amount of gasoline that has the same lower heating value. According to this conversion, 1 kg of hydrogen is approximately equivalent to 1 US gallon of gasoline in heating value.

Several well-to-wheel studies have also evaluated the FE of H_2-FCVs relative to their conventional gasoline ICEV counterparts. Two of these studies are particularly noteworthy because of the extremes in the methodologies employed. The MIT study [1] broadly defines the performance of the FC system by a simple tabulation of the integrated efficiency as a function of the fraction of peak power and makes gross assumptions for the specific power of the FC propulsion system. Other components of the propulsion system and the vehicle power demand were simulated in much greater detail. For a typical US mid-size family sedan, the MIT study estimated the FE of future H_2-FCVs to be 3.5–3.8 times that of the reference 2001 gasoline ICEV. The future H_2-FCV platform in this study was lighter, more aerodynamic, and had lower rolling resistance coefficient and a smaller frontal area than the reference ICEV. With the ICE and FC systems on the same vehicle platforms and with projected advancements in the ICE technology, the study estimated the FE of the future FC vehicle to be 2.2–2.4 times that of the future (2020) gasoline ICE vehicle.

The GM study [2] modeled the complete vehicle architecture and designed the components to meet specified performance requirements such as 0–60 mph acceleration time, passing maneuvers, top speed and gradeability. The baseline vehicle selected for this study was the Chevrolet Silverado full-size pickup truck. Details of the FC system are considered proprietary but the performance was stated to be a projection of experimental data obtained in GM's laboratories. The study conducted a statistical analysis to provide the best estimate and a measure of the uncertainty around the best estimate. At the 50% likelihood point, the GM study estimated the FE of the H_2-FCVs to be 2.1 times that of the ICEV for year 2005 and beyond.

2. FUEL CELL SYSTEM

Pressurized and ambient pressure polymer electrolyte fuel cell (PEFC) systems are being developed for automotive propulsion and other applications. Pressurization permits FC operation at a higher temperature (e.g., 80°C for 3 atm versus 60–70°C for 1 atm), which eases thermal management, improves cell performance, thereby decreasing the required cell active area (hence, volume, weight, and cost), and facilitates water recovery. The scope of the present study is limited to pressurized hydrogen-fueled automotive PEFC systems of the type shown schematically in Fig. 1.

It is assumed in this study that the FC system operates at 2.5 atm (absolute) at the rated power point; the operating pressure decreases at partial loads according to the performance map of the compressor–expander module (CEM). At the rated power point, the average cell voltage is 0.7 V, and the nominal operating

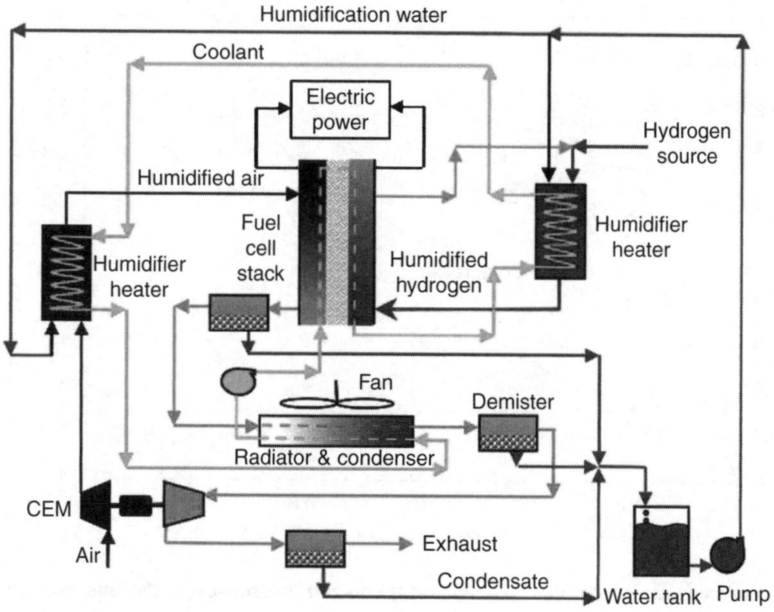

Fig. 1. Schematic diagram of a hydrogen-fueled, PEFC system for automotive applications

temperature of the FC stack is 80°C. At part load, the cell voltage is higher than 0.7 V and it may not be possible to maintain the stack at 80°C. The hydrogen and the air fed to the FC stack are humidified to a relative humidity of 90% at the stack's operating temperature.

The FC power system shown in Fig. 1 may be described in terms of three circuits – one each for the hydrogen, air, and process water. In addition, a coolant circuit is used for thermal management in the system.

Hydrogen from the source (e.g., a compressed gas cylinder) is humidified by using process water and heat from the coolant circuit. The humidified hydrogen is fed to the FC stack; the excess is recirculated to avoid any stagnant zones.

Ambient air is compressed, humidified, and fed to the FC stack at a rate nominally twice that needed for the electrochemical oxidation of the hydrogen to achieve 50% oxygen utilization. During deceleration and at low loads, it may not be feasible to rapidly decrease the air flow rate to achieve 50% oxygen utilization. Exhaust air from the FC stack is cooled in a condenser to recover process water, and then expanded through the turbine to provide some of the compression energy.

Process water is pumped from the tank to the hydrogen and air humidifiers. Inertial separators at the exits from the stack, condenser, and expander recover water from the air stream and return it to the process water tank.

The coolant, which may be water or other fluid, removes waste heat from the FC stack and provides the heat to vaporize water at the gas humidifiers. Excess waste heat is rejected to the ambient air at the radiator–condenser. Although not shown in Fig. 1, the coolant circuit can also provide heat to the passenger cabin, as is done in today's vehicles.

Alternative FC system configurations may also be considered, but they are not discussed in this paper.

3. PERFORMANCE OF FUEL CELL SYSTEM

The performance of the hydrogen FC system shown in Fig. 1 was analyzed with GCtool [3], a systems design and analysis software package developed at Argonne National Laboratory. Although the comprehensive

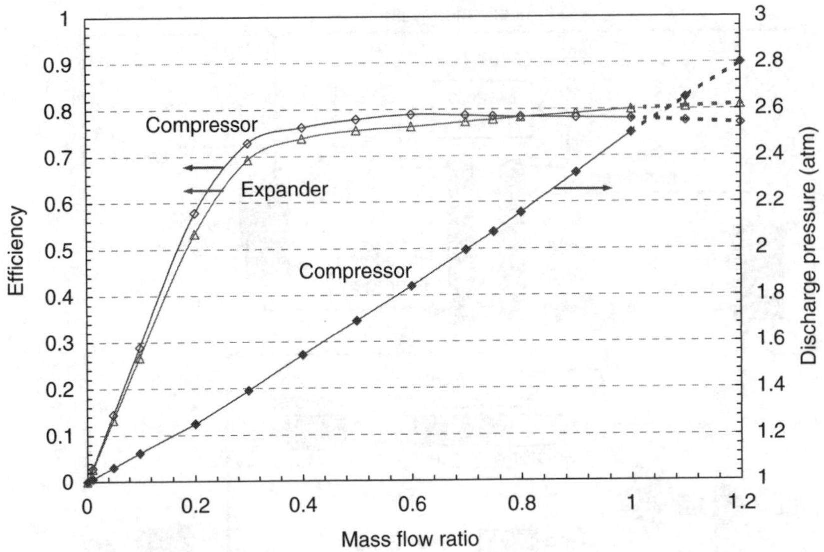

Fig. 2. CEM performance map: isentropic efficiencies of turbo-compressor–expander and compressor discharge pressure

analysis considers heat, mass, and energy flows into and out of each component, only the aspects pertaining to energy efficiency are discussed below.

3.1. Air management system

Developers are pursuing different options for delivery of compressed air to the cathode side of the FC stack. These include a twin-screw compressor, centrifugal compressor and expander on a common shaft, and a toroidal intersecting vane machine [4]. Results presented in this work are based on the projected performance shown in Fig. 2 for a turbo CEM being developed by Honeywell [5]. This module uses a mixed axial and radial flow compressor matched to, and on a common shaft with, an expander with variable inlet nozzle vanes. Because of relatively small flow, less than 200 g/s for a 160 kW FC system, the shaft spins at speeds exceeding 90 000 rpm at the design point. The shaft is supported on air bearings that require a minimum speed of about 36 000 rpm to maintain the air cushion. The module is equipped with a high-speed AC induction motor and a motor controller that also includes a DC/AC inverter so that the module can be directly run by the DC power generated by the PEFC stack. Over the operating range, the motor has an efficiency of about 91% and the motor controller has an efficiency of about 92%. At the design point, the compressor delivers air at 2.5 atm and the compressor and expander have isentropic efficiencies of about 80% and 78%, respectively. With a 20.4 kPa (~3 psi) pressure drop between the compressor and expander, i.e., across the air heater–humidifier, PEFC stack, and the condenser, the module for a 160 kW FC system requires a mechanical power of 7.7 kW at the shaft and a DC power input of 9.1 kW to the motor controller. At part load, the isentropic efficiency of the compressor decreases only gradually from 80% at design point to 72% at one-fourth of the rated flow; the efficiency drops off rapidly as the flow is further reduced towards idling conditions. The expander efficiency behaves similarly. Even though the compressor and expander efficiencies decrease at part load, the net power consumed by the CEM per unit of air flow is actually lower than at rated power because of the drop-off in the compressor discharge pressure.

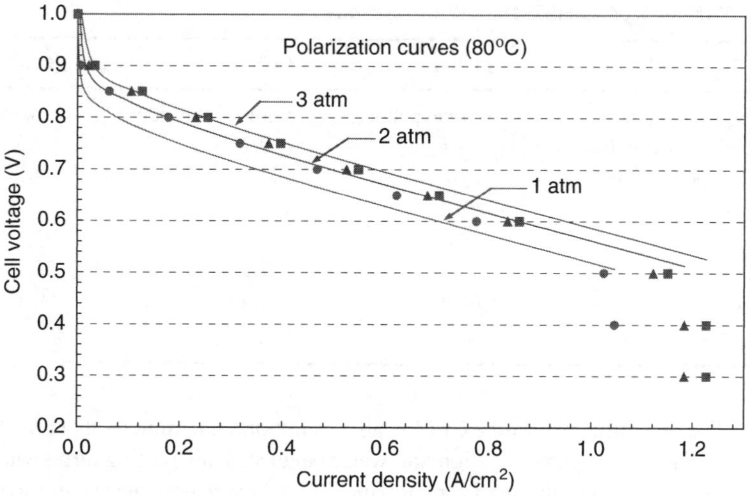

Fig. 3. Polarization curves at 80°C determined from a correlation (solid lines) and a model constructed from laboratory data taken on a small cell (symbols)

3.2. PEFC stack

Figure 3 shows the polarization curves used in this study. The curves are based on a correlation of cell voltage as a function of current density, temperature, and oxygen partial pressure. The experimental data underlying the correlation may be somewhat dated but show the proper trend with respect to the independent variables. Consistent with the reported performances of GM2001 [6] and Ballard stacks [7], PEFC system weights and volumes were estimated by using the published stack power density and specific weight and adjusting them for the design-point cell voltage. For example, the GM2001 stack with 640 cells was stated to have a power density of 1.75 kW/l and specific power of 1.25 kW/kg. Within the simulation model, the current density abscissa in Fig. 3 is renormalized to match the assumed power density at the design point.

3.3. PEFC system performance

Table 1 shows the steady-state performance attributes of three representative stand-alone FC systems. The air management system for FCS-1 includes both a compressor and an expander and the design-point cell voltage is 0.7 V. It has an overall efficiency of 50.6% (based on the lower heating value of hydrogen) at rated power and 61.6% at 25% of rated power.

FCS-2 is similar to FCS-1 except that it does not have an expander. As a result, the compressor in FCS-2 requires an electric motor that is nearly three times the rated capacity of the motor in FCS-1. The FCS-2 has an overall efficiency of 47.0% at rated power and 61.5% at quarter power.

FCS-3 differs from FCS-1 in that the cell voltage at rated power is 0.6 V. FCS-3 has on overall efficiency of only 43.2% at rated power and 60.1% at quarter power.

Also included in Table 1 are the specific powers of the three FC systems. The specific power of FCS-2 is about 10% lower than that of FCS-1 primarily because the stack has to generate additional power to compensate for the power produced by the expander in FCS-1. Because of lower efficiency, the balance of plant is also somewhat larger. The specific power of FCS-3 is about 10% higher than that of FCS-1. The stack for FCS-3 has higher power density because it is sized for a lower cell voltage, but the balance-of-plant is larger due to the lower overall system efficiency at rated power.

Table 1. Attributes of scalable 160 kW FC systems

	FCS-1	FCS-2	FCS-3
Air management system	With expander	Without expander	With expander
Design point cell voltage (V)	0.7	0.7	0.6
CEM motor power (kW)	7.7	22.3	9.0
FCS efficiency			
Rated power (%)	50.6	47.0	43.2
50% power (%)	58.3	57.5	55.5
25% power (%)	61.6	61.5	60.1
Specific power (W/kg)	360	320	400

The maximum allowable turndown of the CEM is also an important parameter that affects the system efficiency at part load as well as oxygen utilization and water balance. This limiting turndown is determined by the minimum idle speed defined as the shaft rpm at which the air management system can provide sufficient cathode air to enable the FCS to generate the power needed by the CEM. This definition is appropriate for a stand-alone FCS in which all FCS accessory loads are provided by the stack. In general, the idle speed may be determined by the power input to the motor controller and by the design of the CEM. For example, in the Honeywell design the air bearings require a minimum speed of 36 000 rpm to support the shaft. Our simulations for the high-speed CEM indicate that with an expander in the system, the idle speed can be as low as 42 500 rpm and the corresponding maximum turndown as high as 20. Without an expander, the minimum idle speed may be as high as 51 500 rpm corresponding to a maximum turndown as low as 5.

Figure 4 illustrates the effect of CEM turndown on dynamic efficiency of FCS-1, the system with an expander, along a simulated urban drive cycle. Simulation results are presented for two variations of FCS-1, one with the theoretically achievable maximum turndown of 20 and the other with a maximum turndown of 5. Differences in efficiencies for the two simulations are clearly evident at low loads. Whereas both give efficiencies in excess of 60% on the urban drive cycle, the peak efficiency can be greater than 70% at low loads for the system with a maximum turndown of 20. However, the scatter in efficiency is also wider at the maximum turndown of 20. This scatter is largely due to acceleration demands that are made with the shaft rpm near the idle speed. At these instances, the CEM motor controller draws large power from PEFC stack to ramp up the shaft speed and increase the cathode flow to meet the surge in demand. The power consumed by the motor is a parasitic loss and contributes to lowering of dynamic efficiency. The frequency and magnitude of the shaft acceleration events are greater at 42 500 rpm idle speed (maximum turndown of 20) than at 51 500 rpm idle speed (maximum turndown of 5). During the acceleration events, the dynamic system efficiency is lower than the steady-state efficiency. Conversely, the dynamic efficiency can be greater than the steady-state efficiency during the deceleration events because some of the CEM parasitic power can be supplied by the inertial power stored in the shaft, compressor, expander, and motor.

For an urban drive cycle, Fig. 5 compares the dynamic efficiencies of FCS-1, the system with an expander, and FCS-2, the system without an expander. In these simulations, FCS-1 is assumed to have the maximum turndown achievable with an expander (20) and FCS-2 the maximum turndown achievable without an expander (5). The differences in efficiencies at high loads are due to the additional power generated by the expander in FCS-1. At low loads, where the parasitic power consumed by the CEM as a fraction of the power produced by the FCS is small, the differences in efficiencies are due to the larger turndown available with the expander in FCS-1. FCS-1 shows a wide scatter in dynamic efficiency at low loads (10–30 kW) whereas FCS-2 exhibits only minor fluctuations. At higher loads (35–60 kW), efficiency fluctuations are damped in FCS-1 but amplified in FCS-2. In general, efficiency fluctuations appear in a narrow band of power demand starting at P/N, where P is the rated power and N the CEM turndown. More importantly, for

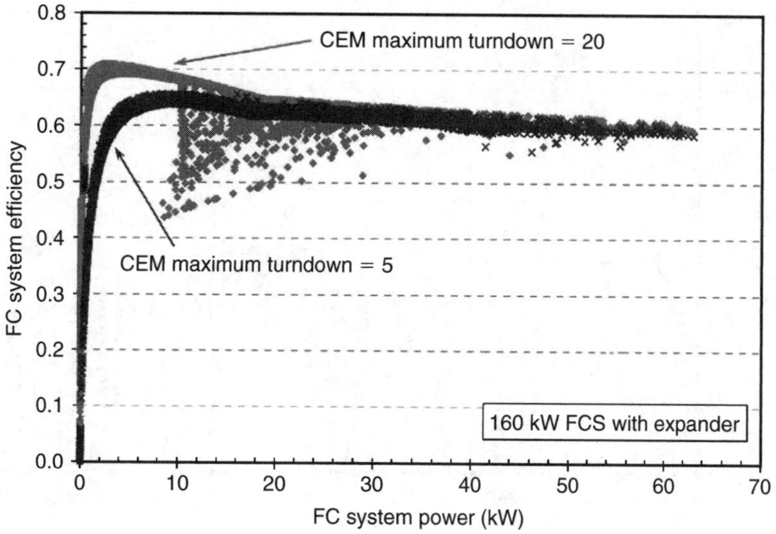

Fig. 4. Effect of CEM turndown on dynamic system efficiency for an urban drive cycle

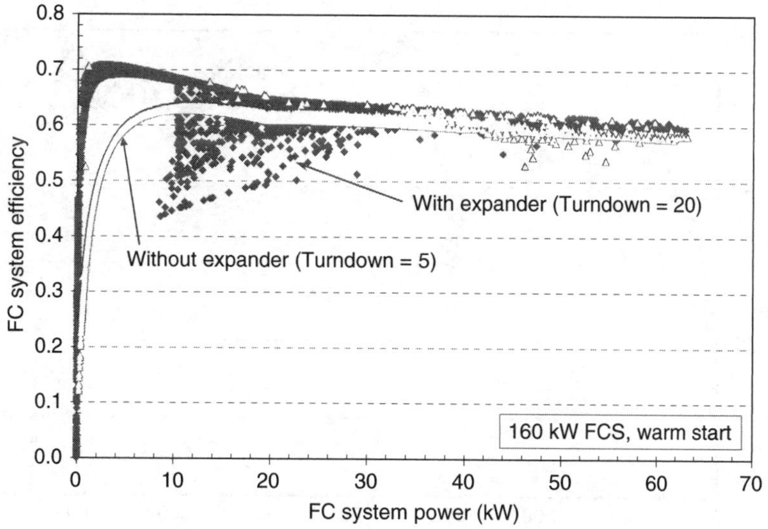

Fig. 5. Effect of expander on dynamic system efficiency for an urban drive cycle

power demand less than *P*/*N*, the oxygen utilization is much lower than the design value of 50% and water recovery becomes an issue.

4. PERFORMANCE OF FUEL CELL VEHICLES

The performance of the hydrogen FCVs was modeled by integrating GCtool with the PSAT vehicle simulation software [8], also developed at Argonne. For this study, we analyzed three US vehicle platforms: the

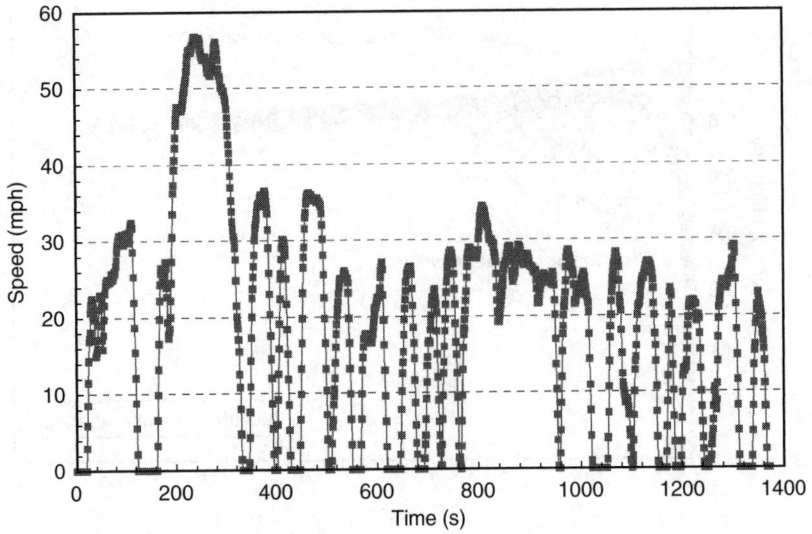

Fig. 6. Vehicle speed versus time for the FUDS

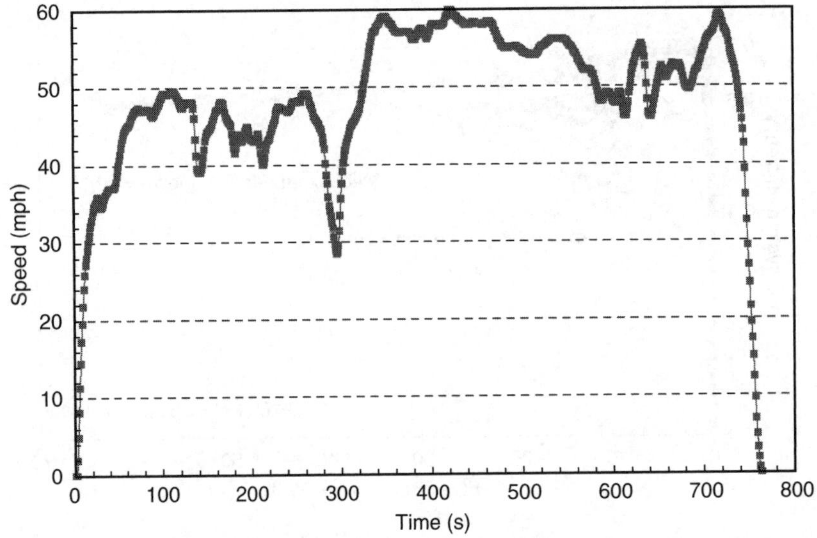

Fig. 7. Vehicle speed versus time for the FHDS

compact Chevrolet Cavalier, the mid-size Ford Taurus, and the sport utility vehicle (SUV) Ford Explorer. We simulated the performance of each vehicle with a gasoline ICE and an H_2-FC power plant. Each vehicle was simulated over the US Federal Urban Driving Schedule (FUDS) (Fig. 6) and the Federal Highway Driving Schedule (FHDS) (Fig. 7). We then determined the FUDS, FHDS, and combined (a weighted harmonic mean of the two) FEs for each vehicle/power train combination.

Table 2 provides values for the major vehicle parameters that affect the FE, including mass, drag coefficient, frontal area, and coefficient of rolling friction. The values for the hydrogen platform are based on FCS-1,

Table 2. Values of key vehicle simulation parameters used in the analyses

	Cavalier		Taurus		Explorer	
Parameter	**ICE**	**H_2-FC**	**ICE**	**H_2-FC**	**ICE**	**H_2-FC**
Mass (kg)	1.214	1.400	1.693	1.850	2.055	2.320
Drag coefficient	0.38		0.32		0.41	
Frontal area (m^2)	1.8		2.2		2.46	
Coefficient of rolling friction	0.009		0.009		0.0084	
Engine power (kW)	86	90	116	120	160	160

the FC system with an expander and PEFC stack at a design-point cell voltage of 0.7 V. For each vehicle, the test mass of the hydrogen version is greater than that of the gasoline version to account for the lower fuel storage energy density and FC system power density. Table 2 also shows the gasoline engine power (mechanical) and the net FC system power (electrical) needed to provide similar drivability characteristics for the FC version of the vehicle. Other vehicle parameters are the same for the two versions of each vehicle.

The modeled electric drivetrain for the FC vehicles consists of a permanent magnet AC traction motor followed by a one-speed transmission and a final drive. The efficiency and torque map of the traction motor includes a DC/AC inverter that is scaled from the data obtained in our laboratory. The electric motor–inverter has a combined peak efficiency of 94%. The one-speed transmission is assigned a constant efficiency of 97% and the final drive a constant efficiency of 93%. It is assumed that the traction motor–inverter is directly connected to the PEFC stack without an intermediate DC/DC converter.

4.1. Fuel economy

Table 3 lists the FE results from base case simulations (i.e., using the FCS-1 FC system configuration) of the six vehicle cases discussed above. The table shows the miles per gallon of gasoline (mpg) or gasoline equivalent (mpgge, based on the lower heating value (LHV) of the fuel) for the different vehicles, power trains, and driving schedules. The US Environmental Protection Agency (EPA) fuel economy ratings for the three conventional gasoline-fueled vehicles are also listed in Table 3 for comparison. In general, there is good agreement between the simulated combined cycle mpg and the values obtained by EPA for the conventional vehicles. The small differences arise from factors such as the engine performance maps in PSAT not being identical to the actual engine performance in the vehicles tested by EPA. The listed FEs have separate correction factors for FUDS and FHDS applied to the calculated values to reflect real-world driving experiences. EPA applies the same correction factors to the FEs of conventional vehicles measured on test tracks; we have assumed that they are applicable to FC vehicles as well.

Table 3 also shows the FE multiplier for the H_2-FCVs versus the ICEVs for the combined urban and highway driving simulations. The multiplier is 2.7 for the compact Cavalier and mid-size Taurus and 2.5 for the SUV. For all three vehicle platforms, the multiplier is higher over FUDS (e.g., 2.6 for SUV) than over FHDS (e.g., 2.3 for SUV). The FHDS to FUDS (FUDS has a lower average speed and power demand than FHDS.) mpgge ratio is higher for ICEVs (e.g., 1.3 for SUV) than for the H_2-FCVs (e.g., 1.1 for SUV). This result simply reinforces our understanding that whereas the efficiency of an ICE decreases at part load that of the H_2 FCS generally increases at part load.

We also examined the sensitivity of these results to variations in several FC and vehicle design and operating parameters. For the FC system, these included the FC polarization curves, the design point cell voltage, and the configuration of and the parasitic losses in the CEM. Vehicle parameters investigated included vehicle mass, drag coefficient, and the coefficient of rolling friction.

Table 3. Calculated FEs of conventional (ICE) and FC (H_2-FC) vehicles

	Cavalier		Taurus		Explorer	
Parameter	ICE	H_2-FC	ICE	H_2-FC	ICE	H_2-FC
Fuel economy (mpgge)						
FUDS	25	73	20	58	18	47
FHDS	32	75	29	69	23	54
Combined	27.6	73.8	23.2	62.4	19.8	49.7
H_2-FC/ICE (mpgge)	2.7		2.7		2.5	
EPA fuel economy combined (mpgge)	26.0		23.7		18.4	

Table 4. Effect of FC system parameters on FE of SUV

FCS configuration	Cell voltage at rated power (V)	CEM maximum turndown	FUDS (mpgge)	FHDS (mpgge)	Combined (mpgge)	FC/ICE (mpgge)	Change in FE (%)
Base case							
FCS-1	0.7	20	46.1	53.0	49.0	2.5	0
More aggressive polarization curve							
FCS-1	0.7	20	49.0	56.1	52.0	2.6	6.1
Effect of cell voltage at rated power							
FCS-3	0.6	20	45.5	52.3	48.3	2.4	−1.4
Effect of removing expander							
FCS-2	0.7	5	41.6	50.4	45.1	2.3	−7.9
Effect of maximum turndown of CEM							
FCS-1	0.7	15	45.7	53.0	48.7	2.5	−0.6
FCS-1	0.7	7	44.2	52.1	47.4	2.4	−3.2
FCS-1	0.7	3	40.6	49.8	44.3	2.2	−9.7
FCS-1	0.7	2.5	38.6	48.4	42.5	2.1	−13.3

4.2. Effects of fuel cell system parameters

The base case analyses discussed above employed reference cell polarization curves based on a published correlation of FC stack data. We also examined the influence of a more aggressive polarization curve (i.e., a higher-performing FC) derived from recent laboratory data on small test cells (see Fig. 3). In the more aggressive polarization curve, the cell voltage at a given current density is less sensitive to system pressure. Table 4 indicates that the FE multiplier for the SUV platform improves by about 6% with the more aggressive polarization curve.

We have also analyzed the effect of selecting a design point cell voltage of 0.6 V (FCS-3) rather than the 0.7 V (FCS-1) used in the base case. At the lower cell voltage, the cell-level power density (in terms of W/cm^2) is almost 40% higher than at 0.7 V per cell. This makes for a smaller, less expensive stack because of the decreased FC active area, but lowers the design point system efficiency. The results are summarized in Table 4, which shows that use of FCS-3 instead of FCS-1 in the SUV platform results in less than 2% decrease in the FE multiplier. Even though FCS-3 is almost 15% less efficient than FCS-1 at the rated power point, the system efficiency is only slightly lower at partial loads. The vehicles' FE is determined

Table 5. Increase in FE of Cavalier and Taurus with improved vehicle parameters

	Cavalier		Taurus	
Parameter	**H_2-FCV**	**H_2-FCV (improved)**	**H_2-FCV**	**H_2-FCV (improved)**
Mass (kg)	1400	1214	1850	1693
Drag coefficient	0.38	0.26	0.32	0.26
Frontal area (m^2)	1.8	1.8	2.2	2.2
Coefficient of rolling friction	0.009	0.006	0.009	0.006
"Engine" power (kW)	90	90	120	120
Fuel economy (mpgge)				
FUDS	73	83	58	62
FHDS	75	101	69	84
Combined	73.8	90.0	62.4	70.4
H_2-FCV/ICEV (mpgge)	2.7	3.3	2.7	3.0

over the FUDS and FHDS, however, which do not require the full rated power at any time during the driving schedules. As shown earlier in Table 1, the difference in efficiencies of FCS-1 and FCS-3 is less than 2% (61.6% versus 60.1%) at 25% of rated power. Consequently, the FE multiplier for the H_2-FCV versus the ICEV decreases very little if the design point cell voltage is 0.6 rather 0.7 V.

The power consumed by CEM represents the largest parasitic loss in the FC system. The FE multiplier can be degraded if the compressor is not as efficient at part load as implied in Fig. 2, does not have the expander, does not have the necessary turndown, or operates as a positive displacement device, and delivers air at constant or high pressure at part load.

We have evaluated the change in FE with removal of the expander, i.e., by using FCS-2 rather than FCS-1 as the FC system configuration for the SUV. Table 4 shows that the effect of removing the expander is to decrease the FE multiplier by about 4% even though FCS-2 is almost 7% less efficient than FCS-1 at the rated power point.

We have also analyzed how the FE of the H_2-FCVs is affected by the maximum turndown of the CEM by varying the minimum "idling" air flow rate and CEM power consumption. Table 4 shows that the FE multiplier decreases by less than 1% if the maximum turndown is 15 rather than the theoretically available value of 20, by about 3% if the maximum turndown is 7, by about 10% at the maximum turndown of 3 and by about 13% at maximum turndown of 2.5.

4.3. Effects of improved H_2-FCV parameters

We have investigated the potential for further improvements in the fuel economies of the two smaller vehicles – the Cavalier and the Taurus – by considering versions of these two vehicles with reduced mass (same as for the ICE version); lower drag coefficient, by improving the vehicles' aerodynamics; and decreased coefficients of rolling friction, by using, for example, advanced tires. The modified parameters, and the resulting FEs, are summarized in Table 5. These results show that significant further increases in FE are possible if these vehicle improvements can be achieved.

The FE multipliers listed in the last row of Table 5 are relative to the conventional ICE vehicles. Of course, with the improved vehicle parameters, the ICEVs' FE would also be improved. In the present discussion, however, we are assessing the potential for improvement in vehicle FE using advanced technologies as compared to current gasoline ICE technology. Thus, the potential improvements in FE offered by lighter, more aerodynamic, and lower-rolling-friction gasoline ICEVs are not discussed.

Table 6. Parameters for compressed hydrogen storage tanks for automotive applications (specifications from quantum [9])

Parameter	Specification			
	Tank 1	Tank 2	Tank 3	Tank 4
Storage pressure (psia)	5000	10 000	5000	10 000
Amount of hydrogen stored (kg)	3	3	7	7
Tank system				
Volume (l)	145	100	320	220
Weight (kg)	45	50	90	100
Tank system				
Energy density (kWh/l)	0.69	1.00	0.73	1.06
Specific energy (kWh/kg)	2.20	2.00	2.60	2.30

Table 7. Hydrogen storage system requirements and parameters for the H_2-FC Cavalier, Taurus, and Explorer to obtain a 320-mile driving range between refuelings

	Cavalier		Taurus		Explorer	
H_2-FC/ICE (mpgge)	2.7		2.7		2.5	
Recoverable H_2 needed (kg)	4.3		5.1		6.5	
Fuel tank						
Pressure (psia)	5000	10 000	5000	10 000	5000	10 000
Volume (l)	205	150	240	165	295	200
Weight (kg)	60	75	75	80	80	95

4.4. Onboard hydrogen storage requirements to yield 320-mile range

A significant issue for hydrogen-fueled vehicles is the amount of hydrogen that must be stored onboard to provide the desired range between refuelings. For the US passenger car market, a minimum driving range of 320 miles is considered essential for customer acceptance. The FEs discussed above are useful for determining the amounts of hydrogen needed for the three vehicles. The corresponding "fuel tank" weights and volumes may be determined by using the characteristics of the specific fuel tank designs or approaches, such as compressed hydrogen, liquefied hydrogen, physical or chemical hydrides, or other hydrogen-storage matrix materials (such as various forms of carbon or complex hydrides).

Table 6 lists the characteristics of representative, current-technology compressed hydrogen storage tanks. We used the energy density (kWh/l) and specific energy (kWh/kg) of the hydrogen stored in the tanks listed in Table 6 as a guide to estimate the volume and weight of compressed hydrogen tanks sized for the H_2-FCVs (Cavalier, Taurus, and Explorer), as shown in Table 7. With the improved vehicle parameters (i.e., lower mass, drag coefficient, and coefficient of rolling friction), correspondingly less hydrogen would be needed onboard the vehicles to yield the same driving range between refuelings.

5. CONCLUSIONS

- For equal LHV energy content of the fuel, the H_2-fueled fuel cell vehicles offer potential mpgge fuel economy multipliers of 2.7, 2.7, and 2.5 for a compact, mid-size, and sport utility vehicle, respectively.

- For a non-hybrid vehicle, the potential improvement in fuel economy over standard urban and highway drive schedules degrades only slightly if the design-point cell voltage is lowered from 0.7 to 0.6 V. Thus, there is little incentive in selecting a higher cell voltage at the rated power point given that the size and cost of the fuel cell stack, likely the most expensive component in the fuel cell system, increase non-linearly with increasing cell voltage.
- The power consumed by the air management system represents the largest parasitic loss in the fuel cell system. To preserve the benefit of the enhanced efficiency of the fuel cell stack at part load, it is important to select an air management system capable of achieving a reasonable turndown and operating at reduced pressures at part load where the CEM component efficiencies are generally lower.
- The compact, mid-size, and sport utility fuel-cell vehicles analyzed in this work will need 4.3, 5.1, and 6.4 kg, respectively, of recoverable H_2 stored onboard to achieve a 320-mile driving range between refuelings (based on the combined fuel economy over the US Federal Urban and Highway Driving Schedules).
- Further gains in the vehicles' fuel economies are possible if their mass, drag coefficient, and/or the coefficient of rolling friction can be reduced.

ACKNOWLEDGEMENTS

This work was supported by the US Department of Energy's Office of Energy Efficiency and Renewable Energy, Office of Hydrogen, Fuel Cells, and Infrastructure Technologies and the Office of FreedomCAR and Vehicle Technologies.

REFERENCES

1. M.A. Weiss, J.B. Haywood, A. Schafer and V.K. Natarajan. *Comparative Assessment of Fuel Cell Cars.* Massachusetts Institute of Technology, Laboratory for Energy and the Environment, Publication No. LFEE 2003-001 RP, February 2003.
2. General Motors. *Well-to-Wheel Energy Use and Greenhouse Gas Emissions of Advanced Fuel/Vehicle Systems–North American Analysis*, Vol. 2. General Motors Corporation, June 2001.
3. H.K. Geyer and R.K. Ahluwalia. GC tool for fuel cell systems design and analysis: user documentation. Argonne National Laboratory Report ANL-98/8, IL, USA 1998.
4. P.B. Davis and F.W. Wagner. Air Management for fuel cell systems: assessment and new activities, SAE Paper 2002-01. *SAE World Conference*, Detroit, MI, 5–8 March 2001.
5. M.K. Gee. Fuel Cell Turbocompressor, DOE Hydrogen, Fuel Cells and Infrastructure Technologies Program 2003 Merit Review and Peer Evaluation Meeting, Berkeley, CA, 19–22 May 2003.
6. GM/Opel Set First Fuel Cell Records with Hydrogen 1, Releases New Stack Design Details, Hydrogen and Fuel Cell Letter, ISSN-1080-8019, June 2001.
7. Ballard sets new standards for automotive fuel cells, News Release, http://www.ballard.com, 26 October 2001.
8. A. Rousseau and P. Pasquier. Validation process of a system analysis model: PSAT, SAE Paper 01-P183. *SAE World Conference*, Detroit, MI, 5–8 March 2001.
9. N. Sirosh. DOE Hydrogen Composite Tank Program, DOE Hydrogen, Fuel Cells and Infrastructure Technologies Program 2003 Merit Review and Peer Evaluation Meeting, Berkeley, CA, 19–22 May 2003.

Chapter 26

PEMFC systems: the need for high temperature polymers as a consequence of PEMFC water and heat management

Ronald K.A.M. Mallant

Abstract

The proton exchange membrane fuel cell (PEMFC) is usually operated at elevated pressure, requiring the use of a compressor. Also the operating temperature is low, generally below 80°C. For many applications, a somewhat higher operating temperature would be preferable. This paper describes the reasons for pressurised operation at, or close to, 80°C in terms of water and heat management issues. It is concluded that a new type of proton conducting material is highly desirable, having many of the good Nafion properties, but based on a proton conduction mechanism that does not require the presence of large amounts of water in the electrolyte.

Keywords: PEMFC; Water management; Heat management; Compressor

Article Outline

1. Introduction . . . 545
2. Theory . . . 546
3. Experimental . . . 547
4. Consequences of high relative humidity . . . 549
   4.1. Humidifier . . . 549
   4.2. Cathode . . . 551
   4.3. Water separator . . . 551
   4.4. Overall heat management . . . 552
5. Discussion . . . 553
References . . . 554

1. INTRODUCTION

It is quite common to operate proton exchange membrane fuel cell (PEMFC) systems at pressures of 2–3 bar (a) and at or below 80°C. Pressurised operation requires that the system includes a compressor, possibly an intercooler and an expander to reclaim part of the energy that was required for compression. A typical air management system is given in Fig. 1. The relevant components are

- a compressor or blower;
- the humidifier, which may be integrated with the cooling circuit;

Fuel Cells Compendium

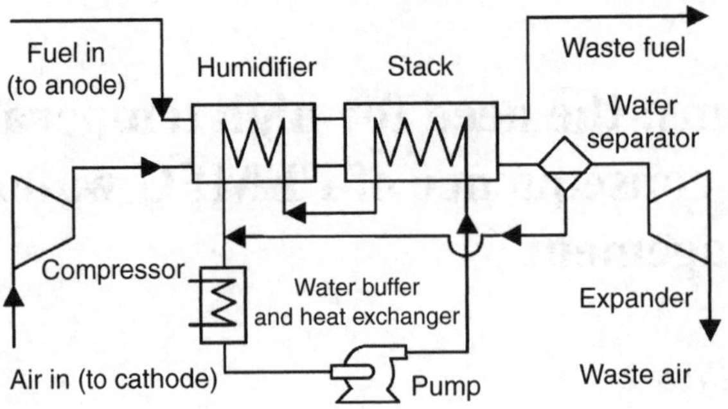

Fig. 1. Simplified layout (intercooler not shown)

- the stack (cathode);
- a water separator;
- a pressure release device (backpressure controller or expander);
- a cooling circuit: pump, buffer and heat exchanger.

In case of pressurisation above about 1.6 bar an air cooler is required. Cooling of the inlet air can also be done by injection of water into the compressor; this option is however not considered here.

For many systems, the use of a compressor is not a viable option. In particular, this is the case for small systems. These systems require low airflows, for which efficient compressors are not available. Also noise and wear may not be acceptable. Such systems are operated close to atmospheric pressure. In this case, the operating temperature will be limited to 60–65°C. This may or may not be acceptable, depending on the application. Some applications however, would benefit from a higher operating temperature in order to make use of the waste heat or to reduce heat rejection problems. This paper addresses why PEMFC systems have difficulty in being operated at both a low pressure and high (>70°C) temperature.

2. THEORY

Application of Faraday's law yields the relation between the current I and the dry air feed (g/s) [1]:

$$q_{\text{air}} = \frac{M_{\text{air}} I \lambda_{\text{air}}}{nNex} \tag{1}$$

where I is the current produced (A), n the number of electrons involved in the reaction (in this case, $n = 2$), e is elementary charge (1.6022E − 19 Coulomb), N the Avogadro's constant (6.022045E ± 23/mol), λ_{air} is the air stoichiometry (the inverse of the oxygen utilisation), x the fraction of oxygen in air (0.20946) and M_{air} the molar mass of air (28.964 g/mol).

For the remainder, all calculations are based on the general gas law

$$\frac{PV}{T} = R$$

and the associated equations

$$P_i = \frac{m_i}{V} \frac{R}{M_i} T$$

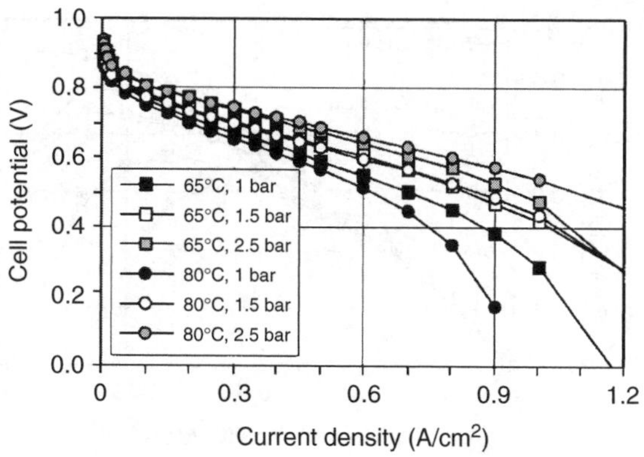

Fig. 2. PEMFC polarisation curves

to calculate the partial pressure for each component in a gas mixture and

$$P = \rho \frac{R}{M} T$$

for calculations on gas mixtures, in which the weighted harmonic mean M of the individual molecular weights M_i of the gases present in masses m_i in the gas mixture is calculated using

$$M = \frac{m_1 + m_2 + \cdots}{m_1/M_1 + m_2/M_2 + \cdots}$$

The calculation of $p_{\max,H_2O}(T)$, the maximal water vapour pressure as function of temperature, was done using a polynomial approximation of water vapour pressure data taken from [1].

3. EXPERIMENTAL

To assess the effect of pressure and temperature on cell performance, a 7 cm^2 cell was operated at 1, 1.5 and 2.5 bar (a), and at temperatures of 65 and 80°C. An air stoichiometry of 2 was maintained by continuously adjusting the cathode flow rate as function of current density. The MEA was made using E-Tek electrodes and a membrane prepared by filling a 50 μm 85% porous substrate (Solupor) with Nafion ionomer [2]. Both anode and cathode gases were fully humidified. Polarisation curves obtained with this cell under various P, T conditions are given in Fig. 2.

Although the PEMFC can be operated over a wide range of the displayed curves, it should be noted that cell efficiency is proportional to cell potential, and is in first approximation given by $\eta_{cell} = 0.8\,V_{cell}$ (V_{cell} in volts, η_{cell} is with respect to LHV, at 100% H_2 utilisation). For that reason, operation at a potential lower than 0.6–0.7 V is hardly attractive. Power density curves are given in Fig. 3. Opting for the highest power density is tempting in order to reduce the power specific size, weight and cost of the stack.

However, compression of air takes energy, and these losses should be accounted for. The polarisation curves given in Fig. 2 can be corrected to account for the losses for compression. A straightforward way to do this is to reduce the cell potential at given current I with a voltage ΔV, such that $I\Delta V$ equals the power required for compression of the air required at that current. The amount of air depends on the operating conditions. At an individual cell potential of 0.675 V and an air stoichiometry of 2, the air consumption of

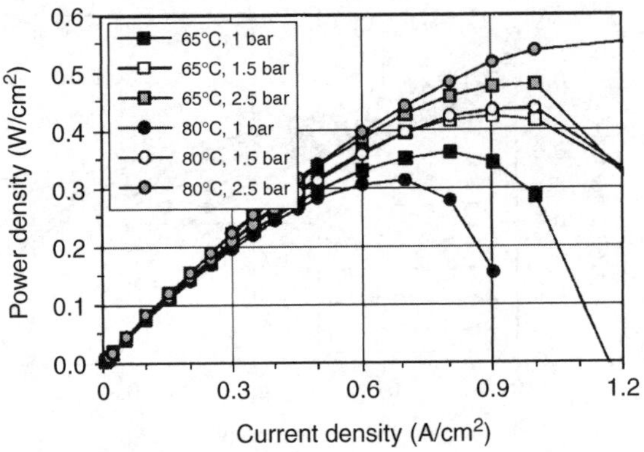

Fig. 3. Power density curves

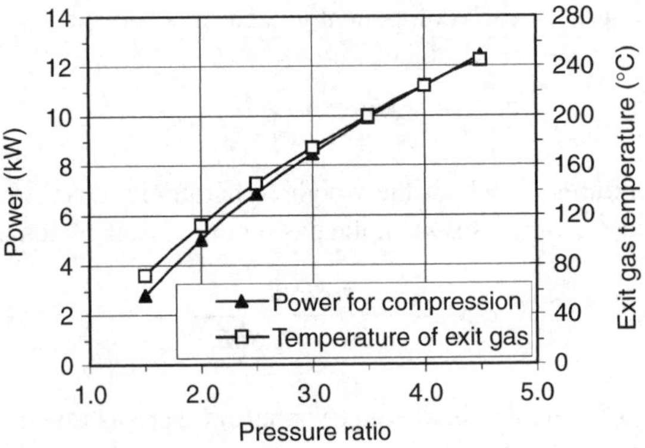

Fig. 4. Calculated shaft power and gas exit temperature for compression of 150 N m³/h air

a 50 kW system is 150 N m³/h. The calculated compressor power as function of pressure ratio is given in Fig. 4 ([3], based on 20°C inlet temperature and 70% compressor efficiency).

Using the calculated energy consumption for air compression, the curves of Fig. 2 were corrected, (see Fig. 5). For the 50 kW (gross power) system referred to in this paper, the compressor uses 8.6 kW, equivalent to 17% of the power produced by the fuel cell. The actual power losses will be slightly higher due to losses in the drive motor and power electronics. On the other hand, for large systems for which a compressor expander can be used, the losses can be reduced. For very small systems (less than several kW), more severe losses must be accounted for, since small air compressors are not likely to attain 70% efficiency.

Ultimately, the two parameters that are of relevance for the system developer and user are the net power density and the system efficiency. For the system dealt with here, the efficiency is mainly determined by cell efficiency and losses for compression. Using the approximate relation between cell potential and cell efficiency, $\eta_{cell} = 0.8\,V_{cell}$, the system efficiency is given by $\eta_{system} = 0.8\,V_{effective}$. The effective cell potential can also be used to calculate the net power density. The resulting relation between net cell power density and system efficiency is given in Fig. 6. From this, it is concluded that the net effect of pressurisation is negligible at preferred operating conditions, that is, at current densities of less than about 0.6 A/cm². So, if for reasons of efficiency, high cell potential/low cell current density operation is preferred, and if no net system power gain is achieved under such conditions, then why would one prefer pressurised operation?

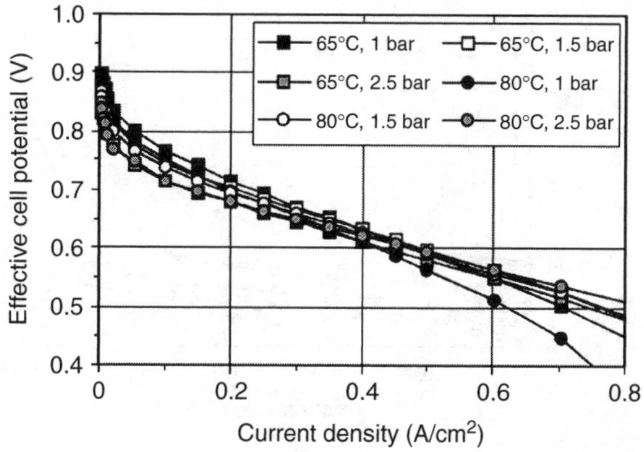

Fig. 5. $V_{effective}$**, based on Fig. 2, corrected for energy consumption for compression**

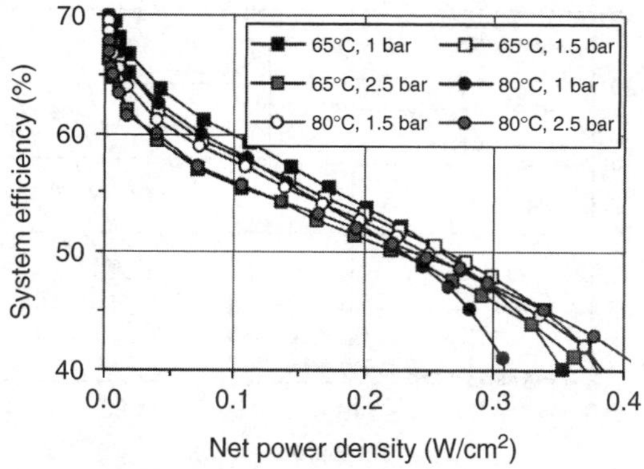

Fig. 6. Relation between system efficiency and net power density

The reason is that at pressurised conditions, the water vapour pressure has less effect on water balance, gas flow rates and gas composition. In the following, this will be shown for the air system only.

4. CONSEQUENCES OF HIGH RELATIVE HUMIDITY

The performance of the PEMFC strongly depends on the conductivity of the polymer [4], which is strongly related to the relative humidity of the gas the cell is in contact with. The requirement of working at saturated gas conditions may perhaps not make it impossible to operate the PEMFC at low P and relatively high T, but such conditions require an adapted design of stack and system. High water vapour pressures lead to strong dilution of reactants by water vapour and large heat exchange surfaces.

In terms of water management, the following occurs in the cathode line.

4.1. Humidifier

The air is humidified and heats up to stack temperature. Using saturation vapour pressure data found in literature [1], the consequences of working at 100% RH are analysed. First of all, the addition of water vapour increases the volume of gas flowing through the system (when pressure is kept constant). Especially at high

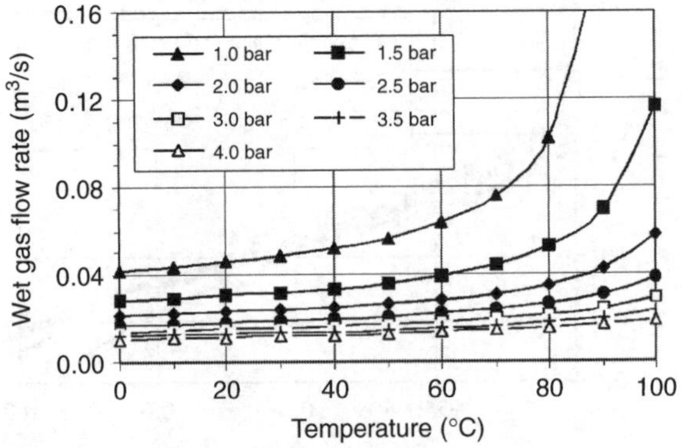

Fig. 7. Calculated wet gas volume at various temperatures and pressures. The original dry gas flow rate used in this calculation is 150 N m³/h, (0.042 N m³/s)

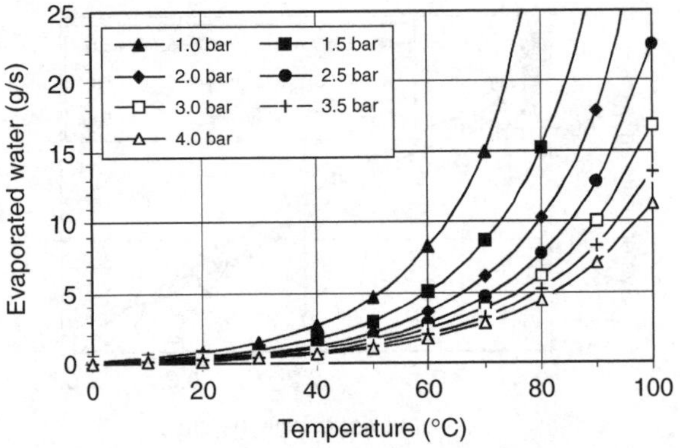

Fig. 8. Amount of water required to humidify 150 N m³/h of initially dry air, as function of temperature and pressure

temperature and low pressure, the increase in gas volume is substantial (Fig. 7). The hydraulic implications of the increase in gas volume have to be taken into consideration in the design of the fuel cell stack and the system. The pressure drop in a tube is given by the equation

$$\frac{dp}{dl} = -\frac{\lambda^{|}}{d}\frac{1}{2}\rho v^2 \tag{2}$$

where dp is the pressure drop over length dl, $\lambda^{|}$ the dimensionless friction factor, d the inside pipe diameter, ρ the average gas density over length dl, and v the average gas velocity. The pressure drop is proportional to the product of the gas density and the square of the gas velocity, although $\lambda^{|}$ is dependent on the Reynolds number, and therefore gas velocity as well. The effect of humidification is that due to the addition of water vapour, the gas density decreases whereas the gas volume increases. The lower gas density results in a lower pressure drop in pipes and bends. The increased volume however will have a much stronger effect, resulting in an increased pressure drop. This can be compensated for by increasing the diameters of flow channels in the fuel cell stack and system piping. This will, however, result in increased system volume and weight.

A trivial consequence of humidification is that water is consumed (Fig. 8). This will particularly become relevant if the water that is produced in the stack is not sufficient to compensate for the water used in the evaporator. The evaporation of water takes energy (Fig. 9). This energy can be taken from the cooling circuit, in which case no extra energy is required. The waste heat of a 50 kW fuel cell stack operating

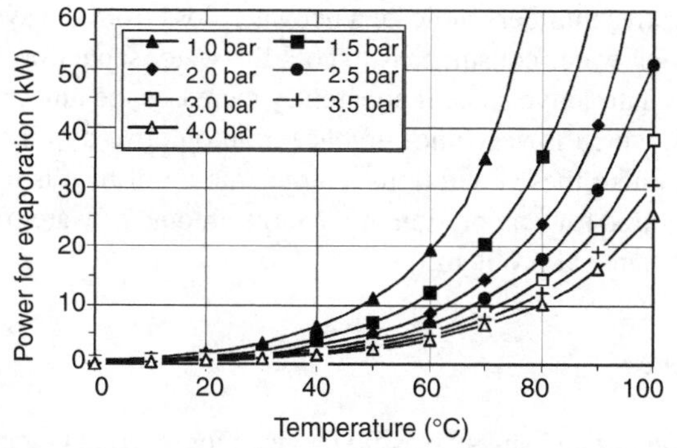

Fig. 9. Power required to fully humidify 150 N m^3/h of initially dry air, as function of temperature and pressure

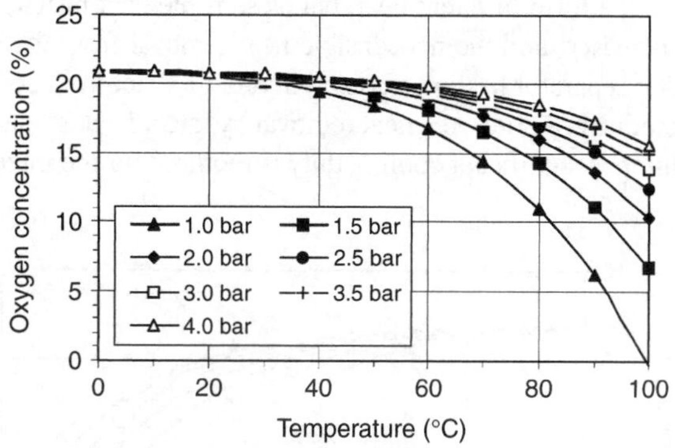

Fig. 10. Calculated oxygen concentration in fully humidified air, in dependence of pressure and temperature

at 0.675 V individual cell potential (the basis for these calculations) under condensing conditions is 61 kW. Comparing this number with the values given in Fig. 9, it can be concluded that there should generally be no problem to fulfil this heat demand.

4.2. Cathode

The addition of water vapour results in a dilution of the oxygen present in air (Fig. 10). The concentration shown here is that at the cathode inlet. It can be concluded that working at low *P*, high *T* conditions leads to high oxygen dilution, which decreases cell performance. Since oxygen is used in the fuel cell reaction, the downstream part of each cell is exposed to substantially lower concentrations.

Under the conditions assumed here (50% oxygen utilisation), half of the oxygen is used by the electrochemical reaction taking place in the fuel cell. Since this reduces the volume of gas, some of the water originally present in the vapour phase will condense out to form liquid water. This produces heat, which has to be removed by the stack coolant circuit. Also, some water is produced, in addition to the water produced by the electrochemical reaction. So, in the cathode, a two-phase flow is established.

4.3. Water separator

The liquid water is separated from the cathode exit gas by a water separator. The amount of water that is collected should match the amount needed for humidification. Fig. 11 shows the water surplus of a system as function of P and T. Negative numbers indicate a net water loss from the system. One way to prevent a system from turning into a net water consumer is to cool the water separator, such that it now becomes a condenser. This may seem unattractive since it apparently increases the amount of heat that has to be discarded to the surroundings, which is very undesirable for automotive applications. Theoretically, this is not the case however since under these conditions a large amount of heat is taken from the stack coolant flow, to provide the heat required for evaporation of the large amount of water for humidification. So, there is merely a shift of the heat removal problem.

4.4. Overall heat management

One can try to match heat sources and sinks, to avoid the need for external heating and to reduce the amount of heat that has to be discarded by a radiator. Even if such matching could be done perfectly, a net cooling duty is still necessary. This is shown in Fig. 12. At low temperatures, the water content of the cathode exit gas is low. Little heat is removed in the form of latent heat, but as sensible heat instead. As temperature increases, cathode exit gas humidity increases, and the net sensible heat removal from the system is reduced. At even higher temperatures, the water separator has to be cooled in order to become a condenser, e.g. at 60°C for the 1 bar situation. For high operating pressures, the heat removal by latent heat is relatively low, since most water is in the liquid form. In addition, a significant cooling duty is required for the inter cooler.

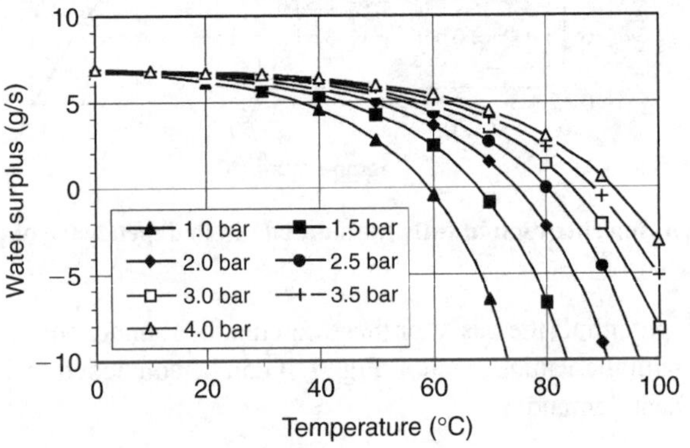

Fig. 11. Water surplus for the system shown in Fig. 1

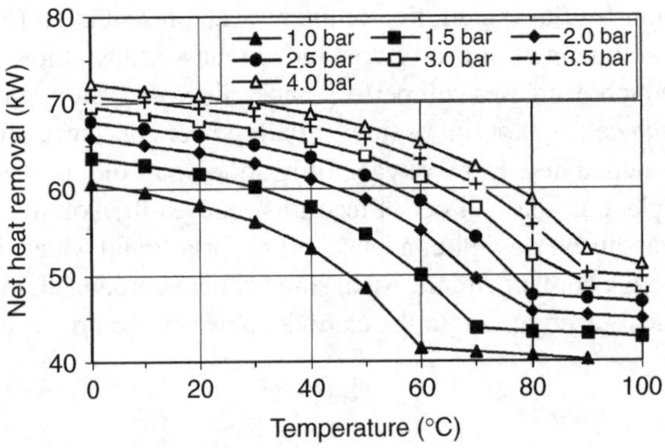

Fig. 12. Calculated amount of sensible heat that has to be discarded to the surroundings

5. DISCUSSION

From Fig. 12 it seems logical to conclude that working at low *P* and high *T* is most attractive. A compressor is not required, and heat rejection to the surroundings is minimised. However, to make such a system work, huge amounts of water have to be evaporated in the humidifier, requiring large amounts of heat. The water condenser has to make up for the water used in the humidifier and therefore has to withdraw large quantities of heat from the humid cathode exit flow. In the calculations used to produce Fig. 12, the heat sources and sinks are coupled. This is however, not realistic at high *T*, low *P* combinations, as can be deduced from Fig. 13. Here, the sum of the duties of the air cooler after the compressor, the humidifier, the condenser and the external heat exchanger (e.g. car radiator) is presented. From Fig. 13, it can be concluded that working at high *T* and low *P* will lead to huge heat duties (hence sizes) for the heat exchanging components. In combination with the large wet gas volumes (Fig. 7), there will be a large effect on system size.

To resume the essentials of all the above, an analysis is made of six system cases, each in the range of 40–50 kW (gross) power (Table 1). The systems are operating at 65 and 80°C and 1, 1.5 and 2.5 bar based on single-cell performances given in Table 1. It is assumed that all the systems operate at a current density of 0.6 A/cm^2. This assumption implies that the systems use the same amount of reactants (H_2: 30 N m^3/h, air: 145 N m^3/h). In the table, a new term is introduced, the "relative pressure drop factor" derived from Eq. (2): $\rho_{T,p}(V_{T,p})^2/\{\rho_{80°C,2.5bar}(V_{80°C,2.5bar})^2\}$. Here, $\rho_{x°C,ybar}$ is the humid gas density at x°C/y bar, whereas $V_{x°C,ybar}$ is the wet gas volume at x°C/y bar. The reference is the 80°C/2.5 bar case. Assuming equal stack design and system piping in all six sample cases, and neglecting the effect of the Reynolds number, the "relative pressure drop factor" is presented as an indicator for the effects of operating conditions on pressure drops in the system.

From Table 1, it can be concluded that the most pronounced differences in the sample systems are in the amounts of heat that have to be rearranged in the systems, and in the relative pressure drop factors. It is also concluded that working at atmospheric pressure and temperatures of 65°C and higher is rather complicated. The 80°C/1 bar case is the most extreme. The high wet gas flow has significant consequences for the pressure drop in the system. The heat required for humidification can be equal to the waste heat of the stack, making it difficult to humidify the gases with the stack coolant water. The oxygen concentration at the cell inlet drops to just 11%. Also, the sum of all heat exchanger duties is increasing significantly, leading to large system dimensions.

It seems that the use of a compressor is inevitable if heat rejection to the surroundings demands for operating temperatures of close to 80°C. For many applications, a different parameter setting would be much more attractive though. Preferably, the temperature should be even higher. That is, close to or above 100°C. The pressure should be close to atmospheric to avoid compressors. For cars, the higher operating

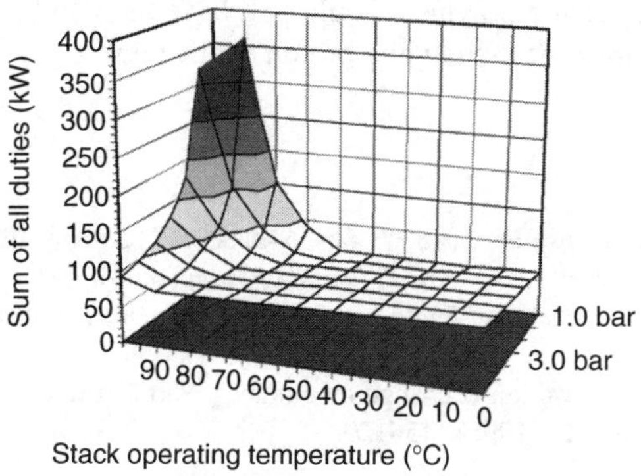

Fig. 13. Sum of duties of air cooler, humidifier, condenser and external heat exchanger as function of stack operating temperature and pressure

Table 1. Overview of essential numbers for six sample systems

	65°C			80°C		
	1 bar	**1.5 bar**	**2.5 bar**	**1 bar**	**1.5 bar**	**2.5 bar**
Power at the rate of 600 mA/cm^2 (kW)	39.6	43.5	46.3	37	43	47.4
Compression losses (kW)	0	2.8	6.7	0	2.8	6.7
Net power (kW)	39.6	40.7	39.6	37	40.2	40.7
Wet gas flow rate (m^3/h)	241	145	80	358	184	93
Oxygen concentration[a] (%)	15.7	17.5	18.9	11	14.3	17
Duty of air cooler (kW)	Not required	0.3	4.3	Not required	Not required	3.5
Heat requirement humidifier (kW)	25.1	15.1	8.4	66.7	34.2	17.3
Heat production stack[b] (kW)	69.8	64.9	61.4	76.8	67.4	61.2
Duty condenser (kW)	6.7	Not required	Not required	44.2	15.1	Not required
Net system cooling duty (kW)	51.4	50.1	57.3	54.3	48.3	47.4
Sum of all duties[c] (kW)	101.6	80.3	74.1	187.7	116.7	82
Relative pressure drop factor[d]	16.4	3.8	0.7	41.9	6.9	1

[a] At cathode inlet.
[b] Includes heat released due to condensation of water as a result of gas volume reflection.
[c] Sum of absolute value of duties, relevant for overall system dimensions.
[d] $\rho_{T,p}(V_{T,p})^2/\{\rho_{80°C,2.5bar}(V_{80°C,2.5bar})^2\}$, Relevant for the pressure drop in the system.

temperature helps to alleviate the problem of heat rejection to the surroundings[1]. For stationary applications, the benefit of a higher operating temperature would be in the more versatile use of the waste heat for space heating and hot tap water.

This paper demonstrated the effect of maintaining 100% relative humidity conditions on operating conditions. The need for operation at such high relative humidity is linked to the characteristics of Nafion and similar proton conductors. It will be clear that as long as such polymers are used in the PEMFC, the problems of pressurised operation and heat and water management will not be solved. This will only be the case if a new type of proton conductor is found, allowing for operation at higher temperatures and low relative humidity. Preferably, such a polymer should still have the good properties of Nafion-type polymers: cold start capabilities, toughness and good proton conductivity.

REFERENCES

1. R.C. Weast, *CRC Handbook of Chemistry and Physics*, 64th edn. CRC Press, Boca Raton, FL, 1984, pp. 83–84.
2. R.K.A.M. Mallant *et al. Electrolytic Membrane, Method of Manufacturing It and Use.* Patent WO98/20063.
3. J.M. Smith *et al. Introduction to Chemical Engineering Thermodynamics*, 4th edn. McGraw Hill, Singapore, 1987, ISBN 0-07-100303-7.
4. A.V. Anantaraman *et al.* Studies on ion-exchange membranes. Part 1. Effect of humidity on the conductivity of Nafion®. *J. Electroanal. Chem.* **414** (1996) 115–120.

[1] The problem of heat rejection is evident when examining current experimental fuel cell vehicles, such as the Daimler Chrysler Necar 4, which have radiators that are substantially larger than those found in the standard ICE powered cars.

Chapter 27
Portable and military fuel cells

K. Cowey, K.J. Green, G.O. Mepsted and R. Reeve

Abstract

Portable power for small electronic devices is the driving force for micro fuel cell development and many prototype systems are being reported with the majority based on direct methanol fuel cells. Several researchers claim to have solved fuel dilution problems and are close to commercialization. Research into micro fabrication techniques for fuel cells promises low-cost, micro-scale power sources for the future.

Keywords: Fuel Cells; Portable Power; Military Fuel Cells; Direct Methanol Fuel Cells; Micro fuel cells; Hydrogen storage; Hydrogen Generation; Hydrides

Article Outline

1. Introduction 555
2. Current developments in PEMFCs for portable applications 556
   2.1. Direct methanol fuel cells 557
   2.2. Hydrogen for portable systems 558
3. Conclusions 559
References 559

1. INTRODUCTION

There is a growing demand for power systems for portable electronic equipment both in the consumer market and for the military. The factors driving the research are fundamentally the same for both, that is power hungry electronics and batteries unable to provide sufficient energy. Fuel cells offer greater energy density coupled with simple and rapid recharging to give extended operation. Portable fuel cells, for this review, are defined as ranging from micro scale for applications such as mobile phones [1] through to small portable generators and generally are based on polymer electrolyte membrane fuel cells (PEMFCs). An additional factor for the military arises from the requirement to have fuel available in the battle area in any part of the world, however remote, hence their interest in reforming diesel or jet fuels as a source of hydrogen.

The portable market is often regarded as secondary after transport and stationary power, but is nevertheless significant in terms of volume of research and the potential size of the market [2–4]. Although much smaller, the military market may well be leading the race to portable fuel cell devices and is a significant source of funding in this area [5].

The main issue facing the portable fuel cell system developer is to produce a compact, lightweight system that is transparent to the user in operation [6]. This means that heat dissipation, water management and fuel supply needs to be developed to the point where the systems can perform reliably in ambient conditions

Fuel Cells Compendium

and be sufficiently rugged (especially for the military) for every day use. All of these must be achieved in a package of similar size and weight to the power sources they aim to replace. Cost is always an issue, but the lithium ion batteries currently in use are expensive and the portable market has potential users such as the military, professional camera operators [7] and even laptop PC users who may be willing to pay a premium for improved power sources.

2. CURRENT DEVELOPMENTS IN PEMFCs FOR PORTABLE APPLICATIONS

The PEMFC in either hydrogen gas or liquid fuelled variant, is by far the most extensively developed for portable applications [8]. Much of the development work for portable electronics is centred on the direct methanol fuel cell (DMFC), discussed separately below. Current PEMFCs research may be divided into the following areas:

- Plate design
- Membranes
- Electrodes
- Gas diffusion layer

The fuel cell plate fulfils a number of functions including gas distribution across, and conduction of electricity from, the membrane electrode assembly and must be corrosion resistant in the fuel cell environment. Development is now focussed mainly on low-cost production materials and techniques, for example, the use of carbon-loaded thermoplastics that can be injection-moulded [9].

Membrane developments are divided into those based on Nafion polymer and those using alternative polymers. Recent developments in Nafion have focussed on large-scale production of durable thin-film membranes [10], which offer superior performance and lifetime. Other membranes receiving attention include low-cost polymers (e.g. polyetheretherketone or PEEK) and high-temperature membranes that improve the fuel cell's tolerance to carbon monoxide.

Electrode fabrication has progressed significantly and currently micro fabrication methods [11,12] are used to provide high performance and good catalyst utilization [13]. These developments have seen catalyst loadings decrease from 4 to $\sim$0.3 mg/cm^2 while the performance has increased. New research is focussed on vacuum deposition (sputtering) techniques for more precisely controlling the catalyst layer, giving high-performance from extremely low catalyst loadings (of the order of 0.01 mg/cm^2) [14].

The gas diffusion layer is the electronic conductor between the fuel cell plate and the catalyst and it serves as a porous media for the passage of reactants and water. The need for high porosity is in opposition to the requirement for high electrical conductivity and research has focussed on optimizing the material and its structure [15,16].

Considerable efforts are also being focussed on the development of micro-fuel cells that could potentially replace battery coin cells for micro-applications [17–19]. These fuel cells are manufactured using adapted semiconductor fabrication techniques since conventional fuel cell components cannot be scaled to this size.

The military are leading many aspects of portable fuel cell development [20] and in the USA, the Department of Defense continues to fund development programs. Manhattan Scientifics has supplied a 40 cell, 700 W NovArs system to the Communications-Electronics Command (CECOM) and the US Fuel Cell Test, and Evaluation Centre has demonstrated an electric transport device (Segway), based on the NovArs stack that effectively doubled its range [21]. Ball Aerospace has developed 50 and 100 systems with the US Defense Advanced Research Projects Agency funding, which utilized an H-Power fuel cell. Researchers at the Case Western Reserve University are working on a prototype fuel cell, the size of a pencil eraser that is produced by screen printing, again funded by the US military [22].

In Europe, Intelligent Energy Ltd has supplied a portable 4 kW PEM generator for the US military and are active in the production of small, low-power stacks (1–100 W) for portable applications. Voller Energy Ltd, another UK-based system producer is bringing a number of fuel cell-based portable power units into the market [23].

QinetiQ has demonstrated a battery-sized 25 W fuel cell system designed specifically to power portable equipment worn by a soldier as part of a UK MOD research programme. Similar systems are under development for the US DoD [24] and the French Ministère de la Defénse.

The big market is in portable civilian electronics such as mobile phones, PDAs and laptop computers. Several companies, principally from the Far East, such as Samsung, Toshiba and Panasonic are developing portable fuel cells based on direct methanol fuel cells.

2.1. Direct methanol fuel cells

Many researchers see DMFCs as one of the best approaches to small fuel cell systems and a considerable amount of research is being conducted in this area [25–28]. DMFC produces power directly from methanol either as a vapor, a liquid or in solution. Liquid solutions are generally preferred due to more convenient thermal management and effective CO_2/methanol separation at the anode exhaust. At a typical working voltage of 0.5 V, the efficiency of the DMFC is up to 41%.

A challenge for the DMFC is to achieve an acceptable power density, currently much lower than the equivalent hydrogen fuelled PEM due to the slow reaction kinetics of the methanol oxidation. Temperatures in excess of 60°C are necessary to raise cell performance to acceptable levels, which can be a problem for miniature fuel cell systems.

A second factor affecting the performance of DMFC arises from the permeability of polymer membranes to methanol that results in a reduction in fuel utilization and a depolarizing effect. Although some work has focussed on methanol-tolerant cathode materials, their intrinsic activity is currently too low to make a significant difference to the overall cell performance [29]. The current approach to minimizing the methanol permeation rate is to limit the methanol concentration to approximately 5 wt.% despite the loss in performance.

One approach to increasing the power density is the application of polymers that allow for higher operating temperature. In particular, acid-doped (PBI) membranes have attracted significant interest due to their low methanol crossover characteristic [30]. Power density as high as 210 mW/cm^2 at 200°C has been demonstrated for PBI membranes operating on fairly concentrated methanol, but questions remain over cell performance at lower temperature and with membrane stability over time.

The primary advantage of the DMFC is the high-energy storage of methanol. Assuming that the oxidant is supplied from the atmosphere, the capacity of methanol is 5019 Ah/kg equating to an energy density of 2509 Wh/kg. In practice the use of dilute fuel solutions reduces this figure significantly, for example a 0.5 M solution equates to only 41 Wh/kg. To improve this energy density figure, in situ methanol dilution systems, using water produced at the cathode, are being deployed.

System control associated with water and methanol management add considerably to the complexity of DMFC systems, particularly those being developed for portable applications [31], since the balance of plant must include fuel/CO_2 separation at the exhaust, in addition to fuel dilution [32].

There is little reported information regarding the projected life of DMFC systems but one example is from Motorola, where a portable air-breathing stack was operated for 3000 h [33].

Development of the DMFC for the portable electronics market represents a major R&D effort with several companies working in this area. To date, there are few commercial products and almost certainly no large-scale manufacturing, although press releases show prototypes at an advanced state of development. Methodologies to eliminate methanol crossover and dilution requirements are claimed but the details remain proprietary [34].

A small number of DMFC systems have been demonstrated publicly from portable electronics manufacturers, including Toshiba [35,36], Hitachi [37], Motorola and Fujitsu. Toshiba has recently announced a range of DMFC-based power sources for handheld electronics, mobile phone recharging and laptop computers. The smallest unit (8.5 g) produces 100 mW and incorporates a concentration gradient to enable it to use neat methanol in the fuel tank. The computer fuel cell (900 g) uses a similar approach to provide an average 12 W (max 20 W) and 11 V and will operate the computer for up to 5 h on a 50 ml cartridge.

An alternative liquid fuel cell based on borohydride solutions is receiving increased interest, although it is still at a very early stage of development [38]. Nevertheless, alkali fuel cells of this type have favorable performance characteristics, but progress is dependent upon the outcome of research into the elimination of the hydrolysis side reaction at the anode.

Medis Energy has developed a proprietary direct liquid fuel cell pack called Power Pack, which uses a sodium borohydride solution, and provides power to portable electronic equipment such as mobile phones [39,40]. They have recently joined with General Dynamics to develop PDAs for the US military.

2.2. *Hydrogen for portable systems*

In very small hydrogen fuel cell systems, the storage and supply of fuel is at least as big an issue as the fuel cell performance. Hydrogen from high-pressure cylinders or hydrides are feasible for portable generators, but are impractical for small electronic devices where the development of liquid fuelled systems using methanol, borohydrides or even primary chemical hydrides are seen as the preferred solution [41].

Micro scale fuel processing is largely the domain of the military who want a single fuel (jet fuel) that is readily available in war zones. Despite the difficulties of processing such fuels, a number of researchers have reported working systems [42–46].

Advanced composite cylinders have been developed using lightweight aluminium [47] or thermoplastic liners [48] that are fiber-wrapped for strength and can store up to 7.5 wt.% hydrogen albeit at pressures of up to 700 bar. Unfortunately, compressed gas storage does not scale-down efficiently, which limits its use in small-scale systems.

Reversible metal hydrides can absorb hydrogen into the structure and desorb it in a controlled way giving high volumetric hydrogen density at ambient temperature and modest pressure (5–20 bar, typically). They are often used for the demonstration of portable fuel cell systems, although their weight probably precludes their use in many applications.

Primary chemical generation offers the highest energy density fuel system with hydrogen released from chemical hydrides by hydrolysis or by thermolysis reactions. Much of the development is being supported by the military who wish to exceed the energy storage capability of batteries.

Hydrolysis of hydrides, such as lithium hydride or sodium borohydride with water, can produce over 8 wt.% hydrogen and has been investigated by a number of workers [49–51]. The US Army funded development of a reactor based upon calcium hydride hydrolysis, but the prototype was unable to control the highly exothermic reaction and tended to overheat. More recently, Millennium Cell has developed a liquid-phase system called "Hydrogen on Demand" [52], where sodium borohydride solution is passed over a metal catalyst to produce hydrogen. Although initially developed for fuel cell vehicles, the approach is now being applied to portable applications by Ballard and Samsung. Another fuel system funded by the US military (CECOM) is based upon the reaction of lithium aluminium hydride with ammonia, rather than water. This reaction theoretically yields 13.2 wt.% hydrogen and workers claim to have produced over 400 Wh of hydrogen, equivalent to 6% (by weight) [53].

Thermolysis involves the thermal decomposition of solid hydride chemicals, but the reactions are extremely exothermic and difficult to control. QinetiQ has developed a portable system for the UK Ministry of Defence using ammonia borane mixed with an exothermic material to give a thermally neutral reaction

[54,55]. Discrete pellets are set off as required and has potential for over 10 wt.% of hydrogen for a complete system.

3. CONCLUSIONS

Fuel cells have considerable potential to power portable and small electronic devices. Academic research into them continues at a high level with many advances, especially in micro fuel cells. However, the commercialization that was widely predicted has not yet occurred. This situation is beginning to change and it is estimated that there are over 1700 prototype and demonstrator programs worldwide in the less than 1 kW power range, many of which have been funded by the military. Portable fuel cell systems now show signs that they are about to go beyond the development stage with a significant number of companies in the US, Japan and Europe claiming to have portable fuel cells systems that will be on the market within a few years.

REFERENCES

1. M. Stefener. Micro fuel cells – an early opportunity. *Proceedings of the 8th Grove Fuel Cell Symposium*, London, 2003.
2. C.K. Dyer. *Sci. Am.* **281** (July) (1999) 88–93.
3. Frost and Sullivan Report D226. *Stationary and Portable Fuel Cells, Developments, Markets and Opportunities*, 2001.
4. Frost and Sullivan Report A162–27. *US Micro Fuel Cell Markets for Mobile Devices*, 2002.
5. Global Information Inc. Report on Evolution of Fuel Cells and Batteries for the Military: Trends and Markets.
6. J. Larminie and A. Dicks. *Fuel Cell Systems Explained*, Wiley, New York, 2000.
7. Website: http://www.jadoopower.com
8. Fuel Cells Review. *J. Electrochem. Soc.* **149** (2002) S59–S67.
9. A. Hienzel, F. Mahlendorf, O. Niemzig and C. Kreuz. *J. Power Sources* **131** (2004) 35.
10. D.E. Curtin, R.D. Lousenberg, T.J. Henry, P.J. Tangeman and M.E. Tisack. *J. Power Sources* **131** (2004) 41.
11. R. Hahn, S. Wagner, A. Schmitz and H. Reichl. Development of a planar micro fuel cell with thin film and micro patterning technologies. *J. Power Sources* **131** (2004) 73.
12. S. Motokawa, M. Mohamedi, T. Momma, S. Shoji and T. Osaka. MEMS-based design and fabrication of a new concept micro direct methanol fuel cell. *Electrochem. Commun.* **6** (2004) 562.
13. Z. Qi and A. Kaufman. *J. Power Sources* **113** (2003) 37.
14. R. O'Hayre, S.J. Lee, S.W. Cha and F.B. Prinz. *J. Power Sources* **109** (2002) 483.
15. E. Antolini, R.R. Passos and E.A. Ticianelli. *J. Power Sources* **109** (2002) 477.
16. M. Glora, M. Miener, R. Petricevic and H. Probstle. *J. Non-Cryst. Solids* **285** (2001) 283.
17. J.P. Meyers and H.L. Maynard. *J. Power Sources* **109** (2002) 76.
18. R. Hahn, S. Wagner, A. Schmitz and H. Reichl. *J. Power Sources* **131** (2004) 73.
19. STMicroelectronics, micro fuel cells for cell phones. *Fuel Cells Bull.* **10** (2003) 10.
20. T.B. Atwater, P.J. Cygan and F.C. Leung. Man portable power needs of the 21st century. I. Applications for the dismounted soldier. II. Enhanced capabilities through the use of hybrid power sources. *J. Power Sources* **91** (2000) 27–36.
21. US Department of Defense Website, Fuel Cell Powered Segway: http://www.dodfuelcell.com/article_010604.html
22. Case Western Reserve University website: http://www.cwru.edu/pubaff/univcomm/fuelcell.htm
23. Website: http://www.voller-energy.com
24. USAF contract for protonex portable soldier power. *Fuel Cells Bull.* **6** (2004) 5.
25. C. Lamy, S. Rouseau, E.M. Belgsir, C. Cooutanceau and J.M. Leger. Recent progress in the direct methanol fuel cell: development of new platinum-tin electrocatalysts. *Electrochem. Acta* **49(22–23)** (2004) 3901.
26. D. Cao and S.H. Bergens. Pt-Pu adatom nanoparticles as anode catalysts for direct methanol fuel cells. *J. Power Sources* **134** (2004) 170–180.

27. E.H. Yu and K. Scott. Direct methanol alkaline fuel cells. Poster presented at *Electrochemical Engineering for the 3rd Millennium, Networking Meeting*, UMIST, November, 2001.
28. UK Department of Trade and Industry publication, DTI/Pub URN 02/592: ETSU F/03/00232/REP.
29. R.W. Reeve, P.A. Christensen, A.J. Dickinson, A. Hamnett and K. Scott. Update on status of direct methanol fuel cells. *Electrochim. Acta* **45** (2000) 4237.
30. J.S. Wainright, R.F. Savinell and M.H. Litt. In: O. Savadogo and P.R. Roberge (Eds.), *Proceedings of the 2nd International Symposium of New Materials for Fuel cell and Modern Battery Systems II*, Montreal, July 6–10, 1997.
31. A. Blum, T. Duvdevani, M. Philosoph, N. Rudoy and E. Peled. Water-neutral micro direct-methanol fuel cell (DMFC) for portable applications. *J. Power Sources* **117(1–2)** (2003) 22–25.
32. R. Luharuka, C. Wu and P.J. Hesketh. Design, fabrication, and testing of a near constant pressure fuel delivery system for miniature fuel cells. *Sensor. Actuator. A Phys.* **112(2–3)** (2004) 187–195.
33. S. Gottesfeld, X. Ren, P. Zelenay, H. Dinh, F. Guyon and J. Davey. Fuel Cell (Seminar Abstracts) (2000) 799.
34. S. Gottefeld. Direct methanol fuel cells for portable power applications. In: *Proceedings of the 8th Grove Fuel Cell Symposium*, London, 2003.
35. Toshiba Corporation website, Press Release March, 2003: http://www.toshiba.co.jp/about/press/2003_03/pr0501.htm
36. Toshiba America Electronic Components, Inc (TAEC) Website, June 24 Press Release, 2004: http://www.toshiba.com/taec/press/dmfc_04_222.shtml
37. Website: http://www.theregister.co.uk/2003/03/07/nec_hitachi_prep_notebook_fuel/
38. R.W. Reeve, I.E. Eweka and G.O. Mepsted. Liquid fuel developments for micro fuel cells. *Proceedings of the 8th Grove Fuel Cell Symposium*, London, 2003.
39. Liquid Fuel Composition for Electrochemical Fuel Cells. US Patent 6 554 877 (2003).
40. Liquid Fuel Composition for Electrochemical Fuel Cells. US Patent 6 562 497 (2001).
41. D. Browning, P. Jones and K. Packer. An investigation of hydrogen storage methods for fuel cell operation with manportable equipment. *J. Power Sources* **65** (1997) 187.
42. J.D. Holladay, D.R. Palo, R.A. Dagle, M.R. Phelps, Y.H. Chin, Y. Wang, E.G. Baker and E.O. Jones. Fuel cell integration and operation with micro-technology based fuel processor systems. *Proceedings of the 8th Grove Fuel Cell Symposium*, London, 2003.
43. D.R. Palo, J.D. Holladay, R.T. Rozmiarek, C.E. Guzman-Leong, Y. Wang, J. Hu, Y. Chin, R.A. Dagle and E.G. Baker. Development of a soldier-portable fuel cell power system. Part I: A bread-board methanol fuel processor. *J. Power Sources* **108(1–2)** (2002) 28–34.
44. InnovaTek wins US Army fuel processor contract. *Fuel Cells Bull.* **2** (2004) 7.
45. Pacific Northwest National Laboratory: http://www.pnl.gov/microcats/fullmenu/minfuelcells.html
46. D.J. Seo, J. Woon, Y. Yoon, S. Park, G. Park and C. Kim. Development of a micro fuelprocessor for PEMFCs. *Electrochim. Acta* (in press).
47 H. Presting, J. Konle, V. Starkov, A. Vyatkin and U. König. Porous silicon for micro-sized fuel cell reformer units. *Mater. Sci. Eng. B* **108(1–2, 25)** (2004) 162–165.
48. Dynetek Industries Ltd, Website: http://www.dynetek.com/frameset.html
49. A.R. Abele and A.P. Niedwiecki. *Proceedings of 2002 Fuel Cell Seminar*, Palm Springs, CA, November 18–21, 2002.
50. Y. Kojima, Y. Kawai, H. Nakanishi and S. Matsumoto. Compressd hydrogen generation using chemical hydride. *J. Power Sources* **135(1–2)** (2004) 36–41.
51. Y. Kojima, K. Suzuki, K. Fukumoto, M. Sasaki, T. Yamamoto, Y. Kawai and H. Hayashi. Hydrogen generation using sodium borohydride solution and metal catalyst coated on metal oxide. *Int. J. Hydrogen Energ.* **27** (2002) 1029.
52. Y. Kojima and T. Haga. Recycling process of sodium metaborate to sodium borohydride. *Int. J. Hydrogen Energ.* **28** (2003) 989.
53. S.C. Amendola, M.T. Kelly, S.L. Sharp-Goldman, M.S. Janjua, N.C. Spencer, P.J. Petillo and M. Binder. *Proceedings of the 39th Power Sources Conference*, Cherry Hill, NJ, June 12–15, 2000, p. 176.
54. Ball Aerospace, Website: http://www.ball.com/aerospace/media/nr09_04_01.html
55. P.B. Jones, D.J. Browning, G.O. Mepsted and D.P. Scattergood. US Patent 2003 0 180 587.
56. D.J. Browning, D.P. Scattergood, G.O. Mepsted, J.B. Lakeman and M.K. Gardner. *Proceedings of the 8th Grove Fuel Cell Symposium*, London, 2003.

Chapter 28

Microfabricated fuel cells

J.S. Wainright, R.F. Savinell, C.C. Liu and M. Litt

Abstract

This paper reports design, materials, and testing of microfabricated fuel cells that can be applied for small power generation. Multiple co-planar fuel cells in series have been built onto ceramic substrates (as well as silicon and polymer, but not reported here) using thick-film fabrication techniques. Testing of these fuel cells gave power densities in the milliwatt (per square centimeter) range for continuous power and tens of miliwatts under pulsing conditions, and have indicated performance limiters and opportunities for higher power densities. These fuel cells are intended to operate on ambient-supplied air, and thus the polymer electrolytes must be capable of operating under a range of humidity conditions. The target conductivity for these low power devices is $>1\,\text{mS cm}^{-1}$ at 15% RH. Although novel polymer electrolytes are being pursued at CWRU for this application, we report here Nafion® films cast from alternate solvents that have adequate conductivity for this application. Several on-board fuel storage approaches are being pursued. We have demonstrated high fuel utilization by a fuel cell operating with on-board source of hydrogen from aqueous sodium borohydride, as well from solid metal hydride systems.

Keywords: PEM; Nafion®; Micropower; Hydrogen storage; Hydride

Article Outline

1. Introduction 562
2. Design principles and fabrication issues 562
   2.1. Current collectors/gas diffusion layers 563
   2.2. Electrolyte issues 564
3. Fuel cell performance 566
4. Hydrogen storage and generation 568
   4.1. Hydrogen generation via decomposition of sodium borohydride solutions 568
   4.2. Hydrogen storage in metal hydrides 569
5. Summary 571
Acknowledgments 572
References 572

Fuel Cells Compendium

1. INTRODUCTION

Efficient small-scale power generation is critical to realizing the full benefit of many portable electronic devices. Wireless electronic devices, e.g. micro sensors and micro electromechanical or microfluidic systems, require electrical power for operation and data transmission. The availability of on-board power with appropriate sized power and energy capabilities, coupled with wireless data transmission, will open numerous possibilities for self-sustaining devices for remote or difficult to access locations. The power and energy needs for small portable devices may vary considerably depending on their design and function. Small microfabricated fuel cells open the door for energy delivery devices with independent sizing of power and energy capacity, while having the capability to supply short pulses of peak power without significant adverse effect on energy utilization. Finally, such fuel cells have the potential for energy densities significantly greater than present and predicted battery technology.

Small fuel cells with various degrees of microfabrication have been reported in the literature. Thin film techniques were used to create flowfields [1–3], some with current collections and catalyst layers [4–8] on a silicon substrate that were attached to a free-standing Nafion® film. Morse et al. [9] did not use a free-standing film, but instead spin-cast a membrane on the microfabricated support components. Individual fuel cell sizes were from 2 to 10 mm on a side, and down to 15 μm on a side. Temperatures above room temperature during testing were maintained by an oven and fuel (hydrogen or methanol) was supplied from an external source.

This paper presents concepts and results arising from a program to develop a small and completely microfabricated polymer-based fuel cell with on-board hydrogen storage. The approach is to co-fabricate a sensor or sensor suite with sufficient analog/digital circuitry for signal conditioning/processing, wireless communications capability (low-power RF) and a power source on a single wafer. The anticipated power requirements are for extended periods of very low (sub-milliwatt) power draws to power the sensor, with random or periodic pulses of 10–100 of milliwatts required for signal processing or data transmission. Potential operational lifetimes are from hours to days (surveillance at a particular location) to years (machinery monitoring). Development of a microfabricated power source is an enabling technology for the autonomous sensor concept. It is envisioned that this device would operate in a passive mode, without active control of temperature, humidity, reactant pressure or flow rate. This greatly simplifies the construction and operation of the fuel cell. The power density achievable in passive operation will naturally be less than that obtained in conventional fuel cells. However, the applications envisioned do not require high power densities.

2. DESIGN PRINCIPLES AND FABRICATION ISSUES

The CWRU microfabricated fuel cell is being developed by using the following guiding principles:

- completely passive device-no fans, pumps, compressors to move reactants,
- ambient temperature operation-no cooling or heating,
- hydrogen fueled at near atmospheric pressure,
- no external humidification-operation at ambient relative humidity,
- fabrication process compatible with silicon-based microelectronics.

The basic design of the fuel cell consists of a planar array of edge collected cells, such that all of the cathodes are exposed to the ambient air, while beneath all of the anodes is a common fuel manifold. Each of the components of the cells; current collectors, gas diffusion layers, catalyst layers, and electrolyte, is deposited in turn using thick film printing techniques. The hydrogen storage/generation capability is similarly fabricated in a second substrate, and bonding the two substrates together creates the final device. A schematic of the final device is shown in Fig. 1 and a photograph is shown in Fig. 2. This fabrication

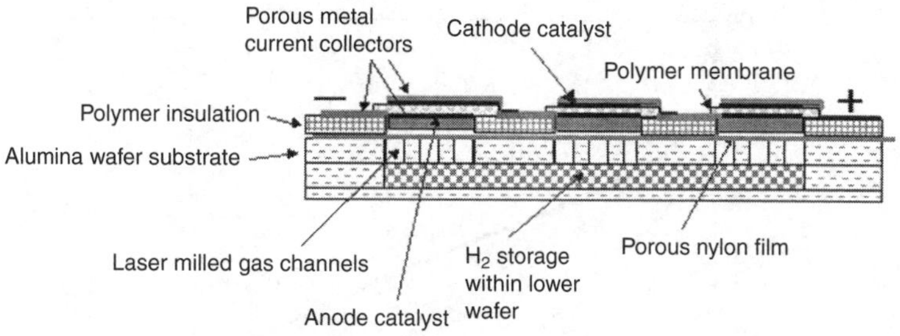

Fig. 1. Schematic cross-sectional view of microfabricated fuel cell with fuel storage. Three series connected cells are shown. Drawing not to scale

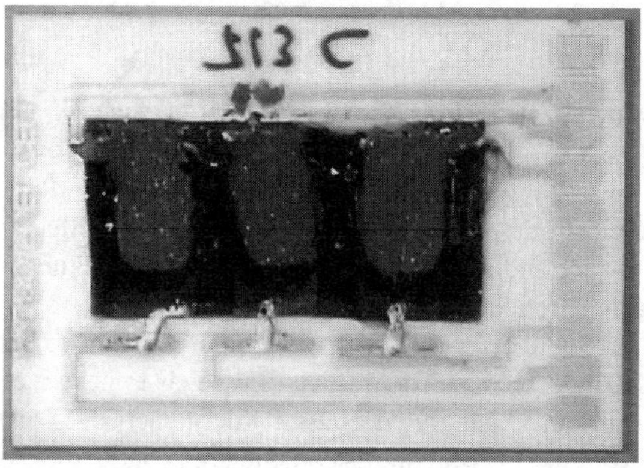

Fig. 2. A photograph of a three cells fabricated onto an alumina wafer with printed current leads that allow cells to be connected in parallel or in series

process is considerably different from the conventional Proton Exchange Membrane (PEM) fuel cell stack construction in which individual components (electrolyte membranes, gas diffusion layers, bipolar plates, gaskets, endplates, etc.) are separately manufactured, and then assembled into the stack. The following two sections will focus on the particular fabrication issues that have been addressed for the current collectors and the polymer electrolyte.

2.1. Current collectors/gas diffusion layers

In order to minimize the number of fabrication steps, the gas diffusion layer (GDL) in the microfabricated fuel cell is also used as the current collector. Typical gas diffusion layers for conventional PEM fuel cells such as carbon cloth or carbon paper have a thickness on the order of 300 μm, porosities on the order of 80%, and conductivities on the order of 50–100 S cm^{-1}. The relatively low conductivity is sufficient given that they are used in conjunction with ribbed flow fields with feature sizes on the order of 1 mm. However, for an edge-collected device, the conductivity required is considerably greater. In addition, with the thick film process, a thickness of 10–50 μm is more realistic, while 300 μm is very difficult to achieve.

The voltage drop along the current collector is given by

$$\Delta\varphi_{ir} = (L^2/2)(i_{avg}/t\sigma)$$

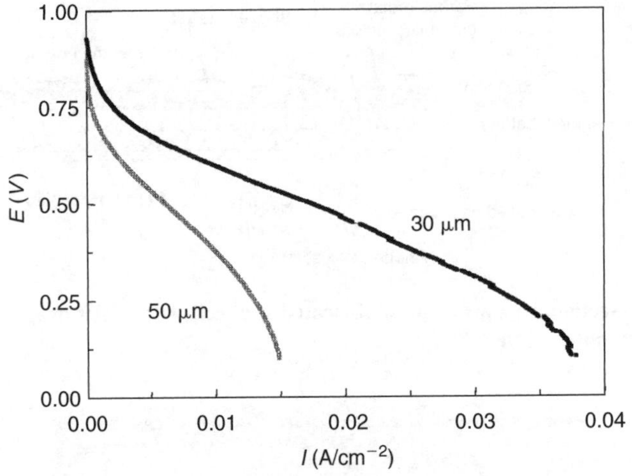

Fig. 3. Air breathing fuel cell polarization curves for two different thickness of a printed gold current collector/gas diffusion layer

where L is the length of the electrode perpendicular from the current-collector edge, i_{avg} the average current density normal to the surface and assumed to be uniform, t the thickness of the current collector and σ the conductivity of the current collector.

Assuming that a maximum voltage drop of 10 mV over a 1-cm-long current collector is desired for a fuel cell current density of 50 mA cm^{-2}, a 25 μm-thick film needs to have a conductivity of 1000 S cm^{-1}. Since the current collector must also be highly porous in order to act as it is clear that carbon-based materials are not ideally suitable for this component. Instead, metallic conductors are preferred. Our experience with attempting to thick film print current collector/GDLs based on metallic conductors has been that the difficulty lies in achieving sufficient porosity. In Fig. 3, fuel cell polarization curves are shown for two cells, one having a 30 μm-thick printed current collector/GDL, the other with a 50 μm-thick film of the same porous material. In both cases there is an appreciable loss due to diffusion of the reactants through the current collector/GDLr, and the loss is considerably greater with the thicker film.

Additional gas permeation resistance can be expected through the substrate that supports the current collector. In practice, the inks are not printed directly onto the substrate because the pores are too large, and the ink from printing the current collector will fill the pores, thus increasing the gas permeation resistance substantially. For the fuel cells tested here, a thin porous nylon film (MAGNA nylon, pore diameter of 0.45 μm, Osmonics Inc.) was secured onto a substrate and then non-porous insulating boundaries and current collectors were printed. The other fuel cell layers were then sequentially printed. The completed fuel cell with nylon backing was then attached to a ceramic wafer with slots for gas access to the anode. The hydrogen gas is then passed through the slots and the porous nylon film to reach the anode surface for oxidation. The observed limiting currents with this cell design were 10–30 mA cm^{-2} with 1 atmosphere hydrogen gas feed.

2.2. Electrolyte issues

There are a number of electrolyte issues that must be considered. Like other macro PEM fuel cells, the electrolyte must be proton conducting, have a favorable environment for electrode reaction kinetics, and must permit gas access to the polymer–catalyst-interface so reaction can occur. In a small microfabricated fuel cell, since the current density will be low under ambient air operation, the conductivity only needs to exceed approximately 1 mS cm^{-1} (125 mV *IR* loss at 50 mA cm^{-2} with a 25 μm film). However, this conductivity

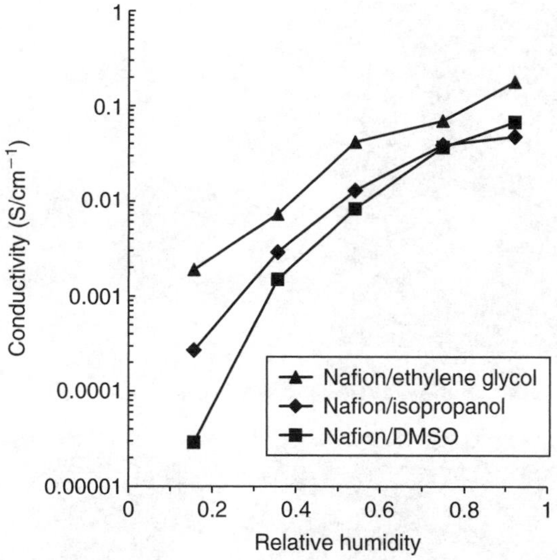

Fig. 4. Conductivity of Nafion® films cast from three different solvents: ethylene glycol, isopropanol, and DMSO. The films were cast, then the solvent subsequently removed in vacuum at 80°C

level must be achievable even at low relative humidities since gas hydration is not feasible in a small passive device. We have established as a working target a conductivity greater than 1 mS cm^{-1} over the relative humidity range of $15 < RH < 100\%$.

A typical 5% Nafion® – alcohol solution has too low of a viscosity to print a well-defined film. In addition it has low proton conductivity at low relative humidities as shown in Fig. 4. We modify the commercially available Nafion® – alcohol solution by adding higher boiling solvents, then concentrating the Nafion® solute. Films were cast and the solvent was evaporated under vacuum at about 80°C. The conductivity of films cast from DMSO and ethylene glycol (EG) are shown in Fig. 4, and the Nafion®/EG film meets our conductivity goals. These films were ca. 25 μm thick and were cast from 10 to 15 wt.% solutions, with the majority of the solvent being the higher boiling component.

Another issue is to deposit electrolyte precisely where it is needed. This is necessary to achieve densely packed cells that make the most use of the available area on the substrate and minimize the length of the current collectors. In addition to minimizing current collector lengths, printing patterned electrolyte layers for each cell, as opposed to having several cells sharing a continuous film of electrolyte, eliminates any shunt currents that may otherwise occur in series connected cells.

In Figs 5a and b are two examples of thick film printing of the modified Nafion® solution. The solution used to produce Fig. 5 ais clearly superior, the electrolyte films are well-defined, and can be closely spaced. The result shown in Fig. 5b is not acceptable, the electrolyte film spreads beyond the desired area when printed, and adjacent cells must be further apart. In addition, the film thickness in Fig. 5a is ≈90 μm and uniform across the area of the cell, while in Fig. 5b, the film thickness varies between 5 and 15 μm. While thinner electrolyte films may be preferred to minimize resistive losses, films less than 25 μm-thick can lead to excessive hydrogen crossover, particularly for airbreathing devices such as these where operating current densities are low as compared to pressurized macro-size fuel cells. The solution used to cast the film in Fig. 5a consisted of about 40 mg of Nafion® and 60 mg of EG with all the alcohol and water of the original 5% Nafion®/alcohol solution removed. The solution used to cast Fig. 5b was similar, but without all of the alcohol and water removed. It was of lower viscosity and Nafion® density. The actual properties (gel content, viscosity, etc.) of the casting solution varied somewhat depending on the processing procedure.

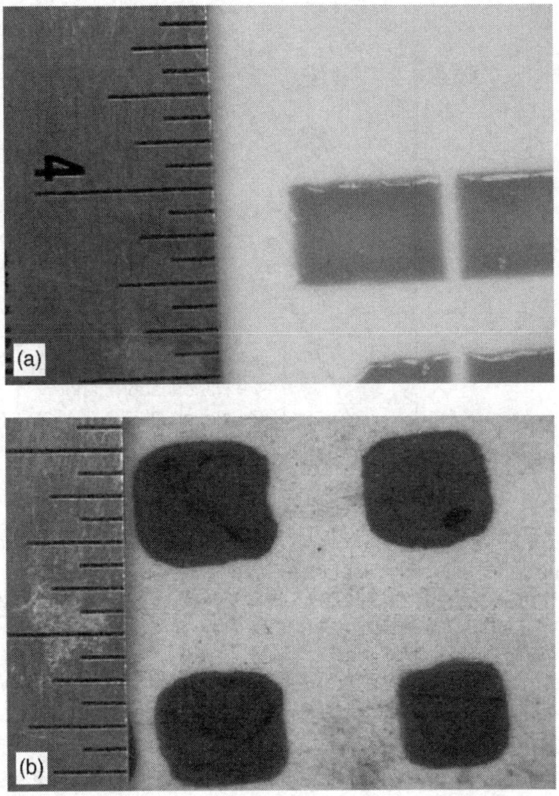

Fig. 5. Two examples of thick film printing of Nafion® electrolyte films on alumina substrates. (b) The black line squares represents the desired film area as defined by the mask used for printing and the overprinting of the film deposit is clearly present. (a) The deposit is the same area as the mask. The smallest division on the ruler shown in the figures is 1/16″ (0.16 cm)

A calculation that estimates the effect of a Nafion® 1100 membrane thickness on cross-over is given in Fig. 6. The cross-over rate is also reflected in a lowering of the open circuit potential because of the mixed electrode reaction (hydrogen oxidation coupled with oxygen reduction) at the cathode. These calculations are based on kinetic parameters taken from experimental data of typical conventional PEM fuel cells operated in our laboratories. The drop in OCV as the membrane becomes thinner will be a strong function of the actual catalyst loading. However, as shown, for a watersaturated Nafion® 1100 film, the cross-over rate increases appreciably for films thinner than 25 μm.

While the Nafion®/EG system appears adequate for microfuel cell applications, it still has its limitations. For example, the conductivity at low RH, although acceptable for low current density operation, will limit pulse power capability. Also, these films will swell excessively at 100% RH and the film will delaminate from the substrate thus resulting in failure. Other electrolyte systems are under development in our laboratories, and progress on these approaches are reported elsewhere [10 and 11].

3. FUEL CELL PERFORMANCE

The performance of representative microfabricated fuel cells are shown in Fig. 7. These fuel cells are all microfabricated by thick film deposition. The current collection/flowfields are 15 μm-thick gold printed

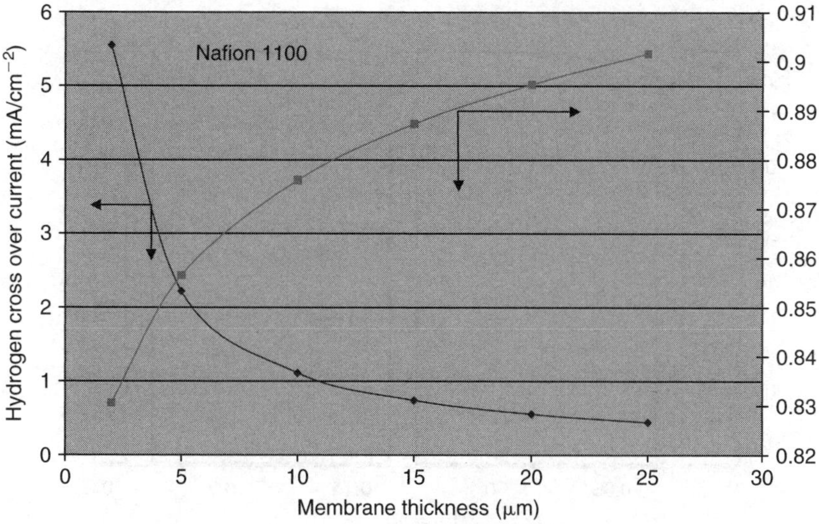

Fig. 6. Calculations that estimate the rate of hydrogen cross-over for a Nafion® 1100 film and that estimate its effect on the open circuit potential based on experimental polarization curves from macro-sized PEM fuel cells. The calculation were based on first calculating the hydrogen cross-over current. A hydrogen permeability of 5.76×10^{-12} was used (e.g. a 10 μm-thick film with one atmosphere hydrogen pressure differential yielded a 1.1 mA cm^{-2} cross-over current). This cross-over current was then used in the cathode potential expression (based on experimental data) to arrive at the cathode potential under the condition of open circuit with only hydrogen cross-over consuming the oxygen. Of course, the cathode potential expression will be strongly related to the platinum catalyst loading and utilization

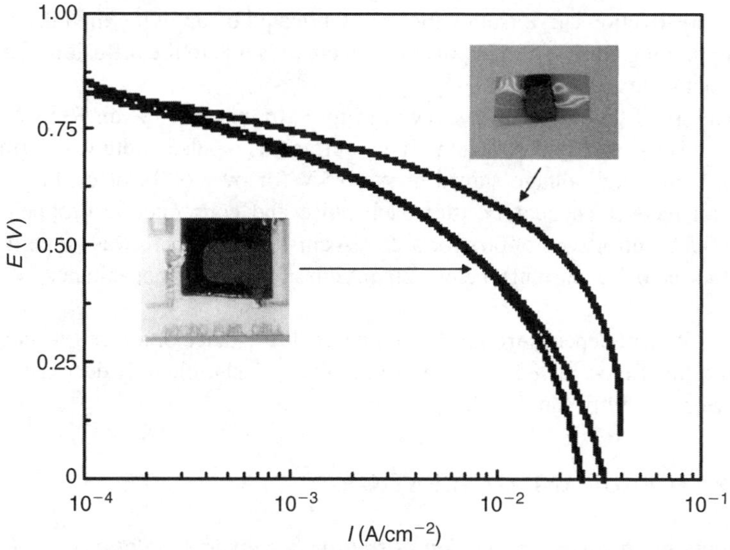

Fig. 7. Polarization curves for two independent microfabricated fuel cells. Also shown is a polarization curve of a fuel cell fabricated by applying the inks used for the microfabricated fuel cells onto a free-standing Nafion® film

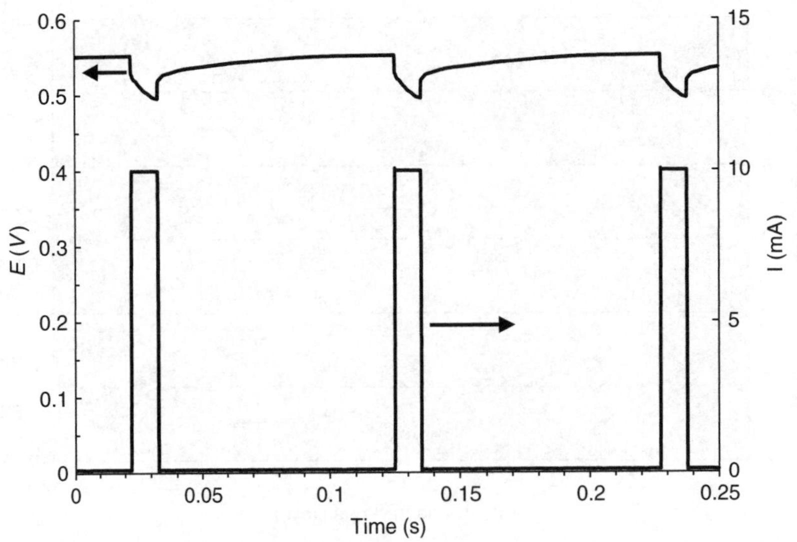

Fig. 8. Example of continuous pulse response of 0.25 cm^2 micro fuel cell. Duty cycle: 10% at 25 mA cm^{-2}

using an Ercon ink. The electrodes were printed from a slurry of particles of C/Pt (40 wt.% Pt on XC-72 from Etek Inc.). It was necessary to have thin electrodes (2–5 μm) to avoid mud-cracking upon drying. The catalyst loading therefore was only 0.2 mg cm^{-2} for both electrodes. The Nafion® was cast from a Nafion®/EG solution. The porous substrate was a nylon porous film. The fuel cell was operated in air at 42% RH at 24°C with humidified hydrogen. The power output at 0.5 V cell was 2 mW cm^{-2} continuous, 5 mW cm^{-2} for 180–300 s durations, and 50 mW cm^{-2} under a 10 ms pulse. Although long term testing was not carried out, there was no noticeable decay in performance over a 20-h period. Also shown for comparison is the polarization curve from a fuel cell fabricated by applying the inks onto a free-standing Nafion® film. The limiting current of this polarization curve is due to the diffusion of oxygen through the porous metal current collector.

The pulse capability of this fuel cell is shown in Fig. 8. In this case the air RH was at 48%. The duty cycle was 90% at 0.5 mA cm^{-2} and 10% at a higher current. The pulse width was 10 ms. When the pulse current was 50 mA cm^{-2} the voltage stayed above 0.5 V for over 600 s of cycling. After about 600 s, the voltage did not recover adequately after each pulse and consequently dropped below 0.5 V. The time to reach the 0.5 V cut-off was 5400 s for a 25 mA cm^{-2} pulse. There was no permanent voltage loss after 50,000 pulses at pulse currents below 25 mA cm^{-2}. The cell capacitance was estimated to be 13.5 mF cm^{-2}.

The pulse capability will depend strongly on the internal resistance of the cell, which is directly related to the relative humidity. Below 40% RH, the pulse capability is significantly decreased while at over 80% RH, the pulse power capability can double.

4. HYDROGEN STORAGE AND GENERATION

4.1. Hydrogen generation via decomposition of sodium borohydride solutions

As described, the goal for these microfabricated fuel cells is passive, ambient temperature operation. Hydrogen storage in the form of stabilized aqueous solutions of sodium borohydride ($NaBH_4$) is one option that meets these criteria. When sufficient base is added to raise the solution pH to ≈14, these solutions are stable at room temperature for several months. However, when brought in contact with a suitable catalyst, they readily decompose according to the reaction.

$$NaBH_4 + 2H_2O \Rightarrow 4H_2 + NaBO_2$$

Sodium borohydride is readily soluble in water, and solutions of over 30 wt.% $NaBH_4$ can be made at room temperature. For a 30 wt.% solution, stabilized with 10 wt.% NaOH, the following theoretical energy densities can be calculated: 2337 mW h cm^{-3} (8.41×10^9 J M^{-3}) and 2091 mW h g^{-1} (7528 J g^{-1}). This is equivalent to ≈7 wt.% H_2. The borohydride system also offers distinct advantages compared to the metal hydride systems discussed below. The primary advantage is that the concentrated solution maintains roughly 35% RH at the fuel cell anode, thus providing a continual source of water vapor to the fuel cell. This has a very distinct benefit in maintaining higher electrolyte conductivity, which is especially important for pulse power delivery.

The simplest design for a hydrogen generator for the microfuel cell consists of a small well (i.e. an opened chamber) etched into a wafer to contain the borohydride solution, which is injected into the well with a syringe when the fuel cell is activated. The decomposition catalyst is printed onto the bottom of the well. The well is capped with the microfuel cell, with the underside of the microfuel cell substrate exposed to the hydrogen that bubbles up from the catalyst at the bottom of the well. A relatively long and small diameter vent through side of the well, above the solution level, allows air to be exhausted when the borohydride solution is injected, and maintains the hydrogen pressure at slightly above ambient, regardless of the relative rates of hydrogen generation by the decomposition reaction and hydrogen consumption by the fuel cell.

Varying the depth of the well for the borohydride solution allows the amount of energy stored to be varied. The obvious drawback of this system is that it is orientation dependent, the fuel cell must be held in an orientation that permits the borohydride solution to contact the decomposition catalyst and that only allows hydrogen and water vapor to be vented to the anode. We are currently working on more advanced designs that would be orientation independent and would incorporate a check valve on the hydrogen vent while remaining completely passive and still be readily manufactured.

A borohydride generator of the simple type was fabricated with a solution volume of 0.6 cm^3. A microfabricated fuel cell bonded to the generator was continually operated at 0.5 V for over 90 h on one charge of borohydride solution. (Polarization curves with the attached hydrogen generator were essentially the same as those generated from hydrogen supplied by a tank.) Over this period, the current produced was monitored and integrated. The integrated current produced of 1822 coulombs was over 67% of the theoretical value for complete decomposition of 0.6 cm^3 of the 20 wt% $NaBH_4$ solution used. The fuel cell was still operating at the end of the 90-h period; however, the power output had dropped to about 25% of the initial value. Considering that the generator is not sealed, but vents excess hydrogen through a small opening, the fuel utilization achieved in this experiment is very promising. Higher utilizations can be achieved by operating the fuel cell at constant current.

4.2. Hydrogen storage in metal hydrides

The advantages of storing hydrogen in metal hydrides include a high volumetric energy density with no operational dependency on orientation. A major goal of this effort was to produce a metal hydride material that is tailored to this micro fuel cell application, with an equilibrium pressure near atmospheric pressure at room temperature, and that can be applied in a thick film fabrication process. Integrated devices with energy densities on the order of 500 mW h cm^{-3} (1.8×10^9 J M^{-3}) for a 50 mW h (180 J) storage capacity should be feasible using this approach.

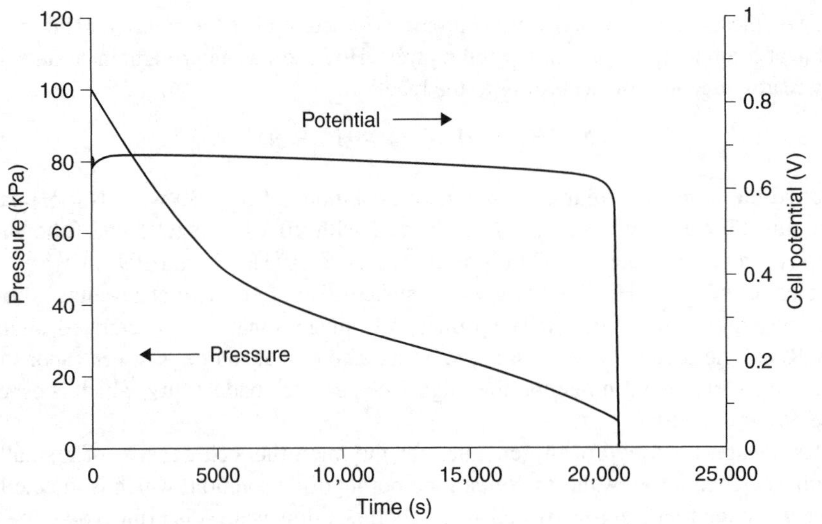

Fig. 9. Small conventional fuel cell potential and pressure above hydride ($LaAl_{0.3}Ni_{4.7}$) storage block while maintaining 20 mA current output. The fuel cell is based on a Nafion® 117 film, with Pt/C printed electrodes of 1 cm^2

$LaAl_{0.3}Ni_{4.7}$ hydride material with an additional surface modification has been formulated into a thick film printable ink with very low binder content. A second member of the AB5 family of hydrides, $CaNi_5$, has also been evaluated. For both hydrides, the binder content was on the order of 1 wt.%, allowing for high volumetric energy density. The surface modification prevents deactivation of the hydride material. Printed inks remained active after over 5 weeks exposure to room air, at which time the tests were stopped. This is an important result because often these materials passivate on exposure to air or water, and can only be re-activated for hydrogen storage by exposure to high hydrogen pressures. As it is impossible to expose material contained within the fuel cell structure to high hydrogen pressures, a non-passivating hydride material is required.

It is also important to show that the hydride material can be repeatedly charged with hydrogen and discharged without loss of capacity. An automated Sievert's apparatus was used for hydrogen sorption/desorption testing of up to six samples simultaneously. Various hydride powders and inks have been subjected to repetitive sorption/desorption testing consisting of 10 minute exposure to ≈133 kPa (1000 torr) H_2 (sorption step) followed by 600 s exposure to vacuum (desorption, PRESSURE = 1.3 Pa (10 millitorr)). SEM micrographs after 1000 cycles showed no change in particle size, and no change in the adhesion of the ink to the substrate. These tests have been extended to over 4000 cycles without any significant change in the total amount of hydrogen stored, or in the rate at which hydrogen is adsorbed or desorbed. Over the same period, the ink based on the $CaNi_5$ hydride has lost ≈15% of its original hydrogen storage capacity.

Printed films of the $LaAl_{0.3}Ni_{4.7}$ ink have been used to provide hydrogen to small, conventionally fabricated fuel cells. In the first experiment, a constant 20 mA current was drawn from the fuel cell, while the cell potential and the hydrogen pressure above the hydride block were monitored, as shown in Fig. 9. Despite the fact that in this experiment, the fuel cell and the hydride block were connected by ≈5 cm of 0.32 cm OD tubing, the fuel cell maintained a constant potential until the pressure dropped to ≈8 kPa (60 torr). The hydrogen consumed during this test represented over 90% of the hydrogen initially stored in the hydride material. With the extremely close coupling of fuel cell and hydride material that will occur in the microfabricated devices, even higher utilizations should be possible. It also is important to note that the potential was essentially constant until the very end of the test; the sloping equilibrium pressure of this

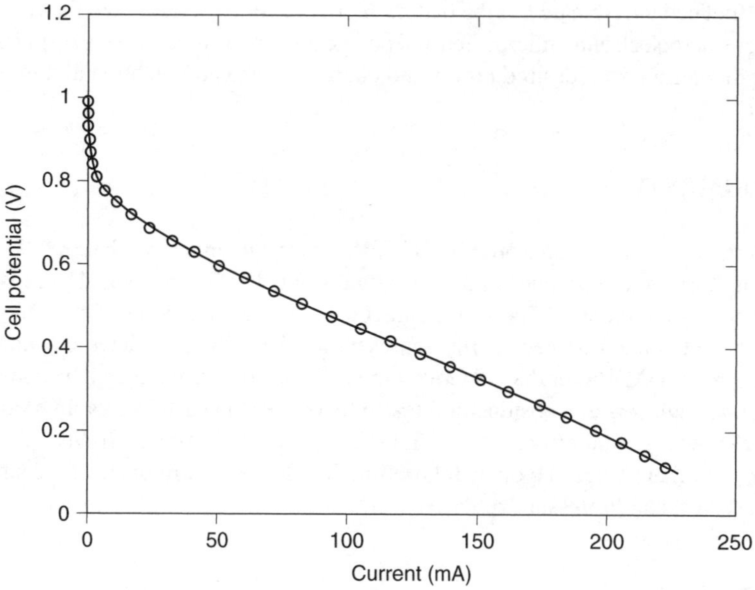

Fig. 10. Polarization curve of the fuel cell described in Fig. 8 coupled to hydride ($LaAl_{0.3}Ni_{4.7}$) storage block

Table 1. Comparison of microfabricated fuel cells and a Li−MnO_2 coin cell

	Micro fuel cell[a]	**Li−MnO_2 coin cell Duracell 2016**
Nominal voltage	3 V	3 V
No. of cells	6	1
Weight	1.7 g	1.8 g
Dimensions	2.7 × 1.8 × 0.2 cm	2 cm dia., 0.2 cm thick
Capacity	60 mA h/cell (216 J/cell)	75 mA h (0.2 mA rate) (216 J)
Energy	200–300 mW h (720–1080 J)	200 mW h (720 J)
Maximum continuous power	3–15 mW (depends on %RH)	6 mW
Peak power (10 ms pulse)	50–100 mW (depends on %RH)	30 mW

[a]Values for micro fuel cell based on single cell and three cell structures scaled to yield a 3 V device.

hydride material was not reflected in a varying fuel cell performance under these conditions. In Fig. 10, a complete polarization curve for the fuel cell is shown. A current output of over 200 mA was achieved without reaching a limiting current, indicating that the hydrogen desorption rates from this hydride ink are more than sufficient for the microfabricated fuel cell, which is expected to operate at currents on the order of 10 mA.

5. SUMMARY

Aspects of the design, materials and fabrication of a microfabricated fuel cell have been presented. An estimate of the performance of these devices is summarized in Table 1, with a comparison to a Li—MnO_2 primary battery of similar voltage, size and energy storage. The interesting advantages of the microfabricated fuel cell as compared to the battery lie in the ability to fabricate the fuel cell on the same substrate as other microelectronic, microelectromechanical or microfluidic devices, in the ability to independently tailor the design for a desired power and energy storage combination and in the ability to provide peak power.

ACKNOWLEDGMENTS

This report is based on research sponsored by DARPA under agreement no. F30602-97-2-0311. The US Government is authorized to reproduce and distribute reprints for Governmental purposes notwithstanding any copyright notation thereon. The views and conclusions contained herein are those of the authors and should not be interpreted as necessarily representing the official policies or endorsements, either expressed or implied of DARPA, or the US Government. This work summarizes the combined efforts of a number of students, engineers and postdoctoral researchers over the past 5 years. In particular the authors would like to acknowledge the efforts of Ms. Laurie Dudik, Dr. Arunkumar Venkatesan, John Staser, Andrew Swann, Ms. Lynn Chen, Hyoung-Juhn Kim, Dr. Snezana Gojkovic, Ms. Liangyuan Chen, Xi Shan and Kenneth Yee, and Professor Joe Payes.

REFERENCES

1. (a) S.J. Lee, S.W. Cha, Y.C. Liu, R. O'Hayre and F.B. Prinz. In: H.Z. Massoud, I. Baumvol, M. Hirose and E.H. Poindexter (Eds.), *Micro-Power Sources*, PV200-3. The Electrochemical Society Proceeding Series, Pennington, NJ, 2000. (b) S.J. Lee, S.W. Cha, A. Chang-Chien, R. O'Hayre and F.B. Prinz. Factorial Design Study of Miniature Fuel Cells with Micromachined Silicon Flow Structures. ECS Abstract No. 452, San Francisco Meeting, September 2001.
2. A. Chang-Chien, S.W. Cha, S.J. Lee, R. O'Hayre and F.B. Prinz. Planar Interconnections of Multiple Polymer Electrolyte Membrane Fuel Cells by Microfabrication. ECS Abstract No. 453, San Francisco Meeting, September 2001.
3. S.J. Lee, S.W. Cha, Y. Liu, R. O'Hayre, A. Chang-Chien and F.B. Prinz. Minature Fuel Cells with Non-Planar Interface by Microfabrication. ECS Abstract No. 241, Phoenix Meeting, October 2000.
4. S.C. Kelley, G.A. Deluga and W.H. Smyrl. *AICHE J.* **48** (2002) 1071.
5. S.C. Kelley, G.A. Deluga and W.H. Smyrl. *Electrochem. Solid State Lett.* **3** (2000) 407.
6. J.P. Meyers and H.L. Maynard. *J. Power Sources* **109** (2002) 76.
7. H. Maynard, J. Meyers and A. Glebov. Silicon Tunnels for Reactant Distribution in Miniaturized Fuel Cells. ECS Abstract No. 60, Toronto Meeting, May 2001.
8. J. Meyers and H. Maynard. Design of Miniaturized Fuel Cells for Portable Power. ECS Abstract No. 64, Toronto Meeting, May 2001.
9. J.D. Morse, A.F. Jankowski, R.T. Graff and J.P. Hayes. *J. Vac. Sci. Technol. A* **18** (2000).
10. H.-J. Kim and M.H. Litt. *ACS Polym. Preprints* **42** (2001) 486.
11. C. Genies, R. Mercier, B. Sillion, R. Petiaud, N. Cornet, G. Gebel and M. Pineri. *Polymer* **42** (2001) 5097.

Chapter 29

Electro-catalytic membrane reactors and the development of bipolar membrane technology

J. Balster, D.F. Stamatialis and M. Wessling

Abstract

Membrane reactors are currently under extensive research and development. Hardly any concept, however, is realized yet in practice. Frequently, forgotten as membrane reactors are electro-catalytic membrane reactors where electrodes perform chemical conversations and membranes separate the locations of conversion. We review first a mature electromembrane technology: the chlor-alkali electrolysis. A second example is the polymer-electrolyte fuel cells (FCs) as an emerging technology, which carries the potential to dramatically change society's energy infrastructure. Finally, we extensively describe a technology where the membrane itself is catalytically effective, splitting water into protons and hydroxyl ions: bipolar ion-exchange membrane technology.

Keywords: Electro-catalytic membrane reactors; Chlor-alkali electrolysis; Polymer-electrolyte fuel cell; Bipolar membrane

Article Outline

1. General introduction . . . 574
2. The chlor-alkali electrolysis . . . 576
   2.1. Introduction – chlor-alkali industry . . . 576
   2.2. Principle – membrane electrolysis cell . . . 576
3. The fuel cell . . . 579
   3.1. Introduction . . . 579
   3.2. Polymer-electrolyte fuel cell . . . 580
      3.2.1. Catalyst in PEFC . . . 580
      3.2.2. Polymeric membrane in PEFC . . . 581
4. Bipolar membranes . . . 582
   4.1. Introduction . . . 582
   4.2. Principle of a BPM . . . 582
   4.3. Principle of BPM electrodialysis . . . 583
   4.4. Preparation of BPMs . . . 584
   4.5. Characterization of BPMs . . . 585
   4.6. Limitations of EDBPM . . . 586
   4.7. EDBPM processes . . . 587

Fuel Cells Compendium

References . . . 588
Nomenclature . . . 591

1. GENERAL INTRODUCTION

The integration of chemical conversion and separation into a single processing step offers opportunities towards process intensification. The later is one of the major forms of sustainable technology development. Membrane reactors are a particular form of the integration of conversion and reaction. A large variety of combinations can be realized in membrane reactors as one easily conceives from the large variety of different reactions as well as membrane separation processes.

Membrane reactors are heavily investigated in the academic and industrial domain. However, only few industrially relevant and realized examples are known in the literature. On large scale, only electro-catalytic membrane reactors succeeded to gain industrial relevance today. In this paper, we will review three different electro-catalytic membrane reactor processes:

- The chlor-alkali process for the production of concentrated caustic and chlorine from concentrated sodium chloride brine.
- The polymer-electrolyte membrane fuel cell (FC) for the production of electricity from hydrogen or methanol.
- The bipolar membrane (BPM) electrodialysis for the simultaneous production of acid and base.

The majority of our review paper focuses on the BMP technology; however, we incorporate the first two as well, since the comparison of the three allows extracting common features of successful membrane reactor development.

We think that there are six essential competencies necessary to guarantee the development of industrially relevant membrane reactor concepts. The six disciplines interact closely with each other and all of them influence each other making membrane reactors development an extremely challenging, but intricate R&D activity. Figure 1 shows the six competencies as they influence each other and hence the ultimate membrane

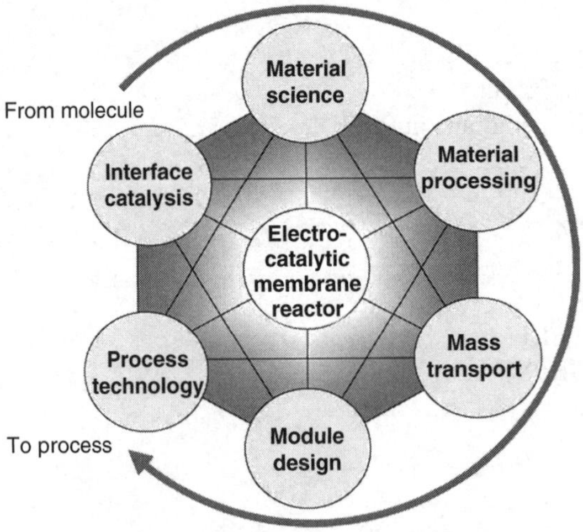

Fig. 1. Visualization of competencies necessary to successfully develop electro-catalytic membrane reactors

reactor concept. In fact, the schematic visualizes the integration of conversion and separation "from molecule to process." *Interface catalysis* addresses all electro-chemical issues occurring at the conversion interface. Most frequently these issues are long-term stability, poisoning of the catalyst and maximizing efficiency. *Material science* focuses on the influence of material properties on conversion and separation. In electro-catalytic membrane reactors, this means in particular the ion transport efficiency. *Material processing* deals with the integration of different materials and layers of such materials into cheap, robust laminates. May be less scientifically challenging, material processing is essential in scaling-up for industrial scale. The competence *Mass transport* is required to understand and describe the mass transport phenomena, microscopically and macroscopically. Continuous models are required to guide the prototyping of membrane reactor module. The interplay of mass transport description and the actual *Module design* leads to viable technology. The integration of the developed membrane reactor modules into existing processes or the design of new processes inspired by the new membrane reactors is the competence called *Process technology*.

The first two examples of electro-catalytic membrane reactors only briefly touch upon each of the disciplines. The example of BPMs, the area that we explored extensively, will be analyzed in more detail with respect to the above-mentioned competencies. Figure 2 shows the historical development of BPMs with respect to the development of regular electrodialysis and chlor-alkali electrolysis. It is surprising to actually realize that the discovery of the chlor-alkali process and the BPM occurred almost at the same time. However, technical realization developed with very different speed. Figure 3 aims to visualize this. It plots the level of maturity as a function of time for the two technologies. As time proceeds and the level of maturity develops, one can distinguish different consecutive phases of technology development. The seed of technology development, R&D, is followed by first successful, frequently even unsuccessful applications. Once first demonstration results convince potential customers, the newly developed technology enters the growth phase and it spreads into the market. At some point in time, the technology must be considered mature and further growth is limited. Figure 3 shows two trajectories of electro-catalytic membrane reactor technologies: the rapidly developed chlor-alkali electrolysis being mature today and the slower BPM process entering the growth phase today.

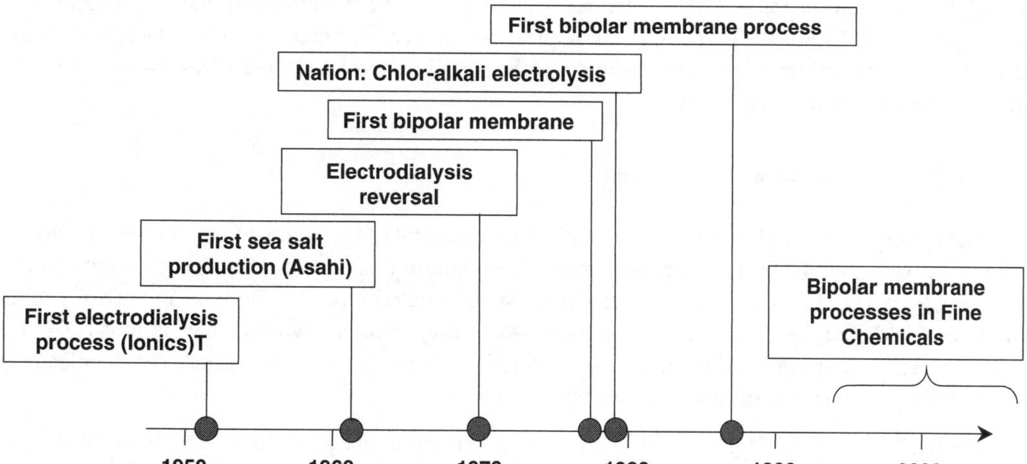

Fig. 2. Time line visualization of ion-exchange membrane-based processes

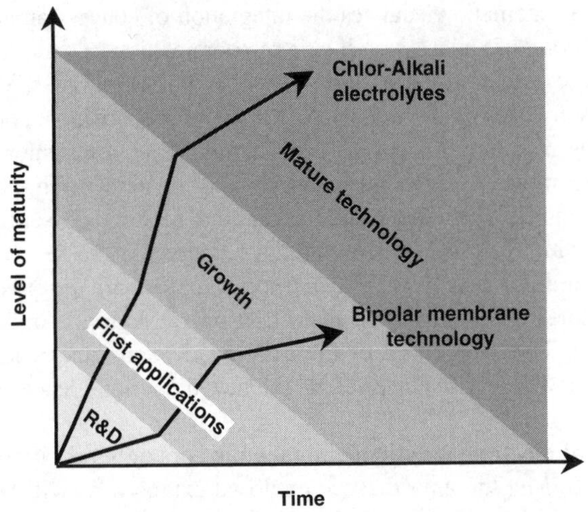

Fig. 3. The level of maturity as a function of time for the chlor-alkali and BPM technology

2. THE CHLOR-ALKALI ELECTROLYSIS

2.1. Introduction – chlor-alkali industry

The chlor-alkali industry is now-a-days one of the largest industries producing annually more than 48 million tons of sodium hydroxide (NaOH) and more than 42 million tons of chlorine (Cl_2). The raw material is sodium chloride (NaCl) and the reaction proceeds according to

$$2NaCl + 2H_2O \rightarrow 2NaOH + Cl_2 + H_2 \quad (1)$$

In this process, two major challenges exist: (i) chlorine reacts explosively with hydrogen and (ii) chlorine dissolves in contact with NaOH to form a hypochlorite solution. It is therefore necessary to separate the products of the reaction. Currently, three different electrochemical cell technologies exist and they are characterized by their means to separate the reaction products: the mercury cell, the diaphragm cell and the membrane cell [1]. For the first two technologies, which were invented in the last decade of the nineteenth century, more information can be found in other publications [1,2]. In this review, our attention is focused on the membrane electrolysis.

2.2. Principle – membrane electrolysis cell

In the membrane reactor cell, the anode and cathode are separated by means of a cation-exchange membrane (CEM) (see Fig. 4). In this case, a saturated NaCl solution is passed through the anolyte compartment and Cl_2 gas is produced at the anode (IrO_2 usually coated with Ti). Sodium ions (Na^+) migrate through the CEM into the catholyte compartment where they combine with hydroxide ions (OH^-) produced in the cathode (usually Ni coated with a catalytic coating) to form hydroxide. All the electrolysis plants and designs share the following characteristics [3]:

- Saturated brine (300–305 g/l) is fed to the anolyte compartment and depleted brine (180–200 g/l) is discharged. The pH of the anolyte is generally kept in the range of 2–4 by addition of HCl. At higher pH, the Cl_2 can react to form hypochloric acid or eventually chlorate following the reactions [4]:

$$Cl_2 + OH^- \rightarrow HClO + Cl^- \quad (2)$$

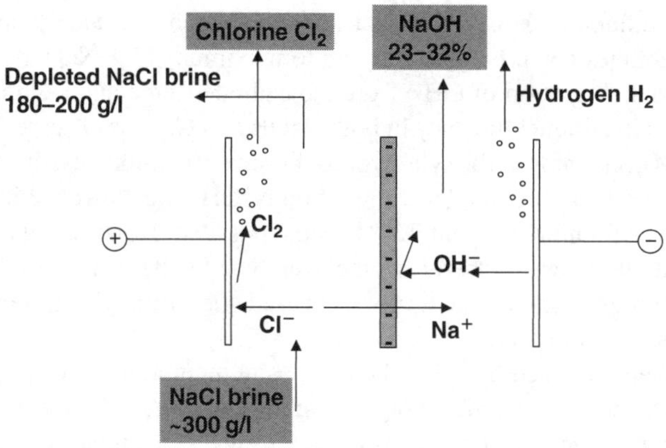

Fig. 4. Membrane electrolysis cell, (adapted from [3])

Table 1. Membrane requirements for the membrane electrolysis process

Physical and chemical stability to withstand NaOH and Cl_2
Low electrical resistance
Transport of only Na^+ from the anyolyte to the catholyte
Operate at high current densities
Immune to interference from impurities (heavy metals) in the solution

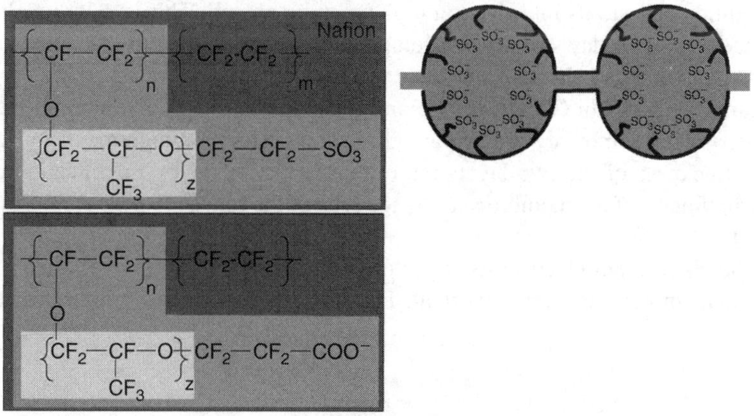

Fig. 5. Perfluorinated polymers containing sulfonic fixed ionic groups (Nafion®)

$$HCl + H_2O - 2e^- \rightarrow 3H^+ + Cl^- + O_2 \tag{3}$$

$$HClO + ClO^- \rightarrow ClO_3^- + 2H^+ + Cl^- \tag{4}$$

However if the pH is kept too low, the competition of H^+ and Na^+ concerning carrying the charge will be high.

- Water is added to the catholyte compartment and concentrated hydroxide overflows from it (Fig. 4).

The basic requirements of the CEM are briefly presented in Table 1 [4]. The first CEM for this application was prepared by perfluorinated polymers containing sulfonate fixed ionic groups (Nafion® polymers, Fig. 5).

The Nafion® consists of a fluorocarbon polymer backbone to which sulfonic groups have been chemically introduced. These membranes could be used only up to maximum 15% NaOH concentration. At higher NaOH concentration, back migration of OH^- from the cathode to the anode compartment occurred and therefore, significant loss in current efficiency in both NaOH and Cl_2 was observed. The situation has been improved by using CEM containing carboxylic groups. For these membranes, the water content was lower than the sulfonic ones, and therefore, the back migration of OH^- was lower. With these membranes, the production of NaOH concentration of about 32% became possible. However, due to the weak carboxylic groups, their conductivity was low and therefore, the overall cell voltage increased in comparison to CEM with sulfonic groups. Moreover when an acid anolyte is used, the carboxylic groups are protonated and the membrane resistance increases further.

Much better results were obtained by using the composite membrane concept (Fig. 6). The composite membranes have a thin layer of carboxylate polymer on the catholyte side of a basic membrane, which is made of sulfonate polymer. The composite construction uses the advantages of both types of materials. Due to the lower water content of the carboxylate polymer, the back migration of the OH^- decrease but because the carboxylate layer is very thin, there is not significant increase of the membrane resistance. In addition, the carboxylate polymer is in contact with the NaOH solution and its protonation can only occur from the high flux of H^+ through the sulfonated polymer [4]. The modern composite membranes can typically work for more than 5 years in a chlor-alkali plant.

The quality of the NaCl feed solution has a significant role on the membrane lifetime. The presence of metal ions such as Mn-$^{(2+)}$, Ca-$^{(2+)}$ and Fe-$^{(2+)}$, $Fe^{(3+)}$ in the NaCl solution can cause severe problems or even failure of the membrane. These metal ions permeate through the membrane and due to the pH gradient in the membrane, precipitate inside it. As a result, the permeation of Na^+ is blocked and the electrical resistance of the membrane increases dramatically. It is therefore necessary to remove the metal ions from the feed NaCl solution prior to its introduction to the membrane cell. This usually involves large volume precipitation process [3,4]. Today's process development, in particular, addresses the optimization of the brine production rather than the electro-membrane reactor.

Several companies (e.g. Asahi Chemical, DuPont, Asahi Glass) are manufacturing membranes for the chlor-alkali industry. The membranes are prepared based on customer requirements by varying the equivalent weight and thickness of the two layers (sulfonic-carboxylic) or by modification of the surface to improve the hydrophilicity of the membrane or by the type of the reinforcement to improve the stability of the membrane [3].

Now-a-days, the chlor-alkali electrolysis is performed on a two-dimensional scale. The surface of the anode, cathode and membrane are very important. The electrodes in the cells can be utilized in two modes,

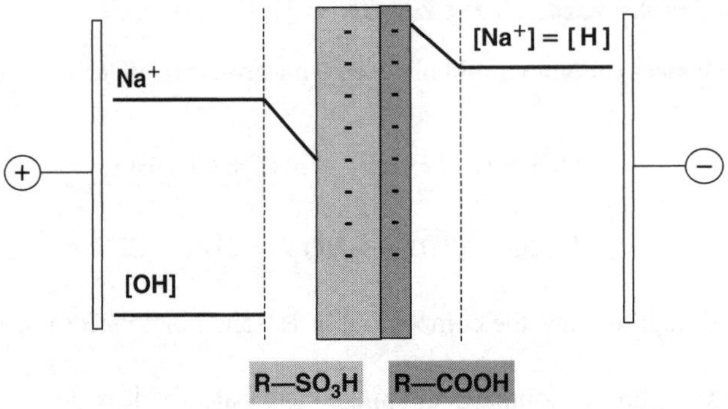

Fig. 6. Composite cation-exchange membrane for membrane electrolysis, adapted from [3]

when both sides serve as either anode or cathode (monopolar configuration) or one side serves as anode and the other side as cathode (bipolar configuration) [3].

3. THE FUEL CELL

3.1. Introduction

The FC is an electro-catalytic membrane reactor where the chemical energy of a fuel and an oxidant is continuously converted into electrical energy [5]. The basic principle of a FC is not different of that of electrochemical batteries. However, in batteries the chemical energy is stored inside and conversion into electrical energy requires either disposal or recharge. In the FC, however, the chemical energy is provided by the fuel and an oxidant, stored outside the cell. Electrical energy can then be continuously produced as long as the fuel and the oxidant are provided. In the FC, the chemical energy is directly converted into electrical energy with efficiencies much higher than the conventional thermo-mechanical systems. In addition, the FC operates without combustion and it is pollution free [6].

The FC is basically composed of two electrodes, the anode and the cathode and an electrolyte. The energy conversion reaction occurs at the electrodes. The fuel is oxidized at the anode; the oxidant moves through the electrolyte and is reduced at the cathode. A typical example of hydrogen–oxygen (H_2—O_2) FC, one of the most important FC types, is presented in Fig. 7. On the fuel-anode-electrode, the hydrogen releases electrons to become hydrogen ions (H^+). The hydrogen ions permeate through the electrolyte and the electrons flow through an external load circuit to reach the air-cathode-electrode. On the cathode, hydrogen ions, electrons and oxygen react to form water (Fig. 7):

$$\text{Anode: } H_2 \rightarrow 2H^+ + 2e^- \tag{5}$$

$$\text{Cathode: } 2H^+ + \tfrac{1}{2}O_2 + 2e^- \rightarrow H_2O \tag{6}$$

$$\text{Overall reaction: } H_2 + \tfrac{1}{2}O_2 \rightarrow H_2O \tag{7}$$

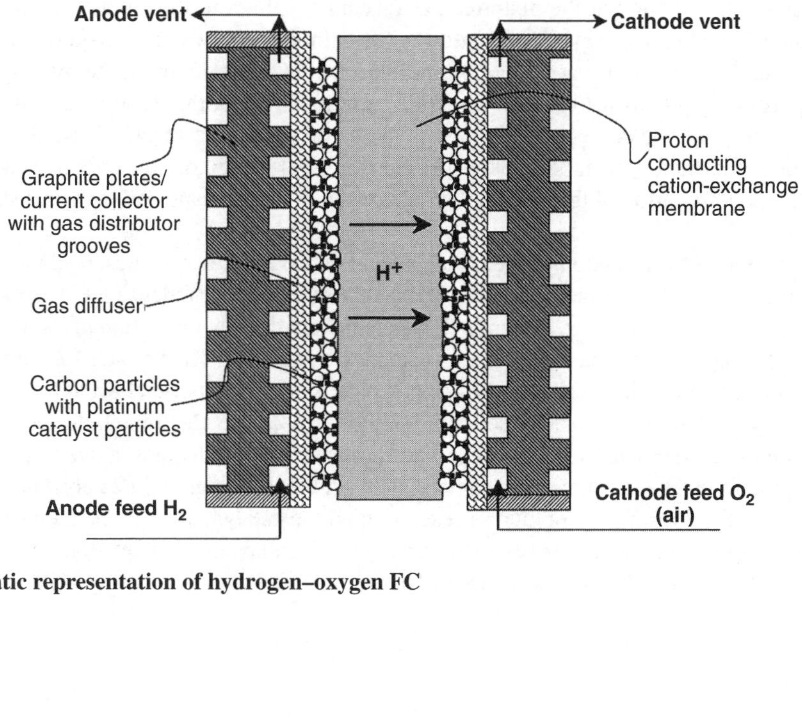

Fig. 7. Schematic representation of hydrogen–oxygen FC

The nature of the electrolyte, liquid or solid, determines the operating temperature of the FC. The electrolyte blocks the electrons and prevents the electrical contact between the electrodes. It can either be a proton contactor or an oxygen ion contactor. The major difference between the two cases is the side of the FC where the water is produced. When a proton (oxygen) conductor is used, the water is produced in the cathode (anode) side of the FC.

The FC systems are commonly distinguished by the type of electrolyte applied, in the following categories: alkaline fuel cells (AFC), phosphoric acid fuel cells (PAFC), molten carbonate fuel cells (MCFC), solid oxide fuel cells (SOFC) and polymer–electrolyte fuel cells (PEFC). In this review, we focus our attention to the PEFC where a polymeric membrane is used as its electrolyte. The reader can find more information about other FC types in other publications [5,7–10].

3.2. Polymer-electrolyte fuel cell

In recent years, great attention has been given to the PEFC because they are considered as the primary power source for vehicles in urban environment. Such vehicles could carry methanol fuel (to be reformed on site) or hydrogen fuel. The cathode could operate in air and the electrolyte could be an ion-exchange polymeric membrane [11].

The PEFC has the following basic structure (from the side to the center): porous backing – gas distributor/catalyst layer/polymeric membrane. The single cell has current collector plates, which usually contain machined flow fields for the effective distribution of the reactant gases along the surfaces of the electrodes (Fig. 7). In a FC stack, the plates are bipolar having flow fields on both sides. The basic part of the FC is the membrane/electrode part. It basically consists of a thin catalyst layer (usually dispersed platinum, Pt) in good contact with the membrane. The backing layers of the catalysts are made of hydrophobic porous material, usually carbon paper or carbon cloth hydrophobized by polytetrafluoethylene (PTFE) of thickness in the range 100–300 μm.

3.2.1. Catalyst in PEFC

In the (H_2—O_2) PEFC, both electrode processes are electro-catalytic in nature. They both require active catalyst sites to break the bond of the diatomic gas reactants. However, the anode and cathode processes do not have the same rate. The oxygen reduction at the cathode requires an over-potential of 400 mV to reach current density of 1 mA/cm^2 at the Pt/polymer interface. On the contrary, the hydrogen oxidation at the anode requires only 30–50 mV to reach the same current density at the Pt/polymer interface. The high over-potential at the cathode is the most important source of loss in the PEFC. Possible solutions to this problem would be the employment of dispersed Pt catalysts, the design of new effective cathode catalysts and the effective incorporation of the catalyst layer between the membrane and the gas distributor/current collectors [11].

The earliest method of preparation of membrane electrode assemblies was presented in a patent of 1967 [12]. In this work, the bonding of metal powder (Pt-black) mixed with a binder as PTFE upon the surface of the membrane is described. However, the limited dispersion and the low utilization of the Pt/PTFE required the application of 4 mg Pt/cm^2 to obtain satisfactory cathode performance. The first step forward in lowering the Pt loading was done by Raistrick [13]. Raistrick impregnated carbon-supported Pt (Pt/C) and PTFE layer on the carbon cloth with Nafion® polymeric solution and hot-pressed the electrode onto the membrane. This approach resulted in a significant increase of FC performance with Pt loading as low as 0.4–0.5 mg/cm^2. Recent developments by Wilson et al. [14–16] have shown a breakthrough in Pt catalyst utilization. In their work, instead of PTFE, the polymer solution is used to bind the catalyst layer to the membrane. Such catalyst layers have very low Pt loading (0.12–0.16 mg/cm^2) in a thin layer (less than 10 μm thick). PEFC with such low catalyst loading exhibited long life (5000 h) and high performance stability [11].

Table 2. Membrane requirements for a polymer-electrolyte FC

High proton conductivity
High chemical stability under PEFC operating conditions
Effective separation of fuel and oxidant gases
Low ohmic losses at the high current densities used in the PEFC

Besides the above concerning the cathode catalyst, an extra problem concerning the anode catalyst should be addressed. The poisoning of the anode catalyst due to the presence of CO in the feed H_2 gas. This problem can be minimized by either applying CO tolerant catalysts like Pt-Ru [17] or air bleeding into the anode [18]. In addition, the Pt-Ru catalyst at low loading (0.1 mg/cm^2) minimize the problem of the in situ generation of CO by reduction of CO_2 within the PEFC cathode.

3.2.2. Polymeric membrane in PEFC

The most important requirements for the membrane in the PEFC are presented in Table 2. The first membranes introduced for PEFC were the poly(styrene sulfonic acid) (PSSA) and sulfonated phenol-formaldehyde membranes. Unfortunately, their stability under PEFC conditions was not sufficient [19]. A breakthrough was achieved by the application of the Nafion® membranes, which were previously developed for the chlor-alkali electrolysis. The Nafion® membrane fulfilled all the above requirements. It is an excellent proton conductor and has high stability in a PEFC for more than 60 000 h [20]. In addition, it has conductivity around 0.2 S/cm at 80°C and 100% relative humidity. Nafion® has extensively been used to PEFC, both hydrogen and direct methanol FC, (typically membranes of thickness between 25 and 175 μm are applied). It has, however, some weak points too. It is expensive and functions as a proton conductor only when it is highly hydrated. Therefore for the hydrogen FC, the gases should be humidified prior entering the FC. In the case of the direct methanol FC, a liquid fuel-water mixture is always in contact with the membrane and keeps it humidified. However, Nafion® is a poor methanol barrier and significant amount of methanol diffuses through the membrane to the cathode [21].

The fact that Nafion® has to be humidified limits its use below 100°C in a direct methanol FC and 80°C in a hydrogen FC, under normal operating conditions. Currently, there is strong tendency to operate the PEFC at higher temperatures (130–200°C) for reasons concerning the improvement of methanol oxidation reaction at the anode of the direct methanol FC or higher tolerance of CO levels in the hydrogen FC etc. Therefore, attention has been drawn to the development of membranes, which either do not require water to maintain proton conductivity or have sufficient proton conductivity at reduced water contents or have resistance to dehydration. Nafion® membranes were doped with molten acidic salts [22] or phosphoric acid [23] to increase its conductivity at higher temperatures. These materials, however, are still based on the expensive Nafion® material and should be tested under FC conditions. In other studies, the sulfonic groups were substituted with phosphoric groups [24]. New polymers based on polymer networks, containing basic sites (ether, amine or imino group) doped with strong acids such as phosphoric or sulfonic acid were prepared. Typical example is doping of polybenzimidazole (PBI) with phosphoric acid [25,26]. These membranes operate at elevated temperatures with very low gas humidification, have good mechanical strength up to 200°C and low methanol permeability. Membranes of sulfonated poly(ether ether ketone) (SPEEK) membranes were prepared as well. The sulfonation level of 60% was found to be good compromise between the conductivity and mechanical membrane properties [27] and their swelling could be reduced by either blending with other polymers [21] or by chemical cross-linking [28]. These materials, however, rely their conductivity to the presence of water-like in the case of Nafion®. Polymer blends were also synthesized by combining polymers containing nitrogen bases with others containing sulfonic groups [29,30].

4. BIPOLAR MEMBRANES

4.1. Introduction

A BPM, a special type of polymeric layered ion-exchange membrane, consists of a cation-and an anion-exchange layer joined together. BPMs allow the electro-dissociation of water into hydroxide ions and protons without generating gases. This membrane water splitting technology is a typical example for process intensification. Electrodialysis with BPMs combines reaction and separation in one unit and allows the design of unique processes like the production and recovery of acids and bases, the variation of the pH of a process stream and the separation of proteins.

4.2. Principle of a BPM

If a CEM and an AEM are placed in an electric field and salt solution is fed in between the two membranes, the salt ions are removed from the enclosed compartment (Fig. 8a). This desalination proceeds until all ions are depleted from the compartment. As a result, a significant increase of the electrical resistance due to the decreased ion conductivity is observed. After the salt ion depletion, current can only be transported by hydroxyl ions and protons generated by water dissociation (Fig. 8b). To minimize the electrical resistance, the AEM and CEM have to be placed closely together. Laminating both on top of each other results in minimum resistance and a BPM is formed. Water diffuses into the BPM from the surrounding solutions and dissociates under the electric field, at the interface between the anion- and the cation-exchange membrane, to generate protons and hydroxyl ions (Fig. 8c).

$$2H_2O \rightleftharpoons H_3O^+ + OH^- \tag{8}$$

The H^+ and OH^- ions further migrate out of the junction layer through the anion- and cation-exchange layers of the BPM. The thickness of the transition region, where the water dissociation takes place, is less than 10 nm [31]. Strathmann et al. [32] reported that the water dissociation in a BPM is accelerated up to 50 million times compared to the rate of water dissociation in aqueous solutions. This acceleration is influenced by the strong electric field across the transition region. The main driving force for the enhanced

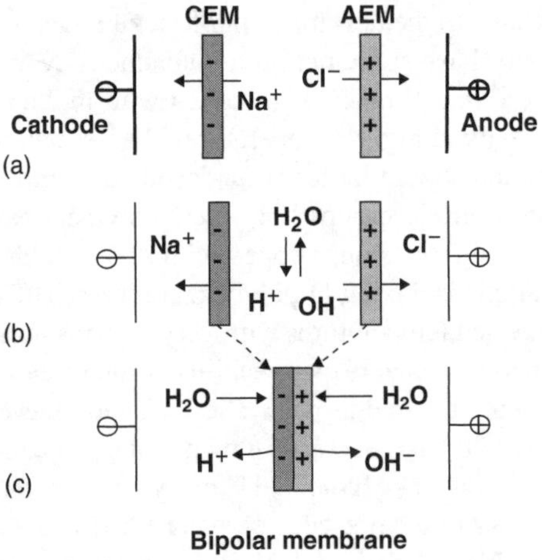

Fig. 8. Schematic drawing of the water splitting function of a BPM, (taken from [33])

water dissociation seems to be the reversible protonation and deprotonation of the functional groups of the ion-exchange membranes, mainly of the tertiary amino groups B of the anion-exchange layer (see Eqs (9) and (10)) [31–34]:

$$B + H_2O \rightleftharpoons BH^+ + OH^- \tag{9}$$

$$BH^+ + H_2O \rightleftharpoons B + H_3O^+ \tag{10}$$

The BPMs can be used not only for the dissociation of water, but also for the splitting of other self-dissociating liquids, like methanol [35].

4.3. Principle of BPM electrodialysis

The BPM is stacked together with monopolar CEM and AEM into a membrane module. A typical application of the BPM electrodialysis (EDBPM), the production of acid and base from salt, is depicted in Fig. 9. The working electrodes establish the electric field as a driving force, salt solution is fed to the central compartment, and the ions migrate out of this compartment into the neighboring ones. Charge compensation because of electro-neutrality occurs due to the water splitting at the interface of the BPM. Other stack configurations (sequence of CEMs, AEMs and BPMs) are also possible as long as electro-neutrality is obeyed

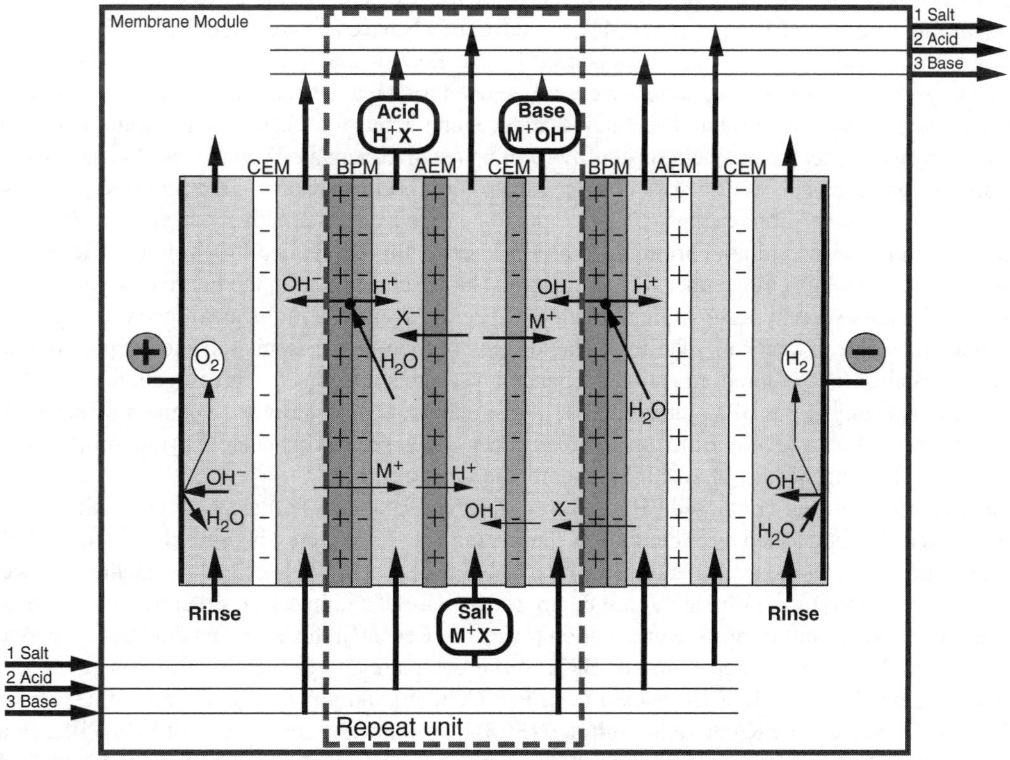

Fig. 9. Schematic drawing of membrane stack with the indication of desired ion migrational fluxes (upper part) and undesired ion migrational fluxes (lower part) (taken from [37])

and counter ions can go through the monopolar membranes to transport the current. The selectivity of the membranes strongly depends on the concentrations of salt, acid and base. According to Donnan equilibrium, more co-ions can migrate through the membranes at higher solution concentrations, which in fact reduces the purity of the produced acids and bases [36,37]. These undesired ion migration fluxes are also visualized in Fig. 9.

4.4. Preparation of BPMs

The performance of a BPM in electrodialysis processes depends on the BPM components. The selection of the membrane materials influences mainly the chemical and mechanical stability, the transport properties of the membrane layers and the strength and topology of the intermediate region. As a result the energy consumption, the product concentration, and quality (extent of salt impurities), the long-term operation of such membranes and their economical feasibility is influenced as well.

The anion- and cation-permeable layers of the BPM consist of materials similar to standard anion- and cation-permeable membranes, which are stable in the environment encountered in acid/base electrodialysis. The two selective layers should allow the selective transport of the water splitting products. Furthermore, the membrane layers should allow a sufficient water flux from the base and/or acid compartment into the membrane junction to replenish the water consumed by the water dissociation reaction (Fig. 8c) otherwise irreversible degradation occurs [38]. In addition, the membrane layers should block co-ions from reaching the membrane junction and the opposite side of the membrane, because these co-ion fluxes are responsible for product contamination. Many BPMs are prepared by using commercial ion-exchange membranes as a precursor for one or both BPM layers [39–44], but novel BPMs have been developed as well [36,37].

The structure of the BPM junction is important for the electro-catalytic water splitting function of the complete system. The water dissociation rate is accelerated for a fixed electric potential across the contact region when a catalyst is present. The catalyst reduces the activation energy of the water dissociation because it provides alternative reaction paths by forming reactive, activated complexes. As catalysts in the bipolar junction, immobile weak acids or bases with an equilibrium constant of the acid/base pair close to that of the water dissociation reaction ($pK_a = 7$) could be used [39]. Alternatively, heavy metals ion complexes, like those of zirconium, chromium, iron or others could be applied [40–46]. The metal ions or complexes are immobilized by either including an insoluble salt of them in the interface layer [44] or by converting a soluble salt by a subsequent treatment. The most suitable multivalent metal ion hydroxides are immobilized due to their low solubility. The location of the water dissociation and the preferred location of the catalyst (to the anion- or cation-exchange layer) are still subject of investigation.

Besides the catalyst, the BPM junction should also have a certain surface roughness to increase the membrane contact area. Several BPMs have already been developed having smooth [41,43,45], corrugated [40,46–48] or heterogeneous [44,46,49–51] membrane junctions.

The preparation of well-composed BPMs requires the establishment of efficient contact of the CEM and AEM. This can be established by either loosely laminating [41–43,45], pressing [44,51], gluing [52] of the two preformed membranes or casting one membrane above the other [47–49,53]. Other methods, however, like co-extruding [54] or modifying [50,55] do not show well-defined interface structures due to the weak influence on these regions during the preparation processes. Recently, the gluing method has been used by Wilhelm et al. [36] for the preparation of BPMs from commercially available AEMs (Neosepta AMX from Tokuyama Corp. and IonClad R4030 from Pall Gelman) and tailor-made cation-permeable layers [36]. The CEM was a SPEEK/poly(ether sulfone) (PES) blend and the adhesive was SPEEK/PES solution. The properties of the CEM could be adjusted through the SPEEK content of the blend with PES. With increase of the SPEEK content, the ion-exchange capacity and the water uptake increase, whereas the resistance and the density of the membrane decrease [36]. Hao et al. [53] have prepared BPMs by casting.

Their membranes consisted of an AEM with cross-linked matrix (prepared by the reaction of chloromethylated polysulfone (PS) with diamine), an interfacial layer (made from chloromethylated PS solution containing cation-exchange resin and amine), and a CEM (made from a cation resin powder dispersed in sulfonated PS).

4.5. Characterization of BPMs

The electrical energy consumption of a BPM process depends on the electrical resistances of the monopolar membranes included in the stack and the current-voltage characteristics of the BPM [56,57]. Current-voltage curves can be obtained with the stack configuration, schematically drawn in Fig. 10. The central BPM in the stack is the investigated one. The other membranes are auxiliary membranes preventing the transport of water dissociation products, formed at the two working electrodes, to the central compartments. During the experiment, the applied current density is increased stepwise and the system is allowed to reach a steady state. The voltage drop across the membrane is measured, followed by the next increase in current density. The experimental current-voltage curve of a BPM recorded in a neutral salt solution (M^+X^-) shows characteristic parts (see Fig. 11) [39]. Below the lower limiting current density (i_{lim1}) the current is only transported by salt ions. At the limiting current density, the electrical resistance is the highest, since all salt ions are removed from the BPM junction. The magnitude of the lower limiting current is a measure for the selectivity of the BPM towards co-ion leakage (the reader is reminded that this lower limiting current density only exists to a membrane in its salt form). Above this current density, water splitting occurs (U_{diss}) and the water splitting products ($J^-_{\mathrm{OH}}/J^+_{\mathrm{H}}$) are also available to transport the current [58]. The operating current density (i_{op}) is chosen as high as possible to reduce the relative salt ion transport and to have a high water splitting efficiency. According to Aritomi et al. [38] and Krol et al. [59], above the upper limiting current (i_{lim2}), the water transport toward the BPM junction is not sufficient to replenish the water split at the corresponding rate.

Another technique commonly used for the characterization of the BPM is chronopotentiometry [60,61]. Chronopotentiometry allows the determination of the actual water splitting voltage, the irreversible energy

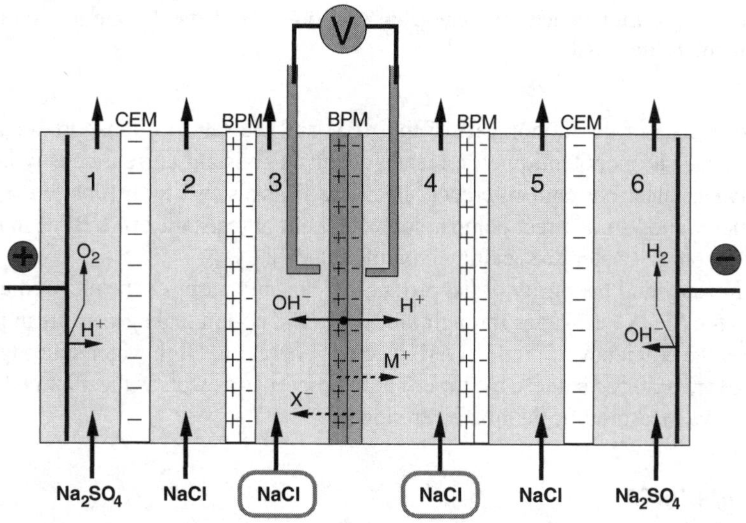

Fig. 10. Schematic representation of a six-compartment measurement module for current-voltage curve and chronopotentiometric measurements, (taken from [36])

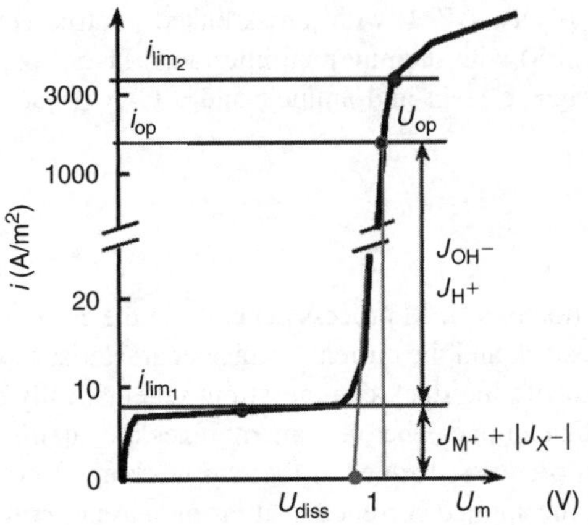

Fig. 11. Schematic current–voltage curve of a BPM in a salt (M^+X^-) solution, (adapted from [36,58])

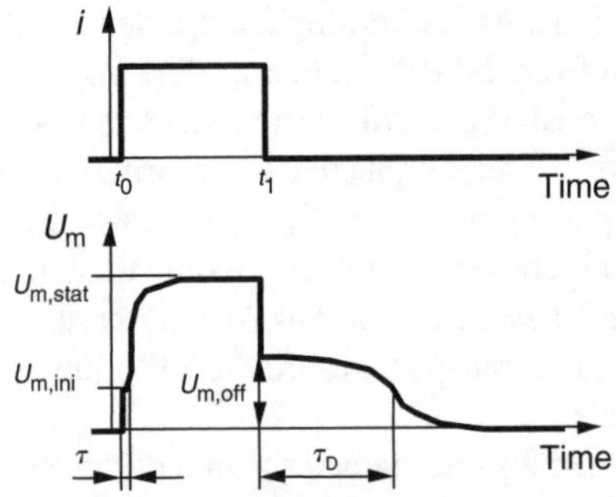

Fig. 12. BPM chronopotentiometry principle (adapted from [61]). A current density i is imposed at time t_0 and switched off at time t_1, while the electric potential difference over the membrane U_m is recorded in time ($U_{m,ini}$ initial; $U_{m,stat}$ static; $U_{m,off}$ after switch off). The transition time τ and the de-charging time τ_D are obtained from the inflection points indicated

loss due to the resistance of the monolayers of the BPM and the salt ion transport behavior through the BPM [57]. During the chronopotentiometric measurements a constant current density is applied and the voltage drop between electrode and reference electrode is measured as a function of time (Fig. 12). Chronopotentiometry allows the direct comparison of the initial resistance of a BPM in equilibrium with the resistance in operation (water dissociation conditions) [57].

The current efficiency and the purity of the produced acids and bases of the electrodialysis process are directly related to the M^+/X^- leakages through the BPM. The co-ion leakage results in product contamination and reduces the efficiency of the H^+/OH^- electro-generation. High water splitting capability and a low co-ion leakage are required at the same time. These properties as well as the maximal obtainable product concentrations are measured by acid/base titration [57].

4.6. Limitations of EDBPM

One of the main limiting phenomena of EDBPM is the water splitting performance [35,62]. Water, which is consumed by the dissociation into protons and hydroxide ions, has to be replenished by diffusion of

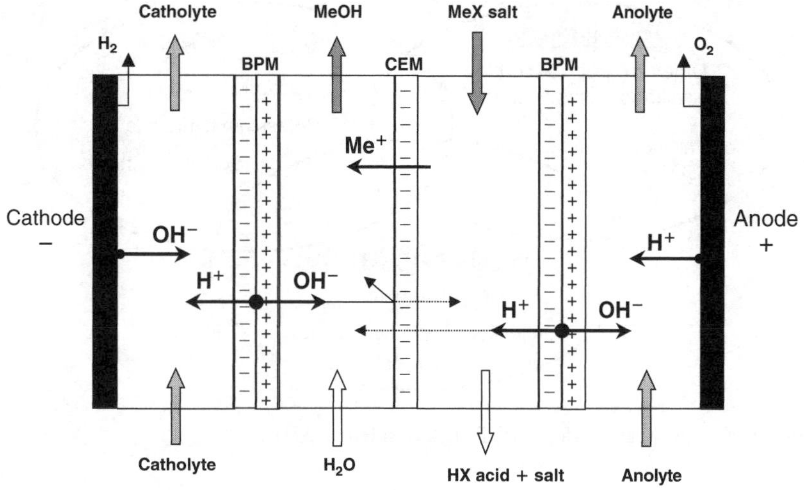

Fig. 13. Cell arrangement configuration for the direct acidification

water from the outer solution of the two monopolar layers into the BPM interphase. When the rate of the water dissociation is faster than the water transport into the BPM transition region, water transport is the limiting step and results in drying out of the BPM, which causes a drastic increase in the resistance [31,63]. This performance is controlled by the permselectivities of the ion-exchange layers of the BPM and by ion diffusive transport. An additional loss of process permselectivity is obtained by the leakage through the monopolar ion-exchange membranes (Fig. 9). Due to these leakages, the current efficiency of the process is reduced and the purity of the products is limited [62].

Another major limitation is the concentration range in which the EDBPM processes can be used. At very low concentrations, the electric resistance of the electrolyte solution in the compartments between the membranes is very high, whereas at high concentrations the selectivity of the BPM and the ion-exchange membranes is very low [64]. Additionally, the chemical stability of the BPM, especially against a concentrated base on the anion-permeable side, is not always sufficient [37].

4.7. EDBPM processes

EDBPM processes can be performed with different stack configurations, depending on the desired application [35,65–67]. For the treatment of concentrated salt solutions and the production of the corresponding acids and bases, a three-compartment cell system consisting of an AEM, a CEM and a BPM as repeat unit is used (Fig. 9). In an industrial-type unit, up to 100 cell triplets could be installed in an electrodialysis stack [68]. In applications where it is not possible to obtain high purity of both products (acid and base) or products with low conductivity are produced, a two-compartment cell is then recommended [35,67]. This configuration can be used together with a CEM for acidification (see in Fig. 13) or with an AEM for alkalization of a salt stream [35,68]. If a high ratio of acid or base with respect to the product salt content is required, a configuration with two monopolar membranes of the same type can be applied. In this configuration, the outlet of the middle loop is recycled again into the acidic or base loop next to the BPM for a more efficient exchange between protons and cations, or between hydroxyl ions and anions [35,68].

Several applications of EDBPM have already been investigated with promising results. Typical examples are the production of acid and base [66,69], the acidification of product streams [70,71] or special

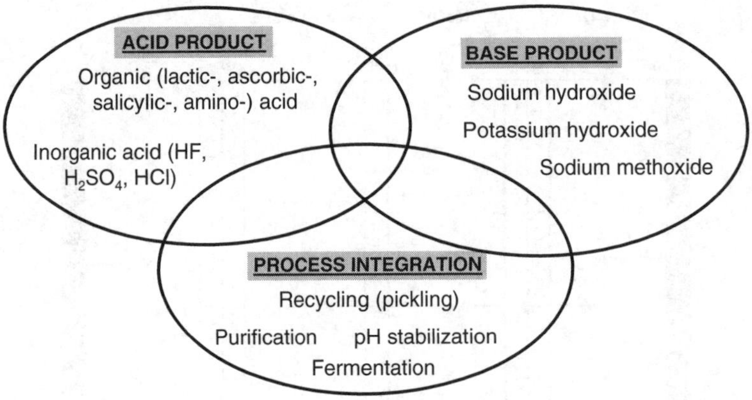

Fig. 14. Possible areas of applications for EDBPM (taken from [37])

separations like the separation of amino acids and proteins on the basis of their isoelectric points [65,66]. Economically, the most interesting applications are those in the overlap of the areas schematically presented in Fig. 14. Designing EDBPM in the overlapping areas provides increasing prospects for its economic feasibility, but it also increases the complexity and the challenges that have to be met [37]. An example of a successful overlap is the conversion of sodium lactate from a fermentation step into lactic acid by EDBPM with a use of the side product, sodium hydroxide, to control the fermentation reaction [4]. Further examples can be found in other reviews concerning EDBPM applications [35,65,67].

REFERENCES

1. T.V. Bommaraju and P.J. Orosz. Brine electrolysis, electrochemistry encyclopedia (http://electrochem.cwru.edu/ed/encycl).
2. L.C. Curlin, T.V. Bommaraju and C.B. Hansson. Alkali and Chlorine Products, *Kirk Othmer Encyclopedia of Chemical Technology*, 4th edn. Wiley, New York, 1991.
3. J.H.G. van der Stegen. The state of the art to modern chlor-alkali electrolysis with membrane cells. Ph.D. Thesis. University of Twente, 2000.
4. T.A. Davis, J.D. Genders and D. Pletcher. *A First Course in Ion Permeable Membranes*. The Electrochemical Consultancy, UK, 1997, p. 199 (Chapter 5).
5. K. Kordesch and G. Simader. *Fuel Cells and Their Applications*. VCH Publishers Inc., New York, 1996.
6. A. Boudghene Stambouli and E. Traversa. Fuel cells, an alternative to standard sources of energy. *Renew. Sustain. Energ. Rev*. **6** (2002) 297–306.
7. K. Joon. Fuel cells – a 21st century power system. *J. Power Sources* **71** (1998) 12–18.
8. G.F. Mc Lean, T. Niet, S. Prince-Richard and N. Djilali. An assessment of alkaline fuel cell technology. *Int. J. Hydrogen Energ*. **27** (2002) 507–526.
9. A. Boudghene Stambouli and E. Traversa. Solid oxide fuel cells (SOFCs): a review of an environmentally clean and efficient source of energy. *Renew. Sustain. Energ. Rev*. **6** (2002) 433–455.
10. G.J.K. Acres. Recent advances in fuel cell technology and its application. *J. Power Sources* **100** (2001) 60–66.
11. S. Gottesfeld and T.M. Zawodzinski. Polymer-electrolyte fuel cells. In: R.C. Alkire, H. Gerischer, D.M. Kolb and C.W. Tobias (Eds.), *Advances in Electrochemical Science and Engineering*, Vol. **5**. Wiley-VCH, Germany, 1997, p. 195.
12. Niedrach. US Patent 3 297 484 (1967).
13. I.D. Raistrick. In: J.W. van Zee, R.E. White, K. Kinoshita and H.S. Burrey (Eds.), *Diaphragms, Separators and Ion-Exchange Membranes, The Electrochemical Society Softbound Proceedings Series*, PV 86-13, Pennington, NJ, 1986, p. 172.

14. M.S. Wilson and S. Gottesfeld. *J. Appl. Electrochem.* **22** (1992) 1.
15. M.S. Wilson. US patent 5 211 984 and 5 234 777 (1993).
16. M.S. Wilson, J.A. Valerio and S. Gottesfeld. *Electrochim. Acta* **40** (1995) 355.
17. M. Iwase and S. Kawatsu. In: S. Gottesfeld, G. Halpert and A. Landgrebe (Eds.), *Proton Conducting Membrane Fuel Cells I.* The Electrochemical Society, Pennington, NJ, 1995, p. 12.
18. S. Gottesfeld and J. Pafford. *J. Electrochem. Soc.* **135** (1988) 2651.
19. J.-P. Randin. *J. Electrochem. Soc.* **129** (1982) 1215.
20. A. Steck. In: O. Savadogo, P.R. Roberge and T.N. Vesinoglu (Eds.), *Proceedings of the First International Symposium on New Materials for Fuel Cell Systems*, Montreal, Canada, 1995, p. 74.
21. C. Manea and M.H.V. Mulder. Characterization of polymer blends of polyethersulfone/sulfonated polysulfone and poly-ethersulfone/sulfonated poly-etheretherketone for direct methanol fuel cell applications. *J. Membr. Sci.* **206** (2002) 443–453.
22. M. Doyle, S.K. Choi and G. Proulx. *J. Electrochem. Soc.* **147** (2000) 34.
23. R. Savinelli, E. Yeager, D. Tryk, U. Landau, J. Wainright, D. Wang, K. Lux, M. Litt and C. Rogers. *J. Electrochem. Soc.* **141** (1994) 46.
24. X. Xu and I. Cabasso. *J. Polym. Mater. Sci.* **68** (1993) 120.
25. X. Glipa, B. Bonnet, B. Mula, D.J. Jones and J. Roziere. *J. Mater. Chem.* **9** (1999) 3045.
26. R. Bouchet and E. Siebert. *Solid State Ionics* **118** (1999) 287.
27. T. Soczk-Guth, J. Baurmeister, G. Frank and R. Knauf. WO Patent 9 927 963 (1999).
28. S.-P.S. Yen, S.R. Narayanan, G. Halpert, E. Graham and A. Yavrouian. US Patent 5 795 496 (1998).
29. L. Jorissen, V. Gogel, J. Kerres and J. Garche. New membranes for direct methanol fuel cells. *J. Power Sources* **105** (2002) 267–273.
30. J. Kerres and A. Ullrich. Synthesis of novel engineering polymers containing basic side groups and their application in acid–base polymer blend membranes. *Separ. Purif. Technol.* **22 –23** (2001) 1–15.
31. H.-J. Rapp. Elektrodialyse mit bipolaren membranen – theorie und anwendungen. Ph.D. Thesis. Universität Stuttgart, Institut für Chemische Verfahrenstechnik, Germany, 1995.
32. H. Strathmann, J.J. Krol, H.-J. Rapp and G. Eigenberger. Limiting current density and water dissociation in bipolar membranes. *J. Membr. Sci.* **125** (1997) 123–142.
33. J.J. Krol. Monopolar and bipolar ion-exchange membranes – mass transport limitations. Ph.D. Thesis. University of Twente, Enschede, The Netherlands, 1997.
34. S. Mafe, P. Ramirez, A. Alcatraz and V. Aquilella. Ion transport and water splitting in bipolar membranes: theoretical background. In: A.J.B. Kemperman (Ed.), *Bipolar Membrane Handbook.* Twente University Press, Enschede, The Netherlands, 2000, pp. 47–78 (Chapter 3).
35. G. Pourcelly. Electrodialysis with bipolar membranes: principles, optimization, and applications. *Russ. J. Electrochem.* **38** (2002) 919–926.
36. F.G. Wilhelm, I. Punt, N.F.A. van der Vegt, M. Wessling and H. Strathmann. Optimization strategies for the preparation of bipolar membranes with reduced salt ion leakage in acid-base electrodialysis. *J. Membr. Sci.* **182** (2001) 13–28.
37. F.G. Wilhelm. Bipolar membrane electrodialysis – membrane development and transport characteristics. Ph.D. Thesis. University of Twente, Enschede, The Netherlands, 2001.
38. T. Aritomi, T. van der Boomgaard and H. Strathmann. Current voltage curve of a bipolar membrane at high current density. *Desalination* **104** (1996) 13–18.
39. F.G. Wilhelm, N.F.A. van der Vegt, M. Wessling and H. Strathmann. Bipolar membrane preparation. In: A.J.B. Kemperman (Ed.), *Bipolar Membrane Handbook.* Twente University Press, Enschede, The Netherlands, 2000, pp. 79–108 (Chapter 4).
40. F. Hanada, K. Hiraya, N. Ohmura and S. Tanaka. *Bipolar Membrane and Method for Its Production*, Tokuyama Soda Ltd. (Japan). US Patent 5 221 455 (1993).
41. R. Simons. *High Performance Bipolar Membranes*. Unisearch Ltd. (Australia). WO Patent 89/01 059 A1 (1989).
42. R. Simons. Preparation of a high performance bipolar membrane. *J. Membr. Sci.* **78** (1993) 13–23.
43. F. Posar and M. Riccardi. *Process for the Manufacture of a Bipolar Membrane and Process for the Manufacture of an Aqueous Alkali Metal Hydroxide Solution.* Solvay S.A. (Belgium). US Patent 5 380 413 (1995).
44. K. Umemura, T. Naganuma and H. Miyake. *Bipolar Membrane*, Asahi Glass Company Ltd. (Japan). US Patent 5 401 408 (1995).

45. F. Posar. *Method for Making a Bipolar Membrane.* Solvay S.A. (Belgium). US Patent 5 849 167 (1998).
46. F.P. Chlanda, L.T.C. Lee and K.J. Liu. Verfahren zur Herstellung bipolarer membranen (Process for preparing bipolar membranes). Allied Chemical Corporation (USA). DE Patent 2 737 131 (1978).
47. R.B. Hodgdon and S.S. Alexander. *Novel Bipolar Membranes and Process of Manufacture.* Ionics Inc. (USA). US Patent 4 851 100 (1989).
48. B. Bauer. Bipolare Membran und Verfahren zu deren Herstellung (Bipolar Membrane and Process for its Preparation). Fraunhofer-Gesellschaft zur Förderung der Angewandten Forschung e.V (FhG) (Germany). EP Patent 563 851 (1993).
49. F.P. Chlanda and M.J. Lan. *Bipolar Membranes and Methods of Making Same.* Allied Chemical Corporation. WO Patent 87/07,624 (1987).
50. K.J. Liu and H.L. Lee. *Bipolar Membranes.* Union Resources and Technology Inc. (USA). EP Patent 0 143 582 A2 (1985).
51. G.J. Dege, F.P. Chlanda, L.T.C. Lee and K.J. Liu. *Method of Making Novel Two Component Bipolar Ion-Exchange Membranes.* Allied Chemical Corp. (USA). US Patent 4 253 900 (1981).
52. H. Müller and H. Pütter. *Verfahren zur Herstellung bipolarer Membranen (Procedure for the Preparation of Bipolar Membranes).* BASF AG, Ludwigshafen, Germany.
53. J.H. Hao, C. Chen, L. Li, L. Yu and W. Jiang. Preparation of bipolar membranes (I). *J. Appl. Polym. Sci.* **80** (2001) 1658–1663.
54. K. Richau, P. Klug, G. Malsch, R. Swoboda, J. Ziegler, Th. Weigel, St. Otto, U. Martens, K. Kneifel and H.H. Schwarz. In: *Proceedings of the Euromembrane on Membranes made from Polyelectrolytes – Preparation and Electrochemical Properties*, Enschede, The Netherlands, 1997, p. 234.
55. R. El Moussaoui and H. Hurwitz. *Single-Film Membrane, Process for Obtaining It and Use Thereof.* Université Libre de Bruxelles (Belgium). US Patent 5 840 192 (1998).
56. J.L. Gineste, G. Pourcelly, Y. Lorrain, F. Persin and C. Gavach. Analysis of factors limiting the use of BPM: a simplified model to determine trends. *J. Membr. Sci.* **112** (1996) 199–208.
57. K. Richau, et al. Bipolar membrane characterization, In: A.J.B. Kemperman (Ed.), *Bipolar Membrane Handbook.* Twente University Press, Enschede, The Netherlands, 2000, pp. 109–153 (Chapter 5).
58. F.G. Wilhelm, I. Punt, N.F.A. van der Vegt, M. Wessling and H. Strathmann. Asymmetric bipolar membranes in acid–base electrodialysis. *Ind. Eng. Chem. Res.* **41** (2002) 579–586.
59. J.J. Krol, M. Jansink, M. Wessling and H. Strathmann. Behaviour of bipolar membranes at high current density: water diffusion limitation. *Separ. Purif. Technol.* **14** (1998) 41–52.
60. F.G. Wilhelm, N.F.A. van der Vegt, M. Wessling and H. Strathmann. Comparison of bipolar membranes by means of chronopotentiometry. *J. Membr. Sci.* **199** (2002) 177–190.
61. F.G. Wilhelm, N.F.A. van der Vegt, M. Wessling and H. Strathmann. Chronopotentiometry for the advanced current–voltage characterisation of bipolar membranes. *J. Electroanal. Chem.* **502** (2001) 152–166.
62. X. Tongwen. Electrodialysis processes with bipolar membranes (EDBM) in environmental protection – a review. *Resour. Conserv. Recycl.* **37** (2002) 1–22.
63. B. Bauer, F.J. Gerner and H. Strathmann. Development of bipolar membranes. *Desalination.* **68** (1988) 279–292.
64. F.G. Wilhelm, N.F.A. van der Vegt and M. Wessling. Bipolar membrane electrodialysis for the production of acids and bases. *Npt-Wetenschapskatern, Npt-Processtechnologie* **1** (2002) 19–21.
65. V. Cauwenberg, J. Peels, S. Resbeut and G. Pourcelly. Application of electrodialysis within fine chemistry. *Separ. Purif. Technol.* **22–23** (2001) 115–121.
66. L. Bazinet, F. Lamarche and D. Ippersiel. Bipolar-membrane electrodialysis: applications of electrodialysis in the food industry. *Trend. Food Sci. Technol.* **9** (1998) 107–113.
67. X. Tongwen and Y. Weihua. Effect of cell configurations on the performance of citric acid production by a bipolar membrane electrodialysis. *J. Membr. Sci.* **203** (2002) 145–153.
68. B. Bauer, H. Holdik and A. Velin. Cell equipment and plant design in bipolar membrane technology. In: A.J.B. Kemperman (Ed.), *Bipolar Membrane Handbook.* Twente University Press, Enschede, The Netherlands, 2000, pp. 155–189 (Chapter 6).
69. G. Pourcelly and C. Gavach. Electrodialysis water splitting – applications of electrodialysis with bipolar membranes. In: A.J.B. Kemperman (Ed.), *Bipolar Membrane Handbook.* Twente University Press, Enschede, The Netherlands, 2000.

70. M. Bailly. Production of organic acids by bipolar electrodialysis: realizations and perspectives. *Desalination* **144** (2002) 157–162.
71. X. Tongwen and Y. Weihua. Citric acid production by electrodialysis with bipolar membranes. *Chem. Eng. Process* **41** (2002) 519–524.

NOMENCLATURE

AEM	anionic exchange membrane
AFC	alkaline fuel cell
BPM	bipolar membrane
CEM	cationic exchange membrane
EDBPM	bipolar membrane electrodialysis
FC	fuel cell
i	current density (A/m^2)
i_{lim}	limiting current density (A/m^2)
i_{op}	operation current density (A/m^2)
J	flux (mol/(m^2(s))
J_{H}^{+}	hydrogen ion flux (mol/(m^2(s))
J_{M}^{+}	cation flux (mol/(m^2(s))
J_{OH}^{-}	hydroxide ion flux (mol/(m^2(s))
J_{X}^{-}	anion flux (mol/(m^2(s))
M^+	salt or base cation
MCFC	molten carbonate fuel cell
PAFC	phosphoric acid fuel cell
PBI	polybenzimidazole
PEEK	poly(ether ether ketone)
PEFC	polymer-electrolyte fuel cell
PES	poly(ether sulfone)
PS	polysulfone
PTFE	polytetrafluoethylene
SPEEK	sulfonated poly(ether ether ketone)
t	time (s)
t_0	starting time (s)
t_1	ending time (s)
U	potential difference (V)
U_{diss}	dissociation potential difference (V)
U_{m}	membrane potential difference (V)
$U_{\text{m,ini}}$	initial membrane potential difference (V)
$U_{\text{m,off}}$	membrane potential difference after switching off (V)
$U_{\text{m,stat}}$	static membrane potential difference (V)
U_{op}	operation potential difference (V)
X^-	salt or acid anion
τ	transition time (s)
τ_{D}	de-charging time (s)

Chapter 30

Compact mixed-reactant fuel cells

Michael A. Priestnall, Vega P. Kotzeva, Deborah J. Fish and Eva M. Nilsson

Abstract

The compact mixed-reactant (CMR) fuel cell is an important new "platform" approach to the design and operation of all types of fuel cell stacks. Amongst several other advantages, CMR has the potential to reduce polymer electrolyte membrane (PEM) stack component costs by around a third and to raise volumetric power densities by an order of magnitude.

Mixed-reactant fuel cells, in which the fuel and oxidant within a cell are allowed to mix, rely upon the selectivity of anode and cathode electrocatalysts to separate the electrochemical oxidation of fuel and reduction of oxidant. A comprehensive review of the 50-year history of mixed-reactant literature has demonstrated that such systems can perform as well as and, in some circumstances, much better than conventional fuel cells.

The significant innovation that Generics has introduced to this field is to combine the concept of mixed-reactant fuel cells with that of a fully porous membrane electrode assembly (MEA) structure. Passing a fuel-oxidant mixture through a stack of porous cells allows the conventional bipolar flow-field plates required in many fuel cell designs to be eliminated. In a conventional PEM stack, for example, the bipolar carbon flow-field plates may block up to half of the active cell area and account for up to 90% of the volume of the stack and of the order of one-third of the materials costs. In addition to all the advantages of mixed-reactant systems, the "flow-through" mode, embodied in Generics' CMR approach, significantly enhances mass-transport of reactants to the electrodes and can reduce reactant pressure drops across the stack. Redesigning fuel cells to operate in a CMR mode with selective electrodes offers the attractive prospect of much reduced stack costs and significantly higher stack power densities for all types of fuel cell.

Initial modelling and proof of principle experiments using an alkaline system have confirmed the validity of the CMR approach and the potential for substantial performance improvements.

Keywords: Mixed-reactant; Fuel–air mixture; Methanol; Sodium borohydride; Fuel cell

Article Outline

1. Introduction 594
2. Significance of the compact mixed-reactant fuel cell 594
3. History of mixed-reactant fuel cells 595
4. Development of CMR systems 600
5. Safety of mixed-reactant and CMR systems 604
6. Conclusion 604

Fuel Cells Compendium

Acknowledgements . . . 605
References . . . 605

1. INTRODUCTION

Conventional fuel cells operate using active but often non-selective electrocatalysts and rely upon a strict segregation of fuel and oxidant to prevent parasitic chemical reactions at the electrodes. Mixed-reactant fuel cells, which have been known for around half a century, rely upon the selectivity of anode and cathode electrocatalysts to separate the electrochemical oxidation of fuel and reduction of oxidant. Without a need for physical separation of fuel and oxidant, there is no longer any need for gas-tight structures within the stack and considerable relaxation of sealing, manifolding and reactant delivery structures is possible.

The compact mixed-reactant (CMR) fuel cell is an entirely new concept in which a mixture of fuel and oxidant is flowed through a fully porous anode – electrolyte – cathode structure. This structure may be a single-cell, series stack or parallel stack and may be in planar, tubular or other geometry. In principle, a CMR cell may be based on polymer, alkaline, phosphoric acid, molten carbonate, solid oxide or any other type of fuel cell chemistry. Whatever the specific geometry or chemistry, for this type of cell to work effectively, the anode and cathode electrocatalysts must be substantially selective – i.e. the anode should be active towards fuel oxidation and tolerant to oxygen, while the cathode should be active towards oxygen reduction and tolerant to fuel.

Selectivity in the electrocatalysts for a mixed-reactant or a CMR fuel cell is needed to minimise mixed potentials at the electrodes which otherwise will reduce the available cell voltage and compromise the efficiency of conversion to electricity. Loss of efficiency reflects the extent of the "parasitic" direct reaction of fuel and oxidant to produce heat rather than electric current. This is directly analogous to the problem of methanol cross-over from the anode compartment to cathode compartment in conventional direct-methanol fuel cells. A range of partially and substantially selective catalysts have become available in recent years, largely as a result of the continuing effort to develop more effective catalysts for direct-methanol fuel cells.

2. SIGNIFICANCE OF THE COMPACT MIXED-REACTANT FUEL CELL

Two key advantages of operating any fuel cell stack in the flow-through CMR mode are as follows:

- flow-field structures can be eliminated from within the stack;
- mass-transport of reactants to electrodes is much higher than in conventional or mixed-reactant stacks.

These key advantages are supported by those more generally of mixed-reactant systems, which include: reduced manifolding and sealing; single feed supply to one side of fuel cell only; reduced quantities of materials and component count; lower tolerance component manufacturing and easier stack assembly.

The consequence of eliminating the bipolar flow-field plates is a huge reduction in volume and cost of a fuel cell stack. For a polymer electrolyte membrane (PEM) stack this volume reduction could be as high as 80–90%. In several semi-commercial designs of planar solid oxide fuel cell, the ceramic or metal bipolar interconnector is the single most expensive component. Removing it could reduce the stack cost by around 30–60%. Increasing the mass-transport of reactants to the electrodes has the immediate effect of increasing cell current density. The reduction in diffusion boundary layers between reactants and catalyst particles may also reduce the amount of catalyst required per kilowatt of power generated by the cell. The combined effects of increased mass-transport and smaller stack volume should be a large increase in volumetric power density for the stack and a reduction in the power required to pump reactants through the stack.

An example of the potential volumetric power density that a CMR stack may deliver is provided by an experimental programme underway at Generics. In this programme, the performance of a small direct methanol fuel cell (DMFC) stack with a mixed vapour feed of humidified methanol and air will be tested in flow-through mode. The stack is designed to consist of a series of closely spaced fully porous membrane electrode assembles (MEAs), with each cell interconnected electrically by a thin and porous carbon gas diffusion layer (GDL) and no flow-field plate. Assuming that each GDL and MEA sheet is around 100 μm thick then a cell pitch of around 50 cells cm^{-1} should be readily achievable. (This compares with a "conventional" cell pitch of around 2–5 cells cm^{-1}.) Using existing selective cathode materials, we anticipate that it should be possible to achieve at least 50 mA cm^{-2} at a cell voltage of 0.4 V in CMR mode. If this moderate cell performance can be achieved at this cell pitch, the stack should achieve a volumetric power density of 1000 W l^{-1}, approximately equivalent to that of a conventional hydrogen-fuelled PEM stack.

With the lower fuel efficiencies resulting from today's generally poorly selective electrocatalysts in CMR mode, one challenge associated with very high stack densities will be cooling. Although cooling is readily achieved in all-liquid fuel cells, this issue may limit the maximum power density of gas-phase CMR systems until improved electrocatalysts are developed specifically for them.

3. HISTORY OF MIXED-REACTANT FUEL CELLS

As has been observed many times elsewhere, the history of the fuel cell goes back more than one and a half centuries. In that period many individuals and organisations have devoted considerable efforts and resources to the invention, development and demonstration of various fuel cell technologies. One such technology that has been investigated for alkaline, solid oxide, polymer and direct-methanol fuel cells is the "mixed-reactant" or "single-chamber" fuel cell [1–33].

Starting in the 1950s, when nuclear fission plants were under intense development, fuel cell engineers examined the possibility of radiolytic splitting of water into hydrogen and oxygen followed by electrochemical recombination of the gas mixture to generate electricity. Fuel cell stack geometries with alkaline electrolytes (anion membrane or potassium hydroxide (KOH) were tested by Grüneberg and others of Varta and Siemens in which the oxygen in a stoichiometric $H_2 + 0.5O_2$ gas mixture was first depleted to below the flammability limit at a selective cathode (e.g. C, Ag, Au) – i.e. a cathode capable of reducing oxygen molecules to oxygen ions but incapable of oxidising hydrogen gas to ions [1]. To complete the cell reaction, the depleted gas mixture was subsequently exposed to an anode electrocatalyst (e.g. Pt, Pd) that promoted hydrogen oxidation. In a later variant, in which the same gas mixture was delivered to each electrode, using a silvered nickel cathode and Pd – Pt anode, Goebel et al. [2], operated a liquid alkaline fuel cell at room temperature on a 9% O_2 + 91% H_2 oxyhydrogen gas mixture. This cell delivered an open circuit voltage (OCV) of 1.0 V and a current density of 4 mA cm^{-2} at 0.4 V.

Around the same time as Grüneberg and Goebel, in 1961, Grimes et al. [3] of Allis Chalmers built and operated a 600 W direct-methanol mixed-reactant alkaline fuel cell. A series-connected 40-cell stack of 25 cm × 25 cm solid bipolar Pt—Ni—Ag electrodes was supplied with a liquid reactant mixture consisting of hydrogen peroxide (0.1–1 wt.%) and methanol (2–10 wt.%) in a 0.5–7 M (KOH) solution and delivered up to 40 A at 15 V. In single-cell tests using a solid anion membrane as electrolyte, an OCV of 0.41 V was measured in the reactant mixture compared to 0.81 V when methanol and hydrogen peroxide were supplied separately in KOH solution to the Pt anode and Ag cathode. Analysis of reaction products determined that the net reaction in the cell was the oxidation of methanol to potassium methanate (formate), possibly either by direct electrochemical reaction for which the theoretical cell voltage is 1.88 V, or via disproportionation of a chemically generated methanol (formaldehyde) intermediate for which the EMF would be 0.94 V. Quantitative comparison of the reaction products with the charge passed by the cell indicated that significant direct chemical reaction occurred between the reactants (as opposed to electrochemical oxidation). Further tests indicated that this chemical short circuiting was primarily catalysed at the platinum electrode. Although not a particularly fuel efficient cell, the Grimes' device was clearly a powerful and early demonstration of the feasibility of liquid-phase mixed-reactant fuel cells.

Also in 1961, Eyraud described another gas-phase mixed oxygen – hydrogen device in which a microporous alumina support flooded with condensed moisture from the humid gas mixture acted as a film electrolyte [4]. In this case the Ni—Al_2O_3—Pd cell was operated as a sensor with OCV varying from −0.35 to + 0.6 V, depending upon gas composition (no power was drawn from this cell). Then, in 1965, extending Eyraud's work on film electrolytes, van Gool, described for the first time, a further variant of the mixed-reactant fuel cell, the "surface-migration" cell [5]. In this geometry, two closely spaced selective electrodes (anode and cathode) are positioned on the same side of an insulating substrate with a film electrolyte between them. While such a geometry could be operated with separate feeds of hydrogen and oxygen, van Gool suggested that the close electrode spacing (~1 μm) required for a film electrolyte would make gas separation impracticable. For the selective anode electrocatalyst he suggested that a metal with a stable or partly stable hydride and unstable oxide (e.g. Pt, Pd, Ir) would be an appropriate starting point for experimentation; for the selective cathode, a metal, such as silver with unstable hydride and stable oxide. van Gool also suggested that the metals W, Ni and Fe which catalyse the dissociation of methane and ethane below ~200°C, might work as effective anodes in a mixed-reactant system based on hydrocarbon fuels, in which oxygen or oxygen ions are available to mitigate carbonisation of the anode. No experimental work was reported by van Gool.

There was a gap of 14 years before Louis et al. [6] of United Technologies tested a single-cell variant of van Gool's [5] mixed-reactant "surface-migration" cell. In the UTC approach, a supported thin-film (3 μm) alumina electrolyte was employed, along with closely spaced (300–400 μm gap) supported Pt anode and $SrRuO_3$ cathode. Each electrode was 5 μm thick and 2.0 cm × 0.5 cm in area. With a single humidified mixed gas feed of 4% O_2, 4% H_2 in nitrogen, an OCV of 0.67 V was obtained at room temperature and, at peak power, a current density of 0.82 μA cm^{-2} at 0.39 V. UTC went on to describe a surface strip-cell geometry in which multiple pairs of surface electrodes are interconnected electrically in series and in which several of such layers are connected in parallel. Alternative electrolyte and selective anode and cathode electrocatalysts (e.g. zirconia, $LaCo_{0.5}Ru_{0.5}O_3$, $LaMnO_3$) were also suggested.

In 1990, Moseley and Williams [7] of AEA Technology described a similar room temperature Pt-oxide—Au surface-migration cell, which they had tested as a sensor in various gas mixtures. Using a porous metal (W, Sn) oxide as substrate for the sputtered metal electrodes and for condensation of an aqueous film electrolyte, they discovered that this cell generated an OCV of up to 0.5 V in humid air alone. Adding small amounts (up to approximately 1%) reducing (fuel) gases, such as H_2, CO, NH_3 or EtOH to the gas mixture, they found that the OCV of the cell increased approximately as the log of the concentration of the reducing gas. The authors reasoned that the OCV response in ambient air was a result of the different mixed potentials established by oxygen reduction and metal oxidation at the respective gold and platinum electrodes. Introduction of an additional fuel gas, they believed, mainly affected the mixed potential at the Pt electrode.

Also in 1990 (although with a Patent priority in 1988), Dyer, who was actually working on hydride batteries at Bell Communications Research, described a quartz-supported thin-film mixed-reactant fuel cell that operated with active but apparently non-selective electrodes [8–11]. As with the systems described by Eyraud et al. [4], van Gool [5] and Louis et al. [6], Dyer used a hydrated alumina film as an electrolyte (0.5 μm). In his system, however, the electrodes, which are positioned on either side of the alumina film, can be identical and are either Pt or Pd. Only one electrode of the thin-film cell is exposed to the fuel (H_2, CH_4, MeOH)-oxidant (air, O_2) mixture, while the other side is supported by an impermeable substrate. Dyer found, in his initial experiments with hydrogen – air gas mixtures, that the outer or first Pt electrode was negative (i.e. was the anode), while the inner Pt electrode was positive (cathode). OCV varied over a range of ~0.2–1.1 V, depending upon gas composition, with OCV > 0.95 V when the gas mixture consisted of at least 50% H_2. Perhaps the most important factor contributing to the magnitude and polarity of the observed OCV is that the inner Pt electrode was treated in boiling water to convert an initial <50 nm coating of Al metal into boehmite phase alumina. It is likely that in this preparation process the Pt surface itself was oxidised. Compared to Pt metal, Pt-oxide has been reported to be a reasonably selective cathode, i.e. promoting

O_2 reduction at the expense of H_2 oxidation [12]. A secondary contribution to OCV may be from the differential oxygen reduction/electrode oxidation reactions occurring in the presence of a hydrous electrolyte film (Moseley and Williams [7] observed potentials up to 0.5 V). A third but possibly weaker contributor to the observed behaviour of Dyer's cell may be that a concentration gradient in the local gas mixture is established either by gettering of hydrogen at the exposed electrode or by differential diffusion rates between oxygen and hydrogen in the pinhole-free alumina layer [9–13]. An alternative explanation, based on the relative ability of the two electrodes to catalyse the formation of hydrogen peroxide (the inner electrode being more active), was given by Ellgen, of Kerr-McGee, which provided the untested basis for a 1991 Patent which claimed a Pt—Pd alloy as a preferred cathode, simultaneously active towards chemical formation and electrochemical reduction of H_2O_2, and an anode that is inhibited towards peroxide formation but active towards hydrogen oxidation [14].

Dyer was able to reverse the polarity of his cell by changing the outer electrode to Ni. This reflects Grüneberg's original mixed-reactant cell of the 1950s which featured selective depletion of oxygen as the gas-phase mixture was exposed first to a hydrogen-inert cathode and then to a hydrogen-active anode [1]. Dyer was also able to draw ~1–5 mW cm^{-2} from his H_2—O_2 Pt—Pt cell with cell voltage falling by ~0.2–0.3 V per decade of current in the range 0.1–3.0 mA cm^{-2}. He proposed that his thin-film fuel cell would be particularly suitable as an integrated power source in planar electronic circuits and, if deposited on a flexible substrate, could also be packaged in a compact stack form (e.g. spirally wound) to replace conventional batteries, potentially with the fuel-oxidiser mixture being supplied in liquid form [10 and 15]. An efficiency improvement to Dyer's cell was suggested by Taylor (also of Bell Communications), involving patterning the outer electrode and coating it with a fuel-permeable, oxygen-impermeable barrier to increase anode selectivity to fuel and the permeability of the cell to oxygen [16]. A later (Motorola, 1996) improvement to the general mixed-reactant cell for battery-type applications included an absorbent material to eliminate water discharge [17].

In 1990, approximately concurrent with Dyer's report [9], Wang et al. [18] of GTE Laboratories, filed Patents on a two-interconnected-chamber mixed-reactant Pt—YSZ—PT solid oxide fuel cell (SOFC). Although a mixed-reactant yttria-stabilised zirconia (YSZ) electrolyte cell was described previously in Louis et al. [6] Patent, this was the first time such an elevated-temperature cell had been demonstrated in practice. The mechanism of operation of Wang's cell is not entirely clear, but seems to be via either differential flow rates of the methane – air mixture to each electrode or via an initial electrical pulse (in the range of several millivolts to volts) that may establish one Pt electrode as anode and the other as cathode. It is stated that the device will operate as a fuel cell only up to 450°C and can be operated on any reducing gas and with alternative electrolytes and with different electrodes (e.g. Pt—Au). At 350°C, Wang's Pt—YSZ—Pt cell provided an OCV of 0.95 V and a current of 70 μA at 0.65 V in a methane (10%) – air mixture. A second embodiment of the device, which was reported to give much improved performance at up to 600°C, appears to supply the fuel (ethane or methane) – air mixture to only one side of the cell, with the other side possibly being exposed to air.

Continuing the development of mixed-reactant SOFCs, in 1993, Hibino and Iwahara reported a high-temperature "single-chamber" solid oxide fuel cell operating on a methane – air mixture [19,20]. At 950°C, Hibino's Ni—YSZ—YSZ—Au single-cell delivered an OCV of 0.35 V and, at peak power, 15 mA cm^{-2} at 0.16 V in a 2:1 CH_4:O_2 mixture. Gas Conversion (GC) analysis of the off-gas from each electrode showed that the Ni—YSZ anode was an effective catalyst for the partial oxidation of methane to hydrogen and carbon monoxide, while the gold cathode was a much less effective catalyst for partial oxidation. The measured gas compositions at each electrode corresponded reasonably well with the Nernstian partial pressure of oxygen at each electrode, calculated from the respective half-cell potentials (−0.9 and −0.2 V, Pt—YSZ reference). This, combined with OCV = 0 V measured when the cell was supplied with air, N_2 or H_2 alone or when both electrodes were identical, led Hibino to conclude that cell EMF is largely due to the local oxygen concentration gradient generated by the difference in catalytic activity (towards partial oxidation of methane) between the two electrodes. Hibino also reasoned that the high cathodic overpotential (0.5 V at 10 mA cm^{-2}), due to gold's poor activity towards oxygen adsorption, diffusion and reduction, could be improved by replacing the oxygen electrolyte with a proton conductor. In subsequent experiments, Hibino was able to increase the peak – specific power of his mixed-reactant SOFC from ~2 mW cm^{-2} (0.5 mm YSZ electrolyte) to 166 mW cm^{-2} (0.5 mm $BaCe_{0.8}Y_{0.2}O_3$) [21,22]. In this latter system, at 950°C, a Pt anode and Au cathode were used, reducing both electrode overpotentials to <50 mV at ~200 mA cm^{-2} and giving an OCV of 0.8 V and current density of 400 mA cm^{-2} at 0.42 V. It was also found that performance of an Au—YSZ—Pt mixed-reactant cell could be significantly improved by surface doping of the YSZ with Pr to render it electronically conducting [23].

Separate measurements by Gödickemeier in 1997 at ETH-Zurich [24] on a Au—TZP—Pt (TZP = 0.3 mol% YSZ) single-cell, over a range of methane – air compositions, indicated an equilibrium oxygen partial pressure, $p(O_2)^{eq}$, at the Pt anode of 10^{-17} atm and a non-equilibrium $p(O_2)^{neq}$ at the Au cathode of 0.18 atm at 700°C in a 3:2 methane:air ratio. OCV for this cell was 0.8 V (corresponding with the Nernstian voltage expected from the measured $p(O_2)$ values), with a current density of ~10 mA cm^{-2} at 0.6 V and ~30 mA cm^{-2} at 0.2 V. These results support Hibino's results and analysis of mixed-reactant SOFCs.

A theoretical analysis of Hibino's and Iwahara [19] and Dyer's [9] results of single fuel cells operating in fuel – air gas mixtures was given by Riess et al. [25]. With a uniform oxygen partial pressure supplied to each electrode, Riess started with the proposition that the asymmetry required to drive an ionic current through the electrolyte must originate from a difference in the catalytic properties of the two electrodes. He showed that an ideal mixed-reactant cell will provide identical OCV and *I–V* behaviour to that of a conventional fuel cell when one electrode is reversible towards oxygen adsorption and completely inert to fuel, and the other exhibits reversible fuel adsorption and is completely inert to oxygen. He argued that the actual OCV measured in Hibino's cell was lower than the theoretical value because imperfect catalyst selectivity promoted direct fuel-oxygen reaction which reduced the chemical potential across the electrolyte. One additional conclusion of Riess' analysis is that the electrolyte does not need to be impervious to fuel oxidant or reaction products.

Logically, Hibino recently tested the practicality of a conventional Ni—YSZ—$La_{0.8}SrMn_{0.2}O_4$ SOFC single-cell in mixed-reactant mode, obtaining an OCV of 0.8 V at 950°C and at peak power, ~300 mA cm^{-2} at 0.4 V with an inlet gas mixture consisting of 19% CH_4, 16% O_2 and 65% N_2 [26,27]. He has also demonstrated that an intermediate-temperature (500°C) SOFC with 0.15 mm $Ce_{0.8}Sm_{0.2}O_2$ (SDC) electrolyte (or with other electrolytes), Ni—SDC anode and $Sm_{0.5}Sr_{0.5}CoO_3$ cathode can operate effectively in "single-chamber" mode when supplied with gas mixtures of methane, ethane, propane or LPG in air [28,29]. In the SDC cell, supplied with an ethane (18%) – air mixture at 500°C, OCV was ~0.9 V and at peak power, a current density of ~800 mA cm^{-2} at ~0.5 V was measured. In all the cases, measurements of electrode off-gas compositions support Hibino's original conclusion (and Riess' later analysis) that cell OCV is determined largely by the oxygen concentration at each electrode which itself is determined by the extent of partial oxidation of the hydrocarbon at each electrode. Hibino uses this as the basis of a Patent [30] on a series-connected surface strip-cell SOFC geometry (exposed to a gas mixture on one side only), similar in many respects to the "surface-migration" cell proposed by van Gool in 1965 [5] and later patented in series-connected form by Louis et al. [6].

In related work at ETH-Zurich, Joerger [31] described a further variant on the strip-cell SOFC geometry. In the ETH design, each side of a zirconia (TZP) electrolyte sheet is coated with a series of alternating and physically separated anode and cathode electrode strips, with each alternated pair interconnected electrically with gold wire. Additionally, the electrodes on the two sides are arranged so that anodes are opposite cathodes and electrically connected anode – cathode pairs are opposite electrically separate cathode – anode pairs. Electrical connection to an eight cell arrangement of this type was made to the outer anode (Pt) and cathode (Au) on one side of the electrolyte (TZP) sheet, giving an OCV of ~4.5 V at 700°C in a methane – air mixture. The connection geometry suggests that the intention in this design is for oxygen ions to be conducted across the thickness of the TZP electrolyte (i.e. between opposite electrodes), rather than along the surface of the TZP, as would be the case in a "surface-migration" strip-cell (i.e. between adjacent electrically separate electrodes). Clearly both ion conduction paths are possible in such a design – risking ionic short-circuiting – with the preferred path being determined largely by electrode separation distances. Recently, Zhu et al. [32] at RIT-Stockholm proposed an essentially identical geometry that solved the issue of ionic short circuiting by replacing the single zirconia electrolyte layer with a double electrolyte sandwich of YSZ (50 μm)—Al_2O_3 (0.5–1 mm)—YSZ (50 μm). The alumina support acts as a barrier to oxygen ion transport, so that each side of the series-connected strip-cell design acts purely in a surface-migration mode. Zhu's design also included stacking of many such layers in a manner similar to that proposed previously by Louis et al. [6].

Very recently, researchers from IFC and U. Texas [33] demonstrated the feasibility of a mixed-reactant direct-methanol PEM cell, showing that performance in mixed-reactant mode with selective electrodes could exceed that in conventional mode when identical rates of fuel and oxidant are supplied to anode and cathode, respectively. They also conducted a design study in which the dimensions of a series of mixed-reactant surface strip-cells were optimised. In the single-cell tests, a two-phase reactant mixture of 1 M methanol solution (3 $cm^3\,min^{-1}$) and air (3 $dm^3\,min^{-1}$) was supplied to both sides of a conventional geometry MEA at 80°C. The 32 cm^2 MEA was a Nafion-117 electrolyte coated on one side with a hydrophobic Pt—Ru C-black (5 mg cm^{-2}) anode, and on the other with iron tetramethoxyphenyl porphyrin (FeTMPP), a methanol-tolerant cathode material. Half-cell experiments were also carried out on these electrodes (with Pt counter electrode) and also on another selective cathode electrocatalyst, Ru—Se—Mo (5 mg cm^{-2}). The half-cell measurements demonstrated that, in this system, there was no significant reaction between oxygen and methanol at the anode and that the main effect of the entrained air (or entrained nitrogen) in the mixed-feed was to impede mass-transport of the fuel to the anode at current densities above 100 mA cm^{-2}. At the cathode, half-cell measurements again showed little significant difference between operation in mixed-reactant mode and conventional mode, a 40 mV drop in OCV being measured for FeTMPP in mixed-reactant mode while a ~20 mV increase in OCV was measured for Ru—Se—Mo.

Cell performance was compared in mixed-reactant mode and in conventional mode (air supplied to cathode; MeOH to anode). For both cathode systems, the OCV (0.5–0.6 V) in mixed-reactant mode was virtually identical to that in conventional mode, while slightly higher current densities were measured for the mixed-reactant systems: at 0.3 V current densities were approximately 12 and 16 mA cm^{-2} for the Ru—Se—Mo cathode, and 23 and 33 mA cm^{-2} for the FeTMPP cathode, in mixed-reactant and selective modes, respectively. One possible explanation for the slightly higher current densities in mixed-reactant mode could be that methanol crossed-over from the cathode side to the anode side by permeation through the electrolyte membrane (driven by depletion of methanol at the anode). If this is the explanation, it is important to note that it is only possible because in their experiments the IFC and U. Texas team supplied the mixed-reactant cell with twice the amount of methanol (3 $cm^3\,min^{-1}$ to both anode and cathode) supplied to the conventional DMFC (3 $cm^3\,min^{-1}$ to anode only).

In the mixed-reactant DMFC, one could argue that methanol cross-over offers a performance advantage – quite the opposite to the situation in a conventionally operated PEM DMFC. In the latter case, methanol leaks constantly from anode compartment to cathode compartment where it reacts at the unselective cathode electrocatalyst (typically Pt), thereby lowering its potential as well as wasting fuel through direct chemical reaction. This has its most significant effect on fuel efficiency (the ratio of electrical energy output of the cell to heat of combustion of fuel entering the cell) when the conventional DMFC cell is being operated at low current densities. The IFC and U. Texas team were able to show that the fuel efficiency of their mixed-reactant DMFC using a selective FeTMPP cathode remained higher than that of a conventional geometry DMFC with non-selective Pt cathode up to a current density of ~100 mA cm^{-2}. Of course, this efficiency improvement would be evident in a conventional DMFC also, if Pt were replaced by FeTMPP. This is one of the primary reasons why selective cathode materials, such as FeTMPP and Ru—Se—Mo were investigated in the first place.

The 50-year history of mixed-reactant fuel cell systems has demonstrated convincingly that they can deliver performance comparable to that of conventional fuel cells, that multiple cells can be stacked in series and that the mechanisms by which they work are well understood in terms of catalyst selectivity and local chemical gradients generated at the electrodes. Moreover, the range of work carried out in this area demonstrates that the mixed-reactant approach is applicable to gaseous and liquid systems, to systems operating over a wide range of temperatures, to alkaline, solid oxide and PEM fuel cell types and, by implication, to all fuel cell types.

The key advantages of mixed-reactant systems identified by the various workers in this field can be summarised as follows:

- more compact designs possible because manifolding simplified;
- surface strip-cell geometry enables series cell connections and exposure to reactant on just one side of structure;
- supported thin-film cell with porous electrolyte and surface electrode enables exposures from just one side;
- higher power densities possible by closer stacking of thin-film and strip-cell geometries;
- lower fabrication costs possible from continuous coating of supported thin-film and strip-cell structures; lower cost and more reliable systems because sealing requirements reduced or eliminated.

Key disadvantages of mixed-reactant systems are as follows:

- selective (as well as active) catalysts required to prevent polarisation losses and fuel inefficiencies due to parasitic side reactions;
- fuel cell exhaust may contain a larger proportion of more dilute unreacted fuel than in conventional anode exhaust;
- fuel (non-reacting) dilutes oxidant concentration (or partial pressure) at cathode (Nernst potential very slightly reduced);
- oxidant (non-reacting) dilutes fuel concentration (or partial pressure) at cathode (Nernst potential very slightly reduced).

At their best, with ideally selective and active anode and cathode electrocatalysts, with identical geometries and fuel and oxidant activities, mixed-reactant single-cells should have identical performance to conventional fuel cells. In certain types of cell, for example, where fuel cross-over depresses performance, or where reaction products may depress electrocatalyst activity, it appears possible for mixed-reactant cells to out-perform conventional separate-reactant cells. At present, ideally selective catalysts are not available for fuel cells and those substantially selective electrocatalysts that do exist are not as active as conventional fuel cell electrocatalysts towards either fuel oxidation or oxygen reduction. Although there are presently no significant efforts to develop selective anode or cathode electrocatalysts for mixed-reactant fuel cell systems, there are significant efforts in the fuel cell community to develop methanol-tolerant cathode materials for DMFCs as one means of helping to alleviate the problem of methanol cross-over. These electrocatalysts are also ideal candidates for the cathodes of mixed-reactant DMFCs and are a starting point for a more substantial development effort for catalysts designed to enable the full benefits of mixed-reactant systems.

4. DEVELOPMENT OF CMR SYSTEMS

Generics has begun initial experimental and theoretical investigations of CMR technology. The approach we are following is first to demonstrate proof of principle in single-cells and small stacks, secondly to develop a valid theoretical understanding of how CMR fuel cells work and thirdly to demonstrate significant performance in pre-prototype devices. This approach, we believe, will enable us to work towards demonstrating the applicability and value of the CMR approach to all the main types of fuel cell (PEM, AFC, PAFC, MCFC and SOFC).

The proof of principle of mixed-reactant fuel cells is already well established by the 50-year history of such systems. Our focus has been to demonstrate that the CMR approach – in which a hydrodynamic flow of mixed reactants moves through a porous cell or stack – also works in principle. At Generics we have used a dissolved-reactant alkaline system as the basis of our initial proof of principle experiments. A series of half-, single- and stacked-cell (series and parallel) experiments were carried out using a commercial anode of Pt—carbon—PTFE(polytetrafluo ethylene)-catalyst on Ni-mesh (EL05&06, Electro-Chem-Technic) and cathode of MnO_2—PTFE on Ni-Mesh (EL01&02, Electro-Chem-Technic – note: backing layer of PTFE was removed). For flow-through experiments, porosity of the electrodes was increased by perforating them with a square array of pinholes. Electrolyte was a 6 M or 10 M solution of KOH. Oxidant was air, either dissolved in the electrolyte, dissolved in water, or entrained as bubbles in the electrolyte. Fuel was either methanol dissolved in electrolyte or 0.008 M sodium borohydride ($NaBH_4$) dissolved in electrolyte.

A three-chamber tubular rig was constructed enabling separate reactant supply to each chamber and also sequential reactant flow through the three chambers. Each chamber was 4 cm in length and were separated by a single anode and single cathode sealed across the diameter of the rig by *o*-rings. External electrical connections were made to the electrodes. In one series of experiments, the performance of an alkaline single-cell in the rig was compared in conventional separated-reactant mode, mixed-reactant mode and in CMR flow-through mode (Fig. 1). In the conventional and mixed-reactant (non-flow-through) modes the electrodes were separated by a 4 cm depth of free 10 M KOH electrolyte, filling the central chamber of the rig.

In the conventional mode, a 0.008 M (3 g l^{-1}) solution of $NaBH_4$ in 10 M KOH, de-aerated with bubbled nitrogen, was fed to the anode chamber while 10 M KOH saturated with dissolved air was fed to the cathode. In non-flow-through mixed-reactant mode the $NaBH_4$—KOH solution was first saturated with bubbled air and then supplied separately to both anode and cathode chambers. In the CMR flow-through mode the same reaction mixture was supplied to the anode chamber, where it flowed through a perforated anode, through the central chamber and then through a perforated cathode before exiting the cathode chamber. In each mode, fluids entirely flooded the three chambers and were pumped through the electrode chambers at a constant rate of 2.5 ml min^{-1}.

Current – voltage data from the three experiments are shown in Fig. 1. At 20°C, the conventional ("classic" in Fig. 1) mode of fuel cell operation gave identical results to that of the non-flow-through mixed-reactant mode ("CMR" in Fig. 1). At 40°C, OCVs and current densities were higher in both cases, with the performance in mixed-reactant mode exceeding that in conventional mode. CMR flow-through mode (which was measured only at 20°C) gave significantly higher performance than either the conventional or mixed-reactant

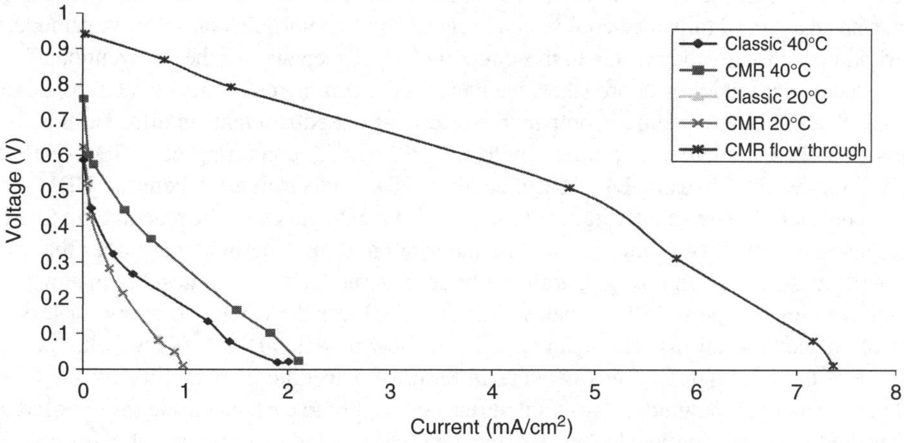

Fig. 1. Performance of flow-through CMR mode (20°C) compared with mixed-reactant (static CMR) and conventional (classic) modes

modes. These results suggest, firstly, that the electrodes in this alkaline system are highly selective towards the respective desired half-cell reactions. Secondly, they suggest that the presence of both reactants at one or other of the electrodes operate in some way to reduce electrode overpotential. Speculative mechanisms could include local decomposition of $NaBH_4$ to H_2 gas at the cathode surface disrupting the local boundary layer and improving oxygen transport to the cathode, or perhaps through local dissolved oxygen reacting at the anode to increase the rate at which adsorbed reaction products are cleared from the platinum catalyst surfaces. Thirdly, and probably the most significant effect, the results demonstrate that mass transport of reactants to the electrode catalysts in a CMR flow-through regime considerably exceeds that in a conventional or in a mixed-reactant "flow-by" regime. In the latter, diffusion boundary layers may be substantial and a smaller volume or surface area of catalyst may be exposed to reactant. Overall, this comparison between the three modes of operation supports the various earlier studies reported in the literature on mixed-reactant fuel cells and clearly shows a potential advantage in operating a single fuel cell in CMR flow-through mode.

As part of the second stage of our approach to CMR systems development, we have developed two simple computer models of direct-methanol CMR behaviour based on initial assumptions of the main electrochemical mechanisms and flow modes that determine CMR cell polarisation performance. We anticipate that these models will become increasingly robust and more sophisticated as they are developed and refined through a process of experimental testing and verification. The purpose of the models is to enable the likely performance of future designs of CMR systems to be predicted with reasonable confidence, for example, to help define measurable characteristics required for suitable active materials and to optimise specific microstructures.

The first of our mathematical models of CMR is based on the Butler – Volmer equation that describes electrode potentials and current density in terms of exchange current density and charge transfer coefficients for the various reactions occurring at each electrode. We initially tested the model by using it to predict the expected performance of a "conventional" DMFC with differing extents of fuel cross-over, populating the model with data extracted from Taffel measurements in the literature [34]. The upper panel in Fig. 2 shows predicted anode (ja) and cathode (jc) polarisation curves for a DMFC with 2% methanol cross-over, where jc is the summed result of the competing oxygen reduction (jcoxid) and methanol oxidation (jcfuel) reactions occurring at the cathode. The lower panel in Fig. 2 shows predicted cell polarisation curves for a range of methanol cross-over, where each cell curve is calculated from the potential difference of curves jc – ja at fixed current. This first electrochemical model assumes that the same reactant mixture is supplied to both electrodes and treats the limits to mass-transport of the reactants simply as a pre-defined maximum flux of each reactant to both electrodes. It does not distinguish, therefore, between the flow-through or flow-across regimes and does not take into account diffusional boundary layers at the electrodes. Despite this gross simplification, with an assumed fully methanol-selective anode and a completely non-selective cathode, the simulated polarisation curves are comparable to those reported experimentally for the "conventional" DMFC.

In the second of our modeling approaches, we have used a commercial finite element package, Femlab (Comsol AB, Stockholm), to begin to compare the effects of mixed-reactants in different flow regimes on cell polarisation. The single-cell is a direct-methanol PEM MEA consisting of a GDL—Pt—Nafion—Pt/Ru—GDL sandwich (120 μm GDL, 10 μm catalyst, 20 μm electrolyte), where the GDL surfaces are 50% blocked on each side by an interdigitated flow-field current collector. The reactant mixture is a hypothetical single-phase 50:50 (v/v) mixture of 1 M methanol and air. The model assumes that in all cases only fuel reacts at the anode and oxygen at the cathode. The model also assumes that in a mixed-reactant environment the cell voltage is 150 mV below standard EMF for the DMFC to allow for less than ideal selectivity of the electrocatalysts. The main basis of the flow model is D'Arcy's law, which describes the rate of flow of a fluid through a porous network in terms of a specific permeability constant for the network and the viscosity of the fluid. Interdigitated reactant supply to each electrode (as opposed to serpentine flow) is used in the conventional separated-reactant (case 1) and mixed-reactant cases (case 2). In the CMR flow-through case (case 3) the cathode catalyst layer is treated as either flooded with liquid (case 3a) or with gaseous air (case 3b) – oxygen in the cathode backing is in gaseous form in both cases. In all cases the reactant flows were driven at a constant pressure differential of 0.5 atm (Fig. 3).

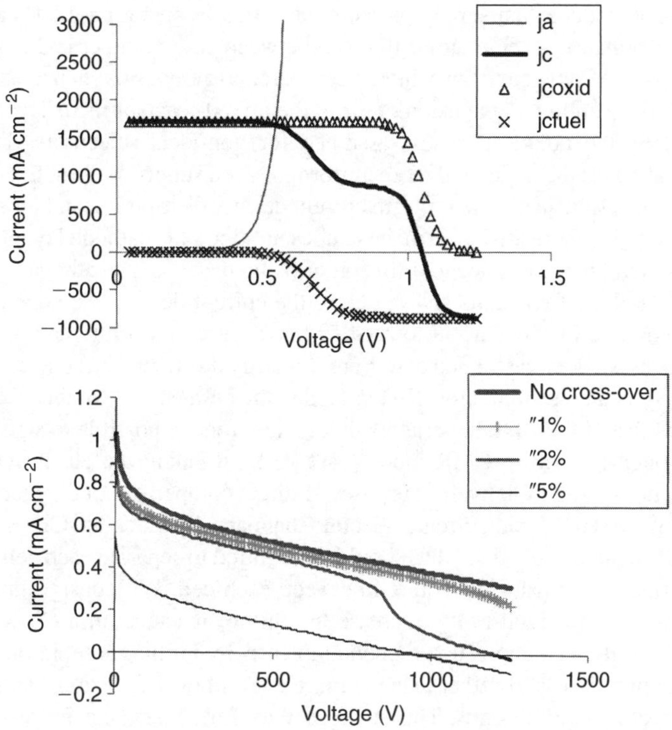

Fig. 2. Simulation of effects on performance of methanol cross-over in a conventional DMFC behaviour using a CMR model. (Upper panel) Simulated half-cell polarisation data for a conventional DMFC with 2% methanol cross-over using a CMR model. Lower panel simulated single-cell polarisation data for a conventional DMFC with 0–5% methanol cross-over using a CMR model

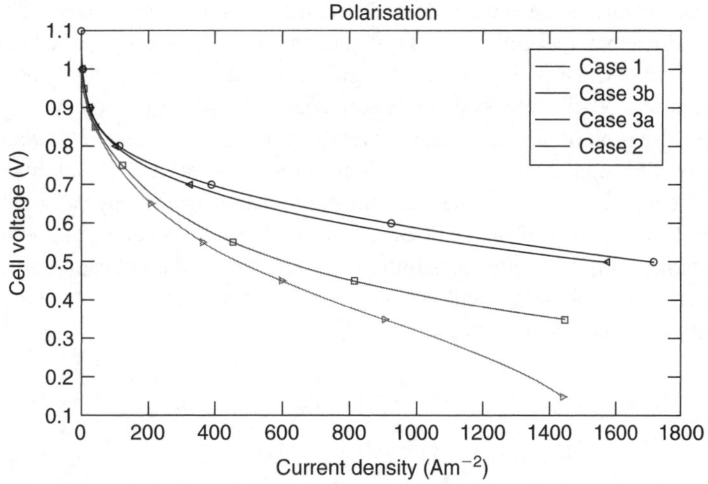

Fig. 3. Predicted-polarisation curves for PEM–DMFC operating in conventional mode (case 1), mixed-reactant (non-flow-through CMR) mode (case 2) and CMR flow-through mode (case 3)

The results show that at a constant supply pressure the different operating modes are likely to result in very different cell performances. The major difference between case 1 and case 2 is accounted for by the much lower flux of dissolved oxygen to the cathode (compared to gaseous oxygen in case 1) and by the 50% dilution of methanol at the anode by inclusion of air in the mixed-reactant feed. The remaining difference is mainly accounted by the polarisation loss assumed for non-ideal selectivity. The major difference between case 2 and case 3a is the larger and more uniform oxygen supply to the active layer (also more fuel reaches the anode). This results in a more uniform current density distribution and thus a lower polarisation of the electrode. In case 3b, where the liquid mixture does not flood the cathode layer (as would be the case in a gas – reactant system), the resistance to diffusion of oxygen across the cathode is much lower (than in case 3a), increasing the flux of reactants and therefore the current density. The decreased mass-transport resistance makes it possible for the current to be distributed more uniformly and also lowers the kinetic polarisation. Differences between case 3 and case 1 are primarily due to the differences in resistance to reactant flow and to the assumed polarisation loss. In the model, the permeability of the electrolyte membrane is assumed to be 10% of that of the GDL material. It should therefore be possible to significantly improve the flux of reactants through the cell in CMR mode (case 3) by thinning the electrolyte layer and also by increasing its permeability. Additionally, for purposes of direct comparison in the model, it is assumed that the MEA in case 3 has two GDLs and current collectors that partially block the GDLs. In what might be an ideal flow-through CMR stack, only one GDL should be required to separate each cell and no other current collector or gas distribution structure is required between each cell. With design improvements, such as these, it should be possible to significantly increase the flux of reactants in a CMR system and thereby increase cell current density well above that modelled in case 3b. Furthermore, in this model an interdigitated forced-flow is applied in case 1 rather than the more conventional diffusional serpentine reactant flow adopted in most current fuel cell designs. This suggests that if the mass transfer resistance of a diffusion boundary layer were included in case 1, an even larger advantage could be identified for the CMR mode.

5. SAFETY OF MIXED-REACTANT AND CMR SYSTEMS

Mixing a fuel and oxidant within a fuel cell raises immediate concerns of potential explosion. In reality, a fuel cell stack is largely filled with electrolyte, electrode and separator materials, which will act as a heat sink, eliminating the possibility of explosion. Analogous systems, such as metal-foams or gauzes are well known as means to prevent explosion in storage or usage of explosive mixtures. The remote possibility of sustained combustion within a mixed-reactant fuel cell will be limited to any open gas supply channels with the structure.

Evidence of the mixed-reactant fuel cells – even high-temperature SOFCs – built and tested in the last 50 years supports the view of safety in these systems. It is, of course, possible to operate with gaseous fuel:oxidant ratios below the explosive limit of the reactants, as several researchers have done. It is also possible to operate with reactant mixtures diluted in an inert carrier, such as nitrogen or water. In liquid mixed-reactant systems, the heat capacity of the liquid medium provides additional protection against any possibility of explosion.

In a CMR fuel cell, where flow of reactants occurs through a porous stacked structure, the elimination of open gas flow channels eliminates any possibility of either sustained combustion or explosion, whatever the mixed-reactant composition. Safety will need to be proved ultimately, of course, by impact, explosion and combustion tests of actual CMR devices.

6. CONCLUSION

For developers of all of today's fuel cell systems, CMR technology offers immediate benefits in cost, size, power density and reliability at the likely expense of some loss in efficiency. For direct-methanol fuel cells,

CMR offers the prospect of an increase in fuel efficiency, as well as these advantages. In the future, with the development of more selective electrocatalysts, the overwhelming advantages of operation in the CMR mode suggest it could displace today's conventional separate-reactant fuel cell in many application areas.

ACKNOWLEDGEMENTS

Thanks are due to Prof. Keith Scott of Newcastle University and Dr. Ed Fontes of Comsol AB, for their assistance in the development of fuel cell models.

REFERENCES

1. G. Grüneberg, W. Wicke and E. Justi. British Patent GB994448 (1961) also equivalent French Patent FR1321373 (1961).
2. G. Goebel, B.D. Struck, W. Vielstich. In: W. Vielstich (Ed.) [English translation by D.J.G. Ives], *Fuel Cells – Modern Processes for the Electrochemical Production of Energy*. Wiley/Interscience, New York, 1965, pp. 374–376, ISBN: 0471906956.
3. P.G. Grimes, B. Fielder and J. Adam. *Proceedings of the Annual Power Sources Conference* **15** (1961) 29–32.
4. C. Eyraud, J. Lenoir and M. Géry. *Acad. Sci.* **252** (1961) 1599–1600.
5. W. van Gool. *Philips Res. Rep.* **20** (1965) 81–93.
6. G.A. Louis, J.M. Lee, D.L. Maricle and J.C. Trocciola. US Patent 4 248 941 (1979/1981).
7. P.T. Moseley and D.E. Williams. *Nature* **346** (1990) 23.
8. C.K. Dyer. US Patent 4 863 813 (1988/1989) and reissued US Patent Re.34248 (1990/1993).
9. C.K. Dyer. *Nature* **343** (1990) 547–548.
10. C.K. Dyer. US Patent 5 094 928 (1990/1992).
11. R. Pool. *Science* **247** (1990) 1034–1035.
12. S. Gottesfeld. *Nature* **345** (1990) 673.
13. T.E. Mallouk. *Nature* **343** (1990) 515–516.
14. P.C. Ellgen. US Patent 5 162 166 (1991/1992).
15. C.K. Dyer. US Patent 4 988 582 (1990/1991).
16. T.M. Taylor. US Patent 5 102 750 (1990/1992).
17. S.M. Scheifers, D.H. Closterman, M.K. Chason and K.W. Wyatt. US Patent 5 723 229 (1996/1998).
18. D.Y. Wang, D.T. Kennedy and B.W. MacAllister. US Patent 5 100 742 (1990/1992); also EU Patent Appl. 91102546 8.
19. T. Hibino and H. Iwahara. *Chem. Lett.* **7** (1993) 1131–1134.
20. T. Hibino, K. Asano and H. Iwahara. *Nippon Kagaku Kaishi* **7** (1994) 600–604.
21. K. Asano, T. Hibino and H. Iwahara. Solid oxide fuel cells IV. In: M. Dokiya *et al.* (Eds.), The Electrochemical Society, *Proceeding Series*, PV 95-1, Pennington, NJ, 1995, pp. 58–66.
22. K. Asano, T. Hibino and H. Iwahara. *J. Electrochem. Soc.* **142**(10) (1995) 3241–3245.
23. K. Asano and H. Iwahara. *J. Electrochem. Soc.* **144**(9) (1997) 3125–3130.
24. M. Gödickemeier, D. Nussbaum, C. Kleinlogel and L.J. Gaukler. *Proceedings of the Joint International 192nd Meeting of Electrochemical Society; Proceedings of the Joint International 48th Meeting of the International Society of Electrochemical*, Paris, France, 31 August–5 September, 1997, Abstract No. 2191.
25. I. Riess, P.J. van der Put and J. Schoonman. *Solid State Ionics* **82** (1995) 1–4.
26. T. Hibino, S. Wang, S. Kakimoto and M. Sano. *Electrochem. Solid State Lett.* **2**(7) (1999) 317–319.
27. T. Hibino. Japanese Patent JP 2000-243412 A2 (1999/2000).
28. T. Hibino, A. Hashimoto, T. Inoue, J.-I. Tokuno, S.-I. Yoshida and M. Sano. *Science* **288** (2000) 2031–2033.
29. T. Hibino, A. Hashimoto, T. Inoue, J.-I. Tokuno, S.-I. Yoshida and M. Sano. *J. Electrochem. Soc.* **147**(8) (1997) 2888–2892.
30. T. Hibino, K. Ushiki and Y. Kuwabara. Japanese Patent JP 02910977 B2 (1995/1998).

31. M. Joerger. *Proceedings of the Joint International 192nd Meeting of the Electrochemical Society Inc.; Proceedings of the Joint International 48th Annual Meeting of International Society of Electrochemical*, Paris, France, 31 August–5 September, 1997, Abstract.
32. B. Zhu, G. Meng and B.-E. Mellander *J. Power Sources* **79** (1999) 30–36.
33. S. Calabrese Barton, T. Patterson, E. Wang, T.F. Fuller and A.C. West. *J. Power Sources* **96** (2001) 329–336.
34. K. Scott, W. Taama and J. Cruickshank. *J. Appl. Electrochem.* **28** (1998) 289–297.

Subject Index

δ-Bi_2O_3 and $Bi_4V_2O_{11}$-based ceramics 200–203

accelerated lifetime
  time to failure 416
acid–base polymer complexes
  polymer electrolyte membranes
    conductivity 401–403
    materials 396–400
    thermal stability 400–401
activation polarization resistance 217, 218, 222, 224, 227, 230
active hydrotreating catalysts 98
active layer *see* catalyst layer
adsorbate-induced changes
  Pt(*h k l*), in UHV
    carbon monoxide adsorption 283–285
    hydrogen adsorption 280–281
    oxygen adsorption 281–283
AES 277, 302, 303, 311, 349
AFC background and development status 119
  corporate activities 121–122
  operating principle
    dry manufacturing method 119
    oxygen reduction reaction 120
  research activity level 121
aging mechanism
  DMFC 503
  PEFC 503
  SOFC
    anode 242–243
    cathode 243–245
    interconnect 245–247
air blower, AFC peripherals 138
air management system 534, 545–546
alkaline fuel cell (AFC) 60, 61, 63
alkaline fuel cell technology, assessment 117
  background and development status 119–122
  cost analysis 133–141
  technical review 122
alkaline peripheral costs 139
alkoxy silane, polymer electrolyte 405–406
alkylsulfonated aromatic polymers
  polymer electrolyte membranes
    materials 389–390
    proton conductivity 393–396
    thermal stability 391–392
    water uptake 392–393
all-polymeric electrolytes 438, 439
aluminosilicate minerals 427
ambient air PEM
  extrapolation 137–138
  peripherals 139–140
anaerobic digester gas (ADG) 161, 162
anode
  aging mechanism
    polarization 242
    triple-phase boundary 242
  impurities, effect 129
  strategies 129
anode materials, development
  CeO_2 (rare-earth doped) anode 226–229
    doped ceria 226, 227, 228, 229
    samaria-doped ceria (SDC) 227, 229
    yttria-doped ceria (YDC) 227
  Ni–$ZrO_2(Y_2O_3)$ cermet 217–226
    anode structure, two-layered 218
    electrocatalyst, nickel 223
    electrochemical reaction process 220–221
    electrochemical vapor deposition (EVD) approach 217, 218
    low sulfur tolerance 225
    methane steam reforming 224, 227–228, 230
    thermal expansion coefficients 222–223
anti-Bragg positions
  thin metal films, surface structures 316
aqueous-phase reforming (APR)
  factors controlling selectivity
    catalyst nature 35–38
    feed nature 39–40
    reaction conditions 38
    reaction pathways 38–39
  heavier alkanes production, favoring factors 41–42
  hydrogen production 42–44
  oxygenated hydrocarbons 29
  process, basis for
    kinetic considerations 33–35
    thermodynamic considerations 31–33
  renewable hydrogen and alkanes 29
  supported metal catalysts 29
area-specific resistance (ASR) 237, 263, 264

Auger electron spectroscopy *see* AES
autothermal reforming
  catalyst formulation 94, 95
  gasoline/hydrocarbon 94
  WGS catalyst 94
azobisisobutyronitrile (ABN) 406

basic research knowledge
  electrocatalysis 180
  electrode kinetics and eletrocatalysis
    intermediate and rate determining steps 179
  methanol oxidation 179–182
beginning-of-life (BOL) performance 504
bipolar membranes (BPM)
  characterization
    chronopotentiometry principle *586*
    current–voltage curves *585, 586*
  EDBPM
    alkalization 587
    limitations 586–587
    processes 587–588
  electrodialysis, principle
    Donnan equilibrium 584
    electro-neutrality 583–584
  ion-exchange membrane 582–583
  preparation 584–585
  principle 582–583
bipolar plate
  design 490
    coated metallic plates 491–493
    composite plates 493
    non-porous graphite plates 491
  manufacturing
    coated metallic plate fabrication 494
    composite plate fabrication 494–496
    non-porous graphite plate fabrication 494
  review and analysis 490
(bi)sulfate adsorption
  butterfly pattern 300
  orientation 301
  radiotracer method 300
Brookfield digital viscometer 414
brownmillerite-like phases 204
Butler–Volmer equation 602

calculated output, model 270
carbon formation, reforming 78
carbon monoxide adsorption
  atop sites 284
  chemisorption 283
  physisorption 283
  strongly adsorbed 283–284
  weakly adsorbed 283–284
carbon monoxide clean-up 112
  acid-doped polydenzimidazole 114
  fuel processor system (FPS) 113
carbon nanotubes (CNTs) 522
carbon powder, influence 464
catalysis, catalyst layer design 474
catalyst deposition, graded 457–458
catalyst layer 446
  design 474
  impregnated 461
  pore formers 452–453
  PTFE-bound 447, 448, 449
catalyst nature
  metal components 35–36
  modified nickel 36–38
  supports 36
catalyst supports
  binary carbon 462
  carbon nanohorn 463
  conducting polymer 463
  Pt/C weight ratio 462
catalytic issues and process conditions, review
  aqueous-phase reforming 29–51
  renewable hydrogen and alkanes 29–51
catalytic WGS 78–79
cathode, aging mechanism
  cell fabrication 243
  pore formation 243–244
cathode, carbon dioxide
  effects 127–128
  strategies 128–129
cathode materials
  coating techniques sol–gel Pechini methods 150
cathode potential expression 505, *567*
cell concepts
  conventional planar *238*
  Jülich membrane–electrode assembly
      (MEA) *238*
cell construction
  tape-casting 148
cell design
  planar cell design *251, 252*
  tubular cell design *255*
CeO_2 (rare-earth doped) anode 226–229
ceramic components 250–254
ceramic materials 150, 151, 169, 192, 245, 254
charged species transport, composite material
  cell resistance control
    mass-transport limitation 430
  percolation threshold 428
chlor-alkali electrolysis
  industry 576
  membrane electrolysis cell, principle

cation-exchange membrane (CEM) 576–578
composite membrane 578
CMR systems, development 600–604
CO, removal 79
CO clean-up, preferential oxidation
in situ DRIFTS 101
low-rate branch vs. high-rate branch 100
PEM fuel cell 100
CO electrooxidation, Pt(*h k l*)
energetics and kinetics 342–346
surface chemistry
on metals 346–348
on Pt bimetallic alloy surfaces 349–353
surface structures 335–342
CO_2-scrubber, AFC peripherals 138
coated metallic plates 491–493
acidic fuel cell environment 493
ionic conductivity 492
PEM fuel cell environment 492
colloidal method 454, 463
combine heat and power (CHP) applications 156, 236
compact mixed-reactant (CMR) fuel cells
development 600–604
Butler–Volmer equation 602
D'Arcy's law 602
history 595–600
fuel–air mixture 597
gas conversion (GC) analysis 597
mixed-reactant systems, advantages and disadvantages 600
surface migration cell 596, 598, 599
principle 594
safety 604
significance 594–595
advantages 594
gas diffusion layer (GDL) 595
membrane electrode assembly (MEA) 595
polymer electrolyte membrane (PEM) 594
complex hydrides 527–528
composite gas diffusion layer 446, 463, 465
composite plates
Micro Molding Technology 493
composite protonic conductors
Nafion® 426, 427
polymer–ceramic
molecular species, permeability 430
protonic, composite 425
thermal and mechanical robustness 430
concentration polarization resistance 217, 218, 222, 264, 265
contact angle 415–416
contact resistance 217, 222, 242, 446
corrosion protection
stainless-steel bipolar plates 151
cost analysis
ambient air PEM, extrapolation 137, *138*
battery range extender 137
consumables, cost 140
gross costs and commercial estimates 134
materials and manufacturing
AFC stack materials 134–135
AFC system costs 136
PEMFC stack costs 136
PEMFC stack materials 135–136
plant balance
AFC peripherals 138–139
alkaline peripheral costs 139
ambient air PEM peripherals 139–140
PEMFC peripherals, compressed 139, 140
production volume, impact 136–137
system cost estimates 140–141
crystal truncation rod *see* CTR
CTR 278, 298, 302, 305, 311, 316, 336, 342

D'Arcy's law 602
design principles and fabrication issues
current collectors
gas diffusion layers (GDL) 563
gas permeation resistance 564
electrolyte issues 564–566
polymer–catalyst interface 564
proton exchange membrane (PEM) 563
differential thermal analysis (DTA) 382, *383*, 400
dimethylacetoamide (DMAc) 389
direct fuel cell (DFC) 2, 8–9
direct methanol fuel cell *see* DMFC
direct oxidation, fuel cell 82
dispersion viscosity 414
DMA 415
DuPont Kapton® polymide film 415
DMFC 14, 21, 175, 426, 594, 595, 599, 602
aging mechanisms and lifetime 503
high-energy storage 557
high humidification 509–512
high temperature 513–515
low humidification
water management 506–509
low reactant flow
degradation 505–506
low temperature 512–513
Power Pack 558
stacks 185
DMFC R&D, international activities
basic research knowledge, status 179
electrode kinetics and electrocatalysis
methanol oxidation 179–182

DMFC R&D, international activities (*contd.*)
oxygen reduction 183–185
methanol crossover 182–183
polymer electrolyte membrane 167
potential applications 168
technologies, status 168
DMFC stacks
conversion efficiency 185
power density output 185
doped ceria electrolytes
gadolinia-doped ceria 199, 200
nanocrystalline-undoped ceria 199–200
Dow membrane 375, 377
dual pathway
formic acid oxidation 353–358
Dupont's new membrane process
advantage 413
dynamic mechanical analysis *see* DMA
DZA model, polarization 270–272

EDBPM designing
effective electrochemical reaction zone (ERZ), 17, 229
electric power plants 56, 65, 67–68, 70, 80
electrical shorts tolerance 416
electrocatalysis, well-defined surfaces
CO electrooxidation
energetics and kinetics 342–346
metals, surface chemistry 346–348
Pt bimetallic alloy surfaces, surface chemistry 349–353
surface structures 335–342
HER/HOR
modified with metals 324–325
modified with pseudomorphic Pd film 325–327
structure sensitivity 319–324
ORR
modified with pesudomorphic Pd film 332
modified with UDP metals 331–332
Pt alloy surfaces 332–335
reaction pathway 327–328
structure sensitivity
oxidation
formic acid 353–358
methnol 358–365
Sabatier principle 318–319
electro-catalytic membrane reactors
bipolar membrane technology
competencies 574
development 573
ion-exchange membrane-based processes *575*
bipolar membranes 582–588
chlor-alkali electrolysis 576–579
fuel cell 579–581
electrochemical oxidation 107
electrochemical vapor deposition (EVD)
approach 217–218
densification sintering 3
plasma spraying 3
electrode catalysis, fuel processing 81–82
electrode designs
PTFE-bound 446
thin-film 446–447
electrode fabrication methods 125–126
electrode performance, improved 250
electrode reaction
hydrogen oxidation, oxygen reduction 566
electrodeposition methods
current control, effect 460
membrane layer 460–461
electrolyte
cathode current collector 149
electrolyte and matrix 158
electrolyte circulation 130–131
electrolyte recirculation loop, AFC peripherals 138
"electron-deficient" molecules 528
electronic power generation system
alternative
solid oxide fuel cells (SOFCs)
electroreduction, oxygen 183–184
end-of-life (EOL) performance 504
energy, sustainable development 57, 58
energy balance, supply-side challenge 55–57
engineered cathodes, high performance SOFCs
development
current distribution estimation 263–264
material properties and microstructural sensitivities 264–266
DZA model, polarization 270–272
microstructural design, polarization 272–274
model parameter 269–270
two-layer cathode, design optimization
inner layer 268
outer layer 267–268
engineered cathode development
current distribution estimation 263–264
material properties and microstructural sensitivities 264–266

fabrication
coated metallic plate 494
composite plate 494–496
non-porous graphite plate 494

feed nature 39–40
field ion microscopy (FIM) 278
forced-deposition chemical method
  thin metal films, surface structures 316–317
formic acid oxidation
  Pt(*h k l*) and bimetallic surfaces
    chemisorbed poison 354
    dual pathway 353
    site blocking species 356
    third-body effect 357
free-standing Nafion® Film 562, *567*, 568
fuel cell applications, processing
  electric power plants 67–68
  fuel options 65–67
  portable power sources 69–70
  residential and commercial sectors 69
  transportation 68–69
fuel cell electric vehicle (FCEV) 377
fuel cell operational flexibility 504
fuel cell performance
  polarization curves *567*
    diffusion of oxygen 568
fuel cell program
  FutureGen 9–10
  molten carbonate fuel cells 8–9
  solid oxide fuel cells 2–8
  US distributed generation 1–11
fuel cell system
  parameters, effects 540–541
  performance 533–537
    air management system 534
    PEFC stack 535
    PEFC system 535–537
fuel cell vehicles, performance
  fuel economy 539
  H_2-FCV parameters, effects 541
  hydrogen storage requirements, onboard 542
fuel cells
  advantages 59–65
  applications 408
  concept 59
  DMFC 503, 504, 510, 512, 515
  efficiency 59–60
  electric power plants 67–68
  experiment 504–505
  fuel options 65–67
  fuel processing 53, 65–70
  PEFC 503, 504, 510
  polymer–electrolyte
    PEFC catalyst 580–581
    PEFC polymeric membranes 581
  portable power sources 69–70
  principle 59–65
  residential and commercial sectors 69
  thermo-mechanical systems 579
  transportation 68–69
  types
    alkaline 63
    molten carbonate 63–64
    phosphoric acid 62–63
    proton-exchange membrane 61–62
    solid oxide 64–65
  water management 506
fuel choice, PEMFC 114
fuel economy
  Environmental Protection Agency 539
  fuel cell vehicles, performance 539
fuel processing
  electrode catalysis 81–82
  fuel cell applications 65–70
  SOFC and MCFC 79–81
fuel processing catalysts, review
  CO clean-up 100–101
  hydrogen production 91–106
  reformer catalysts 92–96
  water–gas shift 96–99
fuel processing research
  challenges and opportunities 70–82
  electrode catalysis 81–82
  fuel processing 79–81
  fuel reforming 74–78
  oxidation, methanol and methane 82
  sulfur removal 73–74
fuel reforming
  PEMFC and PAFC
    alcohol 75–76
    hydrocarbon 76–78
FuelCell Energy (FCE) 2
  DFC technology 8
  proof-of-concept testing 9
FutureGen 2, 9–10

galvanostatic polarization *178, 181*
gas diffusion layer
  design 481
  development
    carbon powder, influence 464
    composite 465
    pore formers 465–466
    PTFE content 464
    thickness 464
gas-processing unit (GPU), design 161–162
Gemini PEFC (Ge PEFC) 377

gradual internal methane reforming 239
gross costs and commercial estimates 134

H_2-FCV parameters, effects 541–542
halide adsorption
  anti-Bragg positions 305, 316
  charge displacements experiments 305
  Gibbs excess analytical method 303
  Pt(*h k l*)–bromide system 303–306
  Pt(*h k l*)–chloride system 302–303
  Pt(*h k l*)–I_{ad} system 306–307
  ring shielding properties 305
heavier alkanes
  alkanes carbon selectivities *41*
  factors favoring production 41–42
HER/HOR
  modified with metals 324–325
  modified with pseudomorphic Pd film 325–327
  structure sensitivity 319–324
high-pressure gas cylinders
  gravimetric and volumetric density 519–520
high-pressure processes
  Nafion® PFSA polymer
  solvent- and water-based 413
high relative humidity, consequences
  cathode
    electrochemical reaction 551
    stack coolant circuit 551
  heat management, overall 552
  humidifier 549–551
  PEMFC performance
    polymer conductivity 549
  water separator 551–552
hybrid membranes, organic–inorganic
  oligoether segments 437
  sol–gel processes 435
hydrocarbon fuels
  direct supply
    direct oxidation 239–241
    reforming 239
  sulfur removal 73–74
hydrocarbon ionomer, polymer electrolytes 406
hydrocarbon polymer, advantages 378
hydrocarbon resources 57–59
hydrocarbons
  alcohols conversion
    carbon monoxide clean-up 112–114
    fuel cells catalysis 107
    fuel choice 114–115
    fuel processor 108, 110, 112, 113
    partial oxidation 111–112
    steam reforming 108–111
hydrogen adsorption 280–281, 291, 293, 298, 317
hydrogen concentrated glucose feeds 44–45
hydrogen evolution/oxidation reaction *see* HER/HOR
hydrogen fuel cell vehicles, fuel economy
  performance
    fuel cell system 533–537
    fuel cell vehicles 537–542
hydrogen–oxygen FC 579
hydrogen peroxide bleeding 474–481
hydrogen portable systems
  hydrolysis 558
  thermolysis 558
hydrogen PEM fuel cells 42
hydrogen physisorption
  adsorbate molecules, monolayer 521
  hydrogen–CNT interaction 522
  Monte Carlo simulation 522
  reversible 523
hydrogen storage
  complex hydrides 527–528
  as gas
    high-pressure gas cylinders 519–520
  hydrogen physisorption 521–523
  liquid 520–521
  metal hydrides 524–527
  via chemical reactions 528
hydrogen storage and generation
  metal hydrides 569–571
    micro fuel cell application 569
  sodium borohydride solution
    decomposition 568–569
hydrogen storage requirements, onboard 542
hydrophilic phase domain 419, 435
hydrophobic phase domain 433

impregnated catalyst layer 461
improved performance and life, PEMFC
  accelerated lifetime 416
  contact angle 415–416
  dynamic mechanical analysis (DMA) 415
  electrical shorts tolerance 416
  Nafion® PFSA
    membranes 412–413
    polymer 412
    polymer dispersions 413, 418–419
    solution-cast membranes 419–421
  Nafion® ST membranes 421–423
  polymer chemical stability 413–414, 416–418
  size-exclusion chromatography (SEC) 415
  surface tension 415
  viscosity measurements 414
infrared reflection/absorption spectroscopy *see* IRRAS

interconnect, aging mechanism
Gibbs energy 246
planar SOFC 245
interconnected materials
ceramic 250, 254, 255, 256
metallic 250, 254
ionomer impregnation 485, 486
ionomer membranes
hydrophilic phase membrane 435
membrane morphology 434
PEMFC 434
proton transport 438
ionomers 434–435, 438
IRRAS 275, 300, 337, 339, 342, 351

Joule–Thompson cycle 520–521

kinetic considerations
C–C bond breaking, catalytic activities 33
Fischer–Tropsch synthesis 34
periodic density functional theory 34

$La_2Mo_2O_9$ (LAMOX), materials
lone-pair substitution (LPS) concept 203
$LaGaO_3$-based electrolytes 194
oxygen vacancy concentration 195
LEED 275, 277, 280, 286, 300, 311, 335
LEIS 277, 278, 285, 286, 287
Life Cycle Assessment (LCA) 498
lifetime and duty cycle information
AFC lifetime, summary 132–133
long-term tests, other 132
Zevco long-term tests 131–132
Linde cycle *see* Joule–Thompson cycle
liquid-hydrogen storage
Joule–Thomson cycle 520
low-energy electron diffraction *see* LEED
low-energy ion scattering *see* LEIS

material, proton-conducting polymer electrolyte membrane
acid–base polymer complexes 396–400
alkylsulfonated aromatic polymers 389–390
sulfonated aromatic polymer 379–382
material properties input, model 270
materials and manufacturing
AFC
stack materials 134–135
system costs 136
PEMFC
stack costs 136
stack materials 135–136

MEA assembly
ionomer impregnation 485, 486
mode 1 485–486, *488*
mode 2 486–487, *489*
MEA design
catalyst layer 474–481
gas diffusion layer 481
membrane 471–474
MEA manufacturing
MEA assembly 485–490
membrane and GDL fabrication 481–485
polymerization process 481
wet-proofing process 485
membrane design 471–474
membrane electrolysis process *577*
membrane layer 460–461, 584
membrane morphology 434–435, 438
metal catalyst
aqueous-phase reforming 35
Fischer–Tropsch reaction 36
hydrogen production 35
metal hydrides
interstitial hydride 524
Vant't Hoff equation 525
metallic materials 237, 250, 254
methanol crossover
DMFC performance, mechanisms and effects
open circuit potential (OCV) 182
inhibition methods 182
methanol oxidation, Pt($h\ k\ l$) and bimetallic surfaces
CO_{ad} coverage 361
FTIR spectroscopy 363
in situ IR spectroscopy 360
potentiostatic oxidation *359*
Pt–Ru alloy surface 362–363
methanol reforming 95–96
microfabricated fuel cells
design principles and fabrication issues 562–566
guiding principles 562
free-standing Nafion® film 562, *567*
fuel cell performance 566–568
hydrogen storage and generation 568–571
microstructural design, polarization 272–274
mixed ionic–electronic conductor (MIEC) 227, 262, 263
mixed-reactant
and CMR systems, safety 604
mixed-reactant fuel cells, history
fuel–air mixture 597
gas conversion (GC) analysis 597
open circuit voltage 595–599
solid oxide fuel cell (SOFC) 597, 598
surface migration cell 596, 598, 599

mixed-reactant systems
advantages 600
disadvantages 600
model parameter 269–270
molecular species, permeability
backscattering 431
ceramic additives 430
molten carbonate fuel cell (MCFC) 63–64, 66, 68, 79–81
anode and cathode materials 150–151
cell construction 148
Direct FuelCell® 8
corrosion protection 151
electrolyte 148–149
FCE 2, 8–9
multiple layer sputtering 458–459

Nafion impregnation 447–448
Nafion loading 448, 449, 451, 456, 459
Nafion® electrolyte films *566*
Nafion® membrane 377, 378, 382, 383, 393, 407–408
solution-casting process *413*
Nafion® PFSA membranes
advantages 413
DuPont's new membrane process 413
solution-casting process 412
Nafion® PFSA polymer
dispersions
manufacture 413
mono-modal distribution 418
SEC–LALLS 418
"second generation" dispersions 413
"third generation" dispersions 413
functional properties 412
industrial applications 412
Nafion® PFSA solution-cast membranes
extrusion-cast membrane 419–420
water back-diffusion 420
Nafion® polymer 412
surface tension 415
Nafion® ST membranes 421–423
MEA performance, single cell *422*
solution-cast and strengthened membranes *422*
Ni–ZrO_2 cermet 217–226
nickel catalysts, modified 36–38
nitrogen purge, AFC peripherals 139
non-porous graphite plates 491
non-precious metal catalysts 97–98
WGS reaction 98

Office of Fossil Energy
molten carbonate fuel cells 2, 8–9
solid oxide fuel cells 2–8
open circuit potential (OCV) 566, *567*, 596, 598
operational parameters 270
organic–inorganic polymer electrolyte 405
organic solvents 451–452
ORR
modified with pesudomorphic Pd film 332
modified with UDP metals 331–332
Pt alloy surfaces 332–335
reaction pathway 327–328
structure sensitivity 328–331
oxygen adsorption
fcc-like sites 283
LEED pattern 282
molecular precursor route 282
oxygen-ion migration 191
oxygen reduction reaction *see* ORR
oxygen storage capacity (OSC) 94, 98

PAFC
applications 155
balance of plant (BoP) 157
components, state of the art 158–159
bipolar plates 159
electrodes 158–159
electrolyte and matrix 158
electrode catalysis 81–82
fuel reforming 74–78
fundamentals 156–158
German case 160–164
anaerobic digester gas, utilization 161–162
fuel cell power plant, overhaul 163
hospitals, energy supply 162
hydrogen operation 160–161
PC25C fuel cell installations 160
service and maintenance 163–164
partial oxidation 93, 94, 95, 111–112
autothermal reformer 111
catalytic and non-catalytic 111
PEFC
aging mechanism and lifetime 503
humidification
high 509–512
low 506–509
hydrogen 504
limiting factor 378
low reactant flow
degradation 505–506
stack 535
system performance 535–537
temperature
high 513–515
low 512–513
water management 506, 508

PEM fuel cell 92, 100 *see also* PEMFC
bipolar plate
design 490–493
manufacturing 494–496
membrane electrode assembly (MEA)
design 471–481
manufacturing 481
PFSA membrane 472, 473
PEM fuel cell electrodes
catalyst supports 461–463
electrodeposition methods 459–461
gas diffusion layer development 463–466
impregnated catalyst layer 461
PTFE-bound methods 447–448
thin-film methods 448–455
vaccum deposition methods 455–459
PEMFC
CO removal 62, 76, 79, 113
current developments
direct methanol fuel cells (DMFCs) 557–558
hydrogen, portable systems 558–559
portable applications 556–559
development, 1970s to new millennium
Ballard Power System 17, 61
International Fuel Cells 16
membrane electrode assembly (MEA) 17, 20
NASA fuel cell 17
perfluorosulfonic acid membrane 16, 17
Perry Energy Systems 17–18
zero-emission bus 18
electrode catalysis 81–82
fuel reforming 74–78
hydrogen 66
key business and supplier alliances
gas diffusion layer 20
proton exchange membrane 21
peripherals, compressed 139
products, needs 22
residential and commercial sectors 69
supplier alliances 20–22
technology development path
compression–ignition engine 14
DuPont's ubiquitous Nafion® membrane 16
four-stroke internal combustion 14
fuel cell effect 14–15
Gemini Space Program 16
polarization curve 15
world needs 22
PEMFC systems
high relative humidity, consequences 549
high-temperature polymers, need
heat management 545, 552–553
water management 545
polarisation curves *547*
power density curves *547–548*
perfluorinated polymer electrolytes 375, 384, 389, 392, 398, 407
perfluorinated polymers
sulfonic fixed ionic groups 577
perfluorosulfonic acid (PFSA)
ionomer impregnation 486
PEM fuel cell 471, 472
polymerization process 481
perovskite 254, 255
perovskite based on $LnBO_3$ 205–206
perovskite-like phases 204
perovskite-type aluminates
applications 205
peroxide stability test 416
PFSA membrane
advantages 472
disadvantages 472–473
possible alternatives *478*
synthesis *482*
phosphoric acid fuel cell *see* PAFC
phosphoric acid group, polymer electrolyte 406–407
planar cell design
anode-supported *252*
electrolyte-supported *251*
plant, balance
AFC peripherals 138–139
alkaline peripheral costs 139
ambient air PEM peripherals 139–140
PEMFC peripherals, compressed 139
platinum group metals (PGM) 94, 97, 98, 100
poisoning and contamination issues
anode
impurities, effects 129
strategies 129
cathode, carbon dioxide
effects 127–128
strategies 128
contamination effect, summary 129–130
poly(benzimidazole) (PBI) 389, 391, 392, 398
polymer–ceramic
composite protonic conductors 425
interaction and microstructure 427–428
interfaces, water retention 427, 431
polymer chemical stability 413–414, 416–418
DuPont's proprietary methods 416–417
measurements 414
Nafion® membranes
fluoride emissions, reduction *417*
peroxide radical attack 414
peroxide stability test 416

polymer electrolyte fuel cell *see* PEFC
polymer electrolyte membranes, proton-conducting
  acid–base polymer complexes
    conductivity 401–403
    materials 396–400
    thermal stability 400–401
  alkylsulfonated aromatic polymers
    materials 389–390
    proton conductivity 393–396
    thermal stability 391–392
    water uptake 392–393
  electrolyte fuel cell
    polymer 376–377
    solid polymer 377–378
  fuel cell applications 408
  hydrocarbon polymer 375–378
  polymers, other 403–408
  sulfonated aromatic polymers
    materials 379–382
    proton conductivity 386–389
    thermal stability 382–384
    water uptake 384–386
  therotical studies 438–439
polymers and oxo-acids, membranes
  dielectric spectroscopy 438
  hopping mechanism 438
  thermogravimetry 437
poly(silamine) (PSA) 397, 398, *399*
polytetrafluoroethylene (PTFE) content 464, 465
pore formers, catalyst layer
  hot spray method 452
  types 453
portable and military fuel cells 555–559
portable power sources, fuel cell 69–70
power, to change the world®
  from curiosity 13–23
  future outlook 22–23
  PEMFC
    development, into the new millennium 16–19
    key business and supplier alliances 20–22
    technology development path 14–16
    world needs 22
power density
  atmospheric pressure cells 123–124
  components, performance
    electrode fabrication methods 125–126
    electrode materials 125
    three-dimensional electrodes 124–125
  PEM, comparison 126
  space applications 122–123
power sources, electric cars
  fuel cells
    direct methanol (DMFCs) 168
    polymer electrolyte membrane (PEMFCs) 168
precious metal catalysts
  gold catalysts 99
  three-way catalysts 98
preferential oxidation (PROX) 100–101
product analyses
  gas or liquid chromatography 180
production volume, impact 136–137, 141
proton-conducting polymer electrolyte membrane
  acid–base polymer complexes 401–403
  alkylsulfonated aromatic polymers 393–396
  sulfonated aromatic polymer 386–389
proton exchange membrane fuel cell *see* PEMFC
proton transport 378, 382, 384, 389, 438, 446, 453
pseudomorphic growth
  thin metal films, surface structures 316
Pt bimetallic surfaces, UHV
  structures and chemistry 285–287
Pt(1 1 1)–H_{upd} system
  entropy of adsorption 296
  Gibbs free energy 295, 296
  interactions 296
Pt(1 1 1)–OH_{ad} system
  chemisorption energy 298
  thermal decomposition 298
  thermodynamic analysis 297
Pt/C weight ratio 450, 462
PTFE-bound methods 447
  Nafion impregnation 447–448
Pt(*h k l*), in electrolytes
  surface structures and energetics
    anion adsorption 299–300
    energetics 294–298
    metal films 316–317
    reconstruction 288–294
    relaxation 298–299
    UPD 307–308
Pt(*h k l*), in UHV
  surface structures and energetics
    adsorbate-induced changes
    clean surface 278–279
Pt(*h k l*) surfaces
  anion adsorption, surface structures and energetics
    (bi)sulfate 300–302
    halide 302–307
    summary 307
  reconstruction
    Pt(1 0 0) 290–291
    Pt(1 1 0) 291–294
    Pt(1 1 1) 288–290
  relaxation

Pt Bragg reflection 298
vibrational spectroscopy 299
visible–infrared SFG 299
UPD and irreversibly adsorbed metals
Bi 313–316
Cu UPD 308–311
Pb UPD 311–313
Pt(*h k l*) system energetics
Pt(1 1 1)–H_{upd} 294–296
Pt(1 1 1)–OH_{ad} system 297–298
Pt–Ni alloys 286–287
Pt–Ru alloy electrocatalyst 285–286
pyrochlores and fluorite-type $(Y,Nb,Zr)O_{2-\delta}$
transition 208

reaction conditions 31, 36, 38, 42, 48, 49, 51
reaction pathways 31, 38–39, 44, 47, 51
ORR 327–328
reflection-high energy electron diffraction (REED) 278
reformer catalysts
autothermal reforming
gasoline/hydrocarbon 94–95
methanol reforming 95–96
partial oxidation
methane and hydrocarbon 93–94
steam reforming
methane and hydrocarbon 93–94
relative pressure drop factor 553
residential and commercial sectors, fuel cell 69
rotating ring disk electrode *see* RRDE
RRDE 305, 308, 311, 328, 331, 346

Sabatier principle 318–319
salvage 131
scanning tunneling microscopy *see* STM
SEC 415, 418
low angle laser light scattering (SEC–LALLS) 418
SECA
auxiliary power unit (APU) 7
SOFC program 7–8
US Hydrogen program 7
selective oxidation, catalyst layer design 474
Siemens Westinghouse Power Corporation (SWPC)
combined heat and power (CHP) generation 3
electrochemical vapor deposition (EVD) process 2–3
SECA program 7
SOFC/gas turbine hybrid system 6
tubular SOFC program 2–7
Sievert's apparatus 570
single SOFC cell 215
single-walled carbon nanotube (SWNT) 522
size-exclusion chromatography *see* SEC

SOFC
activation polarization resistance 217, 218, 222, 224, 227, 230
aging mechanism 241–247
concentration polarization resistance 217, 218, 222
contact resistance 217, 222
interconnect 216
internal resistance 217, 223
manufacturing components
cell design 250
ceramic components 250–254
interconnect materials 254–256
tubular and planar, design 250
operation, benefits 247–248
planar 7, 235, 236, 244, 245
research and development
advances 236–241
hydrocarbon fuels, direct supply 238–241
operation 237–238
Siemens Westinghouse Power Corporation 2–7
single SOFC cell 215
solid state energy conversion alliance 7–8
tubular 2–7, 81, 236
triple-phase boundaries (TPB) 217, 222–223, 225, 226, 242
typology, developers *251*
solid–electrolyte interface 278, 302
solid electrolytes, apatite structure 207–208
solid oxide electrolyte ceramics
δ-Bi_2O_3- and $Bi_4V_2O_{11}$-based ceramics 200–203
brownmillerite-like phases 204–205
doped ceria electrolytes 196–200
$La_2Mo_2O_9$ (LAMOX), materials based 203–204
$LaGaO_3$-based electrolytes 194–196
$LnBO_3$, pervoskite based 205–206
perovskite-like phases 204–205
pyrochlores and fluorite-type $(Y,Nb,Zr)O_{2-\delta}$ 208–209
solid electrolytes
apatite structure 207–208
transport properties
review 189
zirconia-based solid electrolytes 190–194
solid oxide electrolytes 190
electrochemical cells, key components 190
solid oxide fuel cell 60, 64–65, 68, 69, 79–81, 262 *see also* SOFC
solid polymer electrolyte membranes
electric vehicle application 378
Gemini PEFC (Ge PEFC) 377
hydrocarbon polymers, advantages 378
limiting factor 378
Nafion® membrane 377

solid state and materials science
current opinion
PEMFCs, current developments 556–559
portable and military fuel cells 555–559
solid state energy conversion alliance *see* SECA
sputtering techniques
platinum catalyst 172
stack fluidodynamics 177
steam reforming
catalyst deactivation 93
methane and hydrocarbon 93–94
sulphur poisoning 109
water–gas shift (WGS) 108
steam-reforming process 110, 240
STM 278, 284, 288, 291, 303, 331, 335, 365
sulfonated aromatic, polymer electrolyte membranes
materials 379–382
proton conductivity 386–389
thermal stability 382–384
water uptake 384–386
surface structure and energetics
adsorbate-induced changes
carbon monoxide adsorption 283–285
hydrogen adsorption 280–281
oxygen adsorption 281–283
surface tension
Nafion® polymer 415, *418*
Wilhelmy force 415
surface X-ray scattering *see* SXS
sustainable development of energy
energy, sustainable development 57
energy balance, supply-side challenge 55–57
hydrocarbon resources, efficient utilization 57–59
SXS 277, 278, 288, 291, 293, 298, 302, 311, 331, 337
system issues
electrolyte circulation 130–131
salvage 131

technical review, AFC
lifetime and duty cycle information
AFC lifetime, summary 132–133
Zevco long-term tests 131–132
poisoning and contamination issues
anode, strategies 129
anode impurities, effect 129
cathode, carbon dioxide effects 127–128
cathode, carbon dioxide strategies 128–129
contamination effects, summary 129–130
power density
atmospheric pressure cells 123–124
components, performance 124
space applications 122–123
system issues
electrolyte circulation 130–131
salvage 131
technology and potential applications, current status
catalyst layers, fabrication 174
direct liquid methanol (DLM) 175
electrode kinetics 179–182, 183–185
portable power 168–175
LTCC, technology 169
membrance electrode assembly (MEA) 169, 172, 174
sputtering techniques 172
transportation
fabrication techniques 175
proton-conducting ionomer 177
thermal and mechanical robustness
polymer matrices 430
thermal stability
proton-conducting polymer electrolyte membrane
acid–base polymer complexes 400–401
alkylsulfonated aromatic polymers 391–392
sulfonated aromatic polymer 382–384
thermodynamic considerations
steam reforming 32
vapor-phase reforming 32–33
WGS reaction 31
thermogravimetric analysis (TGA) 382, 383, 391, 392, 398, 405, 406
thermoplastic ionomers 453–454
thermoplastic resin, Nafion® 412
thin-film methods
colloidal method 454–455
Nafion loading 451
organic solvents 451–452
pore former, catalyst layer 452–453
self assembly, controlled 455
thermoplastic ionomers 453–454
thin film techniques 562
thin metal films, surface structures 316–317
triple-phase boundary (TPB) 242, 244, 262
activation polarization 217
electrochemical reaction 225
YSZ powder, formulating 219
tubular cell design *255*
two-layer cathode, design optimization
inner layer
small grain diameter 268
outer layer 267–268

ultrahigh vacuum (UHV) surface 278, 280, 281, 285, 291, 302, 306
ultra-shift
hydrogen production
low levels of CO 42–44

vacuum deposition methods
  graded catalyst deposition 457–458
  multiple layer sputtering 458–459
Van der Waals equation 518
Van't Hoff equation 525
viscosity measurement 414

water retention 427
water–gas shift (WGS) 42, 66, 92, 94, 95, 96, 98
  catalytic 78–79
  non-precious metal catalysts 97–98
  precious metal catalysts 98–99
water management, AFC peripherals 138
Westinghouse Electric Corporation *see* Siemens Westinghouse Power Corporation

XPS 277, 282, 311, 313, *314*, 316, 349, 352
X-ray photoelectron spectroscopy *see* XPS

Zevco long-term tests 131–132
zirconia-based solid electrolytes
  dopant–vacancy interactions 191
  electrolytic domain 192
  ionic conductivity
    acceptor-type dopant(s) 190
  oxygen isotopic exchange rate 194

Elsevier Internet Homepage

http://www.elsevier.com

Consult the Elsevier homepage for full catalogue information on all books, journals and electronic products and services.

Elsevier Titles of Related Interest

HYDROGEN AND FUEL CELLS
Bent Sørensen
2005; ISBN: 0126552819
www.elsevier.com/locate/isbn/0126552819

HIGH-TEMPERATURE SOLID OXIDE FUEL CELLS: FUNDAMENTALS, DESIGN AND APPLICATIONS
S.C. Singhal
2003; ISBN: 1856173879
www.elsevier.com/locate/isbn/1856173879

WORLD FUEL CELLS – AN INDUSTRY PROFILE WITH MARKET PROSPECTS TO 2010
G. Weaver
2002; ISBN: 1856173976
www.elsevier.com/locate/isbn/1856173976

Related Journals
The following journals can all be found at: http://www.sciencedirect.com

Fuel Cells Bulletin
Fuel
Journal of Power Sources
International Journal of Hydrogen Energy
Refocus
Electrochimica Acta
Energy Conversion and Management
Renewable & Sustainable Energy Reviews
Energy
Catalysis Today
Chemical Engineering Science
Renewable Energy
Journal of Membrane Science

To Contact the Publisher
Elsevier welcomes enquiries concerning publishing proposals: books, journal special issues, conference proceedings, etc. All formats and media can be considered. Should you have a publishing proposal you wish to discuss, please contact, without obligation, the publishing editor responsible for Elsevier's energy publishing programme:

Heather Johnstone
Publishing Editor, Materials Science & Engineering
Elsevier Limited
The Boulevard, Langford Lane
Kidlington, Oxford
OX5 1GB, UK

Tel.: +44 1865 84 3887
Fax: +44 1865 84 3920
E-mail: h.johnstone@elsevier.com

General enquiries including placing orders, should be directed to Elsevier's Regional Sales Offices – please access the Elsevier internet homepage for full contact details.

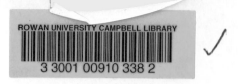